Glimpses of Indian Agriculture

Glimpses of Indian Agriculture

General Editors
Sangeeta Verma and P.C. Bodh

Volume Editors
Vijay Paul Sharma, Parmod Kumar, Nilabja Ghosh,
D.K. Grover, and Usha Tuteja

Ministry of Agriculture and Farmers Welfare
Government of India, New Delhi

Oxford University Press is a department of the University of Oxford.
It furthers the University's objective of excellence in research, scholarship,
and education by publishing worldwide. Oxford is a registered trademark of
Oxford University Press in the UK and in certain other countries.

Published in India by
Oxford University Press
2/11 Ground Floor, Ansari Road, Daryaganj, New Delhi 110 002, India

First Edition published in 2018

ISBN-13: 978-0-19-948883-4
ISBN-10: 0-19-948883-5

Typeset in 10.5/12.7 Adobe Garamond Pro
by Tranistics Data Technologies, Kolkata 700 091
Printed in India by Replika Press Pvt. Ltd

Unless otherwise specified, all tables and figures are part of the Ministry's own research.

Contents

Section II: Technology, Inputs, and Services for Agricultural Development

Section III: Agricultural Marketing, Post-Harvest Management, and Value-Addition

Tables and Figures

TABLES

FIGURES

S.K. PATTANAYAK
SECRETARY

भारत सरकार
कृषि एवं किसान कल्याण मंत्रालय
कृषि, सहकारिता एवं किसान कल्याण विभाग

Government of India
Ministry of Agriculture & Farmers Welfare
Department of Agriculture, Cooperation
& Farmers Welfare

MESSAGE

It is heartening to note that the illustrious research work carried out by various Agro-economic Research Centres under the aegis of the Ministry of Agriculture and Farmers Welfare has found recognition in the form of *Glimpses of Indian Agriculture* published by Oxford University Press. The first *Glimpses of Indian Agriculture* was published in 2008.

The Agro-economic Research Centres, brainchild of Dr VKRV Rao, established in 1953, are entering the 65th year of their existence and have contributed immensely to the corpus of agricultural economic research knowledge derived from field-based diagnostic survey. Divided into five sections, this book has an overview of the content in the first section followed by four sections focused on technology inputs and services; agricultural marketing, post-harvest management and value adition; incentivizing agriculture, prices and food security; and constraints and emerging solutions.

I hope that this publication will be highly useful to policymakers, researchers, and students in sharpening their insights and help in understanding the complexities confronting the farm sector, assess the impact of developmental programmes and help in formulating the policy interventions required to steer agriculture to new heights with a view to ensure farmers welfare and prosperity.

I would like to compliment Dr Sangeeta Verma and Dr P.C. Bodh—the General Editors; Dr Vijay Paul Sharma, Dr Parmod Kumar, Dr Nilabja Ghosh, Dr D.K. Grover, and Dr Usha Tuteja—the volume editors; and the officers and the staff of AER Division of Directorate of Economics and Statistics for making it worthy for publication through Oxford University Press.

8 January 2018
New Delhi.

(S.K. Pattanayak)

Foreword

Glimpses of Indian Agriculture is being published at a time when the country is still grappling with myriad agricultural problems. The period of the research studies appearing in this publication—2008 to 2016—is important as it was during the 11th and 12th Five Year Plans that a massive development policy thrust was witnessed for reviving Indian agriculture from the lows of growth and development and the persistence of agrarian distress since 1995. The significance of this document lies in the fact that the policymakers and academicians get access to a large pool of field-level, evidence-based research and analysis crucial for their working.

The present volume is the outcome of agro-economic research covering the length and breadth of the country, with specific focus on diverse agro-climatic zones, micro and macroeconomic issues, diagnostic insights into problems and solutions emanating therefrom. The publication provides a crisp overview of the emerging trends in agriculture followed by a section on technology, input and services; a section on agricultural marketing, post-harvest management, and value addition; a section on incentives, agricultural prices, and food security; and concludes with a final section on constraints, and emerging solutions. The idea behind this compositional structure of research papers was to provide a complete menu of research-based resources for meeting development policy requirements concerning emerging trends, issues of farm input and services support; imperfections in agricultural produce market and post-harvest management; and, finally, policy regime of incentivizing farming, and replicating success stories.

After the first publication in 2008, which was well received by the academia and the policymakers, it has become an imperative to continue this prestigious publication. It was with the participation of eminent agricultural economists from the agro-economic research network and overall coordination and guidance of Agro-Economic Research Division of the Directorate of Economics & Statistics, Department of Agriculture, Cooperation and Farmers Welfare (DAC&FW) for around two years that this publication became possible. The constant encouragement and support of the Principal Adviser, DAC&FW and the Secretary, DAC&FW were the main motivating factors for this endeavour.

To ensure the worldwide availability of this important research document with high quality touch and at an economic price in the market, Oxford University Press India came to be the preferred publisher and was effective enough to deliver efficiently and strictly according to timelines agreed upon.

K.L. Prasad
Senior Economic & Statistical Adviser
Directorate of Economics & Statistics
Department of Agriculture, Cooperation & Farmers Welfare

General Editors' Note

Glimpses of Indian Agriculture was first published in 2008 by the Directorate of Economics and Statistics, Department of Agriculture, Cooperation and Farmers Welfare, Government of India. This new volume contains a larger number of agricultural economic research works produced by a group of senior economic administrators and career agricultural economists.

Engaged in these research works is a countrywide, grassroots-level agricultural economic research network—the various Agro-Economic Research Centres, which remain devoted to field research in different agro-climatic zones assigned to them by Ministry of Agriculture and Farmers Welfare and allied sector departments and organizations of the union government. Since 1953–54, when the first four centres were set up, they have produced a large number of field research works calling for their availability in the national and global market. This demand was met in 2008 with the initiative of S.M. Jharwal, the then Principal Adviser of DAC&FW. The present issue is in continuation of this endeavour. This issue is based on chapters selected from a large number of studies completed during 2007–08 to 2015–16.

The chapters of this volume share the findings of incisive field evidence-based agricultural economic research. Condensed into these updated research, drawn out of long list of reports brought out during the decade ending 2016, are 65 reports covering as many as subjects. Authored by agricultural economists of fame and experience, the volume talks of the performance of Indian agriculture, crucial issues of technology, inputs, and services; matters of marketing post-harvest management; initiatives for development incentives, price, and food security; and finally of constraints and emerging innovative solutions.

On another level, this volume offers a bird's-eye view of the kind of agro-economic intelligence inputs that go into the formulation of the government's agricultural development policies and programmes; and how the impact on farm production, income, profitability, marketing, soil health, and farmers' welfare is assessed on the basis of thorough field-evidence-based research. Not only this, each of these chapters elaborately dwells on how to improve the effectiveness of agricultural development schemes towards ensuring financial sustainability, ecological effectiveness, and social welfare of the farming communities.

1

Overview of Emerging Trends in Indian Agriculture

Vijay Paul Sharma, Parmod Kumar, Nilabja Ghosh, D.K. Grover, and Usha Tuteja

Agriculture continues to play an enormous role in providing food, nutrition, and livelihood to the teeming millions of India, despite its declining share in the Gross Domestic Product (GDP) of the country. The contribution of the agricultural sector to the national GDP has witnessed a secular decline during the last few decades as a consequence of increase in the shares of other sectors, particularly in that of the service sector. Even at its current share of about 12 per cent in the national GDP, it remains the largest employer and the main source of livelihood to the majority of the rural population (CSO 2014). Consequently, agricultural performance remains crucial for rural livelihood, food security, and reduction of poverty. Indian agriculture, which witnessed a visible deceleration during the 9th and 10th Five-Year Plans, recorded a robust growth during the 11th Plan. The food grains production touched a new peak of 265 million tonnes in 2013–14, with an addition of about 55 million tonnes between triennium ending (TE) 2005–06, and TE 2013–14 (GoI 2015).

While there has been a significant increase in production of food grains in the recent past, the sector has been facing such formidable challenges as the decline in the average size of land holding, dwindling water resources, and inefficient water use, the adverse impact of climate change, shortage of farm-labour, poor and inefficient marketing infrastructure, and increasing costs and uncertainties associated with volatility in international markets. Agriculture production is mainly dependent on the natural resources of land, water, soil, biodiversity (plant, animal, and microbial genetic resources). But these natural resources are rapidly shrinking due to increasing demographic and socio-economic pressures, monsoon disturbances, increasing frequencies of floods and droughts, etc. Overuse of marginal lands, imbalanced fertigation, deteriorating soil health, diversion of agricultural land to non-agricultural uses, misuse of irrigation water, depleting aquifers and irrigation sources, salinization of fertile lands and water-logging continue apace. During the last three decades considerable emphasis has been laid on development of natural resources (land, water, and perennial biomass), attention towards sustainable socio-economic management has reached unprecedented levels. For making agriculture sustainable, to meet country's food requirement, it is a precondition to maintain soil health and water availability at levels that would re-assure farmers to pursue agricultural activities with higher level of productivity. Indian agriculture is also witnessing structural changes with the composition of agricultural output shifting from traditional food grains to high-value products. Agriculture is increasingly being driven by expanding demand for livestock products, fish, and other high-value crops like fresh fruits and vegetables, processed foods, and beverages. The share of high-value commodities/products (fruits and vegetables, livestock products, fisheries) has increased from about one-third in TE 1983–84 to over 50 per cent in TE 2011–12 (Sharma 2011). The composition of export trade has also changed, away from traditional products towards horticulture, meat, meat products, and processed products. The Indian food consumption basket has become increasingly diversified and expenditure on fruits, vegetables, milk, eggs, meat, fish, beverages, and processed food is rising rapidly (Kumar 2014), leading to changes in cropping pattern in the country. Indian agriculture is also becoming increasingly more and more market-oriented and monetized. The proportion of agricultural production marketed by the farmers has increased significantly over the last few decades. In the early 1950s, about 30–35 per cent of food grains output was marketed, which has increased to more than 70 per cent. The marketed surplus remains relatively higher in the case of commercial crops than subsistence crops but the level of marketed surplus for subsistence crops is also on the rise.

Though the share of agriculture in overall economy has been decreasing, share of livestock and fisheries in agriculture is increasing. Within the crop sector there has been significant increase in the cropped area under fruits and vegetables, indicating horizontal and vertical diversification happening in agriculture. The structural changes happening in the agriculture sector require the increasing role of non-farm sector in rural employment and income. The review of studies on rural diversification shows that employment in trade, transport, and business services has increased in the recent period. These activities are dominated by private players. The rural non-farm economy now accounts for more than 60 per cent of rural income though it accounts for only 32 per cent of rural workforce in country. There was a sharp fall in the share of agriculture in the total rural work force after 1999–2000. This fall was more in the case of male workers. Bulk of female workers, that is, 85 per cent rural female workers, is engaged in agriculture, indicating the reality of feminization of agriculture and rural sector. Though manufacturing has traditionally been the most important industry in the rural non-farm sector, its share in employment and income is stagnating. Now construction, transport, trade, and services lead the rural non-farm sector. Women account for around 30 per cent of rural workforce, the disguised unemployment among female workers appears to have reduced marginally following the MGNREGS scheme. The MGNREGS has also helped in increasing rural wages for unskilled workers and also female workers in rural sector (Kumar and Chakraborthy 2016). The rural wages for skilled workers continue to decline in rural sector.

Dependence on agricultural sector, particularly on crop cultivation, has resulted in widespread unemployment and underemployment in the country. The agricultural sector is characterized by ever-declining land–man ratio, predominance of small and fragment land holdings and increasing application of labour saving production technologies. It is thus becoming more and more evident that agricultural sector capacity in itself is not enough to absorb the growing labour force. Not only this, the organized industry sector with its capital-intensive nature does not offer much scope for absorption of additional labour force. Added to these realities is the employment challenges thrown up by liberalization, privatization, and globalization. All these forces have aggravated the un-employment and under-employment situation in India, underscoring the need for discovering alternative avenues for employment generation. This necessitates priority to agro-processing industries development.

Agro-processing industry in India is largely a house of small-scale enterprises. They are highly heterogeneous in terms of capital investment, technology use, scale of operation, quality and quantum of output, composition, and level of employment. More importantly, levels of productivity among tiny and small enterprises are low. There must be a lot of constraining factors covering institutional, technological, and marketing that are holding up productivity of the agro-processing units to low levels. There is therefore need to address these constraints so that productivity of the agro-processing sector may be improved. As a whole, the strength of the agro-based industry is comparatively less than those of non-agro-based industries. Moreover, the growth profile of the number of agro-based enterprises is uneven across the regions of India.

Recognizing the importance of food processing for the country as a whole, the government set up the Ministry of Food Processing Industries in 1988 and the food processing industry was identified as a thrust area for development. Indian food processing industry has become an attractive destination for investors all over the world. The reality, however, is that the food processing sector of India is still at a nascent stage. Processing of the agricultural commodities constitutes a small proportion of the raw material available. In 2004–05, food-processing sector contributed about 14 per cent in the total share of manufacturing in the Gross Domestic Product (GDP). Of this, the unorganized sector accounted for more than 70 per cent of production in terms of volume and 50 per cent in terms of value. On the export front, India contributes only 1.5 per cent (2003–04) to the global agricultural exports despite its leadership in agricultural production (D&B Information Services India Private Limited 2007). The share of this sector in GDP has almost remained same for the last ten years (Kumar et al. 2006). The share of processing of fruits and vegetables is around 2 per cent, around 35 per cent in milk, 21 per cent in meat, and 6 per cent in poultry products. By international comparison, these levels are significantly low. Processing of agriculture produce is around 40 per cent in China, 30 per cent in Thailand, 70 per cent in Brazil, 78 per cent in the Philippines, and 80 per cent in Malaysia. According to one estimate, due to inadequate processing facilities, fruits and vegetables worth Rs 40,000 crore go waste annually. Thus, food processing is necessary for reducing the wastages which normally take place in the post-harvest period or during period of a bumper crop.

Under the Indian Constitution, the state governments have the final say on how marketing of agro-products would operate while the central government can only suggest and advise the states. Existing laws guided by the state's Agricultural Production Marketing Committee (APMC) Regulation Acts provide for regulation of agricultural markets where transactions ideally can take place in a fair and transparent manner. The markets also create self-employment opportunities for a fleet of trades in market chains. Open auction, supervised by democratically elected market bodies would ensure fair practices and competitive prices under idealistic conditions.

The reality is far more complex. Traders who operate in the APMC markets are severely screened and limited through licensing. With nearly no scope of new actors entering the chains, producers enjoy little option in choosing their buyers thus defeating the purpose of the regulation. Moreover, owing to unresolved conflicts in most cases supervision remained poor, bureaucratic, and even corrupt rather than representative. Prices are depressed against the interests of producers owing to the superior power of the traders. Association among traders and alleged collusion between the trader and the state officials make matters worse. Paucity of public funds left little room for modernizing market facilities. However, the states varied widely in the regulated marketing system and the above description may not apply to all regulated markets.

Globalization in the wake of India's formally joining the World Trade Organisation (WTO) made the existence of vibrant and dynamic marketing system compelling. All state governments were asked to amend the state APMC Acts in tune with a model APMC Act finalized and circulated by the central government in 2003. The Model Act was meant to reform the market by allowing more competition and encouraging innovative new marketing methods to evolve. Yet, marketing of agricultural product in India, labouring from the memories of an exploitative past, rural interlocked markets, and food shortages has become a politically sensitive issue. The proposed new Act meant to override the pre-existing legislation and set the reforms in motion has not been accepted by many state governments. Contract farming has in particular been viewed with suspicion drawing from past memories and retail chains calling for external funding and managerial advances have become serious political issues in India.

States vary widely in their nature and pace of reforms in agricultural marketing. In fact, even states that enacted the reforms did not necessarily go by the spirit of the proposal and reforms were at best partial. Contract farming is particularly perceived as 'anti-poor and anti-farmer' in West Bengal but is aggressively promoted in states like Punjab, Haryana, and Andhra Pradesh. Private retail chains are becoming popular with customers and farmers in many states but resistance has been stiff in Jharkhand and Uttar Pradesh. Allowing foreign direct investment (FDI) in multi-brand retail is a critical contention in the political economy of the nation. While West Bengal so far only has shown some interest in amending the Act, Bihar repealed the Act in 2006 but failed to enact a new Law till date. Only partial amendment of the Act is attributed to the two most agriculturally progressive states Punjab and Haryana as well as Madhya Pradesh. Uttar Pradesh amended the Act but, facing serious pressure, withdrew the amendment soon after leaving the old Act still in place.

Certain states are steadfast in averting changes that are in principle administratively possible even under the existing laws. In fact, the older Acts were flexible in principle enabling the States to allow changes through suitable interpretations and 'notifications'. Political will to reform is most important. Regardless of the amendment of the APMC Act early signs of reforms are already apparent in most the states. Development of modern channels such as contact marketing is observable even in states politically averse to these ideas. Entry of new players with modern methods could not be prevented in Uttar Pradesh which did not amend the Act. The central government moved towards a national market using persuasion as a means. Fruits and vegetables followed by cereals, pulse, and oilseeds could be gradually dropped from the APMC schedule of regulated commodities but with all the hesitation shown by the states, Union budget 2015 considered a Constitutional amendment or use of special provisions in the Constitution for setting up a National Common Market for specified agricultural commodities (*Economic Survey 2014–15*). During this period the regular channels, far from being eliminated or phased out were also undergoing transitions in tune with the pressures of competition. Demonstration effects from other states, encouragement from the Centre and overtures of the private companies under the existing regulation are driving the changes on the ground but developments in information and technology were possibly the strongest influence. Flexible Market Intervention Scheme (MIS), e-trading (so far used in trading financial instruments) in the spot and futures market, establishment of derivative exchanges and the ITC e-Choupal deserve mention. Since The Ministry of Agriculture formulated a central sector scheme—Agricultural Research and Marketing Information Network (AGMARKNET) for electronically linking regulated markets spread all over the country. The infrastructure of the APMC markets began to be upgraded.

Integrating the producer with the consumer closely, a modern supply chain can deliver several advantages. By reducing the trader's margin, the farmer's market opportunity and profit can be improved. Besides, managerial efficiency, information dissemination, technologically enabled logistic organization, and the market infrastructure at the producer and retail ends can together help in reducing wastage, saving fuel, curtailing greenhouse gas emission, eliminating unfair practices, maintaining quality standards for consumer safety, improve productivity through private extension, and facilitating convenient marketing and shopping. Yet the reduction of channel length can intensely affect the livelihood of traders whose services remain an unresolved issue. Farmers scattered across underdeveloped rural India are weakly connected with urban and modern society physically. In transacting with poorly informed farmers, traders enjoy the advantage of asymmetric information. Highly maligned for their exploitative practices, the difficult and risky environment in which they operate is given as a rational justification for the low prices they pay producers (Mulky 2008). With lack of competition from the organized sector their methods remained out of tune with the progress of marketing in the outer world. Trading is perceived to be an easy option in employment.

To enhance private investment in agriculture, some important policy measures like Model Act to change the APMC Acts, contract farming, liberalization of agricultural exports/imports and futures trade in agricultural commodities were affected. These were necessitated with the signing of international trade agreements like WTO and Free Trade Area Pacts. It was considered necessary to allow the corporate sector to have captive production, marketing arrangements, storage, processing, transportation facilities, and export and import of agricultural commodities. It was envisaged that with free trade a lot of opportunities will emerge for Indian agriculture in the international markets due to removal of trade barriers, quantitative restrictions, and reduction in tariffs by developed nations, where the demand for competitive commodities from the developing economies was much higher. It was emphasized that once the agricultural subsidies were withdrawn, agricultural trade would expand, leaving export markets open for countries like India (Gulati 1994), where labour costs were less and agricultural production would be cheaper and competitive. The farmers would be able to share benefits if they were linked with international agriculture through trade and industry. If commercial crops replace staple crops threatening food security, it was argued, why to produce everything locally and store for fear of shortages when it could be imported at much lesser costs, even at rates lesser than carry over costs (Jha and Srinivasan 2001). In a counter argument to this Nayyar and Sen (1994) emphatically pointed out that often it might not be possible even to meet the fraction of requirement of India. Whenever a huge player entered the international market as a buyer or seller the international prices significantly shot up or crashed respectively, consequently threatening the food security of the nation and livelihood of the farmers. The insight was that the implementation of these policies would enable corporate sector to enter into farm sector shifting cropping pattern to high value crops like fruits, vegetables and plantation from low margin crops like cereals (Acharya 2001). This would in turn enhance employment opportunities by virtue of these crop's high value items that necessitates more labour, and more investment in transport, processing, and subsidiary activities, like road side small marketing, transport repair, and accessory shops and overall development of infrastructure like roads, warehouses, processing units, banks, etc. Enhancement in employment and production of high value crops will raise income level resulting into food security by easy access to food and nutrition; and eventually help eradicate poverty and hunger. The increased income would mean more demand for industrial goods. Thus reforms in agricultural marketing would result in overall agricultural growth which ultimately will take the growth of the economy to a higher trajectory (Reddy 2001).

With this vision, the central government linked certain grants, like those under National Horticulture Mission, to amendments in APMC Act. The Model Act paved the way for direct purchase by corporate sector from the farmers. Contract farming is seen as a solution to corporate sector's need for assured and high quality farm produced supply; and the farmers' need for assured market with attractive prices.

The contract farming was introduced to help the sector build up its captive supply; and overcome all the shortcomings of extension services in terms of manpower, coverage, quality of services, and to solve the problems of high quality produce need by the food sector industries and to ensure better prices for the farm produce.

The benefits include enhanced income to small and marginal farmers due to assured improved seeds and other inputs and buy back arrangements of produce by corporate sector. To promote regular flow of raw material and to meet trade commitments of processed products uninterruptedly, corporate sector was allowed strategically located huge areas in the form of special economic zones (SEZ); additionally, to cover risks and discover prices future trading was permitted. For that commodity exchanges were established. One of the strongest arguments in favour of future trading was that farmers would benefit from the price discovery as well as by minimizing price risks. Finally to increase share in the international markets trade restrictions were removed and about 1400 product lines were put on open general licence (OGL) in the very first year and substantial reduction in import duties was effected, which was followed by lowering the quantity of buffer foodgrain stocks.

All these developments brought changes in India's pattern of exports and imports. They might have helped achieve higher export-led growth trajectory and higher income through competitive pricing (Bawcutt 1996) but agricultural growth did not see spectacular upward shift. This may be due to the fact that the crucial investment in irrigation, infrastructure, marketing, transportation, and storage was neither made and nor facilitated to the expectations of the corporate sector. Also other administrative barriers like unrestricted movement of agricultural produce throughout the country, choice to sell sugarcane free of area restrictions, etc., were not removed. If higher growth in agriculture could not be achieved, how far cropping pattern, rural income, employment, etc., were affected by these changes are the issues that need serious prudence.

Thus, the analysis of emerging scenarios in the agricultural sector, namely, crop and husbandry sector, agro processing, agricultural marketing, and other various agricultural activities are elementary for the growth of Indian economy as India is one of the major producer and consumer of these items. Further, proper management of agriculture from the policy perspective requires analysis of growth, outcomes, problems, and constraints facing Indian agriculture as inputs for future policy initiatives. This book ponders on these issues and provides policy feedback on various government policies and programmes to present status and prospects of Indian agriculture.

AN OVERVIEW

This book is divided into five sections with each section devoted to one broad theme. These sections are summarized in the following paragraphs.

Section I: Overview and Performance of Indian Agriculture

This section has 20 chapters with a range of subjects. Thematically, these can be divided into three groups. The first group of authors analyse production, economics, marketing problems, and prospects of pulses, oilseeds, basmati rice, and *katrani* paddy. The second category of chapters capture horticultural crops and government interventions for the development of this sub-sector. Third group of authors is focused on allied sector of agriculture, that is, livestock, dairy development, and fisheries.

The first set of chapters in this theme deal with crops and related issues in the case of pulses, rice, and oilseeds. The findings of the first study based on secondary data suggest that the world pulse production grew at a slow rate of 0.48 per cent per annum between 1985 and 2005. A mixed performance across all the producing countries came out to be the reality of during these twenty years. In majority of the cases, poor yield growth was found responsible for slow growth in pulse production. Not only this, the poor yield was characterized by instability in production in most of the pulse producing countries, barring a few cases. The results of secondary data analysis highlight that yield has been a major contributing factor in the growth of pulse production in India. In the case of gram, yield had contributed significantly while in the case of other pulse crops, area, and yield showed more or less equal contribution. The primary data thrown up by the field studies of agro-economic research centres provide findings on the economics of pulse cultivation, adoption of technology, marketing of pulses and farmer's perceptions based in the National Food Security Mission (NFSM) and non-NFSM districts in the surveyed states. The primary field data contained in these chapters suggests that there is an urgent need to enhance availability of improved seeds of pulses at affordable prices and strengthening of other components of technology.

The two field studies on basmati rice in Punjab and Haryana assess cost, profitability, market price, and divergence among farm harvest prices, wholesale prices, retail prices and export prices. The results of these studies suggest that there is a vast untapped potential of basmati in India due to its quality and aroma. As policy prescriptions, improvement in returns from cultivation by yield enhancement through R&D, delivery of required inputs at reasonable prices to producers and provision of physical infrastructure needs to be prioritized. The study on katarani paddy in Bhagalpur, Bihar has interesting findings. It suggests that preservation of traditional katrani paddy seeds and solving agronomical issues would help farmers in realizing its potential.

Another study based on primary data collected from three important states, that is, Punjab, Haryana, and Uttar Pradesh, examined production and procurement pattern, relative economics of paddy vis-à-vis competing crops and highlighted constraints in adoption of alternative crops to paddy in this important region. As policy inputs, the study recommends a favourable price regime, technology for raising the existing level of productivity, financial support, rural infrastructure and, above all, multi-pronged support from the government. Similarly, another chapter assessing spread of new varieties of rice in terms of area, production, and productivity observed that performance of high yield variety (HYV) rice tapered off since the 1990s but hybrid cultivation did not spread sufficiently to compensate it. Thus, adoption of hybrid varieties is possible through dissemination of knowledge and availability of credit to growers. A chapter on potential and prospects of rabi crops cultivation in Assam suggests that consolidation of holdings, supply of institutional credit, strengthening of extension, creation of a single window for input delivery, and supply of HYV seeds would help in improving the profitability and potential of rabi crops in Assam.

That the oilseeds economy is extremely important for India is highlighted in the chapter on important policy implications for improving oilseeds, production, and productivity in India. The chapter states that the strategy for boosting production of edible oilseeds in the country must focus on both price and non-price determinants, such as technological, institutional, and economic factors, that influence supply response of edible oilseeds. In fact, growth in oilseed production in India has been moderate due to yield and price risk despite significant increase in minimum support price (MSP). Therefore, in order to increase the production of edible oil seeds through yield enhancement, a technological breakthrough in terms of suitable high yielding varieties, irrigation as well as acceleration in technology and its dissemination through strengthening extension services should be accorded priority.

In the recent past, production and yield of cotton grew at a commendable rate in India. It was largely due to adoption of Bt cotton which offered a promising solution in terms of yield enhancement. A chapter on cotton examined performance and economics of Bt cotton vis-à-vis non Bt cotton in major producing states of Andhra Pardesh, Gujrat, Maharashtra, and Tamil Nadu. Results show that Bt cotton offers good resistance to bollworms as well as several other pests. The yield of Bt cotton was found higher and statistically significant in the sampled states under irrigated and rainfed conditions and enhanced profitability. Overall, this study, based on a sample of 694 farm households, found a positive impact of Bt cotton on yield and returns. Further, the authors of these chapters suggest that release of area specific Bt cotton varieties, availability of seed at affordable prices and in-time dissemination of information to growers are the keys to farmers' realizing untapped potential. The second study included on Bt cotton in this section examines input use and economic performance of Bt cotton vis-à-vis conventional varieties of cotton in Punjab. The results of the study clearly bring out that yield of Bt cotton can be further increased by use of quality seed, additional human labour and plant protection chemicals. Therefore, farmers need to be facilitated in terms of credit availability from institutional sources and timely release of seeds of the recommended varieties of Bt cotton in Punjab.

Horticulture is emerging as an important sub-sector of agriculture in India. It has immense potential in terms of generating income and employment for small and marginal farmers in rural areas of the country. The present theme contains five chapters on the subject. The first chapter is an evaluation of the National Horticulture Mission (NHM) which assesses the impact on area, production, and yield of these crops covered under NHM, creation of employment opportunities and income for the farmers. Results suggest that there is a large degree of spatial and temporal variation in the yield rate of these crops due to erratic rainfall and periodic occurrence of droughts. Most of the surveyed farmers grew these crops on irrigated land but the quality of irrigation and erratic supply of power were major constraints. Therefore, there is an urgent need to provide a subsidy for adoption of drip irrigation. Improvement in infrastructure and capacity building under the NHM also need strengthening. Another study examined growth, availability of baseline data and other related issues in the case of horticultural crops in north east and Himalayan states. The study concluded that horticulture has emerged as a means of crop diversification and increase in employment for small and marginal farmers in these states. However, non-availability of baseline data on basic parameters at the grassroots level is a serious limitation in designing and planning for improvement in yield through extension, input supply and efficient marketing logistics. Hence, it is essential to collect and publish baseline data regularly on horticultural crops in all states for successful strategy building in the future.

The third chapter on this sub-sector examines the potential and problems of horticultural development in Assam and Meghalaya. The study highlighted that the prospects of these crops in the selected states are bright provided marketing support and infrastructure facilities are available to the farmers. This includes strengthening extension services and input supply, establishment of cooperatives for sale, and processing units to avoid wastage. The fourth chapter on this sub-sector investigates the economics of mushroom cultivation, marketing channels, margins, costs, and constraints in Himachal Pradesh. The authors recommend training, production of compost, and spawn by private units and establishment of processing units at the grower's level. The last chapter in this series evaluated the impact of government intervention in horticultural development in the state of Maharashtra. It stated that government intervention in the state has made possible substantial increase in area under these crops. The production of fruits in the state has increased significantly during the last decade. But, the sector is marked by high wastage due to perishable nature of these commodities. Therefore, it is necessary to invest in post-harvest technology, storage, and processing in order to increase shelf life of the produce.

Animal husbandry sub-sector plays an important role in the agricultural economy of India. This subject is covered in a study on livestock in Maharashtra which evaluate livestock development in the state over the past five decades. The chapter suggests that significant changes in the livestock situation have occurred due to dependence on mechanized sources of farm power. Further, the Indian dairy sector would be competitive only when export subsidies on dairy products are abolished and challenges at the world and domestic level through modernization of the supply chain are taken care of in order to be competitive on the world level. Second study included on live-stock assessed the impact of Integrated Dairy Development Project in the north eastern region in terms of additional employment and income, milk production, and improvement in genetic potential of cattle. The authors suggest that appropriate dairy development policy should pay attention to fodder deficit, establishment of milk cooperatives at the village level, assistance to milk producers in breeding, feeding, animal management, and effective marketing of milk and milk products.

In the recent past, fish farming has assumed special importance for income and livelihood security. A chapter with focus on this sector examines problems and prospects of fish farming in Bihar and Jharkhand. In the light of emerging inputs from primary survey based data, authors recommend professionalism in fisheries through availability of fingerlings, extension backup, quality feed, renovation of ponds, and frequent training to farmers in both the states. The last chapter in this section analysed economic feasibility of fish ponds/raceways in Himachal Pradesh. The authors find that pond fish farming in the state is profitable because it requires low investment, time, and human labour. It is found easier in comparison to trout farming. Authors suggest that availability of fingerlings of required breed, feed, and training related to important aspects to farmers may help in development and realization of potential of fisheries in Himachal Pradesh.

To sum up, the chapters included in this section brought out critical aspects related to subjects covered. Most of the chapters are based on primary survey data and provide crucial insights into the problems and constraints of growth at the grassroots level. The observations and conclusions are useful for future policy initiatives in order to improve performance of agriculture in India.

Section II: Technology, Inputs, and Services for Agricultural Development

This section includes 12 chapters. Under this theme, mainly the role of technology, like machinery, fertilizers, irrigation, and role of services, such as soil testing, crop insurance, and agriculture development. Within this broad sub-sector different sub subsector themes were captured. There are six chapters on irrigation, fertilizer, and farm mechanization; with two chapters each on biotechnology and soil testing; and cop insurance each having one chapter. A brief summary of chapters included in the second theme is presented below.

Biotechnology has tremendous promise and for a country like India that is heavily dependent on agriculture for food security, biotechnology seems to offer the much needed technological innovation. Bt cotton is a significant example of the impact of biotechnology on crop performance, farmer incomes, and profitability. Since the introduction of Bt cotton in 2002, the production growth rate of cotton shot up to 13 per cent and yield growth rate to 10 per cent. Even the area under the crop has grown at 3 per cent. Almost all farmers interviewed indicated that Bt cotton yields more than non-Bt cotton. The results on factors affecting creation of effective demand indicate that the cotton farmers are willing to take risks and be opinion leaders for other farmers. Seed dealers were found the most common and important source of information. Farmers perceive benefits of Bt cotton in the quality and availability of seeds, reduction of the pest incidence, in boll size, staple length, fibre colour, and cotton price. Disadvantages were perceived in seed cost and fertilizer need. No difference was seen in machinery need and irrigation and harvesting cost as well as in marketing and by-product output. The awareness about genetically modified (GM) foods/crops was in general very low. Government and expert intervention to create awareness thorough testing of GM foods, adequate labelling were some of the requirements of the consumers that came out of the survey.

Participatory Irrigation Management (PIM) implies the involvement of irrigation users in different aspects and levels of management of the water resource including planning, design, construction, maintenance, and financing, but particularly in distribution. It generally involves a group of farmers into bodies or institutions, often called Water Users' Associations (WUA) for the purpose of managing a part or more of the irrigation system. The primary objective of PIM is to achieve better utilization of available water through a participatory process that endows farmers with a major role in the management decisions over water in their hydraulic units. The pertinent question is to what extent has PIM resulted in more availability of water for irrigation, greater efficiency in water use, better recovery of water charges, and better operation and maintenance of the irrigation structures? The study indicates that there has been considerable progress in bringing participation and devolution of powers in irrigation management but substantial further efforts are required and will help improve performance. It is found that increased participation commonly brings significant benefits to performance in water resource management. Many of these institutions require not just setting-up but also inputs in institution design, institution building, and training in order to make them strong and sustainable. Greater accountability also needs to be incorporated through proper financial audit, performance evaluation, and social audit, and the financial viability and sustainability of these institutions needs to be enhanced through local resource mobilization as well as external development support.

Micro-Irrigation technique including drip and sprinkler irrigation is considered as water saving technology. It is expected that micro irrigation will help improve the availability of water. However most of the potential area for micro irrigation has not been brought under adoption so far. Various field experiments have shown that this technique increases water use efficiency up to 80–90 per cent depending on the crop and soil type (Sivanappan 1994). The spread of micro irrigation has been restricted to a few pockets in a few states in areas where groundwater withdrawal is high. The lukewarm response is attributed to several causes including lack of access to groundwater, lack of cash, crop specificity and lack of know-how, poor product quality, and absence of credit facilities.

Drip irrigation was seen having positive impact on agriculture through increase in cropping intensity and through increase in yield from existing crops. It also led to reduction of labour use on the farm. The technology also has a direct impact on the prosperity of farmers as majority of adopters confirmed having a positive impact on their income. Drip irrigation also led to the adoption of high value and less water intensive crops. The technology also helped making agriculture more sustainable and profitable through better market prices for the crops grown by the adopters. However, drip irrigation needs a better financial model to ease adoption at the initial stage as only 16 per cent adopters and 12 per cent non-adopters agreed to drip irrigation as a financial proposition without subsidy. Drip irrigation technology has the potential to show conservation effects but these effects are visible only when the adoption is large scale at a cluster level. The government and other promoting agencies need to focus on the cluster approach and on geographic pockets based on hydrology and crop economics to get the best conservation impact.

Crop diversification is a strategy for overcoming food and nutrition insecurity and raising income and employment generation; and maximizing crop production. Water scarcity in agriculture has led to an alternative cropping pattern disfavouring traditional hydrophilic crops such as paddy, banana, and sugarcane; encouraging water conservation measures through a variety of schemes like watershed development scheme. The alternative cropping pattern has been found in diversification of crops towards those crops which require less water and give more yield per unit of water. The farmers facing water deficit adopt several strategies for meeting the water scarcity. Most farmers surveyed under this study have reported that they have left their land fallow during the deficit period. A considerable proportion of them used well irrigation during the deficit period while a few of them borrowed water from other farmers at the time of crop stress. Some farmers used oil-driven pump-sets to irrigate their fields while others reduced their cropped area during deficits. It must be understood that the non-availability of water or the scarcity of water alone cannot determine an alternative cropping pattern. The alternative cropping system needs to be worked out keeping in view the agro climatic zone as the base and keeping in mind the nature of the soils, traditional cropping patterns, and the availability of water resources. State level committees could be constituted with a team of different experts to arrive at a suitable, alternative cropping system.

Tank irrigation is one of the cheapest sources of surface irrigation. This was sustained until 1960s in various states and the country as a whole. After this period, there has been a sluggish growth in irrigated area under tanks in the country more particularly in the state of Karnataka, which has less proportion of irrigated area as compared to national average and southern states. The expert committee, constituted in 1993, suggested Watershed Management in the catchment areas of the tanks, desilting of tanks, modernization of distribution systems under tanks, and initiating conjunctive use of tank and groundwater (GOK, 1993). Under both central state government several watershed development programmes have been implemented. Under these programmes, treatment of catchment areas of tanks was taken up as one of the components with a view to preventing soil erosion and silt deposit in the downstream tanks and arresting rain water run-off for enriching ground water. Under this component, a variety of structures such as farm ponds, check dams, *nala* bunds and percolation tanks, contour bunding, trenching, boulder checks, etc., had been taken up both in arable lands and non-arable lands and across drainages. Some of the structures, particularly, the check dams, farm ponds, percolation tanks had the capacity of storing several cubic metres of rain water. According to farmers, these are not allowing run-off from the catchments leading to non-filling of tanks. Lack of coordination and independent nature of works by each department promoted neither increase in the tank irrigated area nor substantial increase in agricultural production despite the combined efforts of these programmes.

The green revolution, even while achieving the milestones of agricultural development in terms of more area, yield and production through its energy-irrigation-fertilizer approach, bypassed the drought-prone, rainfed, areas of plateau regions of India, confined its developmental boons primarily to the irrigated tracts. This neglect of rainfed areas created unintended agricultural, ecological and socio-economical imbalance between irrigated and rainfed areas. After the 1980s, government realized these disparities and felt that the target of 4 per cent agricultural growth as envisaged in the National Agriculture Policy (NAP) is not possible to achieve without integrated development of rainfed areas on a sustainable basis. The NAP seeks to improve rural livelihood and reinforce farming system/productivity of vast rainfed areas through adoption of holistic participatory watershed development approach. Watershed approach is a low-cost, location-specific technology. It is a vehicle for enhancing the productivity and simultaneously preserving the natural resources of rainfed areas. Watershed comprises of three features namely, arable land, non-arable land, and network of natural drainage lines. From the study it is inferred that The National Watershed Development Project for Rainfed Areas (NWDPRA) holds the key to the development of country's vast rainfed areas. It is observed that the programme improved the groundwater aquifers as well as in-situ soil moisture. Further, in the study it is observed that NWDPRA impacted positively irrigation, cropping intensity, farm employment, fodder and biomass, outmigration, etc. Though, it is difficult to identify a single factor, upsurge in crop yield and farm income owing to improvement in irrigation and in-situ moisture seems to be the key driving force for the noticeable performance of NWDPRA. Though, NWDPRA had an essential component of institutional building, functioning of most of the created institutions (SSGs, UGs) was found weak. The inclusion and support of local non-governmental organizations (NGOs) in the programme will help reducing implementation problems.

The watershed development programme can be considered as an appropriate rural development strategy by implementing all land-based rural development programmes under the concept of watershed development. There should be a holistic approach to rural agriculture development through watershed programme, primarily aiming at integration of several development activities such as soil conservation, land and water management, agriculture, afforestation, and animal husbandry with special emphasis to relate these actions with human issues and to develop the capability of the target population at the micro level befitting the local conditions. The impact evaluation study has demonstrated that watershed development programme to large extent is able to regenerate natural resources including land, forest, and water and play a crucial role in augmenting agricultural growth, productivity, cropping intensity, and cropping pattern.

Chemical fertilizers are a key element of modern technology and have played an important role in the success of Indian agriculture. Fertilizer is assumed to be as important as seed in the Green Revolution (Tomich et. al. 1995), contributing as much as 50 per cent of the yield growth in Asia (Hopper 1993 and FAO 1998). It is also argued that one-third of the world cereal production is contributed by the use of fertilizer and related factors of production (Bumb 1995). There is undisputable need for continuous rapid growth in fertilizer use especially in less-consuming regions in the country in the coming years to increase agricultural production and productivity at the desired rate. In order to meet the additional demand, there is a need to increase fertilizer supplies. For sustained agricultural growth and to promote the balanced nutrient application, it is imperative that an appropriate fertilizer pricing policy is put in place. The current pricing and subsidy schemes generally do not promote the balanced use of nutrients. Therefore, there is a need to keep parity between N, P, and K prices and also address the issue of the deficiency of secondary and micronutrients. A long-term fertilizer pricing policy that promotes fertilizer use as well as production is needed.

Although there is a high degree of inequity in the distribution of fertilizers across states and crops, there is a fair degree of equity in the distribution of these subsidies across different farms sizes. Small and marginal farmers are key beneficiaries of fertilizer subsidies and fertilizer subsidy was not restricted to only irrigated area but had spread to rainfed areas as well. A reduction in fertilizer subsidy is, therefore, likely to have an adverse impact on the income of marginal and small farmers. The increase in fertilizer prices would lead to a reduction in fertilizer use on these farms and consequently lower production and productivity. An increase in prices of fertilizers is also likely to have an adverse impact on agricultural production in low-fertilizer-using regions growing mainly coarse cereals, pulses, and oilseeds. The targeting of fertilizer subsidies is a critical and sensitive issue. Since it is practically not feasible to develop an effective targeting system that reaches poorer households/regions, comprehensive coverage of all farm households is a better alternative than ineffective targeting.

Application of chemical fertilizers depends basically upon soil fertility and previous crop grown and amount of organic manure applied. Due to variations in soil fertility, rainfall, and climatic conditions, a common dose of fertilizer cannot be recommended for all regions. In the sample areas, farmers usually applied recommended dosages of fertilizer. There were no major differences in the adoption of fertilizer use in the sample villages among different categories of farmers. The selected farmers strongly pointed out that the price of fertilizers was the single most important determinant of application of fertilizer. Among the various determinants of fertilizer consumption, the study found that price of urea and rainfall turned out to be the most important variables.

Soil testing is a chemical process by virtue of which requirement of nutrients for plant can be analysed to sustain the soil fertility. The basic objective of the soil testing programme is to provide a service to farmers for better and more economic use of fertilizers and better soil management practices for increasing agricultural production on their farm. There are more than 514 soil-testing laboratories in India with a capacity of about 6.5 million samples per annum (2008). The present infrastructure of soil testing facility is found to be insufficient in different agro climatic regions in Madhya Pradesh. Available infrastructure is not being utlized effectively. Therefore, targets and achievements need to be increased by employing skilled and trained staff in these labs. The quantity and quality of soil samples tested also need to be increased in the State. The awareness about soil testing facility, its need, and significance at the farmers' level is important for the success of the programme.

Given small size of holdings, it is necessary to encourage custom hiring of agricultural machinery especially tractors and combines to raise productivity while maintaining costs of cultivation. This can be done by encouraging panchayats, Primary Agriculture Cooperative Societies (PACS) and agricultural machinery cooperatives/companies, clinics or even private entrepreneurs with loans, training, and subsidies. Easier and wider availability of credit for second hand machines can also come in handy to deepen agricultural mechanization. Since every village now owns 15–20 tractors in northern region, it is necessary to establish agricultural machinery service centres at the village level or the cluster of village's level. Also, contract farming can encourage mechanization as the crops grown under this arrangement are new, more amenable to mechanization and come with good agricultural practices which cut costs or raise yields and productivity.

Combine harvester is a costly machine and requires good and careful usage to attain viability. As seen in Punjab and Gujarat, it is not being used enough (only around 50 days in a year) compared to Maharashtra where it was primarily bought for custom hiring and was used for as much as 90 days. In order to ensure its viability there is a need to exercise caution in its funding and ownership. Banks should insist on definite business plans before approving loans for farmers so that second hand combine markets like tractors do not emerge in India. There is need to provide for collective purchase of combine harvesters by farmers' groups or PACS like tractors so that local availability during times of need especially during peaks of harvesting seasons can be ensured.

In the drip irrigation technique, labour saving on land preparation is a major benefit which farmers have not yet realized. Irrigation companies should highlight this benefit to attract customers for their products. The companies should ensure quality components like drippers, emitting pipes, filters, etc., so that farmers face no major technical problems like system clogging, which affect their motivation on drip irrigation. The decreasing subsidy indicates that low cost would be key factor for survival in the future. Therefore, companies should come up with more cost effective products and business strategies.

The last study in this theme looks into the mechanization in the eastern India. The level of mechanization in the three states covered in the study, the eastern region was very low, and below the national average. Odisha and Bihar were way below the national average, while West Bengal was closer. The cost of machinery was disproportionately high in West Bengal, probably because of the lack of custom hiring facilities. The cost of mechanization appeared lower in Bihar. Machines were mainly used for only three operations—ploughing, irrigation, and marketing. The percentage of farmers using machines for these operations was higher. But the proportion of expenditure was lower for ploughing. Therefore, it might be necessary to encourage more use of tractors and power tillers in ploughing through custom hiring centres. Also, there was a discrepancy between the preference for tools and machines and their actual use for many operations, particularly for ploughing, irrigation, harvesting, threshing, and marketing. Efforts should be made to provide the appropriate tools and machinery to the farmers.

Section III: Agricultural Marketing, Post-Harvest Management, and Value-Addition

As subsistence farming gives way to commercial agriculture in which surplus is generated for sale, marketing becomes important for the viability and success of agriculture. The chapters, based primarily on field surveys, analyse in details the problems and constraints of marketing of different products of Indian agriculture; from crops to livestock and to fodder. Further, in the backdrop of the transitions generated by economic reforms the impacts of various government interventions, new emerging channels, and institutional innovations in Indian agricultural marketing have also been explored. The analysis provides important recommendations for post-harvest management, value addition, and improvement of market efficiency in Indian agriculture which will play an important role in government's initiative of increasing the farmers' incomes. In this section, a total number of 17 chapters related to various aspects of agricultural marketing are included. This theme includes the traditional marketing issues like marketed and marketable surplus of foodgrain crops in various states and pre and post-harvest losses of various crops. The issues related to new marketing system and emergence of new marketing channels like super markets, contract farming, and producer companies; market access and constraints are discussed in various chapters. The theme also includes some interesting chapters on agro-processing and value addition. A brief summary of the chapters included in the third theme is presented in the following paragraphs.

Estimation of marketable and marketed surplus and identification of important factors determining these surpluses will be useful in designing effective food procurement, distribution, and price policy and in understanding the nature of the growth process. The patterns of marketable and marketed surplus of major cereals and pulses in leading producing states in India have been found to be diverse across crops, states, and farm categories in India. Marketed surplus, mostly lower than marketable surplus, was found to be positively related to farm size, output price, distance to the regulated market and to the awareness about MSP wherever there is public procurement. Investments in strengthening infrastructure and improvement of market information will therefore be important in accelerating the growth of agricultural production and generation of marketable surplus which in turn will make farming profitable as an enterprise.

However, not all agricultural products in India are consumed by farmers' households or are marketed. A significant proportion of crops raised is known to be lost even while standing on the field or after harvest due to various reasons. Such losses lead to wastage of valuable resources, compromises of farmer's incomes, and larger dispersions between farmer's cost and the consumer price not attributable to any one's earnings and therefore amount to social losses. Crop losses caused by biotic and abiotic factors are found to be substantial. High cost of inputs, poor quality of seeds, low output price, and pest and disease problems are found to be the important factors responsible for such losses. These losses can be reduced through the dissemination of scientific knowledge on cultivation practices, sensible agronomic practices such as wet and dry system of irrigation and profitable crop rotation along with integrated pest and disease management practices and the use of information and communications technology. Also not just quantity, but quality of public and private investments in rural agricultural infrastructure also needs to be improved. Construction of community storage facilities need to be encouraged with active support from NGOs and gram panchayats.

Considerable post-harvest and processing losses occur to paddy not on the field but during milling. That only a small portion of the entire paddy crop is processed by modern mills is a major problem faced by the rice processing industry in India. Analysing primary data from three major paddy producing States of the eastern, northern, and southern parts of India, revealed to the author noted that the efficiency of traditional huller units is hard to measure and may be undermined because there are several useful by-products which are not separated as in modern mills. The by-products are also not owned because the mills are run under the custom hiring basis. Inefficient traditional hullers are more commonly observed to be discharging the function of milling rice. With custom milling as the practice, the huller units remain very important for the poor farmers even while modern mills have the advantages of large scale and commercial processing. Though traditional and modern mills perform comparably in terms of net revenue, paddy to rice outturn ratio, and cost of production, modernization of the paddy processing mills is advocated for improving the milling ratio, poverty alleviation, employment generation, and entrepreneurship but more training programmes can help in meeting the need for skilled labour.

Agro-processing industry in India consisting largely of tiny and small informal enterprises provide alternative employment to large number of people but the productivity of these units is very low. Constraints to increasing the productivity of agro-processing units in West Bengal, Bihar, and Maharashtra can be overcome through removing infrastructural bottlenecks, ensuring greater availability of raw material, establishment of information Centres, dissemination of technical know-how, and various government schemes meant for promoting agro processing activities. Further, co-operatives and self-help groups need to be strengthened and made instrumental to ensure the supply of raw materials in time at reasonable prices and facilitate marketing of the produces. Further, in West Bengal, fixing prices of the products in advance because of variable prices of raw materials deterred these units from entering into forward contract to sell their products at reasonable prices and hence ensuring marketability of their products. In Bihar, a large scope exists for expanding livestock-based processing activity (milk processing) and for recuperating the handloom and power loom industry specifically for expanding *tussar* and silk units. Setting up of Certification Centres and Laboratories to help in quality standardization of the produces, exhibitions for wide publicity of the products, progressing with technological and infrastructural backup using the PPP model are some of the other measures suggested particularly for the growth of industries in Bihar. In Maharashtra, establishment of co-operative marketing for cashews and fish units was suggested to reduce excessive reliance on agents for marketing of the produce and procurement of raw material. The non-food units mainly reported non availability of labour, absence of governmental support and existence of rivalry as the main problems. Online trading for cashews and aquaculture for supply of fish throughout the year were also suggested to be promoted. Hence, for overcoming the constraints and improving the prospects of agro-processing industries, coordinated efforts among different departments of the government as well as amongst government and non-government agencies are required.

Among non-food agro-processing, silk weaving is a sustainable form of farm-based economic enterprise that supports the rural poor operating in an unorganized sector. Assam is enriched with four varieties of silk products but the average productivity of handloom is very poor. A study on mulberry and *muga* silks weavers of Sualkuchi, known as the *Silk Village of Assam*, suggested that high cost of raw materials which have to be imported from other states was a serious problem of these silk weavers. High prices of real muga yarn and competition from dyed tussar yarn in Muga colour also affected the economics. Dependence on migrant weavers who demand higher wages was also a weakness because of declining trend in the number of local weavers. Therefore, the authors argue that a technological revolution in the age old silk industry of Sualkuchi is needed through the introduction of semi-automatic handlooms and simple power-looms, use of jacquard machine, and computers together with establishment of a research cum training centre for in-plant training. Combining traditional designs with new ideas for making the design patterns more appealing, co-operative societies to ensure that the hired weavers get proper wages and remunerative returns are also suggested. The establishment of a textile museum at Sualkuchi will not only keep alive the unique tradition and culture of silk industry of Assam but will also attract tourists will help in increasing the income of the farmers is suggested.

A persistent 'fodder scarcity' is a challenge in India—a country where dairying is largely the vocation of the poor, especially the poor women. An understanding of the economics of fodder crops is therefore equally important as that of main crops and livestock. The problems faced by fodder growers were noted to be in many ways different across states, Gujarat, Madhya Pradesh, Karnataka, and Punjab, but the biggest problem common faced in all the states was lack of technical knowledge. Other common problems were inadequate and inferior quality of seeds, high expenditure in production, poor access to credit, and labour availability along with non-availability of market information in time. Among the specificities, fodder markets in Gujarat are found to be highly unorganized and unregulated with dry fodder mainly seen as a by-product from cereal crops. In Madhya Pradesh, the progress of fodder cultivation is clearly inadequate. In Punjab, due to heavy pressure of growing wheat and paddy, the area under fodder has been decreasing but the state has abundant roughage derived from wheat and rice crops, which can be used in making silage through processes developed and recommended by Punjab Agricultural University, Ludhiana. Recommendations include organizing fodder banks for Gujarat based on the surplus production that is generated in normal years, regeneration of wastelands, cultivation of fodder trees on marginal land and degraded forest areas, organizational and financial support by government.

An efficient agricultural marketing system is seen as an important means for raising the income level of farmers and for promoting the economic development of a country. An evaluation of spatial price integration of each of the six agricultural commodities; rice, maize, chick pea, pigeon pea, cotton, and groundnut; among 5 major markets in Karnataka under Karnataka State Agricultural Marketing

Board (KSAMB) suggested that there is a degree of pricing inefficiency with regard to the smaller markets of each of the commodities, while the major markets are by and large price efficient. The markets for major cash crops, namely, cotton and groundnut were found better integrated whereas, most of the markets for cereals and pulses that is, for rice, maize, chickpea, and pigeon pea were functioning independent of one another but the distance between two markets was not an important criterion for integration. However, there is a need to improve both the transportation and other infrastructure facilities in these markets and perhaps improve the competition so that price integration is facilitated. Volatility in the markets also undermined the production of these commodities and the operation of the price stabilization fund could be streamlined to ensure stability in prices.

An analysis of integration with international markets and trade competitiveness of Indian agricultural commodities indicated limited trade competitiveness and a small influence of major wholesale markets on international prices. Integration between the domestic markets was observed only in few commodities. Factors affecting trade competitiveness of Indian commodities include future markets, currency exchange fluctuations, low productivity, and poor post-harvest handling. Issues of protection or insulation in many developed countries and allowable export subsidies and domestic support, high tariffs and non-tariff barriers also need to be addressed in future negotiations. A comprehensive plan and strategy for exports accounting for huge domestic demand, modernization of agricultural marketing system, and introduction of information and communications technology (ICT) will help in improving trade competiveness of Indian agricultural commodities.

Market interventions by the government play an important role in protecting the farmers from the volatilities in market prices. An evaluation of MIS for apples in Uttarakhand and Price Support Scheme (PSS) for sunflowers in Haryana found that MSP and PSS will play a major role in improving the cropping pattern, area, production, and yield if measures like crop insurance, credit supply at reasonable rates of interest, modern mills, and equipment for crushing oilseed are developed. Unlike other edible oils, sunflower with no surety of the market price and being susceptible to weather and pests, is not preferred in Haryana and no impact of the two years' intervention was found due to its low coverage. However, in Uttarakhand, the intervention of Mother Dairy and private players like Reliance, Birla, Chirag, Shree Jagdamba Samiti, etc., expanded the horticultural produce market beyond local consumers to distant markets. Like in Haryana, Uttarakhand needs to improve its number of regulated markets, storage, transportation, packing, and processing.

Marketing in agriculture is not limited to field crops alone. Livestock keeping is necessary to compensate for the decreasing income from crops. Rearing of goats provides livelihood and subsistence to millions of small, marginal farmers, and landless farmers across the country because investment and maintenance costs incurred are very low and the gestation period is short. An analysis of market access and constraints in marketing of goats and their products in four states across the country, found that the more numbers of goats were sold at doorstep than in the markets due to the dominance of traders in the markets and the long distances that divide home from the market. Local consumers and hotels were major purchasers of the final products. Local consumers account for over 80 per cent of the purchases. The marketing system of goats was totally dominated by butchers and professional traders across the selected markets of the states with a stronghold of informal financial institutions at the grassroots level. The markets of Uttar Pradesh, West Bengal, and Madhya Pradesh should therefore be governed by APMC Act as in Maharashtra. Infrastructural facilities, basic amenities, communication network, etc., need to be modernized and the credit and delivery system rejuvenated. Weighing machine and slaughter houses have to be modernized for hygienic production of goat meat at par with global standards to facilitate exports. The skin markets should also be regularized to protect skin traders from exploitation of bigger merchants. Better facility for finance, technical expertise, and market intelligence are also necessary to make goat rearing profitable.

New channels of India's agricultural marketing emerging in India did not always involve organized companies in the chain and in some cases even coalesced with the traditional channel but they were always shorter and bypassed the first link which is usually the commission agent. The reforms in agro-marketing have been in process in almost all states, though political will varied in terms of legislation. The new channels provided more options in marketing and cut down marketing costs. They are associated with higher productivity, profit, and returns from farming but when operated by organized companies, they tend to draw participation of the larger farmers indicating implications for disparity. The APMC market still sets the benchmark in pricing and traditional markets, facing competition, are also undergoing transitions. While the emerging channels appear to deliver greater efficiency, reforms accompanied by the strengthening of regulated marketing will also help to provide farmers with options, reduce crop wastage, and support the pricing mechanism, now under challenge.

However, due to the advent of fresh food supermarket retailing and supply chains in India, traditional retailers suffered 20–30 per cent decline in sales. These losses can be reduced through improvement of the functioning of APMC markets, regulated zoning for freeing up the residential localities of cities of supermarket chain outlets, market information and extension to farmers, helping the wholesale sector adjust to the presence of supermarkets, facilitate multi-partite relationships and planning urban development to integrate the interests and concerns of multiple stakeholders.

Another channel of marketing, the producer companies (PCs) which became a relatively new legal entity in India since 2003 to empower the marginal farmers and producers and to improve their bargaining power, net incomes, and quality of life have either been making huge losses or very low profits, enabling members in some ways, like employment generation, income rise, higher market price, dividends, fair and prompt payments and gain of self-respect, and identity for small producers. The PCs are product-focused rather than farmer-focused. Major hurdles for PCs include getting registration and the digital signatures of board of directors who are themselves small farmers and often illiterate villagers with no identity proofs. Being commercial entities, unlike cooperatives, they are not entitled to public grants and have to gain access to the capital market and pay taxes on income. Governmental support and exemption from corporate tax during the initial few years, inclusion of financing agencies on the board of a PC, collateral-free loans, and treatment as a Non-banking Financial Company (NBFC) to provide loans to farmer can help these companies.

Agricultural market reforms through the intervention of Mother Dairy and other such players have shifted area from cereal crops to horticultural crops in both Haryana as well as Uttarakhand although these states are totally divergent in topography, agro-ecology, and production practices. These changes in cropping should be helpful to enhance farmers' income, generate employment opportunities, and increase availability of nutritious food to the consumers. Promotion of horticultural crops at least in states such as Uttarakhand should find some extra support as it most suits the conditions though some concerns about food security does find mention because the demand for cereals in the case of Uttarakhand is being met by buying from the market. Upgrade of marketing facilities, roads, transportation, storage, and onsite processing facility are also recommended. Efforts to increase yield using improved technology and a PPP model are seen as the need of the hour.

As far as the role of establishment of Agri-Export Zones is concerned, on the basis of data from the chickpea growers, village merchants, wholesalers, and processors in the two districts Vidisha and Narsinghpur in Madhya Pradesh, surprisingly no change is observed in developmental activities or employment opportunities even after the declaration of districts under Agri-Export Zones in Madhya Pradesh. To enhance the role of these zones, the steps needed to be initiated include; expansion of area under irrigation, development of water-shed, promotion of varieties resistant to insects, pests and climatic stresses as also the spread of improved technology, involvement of corporate sector and self-help groups in accelerating production, processing, and marketing to meet the global challenges of forward marketing.

Section IV: Incentives, Agricultural Prices, and Food Security

The role of price, food procurement, and distribution and trade policies in promoting agricultural development is important. Agricultural price and trade policies influence the economic incentives necessary to encourage greater efficiency of resource use. There is also increasing recognition that relative price movements create opportunities for institutional change and that institutional innovations cannot be viable unless the economic benefits to individuals or groups in society exceed the costs (Hayami and Ruttan 1985). Beyond the production effects, changes in relative agricultural prices, especially of staple food, have significant implications for income distribution in low-income countries (Mellor 1978). India's experience demonstrates the importance of agricultural price policy in addition to improved technologies, institutional reforms, rural development, and other policies that increase food availability. Technology is a major driving force behind agricultural development. Introduction of HYVs of rice and wheat in the mid-1960s, increased crop yields significantly and helped in achieving self-sufficiency in foodgrain production in the country. Institutional arrangements and other food policies such as agricultural price policy, food procurement and distribution, inputs subsidies, rural development programmes, are also important determinants of India's food production and availability. This theme contains six chapters ranging from food policy transition issues to impact of subsidies and rural development programme on Indian agriculture. The summary of the chapters included in this theme is presented further.

The chapter titled Food Policy Transition in India analyses India's journey towards free food market by hypothesizing the counterfactual of the exit of the State in the recent decades. The chapter examines whether over the period of economic reforms public operations in procurement, distribution, and stocking have grown and whether the food market has become more open to global demand and supply by offering an increasing share of domestic production for international trading. The role of MSP, the main tool for administering food price, in shaping market prices and in deciding the volumes traded in the free market vis-à-vis the administered market is also examined using econometric regressions. The findings suggest that public operations in food market remain unabated, the market has become more open and market prices are coming close to MSPs. The price relative to MSP was higher for rice than wheat but, over time, they converged and while both rice and wheat competed for a place in public channel and trade the market prices and the administered price are currently moving together. Prices of both rice and wheat are strongly influenced by their MSPs. In the case of rice public procurement is strongly determined by MSP and by a public–private competition but production rather than MSP is the driver in the case of wheat. The open-ended procurement policy makes wheat procurement, made mostly in surplus northern states, more responsive to the volume of the production whereas the administered price is a useful tool for directing rice between the two channels.

Over time the share of public sector in food market has not shown any sign of contracting but the market became more open in the international market. Rice and wheat, the two staples of Indian population have gained relative significance in public operations and trade in succession. While the Indian food market is now more open in the international market internally, the traditional food security concerns and misgivings about markets still weigh on the government, sustaining the policy reliant on price administration and public procurement created decades ago. In the current paradigm India needs to do more to work out a mechanism that is responsive to supply and demand forces in the market with a medium term purview for safeguards against short term production fluctuations. A policy for ensuring food security in the current paradigm needs to be less reliant on administration by the state on a regular basis.

Another chapter in this section deals with Food Insecurity and Vulnerability. This on in Himachal Pradesh attempts to identify the farming groups vulnerable to food insecurity and quantify the extent of food insecurity in terms of hunger gap, and coping mechanism adopted by food insecure and vulnerable groups to mitigate the food insecurity. The role of government intervention through policy and programmes for providing food security net to vulnerable groups is also analysed. The study found that there is a scope for cultivation of fruit and vegetables including off-season vegetables in the State but due to lack of irrigation facilities farm families are unable to diversify cropping pattern at the desired extent. Therefore, for alleviating hunger and poverty, it is important to support more public investment in rainfed and backward hilly areas. In hilly topography, there exists a scope for augmenting irrigation facilities through lift and flow irrigation schemes. In order to be effective, the food security policy must evolve as a basic element of a social security policy with proper coordination among the various government departments, private sector, and NGOs. The direct food and nutrition support for the poor through a minimum safety-net should be properly balanced with improvements in the quality of life of local people through investments in education, drinking water, sanitation, and health care. In some categories of households, quantity of cereals supplied through PDS is higher as compared to the quantity produced at home, which indicates that reliance on PDS is increasing steadily leading to neglect of farm production. This is eroding the very production base and pushing local agriculture towards un-sustainability. Considering the overwhelming importance of the rural sector, additional emphasis has to be placed on rural development and non-farm activities to increase income and employment as demand for agricultural goods slows down. The study makes a strong case for a change in the food management policy and, therefore, in the overall agricultural strategy.

Still another chapter on Public Policies and Sustainable Agricultural Development is a case study on commercialized agriculture. It examines changes in public policies towards agriculture and their likely effect on the prices of agricultural commodities; extent of sustainability of agricultural growth in the Northwest India; factors responsible for sustainability; and evaluates responses of alternative policies, such as, price, technology, and institution on various elements of sustainability in a farming system framework. The results indicate that rationalization of farm input and output prices has resulted in decrease of area under paddy and wheat crops and increase in area under maize and potato-based crop rotation. The above reallocation has resulted in loss of farm return due to increase in the cost of agriculture; though there was improvement use of natural resource especially water. The findings also highlighted that too much of emphasis on price or market-based instruments may not encourage sustainable agricultural development; this has to be supplemented with the technological innovations. In the study area, the Integrated Pest and Nutrient Management appeared to be an important option to counter the existing stress on natural resources and simultaneously increase farm income. The chapter suggests that merely rationalization of price would not lead to a sustained growth in agriculture. The sustained growth in fact requires suitable technology, adequate infrastructure facilities, and a set of institutions. The desired institution is not only in the sense of aggregator or facilitator to stem out problems associated with the small and scattered production of agriculture but also effective-enforcer of State laws as in the case of sale of spurious seeds. Institution is also desired in the form of interventionist policies like limiting the use of groundwater aquifer.

The next chapter of this section analyses impact of Diesel/Power Subsidy Withdrawal on Production Cost of Important Crops in Punjab. It was found, after undertaking various simulations regarding change in diesel prices keeping prices of all other inputs at constant level (ceteris-paribus), that the cost of production of paddy increased by 7.06 per cent with the withdrawal of diesel subsidy. Similarly, the increase in cost of production of basmati was by 5.18 per cent; while in other crops it was 3.52 per cent in cotton, 2.63 per cent in sugarcane, 5.83 per cent in maize, 7.90 per cent in wheat, and 8.81 per cent in sunflower. The increase in the cost of production of different crops under various farm categories due to withdrawal of diesel subsidy did not show any specific trend of increase or decline according to size of the farm category. The major impact of power subsidy withdrawal was increase in cost of production of paddy by 25.30 per cent crop due to more number of irrigations applied to this crop; basmati by 21.24 per cent, sunflower by 9.07 per cent, wheat by 6.64 per cent, maize by 3.50 per cent, sugarcane by 3.28 per cent, and cotton by 1.75 per cent. The impact of power subsidy withdrawal was more on semi-medium, medium, and large farm categories as compared to marginal and small farms. The farm category-wise analysis showed that the impact of power and diesel subsidy withdrawal would be more on large and medium farmers as compared to their smaller counterparts. The findings suggest that that Punjab government should emphasize on increase in the minimum support price of paddy and wheat in commensurate with the diesel price hike coefficient. For other crops also, MSP should be enhanced in proportionate to the diesel price hike coefficient, for which MSP is announced but is not actually implemented. In case, power subsidy is withdrawn by the state government, farmers especially marginal and small ones should be compensated according to the electricity usage bill generated for irrigating various crops on their farms. Thus, for keeping marginal and small farmers in farming business, subsidies especially power subsidy should not be withdrawn, however, their form can be changed for the benefit of these farmers in general and farming community in particular.

The chapter, Impact of Mahatma Gandhi National Rural Employment Guarantee Act (MGNREGA) on Wage Rate Food Security and Rural Urban Migration, measures the extent of manpower employment generated under MGNREGA, their various socio-economic characteristics and gender variability in implementing MGNREGA in 16 states. The chapter also analyses effect of MGNREGA on the pattern of migration from rural to urban areas, nature of assets created under MGNREGA, identifies factors determining the participation of people in the MGNREGA scheme and assesses implementation of MGNREGA, its functioning and suggests suitable policy measures to further strengthen the programme. The study found that MGNREGA has not been successful in providing stipulated 100 days employment to all the registered persons. The reasons expressed by the Panchayat and district officials were many, including lack of funds; money not being provided from the Central authorities on time; the gap with which money reaches to the Panchayat officials; and money being provided only for few months and not the whole year. The results of the household survey clearly indicate that unless participants are given work for the stipulated 100 days, MGNREGA shall not be able to make any significant dent on the rural poverty and would fail in its basic objective. Therefore, provision of 100 days employment to all the participants should be made mandatory and strict action should be taken against the Panchayats which fail in fulfilling this target. The issue of timely provision of money to the Panchayats should be looked into so that MGNREGA work does not suffer because of lack of funds with the Panchayats. The study also found big anomaly in the wage rate paid under MGNREGA. Under the MGNREGA, Panchayats are ordained to pay at least equal to the minimum wage determined for the state during a particular period. However, the actual wages paid under MGNREGA were found much lower. The village analysis revealed a conflicting interest between the MGNREGA and the farming community. Farmers across the board are feeling that they are facing labour shortage for agricultural activities because of the diversion of labour caused by MGNREGA activities. A meticulous planning can solve this problem without affecting anyone adversely. There is a need to plan the MGNREGA work at the Panchayat level in such a way that it does not clash with the sowing and harvesting season in agriculture when the demand for agriculture labour is highest. This planning has to be done at the Panchayat/Block and District level, depending upon the cropping pattern of the respective regions. It not only would provide necessary labour force for agricultural operations but also would increase employment and income opportunities for the villagers during the off-season including that of marginal and small farmers who do not have enough work at the farm in the off-season. Proper punishment system should be put up in place for the unscrupulous officials who are found guilty of indulging in corruption and other untoward activities. Similarly, those Gram Panchayats that work efficiently in running the MGNREGA system should be rewarded and felicitated appropriately. The provision of foodgrain at the work place and easy institutional credit can attract more villagers, especially the poor ones towards working in MGNREGA and also ensures better food security to the participants. The study suggests that Unique Identification (UID) should be used for the better functioning of MGNREGA (Anderson et al. 2013). Bank accounts for MGNREGA workers will be linked to the unique biometric ID. As a result, the actual transfer of payments will immediately reach the hands of who it is intended for. This would drastically reduce the alleged inherent corruption in the current system and increase the amounts and reliability of payments to the workers.

Important in this section is another chapter based on a study namely An Evaluation Study of Prime Minister's Rehabilitation Package for Farmers in Suicide-Prone Districts of Andhra Pradesh, Karnataka, Kerala and Maharashtra covers 12 districts from four states, namely, Andhra Pradesh, Karnataka, Kerala, and Maharashtra. Some state governments are implementing parallel programme along with the Prime Minister's Rehabilitation Package (PMRP) to assist distressed farmers in the suicide-prone districts (with identical components). For example, few states have resorted to waiver of loans from cooperative societies, supplying seed with subsidy, promoting micro-irrigation and horticulture with subsidy, construction of farm-ponds, rain water harvesting, etc. The results indicate that parallel implementation of state schemes along with the PM package creates confusion among the beneficiaries. However, it was observed that most of the farmers in the suicide-prone districts benefited from one or the other scheme. Some farmers received benefits from two schemes and a few others from more than even three schemes also. The average amount of ex gratia received by the beneficiaries varied not only across the states but also within the states. Collusion between local leaders and officials is also evident as some well-off farmers got the ex gratia payments whereas poor households were left out. Overdue credit was rescheduled and interest was waived off under the credit component of the PMRP. However, it was observed that many beneficiaries were not aware about the quantum of interest waived off or the yearly instalment of principal they have to repay. Very few borrowers opted for fresh loans. Distribution of certified seeds with 50 per cent subsidy in the identified districts is appreciated by all the farmers. This is one of the important components of the PM's package which helped farmers immensely. Under the micro irrigation scheme, sprinkler and drip irrigation sets are supplied to the farmers at 35 to 50 per cent subsidy with the exception of Andhra Pradesh where subsidy accounted for 90 per cent of the cost of the equipment. It was found that the subsidy amount received by the beneficiaries varied across the districts (Karnataka) as well as within the districts among the beneficiaries (Kerala). There is a need to provide clear-cut guidelines on the subsidy component, so that all the beneficiaries would get the same benefit without any room for leakages or corruption. All the beneficiaries reported that the adoption of micro irrigation system has helped in efficient use of irrigation water leading to modest expansion in the irrigated area. Agriculture Technology Management Agencies (ATMA) is identified as the nodal agency in all the districts to ensure extension support and convergence at district level. A few farmers benefited from the activities identified under extension component in four states. Close monitoring of extension activities is necessary to achieve the desired goals.

There is a provision in the package to provide assistance for feed and fodder for one year for milch animals and also support for rearing a calf. However, none of the respondents has received assistance for feed and fodder (Karnataka). Periodic monitoring and surprise checks by the competent authorities are necessary to control the pilferage of funds. The full potential of milk yield could not be attained in Maharashtra due to heat stress and poor availability of fodder. It was felt that local breeds may be more suitable as they can sustain the heat and require less fodder. Small and marginal farmers benefited the least from the fisheries programme as the initial investment for development of fishponds is very high and the subsidy component is only 40 per cent. Availability of quality seed and infrastructure (availability of ice, transport, markets) are major constraints faced by the beneficiaries in the identified districts. Horticulture has emerged as a sunrise sector having the potential to accelerate the growth of our agrarian economy. Most of the beneficiaries availed subsidies for purchase of micro irrigation equipment, plantation of fruit crops and rejuvenation of old orchards. The benefits from development of minor irrigation are shared by the community as a whole. In Andhra Pradesh, work on minor irrigation projects is not initiated in any of the identified districts. In Karnataka and Kerala, most of the works under the minor irrigation were in various stages of completion, and therefore the actual benefits (expansion of irrigated area, quantity and quality of irrigation water supply, etc.), from these programmes cannot be ascertained at this point of time. It was observed that the quality of civil work done in the rejuvenation/rehabilitation of tanks under the PM's package in Karnataka was of poor quality and the farmers have complained to the authorities. It is suggested that the Gram Panchayat should have the authority to check and supervise the work related to minor irrigation under their jurisdiction.

Accelerated Irrigation Development Programme, Watershed Development Programme, and Micro Irrigation are aimed at increasing the area under irrigation. These schemes did increase the area under irrigation and improved the productivity of all major crops cultivated in the districts. Thus, irrigation projects have to be completed and watershed activities have to be further promoted to recharge groundwater. The farmers will be in a position to face the droughts only when protective irrigation is available. The study shows that farmers benefited from various components of PMRP such as interest waiver, rescheduling of loans and subsidy given under various schemes which enabled them to be eligible for fresh loans and also helped them augment their incomes through subsidiary activities. However, their capacity to cope with drought conditions whenever monsoons failed was still weak and the PM's package through its multiple schemes had a limited impact on this front.

Section V: Constraints and Emerging Solutions

This theme tries to put up major constraints as discussed in the above four themes and tries to provide some emerging solutions to the same. The chapters which speak on the existing constraints include the constraint of climate change; stagnating productivity in agriculture; proper management of agricultural resources and participation of women in agriculture. Some of the brighter side of Indian agriculture likely to emerge in the coming future could be in terms of spread of organic farming at a larger scale; the advent of green revolution in the Eastern India and emerging diversification in rural India. The theme includes total 12 chapters in this section which cover wider areas and topics not touched in the previous four themes. The summary of the chapters covered in this theme is presented below.

The Food and Agriculture Organization of the United Nations (FAO), in collaboration with the International Fund for Agricultural Development (IFAD) has launched a regional programme on 'Pro-poor policy formulation, dialogue and implementation' in the selected countries of Asia-Pacific regions including India. This chapter identified three thematic areas of research for analysing pro-poor policies in India. These were Managing Common Pool Resources (CPR) of Non-Timber Forests Produces (NTFP) in Tribal Areas and small scale culture fisheries, increasing Rural Non-Farm Employment of farm households, Developing Infrastructure for Agricultural Growth and Poverty Alleviation. Findings of these studies and approaches to these policies for formulation of pro-poor policies are discussed in the chapter.

To bring about all-round development of agriculture, a scheme called 'Macro Management of Agriculture' (MMA) became operational from 2001 in all the states and union territories (UTs) by integrating the existing 27 centrally sponsored schemes of agriculture and its related activities. MMA is a significant scheme introduced for the development of the agricultural sector in the country. This scheme has brought about tangible benefits for farmers, especially for the lower-rung farmers. For making the agricultural development programmes much more successful in the states, institutional and infrastructural supports need to be developed. Also, efficient planning, monitoring, and sincere execution of the policies by the government agencies are essential to make the schemes viable and successful. As the success and development of the MMA scheme depends on the involvement of the targeted farmers, more attention should be given to the participation and training of the lower-rung farmers. For disseminating advanced technology to the grassroots farmers under this scheme, their participation in the training provided by the agencies is essential. Unfortunately, the rate of farmers' participation under the scheme is very poor. Proper information has to be disseminated to the farmers and they should be mobilized on time. The seed procurement and the supply of agricultural implements have not met the targets of the scheme satisfactorily. Similarly, most of the implements are received by the higher-rung farmers. More attention needs to be given to the distribution system and delivery mechanism of the scheme.

To discharge the nation's fundamental obligations of food security and sustainable livelihoods adaptation to climatic events needs to be embedded in agricultural policy. Also as a responsible member of the global community, suitable mitigation initiatives through agriculture, compatible with the same obligations, need to be critically and cautiously appraised, taking into view that mitigation can go hand in hand with conservation of resources and sustainable development. Planning and financing of adaptation will reduce farm distress and ensure food sufficiency. Cautious planning of rotations, crops, and fallows, use of cover crops, and mulching can make farming conducive to carbon sequestration turning farm lands to sinks. Mitigation and adaptation, if pursued with prudence, will create new sectors with economic opportunity. India's future agricultural policy needs to be designed seeking balances and synergies among multiple objectives like mitigation, adaptation, food security, sustainability, and livelihood.

The structural changes require the increasing role of non-farm sector in rural employment and income. The review of studies on rural diversification shows that employment in trade, transport, and business services has increased in the recent period. These activities are dominated by private players. The rural non-farm economy now accounts for more than 60 per cent of rural income though it accounts for only 32 per cent of rural workforce in country. Now construction, transport, trade, and services lead the rural non-farm sector. Women account for around 30 per cent of rural workforce. The MGNREGS has helped in increasing rural wages for unskilled workers and also female workers in rural sector. The rural wages for skilled workers continue to decline in rural sector.

Crop insurance, despite its significance for Indian agriculture and the government's active support has proved to be a losing proposition even after several years of the launch of the national level scheme. The reasons would probably be poor risk pooling due to inadequate participation linked to the design. The design of the scheme suffers drawbacks. Besides the uneven ground created by the rigidly constrained uniform prices of insurance for cereal and oilseeds crops despite their vulnerability to risk, the regressive characters of the threshold yield (TY) formulation is a serious weakness. The TY that in principle is linked to the actually revealed uncertainties in nature also serves little purpose by failing to recognize that progressive farmers have to invest to achieve higher yields than experienced before and would take coverage of insurance with the hope that their probable losses will be realistically indemnified. Indeed, it appears that the applied formula could be fixing thresholds at levels way too low to make any meaning for the farmer to pay a price for the insurance. Indeed, they pay for coverage merely mandated for credit.

The programme of Bringing Green Revolution in Eastern India (BGREI) was launched in the year 2010–11 to enhance the agriculture production in the states of Assam, Bihar, Chhattisgarh, Jharkhand, Odisha, Eastern UP, and West Bengal. But there are certain gaps in varying extents between recommended, promoted and implemented strategies across different states due to lack of uniformity in input package and mode of implementation across the states. Though it seems too early to conclude strongly as to the definite impact of the programme nonetheless there are signs towards a positive change. The BGREI programme, as conceived, addressed towards increasing the yield rather than the cropping intensity. Hence, the impact of intervention under block demonstration programmes under BGREI is more prominent in increasing the yield rates for the beneficiary farms as compared to non-beneficiaries. In case of technical backstopping, the scientists of State Agricultural Universities (SAUs), Krishi Vigyan Kendras (KVKs) & Indian Council of Agriculture Researdh (ICAR) (ICAR-SAU system) were identified for providing technical support to the BGREI beneficiaries during 2011–12. Through a regular contact technology dissemination had been quite successful in the BGREI states. Efforts should be made to reduce the gaps between recommended, promoted, and implemented strategies. In course of dissemination of technology, provision of progressive farmers and regular monitoring from State agriculture departments can play vital role. As such, such links between the beneficiaries and State machineries should be encouraged. Interventions through crop demonstrations has helped decline the gap between ecology specific potential and actual yields across beneficiary farms. Hence, such demonstration programmes should be encouraged. Eastern India covered under the BGREI programme has exhibited a glimpse of a high potential for yield enhancement of rice, wheat and rabi pulses through a favourable positive crop response. There is a huge scope to exploit this potential through scientific and technological intervention like BGREI, and hence the programme should continue with greater effort and coordination. An all-round effort should be made to ensure the timeliness of input delivery system prescribed under the recommended technology.

Setting up of organic input units with capital investment subsidy is one of major component under 'National Project on Organic Farming (NPOF)' for encouraging the organic inputs production since 2004. Availability of quality organic inputs is critical for success of organic farming in India. Majority of the sample promoters did not face any problem in establishment of vermi-hatchery units. Very few expressed some difficulties while establishing them. The major problems are: non-availability of quality worms in the vicinity, lack of sufficient raw materials, wild boar attacks on compost units, no proper guidance from NABARD, heavy rains and delay in release of bank loan amounts, etc. The Ministry of Agriculture should introduce favourable governmental policies and strategies for the promotion of organic farming in India. A single authority at national level with a well-defined role should be responsible for the organic sector in the country. The quality organic input production (compost, bio-fertilizers, and bio-pesticides) in the country should be further encouraged with latest technologies and improved way of financial assistance so as to reduce the high dependency on inorganic fertilizers in a phase manner and to save our domestic subsidies. Establishment of organic input marketing channels is the need of the hour for expansion of organic farming in the country.

Jhum cultivation is a primitive method of farming which is practiced by the tribal by slashing and burning the vegetation and then keeping it abandoned for gaining soil fertility during which the jhumias shift their homestead in search of new plot of land for cultivation. The area under jhum cultivation was higher than that of settled cultivation in Mizoram and Meghalaya states of northeast India. Jhumias raise their food crops in the jhum field and the commercial cash crops were raised in the settled cultivation. It was less productive and less remunerative than settled cultivation. It is a continuous socio economic process linked up with religion of the tribals. Jhum cycle was affected mainly due to paucity of land and high population growth. The system of jhuming has not been abolished so far after so much of government plans and programmes. Jhumias must be attracted towards the government programmes which are alternative to jhuming so that these are feasible and acceptable without affecting their religious rites, quality of life of the jhumias can be improved through various measures like improvement of land tenure system, use of common property resources, market development, introduction of small entrepreneurship, food and health-care facilities, good communications, etc.

The basic objective of Tribal Development Authorities is to encourage the tribal women in their income generating non-agricultural activities by providing them various aids by the concerned department. Creation of basic infrastructural facilities such as connecting roads, bridges, transport, marketing facility, electricity, etc., are considered as most essential for development of a region. Investment strategy should be broad-based and the assets and benefits provided by the different departments should be sufficient enough to generate income and employment to help the women in raising their standard of living. Sometimes, fair and exhibitions, etc., should be arranged to promote local products through display or advertisement, as the consumers are less aware of locally manufactured products. This will give an opportunity to the consumers to assess the local products in terms of quality and value. The benefits under the Government sponsored schemes should be offered to economically weaker families and needy lot. Training programme should be arranged in rural areas to encourage the women to adopt modern system of weaving. Training should be of short duration, because women in rural areas cannot stay away from home for a long period because of other commitments. Adoption of modern technologies, especially use of sophisticated tools and equipment as well as machinery in weaving requires reasonable amount of money. Lack of finance in case of the poor weaver families should be mitigated by providing institutional finance to the willing weaver families of the study area.

REFERENCES

Acharya, S.S. 2001. 'Domestic Agricultural Marketing; Policies, Incentives and Integration, Indian Agricultural Policy at the crossroads'. Acharya, S.S. and D.P. Chaudhri (ed.), Rawat.

Anderson, Siwan, Ashok Kotwal, Ashwini Kulkarni, and Bharat Ramaswami. 2013. 'Measuring the Impacts of Linking NREGA Payments to UID'. Working Paper, International Growth Centre, London School of Economics and Political Science, London.

Bawcutt, D.E. 1996. 'Agricultural marketing in a highly competitive and customer responsive food chain-experience in United Kingdom'. *Indian Journal of Agricultural Marketing*, 10(3): 10, September-December.

Bumb, B. 1995. 'Global Fertilizer Perspective, 1980-2000: The Challenges in Structural Transformation'. *Technical Bulletin T-42*. Muscle Shoals, AL: International Fertilizer Development Center.

CSO. 2014. 'National Accounts Statistics 2014'. Central Statistics Office (CSO), Ministry of Statistics and Programme Implementation, Govt. of India, New Delhi, May 2014.

FAO. 1998. *Guide to Efficient Plant Nutrition Management*. Rome: FAO/AGL Publication, FAO.

Ganesh Kumar, A. et al. 2006. 'Reforms in Indian Agro-processing and Agriculture Sectors in the context of Unilateral and Multilateral Trade Agreements'. Report submitted to the Economic Research Service/United States Department of Agriculture, Washington, D.C. WP-2006-011, Indira Gandhi Institute of Development Research, Mumbai, http://www.igidr.ac.in/~agk/AGK_Publications.htm

GoI. 2015. 'Third Advance Estimates of Food Grains, Oilseeds & Other Commercial Crops for 2014–15'. Directorate of Economics & Statistics, Department of Agriculture and Cooperation, Ministry of Agriculture, Government of India, New Delhi, May 2015.

Government of Karnataka. 1993. Stagnation of Agricultural Productivity in Karnataka during 1980's.

Gulati, A. and Tim Kelly. 1994. 'Trade Liberalization and Indian Agriculture'. Oxford University Press, with Anil Sharma: Agriculture under GATT, EPW July 1994; write ups and TV discussions.

Hayami and Vernon W. Ruttan. 1985. *Agricultural Development: An. International Perspective*. Johns Hopkins University Press.

Hopper, W. 1993. 'Indian Agriculture and Fertilizer: An Outsider's Observations'. Keynote address to the FAI Seminar on Emerging Scenario in Fertilizer and Agriculture: Global Dimensions, The Fertilizer Association of India, New Delhi.

Jha, Shikha and Srinivasan, P.V. 2001. 'Taking the PDS to the Poor: Directions for Further Reform'. *Economic and Political Weekly*, September 29.

Kumar, Parmod and Dipanwita Chakraborthy. 2016. *MGNREGA: Employment, Wages and Migration in Rural India.* London and New York: Routledge-Taylor & Francis.

Kumar, Parmod. 2014. *Demand and Supply of Agricultural Commodities in India.* New Delhi: Macmillan Publishers India Ltd.

Mellor, John W. 1978. T*he New Economics of Growth –A Strategy for India and the Developing World.* A twentieth Century Fund Study, Cornell University Press, Ithaca.

Mulky, Avinash. 2008. 'Enhancing Marketing Performance: Academic Perspective'. *IIMB Management review*, 20(4).

Nayyar, D and Sen, A. 1994. 'Agriculture under trade policy regime'. EPW, Vol. 28.

Reddy, Y.V. 2001. 'Advantages of Processing', Address at Conference of Indian Society of Agriculture Marketing at Vizag on February 3.

Sharma V.P. and Dhinesh Jain. 2011. 'High Value Agriculture in India'. Working Paper Number, Indian Institute of Management, Ahmedabad.

Sivanappan R.K. 1994. 'Prospects of micro-irrigation in India: Irrigation and Drainage Systems', 8(1): 49–58.

Tomich, T., P. Kilby, and B. Johnson. 1995. *Transforming Agrarian Economies: Opportunities Seized, Opportunities Missed.* Ithaca, NY: Cornell University Press.

Section I
Overview and Performance of Indian Agriculture

2

Growth and Instability in World Pulse Production and Trade

Usha Tuteja

Economists have extensively investigated growth in production and trade of rice and wheat during the past four decades. It has been widely researched at global, regional, national, state, and household levels. Nonetheless, scant attention has been paid to the study of pulse crops, which play an important role in sustaining crop systems and in maintaining the nutritional security of the poor population in the Asian countries. Inadequate recent information at the macro level has impaired the policy initiatives in the globalized agricultural scenario. Therefore, it is urgent to provide evidence on temporal and spatial dimensions of the world pulse production.

Pulses are the main source of protein for masses in the Asian countries. These are also valuable for the crop systems due to their nitrogen fixing capacity. Pulse crops are cultivated in large number of countries covering around 69 million hectares with a production of almost 60 million tonnes. Area under these leguminous crops remained almost stagnant between 1985 and 2005, but production grew at the slow rate of 0.48 per cent per annum due to positive growth of yield, (0.50 per cent per annum) during this period. Further, around 60 per cent of pulse production is contributed by 10 countries of the world. India is the leading producer with 30 per cent of world area under pulses. World trade in pulses has grown at a significant rate and became 14.07 per cent of total production in 2005. It is largely due to rising demand in countries like India.

This chapter is devoted to the analysis of growth performance and instability in the area, production, and yield of total pulses in the world and in major producing countries between 1985 and 2005. In addition, imports, exports, and their instability have also been examined.

RESEARCH METHODOLOGY AND DATA

In this chapter, the analysis of pulse production and trade is based on the secondary data collected from FAO website for the decades of the 1980s, the 1990s, and up to 2005. Individual countries, which contribute more than one percent of the world pulse production, have been included.

The entire study period is sub-divided into two periods. The first period relates to the 1980s beginning from 1985 to 1995 and the second period extends from 1995 to 2005. These represent pre- and post-WTO periods. The cut-off point of 1995 has special significance since multilateral trade agreements under the aegis of the WTO were signed during this year. Given this framework, two hypotheses are proposed for testing. First, pulse production performance at the world level is poor due to low growth of acreage and yield in the study period. Second, world trade in pulses is low and instable.

The methodology followed for studying each aspect is different. For measuring the growth rates of area, production, yield, exports, and imports of pulses for first, second, and entire study period at the country level, a semi-log equation, log y = a + bt, was used where:

y = area/production/yield/exports/imports of the crop
a = intercept
b = slope
t = time

In constructing an instability index of a parameter, we have used Coppock's (1962) method to estimate instability in area, production, yield, exports and imports of total pulses during the earlier referred three periods at world level and in major producing countries. The magnitude of index exhibits the degree of instability. The formula for calculating the Coppock instability index is as follows:

- Coppock's Instability Index

$$V\log = \frac{\sum(\log\frac{X_{t+1}}{X_t} - m)^2}{N}$$

$$InstabilityIndex = anti\log(\sqrt{V\log} - 1)$$

X_t – variable (area, production and yield of the crop) in year '*t*'
m – arithmetic mean of the difference between the logs of X_t and X_{t+1}
Vlog – logarithmic variance of the series
N – number of years minus one (1)

RESULTS AND DISCUSSION

Growth Performance and Instability in Pulse Production in Important Countries

At the outset, we would examine growth performance and instability in the pulse production in major producing countries of the world.

Table 2.1 provides country wise information on area, production and yield of pulse crops taken as a whole. Pulses were grown on around 69 million hectares of area that produced nearly 60 million tonnes of grain in triennium ending (TE) 2005. It is clear that while pulses are widely grown in India, other countries are not so important as producers of these protein rich foods. China, Canada, Brazil, and Nigeria are important pulse producing countries in that order and accounted together for nearly 25 per cent of the total production in the world. Myanmar and Australia come next, contributing over 7 per cent of total production. Yield levels across the countries show that average yield of pulses in the world (862 kg/ha) is much below the potential yield of 10–15 qtl/ha. This is true for some major countries as well. All-India yield of pulses in TE 2005 was 636 kg/ha and ranked 17th in the world. It was found to be above this level in France (4237 kg/ha), UK (3631 kg/ha), the US (1824 kg/ha) and Canada (1851 kg/ha). The exceptional yield of pulses in France and UK could be attributed to larger share of peas in total pulse cultivation and it is known that peas exhibit the highest yield among pulse crops.

Table 2.1 Share of Important Countries in World Area and Production of total Pulses (TE 2005) Yield: kg/ha

Country	Area (%)	Production (%)	Yield (Kg)	Yield (Rank)
India	30.71	22.66	636	17
China	4.88	8.74	1544	7
Canada	3.25	6.98	1851	3
Brazil	5.76	5.24	783	13
Nigeria	5.90	4.51	660	16
Myanmar	3.70	4.09	952	10
Australia	2.46	3.39	1186	8
France	0.66	3.24	4237	1
Russian Federation	1.61	2.89	1549	6
United States of America	1.36	2.88	1824	4
Turkey	2.22	2.63	1022	9
Mexico	2.53	2.54	865	12
Ethiopia	1.91	1.95	883	11
Pakistan	2.29	1.86	702	14
United Kingdom	0.35	1.48	3631	2
Ukraine	0.60	1.20	1716	5
Iran, Islamic Rep of	1.71	1.16	585	18
Uganda	1.44	1.11	669	15
All (above Countries)	73.36	78.57	862	

Source: http://www.fao.org/faostat/en/#home

The production of pulses in the world has registered a slow growth rate of 0.48 per cent per annum between 1985 and 2005. However, countries such as Canada and Myanmar have shown more than 10 per cent per year growth in pulse production. In addition, Nigeria, Ethiopia, Iran, and Uganda recorded around 3–7 per cent growths in the same period. Area expansion was primarily responsible for production growth in these countries. On the other hand, India, China, the US, and Mexico have exhibited poor growth in pulse production. If we consider the two sub-periods, our conclusions change. In particular, the second period covering 1995– 95 with 0.90 per cent per year growth in pulse production in the world appeared to be better than first period with a growth of 0.48 per cent per annum. The country-wise change in production of pulses in the sub-periods shows that rate of growth of total pulse production in the first period was more than one per cent in 13 countries out of 18 major countries. But, in second period, this reduced to 11 (see Table 2.2).

Table 2.2 Growth Performance of Total Pulses in Important Countries of the World (1985–2005) (per cent per annum)

Country	Area			Production			Yield		
	Period 1	Period 2	Entire Period 3	Period 1	Period 2	Entire Period 3	Period 1	Period 2	Entire Period 3
India	−0.30 (41%)	−1.06 (15%)	−0.78* (0%)	1.17 (13%)	−0.67 (46%)	0.34 (25%)	1.47* (1%)	0.40 (53%)	1.14* (0%)
China	−5.32* (4%)	1.58** (7%)	−0.83 (29%)	−4.14 (18%)	1.96** (6%)	0.20 (83%)	1.25 (27%)	0.38 (50%)	1.04* (0%)
Canada	16.77* (0%)	8.71* (0%)	13.02* (0%)	19.19* (0%)	7.84* (1%)	14.03* (0%)	2.07 (16%)	−0.80 (66%)	0.90 (15%)
Brazil	−1.38 (20%)	−1.61 (16%)	−2.14* (0%)	2.81* (4%)	1.71 (16%)	1.26* (1%)	4.25* (0%)	3.36* (0%)	3.47* (0%)
Nigeria	10.15* (0%)	−0.10 (92%)	5.96* (0%)	11.22* (0%)	4.37* (0%)	6.99* (0%)	0.98 (63%)	4.48* (0%)	0.98 (14%)
Myanmar	9.19* (0%)	5.46* (0%)	9.15* (0%)	5.85** (5%)	9.57* (0%)	10.05* (0%)	−3.06* (1%)	3.90* (0%)	0.82 (12%)
Australia	7.21* (0%)	−2.82* (1%)	2.53* (0%)	7.26* (1%)	−3.78** (8%)	2.81* (1%)	0.04 (98%)	−0.98 (66%)	0.27 (72%)
France	8.21* (1%)	−3.56* (0%)	−0.15 (88%)	10.01* (0%)	−5.27* (0%)	−0.20 (87%)	1.66** (6%)	−1.77** (7%)	−0.06 (88%)
Russian Federation	−7.06* (1%)	−2.86 (22%)	−6.06* (0%)	−20.03 (13%)	2.80 (33%)	−3.63 (15%)	−13.95 (23%)	5.84* (1%)	2.59 (10%)
United States of America	1.70 (12%)	0.37 (79%)	0.98* (4%)	2.36 (13%)	0.12 (94%)	1.39* (2%)	0.65 (31%)	−0.24 (52%)	0.40* (4%)
Turkey	1.35 (33%)	−2.06* (0%)	−1.41* (0%)	−0.03 (98%)	−1.44 (17%)	−1.51* (0%)	−1.36 (18%)	0.63 (35%)	−0.10 (76%)
Mexico	0.55 (73%)	−2.45** (8%)	−0.41 (46%)	2.49 (29%)	0.01 (99%)	1.19 (10%)	1.93** (9%)	2.52* (2%)	1.61* (0%)
Ethiopia	14.62** (6%)	4.44* (0%)	4.83* (0%)	16.81 (49%)	4.82* (0%)	6.37* (0%)	1.91 (93%)	0.36 (63%)	1.46 (10%)
Pakistan	−0.37 (67%)	−1.40* (0%)	−0.76* (1%)	−0.90 (53%)	−0.15 (93%)	0.43 (45%)	−0.53 (53%)	1.27 (33%)	1.19* (1%)
United Kingdom	3.27** (8%)	3.24* (0%)	1.47* (1%)	4.35 (10%)	4.54* (0%)	2.15* (1%)	1.05 (37%)	1.26** (9%)	0.67** (6%)
Ukraine	−5.13* (4%)	−8.60* (0%)	−10.15* (0%)	−18.32 (13%)	−5.67* (3%)	−11.08* (0%)	−13.90 (17%)	3.21 (13%)	−1.03 (54%)
Iran, Islamic Rep of	10.71* (0%)	0.64 (57%)	4.90* (0%)	9.27* (0%)	1.12 (42%)	3.85* (0%)	−1.30 (42%)	0.48 (55%)	−0.99** (5%)
Uganda	4.65* (0%)	3.15* (0%)	3.67* (0%)	4.50* (0%)	5.99* (0%)	2.92* (0%)	−0.15 (82%)	2.75 (11%)	−0.72 (20%)
World***	0.23 (12%)	−0.12 (66%)	−0.02 (82%)	0.69 (12%)	0.90* (1%)	0.48* (0%)	0.46 (24%)	1.03* (0%)	0.50* (0%)

Notes: Period 1: Year 1985–85, Period 2: Year 1995–2005, Entire Period 3: Year 1985–2005.
Source: http://www.fao.org/faostat/en/#home
Significant at below 5% (*) and below 10% (**) level of probability, (brackets show p-values)
***Includes minor producing countries.

The differential growth rates in the pulse production have brought some important changes in the locational pattern of pulse production in the world. The lower growth of production in countries such as India and China implies that growth centres of pulse.

Production is gradually shifting from these countries to countries like Canada and Myanmar. In most of these countries, acceleration in production was primarily due to area expansion. Especially; countries like Canada and Myanmar exhibited an area growth of more than 5 per cent per annum in the study period. Also, yield improvement in these countries was also around one per cent per year. At the world level, whatever little growth has been achieved in pulse production primarily came from yield growth. Contribution of yield growth to production growth was higher in the first period. However, yield growth itself was low. The yield growth of total pulses between 1985 and 2005 was merely 0.50 per cent per annum. Agricultural scientists believe that yield of pulses can be easily raised to above 10 qtl/ha even in rainfed areas. Therefore, efforts should be made to raise yield levels by popularizing improved technology for pulse cultivation through implementation of pragmatic policies.

What could be the plausible explanation for marginal decline in world pulse area. First, pulses are high-risk crops being rainfed and prone to damage due to pests and diseases, often, relegated to marginal and sub-marginal lands. Second, pulses often receive inadequate extension support because priority is not accorded to these crops in foodgrain production system of various countries. Third, no major genetic breakthrough like wheat and rice has yet taken place. Fourth, pulse growers in the largest producing country of India do not get desired price support even if prices are falling below the minimum support price (MSP) level. NAFED is the only agency for purchases under price support and commercial purchases but its operations are extremely limited to a few markets and do not have overall impact. (Tuteja 1999, 2000, 2008).

It is imperative to popularize pulse crops in different regions of the world in lean seasons so that these crops could become part of crop rotation without disturbing existing major crops. It is feasible because pulses are known for low water requirement and adaptability over a wide range of agro-climatic conditions. It would enhance income of the farmers by utilizing the available land in the lean periods and increasing sustainability in agriculture. It would make a significant contribution to total production of pulses and also help evolve a sustainable cropping pattern particularly in the regions with paddy and wheat rotation.

The preceding analysis highlights that growth performance of pulses at the world level has been poor during the reference period (Jain and Singh 1997; Joshi and Saxena 2002; Tuteja 2008). Slow pace of growth could be due to high instability in yield and acreage. Therefore, it is useful to examine instability in area, production, and yield of pulses.

Table 2.3 reveals instability indices in the production of total pulses. It is interesting to note that world production of total pulses was almost stable during the study span. Neither, acreage nor yield was found to be unstable. It could be due to slow adoption of technology and extension services for pulse cultivation in major countries. Surprisingly, even yield instability of pulses was only 3.8 per cent. The differences in the instability in production of total pulses in the two selected periods were narrow despite some fluctuations in area as well as in yield.

The instability indices for the production of total pulses in the world were estimated at 4.6 per cent in the first period, 3.0 per cent in the second period, and 3.9 per cent during the entire study span. Out of the two (area and yield), yield contributed relatively more to instability in the first period but it has decreased during the second period. It may be mentioned that instability around the trend in case of area was relatively low in comparison to yield for the study period. Among important countries, highest uncertainty was found in China and the lowest in Nigeria in the first period but Australia crossed China in the second period. Particularly, instability indices of area as well as yield in Nigeria were found to be less than 20 per cent during the study period. Efforts should be made to reduce production instability, which was found to be more than 10 per cent in most of the cases.

The instability index of pulse production showed high uncertainty at individual country level barring a few exceptions when instability index was below 10 per cent. It was however, low at the aggregate level.

Finally, let us recall our proposed hypothesis that growth in world pulse production is poor due to low growth of acreage and yield. It was confirmed at the world level but Canada, Myanmar, Iran, and Ethiopia experienced high production growth largely due to area expansion.

Growth and Instability in Pulse Trade

After reviewing pulse production performance, we analyse growth of pulse trade during the reference period. In view of large inter-country differences, trade and instability in pulses are analysed at the world and country level. In order to put growth of pulse trade in proper perspective, it would be useful to give a brief idea about trade in pulses at the world level.

Clearly, share of pulse production traded at the world level has more than doubled between 1985–85.

Table 2.3 Country-Wise Instability Indices of Total Pulses (per cent)

Country	Period 1		Yield	Period 2		Yield	Entire Period 3		Yield
	Area	Prod.		Area	Prod.		Area	Prod.	
India	5.0	10.8	6.3	5.7	13.2	10.2	5.3	12.2	8.6
China	25.8	41.7	20.0	7.7	11.6	7.1	18.7	29.5	14.8
Canada	35.6	40.4	20.3	25.4	30.0	18.8	31.6	36.1	19.6
Brazil	17.9	20.6	12.3	17.0	19.2	5.6	17.5	19.9	9.5
Nigeria	11.0	10.4	15.9	14.6	2.2	13.5	13.5	7.9	14.8
Myanmar	16.7	23.1	9.3	9.3	7.5	7.9	13.5	16.9	9.0
Australia	16.1	35.7	38.1	7.6	40.8	46.4	13.8	39.2	42.4
France	16.7	21.3	12.2	9.8	14.5	12.1	15.1	19.7	12.2
Russian Federation	–	–	–	14.8	29.9	25.3	–	–	–
United States of America	14.7	20.7	9.0	18.0	22.2	6.5	16.4	21.5	7.9
Turkey	9.4	19.5	14.3	3.3	8.1	5.8	7.4	14.9	10.8
Mexico	27.8	40.4	15.3	17.5	18.2	12.1	23.3	30.7	13.9
Ethiopia	–	–	–	14.1	13.7	11.3	–	–	–
Pakistan	14.9	21.5	11.7	3.2	17.2	14.8	10.6	19.5	13.4
United Kingdom	17.3	29.6	19.4	12.0	12.6	10.3	14.9	22.4	15.4
Ukraine	–	–	–	17.4	26.0	34.6	–	–	–
Iran, Islamic Rep of	27.7	20.9	25.1	12.9	14.1	12.4	21.8	18.2	19.6
Uganda	6.0	6.4	8.0	0.8	19.6	19.6	4.4	14.4	14.8
World	1.7	4.6	4.7	2.7	3.0	2.6	2.3	3.9	3.8

Source: http://www.fao.org/faostat/en/#home.

It is largely due to rising demand in countries like India at one hand and slow increase is supply owing to low growth in the domestic production. Moreover, pulse exports have experienced high growth between 1985 and 2005 (see Table 2.4).

After presenting the background, we would examine trade scenario for total pulses Tables 2.5 and 2.6 present the country-wise exports and imports of pulses. It may be noted that exporting countries of pulses were only few. In 1985, UK was the leading exporter of pulses. After a decade in 1995, Myanmar, China, and UK acquired around 77 per cent of the world market. In 2005, share of Myanmar was more than 50 per cent in world exports of pulses. India also captured 13.2 per cent share of world pulse exports. The per unit value of pulses rose from Rs 3.4 in 1985 to Rs 7.9 in 2005.

Table 2.4 Share of Pulse Production Traded at the World Level

Year	Exports as percentage of production	Imports as percentage of production
1985	6.16	5.94
1995	12.97	12.20
2005	14.07	14.26

Source: http://www.fao.org/faostat/en/#home.

Table 2.5 Share of Total Pulses Exporting Countries in the World (per cent)

Country	Quantity			Value			Per Unit Value Rs/Kg		
	1985	1995	2005	1985	1995	2005	1985	1995	2005
Afghanistan	7.2	0.0	0.6	9.1	0.0	0.0	4.3	–	0.0
China	0.5	18.2	1.3	1.1	30.3	5.3	7.8	11.0	32.4
India	0.3	9.6	13.2	0.4	33.5	44.8	3.5	23.0	26.9
Kyrgyzstan	0.0	0.0	2.5	0.0	0.0	0.0	–	–	0.0
Myanmar	0.0	48.3	51.6	0.0	0.0	0.0	–	0.0	0.0
Pakistan	0.4	0.1	4.0	0.7	0.3	7.8	6.0	22.0	15.4
United Kingdom	10.2	10.8	21.3	10.2	12.1	24.6	3.4	7.4	9.1
Total	**18.6**	**87.1**	**94.6**	**21.6**	**76.3**	**82.5**	**3.4**	**6.6**	**7.9**

Source: http://www.fao.org/faostat/en/#home.

Table 2.6 Share of Total Pulses Importing Countries in the World (per cent)

Country	Quantity			Value			Per Unit Value Rs/Kg		
	1985	1995	2005	1985	1995	2005	1985	1995	2005
China	4.3	2.4	4.3	2.6	1.9	8.3	3.3	9.8	33.4
Egypt	0.0	0.1	3.7	0.0	0.1	0.0	9.8	10.2	0.0
India	11.6	51.7	61.7	8.2	64.7	63.9	3.7	16.0	17.9
Nepal	0.1	0.0	2.2	0.1	0.0	0.0	2.5	–	0.0
Pakistan	0.4	11.4	6.7	0.3	12.2	5.1	4.0	13.7	13.2
Spain	0.1	0.4	2.4	0.2	0.3	1.3	7.3	10.6	9.2
Turkey	0.0	0.0	1.3	0.0	0.0	1.9	–	16.7	25.4
United Arab Emirates	3.4	13.6	2.3	4.0	0.0	0.0	6.3	0.0	0.0
United States of America	0.0	5.7	3.4	0.0	6.4	7.9	–	14.3	40.7
Yemen	1.0	0.0	0.2	1.1	0.0	0.1	6.2	–	11.7
Total	**20.8**	**85.4**	**88.1**	**16.5**	**85.5**	**88.5**	**5.3**	**12.8**	**17.3**

Source: http://www.fao.org/faostat/en/#home

The number of importing countries of pulses is relatively large. In 1985, India with 11.4 per cent share in world pulse imports was the main importer. After a decade, India's share in world imports of pulses reached 51.7 per cent. The United Arab Emirates and Pakistan were the other two important importers. In 2005, India's share rose further and became almost 62 per cent of world imports. The share of other countries however declined and none of the importing countries crossed even 5 per cent mark except Pakistan. The per unit value of world pulse imports was Rs 5.3 in 1985 and almost trebled in 2005.

A perusal of the growth performance of world pulse exports in Table 2.7 indicates a growth rate of 7.22 per cent per annum for quantity between 1985–2005. However, it was less than half in value terms. The coefficient for the quantity as well as value was significant. Out of the two periods, growth rate in the second period was more than 10 per cent per annum and the coefficient was significant. The growth performance of India in pulse exports with 28.67 per cent per annum increase was found impressive. China and Mynamar were other two important countries indicating more than 20 per cent per year growth in pulse exports.

Growth rate of pulse imports at the world level was 22.67 per cent per year in quantity terms between 1985 and 2005. A contrast was observed between the two reference periods. The growth rate was observed to be as high as 21.64 per cent per year in the second period while it was negative in the first period. India followed by Spain indicated a growth rate of around 20 per cent per annum during the reference period. The growth rate of pulse

Table 2.7 Country wise Growth of Total Pulses Exports (1985–2005)

(per cent per annum)

Country	Exports (Quantity)			Exports (Value)		
	Period 1	Period 2	Entire Period 3	Period 1	Period 2	Entire Period 3
China	77.50* (0.18)	–4.69 (45.38)	24.67* (0.04)	59.78* (0.00)	–4.86 (15.17)	14.41* (0.10)
India	67.40* (0.01)	12.40* (0.00)	28.67* (0.00)	76.43* (0.00)	10.00* (0.19)	29.66* (0.00)
Myanmar	149.24* (4.72)	11.11 (18.39)	23.74* (0.72)	NA	NA	NA
Pakistan	–17.99** (5.75)	118.66* (1.38)	5.19 (61.35)	–19.10** (6.28)	51.45* (1.45)	1.91 (74.97)
United Kingdom	–4.07 (56.49)	25.50* (0.09)	8.35* (0.86)	–7.61 (31.25)	23.46* (0.28)	4.25 (17.62)
World	7.73* (2.76)	11.06* (1.56)	7.22* (0.00)	0.75 (78.21)	8.82* (0.12)	3.06* (0.64)

Notes: Period 1: Year 1985–85, Period 2: Year 1995–2005, Entire Period 3: Year 1985–2005.
Source: http://www.fao.org/faostat/en/#home
Significant at below 5 per cent (*) and below 10 per cent (**) level of probability, brackets show p-values
For Myanmar, value figures are not available

Table 2.8 Country-wise Growth of Total Pulses Imports (1985–2005) (per cent per annum)

Country	Imports (Quantity)			Imports (Value)		
	Period 1	Period 2	Entire Period 3	Period 1	Period 2	Entire Period 3
China	−11.33 (12.51)	8.18* (4.66)	12.38 (18.09)	−10.29 (14.14)	15.07* (3.41)	5.58** (6.06)
Egypt	41.33* (3.88)	31.10* (0.18)	6.22 (22.18)	28.85** (7.05)	94.03 (27.04)	12.54 (42.90)
India	22.30* (0.71)	−4.14 (16.79)	21.64 (12.50)	24.74* (0.28)	10.13 (24.21)	13.53* (0.01)
Pakistan	23.01* (1.35)	5.40 (38.74)	−3.56 (55.38)	23.04* (0.73)	0.38 (91.00)	10.39* (0.01)
Spain	19.04 (12.61)	21.64** (9.40)	19.57 (18.01)	8.96 (25.43)	17.01 (13.83)	15.71* (0.02)
Turkey				−34.06 (12.04)	47.56* (1.16)	43.88* (0.01)
UAE	8.18* (4.66)	12.38 (18.09)	8.57* (3.10)			
USA	31.10* (0.18)	6.22 (22.18)	0.93 (72.73)	18.61* (0.82)	26.85* (3.42)	14.12* (0.70)
World	−4.14 (16.79)	21.64 (12.50)	22.67* (0.03)	−9.22* (0.67)	8.58* (3.69)	1.00 (53.87)

Notes: Period 1: Year 1985–95, Period 2: Year 1995–2005, Entire Period 3: Year 1985–2005.
Source: http://www.fao.org/faostat/en/#home
Significant at below 5 per cent (*) and below 10 per cent (**) level of probability, brackets show p-values.

imports in India was higher in first period in comparison to the second period. Spain showed almost similar growth for both the periods (Table 2.8).

Exports instability behaviour of total pulses diverged from pulse production at the world level. The instability index of world pulse exports was 48.5 per cent for the quantity and 30.8 per cent for the value during the study period. The level of uncertainty was quite high in quantity as well as in value terms during both the periods. Among the important countries, estimated instability index was higher for Pakistan, India, China, and UK. The uncertainty level of imports of total pulses was estimated lower than that of exports at the world level during the study period. It was around the same for the quantity in both the selected periods but in value terms, divergence was noticed across the two selected periods. It is surprising that Pakistan indicated exceptionally high instability index for exports (see Table 2.9).

SUMMARY AND CONCLUSIONS

We have analysed production performance and trade of pulses at the world and country level. Findings suggest that world pulse production grew at a slow rate of 0.48 per cent per annum between 1985 and 2005. It was found to be higher during the second period spanning 1995– 2005 (0.90 per cent per year) in comparison to the first period covering 1985–95 (0.69 per cent per year). A mixed performance has been observed across the developed, developing, and emerging economies. Canada and Myanmar achieved a spectacular growth rate in pulse production (14.03 and 10.05 per cent per annum) during this period. On the other hand, it has been negative in France, Ukraine, Russia, and Turkey. Among the Asian countries, Myanmar has crossed 10 per cent mark in growth of pulse production and performed well. However, India, the leading producer of pulses has exhibited a marginal growth in pulse production (0.34 per cent per annum) during this period. Thus, a considerable diversity has been observed in the growth of pulse production in individual countries. The reasons for varied performance differ from country to country but in majority of the cases, poor yield growth has been responsible for slow growth in pulse production. In India, inadequate adoption of improved technology, low irrigation coverage, and uncertainties related to pulse farming are responsible for slow growth in pulse production between 1985 and 2005. Instability index of pulse production displayed high uncertainty at individual

Table 2.9 Instability in Trade of Total Pulses (1985–2000) (per cent)

Country	Exports (Quantity)			Exports (Value)		
	Period 1	Period 2	Entire Period 3	Period 1	Period 2	Entire Period 3
China	260.8	103.0	194.9	88.2	49.7	78.7
India	218.2	18.3	131.6	72.9	21.7	57.7
Myanmar	–	123.9	–	–	–	–
Pakistan	158.2	4509.0	1551.7	183.3	304.2	250.3
United Kingdom	224.7	67.3	148.3	221.6	59.7	143.7
WORLD	52.5	44.3	48.5	42.6	14.4	30.8
		Imports			Imports	
China	155.4	174.5	166.6	147.5	152.3	153.2
Egypt	258.0	226.0	242.7	201.7	–	–
India	143.8	135.1	139.5	135.3	124.5	130.1
Pakistan	154.0	90.5	125.3	140.0	61.8	106.4
Spain	210.2	292.9	252.0	96.5	238.0	168.9
Turkey	–	262.1	–	–	240.8	–
United Arab Emirates	61.7	61.2	63.1	–	–	–
United States of America	–	296.6	–	–	181.3	–
WORLD	40.3	40.8	41.2	32.7	43.9	39.4

Source: http://www.fao.org/faostat/en/#home

country level barring a few exceptions when instability index was below 10 per cent. It was however, low at the aggregate level.

Major exporters of pulses at the world level were few in 1985 and US was the leading exporter. Among the major exporters, share of Myanmar grew at the phenomenal rate and constituted more than 50 per cent of world pulse exports in 2005. India is the major importer of pulses and its share in world imports of pulses is around 62 per cent. Export growth of pulses at the world level was 7.22 per cent per annum during the reference period. Growth rate in the second period was more than 10 per cent per annum. India, China, and Myanmar indicated impressive growth of pulse exports during this period. Results show that growth of pulse imports at the world level was as high as 22.67 per cent per year between 1985 and 2005. The second period revealed a higher growth rate than the first period. India and Spain registered around 20 per cent per year growth in pulse imports in this period. The uncertainty level in world pulse trade calculated by log variance method suggests that it was 41.2 per cent for exports of total pulses in quantity terms. The country level indices vary significantly. Pakistan, UK, China, and India indicated relatively higher instability in pulse exports.

Worldwide pulse production and trade suggests a fairly complex scenario. The buoyant demand in India and almost constant supply is the main factor in continuous upward pressure in trade and prices of different varieties of pulses in the world. These developments were the result of imbalance in domestic supply and demand in countries with higher consumption. India's share in world Imports was around 62 per cent in 2005. On an average, India imported 2–3 million tonnes per year of pulses in the recent past.

Given the uncertainty of world supply of pulses and rising domestic demand in India, it would be prudent to plan future domestic pulse production in such a way that major share of demand is fulfilled by domestic production. Reducing over dependence on world pulse supply would increase overall welfare of the farmers in rainfed areas and will improve access for consumers. Effective transfer of available technology is the key to narrowing down the gap between potential and actual yield. It has been slow in case of pulses due to non-availability of improved seeds and adequate extension services. For accelerating pulse production in India, providing required inputs for adoption of improved technology and ensuring a high share of prices paid by consumers to the producers can go a long way in solving the problem.

REFERENCES

Coppock J.D. 1962. *International Economic Instability: The Experience after World War II*. New York: McGraw Hill Book Company , 27–48.

FAO website; http://www.fao.org/faostat/en/#home

Government of India. 2003–08. *Agricultural Statistics at a Glance*. New Delhi: Ministry of Agriculture.

Hazell, P.B.R. 1982. 'Instability in Indian Food Grains Production', Research Report 30, Washington, DC: International Food Policy Research Institute.

Hazell, P.B.R. and J.R. Andersen. 1989. *Variability in Grain Yields: Implications for Agricultural Research and Policy in Developing Countries*. Baltimore: John Hopkins University Press.

Jain, K.K. and A.J. Singh. 1991. 'An Economic Analysis of Growth and Instability of Pulses Production in Punjab', *Agricultural Situation in India*. 46(1) April, 3–8.

Joshi, P.K. and Raka Saxena. 2002. 'A Profile of Pulses Production in India; Facts, Trends and Opportunities'. *Indian Journal of Agricultural Economics*, 57(3):.

Tuteja, Usha. 1999. 'Economics of Pulses Production and Identification of Constraints in Raising Production in Haryana', Agricultural Economics Research Centre, University of Delhi.

———. 2000. 'Economics of Pulses Production and Identification of Constraints in Raising Production in Punjab', Agricultural Economics Research Centre, University of Delhi.

———. 2006. 'Growth Performance and Acreage Response of Pulse Crops: A State Level Analysis'. *Indian Journal of Agricultural Economics*, 61 (2): 218–37.

———. 2008. *India's Pulse Production: Stagnation and Redressal*. New Delhi: Pragun Publication.

3

Pulses Production in India

An Empirical Analysis

C.S.C. Sekhar and Yogesh Bhatt

Pulses constitute the major source of protein for a majority of population in India that is predominantly vegetarian in dietary habits. The present study was mainly motivated by the severe stagnation and price rise in the pulses sector in the country during the last few years. India accounts for about 20–3 per cent of the world pulse production—about 93 per cent of the world chickpea (gram) production, and 68 per cent of the world pigeon pea (arhar) production. The high proportion of global production accounted for by India indicates that there are very few import sources in the world market and it is imperative to increase domestic production in order to address the food security concerns. To understand the reasons for stagnation and devise an effective production strategy, it was important to analyse the crop-specific growth trajectory over time and space to unravel the constraints inhibiting the pulses growth in India. The present study aimed:

1. to analyse the temporal, spatial, and crop-specific growth pattern of pulses;
2. to identify the determinants of pulses production (price and non-price factors) and assess their relative importance; and
3. to identify the major constraints and delineate appropriate policy responses.

PULSES SECTOR IN INDIA: AN OVERVIEW

Pulses occupied only about 12–15 per cent of the gross cropped area in the country from 2007 to 2013. Among the foodgrains, pulses occupied about 16–18 per cent of the area and contributed about 6–8 per cent of the foodgrain production during the period, indicating the lower yield levels of pulse crops (Government of India [GoI]). Gram being the predominant pulse crop in India shared about 30 per cent and 40 per cent in the total pulses area and production, respectively. Arhar (16 per cent and 19 per cent), moong (14 per cent and 8 per cent), and urad (13 per cent and 10 per cent) were the other major pulse crops in the country. The major pulse-growing states in the country are listed in Table 3.1.

Pulses production has virtually stagnated over the last 40 years mainly due to two reasons: (i) 87 per cent of the area under pulses is rainfed and (ii) pulses are mainly grown as a residual crop on marginal lands after diverting better irrigated lands to rice and wheat. Farmers are not motivated to grow pulses because of yield and price risk coupled with ineffective government procurement. Pulses face various abiotic (climate-related) and biotic (pest and insect related) stresses. Pulses are more susceptible to pest and insect attacks than cereals like rice and wheat.

Table 3.1 Major Pulse Growing States

Pulse Crop	Major States
Total Pulses	Madhya Pradesh, Maharashtra, Rajasthan, Uttar Pradesh, Karnataka, Andhra Pradesh, Gujarat—Total about 80 per cent
Gram	Madhya Pradesh, Rajasthan, Maharashtra, Uttar Pradesh, Karnataka, Andhra Pradesh—Total about 95 per cent
Arhar	Maharashtra, Karnataka, Andhra Pradesh, Uttar Pradesh, Madhya Pradesh, Gujarat—Total about 90 per cent
Urad	Maharashtra, Andhra Pradesh, Madhya Pradesh, Uttar Pradesh, Tamil Nadu, Rajasthan, Karnataka—Total about 85 per cent
Moong	Rajasthan, Maharashtra, Andhra Pradesh, Karnataka, Odisha, Bihar, Gujarat—Total about 90 per cent
Masur	Uttar Pradesh, Madhya Pradesh, Bihar—Total about 90 per cent

Lower productions as compared to demand, as well as lower stocks in both domestic and global markets have led to price rise in the recent past. Several programmes were launched to boost pulses production including National Pulses Development Programme (NPDP) which was the precursor to the Technology Mission on Oilseeds and Pulses (TMOP) and Integrated Scheme on Oilseeds, Pulses, Oil Palm, and Maize (ISOPOM) 2004. These programmes resulted in some increase in pulse area in rice fallows but limited to regions with irrigation potential. The ISOPOM was replaced in 2008 with a vastly improved National Food Security Mission (NFSM).

BRIEF REVIEW OF LITERATURE

A summary of some of the important literature on pulses sector in India has been presented in Table 3.2.

To sum up, availability of improved technology at affordable prices, input provision particularly pesticides, assured market through procurement were some of the major problems highlighted in the previous work.

METHODOLOGY AND DATA

To address the objectives, a detailed analysis of growth rates and growth acceleration at the national and state levels based on secondary data was presented followed by an econometric analysis to identify the major determinants of production in each state. The period of analysis was 1975–76 to 2007–08. NFSM for pulses was launched in 2008 and has reportedly resulted in changes in the area and production in several states. However, the period after

Table 3.2 Summary of Literature on Pulses Sector in India

S.No	Authors	Major Findings
1	Sharma and Jodha (1982)	Regional specificity in pulses production making uniform policy difficult; farmers' tendency to move away from pulses when irrigation facilities become available; refutes the popular notion that pulses have an inherent yield disadvantage and argues that pulses adapt better to stress along with other positive externalities such as nitrogen fixation.
2	Sharma (1986)	Yield variability, pest and insect problems and non-availability of quality seeds were the major problems; study recommended procurement and distribution by state agencies and improving the extension system.
3	Swarna (1989)	Pulses faced competition from cereals in only five irrigated (or wet) states. In the remaining drier states, where the large proportion of pulse production takes place, pulses do not face any competition from cereals.
4	AERC (2001)	Based on primary data collected in all the pulse-growing states–identified following problem areas: (i) non-availability of high-yielding and short-duration pulses, (ii) lack of extension, training and credit facilities (iii) lower relative profitability (iv) pests and post-harvest losses.
5	Reddy (2004)	Large variability in the production performance of the states and considerable heterogeneity in consumption patterns.
6	Sathe and Agarwal (2004)	Need for further opening up of the Indian markets for pulses imports.
7	Tuteja (2009) and Reddy (2009)	Price support through procurement in the short-run; improved technology and input provision in the long-run were urgently needed

2008 is too small to be treated as a sub-period. Therefore, analysis was restricted until the launch of NFSM. The period has further been divided into three sub-periods—(i) 1975–76 to 1987–88, (ii) 1987–88 to 1997–98, and (iii) 1997–98 to 2007–08. Some of the major pulses development programmes such as the NPDP were initiated in 1985. The effects of the programmes could have been visible in about two to three years. Therefore, 1975–76 to 1986–87 was taken as the first sub-period. There was a widespread slowdown in Indian agriculture since 1997–98 until about 2003–04. Therefore, the second period was from 1987–88 to 1997–98 and the third sub-period from 1997–98 to 2007–08.

Methodology for Computing Growth Rates and Growth Acceleration

For calculating growth rates, the standard method of semi-logarithmic trend equations of the form $\ln y_t = a + bt$ was fitted to the data on area, yield and production. y_t denotes the variable in question and t is the time trend, b gives the exponential growth rate in y_t, and antilog(b)-1 gives the compound growth rate. Growth acceleration (or deceleration) was assessed by determining whether the growth rate calculated in a sub-period was substantially and statistically different from the previous sub-period. Two alternative methods were available in the literature to test this. The first one: to fit two separate functions to each sub-period and examine the break in the trend through a separate test and the second to fit a single function with dummy variable to distinguish the sub-periods. The second alternative was preferred in order to use the entire sample data and also its ease of interpretation. Since there were three sub periods, two regressions have been fitted. The two sub-periods 1975–87 and 1988–96 was denoted as phase1 and sub-periods 1988–96 and 1997–2007 as phase 2.

Phase 1: $\ln y_t = a_0 + a_1D_1 + b_0t + b_1(D_1t)$ Time Period: 1975 to 1996

where $D_1 = 0$ for years 1975 to 1987 and 1 for years from 1988 to 1996

Phase 2: $\ln y_t = a_0 + a_2D_2 + b_0t + b_2(D_2t)$ Time Period: 1988 to 2007

where $D_2 = 0$ for years 1988 to 1996 and 1 for years from 1997 to 2007

To account for the initial output in each sub-period, the intercept dummies have been introduced. The annual rate of growth in the earlier period is given by b_0 and the annual rate of growth of the latter period is given by $b_0 + b_1$ or $b_0 + b_2$ for equations 2 and 3 respectively.

The significance of b_1 in equation 2 and the significance of b_2 in equation 3 directly show if the growth trend in the sub-period was significantly different from the previous sub-period. The positive or negative sign of these coefficients also indicates if the growth rates were accelerating or decelerating over the previous sub-period. Further, determinants of pulses production through an econometric modelling exercise have been identified.

Secondary Data Sources

The secondary data for the study have been collected from the following official publications—(i) *Area, Production, and Yield of Principal Crops in India*, (ii) *Agricultural Statistics at a Glance*, (iii) *Agricultural Situation in India* and *Agricultural Prices in India*, (iv) *Cost of Cultivation of Principal Crops in India*, (v) *Fertilizer Statistics*, and (vi) *All India Reports on Input Survey.*

To derive the crop-specific data on fertilizer consumption and pesticide coverage, the following method has been followed. Crop specific quantities of fertilizers and area under pesticides were collected from various issues of *All India Report on Input Survey* (1981, 1986, 1991, 1996, and 2001). The proportion of fertilizer consumed by pulse crops to total fertilizer consumption was calculated. Suppose this proportion was 0.2 in 1981, this number was assumed to remain constant until the next input survey. Fertilizer consumption for the years 1981 to 1985 was then worked out by multiplying the total fertilizer consumption of the year with 0.2. Similarly, the next input survey data in 1986 was used to compute fertilizer consumption until 1990. This process continued to derive all the yearly figures for fertilizer consumption and area under pesticides for pulses.

Household Surveys

In addition to the secondary-data-based analysis, detailed primary data surveys have also been carried out in seven states to assess the economics of pulses cultivation in these states, constraints faced by the farmers, their suggestions for improving pulses production and the impact of NFSM, if any, on the pulses production in the country. The study was carried out in the following seven states—Uttar Pradesh (UP), Bihar, Haryana, Maharashtra, Rajasthan, Punjab, and Andhra Pradesh (AP). Two districts in each state—one NFSM district and one non-NFSM district were selected. In each district, 50 households belonging to different size-groups of landholding were selected. Therefore, in all, 700 households were surveyed.

RESULTS AND DISCUSSION

Growth Trends in India and Major States

Growth has several dimensions—temporal, spatial, composition across crops, recent performance, etc. It has been attempted to analyse the growth trends along all these dimensions. Yield was the major source of production growth for pulses sector over the period of analysis 1975–2008. Contribution of area was negligible (Table 3.3). It is clear that there was a major stagnation during sub-period 3 due to insignificant growth in area and a decline in yield. The second sub-period, marked by some important programmes, witnessed highest growth rate in yield but contribution of area has been negative even during this period. But it might be noted that some of the trends witnessed at the aggregate level were mainly due to the performance of the major crop—gram.

Gram was the only crop that showed high positive growth rate in yield despite a decline in area. All other crops showed positive contribution from area, with varying degrees of contribution from yield. Since gram occupies about 40–5 per cent of the total pulses production, the positive contribution of yield was reflected at the aggregate level. Similarly, the temporal trends at the all-India level, which did not show any dismal performance in the second sub-period, concealed some of the crop level trends at the state level. There was a general decline or deceleration in production growth of all the pulses in the second sub-period except gram and masur (see Table 3.4). However, growth rate of the pulses sector as a whole did not reflect such deceleration because of gram and masur. Therefore, it is important to recognize that there is a lot of heterogeneity across crops and regions in growth pattern.

The state-level trends during the decade 1997–2007 clearly indicated some troubling trends (Table 3.5). There was general decline or deceleration in area or yield or both in some of the major states for all the pulse crops. For instance, there was a decline in both area and yield of total pulses in Rajasthan and UP. A similar decline is evident

Table 3.3 Crop-wise Growth rates—All India

Crop	Area				Prod.				Yield			
	1975–87	1988–96	1997–2007	1975–2007	1975–87	1988–96	1997–2007	1975–2007	1975–87	1988–96	1997–2007	1975–2007
Total Pulses	−0.03	−0.26	0.10	−0.18	0.81	0.85	0.09	0.67	0.85	1.13	−0.01	0.85
Gram	−1.00	1.08	−0.33	−0.45	−0.46	3.09	−0.16	0.47	0.55	1.99	0.18	0.92
Arhar	2.16	0.04	0.42	1.00	2.94	−0.91	0.73	0.84	0.76	−0.95	0.30	−0.16
Moong	2.22	−1.57	1.34	0.79	4.54	−1.85	0.68	0.97	2.27	−0.29	−0.65	0.18
Urad	3.65	−1.78	0.54	1.05	5.77	−1.77	0.19	2.12	2.05	0.01	−0.34	1.07
Masur	1.29	1.84	0.83	1.81	3.93	1.58	0.64	3.21	2.62	−0.24	−0.16	1.38

Table 3.4 Crop-wise Growth Rates across States 1975–2007

States	Period: 1975–87			Period: 1988–96			Period: 1997–2007			Period: 1975–2007		
	Area	Prod.	Yield	Area	Prod.	Yield	Area	Prod.	Yield	Area	Prod.	Yield
					Growth rates for Gram							
Andhra Pradesh	−3.00	−3.19	−0.20	12.62	20.22	6.75	16.98	26.46	8.10	6.90	11.95	4.72
Bihar	−2.63	1.19	3.92	−3.31	−1.10	2.28	−3.28	−3.30	−0.02	−3.59	−2.03	1.62
Gujarat	2.75	4.06	1.28	6.63	8.40	1.66	5.65	7.98	2.20	1.72	1.93	0.20
Haryana	−5.08	−7.31	−2.35	−4.25	1.35	5.85	−12.09	−13.48	−1.58	−7.30	−6.64	0.72
Karnataka	1.56	1.22	−0.33	7.16	12.39	4.88	5.41	5.69	0.27	4.26	5.37	1.06
Madhya Pradesh	1.62	3.51	1.86	2.57	5.54	2.90	1.23	0.91	−0.32	1.36	3.16	1.78
Maharashtra	1.57	1.67	0.10	2.66	6.48	3.72	3.51	5.64	2.06	2.86	5.25	2.32
Rajasthan	−0.82	−1.25	−0.44	4.18	5.62	1.38	−7.27	−9.11	−1.98	−1.64	−1.74	−0.10
Uttar Pradesh	−1.79	0.70	2.53	−4.61	−2.78	1.91	−2.99	−3.39	−0.42	−3.01	−2.15	0.89
India	−1.00	−0.46	0.55	1.08	3.09	1.99	−0.33	−0.16	0.18	−0.45	0.47	0.92

(Cont'd)

Table 3.4 *(Cont'd)*

States	Period: 1975–87			Period: 1988–96			Period: 1997–2007			Period: 1975–2007		
	Area	Prod.	Yield	Area	Prod.	Yield	Area	Prod.	Yield	Area	Prod.	Yield
Growth rates for Arhar												
Andhra Pradesh	3.26	6.90	3.52	−0.74	6.90	7.70	3.17	8.99	5.64	3.03	6.31	3.19
Bihar	−2.38	3.39	5.91	0.10	−2.30	−2.40	8.28	2.53	−5.31	−0.19	0.59	0.78
Gujarat	12.66	16.51	3.42	1.95	5.49	3.48	−4.31	−2.47	1.92	3.14	4.66	1.47
Karnataka	3.49	1.15	−2.26	−3.58	−2.28	1.35	3.27	7.10	3.71	1.89	1.38	−0.50
Madhya Pradesh	−0.59	2.01	2.62	−2.39	−5.61	−3.30	0.35	−1.47	−1.82	−1.35	−1.28	0.07
Maharashtra	1.76	2.39	0.62	2.99	2.12	−0.85	0.86	2.78	1.91	2.05	2.77	0.71
Odisha	10.09	15.28	4.72	1.35	0.30	−1.03	−0.90	2.09	3.02	2.62	3.43	0.80
Tamil Nadu	2.99	5.70	2.63	−6.07	−7.05	−1.05	−9.39	−9.71	−0.36	−2.15	−1.48	0.68
Uttar Pradesh	−0.25	0.70	0.95	0.54	−2.78	−3.31	−2.24	−4.67	−2.48	−1.07	−2.03	−0.96
India	2.16	2.94	0.76	0.04	−0.91	−0.95	0.42	0.73	0.30	1.00	0.84	−0.16
Growth rates for Moong												
Andhra Pradesh	0.77	4.87	4.07	0.12	2.76	2.63	−0.62	−1.00	−0.38	−0.35	0.53	0.88
Bihar	3.86	7.29	3.30	−1.74	−0.67	1.08	0.63	0.42	−0.20	1.13	2.91	1.77
Gujarat	2.14	−25.19	−26.76	4.74	6.38	1.57	2.01	1.67	−0.33	1.34	0.79	−0.54
Karnataka	6.43	3.75	−2.52	−1.85	−4.70	−2.90	4.86	0.55	−4.11	4.39	2.41	−1.89
Madhya Pradesh	−1.05	−1.06	−0.01	−4.95	−3.28	1.75	−1.85	−2.16	−0.32	−3.31	−2.57	0.76
Maharashtra	−0.34	1.37	1.72	0.48	2.75	2.26	−0.58	−1.19	−0.62	1.06	2.86	1.78
Odisha	3.47	5.72	2.17	−15.34	−22.53	−8.49	1.12	2.73	1.60	−3.94	−7.13	−3.32
Rajasthan	−2.51	−7.24	−4.85	8.89	9.58	0.63	5.17	10.33	4.90	4.12	5.72	1.54
Tamil Nadu	1.26	0.36	−0.89	−1.46	0.32	1.80	1.39	1.13	−0.25	1.46	2.97	1.48
Uttar Pradesh	26.21	28.13	1.52	−1.91	1.90	3.89	−5.24	−5.44	−0.21	2.53	4.12	1.55
India	2.22	4.54	2.27	−1.57	−1.85	−0.29	1.34	0.68	−0.65	0.79	0.97	0.18
Growth rates for Urad												
Andhra Pradesh	4.45	11.38	6.63	2.36	−3.63	−5.86	−0.75	−0.58	0.17	3.84	4.89	1.01
Assam	−0.34	0.04	0.38	−0.11	4.39	4.50	−0.88	−1.17	−0.30			
Bihar	−0.34	1.34	1.68	−3.30	−3.97	−0.70	4.45	3.59	−0.83	−1.08	0.62	1.72
Gujarat	5.12	−1.36	−6.17	4.71	5.81	1.05	−2.82	−4.94	−2.17	1.80	2.55	0.73
Karnataka	4.14	7.02	2.77	6.54	13.66	6.69	−0.56	−6.55	−6.02	4.45	4.41	−0.04
Madhya Pradesh	1.93	1.46	−0.46	−3.97	−0.92	3.17	1.72	2.09	0.36	−0.73	0.60	1.33
Maharashtra	−0.75	−0.73	0.02	1.72	5.71	3.93	−1.43	−1.81	−0.38	0.78	2.55	1.75
Odisha	4.05	10.76	6.45	−16.75	−22.14	−6.47	−0.22	−0.78	−0.57	−6.11	−8.98	−3.06
Rajasthan	5.23	0.72	−4.29	2.16	3.62	1.43	−1.67	−2.51	−0.85	1.30	1.24	−0.06
Tamil Nadu	7.47	8.88	1.31	−4.11	−5.08	−1.02	0.07	−1.77	−1.84	1.93	3.18	1.22
Uttar Pradesh	3.40	1.45	−1.89	2.26	5.65	3.31	6.41	7.42	0.95	4.24	6.03	1.71
West Bengal	−6.60	−5.54	1.13	−1.91	−2.75	−0.86	−4.20	−1.57	2.75	−3.21	−1.82	1.43
India	3.65	5.77	2.05	−1.78	−1.77	0.01	0.54	0.19	−0.34	1.05	2.12	1.07
Growth rates for Masur												
Assam	5.54	1.44	−4.20	−0.65	2.00	2.65	−0.16	0.32	0.47	1.79	2.71	0.90
Bihar	0.96	3.70	2.71	−0.39	0.26	0.64	1.01	−0.90	−1.86	0.53	1.94	1.41
Madhya Pradesh	−0.76	0.02	0.81	5.73	5.81	0.09	0.66	1.13	0.46	2.28	3.24	0.94
Rajasthan	2.94	7.26	4.09	5.59	6.83	1.27	−7.51	−9.46	−2.11	1.39	3.59	2.10
Uttar Pradesh	6.76	10.53	3.54	0.87	1.21	0.35	1.49	1.74	0.28	3.36	5.06	1.67
West Bengal	−4.93	−3.39	1.61	−8.23	−9.54	−1.49	1.74	0.71	−0.90	−2.22	−0.44	1.78
India	1.29	3.93	2.62	1.84	1.58	−0.24	0.83	0.64	−0.16	1.81	3.21	1.38

Source: Authors' calculations based on data from official published sources.

Table 3.5 Growth stagnation in states during 1997–2007

Crop	State	Area	Prod.	Yield	Area Share per cent
Total Pulses	Rajasthan	−1.43	−4.23	−2.86	13.84
	Uttar Pradesh	−0.30	−1.11	−0.82	11.73
Gram	Rajasthan	−7.27	−9.11	−1.98	14.06
	Uttar Pradesh	−2.99	−3.39	−0.42	11.52
Arhar	Uttar Pradesh	−2.24	−4.67	−2.48	10.89
Urad	Maharashtra	−1.43	−1.81	−0.38	17.62
	Andhra Pradesh	−0.75	−0.58	0.17	16.18
Moong	Maharashtra	−0.58	−1.19	−0.62	20.84
	Andhra Pradesh	−0.62	−1.00	−0.38	14.71
	Karnataka	4.86	0.55	−4.11	12.69
Masur	Bihar	1.01	−0.85	−1.86	12.95

Source: Authors' calculations based on data from official published sources.

in area and yield of gram in Rajasthan and UP while in the other major state Madhya Pradesh (MP), there was a decline in yield. In case of arhar, there was a decline in UP (area and yield) and MP (yield). In urad there was a decline in Maharashtra (area and yield) and AP (area) and in the case of moong, there was a decline in Maharashtra (area and yield), AP (area and yield) and Karnataka (yield). Finally, in the case of masur, there was evidence of yield decline in Bihar.

Trends in Growth Acceleration

For pulses sector as a whole, area registered acceleration at least in one phase in few states but yield remained stagnant or has decelerated in both phases in majority of the states, more so in the second phase (Table 3.6). As a result, although some acceleration in production was witnessed in phase 1 (between sub periods 2 and 1) in few states, second phase (between sub periods 3 and 2) was marked by virtual stagnation and decelerating production growth in almost all the states except AP. Overall at the national level, the acceleration (or deceleration) of pulses production remained insignificant in both the phases. Crop-specific patterns show that all the crops, except gram have shown deceleration during phase 1 and slight acceleration in phase 2. However, due to large share of gram, the overall rate of acceleration in both phases turned out to be insignificant.

Table 3.6 Growth Accelerations for Pulses

States	Phase I			Phase II		
	Area	Prod.	Yield	Area	Prod.	Yield
			Growth acceleration for Total Pulses			
Andhra Pradesh	0.01	−0.04***	−0.04***	0.02**	0.06***	0.04***
Gujarat	−0.03**	−0.04	−0.004	−0.04***	−0.05	−0.02
Karnataka	−0.03***	0.02***	0.05***	0.03***	0.01	−0.02**
Maharashtra	0.01***	0.02**	0.01	−0.01	−0.01	−0.01
Madhya Pradesh	0.01***	0.02	0.01	−0.01	−0.03***	−0.02***
Rajasthan	0.06***	0.08***	0.02	−0.05***	−0.1***	−0.05***
Uttar Pradesh	−0.003	−0.02**	−0.01	0.005	−0.01	−0.01***
India	−0.002	0.0003	0.003	0.005	−0.01	−0.01***
			Growth acceleration for Gram			
Andhra Pradesh	0.15***	0.22***	0.07***	0.04***	0.05**	0.01
Gujarat	0.04	0.04	0.004	−0.01	−0.004	0.01
Haryana	0.01	0.09***	0.08***	−0.09***	−0.16***	−0.07***
Karnataka	0.05***	0.1***	0.05***	−0.02	−0.06***	−0.04***
Maharashtra	0.01	0.05***	0.04***	0.01	−0.01	−0.02
Madhya Pradesh	0.01***	0.02	0.01	−0.01**	−0.04***	−0.03***
Rajasthan	0.05***	0.07***	0.02	−0.12***	−0.15***	−0.03***
Uttar Pradesh	−0.03***	−0.04***	−0.01	0.02***	−0.01	−0.02***
India	0.02***	0.03***	0.01**	−0.01	−0.03***	−0.02***

(Cont'd)

Table 3.6 *(Cont'd)*

States	Phase I			Phase II		
	Area	Prod.	Yield	Area	Prod.	Yield
			Growth acceleration for Arhar			
Andhra Pradesh	−0.04***	0.0001	0.04***	0.05***	0.02	−0.02
Gujarat	−0.1***	−0.1***	0.001	−0.06***	−0.08***	−0.02
Karnataka	−0.07***	−0.03**	0.04**	0.07***	0.09***	0.02
Maharashtra	0.01***	−0.003	−0.01	−0.02***	0.01	0.03
Madhya Pradesh	−0.02***	−0.08***	−0.06***	0.03***	0.04***	0.02**
Tamil Nadu	−0.09***	−0.13***	−0.04**	−0.04***	−0.03	0.01
Uttar Pradesh	0.01***	−0.04***	−0.04***	−0.03***	−0.02***	0.01
India	−0.02***	−0.04***	−0.02***	0.004	0.02**	0.01
			Growth acceleration for Moong			
Andhra Pradesh	−0.01	−0.02	−0.01	−0.01	−0.04***	−0.03**
Bihar	−0.06***	−0.08***	−0.02**	0.02***	0.01	−0.01**
Gujarat	0.03	0.35***	0.33***	−0.03**	−0.05	−0.02
Karnataka	−0.08***	−0.08***	−0.004	0.07***	0.05	−0.01
Maharashtra	0.01	0.01	0.01	−0.01	−0.04**	−0.03**
Madhya Pradesh	−0.04***	−0.02	0.02	0.03***	0.01	−0.02***
Odisha	−0.2***	−0.31***	−0.11***	0.18***	0.28***	0.1***
Rajasthan	0.11***	0.17***	0.06	−0.03***	0.01	0.04
Tamil Nadu	−0.03	−0.0004	0.03**	0.03	0.01	−0.02**
Uttar Pradesh	−0.25***	−0.23***	0.02	−0.03***	−0.07***	−0.04***
India	−0.04***	−0.06***	-0.03***	0.03***	0.03***	-0.004
			Growth acceleration for Urad			
Andhra Pradesh	−0.02	−0.14***	−0.12***	−0.03**	0.03	0.06***
Karnataka	0.02***	0.06***	0.04**	−0.07***	−0.2***	−0.13***
Maharashtra	0.02***	0.06***	0.04***	−0.03**	−0.07***	−0.04***
Madhya Pradesh	−0.06***	−0.02***	0.04***	0.06***	0.03***	−0.03***
Odisha	−0.22***	−0.33***	−0.11***	0.18***	0.24***	0.06***
Rajasthan	−0.03***	0.03	0.06	−0.04	−0.06	−0.02
Tamil Nadu	−0.11***	−0.14***	−0.02	0.04	0.03	−0.01
Uttar Pradesh	−0.01	0.04***	0.05***	0.04***	0.02	−0.02**
West Bengal	0.05***	0.03	−0.02**	−0.02**	0.01	0.04***
India	−0.05***	−0.07***	−0.02***	0.02***	0.02***	−0.003
			Growth acceleration for Masur			
Bihar	−0.01***	−0.03***	−0.02***	0.01***	−0.01	−0.03**
Madhya Pradesh	0.06***	0.06***	−0.01	−0.05***	−0.05***	0.004
Uttar Pradesh	−0.06***	−0.09***	−0.03***	0.01	0.01	−0.001
West Bengal	−0.04***	−0.07***	−0.03**	0.10***	0.11***	0.01
India	0.01	−0.02**	−0.03***	−0.01***	−0.01	0.001

Note: *, **, *** denote significance at 10 per cent, 5 per cent and 1 per cent respectively.
Source: Authors' calculations based on data from official published sources.

DETERMINANTS OF PULSES PRODUCTION: ECONOMETRIC ANALYSIS

The pulses sector has been decelerating for a long time and particularly in the last sub-period (1997–2007). Are these declining trends reflective of lower relative profitability which in turn, induces lower input use and land diversion from pulse crops or higher volatility of pulse crop yields or both? To understand this, net returns per unit output[1] (rupees per quintal) of major pulse crops and non-pulse crops in all the pulse-growing states were calculated. The calculations were made for two time points—1997/98 and 2007/08. As an indicator of relative yield variability, the standard deviation of annual growth rates over the period 1997 to 2007 were calculated. The results indicate that the net returns of pulse crops were either equal to or higher than other crops in most of the states. This result was also supported by a number of other primary data-based studies carried out by various Agricultural Economics Research Centres (AERCs) in the country. Also the yield instability did not appear to be higher for pulses than for other major crops. Also, the market prices for pulses had been generally higher and growing in the last few years of the period. Then, what is driving the deceleration in pulses production? It is important to analyse the factors behind stagnation in pulse production in a more systematic way.

MODEL AND RESULTS

A systematic econometric analysis to capture the dynamic effect of major causal factors, after controlling for other relevant factors has been undertaken and the following econometric model in the Nerlovian framework of partial adjustment have been formulated.

Area under ith pulse crop

$$a_{it} = f(a_{it-1}, RP_{it-1}, RF_{it}, GIA, INST_{it}, PEST_{it-1}, z_a, u_a)$$

Yield of the ith pulse crop

$$y_{it} = f(y_{it-1}, RF_{it}, \%Irr_{it}, FERTHA_{it}, PESTHA_{it}, a_{it}, z_y, u_y)$$

Production of the ith pulse crop $q_{it} = a_{it} \times y_{it}$

Notation

a_{it} = Area under ith pulse crop in period t; y_{it} = Yield of ith pulse crop in period t; q_{it} = Production of ith pulse crop in period t; RF_{it}, GIA_{it} = Rainfall and irrigation in year t; RP_{it} = Relative price or relative profitability ith pulse crop vis-à-vis competing crops in year t; $INST_{it}$ = Production Instability measured as the standard deviation of year-on-year growth rates of last three years; $PEST_{it-1}$ = Area under pesticide coverage of the ith pulse crop, as a proxy for pest incidence; $\%Irr_{it}$ = Percentage of irrigated area under the ith pulse crop in period t; $FERTHA_{it}$ = Per hectare fertilizer consumption of the ith pulse crop in period t; $PESTHA_{it}$ = Percentage of area treated with pesticides under the ith pulse crop in period t; $z_{.t}$ and $u_{.t}$ denote the vector of other relevant exogenous variables and error term in the two equations respectively.

Two equations—one each for area cropped and yield were estimated for each of the pulse crops for all the major states. The hypothesized explanatory variables in area function were relative price (or relative profitability subject to the data availability for the state), rainfall, gross irrigated area (GIA), instability (price, yield, or revenue variability), and pesticide use. The GIA was expected to capture the irrigation-induced area shifts away from pulses. There are arguments in the literature that farmers shift away from pulses to more remunerative crops like cereals when irrigation facilities become available. If true, this would lead to a negative relationship in the movements between area under pulse crops and GIA. Thus, the GIA variable was expected to show a negative sign in this equation. The other explanatory variables were the standard ones that are expected to influence area under pulses. All these variables, except instability, were expected to have positive effect on dependent variable.

The hypothesized explanatory variables in the yield function are the rainfall, irrigated area under the crop, fertilizer used, pesticide use, and area under the crop. The inclusion of area under the crop in yield equation is a new feature in our model. This is intended to capture the shift of pulse cultivation to inferior lands (Swarna 1984). The area variable was expected to show negative affect on yield. All other explanatory variables were expected to show positive effect on the dependent variable. Rigorous statistical testing was carried out to assess model adequacy. Breusch-Godfrey Lagrange Multiplier (BG-LM test) for serial correlation, White's test for Heteroscedasticity, Ramsey regression specification error test (RESET), Cusum Q2 test for parameter stability were some of the tests used for the purpose. The results of the econometric analysis are presented in Table 3.7.

Results of the econometric analysis show that the major determinants of area of most of the pulse crops were

[1] Net returns were calculated as the difference between the value of production and cost of production. The costs considered are A2 and B2.

Table 3.7 Econometric Analysis- Summary Results
Time Period: 1975 to 2007 Method of Estimation: OLS

CROP	Area Function	R^2	Yield Function	R^2
Total Pulses	Lagged area, rainfall, relprice/ profitability. Negative irrigation effect negligible	0.75 (RJ) to 0.97 (AP)	Lagged yield, rainfall, fertilizer use, per cent of irrigated area, pesticide use. Negative marginal area effect not present	0.58 (KRN) to 0.97 (AP)
Gram	Lagged area, rainfall, relprice/ profitability, GIA (+ve). Negative irrigation effect negligible	0.79 (GJ) to 0.99 (UP)	Lagged yield, rainfall, fertilizer use. Negative marginal area effect not present but +ve in 3 states	0.58 (RJ) to 0.97 (MHR)
Arhar	Lagged area, relprice/profitability, pesticide, instability. Negative irrigation effect not present	0.94 (TN) to 0.99 (MP)	Lagged yield, rainfall, fertilizer use	0.60 (AI) to 0.92 (AP)
Moong	Lagged area, rainfall, relprice/ profitability. Negative irrigation effect present in 3 states and All-India	0.78 (TN) to 0.99 (MP)	Lagged yield, rainfall, fertilizer use, irrigation, pesticide. Negative marginal area effect only in UP.	0.59 (RJ) to 0.95 (BHR)
Urad	Lagged area, relprice/profitability, rainfall. Negative irrigation effect not present	0.81 (MHR) to 0.99 (KRN, ORS)	Lagged yield, rainfall, fertilizer use, irrigation. Negative marginal area effect only in 2 states (AP, KRN)	0.79 (MHR) to 0.98 (ORS)
Masur	Lagged area and GIA. Irrigation effect is positive.	0.78 (BHR) to 0.99 (AI)	Lagged yield,fertilizer use. Marginal area effect not present	0.86 (WB) to 0.97 (AI)

rainfall and relative price/profitability. The negative irrigation effect, that is, the irrigation-induced area shifts away from pulses was not present except in case of moong. In gram and masur, irrigation has actually contributed positively to area growth in many states. The major determinants of yield were mainly the rainfall, fertilizer use and to a lesser extent irrigation. The marginal area effect, that is, the adverse effect on yield due to cultivation on inferior lands, was present to a small degree only in case of moong and urad. In case of gram, there is some evidence that cultivation appeared to be carried out on better-quality lands in at least three states.

Results Based on Primary Data

Economics of Pulse Cultivation

The net returns per hectare are generally higher for pulses than for other crops. The net returns per quintal (price realized) were higher for pulses in all the districts without exception. Between the NFSM and the non-NFSM districts, the net returns per quintal were lower in the NFSM district for most of the crops and states although the net returns per hectare were higher. This shows that the contribution of area and yield was better in the NFSM district as compared to the non-NFSM district in most of the states, although this could not be attributed to the NFSM programme alone because of a very short period in the study.

Technology Adoption

More than 80 per cent of the farmers in the sampled districts were aware of the improved varieties (IV) of pulses. The level of awareness was generally lower in the non-NFSM districts. The main sources of knowledge about IVs in the NFSM district were *extension agent.* As expected, the role of extension agent was much stronger in the NFSM district as compared to the non-NFSM district. The percentage of households with area under IVs and the percentage of area under IVs were also higher in the NFSM districts than the non-NFSM districts. The percentage of farmers not following even one recommended practice was higher in non-NFSM districts.

Marketing

Majority of the households (>50 per cent) were marketing through the regulated market and majority of production (>50 per cent) was being marketed through regulated market in almost all the states, except Rajasthan. There was no procurement by NAFED in any of the sample districts—NFSM or Non-NFSM in any of the states.

Farmers' Perception

Profitability, lack of irrigation, home consumption, and inferior land quality were the reported reasons for growing pulses in most states. Higher pest incidence and lower yield were reported to be the major problems in growing pulses in majority of the states. Pod Borer was the most serious pest problem, except in UP, where pod fly was the major problem. Moong is the crop affected most by the pest problems followed by gram and arhar. Farmers in most of the states suggested improving irrigation facilities and making high-yielding varieties available as important, showing that non-price factors such as lower yield and yield instability were still important determinants of farmers' willingness to grow pulses.

Impact of NFSM

All the farmers were aware of and had derived benefits from NFSM in Rajasthan, Haryana, and AP. This percentage was slightly lower in Maharashtra and UP whereas in Punjab and Bihar this percentage was very low. The programme was found useful by farmers only in Rajasthan and AP. Assistance in the form of seeds was the most important in most of the states. Farmers in most of the states reported higher yield as the most important benefit derived from the NFSM program followed by increased knowledge and reduced pest attacks. As for impact on area and production, all the states except Haryana, registered an increase in area during 2008–09 of major crops—moong and gram compared to the previous two years. Similar was the case with production. All the crops except arhar showed an increase in production after the NFSM.

POLICY SUGGESTIONS

The overall growth trends of pulses in India showed that yield has been the major contributory factor to production growth. The contribution of area growth has been minimal. There was a deceleration in pulse production during the second sub-period (1987–96) but a major stagnation has set in during the third sub-period (1997–2007). The deceleration in the second sub-period indicated that the major pulses programs such as NPDP and other subsequent programmes launched during this period have not yielded the desired results. A disaggregate analysis at the crop level shows that broad trends noted above concealed the heterogeneity at the individual crop level. Most of the above mentioned trends were mainly due to gram which contributed about 35 to 43 per cent of the total pulse production in the country. For instance, positive production growth rate of 'total pulses' in the second sub-period was only due to gram. All other crops showed a distinct deceleration or decline in the second sub-period. Similarly, the disproportionately large contribution of yield growth to growth in production overall (1975 to 2007) was again mainly due to gram. It was only in case of gram that yield had contributed substantially while in all other crops area and yield showed more or less equal contributions.

State-level analysis of individual pulse crops showed that there was wide variability across states even within the same sub-period. In any crop, the trends at the All-India level result from those from few major states and not because of uniform trends in all the states. There were some worrying trends in the recent past (1997–2007). There was a decline in all three—area, yield, and production in major states like Rajasthan and UP. In case of gram, there was a decline in Rajasthan, UP, and MP (yield). In Arhar there was a decline in UP and MP. In urad, there was a decline in Maharashtra and AP, in moong there was deceleration in Maharashtra, AP, and Karnataka and finally in case of masur there was a major decline in yield levels in Karnataka and Bihar. The pattern in growth acceleration also showed trend similar to that of growth rates. Gram showed distinct acceleration in the first phase (between sub-periods 2 and 1) and deceleration in the second phase (between sub-periods 3 and 2). The trend in the remaining crops was the reverse. As a result of these offsetting trends and the overwhelming weight of gram, the overall acceleration of growth has remained insignificant in both phases. The results of the econometric analysis showed that the major determinants of area of most of the pulse crops were rainfall and relative price/profitability. The negative irrigation effect was generally not present, except in case of moong. In gram and masur, irrigation had actually contributed positively to area growth in many states. The major determinants of yield were mainly the rainfall, fertilizer use and to a lesser extent irrigation. The marginal area effect was present only in case of moong and urad.

Results from the primary data in several states showed that the returns from pulses cultivation were generally higher because of better prices. The farmers in the NFSM districts were aware of and adopted the IVs of pulses widely. Majority of the pulses production was marketed through the regulated market. However, public procurement of pulses by NAFED were not observed in any of the sample district. Higher pest incidence and lower yield are reported to be the major problems in growing pulses. Farmers suggested improving irrigation facilities and making available the improved varieties (pest-resistant) to increase pulses production. Most of the farmers in the NFSM districts were aware of NFSM and also derived benefits from NFSM. Higher yield was reported as the most important benefit derived, followed by increased knowledge and reduced pest attacks. Increase in area and production of pulses were recorded after the start of NFSM programme in most of the study states, except Haryana. Some of the policy implications that emerged from the study include providing assured price support through procurement by NAFED in the short-run; addressing the input supply problems such as provision of improved varieties, irrigation, etc., in the medium term; and developing and disseminating drought-resistant and pest-resistant varieties in the long-run.

Our results showed that one of the major determinants of area under pulses were the relative price/profitability. Therefore, providing assured price support through procurement should be useful in the short run. However, in the medium run, making inputs available at cheaper prices were required in order to increase profitability and induce farmers to devote more area to pulses. The potential for yield improvements also needs to be tapped. Fertilizer use, according to our results, is one of the major determinants of yield. Therefore, timely availability of fertilizer at affordable prices may help improve the yields.

In the long-run, development and dissemination of improved technology is very essential. The results from various primary data surveys showed that there was a need for improved varieties of seeds. The contribution of area to production growth has been minimal so far. Therefore, efforts should be made to increase area under pulses through bringing some of the rainfed rice fallow lands (estimated at 9.4 million ha) in Chhattisgarh, MP, Jharkhand, Odisha, and West Bengal. Lastly, the statistical system needs a great deal of improvement. Data on some of the crucial inputs like fertilizer and pesticide is still not available at the crop level. Efforts should be made to make this data available in order to strengthen the research efforts.

REFERENCES

Agricultural Economics Research Centre (AERC). 2001. 'Economics of Pulses Production and Identification of Constraints in Raising Their Production'. A Consolidated Report of AERC Studies, by S.K. Gupta. Research Study No. 79, Agricultural Economics Research Centre for Madhya Pradesh, J.N.K.V.V.

GoI. *Agricultural Statistics at a Glance* (various issues). Directorate of Economics & Statistics, Ministry of Agriculture, Co-operation & Farmers Welfare, Government of India.

Reddy, Amarender A. 2004. 'Consumption Pattern, Trade and Production Potential of Pulses'. *Economic and Political Weekly*, 39 (44): 4854–60.

———. 2009. 'Pulses Production Technology: Status and Way Forward'. *Economic and Political Weekly*, 44(52): 73–80.

Sathe, Dhanmanjari and Sunil Agarwal. 2004. 'Liberalization of Pulses Sector: Production, Prices and Imports'. *Economic and Political Weekly*, 39(30): 3391–7.

Sharma, D. and N.S. Jodha. 1982. 'Pulses Production in Semi-Arid Regions of India: Constraints and Opportunities'. *Economic and Political Weekly*, 17(52): A135–A148.

Sharma, Rita. 1986. 'Pulses in the Food Economy of India', in J.V. Meenakshi, Rita Sharma, and Thomas T. Poleman (eds), 'The Impact of India's Green Revolution on the Pulses and Oilseeds'. Agricultural Economics Research Report No. 86–22, Department of Agricultural Economics, New York State College of Agriculture and Life Sciences. Ithaca, New York: Cornell University.

Swarna, Sadasivam Vepa. 1984. *Pulses in India*. Unpublished PhD Dissertation, Department of Economics, Delhi School of Economics.

———. 1989. 'Pattern of Pulses Production: An Analysis of Growth Trends'. *Economic and Political Weekly*, 24(51): A167–A180.

Tuteja, Usha. 2009. 'Instability in Production and Trade of Pulses: A Global Analysis' Research Study No. 2009/06. Agricultural Economics Research Centre, University of Delhi.

ANNEXURE 3A

Table 3A The Details of Agro-Economic Research Centres and the Lead Person Involved in the State Report

State	AERC/Unit	Authors
Maharashtra	AERC, Gokhale Institute of Politics and Economics, Pune	Deepak Shah
Andhra Pradesh	AERC, Visakhaptnam, Andhra University	G. Gangadhara Rao, N. Ram Gopal
Bihar and Jharkhand	AERC for Bihar and Jharkhand, T.M. Bhagalpur University, Bhagalpur, Bihar	Rambalak Choudhary, Shambhu Deo Mishra
Uttar Pradesh	AERC, University of Allahabad, Allahabad	Ramendu Roy, S.N. Shukla, H. C. Malviya
Madhya Pradesh	AERC, Jawaharlal Nehru Krishi Vishwa Vidyalaya, Jabalpur	Ashutosh Shrivastava
Delhi	Agro-Economic Research Centre, Delhi University	Usha Tuteja
Punjab	AERC, Department of Economic and Sociology, Punjab Agricultural University, Ludhiana	D.K. Grover, J.M. Singh
Rajasthan and Gujarat	AERC, Sardar Patel University, Vallabh Vidyanagar, Gujarat	Rajeshree A. Dutta, Kalpana M. Kapadia

4

Problems and Prospects of Oilseeds Production in India

Vijay Paul Sharma

Indian agriculture has made considerable progress, particularly in respect of food crops such as wheat and rice in irrigated areas. However, performance has not been so good in case of other crops such as oilseeds, pulses, and coarse cereals. Therefore, after achieving self-sufficiency in foodgrains, the government is focusing attention on these crops. The oilseed sector has been an important area of concern and interventions for Indian policymakers in the post-reform period when India has moved from near self-sufficiency in the early 1990s to one of the largest importer of edible oils in the world, importing more than half of total domestic requirement (Government of India [GoI] 2015 and Directorate General of Commercial Intelligence & Statistics [DGCIS] 2015).

On the oilseeds map of the world, India occupies a prominent position, both in regard to acreage and production. India is the fourth largest edible oil economy in the world and contributes about 10 per cent of the world oilseeds production, 6–7 per cent of the global production of vegetable oil, and nearly 7 per cent of protein meal (SEA 2014). This sector has also an important position in the Indian agricultural sector covering an area of about 28.7 million hectares, with total production of about 31.2 million tonnes in triennium ending (TE) 2012–13 (GoI 2015, 2015a). This constitutes about 14.7 per cent of the gross cropped area (GCA) in the country. India was self-sufficient in edible oilseeds and oils until the mid-1960s and was a substantial export earner through exports of oilseeds, meals, extractions, and edible oils. With stagnation in production as well as a rise in population and demand for edible oils, the oilseed production fell short of its demand in the early 1970s. By the mid-1980s, edible oils was the most imported item, constituting about 30 per cent of the total imports, next only to petroleum products despite the fact that India had the world's second largest area under oilseeds. This concerned the policy planners, and government took a decision to achieve self-sufficiency in edible oilseeds. The initial strategy to overcome stagnant oilseed production was to promote new technologies in oilseed production and processing through Centrally Sponsored Schemes. The National Oilseed Development Project (NODP) was initiated in 1984–85 and launched in 1985–86 by reorienting various centrally sponsored schemes for oilseeds development. In May 1986, the GoI launched Technology Mission on Oilseeds (TMO) to increase production of oilseeds, reduce the import of oilseeds/oils and achieve self-sufficiency in edible oils. Oil Palm Development Programme (OPDP) was launched during 1991–92 under the TMOP with a focus on area expansion in Andhra Pradesh, Karnataka, Tamil Nadu,

Odisha, Gujarat, and Goa. As a result of concerted policy initiatives like TMOP in the mid-1980s and protection to domestic industry from imports, there was a significant progress in the production of oilseeds from the mid-1980s to the early 1990s. Between TE 1985–86 and TE 1993–94, production of oilseeds increased from 12.1 million tonnes to over 20 million tonnes, largely due to improved yields. Average yields increased from 644 kg/ha to 772 kg/ha during the corresponding period (GoI 2015b).

However, in pursuance of the policy of liberalization and globalization in the early 1990s, there were progressive changes in the trade policy in respect of edible oils. Edible oils, which were in the negative list of imports, were first decanalized partially in April 1994 with permission to import edible vegetable palmolein under open general license (OGL). This was followed by increasing the basket of oils under OGL in March 1995 when all edible oils (except coconut oil, palm kernel oil, RBD Palm Stearin), were brought under OGL. With decanalization and placing imports of edible oils under OGL in 1994–95, imports of edible oils particularly palm oil and soybean oil increased substantially during the subsequent years due to reduction in import duty, removal of quantitative restrictions (QRs) and other non-tariff barriers on all edible oils. Due to opening up of domestic markets and lack of appropriate technologies, the production of oilseeds in the country remained stagnant at about 20 million tonnes during the 1990s.

During the Tenth Five Year Plan, the Integrated Scheme on Oilseeds, Pulses, Oil Palm, and Maize (ISOPOM) was implemented by converging earlier schemes like Oilseeds Production Programme (OPP), OPDP, NPDP, and Accelerated Maize Development Programme (AMDP). In April 2010, pulses component of ISOPOM was with National Food Security Mission (NFSM) to intensify efforts for the production of pulses. In order to accelerate production of vegetable oils from oilseeds, oil palm and Tree Borne Oilseeds (TBOs) government launched National Mission on Oilseeds and Oil Palm (NMOOP) by merging ISOPOM, Oil Palm Area Expansion Programme (OPAE), and Integrated Development of TBOs (GoI 2014).

Given the competing demands on agricultural land from various crops and enterprises, the production of oilseeds can only be increased if productivity is improved significantly and farmers get remunerative prices and assured market access. However, farmers face various constraints in oilseeds production. As most of the oilseed crops are grown under rainfed conditions, several biotic, abiotic, technological, institutional, and socio-economic constraints also inhibit the exploitation of the yield potential of crops. Therefore, there is a need to identify and analyse important constraints that oilseeds farmers face and help them coping with these constraints. The present study attempts to analyse performance and potential of Indian oilseeds sector, identify major constraints in oilseeds production and suggest options for increasing oilseeds production and productivity in the country.

OBJECTIVES

The specific objectives of the study are:

1. to examine trends and pattern of growth of different edible oilseeds over time and across states and identify the sources of growth in edible oilseeds output in India, and
2. to identify major constraints in the edible oilseeds cultivation and suggest policy options to increase oilseeds production and productivity in the country.

RESEARCH METHODOLOGY

This study is based on both primary and secondary data pertaining to major edible oilseeds, namely soybean, groundnut, rapeseed and mustard, sesamum, and sunflower, grown in the country. In order to provide an overview of oilseed economy of the country, secondary data related to area, production and productivity of major oilseeds have been collected from different published sources such as *State-wise Area, Production and Yields Statistics*, *Agricultural Statistics at a Glance*, *Land Use Statistics at a Glance*, *Report on Price Policy for Kharif and Rabi Crops*, etc., published by Ministry of Agriculture and Farmers Welfare, Government of India. In order to study the growth trends and patterns, the study analysed a disaggregated time series data for major oilseeds in important states and the country. The study covered five major oilseeds, soybean, rapeseed and mustard, groundnut, sesamum, and sunflower, which account for over 90 per cent of total acreage and production of nine oilseeds in the country.

In order to identify constraints in edible oilseeds production in the country, primary data from the households growing oilseeds in selected states were collected. The states were selected based on share in production/acreage of respective oilseeds. Madhya Pradesh and Maharashtra for soybean, Rajasthan, Madhya Pradesh and Uttar Pradesh for rapeseed-mustard, Gujarat and Andhra Pradesh for groundnut, West Bengal for sesame, and Karnataka and Andhra Pradesh for sunflower crop

Table 4.1 List of Selected Crops, States, and Farm Category-wise Sample Size

Oilseed	Selected State	Marginal	Small	Medium	Large	Total
Soybean	Madhya Pradesh	62	47	93	38	240
	Maharashtra	110	70	69	1	250
	Total	172	117	162	39	490
Rapeseed and Mustard	Rajasthan	19	38	116	27	200
	Madhya Pradesh	23	34	46	17	120
	Uttar Pradesh	55	68	61	12	196
	Total	97	140	223	56	316
Groundnut	Gujarat	15	66	161	8	250
	Andhra Pradesh	31	78	130	11	250
	Total	46	144	291	19	470
Sesamum	West Bengal	165	43	42	–	250
Sunflower	Karnataka	72	110	66	72	320
	Andhra Pradesh	9	37	91	13	150
	Total	81	147	157	85	470

Source: Field Survey.

were selected for present study. The data on the socio-economic profile, operational holding, cropping pattern, area, production and yield of oilseeds and their cultivation aspects, and major constraints in cultivation of oilseeds were collected from oilseed producers in selected States (Table 4.1).

In order to meet the first objective of the study, secondary data on state-wise area, production, and yield of major edible oilseed crops/crop groups were analysed using compound annual growth rates (CAGR), averages, the coefficient of variations, etc. The analysis of trends and patterns of growth of different edible oilseeds over time and across states was done for the last three decades with a special focus on post-reform period. In order to identify major constraints in oilseeds production, farmers' responses were rated on a Likert scale ranging from 4 (severe) to 1 (not important).

RESULTS AND DISCUSSION

Trends in Area, Production, and Yield of Oilseeds

Indian agriculture has witnessed important changes over the last 3–4 decades and the most significant change has been a shift of acreage from coarse cereals to rice, wheat, and commercial crops, mainly fruits and vegetables and crop intensification. In relative terms, the share of total cereals in the GCA has declined from about 59.6 per cent in TE 1983–84 to about 51.7 per cent in TE 2010–11, indicating that increase in share of area under rice and wheat was less than the decline in area under coarse cereals. The share of oilseeds in GCA has increased from around 10.5 per cent in TE 1983–94 to 14.8 per cent in TE 2010–11 (Sharma 2014).

Average area under oilseeds, which was estimated at 12.4 million ha during the 1950s, increased to about 26 million ha during the last decade (Table 4.2). Annual production, which was about 6.1 million tonnes during the 1950s registered a rapid rise and reached a level of 26.6 million tonnes during the 2000s. The average productivity also increased from 488 kg/ha to 1015 kg/ha during the same period.

The area, production, and productivity of oilseeds grew at an annual compound growth rate of 1.48 per cent, 3.07 per cent, and 1.56 per cent, respectively during the period 1951–52 to 2013–14 (Table 4.3). Instability in area, production, and productivity of oilseeds computed using coefficients of variation, showed that the highest variability has been observed in case of production (56.8 per cent), followed by productivity (31.9 per cent), and the lowest in area (27.2 per cent) of oilseeds during the period 1951–2013. However, the performance of oilseeds during different decades shows quite interesting trends. As is evident from Table 4.3, oilseeds production recorded the highest growth rate (5.8 per cent) during the 1980s, followed by 2000s (4.42 per cent) and the lowest (0.57 per cent) during the 1990s. Almost a similar trend was observed in the case of variability in production. Yield variability has been a major factor for production variability during all decades, which is an indication of high yield risks associated with oilseeds.

Table 4.2 Trends in Average Area (million ha), Production (million tonnes), and Yield (kg/ha) of Oilseeds in India

	1951–52 to 1960–61[1]	1961–62 to 1970–71	1971–72 to 1980–81	1981–82 to 1990–91	1991–92 to 2000–01	2001–02 to 2013–14
Area	12.4	15.2	17.0	20.1	25.5	26.0
Production	6.1	7.6	9.2	13.6	21.3	26.6
Yield	488	497	538	671	836	1015

Source:

Table 4.3 Trends in Compound Annual Growth Rates (per cent) and Variability in Area, Production, and Yield of Oilseeds in India

Period	CAGR (per cent)			Coefficient of variations (per cent)		
	Area	Production	Yield	Area	Production	Yield
1950s	2.41	4.24	1.78	8.0	13.8	8.3
1960s	0.47	1.55	1.08	4.0	13.4	10.5
1970s	0.51	1.22	0.70	3.3	10.4	8.5
1980s	3.02	5.80	2.70	10.6	22.8	12.9
1990s	−0.87	0.57	1.45	4.7	9.8	7.8
2000s	1.45	4.42	2.88	8.0	18.9	14.1
All Periods	1.48	3.07	1.56	27.2	56.8	31.9

Source: Author's calculations using *State-wise Area, Production and Yield Statistics* (GoI 2013, 2015b).

The results of relative contribution of area expansion and yield improvement towards total change in oilseeds production using decomposition analysis showed that acreage expansion was a more important source of growth (55.7 per cent) in oilseeds output than yield improvement (31.4 per cent) between TE 1983–84 and TE 1993–94. However, the contribution of improved yield was higher (60.3 per cent) than area expansion (31.1 per cent) during the TE 2001–02 and TE 2011–12. These trends clearly show that yield has been a major contributor to increased output during the last decade, which is a healthy trend for oilseeds sector.

Regional Variations in Oilseeds Acreage and Production

The relative position of various oilseeds in total area and production of oilseeds is given in Figure 4.1. As is evident, soybean enjoys a dominant position both in terms of area and production. Its share in the output of oilseeds is over 40 per cent and in respect of total oil production, 29.4 per cent during the TE 2011–12. Rapeseed and mustard is the second important crop, its share being 24.5 per cent of oilseeds output and about 22.8 per cent of the acreage. It is interesting to note that rapeseed-mustard oil contributes a significant share to domestic supply, ranking number one, and its share in oil production being 35 per cent. Groundnut, which was predominant crop during the 1980s and the early 1990s, lost its share and accounted for 23.7 per cent of total production and 20.6 per cent in acreage during TE 2011–12. The share of kharif oilseeds was about 67 per cent and rabi oilseeds 33 per cent. The share of kharif oilseeds has increased during the last two decades.

The top four oilseed producing states, namely, Madhya Pradesh, Rajasthan, Gujarat, and Maharashtra accounted for over three-fourth of the total production and about two-third oilseeds acreage in the TE 2011–12 (Figure 4.2). Madhya Pradesh alone accounted for 27.5 per cent of the total oilseed production in India, with other three states contributing 48.3 per cent. Madhya Pradesh, Rajasthan, Gujarat, and Maharashtra have increased their share in oilseeds production during the last two decades while all other States have lost their share. Between TE 1991–92 and TE 2011–12, Madhya Pradesh recorded the highest increase (11.7 per cent) in its share, followed by Rajasthan (6.4 per cent), and Maharashtra (5.3 per cent). In case of acreage shares, the situation is slightly different. Andhra Pradesh, which is the 5th largest producer of oilseeds in the country, accounted for 12.9 per cent acreage (second largest acreage) during TE 1991–92, which declined to 8 per cent (5th position) during the TE 2011–12. Madhya Pradesh gained share in the area between TE 1991–92 and TE 2011–12 (from 16.4 per cent to 27.6 per cent). Other states like Rajasthan, Karnataka, Uttar Pradesh, Tamil Nadu, Odisha, and Haryana lost their share in oilseeds acreage. Area expansion in Madhya Pradesh and Maharashtra has been primarily driven by soybean cultivation. Among the major states, Maharashtra, Rajasthan, Madhya Pradesh,

[1] Data for 1951–52 to 1969–70 relates to total of five major oilseeds namely, groundnut, castor seed, sesamum, rapeseed, and mustard, and linseed.

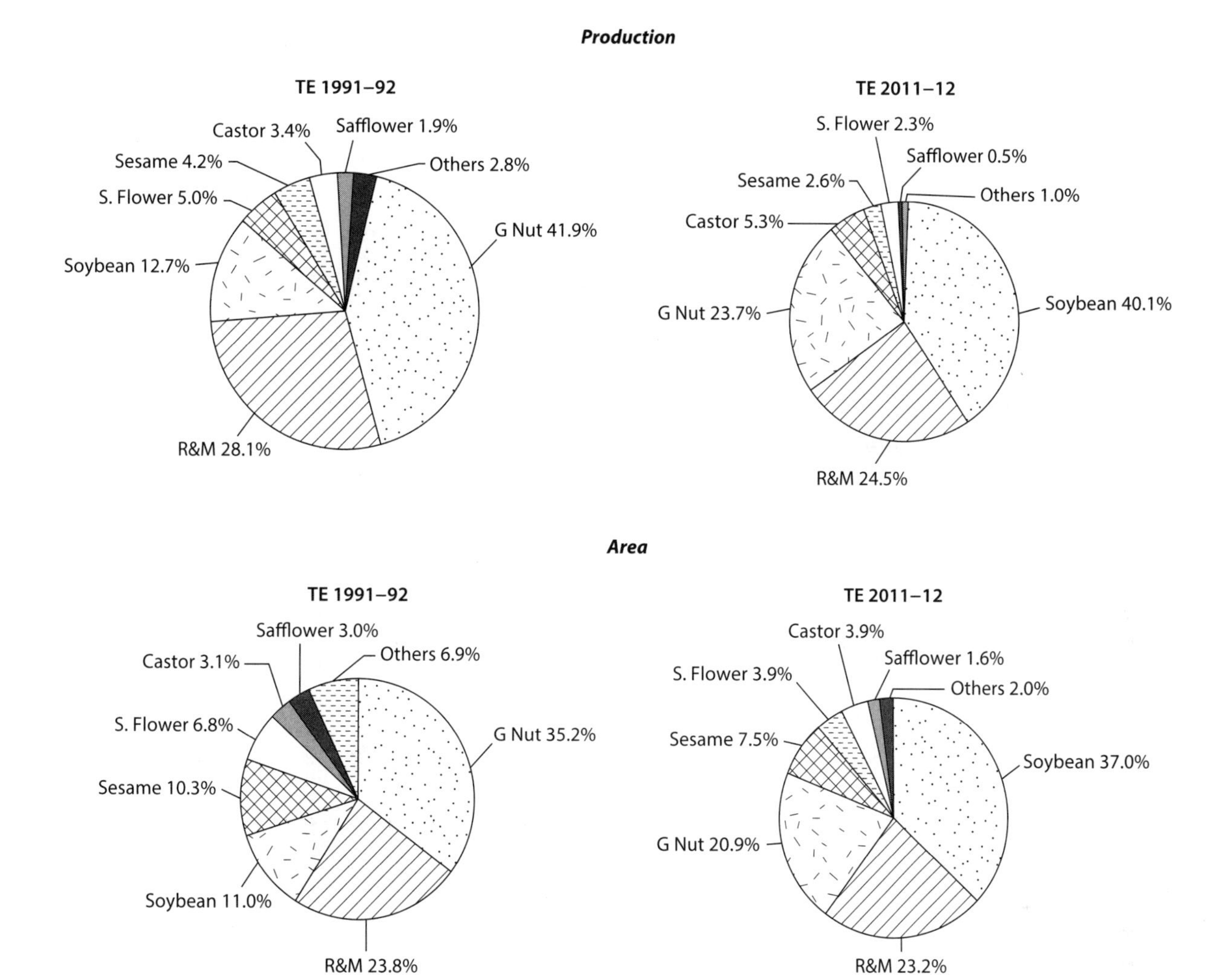

Figure 4.1 Changing Shares of Major Oilseeds in Total Acreage and Production of Oilseeds in India, TE 1991–92 to TE 2011–12
Source: Author's calculations using *State-wise Area, Production and Yield Statistics* (GoI 2013).

Gujarat, and West Bengal exhibited healthy growth rates in the area, production, and productivity during 1991–2011. However, there are wide variations in performance of different states during different time periods.

On a regional basis, Indian oilseed production (soybean, sunflower, and safflower), is highly concentrated (Sharma 2014). Soybean production is concentrated in three states, namely, Madhya Pradesh, Maharashtra, and Rajasthan, accounting for about 96 per cent of total production. Maharashtra and Rajasthan have increased their share in production while the share of Madhya Pradesh, the largest producer, has declined during the last two decades. Compared to soybean, the other major oilseeds are widely distributed and grown in many states. The main producers of rapeseed-mustard are Rajasthan (48.1 per cent), Madhya Pradesh (12.3 per cent), Haryana (11.9 per cent), Uttar Pradesh (10 per cent), West Bengal (5.8 per cent), and Gujarat (4.8 per cent). During the last three decades, share of Rajasthan in total production has increased significantly while Uttar Pradesh, which used to be the largest producer, has lost its share from 38 per cent in the early 1980s to about 10 per cent. About 85 per cent of groundnut production in concentrated in five states, namely, Gujarat, Andhra Pradesh, Tamil Nadu, Rajasthan, and Karnataka. Gujarat and Rajasthan have increased their share in national production while all other major producers like Andhra Pradesh, Tamil Nadu, Karnataka, and Maharashtra lost their share in total production during the last 2–3 decades. Groundnut area has been replaced by cotton due to the popularization of Bt cotton and higher income from Bt cotton in Gujarat and Andhra Pradesh.

Karnataka is the largest producer of sunflower seed in the country and has maintained its leadership during the

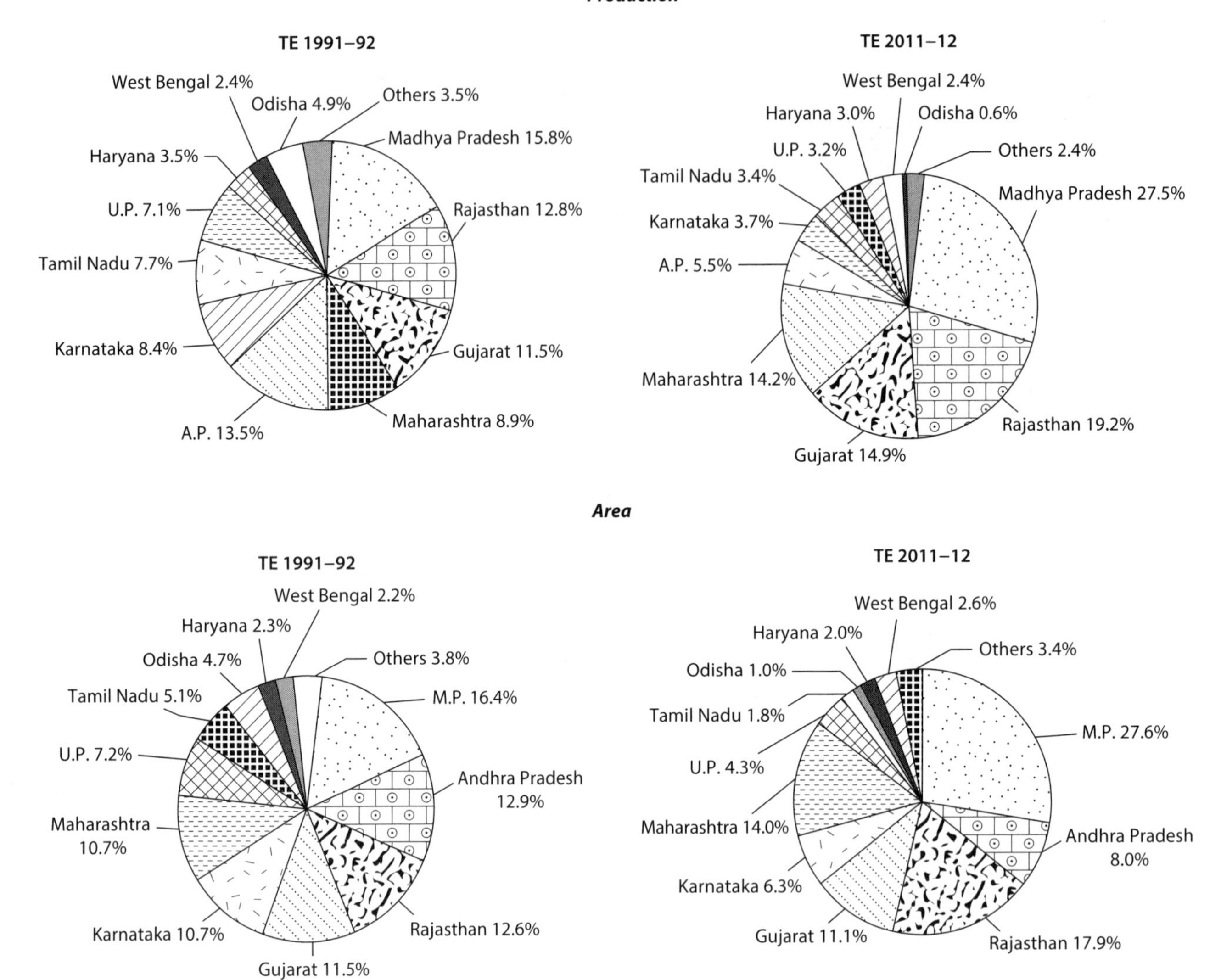

Figure 4.2 Changing Shares of Major States in Production of Oilseeds in India,[2] TE 1991–92 and TE 2011–12

Source: Author's calculations using *State-wise Area, Production and Yield Statistics* (GoI 2013).

last two decades. The other two major producers, Andhra Pradesh (27.2 per cent) and Maharashtra (14.6 per cent) account for over 40 per cent of the total production. Maharashtra has lost its share in sunflower production to other oilseeds, particularly soybeans while Andhra Pradesh has increased its share during the last three decades. Sesamum is grown in a number of states, but West Bengal and Rajasthan are major producers accounting for over 40 per cent of total production in the country. Top five producers account for over 80 per cent of production.

Among the major oilseeds, the performance of soybean has been much better than other oilseeds. Soybean production recorded the highest growth rate (6.47 per cent), followed by rapeseed and mustard (1.68 per cent) during the last two decades (Table 4.4). Groundnut and sunflower production had a negative growth in production. However in terms of productivity, rapeseed and mustard has performed better than soybean and groundnut. The performance of oilseeds sector, in general, has improved during the last decade. Groundnut, which had negative growth in production (–2.26 per cent) during the 1990s, recorded 1.63 per cent growth rate in production during the last decade and it was primarily driven by yield improvement (2.92 per cent) as groundnut acreage had negative growth rate (–1.26 per cent). Similarly, the

[2] In order to compare figures between two time periods, production/area data for Madhya Pradesh includes both for Madhya Pradesh and Chhattisgarh combined, for Uttar Pradesh both Uttar Pradesh and Uttarakhand, and for Bihar both Bihar and Jharkhand.

Table 4.4 Compound Annual Growth Rates of Area, Production, and Yield of Major Oilseeds in India: 1991–92 to 2011–12

	1990s	2000s	All Periods
	Soybean		
Area	8.08	5.53	5.18
Production	9.85	8.88	6.47
Yield	1.64	3.17	1.22
	Rapeseed and Mustard		
Area	−1.78	1.82	−0.12
Production	−1.15	3.71	1.68
Yield	0.63	1.85	1.80
	Groundnut		
Area	−2.75	−1.26	−2.12
Production	−2.26	1.63	−0.78
Yield	0.51	2.92	1.36
	Sunflower		
Area	−6.97	−5.44	−2.61
Production	−6.99	−2.53	−1.72
Yield	0.05	3.07	0.87
	Sesamum		
Area	−3.73	2.31	−0.19
Production	−3.48	2.56	1.36
Yield	0.27	0.24	1.56

Source: Author's calculations using *State-wise Area, Production and Yield Statistics* (GoI 2013, 2015b).

rapeseed and mustard production also increased at a faster rate (3.71 per cent) and was driven by both area expansion and yield improvements. Soybean witnessed the highest growth rate in production (8.88 per cent) among all oilseeds during the last decade but was slightly lower than the 1990s (9.85 per cent). Soybean production has been mainly driven by area expansion while yield improvement has been marginal. Therefore, efforts are needed to improve crop yields as scope for area expansion is limited. The above results clearly show that oilseeds sector, which had poor performance during the 1990s, has gained momentum during the last decade. In order to maintain the pace of growth in oilseeds production, there is a need to address technological, institutional, and socio-economic factors limiting oilseeds production in the country, which are discussed in the next section.

Factors Constraining Oilseeds Production

Given the rising demand for edible oils and increasing dependence on imports, there is a need to increase edible oilseeds production in the country. However, there are competing demands for agricultural land from various crops and scope for increasing area under oilseeds is very limited. Therefore, production of oilseeds can be increased if productivity is improved significantly and farmers get remunerative prices, better market access, technology, and other infrastructure facilities. However, oilseeds farmers face various constraints as most of oilseed crops are grown under rainfed conditions, and these constraints inhibit the exploitation of the yield potential of crops. Therefore, for improving crop yields, the first point to be emphasized is the magnitude of the yield gap and its main causes.

Table 4.5 Technological Gap and Extension Gap (in per cent) for Major Oilseeds-producing States

Crop/State	Technological Gap	Extension Gap
Soybean		
Madhya Pradesh	16.5	29.6
Maharashtra	41.9	21.0
Rapeseed and Mustard		
Rajasthan	1.8	9.0
Madhya Pradesh	12.5	22.4
Uttar Pradesh	–	11.7
Sunflower		
Karnataka	31.8	21.4
Andhra Pradesh	31.9	16.5
Safflower		
Maharashtra	28.8	23.0
Karnataka	49.9	19.9

Source: Field Survey.

The results of the yield gap analysis showed that significant gaps exist between actual and potential yields for different oilseeds crops (Table 4.5). The yield gap for safflower, sunflower, and soybean is higher than rapeseed and mustard. In case of soybean, Maharashtra has higher technological gap than extension gap, while, in Madhya Pradesh, the largest producer of soybean, extension gap is higher than the technological gap. In case of rapeseed-mustard, extension gap is higher compared with a technological gap. The yield gap estimates clearly show that there is a vast potential to expand oilseeds production in the country if farmers can access and efficiently use the available knowledge and technologies. The yield gap for most crops can be reduced to obtain yields closer to the potentially achievable yield by using improved crop varieties, the recommended levels of inputs, and better management of water, insects pests and diseases. But there are several questions which need to be addressed. Are these technologies and knowledge really available to the farmers? Are our institutions equipped to transfer the technologies and knowledge?

Narrowing yield gaps not only increases oilseeds yield and production, but also improves the efficiency of input use, reduce production costs, and increase sustainability. Exploitable yield gaps are caused by various factors, such as physical, biological, socio-economic, and institutional constraints, which can be effectively improved through identification and prioritization of major constraints affecting oilseeds production, appropriate government policy support, effective transfer of technologies, adequate and timely supply of quality inputs and farm credit, reduction of post-harvest losses, and strong linkages among research, extension, and farmers.

The results of constraint analysis show that at the national level, economic factors were the most important constraints in oilseeds production, followed by institutional factors, technological constraints, and agro-climatic constraints (Table 4.6). Among technological constraints, the incidence of insect pests and diseases and poor crop germination are the main problems for oilseeds production in the country. Policy-related impediments include unfavourable policies such as high costs of inputs, low and fluctuating crop prices, non-availability of timely and quality seeds and other inputs, and poor extension services. Lack of access to markets, exploitation by market intermediaries, lack of processing facilities in the region, and high transportation costs were major post-harvest management and market-related constraints. Most rural areas are inaccessible largely due to poor roads, which often restrict their access to market and prevent them from getting technologies and extension services.

The results showed that economic constraints constitute the major obstacles to the soybean and groundnut production while, in case of rapeseed-mustard, institutional constraints were the most important. Technological constraints ranked number two in case of soybean, groundnut, and sesamum cultivation. In case of sunflower, post-harvest management and value-addition were the most important constraint. Agro-climatic factors turned out to be the 3rd important constraint in oilseeds cultivation in the study states.

Figure 4.3 shows main constraints experienced by oilseeds farmers. The high cost of inputs was perceived to be the most important constraint faced by oilseeds producers, followed by low and fluctuating prices and incidence of diseases and pests. Other important constraints included shortage of human labour, timely availability of quality seeds, poor crop germination, exploitation by middlemen, poor extension services, higher risks and low profitability compared to competing crops, lack of irrigation facilities, access to institutional credit, poor extension services leading to lack of awareness about improved technologies, lack of processing facilities, and lack of information about prices and markets.

CONCLUSIONS AND POLICY IMPLICATIONS

The chapter presents important policy implications for improving oilseeds production and productivity in the country. The findings suggest that the strategy for boosting edible oilseed production in the country should lay emphasis on both price and non-price factors because technological, institutional, and economic factors influence the supply response of edible oilseeds. However, while recognizing the importance of price policy in accelerating the edible oilseed production, it is non-price factors like technology (crop varieties, irrigation) and institutional infrastructure (access to markets and market information), which are more important in influencing the crop area allocation decisions. There is a general perception that unfavourable prices for oilseeds is a main constraint to increasing oilseed production, however, there has been a conscious attempt in recent years to improve price parity of oilseeds vis-à-vis other competing crops through significant increase of minimum support price (MSP) to encourage cultivation of

Table 4.6 Ranking of Major Constraints in Oilseeds Production in India

Crops	Technological	Agro-climatic	Economic	Institutional	Post-harvest Management & Marketing
Soybean	2	3	1	4	5
Rapeseed and mustard	3	5	2	1	4
Groundnut	2	4	1	3	5
Sunflower	4	5	2	3	1
Sesamum	2	1	3	4	5
All Crops	3	4	1	2	5

Source: Field Survey.

Figure 4.3 Major Constraints Faced by Oilseed Growers in Production of Oilseeds

Source: Field Survey.

oilseeds crops. The trends in procurement prices of edible oilseeds during the last decade indicate that there has been a substantial increase (10–17 per cent per annum) in prices of edible oilseeds, much higher than main competing crops. It's also true that government procurement of oilseeds has been negligible as a major focus of the procurement has been on rice and wheat. Despite such increase in procurement prices, the growth in oilseeds production has been moderate because per hectare profitability of major oilseeds are much lower than competing crops mainly due to low yields and high risks. Therefore, price support policy alone cannot encourage oilseeds production. The increase in the MSP of oilseeds leads to an increase in the market price of edible oils and other by-products, which hurts consumers and processors. A significant increase in MSP of oilseeds may result in a rise in import of relatively cheaper edible oils and have an adverse effect on domestic producers and processors. Therefore, in order to increase edible oilseeds production and yields, a technological breakthrough in terms of suitable high-yielding varieties, irrigation, as well as accelerating technology dissemination through strengthening of extension services is required. It is also necessary to mention that there should be a regional approach to boost edible oilseeds output taking into account regional diversities in the trends and patterns of growth of different edible oilseeds.

The technological gap (difference between experimental and frontline demonstration yield) is quite high for most oilseeds and is caused mainly by factors that are generally non-transferable including environmental conditions. It is, therefore, difficult to economically narrow this gap. This calls for a review and refinement of production technologies developed to bridge this gap. The extension gap (difference between frontline demonstration yield and actual farm yield) is also high for most crops and is mainly caused by lack of proper management practices, suboptimal use of inputs and institutional bottlenecks. The lack of availability of quality seed of improved varieties and other inputs and services is perceived to be a major concern for oilseed cultivators. Ensuring the availability of key inputs such as quality seed, fertilizers, pesticides, credit, risk management tools including crop insurance, and extension services in oilseeds producing regions would help in increasing productivity and production. The Research–Extension–Farmer–Industry linkages should be strengthened to reduce yield gap.

Over the last two decade, Indian edible oil sector has become more liberalized but there are two major problems with the edible oil import duty structure in India: low bound rate of duty (45 per cent) on soybean oil, which has the second largest share in edible oil imports and low applied duty rates and high variability in import duty structure with frequent changes in tariffs. The bound rate of duty for soybean is not sufficient to protect domestic producers/processors when world prices are low (for example, in the first half of last decade) as there is considerable substitution among various oils based on prices. Second, import duty on edible oils has been very low since April 2008 when import tariff on crude palm oil was reduced to zero and on refined palm and soybean oil to 7.5 per cent.

Low import duties on edible oils adversely affect the oilseeds farmers. Moreover, high dependence on world market for large quantity of oils is risky given the fact that world oilseeds production has high fluctuations due to dependence on weather in major exporting countries and demand in some importing countries may go up for non-edible purpose like biodiesel. Therefore, there is a need to have consistent trade policy which protects the interests of both producers and consumers and helps in making India self-sufficient in edible oils in the long run.

The long-term strategy to make India self-sufficient in edible oilseeds/oils should focus on technology by evolving new location-specific high yielding varieties, more coverage under assured irrigation and better water use efficiency, appropriate pricing incentives, and trade policy and ensure timely availability of quality inputs such as fertilizers, pesticides, credit facilities, crop insurance, and assured market access. Investment in research and development of oilseeds complex is a key element and should be stepped up. The dissemination of technology is equally important and needs to be strengthened through effective agricultural extension system. Extending oilseed cultivation to non-traditional areas and as mixed cropping system is worth considering. The potential of non-traditional edible oils like rice bran oil, corn oil, cottonseed oil, needs to be exploited to boost India's edible oil output and reduce the dependence on imports.

REFERENCES

Directorate General of Commercial Intelligence & Statistics (DGCIS). 2015. *Monthly Statistics of Foreign Trade of India*, Vol. II: *Imports*. Kolkata: Directorate General of Commercial Intelligence & Statistics.

Government of India (GoI). 2013. *State-wise Area Production and Yields Statistics (Major Crops, 2012–13).* New Delhi: Ministry of Agriculture, Government of India.

———. 2014. 'National Mission on Oilseeds and Oil Palm (NMOOP) Operational Guidelines', National Mission on Oilseeds and Oil Palm (NMOOP), Ministry of Agriculture, Government of India. Available at http://nmoop.gov.in/Guidelines/NMOOP20114.pdf.

———. 2015. 'Third Advance Estimates of Food Grains, Oilseeds & Other Commercial Crops for 2014–15', Ministry of Agriculture, Government of India, New Delhi.

———. 2015a. *Land Use Statistics at a Glance 2003–2004 to 2012–2013.* New Delhi: Ministry of Agriculture, Government of India.

———. 2015b. 'Agricultural Statistics at a Glance 2014'. Directorate of Economics and Statistics, Department of Agriculture and Cooperation, Ministry of Agriculture, Government of India, New Delhi, 103.

Sharma, Vijay Paul. 2014. *Problems and Prospects of Oilseeds Production in India,* Report Submitted to the Ministry of Agriculture. Ahmedabad: Centre for Management in Agriculture, Indian Institute of Management.

ANNEXURE 4A

Table 4A The Details of Agro-Economic Research Centres and the Lead Person Involved in the State Report

State	AER Centre/Unit	Authors
Andhra Pradesh	AERC Waltair	M. Nageshwara Rao
Gujarat	AERC Vallabh Vidyanagar	Mrutyunjay Swain
Karnataka	ISEC, Bangalore	Komol Singha, Parmod Kumar and Kedar Vishnu
Madhya Pradesh	AERC Jabalpur	Hari Om Sharma, Deepak Rathi
Maharashtra	GIPE, Pune	Jayanti Kajale and Sangeeta Shroff
Uttar Pradesh	AERC Allahabad	Ramendu Roy
West Bengal	AERC Shantiniketan	D. Roy, F.H. Khan

5

Production and Marketing of Basmati Rice in Punjab

D.K. Grover, Jasdev Singh, Sanjay Kumar, and J.M. Singh

Indian farmers are vulnerable to both production and market risk throughout their life. Vagaries of monsoon, occurrence of pest and diseases, etc., marked the risks associated with the production. The output variations in agricultural commodities lead to wide fluctuations in their prices exposing the growers to considerable marketing risk. Inadequate market infrastructure and prevalence of too many intermediaries between producer and consumer result in higher marketing costs thus lowering the share of producer in the consumer's rupee. Changes in domestic and global fundamentals, poor market intelligence, poor infrastructure facilities, large number of intermediaries, lack of awareness about quality standards, etc., are the main causes for market-related risks. Price volatility led to disadvantage to producers and final consumers but benefitted to the large number of intermediaries. The nature of markets of agricultural commodities and imperfections in these markets also influence the price transmission and the final consumer prices. Instability in commodity prices has always been a major concern of producers, processors, traders as well as the consumers in an agriculture dominated country like India. Farmers' direct exposure to price fluctuations, for instance, makes it too risky for many farmers to invest in otherwise profitable activities (Sahadevan 2008). Considering the above observations, it is necessary to study the farm level crop profitability and the relationship between movements in market prices of important agricultural commodities along with the analysis of price relationships in movement of these commodities from farm gate to the retail level. In current study this inquest was applied to the basmati rice in Punjab state.

Basmati is long-grain aromatic rice grown for many centuries in the specific geographical area at the Himalayan foot hills of Indian subcontinent. India being the world's largest producer contributes more than 70 per cent of the total world basmati rice production Annual production of basmati rice in India hovers around 4 million tonnes and it was estimated to be 4.7 million tonnes in 2012. Export of basmati rice from India has increased from about 8.51 lakh tonnes during 2000–1 to 37.57 lakh tonnes during year 2013–14. During 2012–13, the country earned crucial foreign exchange having worth of Rs 19.40 thousand crore contributing about 8.4 per cent towards the total earnings from the export of agricultural commodities (Adhikari and Sekhon 2014). Basmati exports from India are highly concentrated to few countries. During 2007–08, the share of Saudi Arabia in the total export from India was about 47 per cent followed by UAE (15.87 per cent), Kuwait (9.24 per cent), UK (6.69 per cent), and USA (3.27 per cent). Since 2008, after Pusa Basmati 1121 variety of rice got notified as basmati, Iran became the important buyer of this rice variety (Sidhu, Singh, and Kumar 2014).

Haryana, Punjab, Uttar Pradesh (UP), Uttarakhand, and Jammu and Kashmir (J&K) are the basmati growing states in India. The share of Haryana in total Indian basmati production during 2011–13 was about 37.18 per cent followed by Punjab with 36.47 per cent, and UP with 23.65 per cent (see www.airea.net). During 2000–01, area under basmati rice in Punjab was 1.04 lakh hectares and production 1.61 lakh tonnes which increased to 5.50 lakh hectares with production of 13.29 lakh tonnes in 2010–11. During 2013–14, area and production of basmati rice in Punjab was estimated at 5.59 lakh hectares and 14.73 lakh tonnes respectively (Directorate of Agriculture, Punjab 2014). The present study is an attempt to analyse the profitability of basmati cultivation, behaviour of prices of basmati, both over years as well as across the selected markets along with enquiry into the relationship among farm harvest prices, wholesale prices, retail prices, and export prices. The specific objectives of the study are:

1. to study the cost structure and profitability of basmati cultivation in state;
2. to study the relationship between movements in market prices at important markets; and
3. to study the divergence among farm harvest prices, wholesale prices, retail prices, and export prices.

METHODOLOGY AND DATA

The study has been based on both primary as well as secondary data. Among the different districts of Punjab, major proportion of area under basmati is concentrated in its traditional belt comprising Gurdaspur, Amritsar, and Tarntarn districts. However, recently, the area under basmati cultivation in non-traditional districts of state such as Ferozepur, Sangrur, Patiala, and Mukatsar, has also been picked up in a major way. Multistage sampling procedure was adopted for selection of sample basmati growers from three major basmati growing districts (that is, Gurdaspur, Amritsar, and erstwhile Ferozepur) accounting for nearly 46 per cent of total basmati production of state during 2012–13. Further, from each of the selected district, two blocks with highest area under basmati were selected. Thus overall six blocks from the sample districts were selected. At next stage of sampling a cluster of two to three villages with concentration of basmati rice was chosen randomly from each of the selected block for the farm household survey. Finally from each of the selected village cluster, 25 basmati growers, in proportion to their respective proportionate share in different categories of operational holdings in state, namely, marginal (<1 ha), small (1 to 2 ha), medium (2 to 4 ha) and large (>4 ha) were selected randomly. Thus, overall from state, total sample of 150 households growing basmati, comprising 18 marginal, 30 small, 48 medium, and 54 large farmers forms the basis for the present enquiry. To study the market channels and price spread, the information was collected from ten wholesalers and ten retailers selected randomly from the local markets. Similarly five basmati exporters/millers were randomly selected to collect the data. The comprehensive survey was conducted in the sample villages at end of crop year 2013–14 (Reference year). Secondary data on monthly market arrivals and prices of basmati were collected from the offices of Agricultural Produce Market Committees (APMC) of the four major basmati markets namely, Amritsar, Tarntarn, Jallalabad, and Fazilika. Simple statistical tools like averages and percentages were used to compare and interpret results properly. Time series analysis was conducted to examine the patterns in basmati paddy prices. To measure inter- and intra-year price variability coefficients of variation were also worked out. Correlation coefficients were worked out to examine the paddy price relationship of different sample markets.

Time series analysis was conducted to examine the patterns in basmati paddy prices. Seasonal behaviour was studied by constructing the seasonal indices. The seasonal indices were worked out by using ratio to moving average decomposition method. Time series is very likely to show a tendency to increase or decrease. This illustrates the trend which usually predominant in time series. To separate out the trend factor, straight line trend was fitted on the moving average data of the type: $Yt = a + bT + Ut$, which gives us the trend values (T). To calculate cyclical and irregular movements the residual method was used. This method consists of eliminating seasonal variation and trend from the original series as following:

$$T \times C \times I = (T \times S \times C \times I)/S$$
$$C \times I = (T \times C \times I)/T$$

where, T = Trend component; S = Seasonal component; C = Cyclical element, and I = Irregular fluctuations.

The trend cycle components were plotted against the time to examine the cyclical behaviour of prices. To check inter-year and intra-year instability of prices, the coefficients of variation (CV) were worked out for de-seasonalized price data. Market integration among spatially separated markets of state was evaluated by constructing the correlation coefficients (r). The methodologically correct way of using correlation coefficients in spatial price analysis is to calculate the correlation coefficient of the residuals (Croxton and Cowden 1955). Trend and seasonal components reflect generally pervasive influences throughout an agricultural region. The trend due to rising demand occasioned by population increase, for instance apt to affect all the regions; change in supply during the year reflects in part a common climatic pattern (Blyn 1973). Thus, in this chapter, market integration among state markets was examined through calculating the correlation coefficients of the residuals. For this, trend and seasonal components from the raw price data of the study markets were eliminated through decomposition analysis (as discussed above). The significance of 'r' was tested using t test. A higher degree of the correlation coefficient indicates a greater degree of integration.

RESULTS AND DISCUSSION

The overall average operational farm size on sample farms in state was 4.82 ha. The average operational area on marginal, small, medium, and large category farms was 0.78, 1.61, 3.29, and 9.31 ha, respectively. There was adequate availability of irrigation water as entire (100 per cent) operational area on the sample farms was under assured irrigation Wheat and basmati were major crops on all the farm size categories and on average accounted for 44.77 and 32.00 per cent of the gross cropped area (GCA) on the sample farms in state followed by rice, other major crops (primarily fodder) and sugarcane. Cropping intensity on overall farms was 194 per cent. On overall sample farms, Pusa Basmati 1121 was the most dominating variety of basmati which accounted for nearly 90 per cent of the total area under basmati crop. This was followed by Traditional Basmati and Pusa Punjab Basmati 1509 varieties accounting for 6.96 and 3.42 per cent of the total area under basmati, respectively. On sample farms basmati varietal diversification showed a positive relationship with the farm size, which meant that larger size categories of farms had sown more varieties.

The results have been discussed under the following sub heads:

1. Production and marketed surplus of basmati
2. Economics of basmati cultivation
3. Marketing of basmati
4. Price patterns and price spread of basmati rice
5. Stakeholders' perceptions on basmati production and trade

Production and Marketed Surplus of Basmati

Status of basmati area on different categories of farmers revealed that total area under basmati cultivation on average farm under basmati was 3.00 ha (Table 5.1). The average per farm basmati production on overall farms was 123.54 quintals. Category-wise, per farm production of total basmati on marginal, small, medium, and large farms was 24.28, 42.14, 87.49, and 233.90 quintals respectively. Average per farm and family requirements/ consumption of basmati was 1.34 quintals which accounted for 1.01 per cent of the total production on overall farms. Proportionate share of per farm retention of total basmati for other uses and wastage was 0.90 quintal (0.73 per cent) and 0.57 quintal (0.46 per cent) respectively. While the proportionate share of consumption in total production was inversely related to the farm size, share of wastage to total production varied directly with the farm size. The marketed surplus of basmati collectively for all varieties

Table 5.1 Area, Production, Consumption, and Marketed Surplus of Basmati (All Varieties) on Sample Farms, Punjab, 2013–14 (Per farm)

Farm category	Area (ha)	Production (qtls)	Consumed (qtls)	Retained/stocked for future use (qtls)	Wastage	Sold (qtls)	Price (Rs/qtl)
Marginal	0.60	24.28 (100.00)	0.60 (2.47)	0.24 (0.97)	0.00 (0.00)	23.44 (96.56)	3605
Small	1.08	42.14 (100.00)	0.83 (1.98)	0.44 (1.05)	0.10 (0.24)	40.77 (96.73)	3983
Medium	2.10	87.49 (100.00)	1.45 (1.65)	0.78 (0.89)	0.31 (0.35)	84.96 (97.10)	3922
Large	5.67	233.90 (100.00)	1.76 (0.75)	1.49 (0.64)	1.26 (0.54)	229.39 (98.07)	3917
Average	3.00	123.54 (100.00)	1.34 (1.08)	0.90 (0.73)	0.57 (0.46)	120.73 (97.73)	3915

Note: Figures in the brackets indicate the per cent share in production.
Source: Field survey data.

on the overall sample farms increased both in absolute as well as proportionate terms with the increase in farm size. The marketed surplus on overall farms accounted for 97.73 per cent of total basmati production. The average price received by farmers for their marketed surplus was Rs 3915 per quintal.

Economics of Basmati Cultivation

The total variable cost of cultivation of overall basmati (all varieties) on the sample farms including cost involving the post-harvest operations of basmati paddy was Rs 30256 per ha (Table 5.2). The variable input cost accounted for

Table 5.2 Economics of Basmati Cultivation (All Varieties) on Sample Farms, Punjab, 2013–14 (Rs per ha)

Particulars	Farm Category				
	Marginal	Small	Medium	Large	Total
Input costs					
Seed	660 (2.69)	712 (2.68)	707 (2.55)	761 (2.56)	743 (2.57)
Irrigation	978 (3.99)	422 (1.59)	707 (2.55)	445 (1.50)	515 (1.78)
Manure and fertilizer	2400 (9.79)	2890 (10.88)	3200 (11.53)	3890 (13.08)	3627 (12.54)
Labour (bullock + human)	11210 (45.73)	13958 (52.56)	13559 (48.85)	15074 (50.71)	14562 (50.33)
Machinery hire charges	6078 (24.79)	5250 (19.77)	5709 (20.57)	5385 (18.12)	5465 (18.89)
Pesticides/weedicides	3188 (13.00)	3325 (12.52)	3877 (13.97)	4170 (14.03)	4020 (13.89)
I) Total input cost	24513 (100.00)	26557 (100.00)	27759 (100.00)	29726 (100.00)	28931 (100.00)
Storage, transportation, and marketing cost					
Storage	0 (0.00)	17 (1.19)	64 (4.96)	76 (6.04)	67 (5.07)
Transportation	535 (43.19)	486 (34.07)	266 (20.58)	278 (22.09)	345 (26.03)
Marketing losses	148 (11.93)	206 (14.47)	186 (14.40)	132 (10.49)	150 (11.30)
Marketing (unloading, cleaning, etc.)	523 (42.22)	506 (35.43)	543 (41.97)	537 (42.71)	535 (40.40)
Kind payments	33 (2.65)	207 (14.52)	205 (15.83)	180 (14.33)	184 (13.88)
Misc.	0 (0.00)	5 (0.32)	29 (2.26)	55 (4.35)	44 (3.32)
II) Total storage, transportation, and marketing cost	1239 (100.00)	1427 (100.00)	1293 (100.00)	1256 (100.00)	1325 (100.00)
Percentage share to the total cost (I + II)					
Total input cost (I)	95.19	94.90	95.55	95.94	95.62
Total storage, transportation, and marketing cost (II)	4.81	5.10	4.45	4.06	4.38
Total variable cost (I + II)	25752 (100.00)	27985 (100.00)	29052 (100.00)	30982 (100.00)	30256 (100.00)
Profitability					
Productivity (qtl/ha)	40.24	38.90	41.75	41.28	41.18
Gross returns (Rs/ha)	148697	158261	167057	165114	164658
ROVC (Rs/ha)	122945	130276	138005	134132	134402

Note: Figures in parentheses indicate per cent to the respective total cost.
Source: Field survey data.

about 96 per cent where as the rest about 4 per cent constituted the post harvest/marketing costs. Category-wise, the lowest cost (Rs 25752 per ha) was observed to be on the marginal category farms and the highest (Rs 30982 per ha) on the large category farms. In overall variable input cost of basmati cultivation, the major share was contributed by labour cost to the tune of about 50 per cent (varying from 45.73 per cent on marginal to 52.56 per cent on small farms) followed by the machine cost at about 19 per cent (varying from 18.12 per cent on large to 24.79 per cent on marginal farms). The other cost components like expenses on pesticides/weedicides accounted for about 14 per cent (varying from 12.52 per cent on small to 14.03 per cent on large farms) and expenditure on fertilizers and manures at about 13 per cent (varying from 9.79 per cent on marginal to 13.08 per cent on large farms) of the total variable input cost. On account of free electricity supply to irrigation pumps the expenditure on irrigation component accounted for only 1.78 per cent of total variable input cost on overall farms and it varied from 1.50 per cent on large farms to 3.99 per cent on marginal farms. Total cost on post harvest/marketing operations on overall sample farms was Rs 1325 per ha which varied from Rs 1239 per ha on marginal farms to Rs 1427 per ha on the small farms. Among different post-harvest/marketing costs the unloading and cleaning accounted for the major proportion followed by transportation, kind payments, marketing losses, and storage cost.

The per hectare gross returns and returns over variable cost (ROVC) on overall sample farms were Rs 164658 and Rs 134402 respectively. Among different categories the gross returns were found to be highest on medium size farms (Rs 167057/ha) and lowest on the marginal farms (Rs 148697/ha). ROVC were also observed to be highest on medium size farms (Rs 138005/ha) and lowest on the marginal farms (Rs 122945/ha). Lowest returns on marginal farms were particularly due to the relatively low price received by them for their produce which was significantly lower at Rs 3605 per quintal against overall average price received by the sample farms at Rs 3915 per quintal.

Marketing of Basmati

On overall sample farms, per farm marketed surplus of basmati (all varieties) was recorded at 120.73 quintals which was sold on an average price of Rs 3915/qtl. Market-wise, 96.33 per cent of marketed surplus was sold in the regulated market at price of Rs 3935/qtl. The basmati sold in other channels like village market and to others comprised only 2.83 and 0.83 per cent of total marketed surplus. The average price received in these markets was Rs 3062/qtl and Rs 4500/qtl respectively. Category-wise, marginal farmers sold their 91.94 per cent marketed surplus in regulated markets at price of Rs 3636/qtl and 8.06 per cent in the village market at price of Rs 3247/qtl. The large farmers had sold their 94.91 per cent proportion of total marketed surplus in the regulated market at price of Rs 3945/qtl followed by 3.87 per cent in village market at price of Rs 3048/qtl and other 1.22 per cent to others at price of Rs 4500/qtl. Small and medium farmers had sold their entire marketed surplus in regulated market at price of Rs 3983/qtl and Rs 3922/qtl respectively. Category-wise, for overall marketed surplus the average price received by small, medium and large farmers had very small difference and it varied from Rs 3917/qtl to Rs 3983/qtl received by the large and small farmers respectively. Lack of bargaining power as well as selling relatively largest proportion (60.19 per cent) of marketed surplus during October, month in which the prices remained significantly low as compared to other months of marketing year seems to be the reason for relative overall low price (Rs 3605/qtl) received by marginal farmers.

It was observed that sample farmers had sold 82.43 per cent of their marketed surplus during the crop harvesting period consisting months of October and November of marketing year 2013–14. During October, 34.44 per cent of the total marketed surplus was sold, however the price received during this month was Rs 3463/qtl which on an average was 11.55 per cent lower than the average annual prices (Rs 3915/qtl) received by the farmers. As harvesting season starts in October, the high moisture content of output during this month results in the low prices. The proportion of marketed surplus sold during month of November was 47.99 per cent which on average fetched price of Rs 4007/qtl which was 2.35 per cent higher relative to the annual average price. During December, January, February, and others months (March onwards) of the marketing year 2013–14, the overall quantity sold constituted 6.65, 3.57, 3.05, and 4.31 per cent of the marketed surplus respectively. Respective prices received during these months were Rs 4820, Rs 4378, Rs 4436, and Rs 4356/qtl which were 23.10, 11.81, 13.30, and 11.26 per cent higher as compared to the annual average price received by the farmers. During December month nearly 78 per cent of total marketed surplus of the traditional aromatic basmati fetching significant higher price was sold, which is reflected in the price received for overall basmati during this month. Category-wise it was found that marginal farmers had sold their entire marketed surplus during harvesting months. The small farmers besides selling 89.27 per cent of their marketed surplus in harvesting months had also sold 10.73 per cent of it during the month of December. Medium farmers besides selling major proportion (86.99 per cent) of marketed surplus in months of October and November had also sold during the months of December (7.82 per cent), and January (5.18 per cent). It was the large farmers who on account of their output holding capacity had sold their marketed surplus even after month of January. Besides selling major proportions (79.64 per cent) of marketed surplus during October and November months, farmers of this category had also sold significant proportions, that is, 6.09, 3.51, 4.46, and 6.30 per cent during the months of December, January, February, and March onwards respectively.

The promotion of basmati paddy cultivation is part of the state government's diversification plan aimed at reduction of area under water consuming common paddy crop in Punjab. Although government had not provided any direct benefit to the basmati farmers, but major relief provided to basmati millers in the form of market tax exemptions on basmati purchase from Punjab markets is primarily aimed at ensuring remunerative prices to basmati farmers. During 2013–14 marketing year, the Punjab Government had waived the market fee (2 per cent), rural development fee (2 per cent), and infrastructure development fee (3 per cent) on purchase of basmati paddy for milling in Punjab. Besides, Value Added Tax (VAT) is zero on all basmati exports which are sourced from Punjab.

All of the respondent wholesalers procured their major part of supplies from the rice millers. Second important source of supply for wholesalers was observed to be the other wholesalers. The major source of supply of basmati for the retailers was the wholesalers. Besides, second most important source of supply for them was either processors or other retailers in the market. It has been found that most of the exporters dealing with basmati rice had their own rice milling units established in producing areas of state and their major source of supply was the farmers from whom they purchase large quantities of basmati paddy. Almost entire supply from farmers was purchased through commission agents in the regulated markets of state. Other rice millers were the second important source of supply for exporters.

Price Patterns and Price Spread of Basmati Rice

The decomposition of time series data of basmati paddy prices revealed the presence of considerable seasonality in study markets and prices remained above the annual average prices during the lean period consisting months of April to July and highest prices prevailed during the months of April to July. Seasonality in basmati prices in Punjab markets revealed that in comparison to annual average prices,

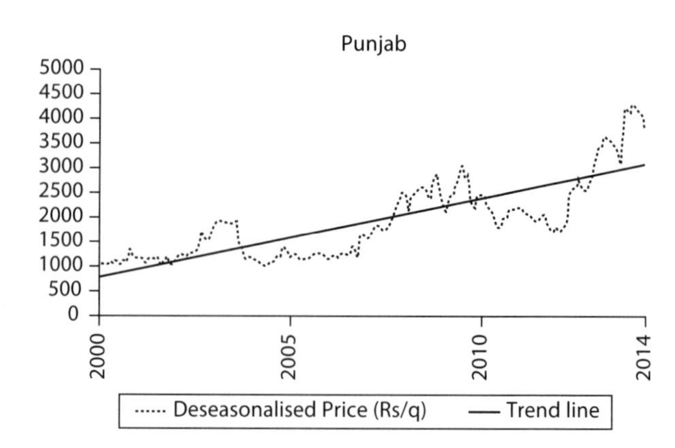

Figure 5.1 Movement of Deseasonalized Prices of Basmati Paddy around Trend Line in Punjab

Source: Field survey data.

monthly prices in month of October remained about 8 per cent lower whereas the same were observed by higher by 6 per cent during the month of May.

As the harvesting season starts in October, the basmati paddy prices in the state markets remained relatively low from October to January. Movement of deseasonalized prices around trend line exhibited the presence of oscillatory movements pointing towards existence of the cyclical patterns in basmati paddy prices during the past years (Figure 5.1). Price linkages of different Punjab revealed that price behaviour of basmati paddy in different markets were similar and the markets of state were well integrated as the estimated correlation coefficients in this regard among all pairs of markets were more than 0.85 and were significant at one per cent level of probability. High integration of rice markets was also reported by Pandit, Samal, and Mishra (2012).

An average wholesaler had monthly traded 28.65 quintal of total basmati rice comprising 96.12 and 3.88 per cent of Pusa Basmati 1121 and traditional basmati respectively. Wholesale of basmati of both of these varieties were found to be highest in month of December. Overall for both varieties taken together percentage mark-up in wholesale trade turns out to be 4.42 per cent. The monthly basmati trade of an average retailer in local market was 0.59 quintal comprising 78 per cent of Pusa Basmati 1121 and 22 per cent of traditional basmati. The retail sales of both varieties were relatively more during October–December months and low during the January–February months. Retailer's overall absolute mark up and percentage mark-up in both varieties taken together turns out to be Rs 1073/qtl and 11.90 per cent respectively. Pusa Basmati 1121 and Traditional Basmati accounted for 96.59 and 3.41 per cent respectively of the total quantity exported by the respondent exporters. Over the study period from October to February, monthly export was found to be the highest in month of February and least during the month of October. The exporter's overall absolute weighted average absolute mark-up and percentage mark-up for both varieties taken together was Rs 807/qtl and 10.76 per cent respectively.

In domestic market the farmer's net price accounted for about 68 and 73 per cent of the consumer's purchase price of Pusa Basmati 1121 and Traditional Basmati respectively (Table 5.3). The rice miller's net margin for Pusa Basmati 1121 and Traditional Basmati varieties was 8.06 per cent and 4.31 per cent of the consumer's price respectively. While the wholesaler's net margin as percentage to the final price remained more or less equal in both of the basmati varieties (3.40 per cent and 3.43 per cent), retailer's net margin in sale of Traditional Basmati was relatively low (8.40 per cent) when compared to that in the Pusa Basmati 1121 (9.47 per cent).

Although there was no difference with regard to the net sale price received by the farmers in export market channel as compared to local market channel for both of the varieties, however farmer's net margin in export price was significantly higher at 77.47 to 79.52 per cent of the export price as compared to that in the consumer price in domestic local market channel (Table 5.4). Relatively low total cost of miller in this channel as compared to the domestic market channels was on account of the exemption/refund of Value Added Tax (5 per cent) by state government under the basmati export promotion scheme. The net margin of miller accounted for 4.86 and 4.34 per cent of export price of Pusa Basmati 1121 and Traditional Basmati respectively. The exporter's net margin was about 5.92 per cent (Rs 492/q) and 5.01 per cent (Rs 542/q) for Pusa Basmati 1121 and Traditional Basmati respectively.

Stakeholders' Perceptions on Basmati Production and Trade

Farmers' responses regarding motive of basmati cultivation revealed that profitability along with usefulness of its by-product and the proper adjustment of basmati in the adopted crop rotation were the principal reasons for cultivation of basmati in Punjab. According to farmer's perceptions, the marketing-related constraints or problems in basmati cultivation, which are: lack of minimum support price (MSP) and government procurement, lack of remunerative price, high fluctuations price, lack of market information, and collusion among traders were the major problems being faced by them. The main production

Table 5.3 Price Spread for Basmati Rice, Punjab, 2013–14
(Channel: Farmer → Rice Miller (through commission agent) → Wholesaler (Local Market) → Retailers → Consumer)
(1 qtl of rice = 1.667 qtl of basmati paddy)

S. No.	Particulars	Pusa Basmati 1121		Traditional Basmati	
		Rs/qtl	Percentage share in Consumer's rupee	Rs/qtl	Percentage share in Consumer's rupee
A	Net price received by the farmer	6438	67.53	8595	72.85
	Expenses borne by the farmer	53.87	0.57	54.58	0.46
B	Miller's purchase price/Farmer's sale price	6492	68.10	8650	73.32
	Miller' expenses on processing and selling	878	9.21	1126.82	9.55
	Total cost to the miller	7370	77.31	9667	82.87
	Miller's net margin	757	7.94	508	4.31
C	Wholesaler's purchase price/miller's sale price	8127	85.25	10285	87.18
	Expenses borne by the wholesaler	37.62	0.39	37.62	0.32
	Wholesaler's net margin	324.58	3.40	405.18	3.43
D	Retailer's Purchase price/Wholesaler's sale price	8489.20	89.05	10666.00	90.41
	Expenses borne by the retailer	141.16	1.48	141.16	1.20
	Retailer's net margin	902.64	9.47	990.84	8.40
E	Consumer's purchase price/retailer's sale price	9533	100.00	11798	100.00

Source: Field survey data.

Table 5.4 Price Spread for Basmati Rice, Punjab, 2013–14
(Export Channel: Farmer → Rice Miller (through commission agent) → Exporter)
(1 qtl of rice = 1.667 qtl of basmati paddy)

S.No.	Particulars	Pusa Basmati 1121		Traditional Basmati	
		Rs/qtl	per cent to export price	Rs/qtl	per cent to export price
A	Net price received by the farmer	6438	77.47	8595	79.52
	Expenses borne by the farmer	53.87	0.65	54.58	0.50
B	Miller's purchase price/Farmer's sale price	6492	78.12	8650	80.02
	Miller's expenses	444	5.34	670.82	6.21
	Total cost to the miller	6936	83.47	9321	86.22
	Net margin of miller	404	4.86	469	4.34
C	Exporter's purchase price/Miller's sale price	7340	88.33	9790	90.56
	Expenses borne by the exporter	477.97	5.75	477.97	4.42
	Total cost of exporter	7817.97	94.08	10267.97	94.99
	Exporter's margin	492.03	5.92	542.03	5.01
D	Export price	8310	100.00	10810	100.00

Source: Field survey data.

related problems in order of their degree of severity were the lower yield, yield instability, problem of diseases, problem of weeds, and lack of technical knowhow or extension related problem. Major problems of wholesalers were regarding the competition from other wholesalers, problem of high marketing charges/taxes and poor road network. The major problems perceived by sample retailers in order of severity were competition from large organized retail chains, competition from other retailers, and the poor infrastructure. Infrastructure-related problem was the most severe problem faced by the exporters. The chemical residue on account of heavy use of insecticides and fungicides in basmati cultivation was a high and severe problem faced by them in basmati exports. The other problems of severe and high nature perceived by the exporters were the uncertainty of the government's export policy, poor quality supply, competition from wholesalers, and poor road network and port facilities.

CONCLUSIONS AND POLICY IMPLICATIONS

Recently, basmati replaced cotton as the third most important crop after wheat and non-basmati rice in Punjab state. During 2008, Government of India's notification declaring Pusa Basmati 1121 variety as basmati played major role in this regard and the area under basmati in state increased in a major way. This particular variety with high demand in gulf countries had broken the long-time genetic yield barrier of basmati varieties. Unlike non-basmati paddy, basmati does not get price support/procurement by the government, thus marketing-related problems faced/perceived by basmati growers, that is, high fluctuations in price, lack of remunerative price, lack of MSP, and Government procurement, lack of market information and collusion among traders need to be addressed through various institutional and policy instruments. As basmati cultivation is important from ecological as well as farm incomes point of view; both state as well as central governments should frame a policy to save farmers from price fluctuations through ensuring some minimum income support to its cultivators. As significant proportion of farmers still depend upon the traders for basmati price information, there is need of providing wider coverage to collecting and dissemination of agricultural market intelligence/information so that prices prevailing in each and every market is available to them for making adequate marketing decisions. Study revealed that basmati paddy prices remained generally low during harvesting/peak period and high during the lean period. However, farmers had disposed of more than 80 per cent of marketed surplus immediately after harvesting at relatively low prices. Especially, small and marginal farmers lack facility or money to sell where they want or keep their produce and sell when price is suitable/high. Proper organizations, both collective and government organizations may tackle these issues in order to ensure good price of farmer's output. There is need to encourage the farmers to opt for farm level storage through creation of efficient storage structures at farm level.

The production-related problems can be handled through improvement in the knowledge of farmers regarding new techniques and technologies. Even though the research and extension linkages in Punjab are fairly strong, the qualitative improvement through involvement of crop scientists and extension personnel is very much needed. Improvement in road and other infrastructure in markets can solve the problems being faced by different market intermediaries. The problem of chemical residue on account of heavy use of insecticides and fungicides in basmati cultivation needs immediate intension to address the exporter's quality concerns. To be credible supplier of basmati rice in world market, the country should formulate a long run export policy. Although, short-run benefits to basmati rice milling industry/exporters provided by state government had helped the industry in competing in the world market, urgent steps should be taken to frame the long run policy in this regard.

REFERENCES

Adhikari A. and Sekhon, M.K. 2014. 'Export of basmati rice from India: Performance and Trade direction'. *Journal of Agricultural Development and Policy* 24 (1): 1–13.

Anonymous. 2014. Agriculture at a Glance, Information service, Directorate of Agriculture, Punjab, Chandigarh.

Blyn, G. 1973. 'Price Series Correlation as a Measure of Market Integration'. *Indian Journal of Agricultural Economics* 28: 56–9.

Croxton, F.E. and Cowden, D.J. 1955. *Applied General Statistics*, Second Edition. New York: Prentice Hall.

Pandit A., Samal, P., and Mishra, J.R. 2012. 'An Analysis of Price Behaviour of Rice in Eastern Indian Markets'. *Indian Journal of Agricultural Marketing* 26 (2): 102–14.

Sahadevan, K.G. 2008. 'Mentha Oil Futures and Farmers'. *Economic and Political weekly* 43 (4): 72–6.

Sidhu J.S, Singh, J., and Kumar, R. 2014. 'Role of Market Intelligence in Agriculture: A Success Story of Basmati Cultivation in Punjab'. *Indian Journal of Economic Development*, Seminal Issue, 10: 26–31.

6

An Economic Analysis of Basmati Cultivation in Haryana

Usha Tuteja

India is the second largest producer of rice after China in the world. It grows a large number of varieties across the regions. Basmati is very special and regarded as the gold standard of rice. It is one of the India's great national treasures, at par with saffron from Kashmir, pepper from Kerala and tea from Darjeeling. What makes basmati so special? After all, there are thousands of rice varieties in India. Why does basmati deserve special attention? First of all, basmati is the Indian rice that we have grown in the foothills of the Himalayas for many centuries. Secondly, the best kinds of basmati have long grains that stay separate and distinct, even when they are cooked. The third reason is fragrance. So basmati is one of the world's most special rice varieties. It is not just the flavour and the shape of the grain. It is also that distinctive aroma that few other rice breeds can hope to match.

India produces about 7–8 million tonnes of basmati rice (12 million tonnes of paddy at 66 per cent conversion ratio) primarily in three states, namely Haryana, Punjab, and Uttar Pradesh (UP). It is one of the major export items from India. The exports of basmati rice touched about 3 million tonnes, equivalent to Rs 15336 crore during 2011–12. Pusa Basmati 1121 and 1509 which are hybrid varieties and yield higher than traditional basmati have become popular in Iran and other export markets of West Asia.

Although, we have large number of studies on rice (David and Huang 1996; Kumar and Sharma 2003; Sekher 2003, 2008, among others), the study of basmati rice (Dwivedi, Dwivedi, and Singh 2011; Ghani, Metzel, and Salinger 1993; Grover 2012; Sidhu 2014) has received scant attention of scholars despite its importance in future development of agriculture in irrigated areas of Punjab and Haryana. This chapter aims to analyse economics of major basmati varieties grown by the farmers in Haryana.

RESEARCH METHODOLOGY AND DATA

The scope of the study is confined to two most popular varieties (Pusa Basmati 1121 and 1509) of basmati rice grown by the farmers in Haryana. Three districts, namely Kaithal, Jind, and Sonipat with highest share of area under basmati rice in Haryana were selected for in-depth study. The selection of respondents is based on multistage sampling design. At the first and second stages, basmati rice producing districts and blocks in these districts were selected. At the third stage, villages were selected on the same criterion. A questionnaire was canvassed to the farmers growing basmati rice. All farm size categories—marginal, small, medium, and large—were covered in the sample. The number of farm households in each category was decided according to their proportion at the district level. The primary data pertaining to the year 2013–14 were collected from 150 farmers.

RESULTS AND DISCUSSION

Basmati Production in India

During the past two decades, area, production, productivity, availability, and exports of basmati rice from India increased manifolds which provided ample opportunities to producers and exporters in major basmati growing states such as Haryana and Punjab.

Traditionally, basmati rice is a crop of north-west Himalayas in India. This area is blessed with producing extra-long, slender, aromatic grain that elongates at least twice its original size, with a soft and fluffy texture upon cooking, and has a delicious taste. Also, known as king of rice, basmati uses less water and fertilizer, has high export potential and its straw is used for livestock feed, rather than burning in the field and creating atmospheric pollution.

Production of basmati rice is concentrated in north-west Indian states—Haryana, Punjab, Western UP, and to a limited extent in Uttarakhand, Himachal Pradesh, and Jammu & Kashmir. Currently, Haryana is the leading producer of basmati in India. The production of basmati in India was 6616 thousand tonnes in 2013 (see Table 6.1). The share of Haryana in total basmati production was about 43.8 per cent followed by Punjab with 34.7 per cent and Uttar Pradesh with 19.2 per cent. Haryana and Punjab together constituted more than 75 per cent of basmati rice produced in India. It may be noted that production of basmati rice has increased by 32.61 per cent in 2014 over 2013. The highest increase may be observed in Haryana and Punjab. The yield rate of basmati rice was 3944 kgs/ha which rose to 4110 kgs/ha in 2014. It is worth recording that Haryana was leading in productivity.

Area, Production, Consumption, and Marketed Surplus

The quantum of marketed surplus for disposal in the market depends on the level of production and retention. Normally, farmers retain a part of output for consumption of family, seed requirement, animal feed, and other purposes. The pattern of area, production, consumption, retention for future use, wastage, quantity sold, and price realized of Pusa Basmati 1121 is presented in Table 6.2. It may be noticed that sampled farmers devoted around 2.81 ha to

Table 6.1 Area, Production, and Yield of Basmati Rice in Major Growing States of India (2013 and 2014)
Area: '000 ha, Production: '000 Tonnes, Yield: Kgs/Ha

State	2013			2014		
	Area	Production	Yield	Area	Production	Yield
Punjab	590.01	2292.75	3885	857.68	3498.88	4079
	(35.17)*	(34.65)		(40.18)	(39.88)	
Haryana	711.11	2898.98	4077	832.54	3701.88	4446
	(42.39)	(43.82)		(39.00)	(42.19)	
Uttar Pradesh	318.75	1270.09	3985	354.39	1260.69	3557
	(19.00)	(19.20)		(16.60)	(14.37)	
Uttrakhand	18.30	54.16	2960	20.34	66.41	3265
	(1.09)	(0.82)		(0.95)	(0.76)	
Jammu & Kashmir	37.28	92.66	2486	68.45	240.77	3517
	(2.22)	(1.40)		(3.21)	(2.74)	
Himachal Pradesh	1.00	3.40	3400	0.45	2.15	4777
	(0.06)	(0.05)		(0.03)	(0.03)	
Delhi	1.00	4.09	4090	0.70	3.00	4286
	(0.07)	(0.06)		(0.03)	(0.03)	
Total	1677.45	6616.13	3944	2134.55	8773.78	4110
	(100.00)	(100.00)		(100.00)	(100.00)	

Note: *Percentage of total.
Source: Rice Exporters Association, New Delhi.

Table 6.2 Per Farm Production, Consumption, Marketed Surplus, and Price Realized by Sampled Farmers during 2013–14

Farm Size	Area (ha)	Production (qtls)	Consumption (qtls)	Retained/stocked for future use (qtls)	Wastage (qtls)	Marketed surplus (qtls)	Price (Rs/qtl)
				Pusa Basmati 1121			
Marginal	0.43	18.11	1.34	0.12	0.09	16.56	3824
Small	1.11	41.12	1.25	0.27	0.08	39.52	3511
Medium	1.93	74.38	1.79	0.16	0.67	71.76	3627
Large	4.05	165.59	2.36	0.93	1.00	161.32	3608
Total	2.81	113.45	1.97	0.60	0.70	110.20	3607
				Pusa Basmati 1509			
Marginal	0.40	16.00	0.30	0.25	0.00	15.45	3200
Small	1.21	66.00	1.00	0.50	0.00	64.50	2900
Medium	1.19	44.67	1.00	0.67	0.00	43.00	3567
Large	1.65	71.86	1.34	0.68	0.17	69.66	3357
Total	1.52	65.89	1.24	0.66	0.14	63.85	3354
				Total			
Marginal	0.46	19.25	1.36	0.14	0.09	17.66	3785
Small	1.15	43.76	1.29	0.29	0.08	42.10	3474
Medium	2.05	79.00	1.89	0.23	0.67	76.21	3624
Large	4.44	182.24	2.67	1.09	1.04	177.46	3585
Total	3.06	123.99	2.17	0.70	0.72	120.41	3586

Source: Field Survey.

this crop and produced 113.45 qtls per farm. Out of total produce, they consumed 1.97 qtls and retained around 0.60 qtl per household for future use. They also incurred wastage of approximately 0.70 qtl of basmati production. The remaining produce of 110 qtls was disposed of in the market. They realized a price of Rs 3607 per quintal after selling the produce. As expected, production and marketed surplus were several times higher in case of large farmers when compared to marginal and small farmers. It may be noticed that marginal farmers disposed of 16.56 qtls per farm of Pusa Basmati 1121 but the price realized by them was higher than other categories. It could be due to better quality of their produce.

Table 6.2 also suggests that sampled farmers produced 65.89 qtls per farm of Pusa Basmati 1509. On an average, they retained 1.24 qtls for domestic consumption and 0.66 qtl for future use. A marginal quantity of 0.14 qtl per farm was wasted in the process. The remaining quantity of 63.85 qtls of Basmati 1509 per farm was disposed in the market. It may be pointed out that each category of farmers retained a part of produce for self-consumption. One may observe class disparities in the production as well as in the consumption. The marginal farmers sold only 15.45 qtls per farm against 69.66 qtls by the large farmers. This result is on the expected lines.

It would be useful to combine Pusa Basmati 1121 and Pusa Basmati 1509 for examining the overall scenario. This information is presented in Table 6.2. The area per farm under these varieties was 3.06 hectares and production was 123.99 qtls. Each household retained 2.17 qtls for domestic consumption and stocked 0.70 qtls for future use. A small wastage of 0.72 qtl per farm was recorded. After retaining a part of produce for consumption and future use and accounting for wastage, each farmer sold around 120 qtls of this high value grain in the market. The farm size variations were significant. The large farm category sold 177 qtls per farm against around 18 qtls by marginal and 42 qtls by small farmers during 2013–14.

Cost of Cultivation

The utilization of high yield variety (HYV) seeds, fertilizer, pesticides, tractor, and tube wells play an important role in boosting the agricultural development of a region. Haryana is using these inputs for a long time. The consumption of fertilizer in the state was 386 kg/ha during 2010–11. The nitrogenous fertilizers were preferred over phosphatic and potassic fertilizer. The state of Haryana has already moved towards agricultural mechanization. Use of tractors, tube wells, and pumping sets is common in the state. It may be pointed out that Haryana is ahead of other states in the production as well as distribution of HYV seeds. These were used on 98.5, 66.7, and 97.6 per cent of cultivated area in case of wheat, rice and bajra, while for maize it was 70.0 per cent during 2009–10.

With this brief introduction, we analyse cost of cultivation of Pusa Basmati 1121 and Pusa Basmati 1509 during the reference year. As a part of cost, we have also included storage, transportation and marketing cess, etc.

We have provided details of cost of cultivation of Pusa Basmati 1121 in Table 6.3. The per hectare cost of the Pusa Basmati 1121 cultivation was Rs 39850 at the aggregate level. Evidently, the maximum proportion of cost was incurred on human labour followed by fertilizer including manure and pesticides. These items constituted 47.37, 14.91, and 12.90 per cent of total cost of cultivation of Pusa Basmati 1121. The share of these items in total cost was more than 75 per cent. In the array, expenditure on irrigation was the next item of the cost and constituted approximately 5 per cent of total cost.

The sampled farmers also incurred Rs 148 and Rs 1184 per hectare as a cost of storage and transportation. Foodgrains, including paddy, are bulky in nature and require higher space both in storage and transportation. This causes relatively higher cost of storage and transportation per unit of produce. Often, higher cost of transportation restricts the movement from surplus to deficit areas. This also results in lower price of the produce in growing states and higher price in the deficit states. Owing to these reasons, marginal and small farmers sell their produce immediately after the harvest and realize low price due to higher supply in the harvesting months.

The surveyed farmers spent Rs 2685 per hectare on storage, transportation, and marketing cost. It may be noticed that cost of cultivation of Pusa Basmati 1121 varied across the farm sizes. In case of human labour, small farmers incurred lower cost in comparison to other categories. Surprisingly, cost of human labour per hectare on marginal farms was higher than the large farms. Also, cost of cultivation of Pusa Basmati 1121 could be observed maximum on marginal farms. It was largely due to higher expenditure on irrigation and some other items.

The information related to expenditure incurred by the growers of Pusa Basmati 1509 on various inputs used by them and associated cost in terms of storage, transportation and marketing cost is presented in Table 6.4. Clearly, cost of cultivation of this variety on sampled farms at the aggregate level was Rs 35278per hectare during the reference year. It may be observed that cost of cultivation of Pusa Basmati 1509 was lower in comparison to

Table 6.3 Per Hectare Cost of Production of Pusa Basmati 1121 on Sampled Farms during 2013–14

Item	Marginal	Small	Medium	Large	Total
Area (ha)	6.03	27.64	55.85	332.46	421.97
Input Costs (Rs/ha)					
Seed	1327	1148	1225	969	1020
Irrigation	3279	2291	2197	1908	1991
Manure & Fertilizer	6099	5973	6365	5869	5945
Labour (bullock+manual)	19042	18690	19617	18765	18877
Machinery hire/owned charges	5260	5693	5098	3893	4190
Pesticides/Weedicides	4826	6358	6294	4852	5141
Any other cost (specify)	0	0	18	0	2
(I) Total input cost (Rs)	39833	40153	40813	36256	37166
Storage, transportation, and marketing Costs (Rs/ha)					
Storage	116	105	179	147	148
Transportation	1321	1298	1202	1169	1184
Marketing and other (market fees, cess, if any etc)	2665	1747	1253	1313	1353
Any other cost (specify)					
(II) Total storage, marketing, cost (Rs)	4102	3150	2634	2629	2685
Production (qtl)	42	37	39	41	40
Total Cost (I+II)	43935	43303	43447	38885	39850

Source: Field Survey.

Table 6.4 Per Hectare Cost of Production of Pusa Basmati 1509 on Sampled Farms during 2013–14

Item	Marginal	Small	Medium	Large	Total
Area (ha)	0.40	1.21	3.56	31.26	36.44
Input Costs (Rs/ha)					
Seed	890	1112	1178	1160	1157
Irrigation	1730	1829	1794	1956	1934
Manure & Fertilizer	4806	4794	4859	5279	5217
Labour (bullock+manual)	15567	14332	14158	14364	14356
Machinery hire/owned charges	4942	4942	4201	4246	4273
Pesticides/ Weedicides	3707	3954	3370	3903	3851
Any other cost (specify)	0	0	0	0	0
(I) Total input cost (Rs)	31641	30962	29559	30909	30787
Storage, Transportation, and Marketing Costs (Rs/ha)					
Storage	0	0	174	71	78
Transportation	3163	1631	977	1662	1611
Marketing and other (market fees, cess, if any etc.)	2530	3153	2686	2805	2802
Any other cost (specify)					
(II) Total storage, marketing, cost (Rs)	5693	4784	3836	4539	4491
Production (qtl)	40	54	38	44	43
Total Cost (I+II)	37334	35745	33395	35447	35278

Source: Field Survey.

Pusa Basmati 1121. Like Pusa Basmati 1121, marginal farmers incurred higher cost per hectare in comparison to other categories. The expenditure on human labour was the highest irrespective of farm category. In case of marginal farmers, around 42 per cent of the cost was spent on this item alone. Other categories of farmers also incurred around 40 per cent of total cost on human labour. The high cost of human labour was due to shortage which resulted in higher wages in turn increasing the cost on this item. Further, cost of fertilizer and machinery were other major items which constituted sizeable proportion of the total cost. None of the farm categories spent less than Rs 4000 per hectare on these items. We could not find a clear cut advantage of family labour on marginal and small farms. The expenditure on pesticides and weedicides ranged between Rs 3370 and Rs 3954 per hectare. All these input items constituted around 87 per cent of total cost of cultivation at the overall level. The remaining 13 per cent of cost was incurred on storage, transport and marketing.

The combined results of Pusa Basmati 1121 and 1509 on cost of cultivation are presented in Table 6.5. The cost of cultivation per hectare was Rs 39485 at the aggregate level during the reference year. Like separate results of Pusa Basmati 1121 and Pusa Basmati 1509, marginal farmers incurred higher cost in comparison to other categories of farmers. Among the included items, human labour, fertilizer, and pesticides constituted 46.90, 14.91, and 12.76 per cent of the total cost.

Thus, these three items alone formed around 75 per cent of cost. The next item was machinery with an expenditure of Rs 4197 (10.63 per cent). The expenditure on these four items across farm sizes ranged between Rs 4197 and Rs 19289. The cost of human labour was high due to shortage and escalating wages. In addition to cost of inputs, sampled farmers spent on storage, transportation, and marketing. All these associated costs together formed around 7 per cent of the total cost. Farm size variations are a common phenomenon in the expenditure incurred by the sampled farmers on various items in cultivation of basmati paddy. Surprisingly, cost of several items on marginal and small farms was higher than the overall level.

Economics of Basmati Cultivation

Having analysed cost of Pusa Basmati 1121 and Pusa Basmati 1509 cultivation, we now discuss economics of cultivation on the sampled farms during the reference year. Table 6.6 presents gross returns and net returns per hectare and per quintal and per farm value of marketed surplus of

Table 6.5 Per Hectare Cost of Production of Pusa Basmati 1121 and Basmati 1509 on Sampled Farms during 2013–14

Item	Marginal	Small	Medium	Large	Total
Area (ha)	6.43	28.85	59.41	363.72	458.42
Input Costs (Rs/ha)					
Seed	1299	1146	1222	985	1031
Irrigation	3181	2272	2173	1912	1986
Manure and Fertilizer	6017	5924	6275	5819	5887
Labour (bullock+manual)	18824	18507	19289	18387	18517
Machinery hire/owned charges	5240	5661	5044	3924	4197
Pesticides/Weedicides	4756	6257	6119	4770	5039
Any other cost (specify)	0	0	17	0	2
(I) Total input cost (Rs)	39318	39767	40122	35796	36656
Storage, transportation, and marketing Costs (Rs/ha)					
Storage	109	101	179	140	142
Transportation	1437	1312	1189	1211	1218
Marketing and other (market fees, cess, if any, etc.)	2657	1806	1339	1442	1468
Any other cost (specify)					
(II) Total storage, marketing, cost (Rs)	4202	3218	2706	2793	2828
Production (qtl)	42	38	39	41	41
TOTAL COST (I+II)	43520	42985	42828	38590	39485

Source: Field Survey.

Table 6.6 Returns from Basmati Cultivation on Sampled Farms during 2013–14

Farm Size	Gross Returns per ha (Rs)	Net Returns per ha (Rs)	Gross Returns per qtl (Rs)	Net Returns per qtl (Rs)	Value of Marketed Surplus Rs/Farm
Pusa Basmati 1121					
Marginal	164167	120232	3905	2860	63310
Small	134376	91074	3613	2449	138756
Medium	143372	99924	3712	2587	260030
Large	150574	111689	3687	2735	582374
Total	148754	108903	3689	2700	397672
Pusa Basmati 1509					
Marginal	129678	92344	3280	2336	49440
Small	161999	126253	2980	2322	187050
Medium	136725	103330	3634	2746	152800
Large	149965	114517	3434	2622	233956
Total	148847	113569	3430	2617	214169
Pusa Basmati 1121 + 1509					
Marginal	161998	118478	3868	2829	112750
Small	135539	92554	3575	2441	325806
Medium	142973	100145	3708	2597	412830
Large	150521	111932	3664	2724	816330
Total	148761	109276	3667	2693	611841

Source: Field Survey.

Pusa Basmati 1121. One may notice from the table that the gross and net returns per hectare from cultivation of Pusa Basmati 1121 were Rs 148754 and Rs 108903 respectively. Like per unit price, marginal farmers reaped higher gross and net returns per hectare. Further, wide variations may be noticed in the gross returns and net returns per qtl from cultivation of Pusa Basmati 1121 during 2013–14. Obviously, marginal farmers emerged greater beneficiaries than other categories of surveyed farmers. The value of marketed surplus disposed of by the farmers was Rs 397672 per farm during the reference year. Since, production of large farmers was higher than other categories; their marketed surplus was also recorded maximum. As expected, large group followed by medium farmers indicated higher marketed surplus in comparison to small and marginal farmers who allocated low area to Pusa Basmati 1121 due to tiny pieces of their land holdings. This implies that marketed surplus of Pusa Basmati 1121 is primarily concentrated in the hands of large land owning classes who constitute low proportion in numbers.

Table 6.6 also presents the status of gross and net returns per hectare and per qtl and marketed surplus of Pusa Basmati 1509 on sampled farms during 2013–14. The gross and net returns per hectare from cultivation of basmati 1509 worked out Rs 148847 and Rs 113569 respectively. We could not ascertain any relationship between returns per hectare and farm size. However, these could be noticed maximum on small farms. Further, an examination of gross and net returns per per quintal from cultivation of Pusa Basmati 1509 at the overall level were computed Rs 3430 and Rs 2617 respectively during 2013–14. The medium category of farmers could reap higher returns in comparison to other categories. The value of marketed surplus per farm was Rs 214169 at the aggregate level. As expected, it was much higher on large farms in comparison to other categories of surveyed farmers.

Finally, we present economics, returns, and marketed surplus on the sampled farms by combining the results of Pusa Basmati 1121 and Pusa Basmati 1509 during the reference year. Table 6.6 points out that the gross and net returns per hectare from cultivation were estimated Rs 148761 and Rs 109276 respectively. It is worth mentioning that returns per hectare were found highest on marginal farms. The gross and net returns per quintal also followed the same pattern. As a result, these were also maximum on the marginal farms. The gross and net returns per quintal were worked out Rs 3667 and Rs 2693 respectively. The value of marketed surplus per farm was Rs 611841 at the overall level. The share of marginal, small, medium, and large category farms was 1.44, 5.55, 12.43, and 80.48 per cent respectively. Evidently, large category farmers emerged as a dominant group due to concentration of land in their hands.

CONCLUSIONS AND POLICY IMPLICATIONS

Findings regarding production, retention, and disposal of Pusa Basmati 1121 and Pusa Basmati 1509 grown by the farm households during the reference year revealed that production of Pusa Basmati 1121 was around 113 qtls per farm during 2013–14. Farm size variations were found wide. The sampled households retained a part of production (1.97 qtls) for domestic consumption. In retention, self-consumption dominated whereas other requirements were found marginal. The quantity of marketed surplus of Pusa Basmati 1121 was around 110 qtls per farm whereas, a relatively smaller quantity of 64 qtls per farm of Pusa Basmati 1509 was disposed during the reference year. Since, large farm category produced higher quantity than other categories, they also dominated in sales. The price of Pusa Basmati 1121 realized by the farmers was Rs 3607 per qtl while Rs 3364 per qtl were received for Pusa Basmati 1509.

The sampled farmers incurred cost on human labour, seed, irrigation, fertilizer, and manure and pesticides used by them in cultivation of basmati paddy in kharif season. They also incurred expenditure on storage, transportation, and marketing. The per hectare cost of cultivating Pusa Basmati 1121 was Rs 39850 on sampled farms and the maximum proportion of cost was incurred on human labour followed by chemical fertilizer and pesticides. Findings show that per hectare cost of cultivating Basmati 1509 on sampled farms was Rs 35447 during 2013–14. The human labour and fertilizer were found the major components of cost. Thus, human labour, machine labour, fertilizer, and plant protection were the major items in cost composition in cultivation of study crops.

The per hectare yield of Pusa Basmati 1121 on sampled farms was 40.32 qtls. Farm size and productivity were found related. Thus, productivity on marginal farms was higher than large farms. After deducting the cost from gross returns, producers earned a profitability of Rs 109903 per hectare during 2013–14. As expected, marketed surplus in terms of value was much higher in case of large farmers in comparison to other categories. The net returns per quintal from Pusa Basmati 1121 were Rs 2700 and these were found highest on marginal farms.

The results of economics of Pusa Basmati 1509 revealed that per hectare input cost of cultivation was Rs 30787 on sampled farms during the reference year. The major cost items were human labour followed by fertilizer and machinery. Other costs such as storage, transportation, marketing cess, etc., were estimated Rs 4491 per hectare. Thus, total cost of cultivation for Pusa Basmati 1509 was Rs 35278 per hectare on sampled farms. In particular, marginal farmers incurred higher cost in comparison to other categories. The net returns per hectare after deducting the cost from gross returns were computed Rs 113569 during 2013–14. The net returns per quintal were estimated Rs 2617. Like Pusa Basmati 1121, marketed surplus was recorded higher on large farms in comparison to other categories.

After combining the results for cost of cultivation and net returns from Pusa Basmati 1121 and 1509, it was found that sampled producers earned a profit of Rs 109276 per hectare and Rs 2693 per quintal during 2013–14. The share of marginal, small, medium, and large categories of farmers in marketed surplus of basmati paddy was positively related to farm size.

POLICY IMPLICATIONS

Basmati rice is a great strength of India since its quality in terms of grain length and aroma can hardly match any other variety of rice in the world. However, a huge potential still remains to be realized. Haryana is the leading producer of basmati rice in India. The production can be further improved through pragmatic policy initiatives. The following policy measures are recommended for achieving this objective.

1. Improvement in returns from cultivation of basmati paddy by yield enhancement through research on improved varieties and its transfer at the farm level.
2. In addition to technology generation for improved yield of basmati paddy, timely delivery of required inputs at reasonable price to the farmers should be prioritized.
3. Provision of necessary physical (storage, credit, etc.), and marketing infrastructure.

REFERENCES

David, C. Cristina and Jikun Huang. 1996. 'Political Economy of Rice Price Protection in Asia', *Economic Development and Cultural Change*, 44(3): 463–83.

Dwivedi, Sudhakar, M.C. Dwivedi, and Tarunvir Singh. 2011. 'An Economic Analysis of Basmati Rice Production in Jammu District of Jammu and Kashmir', *Journal of Research, SKUAST-J*, 10(1): 93–9.

Ghani, Abdul, Jeffrey C. Metzel, and B. Lynn Salinger. 1993. 'Diversification within Rice: Production Opportunities and Export Prospects of Specialty Rice in Bangladesh', *The Bangladesh Development Studies*, 21(3): 111–23.

Grover, D.K. 2012. 'Basmati Rice Cultivation for Resource Conservation and use Efficiency in Context of Sustainable Agriculture in Punjab', *Indian Journal of Economic Development*, 8(2): 11–26.

Kumar, Parmod and R.K. Sharma. 2003. 'Spatial Price Integration and Pricing Efficiency at the Farm Level: A Study of Paddy in Haryana', *Indian Journal of Agricultural Economics*, 58(2): 201–17.

Sekhar, C.S.C. 2008. 'World Rice Crisis: Issues and Options', *Economic and Political Weekly*, 43(26/27): 13–17.

———. 2003. 'Agricultural Trade Liberalisation: Likely Implications for Rice Sector in India', *Indian Journal of Agricultural Economics*, 58(1): 42–63.

Sidhu, J.S., Jasdev Singh, and Raj Kumar. 2014. 'Role of Market Intelligence in Agriculture: A Success Story of Basmati Cultivation in Punjab', *Indian Journal of Economic Development*, 10(1a): 26–31.

7

Problems and Prospects of Katarni Paddy Production in Bihar

Rajiv Kumar Sinha and Rosline Kusum Marandi

With about 43 million hectares of rice area, India is the second largest producer of rice in the world after China. Rice production in India reached 104.40 million tonnes during 2012–13. Of these, the three states: West Bengal, Uttar Pradesh (UP), and Andhra Pradesh account for about 42 per cent of the total production. India also produces some of the best quality rice in the world. These include the long grained export quality Basmati and a host of locally adapted small- and medium-grained scented rice varieties known for their excellent cooking and eating qualities. While the Basmati is found in Punjab, Haryana, and western UP, the small- and medium-grained scented rice varieties like: (i) kalanamak, (ii) shankarchini, and (iii) Hansraj are found in UP, (i) Dubraj, (ii) Chinoor in Chhattisgarh, Kalajoha in northeast Ramdhuni Pagal in Odisha, Ambemohar in Maharashtra, and so on. The long-grained Basmati rice is generally exported and has assured markets, whereas the small- and medium-grained non-basmati rice is consumed locally.

In Bihar, although aromatic rice is grown all over the state, it is mainly concentrated in Bhagalpur and Magadh divisions (Table 7.1). Bhagalpur has been a traditional aromatic rice-growing area, where the varieties, such as: (i) Katarni, (ii) Tulsi Manjari, (iii) Badshahbhog, (iv) Br-9, and; (v) Br-10 are mostly common. These are photoperiod-sensitive, tall and, hence, susceptible to lodging and several diseases and pests. Their yield vary from 2.0 to 2.5 t/ha (Katarni is the most prevalent variety of the region). However, over the period, a large variation has occurred, which has resulted into various types, such as (i) Bhauri katarni, (ii) Deshla katarni, (iii) Sabour katarni and; (iv) Ghorayiya katarni. In Magadh region, which is the main rice growing tract of Bihar, farmers grow Karibank, Marueya, Mehijawain, Shyamjira, Tulsiphool, Sonachur, and Shah Pasand. Overtime, the areas under these varieties has drastically reduced, although farmers still grow Karibank and Marueya, but on a small scale only. The Tarai region of West Champaran was, at one time, known for its good quality aromatic rice varieties, that included (i) Lal champaran basmati, (ii) Bhuri champaran basmati, (iii) Kali champaran basmati, (iv) Baharni, (v) Badshahbhog, (vi) Chenaur, (vii) Dewtabhog, (viii) Kesar, (ix) Kamod, (x) Kanakjeera, (xi) Marcha, (xii) Ram Janwain, (xiii) Sonalari, and; (xiv) Tulsi Pasand. Most of these varieties have either already lost, or are at the verge of extinction (Singh et al. 2003).

Local varieties have yield potential ranging from 15 to 30 qtls/ha, and are tall, possessing short grains. Many of them are highly susceptible to various insect pests and diseases, like: (i) stem borer and (ii) bacterial blight.

Table 7.1 Local Scented Rice Varieties and Landraces of Bihar

S. No.	Location	Scented Rice Varieties/Landraces	Land Type	Important Characteristics
1.	Patna	Basmati-3, Karibank-2, Mohin Dhan, Sagarbhog, and Hansraj	Medium and Low lands	Tall, late duration, small to medium fine grain, aromatic and photoperiod sensitive
2.	Bhojpur	Sonachur, Karibank, Basmati, Badshahbhog, and Kanakjeera	Medium and Low lands	Tall, late duration, small to medium fine grain, aromatic and photoperiod sensitive
3.	Rohtas	Sonachur, Shyamjeera, Basmati-3, Shahpasand, and Thulsiphool	Medium and Low lands	Tall, late duration, small to medium fine grain, aromatic and photoperiod sensitive
4.	Gaya	Basmati and Kanehonehur	Medium and Low lands	Tall, late duration, small to medium fine grain, aromatic and photoperiod sensitive
5.	Aurangabad	Shyamjeera and Mehijawain	Medium and Low lands	Tall, late duration, small to medium fine grain, aromatic and photoperiod sensitive
6.	Bhagalpur	Tulsimanjari, Katarnibhog, Badshahbhog, and Br-9 & Br-10	Medium and Low lands	Tall, late duration, small to medium fine grain, aromatic and photoperiod sensitive
7.	Munger	Tulsimanjari, Shyamjeera, Karibank, Marueya, and Lakhisar	Medium and Low lands	Tall, late duration, small to medium fine grain, aromatic and photoperiod sensitive
8.	North Bihar	(i) Badshahbhog, (ii) Badshahpasand, (iii) Baharni, (iv) Basmati03, (v) Br-9, (vi) Br-10, (vii)Bhuri Champaran Basmati, (viii) Chenaur, (ix) Dewtabhog, (x) Hansraj, (xi) Kamod, (xii) Katarnibhog, (xiii) Kali Champaran Basmati, (xiv) Karibank, (xv) Karibank-2 (xvi) Kanakjeera, (xvii) Kesaurbani, (xviii) Kesar, (xix) Lal Champaran Basmati, (xx) Lakhisar, (xxi) Marcha, (xxii) Marueya, (xxiii) Malbhog, (xxiv) Mehijawain, (xxv) MohinDhan, (xxvi) Ram Jawain, (xxvii) Sagarbhog, (xxviii) Sonalari, (ixxx) Sonachur, (xxx) Shyamjeera, (xxxi) Shahpasand, (xxxii) Tulsipasand, (xxxiii) Tulsimanjari, and; (xxxiv) Tulsiphool.	Medium and Low lands	Tall, late duration, small to medium fine grain, aromatic and photoperiod sensitive

Source: Singh and Singh (2003).

Since they have excellent cooking quality and aroma, they are still grown by farmers on small scale particularly in case of three varieties: (i) Kamini (Katarni), (ii) Mircha, and; (iii) Malida. Each of these three has its own speciality: Katarni for cooked rice, Mircha for Cheura, and Malida for its adaptability in low-land deep water. All landraces of aromatic types grown in Bihar have fine, but short grains and consequently of low export values.

In view of the lower yield of traditional aromatic rice varieties in Bihar declining area under Katarni paddy over the years and most of the good quality aromatic rice varieties either being lost or facing the threat of extinction the study has been undertaken with the objectives noted below:

i. to assess the potential area of Katarni paddy in the study area;
ii. to find out socio-economic characteristics of the farmers, who cultivate Katarni paddy;
iii. to study the economics of Katarni paddy in the study area;
iv. to identify the marketing channels of Katarni paddy in the study area;
v. to identify the constraints in raising the area under Katarni paddy in the study area; and
vi. to suggest suitable measures for the development of Katarni paddy in the study area.

METHODOLOGY AND DATA

This chapter is mainly based on primary data collected from 30 Katarni-paddy-growing cultivators each from Bhagalpur and Banka districts. Multi-stage random sampling method was followed to select respondents.

At the first stage of sampling, the two districts, Bhagalpur and Banka, were purposively chosen, as the specific variety of Katarni, to which this study is devoted are grown only in particular areas of these two districts. At the second stage of sampling, one block, in from each district was selected on the basis of larger area under Katarni paddy and potential. On this basis, Jagdishpur block and Amarpur block were selected from Bhagalpur and Banka Districts, respectively. At the third stage of sampling, maintaining the harmonious basis of choosing potential villages, in regard to cultivation of Katarni paddy, two villages each from the two selected blocks of the concerned districts were identified. Thus, two villages, namely: Bhawanipur-Deshari and Jagdishpur cluster of villages under Jagdishpur block were selected. Similarly, (i) Tardih-Lakshmipur, and; (ii) Ramchandrapur-Bhadariya villages were selected from Amarpur block of Banka district. At the fourth stage of sampling, enlistment of Katarni paddy growers in the selected villages was made. In Bhawanipur-Deshari and Jagdishpur cluster of villages under Jagdishpur block of Bhagalpur district, the number of marginal, small, medium, and large farmers growing katarni paddy also, were 40, 50, 55, and 21 respectively. Number of katarni paddy growers, who belonged to marginal, small, medium and large farm size classes of 'Tardih Lakshmipur' and 'Ramchandrapur Bhadaria' villages in Amarpur Block of Banka district were 45, 40, 60, and 18, respectively. At the fifth stage of sampling, indispensable classification of farmers from out of the enlisted growers was done based on farm size owned by them. All the enlisted growers were broadly kept in four categories: (i) Marginal—owning land up to 1 hectare, (ii) Small—1.01 to 2 hectare, (iii) Medium—2.01 to 4 hectare, and; (iv) Large—> 4 hectare. At the sixth stage of sampling, 15 farmers from each of the selected villages (if required number of Katarni paddy growers was not found in a particular village, then cluster of adjoining villages was also considered) were selected for detail study. The selection of farmers was done on probability proportion method. Further, with the view to maintain discreet selection of respondents, due emphasis was given on social composition of the enlisted growers.

In this way, the selection of sample can be illustrated as below: 2 districts × 1 block each (= 2) × 2 villages each (=4) × 15 farmers = 60 Katarni paddy growers. Simple tabular and percentage methods have been followed to analyse the data and interpretation of observed facts. Reference year of the primary data collection is 2010–11. However, the secondary data pertain to the latest data available in the Department of Agriculture, Government of Bihar, Bhagalpur, and Banka districts.

RESULTS AND DISCUSSIONS

Bhagalpur district comprises Bhagalpur sadar and Kahalgaon that lie on the southern bank. The district was spread over 2.54 lakh square hectares and divided into 16 blocks and 242 gram panchayats. The city of Bhagalpur is the headquarters of Bhagalpur division as also of the district and sadar sub-division. The district had 1519 revenue villages. Out of it, 923 villages (60.76 per cent) are inhabitated and 596 (39.24 per cent) un-inhabitated.

However, as per the Census 2011, Series–II, the population of Bhagalpur district was 3032226 which accounted for 2.92 per cent of the state's total population. Sex ratio of the district was distressing (879 females/1000 males). The population density was found 1180/sq. km. The literacy rates of males and females were 72.30 and 56.50 per cent respectively. It revealed that the gender gap in literacy was 15.80.

The data classification of workers reveals that 48.39 per cent were agricultural workers followed by 19.63 per cent cultivators. 7.43 per cent workers were engaged in household industries and 24.55 per cent constituted other workers. The work participation rate in the district was 35.37 per cent with only 21.34 per cent in case of females. Data on sector wise employment pattern revealed that 68.10 per cent workforce was employed in primary sector followed by 24.50 per cent in tertiary sector and only 7.40 per cent in secondary sector.

As regards the pattern of land utilization in the district of Bhagalpur, out of the total geographical area, net sown area was 1.53 lakh hectares i.e., 61.91 per cent. While forest coverage had remained at 0.10 per cent, permanent pasture land was 0.90 per cent. Current fallow land came to 20.87 per cent. Cultivable waste land (3.30 per cent) and land under non-agricultural uses (18.33 per cent). The cropping intensity was 124.42 per cent only, which was slightly less than the state figure of 132.78 per cent.

Irrigation is one of the major inputs of agricultural development. Though, several measures have been taken to enhance the scope of irrigation ever since India became independent, however, things have not undergone metamorphosis change so far. Consequently, various sources of irrigation were taken recourse depending also on the status of the farmers. Notwithstanding these efforts, Bhagalpur lagged behind in terms of irrigational base and irrigational intensity compared to the state average.

The Census 2011, Series –II report reveals that the population of Banka district was 2029339, which accounted for 1.96 per cent of the state's total population. Sex ratio of the district was distressing (907 females/1000 males). The population density was 672/sq. km. The literacy rates of males and females were 69.80 and 49.40 per cent respectively. It revealed that the gender gap in literacy was 20.40 per cent.

The number of total workers in the district was 6.39 lakh, which accounted for 39.74 per cent of the total population. The data on classification of workers reveals that 51.71 per cent were agricultural labourers followed by 33.74 per cent cultivators, 4.62 per cent workers engaged in household industries, and 9.93 per cent constituted other workers. The work participation rate in the district was 39.70 per cent with only 28.16 per cent in case of female.

While changes in land utilization pattern were very slow and marginal throughout the state, there had been a marginal increase in the non-agricultural use of geographical area across the state. As regards the pattern of land utilization in the district of Banka, out of the total geographical area, net sown area was 1.52 lac hectares i.e., 49.86 per cent. While forest coverage had remained at 14.18 per cent, permanent pasture land was 0.56 per cent. Current fallow land came to 1.25 per cent, cultivable waste land (2.61 per cent) and land under non-agricultural use (13.35 per cent). The cropping intensity was 106.00 per cent only, which was much lower than the state figure of 132.78 per cent.

Banka lagged behind in terms of irrigational base and irrigational intensity compared to the state average. As per latest data, out of the net sown area, only 83.72 per cent of land had the scope of irrigation and the rest either remained rainfed or faced the worst. The data on source-wise distribution of irrigated area revealed that canal (70.57 per cent) was the major source followed by bore well, open well (5.67 per cent), tank (2.34 per cent), and others (2.68 per cent). A glance on the table containing data related to socio-economic features of sample respondents reveals highest number of respondents (growing Katarni paddy) to be in the age group of 36–60 years in both Bhagalpur (93.33 per cent) and Banka districts (83.33 per cent). Social group wise composition of the surveyed respondents reveals that the highest number of Katarni paddy growing farmers belonged to OBC group 23 (76.67 per cent) in Bhagalpur and 17 (56.67 per cent) in Banka district.

All of the surveyed Katarni paddy growers in both the selected districts were male (100 per cent each). In regard to main occupation and other activities of the surveyed farmers, agriculture was evinced to be the main source of livelihood for all the surveyed farmers of both the districts—Bhagalpur and Banka 30–30 (100 per cent each). On overall level, share of the sources of livelihood for the surveyed katarni paddy growers were: (i) Agriculture (as main occupation) 100 per cent, (ii) Business/Trade 8.34 per cent, (iii) Service (Public/Private Sector 10.00 per cent, and; (iv) Agricultural Labourers 6.66 per cent).

On an average, the surveyed Katarni paddy growers of Bhagalpur and Banka districts both fell under the broad category of medium land holdings. Data further furnishes that Katarni paddy growing not being a much remunerative exercise, mostly the medium and big farmers preferred to undertake its cultivation.

It is interesting to note that in Bhagalpur 4.54 ha and Banka 25.14 ha of leased in land areas were from irrigated conditions, that is, higher than unirrigated ones 3.48 ha and 10.71 ha) respectively.

Having groped the reason, most probably responsible for larger gross areas on aggregate level in Banka district (100.93 ha) than Bhagalpur (79.16 ha), the revealed factor may be attributed to much higher leased-in area of 35.85 ha actually leased out by the big and prosperous Rarhi Kayastha caste landowners of the district.

It was found that average sizes of leased-in land were higher in cases of marginal farmers on overall level (0.53 ha) in Bhagalpur district, while small farmers (2.10 ha) in Banka district. It is interesting to note that no medium and/large farmer leased out their land in Bhagalpur district, while Banka district witnessed as per normal belief large (1.16 ha) and medium farmers (0.21 ha) to have leased out their cultivable land. Small and medium farmers of Banka district were clearly ahead in leasing in lands followed by marginal farmers in terms of aggregate total (14.65 ha, 14.48 ha, and 6.72 ha), respectively. In Bhagalpur district, marginal farmers were ahead 3.67 ha followed by small and medium (3.35 and 1 ha), respectively.

In nutshell, due to 'larger total' and 'average land areas' leased out by big farmers of Banka district (majority of them belonging to prosperous R K caste, 3.47 ha, 2.53 ha, 1.16 ha, and 0.21 ha), respectively, the sample Katarni paddy growers of this district were at more privileged stage having taken larger areas as leased in land.

In Bhagalpur district, larger areas under unirrigated conditions were used by the sample respondents for growing cereal mainly paddy 34.25 ha pulses, mustard 7 ha, and orchards 3.50 hectares. In case of Banka district, areas under pulse crop masoor (lentil) 12.50 ha, Khesadi 7 ha, mustard 8 ha, Tisi 4.92 ha, and orchards 5.55 ha under unirrigated land areas were higher. Paddy in Banka district got greater share under irrigated condition 38.13 ha. Larger areas under Katarni paddy were seen to be devoted/used in irrigated conditions in Bhagapur and Banka districts both, when compared to areas under unirrigated conditions (6.35 ha, 10.20 ha, and 2.08 ha and 9.15 ha), respectively.

As for cultivation of vegetables is concerned, larger areas devoted towards onion (in irrigated condition) only was revealed in both the surveyed districts in comparison to brinjal and potato also in Banka district (0.56 ha, 1.00 ha, 0.50 ha, 0.25 ha, and 0.65 ha), respectively.

Data contained in concerned table captivates towards largest areas under Katarni paddy devoted by big large farmers of Bhagalpur and Banka districts both (4.22 ha and 10.00 ha, that is, 50.06 per cent and 51.68 per cent of the total operational areas of the concerned districts) respectively. Having a glance on data, it can also be framed that there is direct and positive relationship between farm size and areas devoted for growing Katarni paddy, that is, with the increase in the farm size, there was increases in land areas used for growing Katarni paddy.

On overall level, under variable cost, maximum expenditure, that is, in percentage terms, was incurred on labour (both hired and imputed family labour) Rs 9070 per hectare (24.68 per cent), and the minimum being in case of transplantation Rs 1000 (2.72 per cent). While the amount paid as wages to labourers was higher in Banka district (Rs 9,100/-), the transplantation cost was higher in Bhagalpur (Rs 1050). As the quantum of net returns were quite higher in Bhagalpur district (Rs 12986.72) in comparison to that of Banka district (Rs 9,552.70), so cost–benefit ratio (CBR) of Bhagalpur district (1:1.36) was found to be genuinely higher than that of the later district (1:1.26) (Table 7.2).

Data in table made it crystal clear that labour had remained the major items of expenditure (Rs 9070 per ha (24.68 per cent) followed by irrigation (12.24 per cent), harvesting (8.59 per cent), ploughing (8.30 per cent), manure (6.87 per cent), fertilizers (5.96 per cent), seeds (5.17 per cent), transplantation (2.72 per cent), and interest on working capital (1.80 per cent).

Farm size wise areas under Katarni paddy in both the selected districts leads us to crunch for revealing that out of the total operational area owned by different size groups of respondents in Bhagalpur and Banka districts, in physical and percentage terms, large farmers were much ahead (4.22 ha or 50.06 per cent and 10 ha or 51.68 per cent) respectively.

In case of marginal farms, on overall level, highest expenditure was incurred in labour Rs 8750 (26.22 per cent) of the total, that is, fixed cost and variable costs. Gross return

Table 7.2 Per hectare Cost of Cultivation of Katarni Paddy (Rs)

SN		Particulars	Bhagalpur	Banka	Overall (In per cent)
Average Area Under the Crop (In ha)			0.281	0.645	0.463
A		**Fixed Cost**			
	i.	Value of Land (Lakh/ha)	3.71	3.46	—
	ii.	Interest on Fixed Capital	6183.33	5766.66	—
	iii.	Land Revenue Paid	45.28	50.00	47.64 (0.13)
	iv.	Rental Value of Land	8251.50	7757.69	8004.59 (21.77)
B.		**Variable Cost**			
	i.	Ploughing	3100.00	3000.00	3050.00 (8.30)
	ii.	Transplantation	1050.00	950.00	1000.00 (2.72)
	iii.	Seeds (both farm produced and purchased)	1800.00	2000.00	1900.00 (5.17)
	iv.	Fertilizer	2187.50	2193.25	2190.38 (5.96)
	v.	Manure (owned and purchased)	2500.00	2550.00	2525.00 (6.87)
	vi.	Labour (hired and imputed family labour)	9040.00	9100.00	9070.00(24.68)
	vii.	Irrigation	4000.00	5000.00	4500.00 (12.24)
	viii.	Harvesting	3318.00	3000.00	3159.00 (8.59)
	ix.	Interest on working capital	630.00	692.52	661.26 (1.80)
	x.	Depreciation on implements and Farm Buildings	600.00	700.00	650.00 (1.77)
		Total	**28225.50**	**29185.77**	**28705.64**
		Total Cost (A (iii, iv) + B)	**36522.28**	**36993.46**	**36757.87 (100.00)**
C.		**Gross Return** (including price of straw @ Rs 5000 - per ha)	49509.00	46546.16	48027.58
	i.	Rates (@ Rs/Qtl)	2350.00	2280.25	2315.13
	ii.	Yield rate (Qtl/ha)	18.94	18.22	18.58
D.		**Net Return (in Rs)**	12986.72	9552.70	11269.71
E.		**Cost of Production (in Rs)**			
	i.	Per quintal (Rs)	1928.31	2030.37	1979.34
F.		**Cost-Benefit Ratio**	1:1.36	1:1.26	1:1.31

Note: The Calculated Value of Rent is meant for 5- month period from sowing to harvesting of Katarni Paddy (already paid before).
Source: Field survey data.

(including straw) was found lower in Bhagalpur district (Rs 44050). Besides total cost of production (Rs 26427.32) being higher, the rate of sale of katarni paddy, and yield rate were found lower in Bhagalpur district. Hence, CBR was lower (1:1.30) because net return was also calculated quite lower at Rs 10258.39 than that of Banka district Rs 12089.90.

Like marginal farmers, the surveyed growers belonging to small farm size on overall level, incurred maximum expenditures in labour (24.78 per cent), irrigation (12.74 per cent), ploughing (8.78 per cent), manure (7.36 per cent), and harvesting (6.16 per cent). Gross and net returns were higher in Banka district Rs 47550 and Rs 11541.24 respectively. So, cost benefit ratio was marginally higher in Banka district (1.1:32) than that of Bhagalpur district (1:1.31). On overall level, it was estimated at 1:1.32.

Maintaining unchanged scenario, at the overall level, the surveyed farmers (belonging to medium farm size group) evinced highest expenditures to have made in items of labour (24.56 per cent) followed by irrigation (11.04 per cent) and ploughing (8.30 per cent) like that of marginal and small farmers. The CBR in Bhagalpur district (1:1.35) was higher than that of Banka district (1:1.30). Dwelling upon the reasons for it, some of the factors could be higher rate sale price of Rs 2350 per qtl and yield rate 19.10 qtls/ha in the former district. On overall level, in percentage terms, elucidates similar scenario/trend of highest expenditures was elucidated to have been made by large farmers in the heads of labour (25.00 per cent) and irrigation (10.25 per cent) as could be seen in case of marginal, small and medium farmers surveyed. It was followed by harvesting (8.78 per cent), ploughing 8.56 per cent), manure (6.41 per cent), fertilizers (5.39 per cent), and seeds (4.73 per cent). CBR in Bhagalpur and Banka districts could be found almost same (1.1:30).

It can also be interpreted that higher share of irrigation expenditure incurred by all farm size classes is one of the reasons for farmers having developed indifferent attitude towards growing Katarni paddy in larger areas. So, if irrigation facility is strengthened and expanded in the region, farmers may be encouraged to jump towards growing Katarni paddy. While looking at data in the table, it can be observed that out of total production of Katarni paddy by the surveyed farmers, highest quantum in percentage terms in both the districts, namely, Bhagalpur and Banka districts were meant for marketable surplus (67.74 per cent and 65.20 per cent) followed by home consumption (32.26 per cent and 34.80 per cent), respectively. Further, out of the total marketable surplus lower quantum were found to have been retained for further sale/home consumption in both the districts (12.09 per cent and 5.98 per cent) respectively.

Four marketing channels have been identified here:

Channel I:	Producer—Consumer (PC)
Channel II:	Producer—Itinerant trader—Consumer (PITC)
Channel III:	Producer—Wholesaler—Retailer—Consumer (PWRC)
Channel IV:	Producer—Retailer—Consumer (PRC)

Table 7.3 clearly reveals Channel III to be the most prominent one for selling maximum quantities of Katarni paddy by the surveyed growers of Bhagalpur and Banka districts (59.38 per cent and 54.34 per cent) respectively.

Table 7.3 Quantity Sold by Different Marketing Channels (In qtls)

Districts	Particulars	Zero level Producer-consumers (PC)	One level producers-itinerant-traders consumers-(PITC)	Third level producers-whole-sellers—retailer-consumer	Fourth level producers-retailer consumer-(PRC)
Bhagalpur	Qty	7.00	12.00	57.00	20.00
	(in qtls)	(7.29)	(12.50)	(59.38)	(20.83)
	Selling Price	2100.00	2250.00	2400.00	2200.00
	(in Rs/qtls)				
Banka	Qty	10.50	50.00	119.61	40.00
	(in qtls)	(4.77)	(22.72)	(54.34)	(18.17)
	Selling Price	2100.00	2250.00	2400.00	2150.00
	(in Rs/qtls)				

Note: Figures in brackets indicate percentages of the Marketed Surplus of the Districts concerned, that is, out of the total quantities sold—estimated at 96 qtls for Bhagalpur and 220.11 qtls for Banka.
Source: Field survey data.

The channel of sale through which lowest quantities were sold, was channel – I for both the districts (7.29 per cent and 4.77 per cent) respectively. In Bhagalpur district as it is revealed, the marketing margins at Channel II, Channel III, and Channel IV could be calculated at Rs 150, Rs 300, and Rs 100, respectively. It means that from Producer and Consumer points of view, Channel IV is the most advantageous, desirable, and effective channel. In Banka district, almost similar scenario of marketing margins could be seen, except the amount of margin in case of Channel IV being Rs 50 lower than that of Bhagalpur district. It was Rs 2,150 per quintal in PRC Channel.

It can thus be concluded that if the number of intermediate traders are reduced, then the growers may earn higher sale price, which is urgently desired with the view to encourage the farmers to undertake cultivation of Katarni paddy in more areas.

CONCLUSIONS AND POLICY IMPLICATIONS

It can be induced that if suitable and stringent measures are taken for stopping uncontrolled excavation of sand from river Chandan and original/certified seeds of Katarni paddy are made available to farmers by any government agency, then a remarkable increase in areas under Katarni paddy in both the districts can be realized.

However, it may be taken as a matter of encouragement that now cultivation will be done by using foreign techniques. Blueprint was being prepared for this. In Bihar Agriculture University (BAU), Sabour, (Bhagalpur) also, research is being conducted to save the fragrance of Katarni paddy. As stated by the scientists from Philippines, productivity will be doubled by using foreign techniques/ technology. This will not only help farmers in being more prosperous, but, Katarni's fragrance will smell in foreign countries also. As per a rough estimate, about 20,000 qtls of Katarni rice is required. To meet this high demand, modification in the variety of Katarni paddy (into dwarf variety is desired).

Katarni paddy is a specific area-based variety of rice. It has high potential with vast untapped opportunities of marketability. In true sense, basmati cannot be produced in Bihar despite willingness of the farmers. Agricultural commodities do possess their area specific characteristics. Marketing of any good or service involves already existing features of competition. Having classified under hard aroma group (Katarni rice) is mainly termed as kheer and khichadi rice. On the other hand, basmati rice is soft aroma rice, and it is known as polao rice. Actually Katarni is ceremonial rice, so there is no option to it, particularly on special occasions. Bhagalpur and Banka districts are the main areas for growing Katarni paddy. It is worth mentioning that fine scented rice is the strength of Bihar. Patna rice was also the main source of economy in Bihar, but now it has fully extinguished. It is dismal to note here that none of the public or private agencies in Bihar did take any endeavour to produce/develop or did make any effort to promote production of Katarni paddy seed. The same observation-based feelings were genuinely expressed by sample and other Katarni paddy growers as well. Names of such public/private agencies or departments may be noted as below:

(i) Bihar Rajya Beez Nigam (BRBN), (ii) Tarai Development Corporation (TDC), (iii) State Food Corporation (SFC), (iv) Food Corporation of India (FCI), (v) Rajendra Agricultural University (RAU), (vi) National Seed Corporation (NSC), and; (vii) Private Companies. Although, these agencies use to provide or promote production of Hybrid Varieties (HYVs) of all other types of paddy, they didn't take pain of producing and preserving traditional seed of Katarni paddy.

Some varieties of fine scented rice, namely: (i) kalanamak, (ii) adam chini, (iii) Mirzapur (UP) do belong to Varanasi and Gorakhpur regions, while Mircha variety belonged to Motihari district of Bihar state. There is 2 to 3 per cent chance of cross-pollination. At the same time, the threat of deterioration in quality is also involved in it. It was also embodied that due to exchange of seeds from, in some cases, non-real Katarni seeds, the originality of the product is being vanished. Besides the above threat, grains are also being deteriorated. Another one of the challenges before the prospects of Katarni paddy production in Bihar is its lower demand.

Following policy suggestions have emerged from the analysis of the study:

i. Katarni paddy should be improved agronomically.
ii. Causation endeavours may be made by the Agricultural Scientists' and the farmers to maintain the natural and unique fragrance of Katarni Paddy and not to increase its yield only.
iii. Agronomically developed seeds need to be innovated.
iv. All the public and private agencies (directly or indirectly involved in the production/manufacturing of seeds), should emphasize its activities towards preserving traditional Katarni paddy seeds. These agencies need to include preparing and preservation of traditional Katarni paddy seeds in their programme.
v. Emphasis has to be given on preserving aroma of Katarni paddy. For this, National Bureau of Plants

and Genetic Resources, New Delhi needs to be invited to look into the problem/matter in Bhagalpur and Banka districts of Bihar.

vi. Scientific research needs to be conducted to ascertain as to which particular micronutrients are responsible for maintaining aroma in Katarni paddy.

vii. By mutation (gama radiation), genetical improvement, yield increase, and quality improvement can be obtained.

viii. With the view to enhance yield of Katarni paddy, organic farming should be propagated and encouraged.

ix. With the objective to deter the problem of adulterated seed, nucleus seed has to be maintained. After every 3–4 years, practice of using new seeds needs to be promoted. If the breeder does not have nucleus seed, arrangements may be made to bring it from National Bureau of Plant Genetic Research (NBPGR), New Delhi, so that its originality and quality could be maintained.

x. Genetic purity (True to Type) of katarni paddy should be maintained.

xi. Drive to popularize Katarni paddy should be launched with the objective to check the decline of area under it.

xii. Threat of eroding natural taste/fragrance of Katarni paddy as a result of excessive use of chemical fertilizers needs to be countered by creating awareness among the farmers to cultivate it by using bio-fertilizers or natural manure only.

xiii. Check dams in River Chandan in quite a few numbers should be constructed to uplift and retain water level at a height that could facilitate easy cheaper and assured irrigation.

xiv. Unchecked excavation of sand from different points of River Chandan in large quantum everyday has caused deepening of river bed. It has been causing outflow of water from such a depth, from where irrigation is too costly and cannot be done at required intervals. Sand-less bed of River Chandan has lost water-retention capacity making farmers indifferent towards cultivation of Katarni paddy on a large scale.

xv. As Basmati, the aromatic rice grown in Northern India, has seen phenomenal growth in regard to its processing industries in the past five years (2006–07 to 2010–11), due to the result of more than doubled demand, similarly environment for increase in production of Katarni paddy could be created by promoting and carrying out branding exercise.

REFERENCES

Bhagalpur representative, 'Ab videshon me bhi gamkegi Bhagalpuri katarni ki sugandh: Delhi ki multi-national company ne dikhai dilchaspi'. *Dainik Jagran*, Hindi daily, 2011, pp. 1, 9.

Correspondent. 'Chamkegi Kismat: Katarni ko milegi videshi taknik-wideshiyon ko bhi bhayi khushboo' *Prabhat Khabar*, Hindi daily, Bhagalpur edition, 2011 p. 6.

Kumar, Prateek. 2012. 'Sukhe ki aashanka se kam huie katarni ki khushboo'. *Hindustan*, Hindi daily.

Kumar, Roop. 2011. 'Jagdishpur banega kisano ka s'. *Dainik Jagran*, Hindi daily, p. 8.

Legal Correspondent. 'Bhagalpur ke Commissioner se baloo (sand) khanan par report talab'. *Dainik Jagran*, Hindi daily, 2012, p. 3.

Mishra, Lalit Kishore. 2012. 'Katarni ki khushboo phir failegi'. *Prabhat Khabar*, Hindi daily, p. 1, 17

Representative. 'Katarni khushboo bachane me jute Waigyanik'. *Prabhat Khabar*, Hindi daily, 2011, p. 3.

Senior Correspondent, 'Bahoot jald loutega katarni ka purana swad'. *Hindustan*, Hindi daily, 2011.

Singh, A.N. and V.P. Singh. 2003. 'Extent, Distribution and Growing Environments of Aromatic Rices in India', in R.K. Singh and U.S. Singh (eds), *A Treaties on the Scented Rices of India*, pp. 211–29. Ludhiana: Kalyani Publishers.

Singh, R.K. and U.S. Singh. 2003. 'Scented Rices of India: An Introduction', in R.K. Singh and U.S. Singh (eds), *A Treaties on the Scented Rices of India*, pp. 1–4. Ludhiana: Kalyani Publishers.

Singh, R.K., S.C. Mani, Neelam Singh, G. Singh, H.N. Singh, Rashmi Rohilla, and U.S. Singh. 2003. 'The Aromatic Rices of Uttar Pradesh and Uttaranchal', in R.K. Singh and U.S. Singh (eds), *A Treaties on the Scented Rices of India*, pp. 403–20. Ludhiana: Kalyani Publishers.

Thakur, Pran Mohan, Ex-Commissioner, 'Katarni ki kheti ka dayara badhane par Aayukta ka jor' (emphasis), Bhagalpur. 2010. *Hindustan*, Hindi daily.

Tiwari, Amarendra Kumar. 2011. 'Swad achha, sugandh badhiyan-Par Nahin Mili Pahchan'. *Dainik Jagran*, Hindi daily, p. 2.

8

Spread of New Varieties of Hybrid Rice and Their Impact on the Overall Production and Productivity

Pranab Kanti Basu and Debajit Roy

India has a large agrarian economy with majority of its rural population subsisting on farming. Over the decades since independence, Government of India has made concerted efforts to improve the lot of the farmers. By the mid-1960s it was realized that for India to achieve self-sufficiency in foodgrains, there was no alternative to technological change in agriculture. The spread of high yield variety (HYV) technology resulting in the green revolution in India in the last decades and achievement of self-sufficiency in foodgrains represent a success story for the science and technology sector. The most widely debated issue about the green revolution was the growing disparities in income between the different regions and the different classes of farmers. This was observed in the early phase of the green revolution, that is, until about the mid-1970s. These trends however got reversed after the mid-1970s, typical of a diffusion process characterized by the spread of green revolution to new areas, and the increasing adoption of new technology by the small farmers. The achievements so far in respect of raising yields and reducing variability in the unfavourable agro-climatic regions are not comparable with those realized for the favourable environments. The limited spread of the green revolution can be explained partly by the nature of available technology itself and partly by the uneven development of infrastructure, physical as well as institutional which is pre-requisite for the adoption of improved practices. It was also observed that the rise in yield as a result of green revolution had reached a plateau. Further there was apprehension that it might take a downturn because of inherent drawbacks in the bio-technological character of the revolution. For example it was causing a much faster rate of depletion of ground water resources.

Against such a background, it is necessary to examine the needed changes in agricultural research strategy. Minimizing regional imbalances in growth, imparting stability to agricultural output, and bringing the benefits of agricultural research technology to the resource poor farmers are the three major concerns that motivate research. The research scientists considered hybrid rice technology as a readily available option to shift the yield frontier upward in the face of declining trend of the yield potential of the existing varieties. It was projected that hybrid rice technology would bring about another rice revolution in the country. However, although a number of varieties of hybrid rice are released by the Government, the extent of adoption of hybrid rice varieties in the country is too meagre to make an impact on rice production. Against this backdrop, the present study is conceptualized and undertaken at the instance of the Directorate of Economics and Statistics, Ministry of Agriculture and Farmers Welfare, Government of India, with a view to assessing the actual spread of hybrid rice varieties replacing the conventional HYVs to make an overall impact of rice production.

Encouraged by the success of hybrid rice technology in enhancing the rice production and productivity in China, the Indian Council of Agricultural Research (ICAR) initiated a national programme for development and large-scale adoption of hybrid rice in the country in December 1989. The project was implemented through a National Network comprising research, seed production, and extension networks. The hybrid rice research network consisted of 11 research centres and many voluntary centres spread across the country. The seed production network consisted of public sector seed production agencies such as National Seed Corporation, State Farms Corporation of India, and the State Seed Development Corporations in addition to many private sector seed companies. The extension network consisted of state departments of Agriculture, extension wings of the State Agricultural Universities (SAUs), Krishi Vigyan Kendras (farm science centres), and the NGOs. Effective linkages were established within the different sub-components of the network. The entire project was co-ordinated and implemented by the Directorate of Rice Research (DRR), Hyderabad. The project initiated by the ICAR, was strengthened by the technical support from IRRI Philippines, FAO, the financial support from the UNDP, Mahyco Research Foundation (MRF), World Bank funded National Agricultural Technology Project (NATP), and IRRI/ADB Project on Hybrid Rice.

Hybrid rice technology is likely to play a key role in increasing the rice production. During the year 2008, hybrid rice was planted in an area of 1.4 million ha and an additional rice production of 1.5 to 2.5 million tonnes was added to our food basket through this technology. More than 80 per cent of the total hybrid rice area is in eastern Indian states such as Uttar Pradesh, Jharkhand, Bihar, Chhattisgarh, with some little area in states such as Madhya Pradesh, Assam, Punjab, and Haryana. As rice is a key source of livelihood in eastern India, a considerable increase in yield through this technology will have a major impact on household food and nutritional security, income generation, besides an economic impact in the region. In view of this, hybrid rice has been identified as one of the components under the National Food Security Mission (NFSM) launched by the Government of India (GoI) with the aim to enhance rice production by 10 million tonnes by 2011–12. Under the scheme it has been targeted to cover 3 million ha area under hybrid rice by the year 2011–12. The approach is to bridge the yield gap in respect of rice through dissemination of improved technology and farm management practices. Similarly, added emphasis is being given for adoption of hybrid rice under the special scheme (BGREI) of GoI to bring green revolution to eastern India.

As a result of concerted efforts for over two decades, a total of 46 hybrids have been released for commercial cultivation in the country. Among these, 29 have been released from the public sector while remaining 17 have been developed and released by the private sector. Though 46 hybrids have been released in the country so far, some of them have been outdated, and some are not in the production chain. Such hybrids related to production chain and available for commercial cultivation are listed in Table 8.1.

The farmers of the country are growing mostly the varieties bred by the research system such as ICAR, SAUs, and other research institutions connected to agriculture. The varieties are normally bred taking into consideration, various characters like yield potential, resistance to biotic and abiotic stress of the existing popular variety/varieties. The new varieties are bred by the research institutions and screened for their performance at different locations through initial evolution trial and advance varietal trial. A Technical Committee finally considers these varieties and releases only those varieties which are found superior over the existing best varieties. While releasing these varieties the Technical Committee also specifies the ecology, that is, the state area within State, season in which the varieties are to be grown. The newly released varieties normally have edge over the existing varieties in yield, resistant to serious pest and diseases, resistant to the abiotic stresses, that is, water-related problems such as droughts, etc. Although a number of varieties are being released by the government to meet the demand of the farmers, the spread of these newer varieties in place of the conventional varieties that

Table 8.1 Mean Yield Levels of Hybrids and HYV 2010–11 for Selected States (Pooled)

Farm Size Classes (ha)	Mean Yield (Kg/ha)		% Difference
	Hybrid	HYVs	
Below 1	6173	4259	31.00
1–2	6212	4413	28.96
2–4	6462	4401	31.90
4–10	6443	4207	34.70
10 ha and above	5053	3034	39.95
All sizes	6331	4379	30.83

Source: Field Survey.

are grown by the farmers for a longer period has not been assessed properly. There is no comprehensive evaluation study to document farm-level insights into hybrid rice performance except very few studies citing the instance of yield superiority of hybrid rice but less profitable than the inbred varieties i.e. conventional HYVs (Chengappa, Janaiah, and Gowda 2003; Janaiah 2002).

The spread of the newer varieties replacing the older varieties need to be closely monitored to take advantage of the superior characters of these newer varieties released by various Research Institutions. This will help to break the yield plateau in rice production of the recent past. Though the government is taking a number of steps to popularize these varieties like frontline demonstration, mini-kit supply, organizing training programmes (1–21 days) for farmers, farm women, seed growers, seed production personnel of public and private seed agencies, extension functionaries of state departments of agriculture, officials of SAUs, and NGOs, there is no concrete evidence that the newer varieties of rice are spreading faster and replacing the older ones. Therefore, it is essential to conduct a study to assess the actual spreading of these newer varieties in terms of area. This will help GoI to draw a plan for augmenting the spread of the superior newer varieties. The specific objectives of the present study are:

1. to indicate the extent of adoption and the level of participation by the different categories of farmers in the cultivation of hybrid rice;
2. to assess the overall impact of hybrid rice cultivation on rice production and productivity;
3. to study the economics of cultivation of hybrid rice varieties vis-à-vis inbred varieties;
4. to identify factors determining the adoption of hybrid rice varieties;
5. to determine constraints and outline the prospects for increasing hybrid rice cultivation; and
6. to suggest policy measures for expansion of hybrid rice cultivation.

METHODOLOGY AND DATA

This chapter covers five states, namely West Bengal, Uttar Pradesh, Bihar, Madhya Pradesh, and Andhra Pradesh. The study is based on both secondary and primary data. Secondary data was obtained from different state government publications relating to area, production, and productivity of rice.

Keeping in mind that the first hybrids were developed and released for commercial cultivation in India in 1994, the study period was divided into three sub-periods namely, 1984–85 to 1993–94, 1994–95 to 2003–04, and 2004–05 to 2009–10. The period I namely, 1984–85 to 1993–94 refers to the pre-introduction period of hybrid rice while other two period's namely, period II and period III correspond to post-introduction periods.

Primary survey was confined to the NFSM districts in the states. The two districts having relatively higher concentration of hybrid seeds cultivation within the group of NFSM districts were chosen for the study. In each of the district, two representative blocks were taken and within each block two villages are selected. In each village, a complete list of cultivating households growing hybrid rice varieties and inbred varieties were prepared and stratified according to four standard land size groups such as marginal (less than 1 hectare), small (1 to 2 hectares), medium (2 to 4 hectares), and large (more than 4 hectares) farmers. In each district, 40 hybrid rice growers from the list of hybrid rice growing cultivators were drawn at random from different land size groups on the basis of their proportion in the universe. In addition to this sample, 10 inbred variety (traditional HYVs) rice growers but non-adopters of hybrid rice were selected randomly from the different land size groups amongst inbred rice growing cultivators following the same procedure. Thus altogether, 50 rice growing cultivators were chosen from each selected district. In all, 100 rice growing cultivators in each state equally spread over two selected districts constituted the size of the sample in the study. Thus the total sample size consists of 500 farmer households (400 beneficiaries and 100 non-beneficiaries) across the five states covered in the study. The primary survey was conducted over 2009–10 and 2010–11. Some state surveys covered different seasons over these years. However, this consolidated report does not present the seasonal patterns in the body because of non-comparability over states.

RESULTS OF THE STUDY

Yield and productivity under paddy in all states together increased in all the periods (Figure 8.1). Area fluctuated and there was no upward trend. In fact the area under paddy at the end of the entire study period was lower than at the beginning. This indicates that the scope of increasing output through extension of area has been exhausted and it is imperative to concentrate on yield improvement, through Hybrid seeds, etc. It is also noticeable that yield and productivity performed substantially better during the pre-hybrid period (1984–85 to 1993–94). This probably indicates the fact that HYV performance tapered off since the 1990s. Hybrid cultivation did not spread sufficiently so as to compensate.

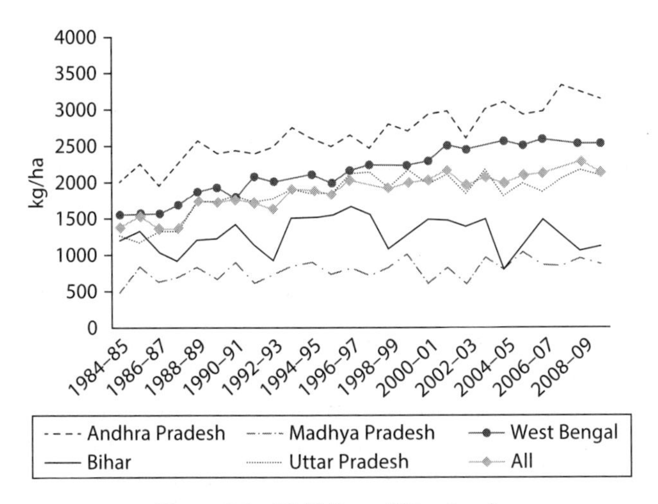

Figure 8.1 Yield Rate of Rice (total)
Source: Agricultural Statistics at a Glance, Various Issues.

It can also be observed that the increase in production can be attributed more to gain in productivity than to increase in area under crop, which in fact declined, as we have already indicated. Both yield and production showed similar and substantial gains.

For both years surveyed, the receptivity by size class to hybrid cultivation takes the form of a U, with the size class 2 to 4 ha being the least receptive. This suggests that there is a conflict between equity and efficiency in the case of hybrid cultivation. In striking contrast the receptivity to HYV takes the form of an inverted U, with the same size class being most receptive.

Further apart from the largest farms, area under hybrid cultivation has increased between 2009–10 and 2010–11. Correspondingly, there has been a decline in area under HYV (Figure 8.2). Though the time span is too short, the result is intuitively expected. With time information about and confidence in hybrid cultivation is likely to increase.

It was observed that a significantly higher proportion of head of households adopting hybrid farming belong to the younger generation. At the same time, the ability to read literature on hybrid cultivation is sufficient for adoption of new technology and that higher formal education is unnecessary. It has also been found out that a significantly larger proportion of scheduled caste (SC) and scheduled tribe (ST) farmers compared to general caste cultivators go in for hybrid cultivation.

The survey reveals that state plays predominant role in dissemination of information of new agricultural technology mainly through extension workers and, next through training programmes. So the spread of this technology cannot be entirely entrusted to the private sector.

Training programmes have to be toned up, as the extension workers are more effective in persuading farmers to adopt appropriate input mix, while participation in training programmes yields much poorer results. Participation in demonstration programmes is even less effective for disseminating knowledge about proper input mix. There is also great regional variation in effectiveness of government servants and programmes in disseminating information. This suggests that some monitoring devise has to be positioned.

Hybrid technology is substantially more productive compared to HYV across farm sizes. It is noticeably more productive in the largest farm size (Table 8.1). This suggests that the spread of the technology may have regressive impact on distribution. At the same time, Hybrid cultivation is more labour intensive than HYV cultivation. Hybrid rice cultivation also involves greater use of female labour. Hybrid rice cultivation is thus likely to generate additional employment opportunities for workers in general and specially for female labour rural areas.

Area-wise, the cost of hybrid cultivation was significantly higher. But the higher productivity compensated. Thus the

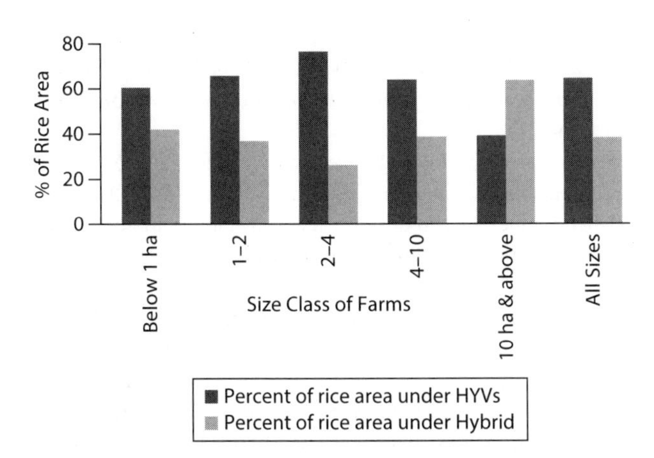

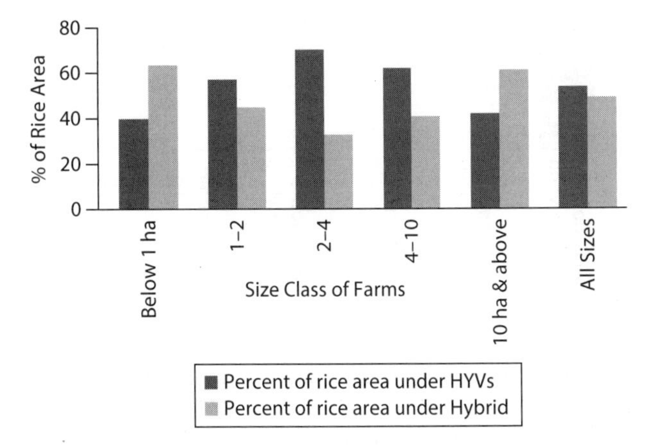

Figure 8.2 Extent of Adoption of Hybrid Rice Technology by Farm Size in 2009–10 and 2010–11 for All States
Source: Agricultural Statistics at a Glance, Various Issues.

cost per quintal was lower for hybrid. This suggests that to popularize hybrid cultivation credit needs have to be addressed. The average rate of return on working capital was higher for hybrid cultivation, though in some states the opposite was obtained (Figure 8.3).

It was also observed that grain quality of hybrid rice, in terms of hulling and milling ratio is inferior to HYV rice. This suggests that research must concentrate on improving this aspect of hybrid rice. In case of marketing of paddy, it comes out that a greater percentage of hybrid output is marketed compared to HYV. This suggests that hybrid cultivation is suitable to the expansion of grain markets. The price of hybrid rice is lower than that of HYV rice, on an average.

Although government is the main source of hybrid seeds, there is great regional variation in the proportion of seeds supplied by government sources. There is, therefore, scope for improving government intervention in this area. Also seeds are not often supplied in time. This needs to be looked into. However, it seems that there is a perception of poor quality of seeds supplied among the farmers. The reasons for this are not clear. This needs investigation.

Importantly enough, it is observed that hybrid cultivators are often using inputs in incorrect proportion. Though lack of financial ability has been indicated as a reason, lack of knowledge has also played a significant role. Thus the government needs to improve the quality of knowledge dissemination and also provide sufficient credit. The need for proper credit provision is more pronounced because hybrid cultivation is costlier.

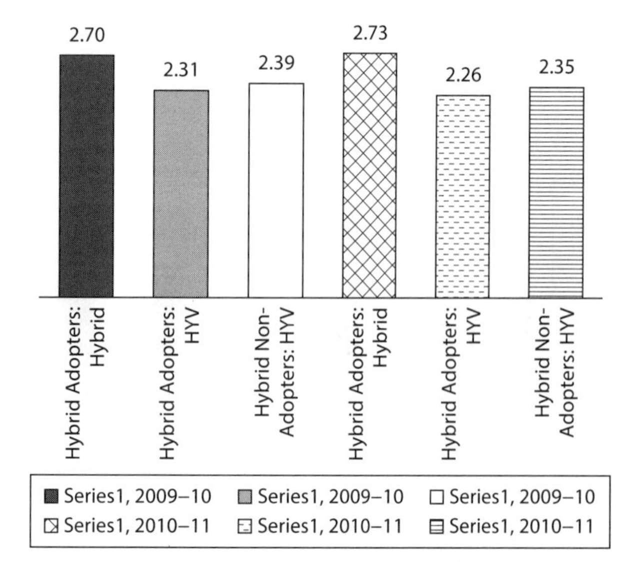

Figure 8.3 Rate of Return on Working Capital for Hybrid Adopters and Non-adopters in 2009–10 and 2010–11

Source: Field Survey.

The quality of hybrid rice, in the perception of the consumer, is poorer than HYV rice. This makes marketing difficult. This suggests that research should concentrate on improving quality like decreasing stickiness of cooked hybrid rice. The rate of degeneration or 'keeping quality' also needs to be improved.

CONCLUSIONS AND POLICY IMPLICATIONS

The spread of the newer varieties replacing the older varieties need to be closely monitored to take advantage of the superior characters of these newer varieties released by various Research Institutions. This will help to break the yield plateau in rice production of the recent past. However, there is no concrete evidence that the newer varieties of rice are spreading faster and replacing the older ones. It is here that the present study assesses the actual spreading of these newer varieties in terms of area, production, and productivity while at the same time tries to find out the impediments in the spread of new varieties of hybrid rice. On the whole, the study finds that though that performance of HYV rice tapered off since the 1990s, hybrid cultivation did not spread sufficiently so as to compensate. The present study, based on its findings, thus suggests that the government needs to improve the quality of knowledge dissemination and also provide sufficient credit. The need for proper credit provision is more pronounced because hybrid cultivation is costlier. Training programmes have to be toned up to popularize hybrid cultivation, particularly among the youth. However, research should concentrate on improving quality like decreasing stickiness of cooked hybrid rice and improve the rate of degeneration of hybrid rice.

REFERENCES

Chengappa, P.G., Aldus Janaiah, and M.V. Srinivasan Gowda. 2003. 'Profitability of Hybrid Rice Cultivation: Evidence from Karnataka'. *Economic and Political Weekly*, 38(25), 21 June.

Janaiah, A. 2002. 'Hybrid Rice for Indian Farmers: Myths and Realities'. *Economic and Political Weekly*, 37(42) October.

Mahendra Dev, S. 1987. 'Growth and Instability in Food Grains Production: An Inter-state Analysis'. *Economic and Political Weekly*, 22(39), September.

Rawal, Vikas and Madhura Swaminathan. 1998. 'Changing Trajectories: Agricultural Growth in West Bengal, 1950 to 1996'. *Economic and Political Weekly*, 33(40), July.

Saha, Anamitra and Madhura Swaminathan. 1994. 'Agricultural Growth in West Bengal in the 1980s: A disaggregation by Districts and crops'. *Economic and Political Weekly*, 29(13), March.

9

Possibilities and Constraints in Adoption of Alternative Crops to Paddy in Green Revolution Belt of India

D.K. Grover, Sanjay Kumar, J.M. Singh, and Jasdev Singh

India has made tremendous progress in agriculture over the past decades. Technological change with the introduction of short-duration high yielding varieties (HYVs) of wheat and rice in the 1960s increased productivity of these crops manifold. The effective price policy coupled with relatively better technology has resulted in the emergence of paddy in kharif and wheat in rabi as the most secured and profitable crops in several states (Bhatia 2002). Consequently, production of wheat and rice in India has increased from 23.8 and 42.2 million tonnes in 1970–71 to 95.8 and 106.3 million tonnes in 2013–14. This translates into a growth rate of 2.82 and 1.86 per cent per annum for wheat and rice during this period. The output of wheat and rice in the country has reached a saturation point; still farmers in agriculturally advanced states such as Punjab, Haryana, and western Uttar Pradesh still prefer to grow rice despite being aware of problems created by this crop rotation in terms of deteriorating soil health and depleting water table. The emerging scene of Punjab, Haryana, and western UP agriculture is not free from some serious concerns. Overtime, other states are becoming self-sufficient as they are also emerging as the potential contributor towards the production and procurement of rice and wheat in India, resulting into the declining demand of these crops grown in these states (Rao, Birthal, and Joshi 2006). In wake of the emerging scenario in country, the union government is advising these states to shift some area towards production of other crops. The state cropping pattern of these states dominated by wheat–rice rotation is also causing a serious damage to the natural resource base. Paddy in particular, a water-intensive crop is blamed for water-table depletion in tube-well-irrigated areas and water-logging in canal-irrigated areas. Due to unregulated use and heavy subsidies on power, there has been a tendency of excess withdrawal of this precious resource. Decline in water level is observed mostly in northern, north-western, and eastern parts of the country including the states of Punjab, Haryana, and western UP. The monoculture of paddy and wheat has also resulted into increasing incidence of nutrient deficiency in the soils, including micronutrients and insect/pest attacks on the crops are also posing major threats to productivity, foodgrain production, and sustainability of agriculture in the long run. Diversification of cropping pattern especially from paddy towards environment friendly crops with emphasis on quality output and promotion of agro-processing industry is the need of hour. There is need to examine biotic as well as abiotic constraints inhibiting the farmers to replace such a water galloping and environmentally unsustainable crop. The present study is an attempt in this direction with specific objectives:

1. to examine the production and procurement pattern of paddy in selected states of India;
2. to work out the relative economics of paddy vis-à-vis competing/alternative crops;
3. to bring out the constraints in adoption of alternative crops; and
4. to suggest policy measures to overcome the constraint in adoption of alternative crops to paddy in the selected region of India.

METHODOLOGY AND DATA

The study was conducted in Punjab, Haryana, and UP (western) states of India and coordinated by Agro Economic Research Centre (AERC), Punjab Agricultural University, Ludhiana. The important crops competing with paddy during kharif season, namely basmati paddy, maize, cotton, guar, sugarcane, bajra, and urad were selected for the in depth analysis. Amongst different districts of the state, six districts with the highest area/production of crops/highest diversification in the state (depending upon the coverage of the crops to be selected) were taken purposively. Amongst the selected districts, one block was selected randomly from each district. From each block, a cluster of 3 to 5 villages was randomly chosen. Finally, a sample of 35 farmers was selected randomly from each selected cluster, making a total sample of 210 farmers in each state. The primary data were collected by the respective AERCs through personal interview method for the reference year 2012–13. The data were suitably tabulated to draw logical interpretation.

RESULTS AND DISCUSSION

The results have been discussed under the following sub-heads:

1. Production status for major kharif crops
2. Agro-socio-economic characteristics of sample households
3. Economics of production of paddy vis-à-vis competing crops
4. Constraint analysis for various Alternative crops

Production Status for Major Kharif Crops

An examination of area under important crops in kharif season in Haryana, Punjab, and UP states of India shows that in the post green revolution era, these states have witnessed a considerable change in its cropping pattern. Traditionally, these states were predominantly a wheat-growing area. Rice stormed in the cropping pattern since the mid-1970s as a commercial crop and made a major impact on the agriculture in these states. The area under paddy in Haryana, Punjab, and UP which occupied only about 5, 6, and 13 per cent of gross cropped area (GCA) during triennium ending (TE) 1970–71 jumped to around 19, 36, and 23 per cent in TE 2011–12 in these states, respectively (Table 9.1). The area under rice has increased by replacing crops such as kharif pulses, maize, jowar, bajra, and kharif oilseeds. In Haryana, during the TE 1970–71, bajra was the dominant crop which occupied more than 18 per cent of GCA. In ranking, paddy and jowar were next and each one was allotted more than 4 per cent of GCA. Cotton and pulses received 4 per cent of GCA. In nutshell, pattern of area allocation in kharif season during the TE 1970–71 in Haryana was found skewed towards bajra, which requires minimum irrigation. Recently, the acreage allocation under kharif crops in TE 2011–12, in Haryana indicates significant increase in acreage under paddy and cotton which occupied almost 19 and 8 per cent of GCA, respectively. Maize, a multipurpose crop became almost negligible. However, bajra still occupied around 9 per cent of total kharif area. It is important to note that horticultural crops, that is, fruits and vegetables gained significantly but latter was an important beneficiary by indicating around 6 per cent of GCA during TE 2011–12 in Haryana. This trend could be due to increasing demand for high-value crops arising out of changes in consumption pattern of population. In Punjab, the area under rice has increased eightfold during last four decades by replacing crops such as cotton, kharif pulses, maize, jowar, bajra, and kharif oilseeds. Similarly, the production of rice during this period increased by about twelve times from 0.56 million tonnes to 10.9 million tonnes that is at CAGR of 6.07 per cent per annum. The area under rice which occupied only 6.43 per cent of gross cropped area during TE 1970–71 jumped to around 35.73 per cent in TE 2011–12. The state has extreme specialization of paddy-wheat cropping system which may be attributed to effective implementation of agricultural price policy with MSP and relative profitability of these crops as compared to other crops. Cotton is ranked third in the cropping pattern of the state. The area under this crop in TE 1970–71 was about 5 per cent of gross cropped area (GCA) and its share in GCA went down to

Table 9.1 Change in Cropping Pattern, Selected States, India, 1970–71 to 2011–12
(Per cent to gross cropped area)

Crops	Haryana				Punjab				UP			
	TE 1970–71	TE 1985–86	TE 2000–01	TE 2011–12	TE 1970–71	TE 1985–86	TE 2000–01	TE 2011–12	TE 1970–71	TE 1985–86	TE 2000–01	TE 2012–13
Paddy	4.97	10.05	17.46	19.10	6.43	22.82	31.83	35.73	13.13	20.77	22.24	23.12
Jowar	4.35	2.48	1.90	1.11	–	–	–	–	–	–	–	–
Bajra	18.05	13.21	9.79	9.44	3.59	0.64	0.06	0.04	4.57	3.79	3.19	3.62
Maize	2.11	1.01	0.30	0.17	8.81	4.05	1.99	1.69	6.24	4.46	3.56	2.87
Total Kharif Cereals	29.48	26.74	29.45	29.81	18.83	27.50	33.88	37.45	–	–	–	–
Urad	–	–	–	–	–	–	–	–	0.58	0.75	1.10	2.03
Mash	0.19	0.11	0.04	0.03	0.46	0.18	0.05	0.04	–	–	–	–
Arhar	–	–	–	–	–	0.58	0.11	0.05	–	–	–	–
Moong	0.36	0.10	0.20	0.30	0.09	0.54	0.46	0.09	–	–	–	–
Total Kharif Pulses	0.55	0.22	0.24	0.33	0.53	1.30	0.62	0.18	–	0.83	0.94	2.19
Groundnut	0.25	0.15	0.01	0.03	3.42	0.69	0.06	0.03	–	0.00	0.00	0.00
Sesamum	–	–	–	–	0.21	0.20	0.20	0.08	–	33.14	32.15	32.61
Total Kharif Oilseeds	–	–	–	–	3.63	0.89	0.26	0.10	–	–	–	–
Cotton	4.03	6.16	9.11	8.30	7.03	7.92	6.22	6.38	0.22	0.07	0.02	0.02
Sugarcane	–	–	–	–	2.55	1.13	1.37	0.89	5.51	6.11	7.38	8.64
Total Foodgrains	–	–	–	–	19.36	28.81	34.50	37.63	–	33.14	32.15	32.61
Total Area under kharif crops	34.31	33.27	38.81	38.48	–	–	–	–	–	–	–	–
Fruits	0.21	0.22	0.20	0.70	–	–	–	–	–	–	–	–
Vegetables	0.48	0.72	0.82	5.21	–	–	–	–	–	–	–	–
Total Area under kharif crops	35.00	34.21	39.83	44.38	–	–	–	–	–	–	–	–
Gross Cropped Area ('000 ha)	4957	5647	6155	6428	5678	7068	8099	7879	23297	25074	26769	25074

Source: States Statistical Abstracts – Various Issues.

5.97 per cent in TE 2000–01. With the introduction of Bt varieties, area under cotton again rose to 6.54 per cent in 2011–12. The proportionate area under maize kept on declining since TE 1970–71 from 9.77 per cent to 1.69 per cent in TE 2011–12. The area under sugarcane decreased by 0.28 per cent per annum. On the other hand, the production of kharif pulses and oilseeds went down drastically over this period. The reason of decline of production of these crops was the drastic decline of area under these crops due to encroachment by paddy and wheat. On the other hand the area under kharif oilseeds have shown the continuous decrease since the TE 1970–71 and the area has declined from 3.63 per cent to meagre 0.1 per cent of GCA in the TE 2011–12. It can be concluded that imbalance in favour of two main cereals, namely rice and wheat in the cropping pattern has further sharpened despite all efforts of diversification in the state agriculture. In Uttar Pradesh, the area under paddy was 3059.70 thousand hectares in TE in 1970–71 which has increased to 5797.30 thousand hectares in TE 2012–13, showing 189.47 per cent increase over the period. The production of paddy has been continuously increasing from TE 1970–71 to TE 2012–13 in UP. The production of paddy was 2362.32 thousand metric tonnes in TE 1970–71 which has gone up to 13418.24 thousand metric tonnes in TE 2012–13, showing 568 per cent increase over the period. The yield of paddy was 7.67 qtls per ha in TE 1970–71 which has increased to 23.12 qtls per ha in TE 2012–13 thereby showing 301 per cent increase over the period. It shows that area, production, and yield of paddy have positive rate of growth during the study periods. The area and production of maize have maintained the decreasing trends during corresponding periods. The area under maize was 1454.78 thousand hectares in TE 1970–71 which has decreased to 720.44 thousand hectares in

TE 2012–13, thereby showing 150.48 per cent decrease over the period. It shows that area under maize was shifting to another kharif crops in UP, despite its better yield per ha. The area under bajra was 1064.90 thousand hectares in TE 1970–71, which has decreased to 907.93 thousand hectares in TE 2012–13, showing 114.74 per cent decrease over the period. Urad is most important kharif pulse in Uttar Pradesh. The area, production, and yield of urad have also maintained increasing trends over a period of 42 years. The area under urad was 51.96 thousand hectares in TE 1970–71 which has gone up to 508.65 thousand hectares in TE 2012–13 showing 878.93 per cent increase over the period, while yield rate of urad varied from 1.49 qtls per ha in TE 1985–86 to 6.69 qtls per ha in TE 2012–13. As far as other kharif crops are concerned, area, production, and yield of sugarcane and pulses have significantly increased in UP in the study period. It may be concluded with this impression that among Kharif crops of UP the maximum decline in area under maize was witnessed during the study periods. The area, production, and yield of most of other important kharif crops of UP have maintained rising trends in UP during the study periods.

Agro-socio-economic Characteristics of Sample Households

Majority of the sample households had young head with age above 31–50 years except in western UP, where about 61 per cent of sample households had a family head of more than 50 years. The data showed that heads of more than 79 per cent sample households were literate in the selected states. Illiteracy was found to be significantly higher among the Punjab farmers, that is, about 21 per cent. The family size was found to vary between 6 for Punjab farmers to 9.21 for western UP households. Agriculture was found to be the main occupation of head of the family as revealed by more than 96 per cent of the heads of the family. The average land holding was the highest for Punjab farmers (7.97 ha) and the least for western UP farmers (3.04 ha). The average asset value per farm was the highest for Haryana farmers (4.03 lakh) and the least for western UP farmers (1.49 lakh). Paddy and wheat were the major kharif and rabi crops in all the selected states of India. In Haryana, cotton followed by fodder and maize was observed as the important crops in terms of Net Area Sown (NAS) devoted in kharif season by the sampled households. Bajra and sugarcane were grown on 4.86 and 4.07 per cent of NAS, respectively. In Punjab, paddy and wheat were grown over about 47 and 79 per cent area respectively. Sugarcane, basmati paddy, maize, cotton, and guar were the other important crops grown during the season. In western UP, wheat, paddy, bajra, vegetables, urad, mustard, and maize were important crops on the sample farms which accounted for 56.64 per cent, 41.02 per cent, 17.44 per cent, 12.42 per cent, 11.24 per cent, 10.69 per cent, and 10.67 per cent to net cropped area, respectively. Paddy, bajra, maize, urad, and sugarcane were important crops in kharif season.

Economics of Production of Paddy vis-à-vis Competing Crops

The operational cost of cultivation of paddy was compared with the most important competing crops during the kharif season in the selected states of India namely, Haryana, Punjab and western UP and the results are presented in Table 9.2. During the kharif season, paddy is the most important crop of these states. The results showed that the total variable cost on per hectare basis for paddy crop was found to vary between Rs 27465 for Punjab to Rs 38132 for the Haryana state, which was due to the highest cost of machine labour in Haryana during the reference year. Further, the total variable cost on per hectare basis for basmati paddy crop was found to vary between Rs 26704 for Punjab to Rs 38912 for western UP, which was due to the per hectare highest cost incurred on human labour, seed and fertilizer for basmati paddy crop in western UP. The total variable cost on per hectare basis for bajra was found to be higher (Rs 17088) for western UP as compared to Rs 11039 for the Haryana state due to the higher cost of human and machine labour. For maize, the total variable cost on per hectare basis for maize crop was found to vary between Rs 20829 for western UP to Rs 32179 for Punjab as the Punjab farmers incurred almost double expenses on the use of human labour per hectare as compared to the other states. In western UP, the average cost of cultivation of urad was estimated at Rs 15437 per hectare and the per pectare net income of urad was estimated at Rs 27117 against Rs 76254 per hectare net income of paddy and the per hectare yield of urad was worked out to be 12.04 qtls. The total variable cost on per hectare basis for cotton crop was found to marginally higher (Rs 38999) for Haryana as compared to Rs 36523 for the Punjab state due to the higher cost of human and machine labour in Haryana state. The total variable cost on per hectare basis for guar crop in Punjab was found to be Rs 9943.

The economics of paddy was compared with the most important competing crops during the kharif season in the selected states of India, namely Haryana, Punjab, and western UP and the results are presented in Table 9.3. It is to be noted that the economics of sugarcane, being the annual crop, cannot be compared with other kharif crops and only the benefit-cost analysis has been utilized for the purpose.

Table 9.2 Operational Cost of Cultivation of Paddy vis-à-vis Competing Crops, Sample Households, Selected States, India

(Rs/ha)

Crop	Haryana	Punjab	Western UP
Paddy			
Human labour	15011	12279	18956
Machine labour	14397	6480	3027
Seed	1569	670	788
Farm Yard Manure (FYM) and Fertilizer	5255	4649	4670
Plant protection measures	1284	2915	646
Interest on working capital	616	472	515
Misc	–	–	1345
Total variable cost	38132	27465	29947
Basmati-paddy			
Human Labour	15113	15077	18248
Machine labour	10418	3618	9978
Seed	990	763	2070
FYM and Fertilizer	5517	3500	5878
Plant protection measures	2783	3287	712
Interest on working capital	586	459	669
Misc	–	–	1358
Total variable cost	35407	26704	38912
Bajra			
Human Labour	5730	–	7822
Machine labour	2438	–	6005
Seed	691	–	786
FYM and Fertilizer	1853	–	1708
Plant protection measures	153	–	77
Interest on working capital	174	–	294
Misc	–	–	396
Total variable cost	11039	–	17088
Maize			
Human Labour	8097	15991	8286
Machine labour	3713	3963	5406
Seed	5076	4645	2493
FYM and Fertilizer	4145	5119	3711
Plant protection measures	1220	1908	294
Interest on working capital	360	553	358
Misc	–	–	280
Total variable cost	22613	32179	20829

(Rs/ha)

Crop	Haryana	Punjab	Western UP
Urad			
Human Labour	–	–	6582
Machine labour	–	–	5355
Seed	–	–	1298
FYM and Fertilizer	–	–	1654
Plant protection measures	–	–	119
Interest on working capital	–	–	250
Misc	–	–	178
Total variable cost	–	–	15437
Cotton			
Human Labour	16392	14912	–
Machine labour	7345	5184	–
Seed	4562	6249	–
FYM and Fertilizer	5504	4403	–
Plant protection measures	4554	5147	–
Interest on working capital	643	628	–
Total variable cost	38999	36523	–
Guara			
Human Labour	–	5316	–
Machine labour	–	2431	–
Seed	–	1560	–
FYM and Fertilizer	–	105	–
Plant protection measures	–	316	–
Interest on working capital	–	171	–
Total variable cost	–	9945	–
Sugarcane			
Human Labour	–	38983	–
Machine labour	–	5920	–
Seed	–	19516	–
FYM and Fertilizer	–	9498	–
Plant protection measures	–	4239	–
Interest on working capital	–	2735	–
Total variable cost	–	80891	–

Source: Field Survey Data.

During the kharif season, paddy is the most important crop of these states. The results showed that the ROVC) fetched from paddy on per hectare basis were the highest for Punjab (Rs 60113) and the lowest for western UP (Rs 19138) as the productivity of paddy was the highest (68.42 qtl /ha) in Punjab. Further, the returns over variable cost fetched from basmati paddy on per hectare basis were found to vary between Rs 122276 for Punjab to Rs 76714 for western UP, which was mainly due the highest average price of basmati paddy (Rs 3673/qtl) fetched in Punjab

Table 9.3 Economics of Paddy vis-à-vis Competing Crops, Sample Households, Selected States, India

(Rs/ha)

Crop	Haryana	Punjab	Western UP
Paddy			
Yield (qtl/ha)	51.40	68.42	35.89
Price (Rs/qtl)	1280	1280	1368
Gross returns (main product + by-product)	65855	87578	49085
Total Variable cost	38132	27465	29947
Returns over variable cost	27723	60113	19138
Benefit–cost ratio	1.73	3.19	1.64
Basmati-paddy			
Yield (qtl/ha)	43.90	40.56	45.72
Price (Rs/qtl)	3421	3673	2529
Gross returns (main product + by-product)	150230	148980	115626
Total Variable cost	35407	26704	38912
Returns over variable cost	114823	122276	76714
Benefit–cost ratio	4.24	5.58	2.97
Bajra			
Yield (qtl/ha)	20.70	–	32.89
Price (Rs/qtl)	1114	–	1104
Gross returns (main product + by-product)	23003	–	36311
Total Variable cost	11039	–	17088
Returns over variable cost	11964	–	19223
Benefit–cost ratio	2.08	–	2.12
Maize			
Yield (qtl/ha)	27.3	50.08	32.44
Price (Rs/qtl)	1267	860	1345
Gross returns (main product + by-product)	34563	45781	43632
Total Variable cost	22613	32179	20829
Returns over variable cost	11950	13602	22803
Benefit–cost ratio	1.53	1.33	2.09

(Rs/ha)

Crop	Haryana	Punjab	Western UP
Urad	–	–	
Yield (qtl/ha)	–	–	12.04
Price (Rs/qtl)	–	–	3551
Gross returns (main product + by-product)	–	–	42754
Total Variable cost	–	–	15437
Returns over variable cost	–	–	27317
Benefit–cost ratio	–	–	2.77
Cotton			
Yield (qtl/ha)	20.8	20.46	–
Price (Rs/qtl)	4955	4150	–
Gross returns (main product + by-product)	103051	86930	–
Total Variable cost	38999	36523	–
Returns over variable cost	64052	50407	–
Benefit–cost ratio	2.64	2.38	–
Guar			
Yield (qtl/ha)	–	12.13	–
Price (Rs/qtl)	–	5525	–
Gross returns (main product + by-product)	–	67018	–
Total Variable cost	–	9943	–
Returns over variable cost	–	57075	–
Benefit–cost ratio	–	6.73	–
Sugarcane			
Yield (qtl/ha)	–	706.26	–
Price (Rs/qtl)	–	270	–
Gross returns (main product + by-product)	–	196303	–
Total Variable cost	–	80891	–
Returns over variable cost	–	115412	–
Benefit–cost ratio	–	2.43	–

Note: The economics of sugarcane, being the annual crop, cannot be compared with other kharif crops and only the benefit–cost analysis has been utilised for the purpose.
Source: Field Survey Data.

during the reference year under the study, although the productivity for the crop was the lowest in the state. The production scenario of basmati paddy in these states has significantly improved in the recent years due to the adoption of Pusa Basmati 1121 and Pusa Basmati 1509 varieties, which have become popular in the state due to their better yield. The total variable cost on per hectare basis for bajra was found to be higher (Rs 17088) for western UP as compared to Rs 11039 for the Haryana state. Similarly, ROVC was also higher for western UP. For maize, the returns over variable cost fetched on per hectare basis were the highest for western UP (Rs 22803) and the lowest for Haryana (Rs 11950) due the highest average price of maize (Rs 1345/qtl) fetched in western UP during the reference year under the study. In western UP, the average cost of cultivation of urad was estimated at Rs 15,437 per hectare and the per hectare net income of urad was estimated at Rs 27,317 and the per hectare yield of urad was worked out to be 12.04 qtls. The returns over variable cost on per hectare basis for cotton crop was found to be higher (Rs 64052) for Haryana as compared to Rs 50407 for the Punjab state, which was due the highest average price of cotton (Rs 4955/qtl) fetched in Haryana. The guar growers got handsome returns over variable cost (Rs 57075/qtl) in Punjab due to remunerative prices in the market. In the long run, the yield of the crop would have to be increased to make guar crop remunerative.

Constraint Analysis for Various Alternative Crops

The important problems faced by the growers of important crops in the kharif season competing with paddy crop which are related to production, environment, inputs, and marketing for the selected states of India namely, Haryana, Punjab, and western UP have been analysed and presented in Table 9.4. More than 54 per cent of the basmati paddy farmers in Punjab pointed out the prevalence of insect pest and 45 per cent of Haryana farmers reported the attack of diseases as the most prevalent problems during production of the crop. More than 29 per cent of the basmati paddy growers of the region were reported to be facing the shortage of labour for performing various cultivation operations especially during transplantation and harvesting of the crop, which is highly labour intensive. Lack of market intelligence and low price in the market were reported as the major marketing problems confronted by growers of the study area. The price in market abruptly changes with the arrival of crops in the market. Whenever there is glut in the market, the prices comes down and farmers find it very difficult to dispose of the produce at the remunerative prices in the market. The bajra growers revealed that marketing problems were among the most important constraints in the cultivation of bajra on their farms during the reference year. The weeds and environment were also important problems in the cultivation of bajra as had been revealed by 40 per cent and 45 per cent of sample growers of bajra, respectively. For maize, lack of market intelligence and low price in the market were reported as the major problem confronted by maize growers of the study area. The price in the market for most of the time remains lower as compared to MSP and as the prices comes down the farmers find it very difficult to dispose of the produce at the remunerative prices in the market due to its short shelf life. More than three per cent of the maize growers also reported the problem of destruction by the stray animals and monkeys when the crop is ready for the harvest. More than 39 per cent of the maize growers were reported to be facing the shortage of labour for performing various cultivation operations especially during harvesting of the crop, which is highly labour intensive. The prevalence of weeds was the important problem faced by sample growers during the cultivation of urad on their farms. Although, with the introduction of Bt cotton, the attack of American bollworm has decreased substantially but still the farmers felt that it does not provide effective control of sucking pests and tobacco caterpillar and attack of insects/pests is most prevalent in the crop as revealed by about 72 per cent of the cotton growers. The crop requires huge human labour during the harvesting/picking season and availability of labour at the peak season was also one of the problems confronted by more than 41 per cent of the growers. Lack of market intelligence and fluctuation of prices in the market were reported as the major marketing problems confronted by cotton growers of the study area. The guar is grown in south western districts of the Punjab state where about 26 per cent of the farmers confronted the problem of water logging on their farms. Amongst other problems, attack of insects/pests and diseases were also prevalent in the crop as revealed by more than 16 per cent of the guar growers. Lack of market intelligence and low prices in the market were reported as the major marketing problems confronted by guar growers of the study area. The attack of insects/pests was the most prominent problem faced as revealed by more than 32 per cent of the sugarcane growers in Punjab. The crop requires huge human labour during the harvesting season and availability of labour at the peak season was also one of the problem confronted by more than 52 per cent of the growers Delay in payment by the sugar mills was also observed to be the problem felt by more than 65 per cent of the growers. The farmers reported that their payments were not made even after six months of sale to the sugar mills.

CONCLUSIONS AND POLICY IMPLICATIONS

Crop diversification away from paddy towards alternative crops in the kharif season in the selected states requires a favourable price regime, technology for raising the existing levels of productivity, financial support, rural infrastructure, and above all, multi-pronged government support. The results showed that the returns over variable cost fetched from basmati paddy were the highest on per hectare basis even more than the fine varieties of paddy, which was mainly due the higher average price of basmati paddy fetched during the reference year under the study. Otherwise, the last year average price received by the farmers was almost 60 per cent. Therefore, ensuring profitability of alternative crops on sustainable basis through suitable

Table 9.4 Main Problems Faced by Sample Households, Selected States, India (Per Cent Multiple Response)

Reason	Basmati–paddy				Bajra				Maize				Urad			
	Haryana	Punjab	Western UP	Overall	Haryana	Punjab	Western UP	Overall	Haryana	Punjab	Western UP	Overall	Haryana	Punjab	Western UP	Overall
Production problems																
Diseases	45.20	22.52	10.95	23.22	–	–	–	–	7.10	5.10	3.39	5.20	–	–	–	–
Insects/ pests	44.80	54.97	–	33.25	2.35	–	–	1.17	1.60	6.12	–	2.57	–	–	–	–
Weeds	28.60	1.32	24.76	18.23	26.52	–	53.61	40.07	–	–	32.55	10.18	–	–	61.25	61.25
Destruction by Animals	–	–	–	–	–	–	–	–	43.10	55.10	–	32.73	–	–	–	–
Environmental problems	–	–	43.34	14.45	29.10	–	61.86	45.48	–	–	45.35	15.12	–	–	57.50	57.50
Non availability of inputs	–	–	37.14	12.38	25.70	–	17.53	21.62	–	–	–	–	–	–	–	–
Seed	1.24	0.66	–	0.63	–	–	–	–	4.10	1.20	–	1.77	–	–	–	–
Labour	35.70	53.63	–	29.77	–	–	–	–	53.18	59.18	4.70	39.02	–	–	–	–
Marketing problems	–	–	–	–	26.70	–	67.00	46.85	–	–	–	–	–	–	–	–
Prices	41.79	4.39	–	15.09	–	–	–	–	73.25	62.25	77.91	71.14	–	–	–	–
Information	84.76	81.45	74.77	80.33	–	–	–	–	63.67	71.67	–	45.11	–	–	–	–
Transport	–	–	–	–	–	–	–	–	3.10	1.20	–	1.11	–	–	–	–
Demand	4.00	2.65		2.32	–	–	–	–	–	–	–	–	–	–	–	–

Reason	Cotton				Guara				Sugarcane			
	Haryana	Punjab	Western UP	Overall	Haryana	Punjab	Western UP	Overall	Haryana	Punjab	Western UP	Overall
Production problems												
Diseases	47.60	54.00	–	50.80	–	19.00	–	19.00	–	3.00	–	3.00
Insects/ pests	47.70	96.00	–	71.85	–	16.67	–	16.67	–	5.00	–	5.00
Weeds	27.50	–	–	13.75	–	–	–	–	–	31.25	–	31.25
Environmental problems	–	–	–	–	–	26.19	–	26.19	–	–	–	–
Non availability of labour	77.00	6.00	–	41.50	–	38.10	–	38.10	–	52.50	–	52.50
Marketing problems												
Prices	36.00	54.00	–	45.00	–	56.81	–	56.81	–	7.50	–	7.50
Information	63.00	74.00	–	68.50	–	66.66	–	66.66	–		–	
Transport	29.60	36.00	–	32.80	–		–		–	36.25	–	36.25
Demand	4.00	2.00	–	3.00	–	2.38	–	2.38	–	12.50	–	12.50
Delay in Payment	–	–	–	–	–	–	–	–	–	65.00	–	65.00

Source: Field Survey Data.

policy reforms appears to be a pre-requisite for successful crop diversification in the region. In order to reduce area under paddy, there is an urgent need to ensure parallel facilities for alternative crops including research and development to augment yield levels and its effective dissemination at the grass root level. The degree of production and price risks in alternative crops is higher than paddy. Climate change is further aggravating the yield risk. The first risk can be reduced by development of suitable technology and second, by favourable price policy, credit and insurance facilities, investment in creating nearby markets and rural infrastructure. This is possible through wholehearted support of the government and participation of the private sector. In Punjab, the adaptation to the soil/climate type and attractive price of the basmati paddy, cotton, and guar has been the major reasons for attraction of the farmers to the crop. The prevalence of insect pest and diseases and shortage of labour for performing various operations were the most prevalent problems during production of the various kharif season crops. Low/fluctuation of prices in the market and lack of market intelligence were reported as the major marketing problems confronted by growers of kharif crops in the study area. Delay in payment by the sugar mills to sugarcane growers and the problem of damage by the stray animals and monkeys as revealed by the maize growers were some other constraints. Cotton and guar crops growers in south western districts of the Punjab state confronted the problem of water logging on their farms. The attack of insects/pests was found to be more prevalent in the kharif crops as compared to diseases and weeds. The productivity of cotton and guar crops was found to be affected due to water logging problem in some areas. The farmers felt the need for effective procurement of produce by government agencies at MSP and better market intelligence so that the farmers may get the remunerative prices for their produce. The sugarcane growers were advocating that the government should work as regulatory in ensuring the timely availability of payments for the produce by the sugar mills. To improve the yield of these alternative crops, the researcher should develop disease-resistant, excess-moisture-tolerant varieties for guar and cotton and drought-tolerant varieties for sugarcane. On the production front, application of the irrigation at the right time, timely sowing and transplanting schedule, monitoring of the insect/pest population/damage and use of recommended control measures and seed treatment to avoid seed borne diseases were the secrets of success for the alternative crops. Most of the growers felt the need to improve the extension activities through increase in number of training camps or field visits by the experts and providing the information particularly regarding the high yielding recommended varieties of the crop particularly hitherto neglected guar crop.

In western UP, in order to popularize the basmati paddy, maize, bajra, and urad in cropping pattern in place of paddy crop, there is a need to increase the production and marketing efficiency of these competing crops to provide the best scientific techniques to the farmers. The coarse cereals, pulses, and oilseeds cannot be grown in low lying areas and flooded fields. Therefore, drainage networks should be expanded in low lying and flooded areas to enable the farmers to grow maize and bajra in place of paddy during kharif season. In order to promote competing crops of paddy, high yielding variety seeds of maize, bajra, and urad should be made available at reasonable prices and well before sowing times. The Seed Replacing Rates (SRR) of maize, bajra, and urad is much less as compared to SRR of paddy. Therefore, an effort should be made by the research scientists to make available certified seeds of maize, bajra, urad at par with certified seeds of paddy. The researchers should take sincere efforts to evolve short duration drought and excess moisture tolerant varieties of alternative crops at par with evolved varieties of paddy. There is a need to improve the efficiency of government procurement agencies so as to purchase maximum quantities of produce of competing crops of paddy at MSP from the farmers.

REFERENCES

Bhatia, M.S. 2002. 'Appraisal of Agricultural Price Policy with Special Reference to Punjab', in S.S. Johl and S.K. Ray (eds), *Future of Agriculture in Punjab*. Chandigarh: Centre for Research in Rural and Industrial Development.

Rao, Parthasarathy P., P.S. Birthal, and P.K. Joshi. 2006. 'Diversification towards High Value Agriculture: Role of Urbanization and Infrastructure', *Economic and Political Weekly*, 41(26): 2747–53.

ANNEXURE 9A

Table 9A The Details of Agro-Economic Research Centres and the Lead Person Involved in the State Report

S. No	State	Authors
1	Haryana	Usha Tuteja
2	U.P.	Ramendu Roy

10

Economics of Bt Cotton vis-à-vis Non-Bt Cotton in India

A Study across Four Major Cotton-growing States*

Vasant P. Gandhi and N.V. Namboodiri

There have been major advances in biotechnology in the recent years and these have made it possible to directly identify genes, know their functions, isolate them, and transfer them from one organism to another. These developments have spanned the entire biological sciences. An important outcome of this is the development of Bt cotton. Bt cotton was developed by Monsanto and it is now one of the most widely grown transgenic crops, currently grown in many countries including United States of America, China, India, Australia, Argentina, South Africa, and Indonesia. The adoption of Bt cotton has been rapid, from an estimated 0.8 million ha in 1996 to over 6 million ha in 2004 globally. Between 1996 and 2003 the global area under all transgenic crops has increased 25-fold from 1.7 million ha to 68 million ha. In the year 2005, which marked the tenth anniversary of commercialization of transgenic or biotech crops, the global area was estimated to be around 90 million ha. This came from 21 countries, 11 developing, and 10 industrialized countries (Table 10.1).

This chapter has examined the performance and economics of Bt cotton vs Non-Bt cotton in the major cotton states of Andhra Pradesh, Gujarat, Maharashtra, and Tamil Nadu which together account for about 70 per cent of the cotton production in the country. The study has been undertaken at the request of Ministry of Agriculture and Farmers Welfare, GoI. It is undertaken as a coordinated research project with the respective state AERCs to objectively examine performance of Bt cotton, in light of various conflicting views and opinions.

The reported advantages of Bt cotton include agronomic, economic, and environmental benefits. The major agronomic advantage of Bt cotton over the conventional cotton is the resistance to the bollworm pest. The major economic benefits are reduced need for pesticides and, yield superiority through the resistance over non-Bt cotton

* The authors wish to gratefully acknowledge the contribution of V.D. Shah, Agro-Economic Research Centre (AERC) Vallabvidyanagar, Gujarat; S.S. Kalamkar, AERC Pune, Maharashtra; N. Ramgopal, AERC Vishakapatnam, Andhra Pradesh; A. Pushpavalli, AERC Chennai, Tamil Nadu; as well as the various AERC heads and staff. Support of the Ministry of Agriculture and Farmers Welfare, Government of India (GoI), and CMA-IIM, Ahmedabad is also gratefully acknowledged.

Table 10.1 Adoption of Bt Cotton in Major Cotton-growing Countries

Country	1996	1997	1998	1999	2000	2001	2002	2003	2004	2005
USA	√	√	√	√	√	√	√	√	√	√
Australia	√	√	√	√	√	√	√	√	√	√
China		√	√	√	√	√	√	√	√	√
India							√	√	√	√
Indonesia							√	√		
Mexico	√	√	√	√	√	√	√	√	√	√
Argentina			√	√	√	√	√	√	√	√
Colombia							√	√		
South Africa		√	√	√	√	√	√	√	√	√
Costa Rica									√	√
Pakistan										√
Paraguay										√

Source: www.grain.org/research

varieties. Even though there are some potential environmental risks, the major environmental benefits include reduction in number of pesticides sprays, less exposure to pesticides for human beings and animals, and less pesticides in the water and soil.

Many countries have reported positive experiences with Bt cotton. India entered relatively late after much hesitation. The GoI allowed the cultivation of three genetically modified Bt cotton hybrids initially for three years from April 2002 to March 2005. Even though the performance of Bt cotton has been projected to be satisfactory, there is great discontent in some quarters. Those in favour indicate reduction in the use of insecticides, better yield per unit of input use, equal or better quality, and lesser residue of pesticides in the fibre. Those against indicate concerns such as: the gene spread may have adverse impact in the ecosystem, Bt cotton seed is expensive compared to non-Bt seeds, inadequate resistance so the farmers may still require to use insecticides, and other issues.

COTTON IN INDIA AND THE STUDY STATES

Though India ranks first in area cultivated of cotton in the world, it occupies the third position in production after China and US because of low ranking in yields. About 65 per cent of the cotton cultivation in India is unirrigated and therefore less productive and subject to vagaries of monsoon. The cotton crop is highly susceptible to insects/pests and about 166 different species of insects/pests are reported to attack cotton at various stages of its growth. It is estimated that the pests and diseases cause over 50 per cent damage to cotton in India, compared to 24.5 per cent world over. Of about 96,000 metric tonnes of technical grade pesticides produced in the country, about 54 per cent is estimated to be used on cotton.

Area under cotton in India is about 9 million hectares which is about 5 per cent of the total cropped area in the country. Large variation in the area under cotton is observed from year to year due to the vagaries of rainfall, as well as prices and profitability of cotton. The cotton yield in India is one of the lowest in the world and it stagnated or declined during the 1990s. However, there is significant growth after 2002–03 in the wake of Bt cotton. The estimated production of cotton in 2004–05 was a record in the history of cotton cultivation in the country at 21.3 million bales (1 bale = 170 kg), and this has further risen to 22.9 million bales by 2007/08. The cotton yield from 1990–91 to 2000–01 shows an annual growth of merely 1.45 per cent, but taken between 1990–1 and 2007–08, it is more than double at 3.01 per cent per annum.

The Bt cotton was approved for commercial cultivation in India in 2002. In March 2002 the Genetic Engineering Approval Committee (GEAC), the regulatory authority for transgenic crops in India, approved the commercial cultivation of three Bt cotton varieties namely, Bt Mech 12, Bt Mech 162, and Bt Mech 184. This remained and only after several years in 2005, the GEAC approved large scale field trials and seed production of 12 more varieties of Bt hybrids. Gujarat and Maharashtra were the early adopters of Bt cotton on a large scale that commenced from 2002, followed by Andhra Pradesh, Karnataka, Tamil Nadu, and Madhya Pradesh.

Based on cotton production during the recent triennium ending 2007–08, Gujarat ranks at the top with a share of 36 per cent, followed by Maharashtra with 17.8 per cent and Andhra Pradesh with 13.2 per cent. Tamil Nadu has a share of only 1.86 per cent in the national production. Together, Gujarat, Maharashtra, Andhra Pradesh, and Tamil Nadu accounted for 69 per cent of the cotton production in India, in the triennium ending (TE) 2007–08. In terms of area under cotton, Maharashtra occupies the top position with a share of 33.2 per cent in the 9.2 million hectares of area under cotton cultivation in the country, followed by Gujarat with 25.36 per cent and Andhra Pradesh with 11.3 per cent during triennium ending 2007–08. However, the average yield of cotton is among the lowest in Maharashtra at 273 kg per hectare as against 514 kg per hectare for the country as a whole.

DATA AND METHODOLOGY

This study is based on the primary data collected from four states namely Andhra Pradesh, Gujarat, Maharashtra, and Tamil Nadu. Effort was made to adopt similar methodology, content, and survey instruments in all these states, as far as possible. Effort was made to have nearly an equal number of Bt and non-Bt farmers in the sample. Effort was also made through stratification to cover both irrigated and unirrigated farms under Bt and non-Bt cotton, as well as small, medium and large farmers. The primary data collected pertains to the agricultural year 2004–05.

The target sample size for each state was 180 cotton farmers with 90 Bt and 90 non-Bt farmers (Table 10.2). Maharashtra was an exception where the coverage was 85 Bt and 69 non-Bt farmers. The Gujarat sample did not have unirrigated cotton as a sufficient number of such farmers were not available in the sample districts. The number of sample farmer households under unirrigated Bt cotton was relatively less in Andhra Pradesh and Maharashtra, but were relatively more in Tamil Nadu. The study had an overall sample size of 694 farm households.

Within the states, the study sampled districts which were important for cotton growing, and provided some variety in the location type. The following districts were sampled based on this information: Gujarat—Rajkot and Vadodara districts; Maharashtra—Jalgaon and Buldhana districts; Andhra Pradesh—Guntur and Warrangal districts; Tamil Nadu—Salem and Perambalur districts.

Table 10.2 Sample Size

	Bt Cotton	Non-Bt Cotton	Total
Gujarat	90	90	180
Maharashtra	85	69	154
Andhra Pradesh	90	90	180
Tamil Nadu	90	90	180
Total	355	339	694

Source: Filed survey.

RESULTS AND DISCUSSION

Varieties Grown

Many Bt and non-Bt varieties were reported. The use of non-genuine Bt seeds was reported by both Gujarat and Andhra Pradesh sample households. This accounted for 56 per cent of the Bt sample in Gujarat and 20 per cent in Andhra Pradesh. The genuine Bt seeds that were used by the sample farmer households were of RCH and MAHYCO varieties—MECH 184, MECH 12, MECH 162, and Rasi (RCH) 2, in Andhra Pradesh and Gujarat besides the use of non-confirmed Bt seeds. While in Maharashtra both Mahyco and Rasi Bt seeds were used by the sample households, in Tamil Nadu only Rasi Bt seeds were used. A large number of non-Bt hybrid seeds were also reported among the sample households such as Bunny, Super Bunny, Brahma, Satya, Attara, JK, Tagore, Bindu, Sankar, Vikram, Navbharat-Deshi Ankur, Banny, and Ajit.

Cost of Seeds

The average cost per hectare of Bt cotton seeds used by the sample households varied substantially across the states. It was the least in Gujarat at Rs 3079 per hectare followed by Andhra Pradesh with Rs 3313. On an average the cost of Bt seeds per hectare was more than double than the non-Bt seeds in all the four states.

Application of Pesticides

The average number pesticides sprays as well as its cost per hectare was higher on non-Bt cotton in all the states. But the difference across states was very high. The average number of spray on non-Bt cotton in Gujarat was 0.52, but it was as high as 3.84 in Andhra Pradesh. Similarly the cost of pesticides was higher on non-Bt cotton by Rs 436 per hectare in Gujarat, and Rs 2749 in Andhra Pradesh.

Yields of Bt and Non-Bt Cotton

In Andhra Pradesh the Bt yields were higher but the difference in the yield levels under irrigated and unirrigated was not consistent. The yield of Bt cotton irrigated was significantly higher in Maharashtra and Tamil Nadu. There was no consistent yield difference across farm sizes. The average yield of Bt cotton over non-Bt cotton was higher in all the states in the study and it was higher by 18.2 per cent in Andhra Pradesh, 28.4 per cent in Gujarat, 46.4 per cent in Maharashtra, and 28.5 per cent in Tamil Nadu.

Cost of Production, Value of Output, and Profit

The total cost of cultivation under irrigated condition was in general higher than under unirrigated conditions both for Bt and non-Bt cotton. The cost of cultivation per hectare of Bt cotton exceeded that of non-Bt cotton in all the states and the difference was the highest in Maharashtra exceeding Rs 8250.

The average per-hectare seed cost for Bt cotton was higher compared to non-Bt cotton and the difference ranged from Rs 1600 per ha in Gujarat to over Rs 2700 per ha in Tamil Nadu. The lower cost of Bt cotton seeds in Gujarat and Andhra Pradesh as opposed to Maharashtra and Tamil Nadu could be due to use of non-genuine Bt seeds in Gujarat and Andhra Pradesh. The share of seed cost of Bt seeds in total cost of production was about 10 to 17 per cent in the selected states, whereas it varied from 4 to 6 per cent for non-Bt cotton seeds

The per-hectare cost of pesticides under Bt cotton varied from Rs 2732 in Gujarat to Rs 7806 in Andhra Pradesh. This is as against Rs 3168 in Gujarat to Rs 10878 in Andhra Pradesh for non-Bt. While the share of pesticide cost in total cost of cultivation ranged from 8.29 per cent in Tamil Nadu to 24.29 per cent in Andhra Pradesh for Bt cotton, it varied from 11.96 per cent in Gujarat to 35.73 per cent in Andhra Pradesh for non-Bt cotton.

The value of output of Bt cotton per hectare exceeded that of non-Bt cotton by 30.84 per cent in Gujarat, 42.1 per cent in Andhra Pradesh, 44.74 per cent in Maharashtra, and 47.06 per cent in Tamil Nadu. The net profit per hectare from Bt cotton ranged from Rs 15242 in Tamil Nadu to Rs 34199 in Gujarat as opposed to Rs 5772 in Tamil Nadu to Rs 21880 in Gujarat for non-Bt cotton. The benefit–cost ratios of Bt cotton are higher than that of non-Bt cotton.

Econometric Analysis

The results of statistical/econometric analysis of the whole sample indicate that the positive impact of Bt cotton on the yields has strong statistically significance (Table 10.3). It is significant at the 99 per cent level and the estimates indicate that Bt cotton yields are 30.71 per cent higher. The impact of the value of output is also highly significant and estimates show that this is boosted by 33.35 per cent. However, the cost also rises significantly, and this rise is estimated to be 6.69 per cent. The pesticide cost is reduced by 23.98 per cent, but the seed cost rises by 168.77 per cent.

Table 10.3 Regression Results: Impact of Bt Cotton

Dependent Variable		Independent Variables		N = 515
		Constant	Bt	Per cent Impact of Bt
Yield	Coefficient	2212.25	679.45	30.71
	t-stat	47.05	10.37	
	Signifi.	***	***	
Value of output	Coefficient	41861	13960	33.35
	t-stat	45.2	10.81	
	Signifi.	***	***	
Total cost	Coefficient	28066	1878.56	6.69
	t-Stat	71.5	3.43	
	Signifi.	***	***	
Pesticide cost	Coefficient	7387.95	−1771.47	−23.98
	t-Stat	33.01	−5.68	
	Signifi.	***	***	
Seed cost	Coefficient	1296.12	2187.41	168.77
	t-Stat	28.71	34.76	
	Signifi.	***	***	
Price	Coefficient	19.04	0.28679	1.51
	t-Stat	140.45	1.52	
	Signifi.	***	NS	
Profit	Coefficient	13795	12081	87.58
	t-Stat	16.1	10.11	
	Signifi.	***	***	

Note: *** = significant at 99 per cent, ** = significant at 95 per cent, * = significant at 90 per cent, NS = not significant
Source: Author's computations.

The difference in the output price between Bt and non-Bt cotton is positive but not statistically significant. The results indicate that the profit rise is also highly significant and the increase is estimated to be 87.58 per cent. The results explain the popularity of Bt cotton, at the same time, the opposition to the high seed cost.

The performance varies from state to state. The results for Gujarat indicate that the positive impact on yield and value of output is greater than the combined results, but the cost increase is also greater. The reduction in the pesticide cost is somewhat lower, but the increase in the seed cost is also lower. The price increase is statistically significant but small, and the profit increase is 73.81 per cent. In the case of Maharashtra indicate that the impact on the yield and the value of output at 42.67 and 42.79 per cent respectively are the highest among the three states, and the impact on the total cost is relatively low at 5.18 per cent. Pesticide cost is reduced by 22.38 per cent, and the profit increase is the highest at 120.08 per cent.

In the case of Andhra Pradesh, the impact on the yields as well as the value of outputs is the lowest at about 21.33 per cent, but the rise in total cost is also lower. The fall in the pesticide cost is the highest in Andhra Pradesh at –28.17 per cent, but the rise in the seed cost is also the highest at 192.53 per cent. This is supports the great opposition to the seed prices in Andhra Pradesh. The rise in the profits is statistically highly significant and amounts to a 78.18 per cent, which is in between Gujarat and Maharashtra. The absolute level of profitability of cotton in Andhra Pradesh is lowest amongst the three states.

The model relating the performance to all the covered inputs/factors including Bt, pesticide, seed, fertilizer, irrigation, and state of location together, though affected by multicollinearity, indicates that Bt alone is still statistically highly significant as a determinant of the yield, value of output and profitability. Its exclusive impact on yield is estimated to be about 22 per cent and the impact on profitability about 35 per cent. Profit is negatively related to pesticide cost and positively related to seed cost (reflecting use of Bt seeds) and irrigation. The results also indicate that while profits are significantly higher in Maharashtra as compared to Gujarat, there is no statistically significant difference in the profitability between Gujarat and Andhra Pradesh.

Observations of Farmers

The sample farmer households of Andhra Pradesh expressed that the major advantages of Bt cotton over non-Bt cotton varieties were lesser need for pesticides, better yields, and profits. No major differences were expressed in terms of the use of inputs other than pesticides, and price of cotton. Two major disadvantages expressed by the sample households were the availability of Bt seeds and its high price. 96 per cent of the sample households reported that Bt cotton has more resistance to bollworm attack. Over 90 per cent of the sample households expressed their willingness to continue with Bt seeds considering better yields and profitability. About 10 per of the sample households expressed their unwillingness to continue with Bt cultivation due to high cost of seeds and its unsuitability to their land.

The sample farmer households in Gujarat expressed that the major advantages of Bt cotton over non-Bt cotton varieties were better yields and profits, less pest incidence and pesticides cost, and suitability for early sowing. No major differences were expressed in terms of the use of inputs other than pesticides, and price of cotton. Two major disadvantages expressed by the sample households were high seed price and the availability of Bt seeds.

Observations on the pest incidence available from the Maharashtra Bt growers indicate that for boll worms, including American, pink and spotted bollworms, no infestation on Bt cotton is indicated by over 70 per cent of growers, whereas no infestation is reported by only 2–30 per cent on non-Bt cotton. Only about 4–6 per cent report moderate to heavy infestation on Bt, whereas this number is as high as 20–60 per cent on non-Bt. Surprisingly, there is also a difference in the infestation of sucking and foliage feeding pests, for which the incidence is none to light in the case of

Table 10.4 Regression Results: Impact of Bt Cotton and Other Determinants

Dependent Variable	Coefficient /t-stat/ Signifi	Independent Variables (N = 515)							
		Constant	Bt	Pesticide Cost	Seed Cost	Fertilizer Cost	Irrigation Status	Maharashtra Dummy	Andhra Pradesh Dummy
Yield	Coefficient	1912.78	428.03	0.0318	0.1469	−0.0819	475.07	−333.41	−37.70
	t-stat	16.36	3.78	2.27	3.44	−4.43	6.22	−2.94	−0.3
	Signifi.	***	***	**	***	***	***	**	NS
Value of output	Coefficient	35854	7568.29	0.6136	3.3843	−1.2864	8440.21	−2875.29	−6810.50
	t-stat	15.08	3.28	2.16	3.9	−3.42	5.43	−1.25	−2.67
	Signifi.	***	***	**	***	***	***	NS	***
Total cost	Coefficient	15637	392.30	1.2683	1.6172	1.2548	1117.76	−9105.81	−6689.49
	t-stat	20.7	0.54	14.03	5.86	10.49	2.26	−12.43	−8.25
	Signifi.	***	NS	***	***	***	**	***	***
Profit	Coefficient	20217	7175.99	−0.6547	1.7671	−2.5412	7322.45	6230.52	−121.014
	t-stat	9.85	3.61	−2.67	2.36	−7.82	5.46	3.13	−0.05
	Signifi.	***	***	***	**	***	***	***	NS

Note: *** = significant at 99 per cent, ** = significant at 95 per cent, * = significant at 90 per cent, NS = not significant.
Source: Author's computations.

Bt, whereas it is moderate to heavy in the case of non-Bt. Thus, Bt cotton appears to provide considerable resistance to boll worms, and even shows resistance for other pests.

Over 50.6 per cent of the Maharashtra sample households adopted Bt seeds with the recommendation of fellow farmers, and the next major sources of information was seed company agents/dealers. Government extension agencies did not play much role. The main communication from the agents was about its superiority in terms of better profits and lesser pesticides sprays, more bolls per plant, and no bolls shedding. The Bt cotton growers indicate that no government agencies had approached them for inspection of Bt cotton, and none of the sample households had problems with respect to the marketing of Bt cotton. None of the Bt growers had observed any adverse environmental impact as a result of the cultivation of Bt cotton. They also do not indicate any increase in the pest attack on other crops as a result of the cultivation of Bt cotton.

Almost 98 per cent of the sample farmers did not face any difficulty in getting quality Bt seeds in time. As high as 94.1 per cent of the sample farmers were positive on continuing with Bt cultivation in the future. The major advantages of Bt cotton that are expressed by majority of the Maharashtra sample farmer households were yield superiority, more profit, lesser need of pesticides, better quality, and its suitability for early sowing. On the other hand a common disadvantage expressed was the high cost of seed. To improve the improve the use of Bt technology, the most frequent suggestion given by the Maharashtra farmers was to reduce the cost of Bt cotton seed. Other suggestions were: arranging field demonstrations, seed packages with smaller quantities, and assurance of seed quality.

Majority of the sample farm households from Tamil Nadu said that the plant size of Bt cotton is shorter than non-Bt cotton but the boll size of Bt cotton is bigger. 92 per cent of the sample farmers indicated that the number of pickings is same for Bt and non-Bt cotton. Majority of the sample households did not observe any major difference in terms of the flowering time between Bt and non-Bt cotton either under irrigated or unirrigated conditions. None of the sample farmers had observed any adverse impact on the environment caused by the cultivation of Bt cotton. Major measures suggested by the Tamil Nadu sample households to help Bt cultivation are reduction in the seed price, guidance from extension agencies, and prevention of the sale of spurious Bt cotton seeds.

CONCLUSIONS AND POLICY IMPLICATIONS

Biotech crops, which made their appearance in the world in the mid-1990s, have gained substantial popularity and acceptance in many parts of the world. However, their introduction in India has been relatively late. Cotton is a very important commercial crop in India but has had substantial problems particularly from extensive pest damage and poor yields. In light of this, Bt cotton offered a very promising solution to these serious problems. The study covered the important cotton states of Gujarat, Maharashtra, Andhra Pradesh, and Tamil Nadu. The study finds that Bt cotton offers good resistance to bollworms as well as several other pests. The incidence of these pests is found to be considerably lower in Bt cotton versus non-Bt cotton. The yields of Bt cotton are found to be higher and the yield increase statistically significant in all the states under both irrigated and rain-fed conditions. As a result, given the good market acceptance of the product, the value of output is substantially higher in all the states and conditions sampled. The cost of cultivation is generally higher because of higher seed cost and not a commensurate reduction in pesticide cost. However, the profits/returns are found to be higher in all the states. The returns are highest in Maharashtra followed by Gujarat and then Andhra Pradesh in value terms. Responses indicate that farmers find advantage in lower pest incidence and pesticide cost, and better cotton quality, yields and profits. Almost all farmers indicate that they plan to plant Bt cotton in the future.

Overall, the study, with a sample of 694 farm households across the major cotton states of Andhra Pradesh, Gujarat, Maharashtra, and Tamil Nadu, finds a consistent positive impact of Bt cotton on the yields with strong statistically significance. Bt cotton yields are found to be 30.71 per cent higher. The pesticide costs are reduced, but the seed costs rise, and as a result, the overall cost increases by 6.69 per cent. The profit rise is statistically highly significant and the increase is estimated to be 87.58 per cent. No adverse environmental impacts are reported by farmers. Measures indicated for improving the impact of the technology include release of more Bt varieties for different areas, better seed availability and lower prices of seeds, prevention of sale of spurious seeds, and thorough extension/information dissemination efforts from both public and private agencies for conveying to the farmers the correct package of practices to follow with Bt cotton.

REFERENCES

Chaturvedi, Sachin. 2002. 'Agricultural Biotechnology and New Trends in IPR Regime: Challenges before developing Countries', *Economic and Political Weekly*, 37(13): 1212–22.

Despande, R.S. 2002. 'Suicide by Farmers in Karnataka: Agrarian Distress and Possible Alleviatory Steps', *Economic and Political Weekly*, 37(26): 2601–10.

Dong, Hezhong, W. Li, W. Tang, and D. Zhang. 2004. 'Development of Hybrid Bt Cotton in China—A Successful Integration of Transgenic Technology and Conventional Techniques', *Current Science*, 88(6): 778–82.

Gandhi, Vasant P. and N.V. Namboodiri. 2006. 'The Adoption and Economics of Bt Cotton in India: Preliminary Results from a Study', Paper presented at the IAAE 2006 Conference Symposia: 'The First Decade of Adoption of Biotech Crops–A World Wide View', at the Conference of the International Association of Agricultural Economist (IAAE), Gold Coast, Australia.

Ghosh, P.K. 2000. 'Genetically Modified Crops in India with Special Reference to Cotton', Indian Society of Cotton Improvement, Central Institute of Research on Cotton technology, Mumbai.

Indian Express Bureau. 2003. 'Government Defends Bt Cotton', *The Indian Express*, New Delhi.

Iyengar, S. and N. Lalitha. 2002. 'Bt Cotton in India: Controversy Visited', *Indian Journal of Agricultural Economics*, 57(3): 459–66.

James, C. 2002. 'Global Review of Commercialized Transgenic Crops: 2001', ISAAA Brief No.26-2001. Ithaca, New York: ISAAA.

Joseph, Reji K. 2005. 'Is GM Technology Desirable?', *Economic and Political Weekly*, 40(49): 5210.

Kranthi, K.R. 2005. 'Is Bt cotton unsustainable?'. *Hindu Online*.

Mayee, C.D. 2002. 'Cotton scenario in India. Challenges and approaches for increasing productivity', Paper presented in Workshop, Tamil Nadu Agricultural University, Coimbatore.

Naik, Gopal M. Qaim, A. Subramanian, and D. Zilberman. 2005. 'Bt Cotton Controversy: Some Paradoxes Explained', *Economic and Political Weekly*, 40(15): 1514–17.

Narayanamoorthy, A. and S.S. Kalamkar. 2005. 'Economics of Bt Cotton Cultivation in Maharashtra', Gokhale Institute of Politics and Economics, Pune.

Nelson, A.C. and ORG MARG Study. 2004. 'Benefits of Bollgard Cotton', MAHYCO and MONSAN to Biotech, Mumbai.

Pray, Carl E., J. Huang, R. Hu, and S. Rozelle. 2002. 'Five Years of Bt Cotton in China–the Benefits continue', *The Plant Journal*, 31(4): 423–30. DOI: 10.1046/j.1365-313X.2002.01401.x.

Pray, D. Ma, J. Huang, and F. Qiao. 2001. 'Impact of Bt Cotton in China', *World Development*, 29(5): 813–25.

Purcell, John P. and Frederick J. Perlak. 2004. 'Global impact of insect resistant (Bt) cotton', *AgBioForum* 17 (1 and 2): 27–30.

Pushpavalli, A. 2004. 'Returns to Bt Cotton vis-à-vis Traditional Cotton in Tamil Nadu', Agro-Economic research Centre, University of Madras, Chennai.

Qayum, Abdul and Kiran Sakkhari. 2003. 'Did Bt Cotton Save Farmers in Warrangal?', Deccan Development Society, Hyderabad.

Ramgopal, N. 2006. 'Economics of Bt Cotton vis-à-vis Traditional Cotton Varieties', Study in Andhra Pradesh, Agro-Economic Research Centre, Andhra University, Visakhapatnam.

Research Foundation for Science, Technology and Ecology (RFSTE). 2002. 'Failure of Bt Cotton in India', Press Release, Research Foundation for Science, Technology and Ecology.

Sahai, Suman and S. Rahman. 2003. 'Performance of Bt Cotton: Data from First Commercial Crop', *Economic and Political Weekly*, 38(30): 3139–41.

Shah, V.D. 2007. 'Returns to Bt Cotton vis-à-vis Traditional Cotton Varieties in Gujarat State', Agro Economic Research Centre, Sardar Patel University, Gujarat.

Shiva, Vandana, A. Emani, and A.H. Jafri. 1999. 'Globalization and Threat to Seed Security: Case of Transgenic Cotton Trials in India', *Economic and Political Weekly*, 34(10 and 11): 601–13.

Shourie, David G. and YVST. Sai. 2002. 'Bt Cotton Farmer's Reactions', *Economic and Political Weekly*, 37(47).

Smetacek, Ranjana. 2003. 'Performance of Bt Cotton', *Economic and Political Weekly*, 38(33).

ANNEXURE 10A

Table 10A The Details of Agro-Economic Research Centres and the Lead Person Involved in the State Report

State	AER Centre/Unit	Authors
Andhra Pradesh	AERC Vishakhapatnam	N. Ramgopal
Gujarat	AERC Vallabh Vidyanagar	V.D. Shah
Maharashtra	AERC Pune	S.S. Kalamkar
Tamil Nadu	AERC Chennai	A. Pushpavalli

11

Performance Evaluation of Bt Cotton Cultivation in Punjab

D.K. Grover

OBJECTIVES

With cultivation on around 9 million hectares, India's cotton acreage is the largest in the world and India is the third largest cotton producer after US and China. Cotton, being a major cash crop of India is grown under rain fed as well as irrigated conditions. In India, cotton is grown mainly in nine states spread over three zones, the north, central and South. The major cotton-producing states in India are Maharashtra, Gujarat, Andhra Pradesh, Punjab, Karnataka, and Madhya Pradesh. The productivity of cotton in India is, however, very low. The pest problem in cotton is one of the worst among all crops. The main pest is boll worms and the largest quantity of pesticides among all crops is applied to control pests in cotton. Outbreak of American bollworm in epidemic proportion during crop season of 2001 resulted in very heavy damage to cotton crop, especially in the north zone, which recorded as much as 20–50 per cent reduction in yield compared to the previous year. Cotton cultivation had recently become uneconomic in many parts of the country due to the high cost of pesticides and the low yields. It is under this background, and after much government hesitation, that the introduction of Bt cotton took place in India in 2002.

The Government of India (GoI) allowed the production of three genetically modified Bt cotton hybrids for three years from April 2002 to March 2005. According to official estimates, the area under Bt cotton in India is about 1 million hectare, or about 11 per cent of the total area under cotton in the country. During 2005, the share of area under Bt cotton to total area under cotton was over 27 per cent in Madhya Pradesh, and about 18 per cent in Maharashtra. The performance of Bt cotton in India, during 2005–06, the fourth consecutive season and in Punjab, the only first year, appears to be very encouraging. The most important reason for the adoption of Bt cotton is its resistance to pests, particularly boll worms, which can be a devastating problem for cotton. The area under Bt cotton varieties has consistently increased. Starting with 12 Bollgard varieties on 28800 hectares in the 2002–03, Bt cotton cultivation rose to 92000 hectares in the 2003–04, 5,20,000 hectares in 2004–05, and during 2005–06 there were 20 Bollgard varieties planted on 12,50,000 hectares. Punjab, Haryana, and Rajasthan, taken together produce nearly a quarter of the country's cotton. In the year 2003–04, the overall crop damage caused by the American bollworm in these three states was estimated at 15.50 lakh bales of 170 kg each. Considering the sheer extent of crop losses suffered, officially valued at about Rs 1364 crore, a lot of hope has been generated over Bt cotton amongst the lakh bales of 170 kg each. Considering the sheer extent of crop losses suffered, officially valued at about Rs 1364 crore, a lot of hope has been generated over Bt cotton amongst the region's cotton growers. This enthusiasm has also been echoed by planners and policymakers, keen to divert surplus paddy growing areas in Punjab and Haryana to less water requiring crops in kharif season. Resultantly, during the kharif season of 2005–06, the States of Punjab, Haryana, and Rajasthan have also officially came under Bt cotton cultivation (www.fbaeblog.org) Punjab Agricultural University has recommended cultivation of six Bt cotton varieties, namely RCH 134, RCH 317, MRCH 6301, MRCH 6304, Ankur 691, and Ankur 2534 in Punjab with the approval of GEAC from kharif season of 2005. A study of the adopted villages has revealed that Bt cotton brought down significantly the number of sprays, slashing the cost of spray on one hand and improving ecology on the other. Obviously, there are some areas where Bt cotton's performance was below optimal, for reasons that may largely lie outside Bt technology, such as growing cotton in unsuitable areas and faulty management practices. This is what we have to seriously look into and its remedy. Therefore, it becomes pertinent to conduct a comprehensive study on the performance evaluation of boll worm resistant Bt cotton varieties and its various advantages over the conventional cotton varieties in the state. The study focused on the following specific objectives:

1. to study the input use pattern for cultivation of Bt cotton vis-à-vis conventional cotton varieties;
2. to study the economic performance of Bt Cotton over the conventional cotton varieties;
3. to study the input use efficiency in cultivation of Bt cotton and conventional cotton varieties;
4. to identify the various production and marketing constraints of Bt as well as non-Bt cotton; and
5. to evaluate the potential and suggest policy measures for expanding area under various Bt cotton varieties.

METHODOLOGY AND DATA

To accomplish the various objectives of the study, both secondary as well as primary data were required. The secondary information such as time series data on area, production, yield of cotton (American and desi) in India and its different states, in Punjab along with its different districts, etc., were collected from the secondary sources such as, various publications of Ministry of Agriculture and Farmers Welfare and Ministry of Commerce and Industry, data websites, *Statistical Abstracts of Punjab, Agricultural Statistics at a Glance*, and others. In Punjab, Cotton is grown over 6 lakh hectares, largely in the geographically contiguous tract of south-west Punjab, that is, Bathinda, Muktsar, Mansa, and Ferozepur districts. Hence, the study has been concentrated in these cotton growing districts of Punjab. Based on the concentration of cotton cultivation, the study covered two districts namely, Bathinda and Ferozepur which taken together constituted over 50 per cent of total cotton cultivation in the state. From each selected district, two blocks with maximum concentration of cotton cultivation was selected for the field-level base survey data. These blocks were Bathinda and Talwandi Sabo from Bathinda district and Abohar and Fazilka from Ferozepur district. From each block, two clusters of three villages each were selected for intensive survey. To study the relative performance of Bt and non-Bt cotton, the planned sample size was 120 Bt cotton growers and 60 non-Bt cotton growers. Owing to non-availability of farmers in the latter category, the study has been based on 121 Bt cotton growers and 29 non-Bt cotton growers spreading over 8 village clusters (24 villages), 4 blocks and 2 districts of Punjab. The required information and data pertaining to the performance related parameters of Bt cotton and non-Bt cotton cultivation such as input use pattern, yield, price, biotic/a biotic constraints, cost and return, and production constraints/perception, etc., were collected from these 150 sampled farmers with the help of an especially designed schedule for the purpose following random sampling technique. The study was based on two production year's experience of Bt cotton non-Bt growers in Punjab during 2006–07 and 2007–08. The simple statistical techniques like averages, percentage, tabular analysis, frequency distribution, and ranking analysis, compound growth rates (CGR) and coefficients of variation (CV) have been applied for better explanation and interpretation of the results. The graphical/diagrammatical presentation was also made in the analysis for better understanding of the trends. To identify the determinants of cotton yield, Cobb-Douglas type production function, and linear production function were tried with different combinations of independent and dependent variables.

RESULTS AND DISCUSSION

In recent years especially during 2004–05 to 2006–07, cotton performance in the state have been quite encouraging as its per cent share in India's cotton production was much higher in relation to its share in India's total cotton acreage e.g. during 2005–06, Punjab shared only 6.42 per cent of country's cotton area and contributed as high as 12.95 per cent of total cotton production in India, reflecting much higher production efficiency of cotton per unit of area in the state. Results have been discussed under the following sub heads:

1. State-wise trends in area, production, and yield of cotton in India
2. District-wise trends in area, production, and yield of cotton in Punjab
3. Agro-socio-economic distinctiveness and Agronoic practices of sample households
4. Economics of cotton cultivation
5. Yield variations and response function
6. Constraint Analysis

State-wise Trends in Area, Production, and Yield of Cotton in India

Maharashtra state has maximum area under cotton in the year 2005–06, that is, 2875 thousand hectares constituting 33.13 per cent of total cotton acreage in the country, followed by Gujarat (21.97 per cent), Andhra Pradesh (11.90 per cent), Madhya Pradesh (7.15 per cent), Haryana (6.72 per cent), and Punjab (6.42 per cent). The cotton acreage increased in states such as Punjab, Gujarat and Madhya Pradesh over 2000–06 while it declined in Rajasthan, Maharashtra, Karnataka, and Tamil Nadu. Only marginal increase has been noticed in the cotton acreage in India from 8576 thousand hectares to 8677 thousand hectares during this period. Gujarat state produced maximum cotton during 2005–06 i.e. 6772 thousand bales constituting 36.61 per cent of total cotton production in the country, followed by Maharashtra (17.08 per cent) and Punjab (12.95 per cent). The cotton production increased in states like Punjab, Haryana, Rajasthan, Gujarat, Maharashtra, Madhya Pradesh, and Andhra Pradesh over 2000–06 while it decreased in Karnataka and Tamil Nadu. The production of cotton in India has almost doubled over this period (2000–06), that is, from 9652 thousand bales in 2000–01 to 18499 bales during 2005–06. As far as yield of cotton is concerned it was the highest in Punjab state (731 kg/ha) followed by Gujarat (604 kg/ha) and Haryana (437 kg/ha) during 2005–06. The states with the lowest yield cotton was Maharashtra (187 kg/ha) followed by Madhya Pradesh (204 kg/ha), Karnataka (228 kg/ha), Tamil Nadu (258 kg/ha), Rajasthan (318 kg/ha), and Andhra Pradesh (347 kg/ha). The overall cotton yield for the country as a whole was 362 kg/ha during 2005–06 as compared to only 192 kg/ha in 2000–01. The states which experienced rapid cotton yield increase over this period were Punjab (430–731 kg/ha), Gujarat (122–604 kg/ha) and Madhya Pradesh (80–204 kg/ha). The states with marginal yield increase were Haryana (423–37 kg/ha), Rajasthan (269–318 kg/ha), and Andhra Pradesh (277–347 kg/ha). The states with lower cotton yield were Karnataka (298–228 kg/ha) and Tamil Nadu (286–258 kg/ha).

District-wise Trends in Area, Production, and Yield of Cotton in Punjab

Relative share of American cotton and cotton (desi) in Punjab: The relative share of American Cotton and Desi cotton in the total cotton acreage and production in the state over the years has been calculated. The American cotton constituted the big chunk of total cotton cultivation in the state. Of the total cotton acreage in Punjab, American cotton is grown over 96 per cent area. The share of Cotton desi used to be as high as 24 per cent during 2000–01 has drastically declined to only 4 per cent during 2007–08. Similarly, the share of production declined from 23 per cent to only 3 perc ent in the state over the years. Trends in American cotton: Bathinda district has the highest area under American cotton during 2007–08 (160 thousand hectares) and 115 thousand hectares in 2000–01, accounting for 28 and 32 per cent of the total area under American cotton in the state during these years. The compound annual growth rate (CAGR) of area under American cotton (4.57 per cent) turned out to be significant at 5 per cent level. The CAGR in the area under American cotton has been noticed as positive in all the districts of Punjab though it was significant only in four districts namely Ferozepur, Bathinda, Mansa, and Muktsar. The total production of American cotton in the state increased from 922 thousand bales in 2000–01 to 2280 thousand bales in 2007–08. The annual growth rate of production has been as high as 17.26 per cent, significant at 5 per cent probability level. Bathinda district alone produced 645 thousand bales sharing around 28 per cent of total American cotton in the state. As far as yield of American cotton in the state is concerned, it increased from 437 kg/ha in 2000–01 to 668 kg/ha in 2007–08 with highest yield level of 763 kg/ha in 2006–07. The decline in the yield during 2007–08 was mainly attributed to appearance of a cotton pest 'mealy bugs' in the state in a big way. During 2007–08 Mansa district enjoyed the highest yield of American cotton cultivation i.e. 753 kg/ha followed by Sangrur (749 kg/ha) and Bathinda (685 kg/ha). The lowest yield of American cotton during this year was recorded as 577 kg/ha in Ferozepur district of Punjab. The high value of CV indicates the large fluctuations in yield of American cotton in various districts of Punjab. The annual growth rate in the yield of American cotton has been observed positive in all the districts while remained significant only in Faridkot, Bathinda, Sangrur, Mansa, and Muktsar.

Trends in desi cotton: As far as the area under desi cotton in the state is concerned, it shows a declining trend. The area under desi cotton in the state decreased from 115 thousand hectares in 2000–01 to merely 24 thousand hectares during 2007–08. The annual growth rate of area under desi cotton has been estimated as –19.47, highly significant. At present the maximum area under desi cotton is in Bathinda district (5 thousand hectares) which was 32.3 thousand hectares in 2000–01. The negative and significant CAGR in the entire districts showed declining trend in the area under desi cotton due to substitution of American cotton in the state. Due to fast decline in its acreage, production of desi cotton in the state also shows declining trend. The production of desi cotton in the state was 278 thousand bales in 2000–01 which declined to 76 thousand bales in the year 2007–08. In the year 2007–08 maximum production of desi cotton was in the Muktsar district (17 thousand bales) followed by Bathinda (15 thousand bales), Ferozepur (14 thousand bales), Mansa (8 thousand bales), Sangrur (7 thousand bales), Faridkot (6 thousand bales), and Moga (4 thousand bales). Similar to area, the annual growth rate of production of desi cotton in various districts was also found to be negative, implying declining trend of production of desi cotton. Contrary to area and production of desi cotton in the state, the yield of desi cotton showed a little improvement over the years. In 2007–08 average yield of desi cotton in the state was 541 kg/ha as against only 408 kg/ha during 2000–01. The yield of desi cotton was maximum in Muktsar district (704 kg/ha) and minimum in Sangrur (389 kg/ha). The CAGR of yield of desi cotton in the state was recorded as 8.80 per cent over a period of 2000–01 through 2007–08. Though the growth rate turned out to be positive in all the districts, yet it was positive and significant in Faridkot and Bathinda. Higher value of CV denotes a fluctuating trend in the yield of desi cotton in the state as well its major cotton producing districts.

Agro-socio-economic Distinctiveness and Agroeconomic Practices of Sample Households

The overall family size of Bt cotton growers was a little smaller but the number of family members in the productive age group (18–60 years) was comparatively higher as compared to non-Bt cotton growers. Though not much difference with regard to education level of family members of Bt cotton and non-Bt cotton growers have been observed, yet the analysis indicates that Bt cotton growers' families were better literate as compared to their counterparts, that is, non-Bt cotton growers The Bt cotton growers possessed relatively better quality/shaped agricultural machinery on their farms as compared to non-Bt cotton growers. The average number of agricultural machinery like tractor, trailer, Tube well/pump set, seed—fertilizer drill, etc., possessed by Bt cotton growers were also more as compared to non-Bt cotton growers on sample farms. There was not much difference in the average operational land holding of Bt cotton growers (8.7 ha) and non-Bt cotton growers (8.6 ha). In kharif season (2007) area under cotton crop, was about 80 per cent for both the Bt cotton growers and non-Bt cotton growers' category, and in rabi season the area under wheat in both the cases, was 87 to 89 per cent. Hence, in the study area, mainly cotton wheat rotation was followed by the sample farmers both Bt cotton growers as well as non-Bt cotton growers. During 2007, production of Bt and non-Bt cotton declined due to the severe appearance of mealy bugs—a cotton pest. The produce was sold by both the categories and not any quantity was kept for home need indicates the absolute commercial nature of the crop. The year 2007 was favourable for the cotton output prices, which has to some extent compensated the profitability of the farmers caused due to yield decline by appearance of mealy bugs. In this year due to shortage of cotton production, prices remained almost same across various varieties of Bt as well as non-Bt cotton. During the year 2006, price of Bt cotton (Rs 2134/qtl) ruled marginally higher than the non-Bt cotton varieties, that is, Rs 2112/qtl. Major factors responsible for adoption of Bt cotton varieties were resistant to attack of pests and higher yield of crop and in case of non-Bt cotton varieties, low cost of cultivation and low water requirement were the major reasons for adoption of the varieties.

Agronomic practices: All the land preparation practices such as ploughing and levelling were similar for Bt cotton and non-Bt cotton cultivation. The practice of seed treatment was not found to be prevalent among Bt cotton growers, however, among non-Bt cotton growers 28 per cent farmers adopted seed treatment practices. The seed rate/ha was only 2.59 kg in case of Bt cotton cultivation as against as high as 14.35 kg/ha for non-Bt cotton cultivation. Consumption of fertilizers was much more in Bt cotton varieties. The extent of various fertilizers used in Bt cotton were to urea (262 kg/ha) followed by DAP (116 kg/ha), super (132 kg/ha), and potash (77 kg/ha) against the consumption of urea (184 kg/ha), DAP (94 kg/ha), super (124 kg/ha) and potash (37 kg/ha) in non-Bt cotton varieties. Consumption of micro nutrient was almost same for Bt cotton and non-Bt cotton varieties (20 kg/ha to 21 kg/ha). Bt cotton varieties need more number of irrigations as compared to non-Bt varieties which are somewhat drought resistant. On an average, farmers applied 14 irrigations for Bt cotton varieties but in case of non-Bt cotton varieties, only 8 irrigations were applied by the farmers. Inter-culture and weeding operation was also more frequent in Bt cotton varieties (5 and 4 times as against 4 and 3 times in non-Bt cotton varieties). Both varieties yielded maximum cotton at second picking, which varied between 33 per cent and 37 per cent of the total production, followed by the third picking, yielding 22 to 27 per cent of the total production. The production was minimum at fourth and last picking that yielded only 14 per cent in case of Bt cotton and 11 per cent in case of non-Bt cotton of the total production per hectare in the study area. During 2006, majority of the farmers (96 per cent) applied 2 to 4 number of sprays to Bt cotton varieties whereas in the case of non-Bt cotton varieties frequency of chemical sprays was more, over 83 per cent farmers used 4 to 8 sprays and 14 per cent of the total farmers even applied more than 8 sprays. Contrary to 2006, in the year 2007 number of chemical application increased in both the Bt and non-Bt cotton varieties due to appearance of 'mealy bugs' pest. In this year, 60 per cent of the Bt cotton grower's applied 6–8 number of sprays to Bt cotton varieties. In case of non-Bt cotton varieties 76 per cent of the total farmers applied more than 8 sprays to non-Bt cotton varieties.

Economics of Cotton Cultivation

Cost–benefit analysis of cotton cultivation: Detailed comparative analysis of cost of cultivation of Bt cotton and non-Bt cotton crop was made and presented in the Table 11.1. Table clearly indicates that the cost of cultivation was found to be more in case of Bt cotton as compared to non-Bt cotton varieties in both the years 2006 and 2007. It was Rs 28529/ha for the Bt cotton in 2006 as compared to Rs 23551/ha for the non-Bt cotton crop which was 21.14 per cent higher than non-Bt cotton. Similarly in 2007, cost of cultivation of the Bt cotton (Rs 34607/ha) was higher by 5.00 per cent than non-Bt cotton crop (Rs 32960/ha). In year 2007, the various costs of production like seed (Rs 2876/ha), irrigation (Rs 1552/ha), fertilizers (Rs 5985/ha), and manures and human labour (Rs 10943/ha) were higher than non-Bt cotton. Costs of these inputs were higher by 598, 71, 15, and 9 per cent than non-Bt cotton. But cost of some other inputs like insecticides and pesticides was low in Bt cotton. In 2007 farmers incurred Rs 8264/ha on plant protection chemicals for Bt cotton, which were 25 per cent lower than such expenses on non-Bt cotton (Rs 11027/ha). The corresponding figure for the year 2006 was Rs 709/ha for Bt cotton as against Rs 2033/ha

Table 11.1 Economics of Bt/Non-Bt Cotton on the Sample Bt/Non-Bt Growers in Punjab

Particulars	2006			2007		
	Bt Growers	Non Bt Growers	% Change	Bt Growers	Non Bt Growers	% Change
Variable Cost (Rs/ha)						
Human labour	10943	10027	9.14	10943	10027	9.14
Machine expenses	4143	4346	−4.67	4143	4346	−4.67
Seed treatment	0	247	−100.00	0	247	−100.00
Insecticides and Pesticides	709	2033	−65.13	8264	11027	−25.06
Cost of seed used	4500	425	958.82	2876	412	598.06
(a) Recommended	3599	30	–	2077	13	–
(b) Un-recommended	901	395	–	799	399	–
Manures and Fertilizers	5986	5187	15.40	5985	5187	15.38
Irrigation	1552	859	80.67	1552	910	70.55
Interest on working capital	696	427	63.00	844	804	4.98
Total variable cost	28529	23551	21.14	34607	32960	5.00
Yield (qtl/ha)	29.38	23.36	25.77	25.2	19.38	30.03
Price (Rs/qtl)	2134	2112	1.04	2419	2429.7	−0.44
Value of Main Product (Rs)	62697	49336	27.08	60959	47088	29.46
By-Product (qtl/ha)	101.82	121.8	−16.40	86.66	126.81	−31.66
Price (Rs/qtl)	27.4	39.49	−30.62	26.94	35.17	−23.40
Value of by-product (Rs)	2790	4810	−42.00	2335	4460	−47.65
Gross Returns/ha	65487	54146	20.95	63294	51548	22.79
Net Returns Over Variable Cost/ha	36958	30595	20.80	28687	18588	54.33

for non-Bt cotton crop. The huge difference in the plant protection expenses in both the year was due to severe attack of 'mealy bugs' pest in the year 2007, by which the cost of cultivation increased considerably. There are several reasons for the higher cost of cultivation of Bt cotton varieties. First, the seed cost of the Bt cotton varieties is quite higher than that of non-Bt cotton varieties which accounts 8 per cent of the total variable cost in the year 2007. Secondly, due to higher yield the cost of picking was also higher in Bt cotton. Thirdly, the irrigation costs were more in Bt cotton varieties due to more water requirement of the crop as compared to non-Bt cotton varieties, irrigation charges were higher by 70.55 per cent in Bt cotton crop. Hence the cost of cultivation is substantially higher in Bt cotton except the cost of plant protection measures.Further, there was only marginal difference in the output prices of Bt and non-Bt cotton varieties. Regarding the gross and net returns in Bt cotton and non-Bt cotton varieties, both the gross returns and net returns were higher in Bt cotton in both the years. The gross returns in Bt cotton were Rs 65487/ha in 2006 and Rs 63294/ha in the year 2007, corresponding values for the non-Bt cotton were Rs 54146/ha and Rs 51548/ha. As far as net returns are concerned, despite the higher cost of cultivation in Bt cotton varieties, but due to resistant against the pests and diseases which resulted into higher productivity of Bt cotton varieties. In year 2006, the net returns in Bt cotton was higher by 20.80 per cent than non-Bt cotton. In Bt cotton per hectare net returns over variable costs were Rs 36958/ha as compared to Rs 30595/ha in non-Bt cotton. Surprisingly in the year 2007, with the high intensity of mealy bugs attack the Bt cotton varieties yielded 54.33 per cent more net returns than non-Bt cotton varieties. The relative figures for the Bt cotton and non-Bt cotton were Rs 28687/ha and Rs 18588/ha. This enhanced performance of Bt cotton varieties may be attributed to higher yield of Bt cotton varieties in the respective years 2006 and 2007, that is, 29.38 qtl/ha and 25.29 qtl/ha in Bt cotton as compared to 23.36 qtl/ha and 19.38 qtl/ha in non-Bt cotton varieties. It indicates that the yield of Bt cotton was higher by 25–30 per cent than that of non-Bt cotton.

Yield Variations and Response Function

Though, the overall yield of Bt cotton varieties was found to be much higher (25–29 qtl/ha) during the years of study, that is, 2006 and 2007, yet acute yield variations have been observed from farmer to farmer in the sample study area as shown in Figure 11.1. Amongst the sample Bt cotton growers, the yield obtained has been as high as about 38 qtl/ ha on some farms and as low as about 10 qtl/ha on the others. These large variations may majorly be attributed to the crop management skill of Bt cotton growers, as the agro climatic conditions were more or less the same in all sites of study area.

Yield response function: The log-linear regression model was selected to determine the factors affecting value productivity per hectare in Bt cotton cultivation. The R^2, coefficient of determination was 0.25 showing that 25 per cent value productivity was explained by the independent variables included in the model. The overall model was statistically significant as shown by the F value given in Table 11.2. So far the identification of variables having significant effect is concerned, it was seen that regression coefficients for seed, human labour, and plant protection chemicals turned out to be positively significant at 5 per cent probability level. The coefficient of operational holding turned out to be negatively significant at 5 per cent level of significance. In terms of magnitude, 1 per cent increase in area would decrease the value productivity of Bt cotton crop by 0.15 per cent. If 1 per cent expenditure is increased on seed, plant protection chemicals and human

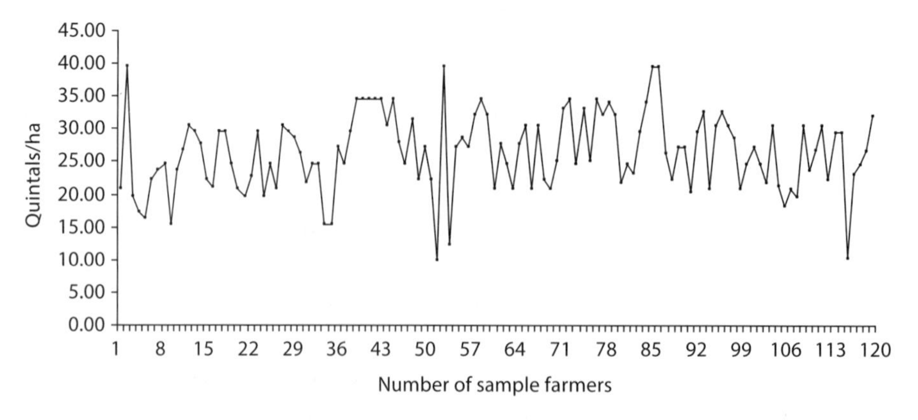

Figure 11.1 Variation in Bt Cotton Productivity among the Sample Farmers, 2007

Table 11.2 Yield Determinants of Bt Cotton Cultivation: Log-Linear Equation)

Variable	Unit	Regression Coefficients
Intercept		9.03* (1.16)
Education (X_1)	No. of years	−0.00036 (0.01)
Operation holding (X_2)	Hectares	−0.15* (0.033)
Seed (X_3)	Rs/ha	0.095* (0.04)
Human labour (X_4)	Rs/ha	0.15* (0.67)
Machine labour (X_5)	Rs/ha	−0.13 (0.09)
Plant protection chemicals (X_6)	Rs/ha	0.05* (0.02)
Manures & Fertilizers (X_7)	Rs/ha	0.07 (0.09)
Irrigations (X_8)	No.	0.06 (0.09)
R-SQ (F-Ratio)		0.25 (4.77)
No. of observations		121

Notes:

(a) The figures in the parentheses show the respective standard error of the coefficient.

(b) * shows the significance at 5 per cent level.

labour, the value productivity of cotton would increase by 0.095, 0.048, and 0.15 per cent, respectively. The regression coefficients of education, machine labour, manures and fertilizers and irrigation were found non-significant.

Constraint Analysis

Spotted/spiny bollworms, American bollworms, Tobacco caterpillar, kali sundi (Black Larva), mealy bugs, and White Fly were some of the insects/pests of cotton belt in the state. Of these insects, mealy bugs, and American bollworm were the major enemies of Bt and non-Bt cotton varieties respectively that requires the immediate attention of the scientists to avoid any epidemic loss of the crops in coming years. During 2006–07, there was only slight attack of leaf curl disease as perceived by 8 per cent Bt cotton growers and 34 per cent non-Bt cotton growers. Due to this disease farmers reported only 1 to 3 per cent yield decline in Bt and non-Bt cotton varieties, respectively. High temperature was considered by 21 per cent farmers as a severe problem among Bt cotton growers as compared to 7 per cent of non-Bt cotton growers. Low temperature was almost not considered as a problem in both the categories as only 1 per cent Bt cotton growers perceived it without any yield losses. Production constraints like poor quality of underground water, inadequate irrigation facilities, high incidence of pest/diseases, high interest on credit, and lack of credit availability from institutional resources were highly intensive both in Bt cotton growers and non-Bt cotton growers. The production constraints whose intensity was low but higher percentage of farmers faced were, availability/higher prices of good quality seed, lack of knowledge about package of practices, availability of appropriate machinery, low yield, and non-availability of disease resistant varieties. Price variability, losses during storage, high labour needs for sorting/packaging and low market demand were the important marketing constraints, as perceived by both Bt and non-Bt cotton growers. Cotton being a kharif crop its competitiveness for irrigation with other crops like paddy is a very important factor for the adoption of various varieties of crop. Regarding water availability from the tube well, 84 per cent of the farmers reported it was sufficient at various stages of crop growth like sowing and flowering, etc. Use proper plant protection measures, provide irrigation at right time and timely application of fertilizers were the most desired practices at the farmers own level for yield improvement as perceived by Bt cotton as well as non-Bt cotton growers. Some other practices in order of importance were use more fertilizers and timely planting for Bt cotton growers and timely planting and use more fertilizers for non-Bt cotton growers category. Of the total area under cotton in the sample villages, around 82 per cent was under Bt cotton, both recommended as well as non-recommended varieties and the remaining 18 per cent was under non-Bt cotton recommended and non-recommended varieties during the year 2006. The most important recommended Bt cotton varieties were RCH 134 and MRC 6304. The major non-recommended Bt cotton varieties were Gujarat and Om 3. Most of the total area under non-Bt cotton was found under non-recommended varieties such as SLM 8 and 104. During 2007, the per cent area under Bt cotton increased though the varietal distribution of Bt cotton remained almost the same as during previous year.

CONCLUSIONS AND POLICY IMPLICATIONS

Based on the farmers' experience of two consecutive production year, following suggestions may be put forwarded to revive the cotton cultivation and hence cotton industry in Punjab state.

1. Though the overall experience of Bt cotton in the state has been quite encouraging, yet acute yield variations have been observed from farmer to farmer in the sample study area. Amongst the sample Bt cotton growers, the yield obtained has been as high as about 38 quintals/ha on some farms and as low as about 10 quintals/ha on the others. There were some farmers where Bt cotton's performance was below optimal, for reasons that may largely lie outside Bt technology, such as growing cotton in unsuitable areas and faulty management practices. This is what we have to seriously look into and its remedy.
2. The high-value of coefficient of variations in the yield and production of cotton at district as well as state level indicates that the crop could not be stabilized even after a long period of its cultivation. Much more research efforts need to be put by the scientists so that yield variability may be minimized. Crop insurance scheme must be effectively introduced in such risky crops helping the farmers to minimize the yield risk and hence the profitability.
3. Spotted/spiny bollworms, American bollworms, Tobacco caterpillar, Black Larva, Mealy bugs, and White Fly were some of the insects/pests of cotton belt in the state. Of these insects, mealy bugs and American bollworm were the major enemies of Bt and non-Bt cotton varieties respectively. Due to appearance of severe mealy bugs in the state, the Bt cotton yield has declined from 763 kg/ha in 2006–07 to 668 kg/ha in the subsequent year of 2007–08. Therefore, this requires the immediate attention of the scientists to avoid any epidemic loss of the crops in coming years
4. The regression analysis brings out that yield of Bt cotton can significantly be increased by applying more seed, human labour and plant protection chemicals. Hence farmers need to be educated and facilitated on these fronts through appropriate policies to further improve upon the cotton productivity in the state and also to better farmers' return from cotton cultivation.
5. Production constraints like poor quality of underground water, inadequate irrigation facilities, high interest on credit, and lack of credit availability from institutional resources were highly intensive both in Bt cotton growers and non-Bt cotton growers. Hence, suitable policies need to be evolved to tackle these problems being confronted by the sample farmers in the study area.
6. Use proper plant protection measures, provide irrigation at the right time and timely application of fertilizers were the most desired practices at the farmers own level for yield improvement as perceived by Bt cotton as well as non-Bt cotton growers.
7. The study highlighted that of the total area under cotton in the sample villages, around 82 per cent was under Bt cotton, both recommended as well as non-recommended varieties and the remaining 18 per cent was under non-Bt cotton recommended and non-recommended varieties. Around 45 per cent of the total Bt cotton area on the sample holdings were under non-recommended varieties posing a serious risk to their yield. Therefore, farmers should be convinced to sow only the varieties as recommended by the scientists of state agricultural university and also follows other package of practice as provided by them for better management of the crop.

REFERENCES

Bose, Ashish. 2000. 'From Population to Pests in Punjab: American Boll Worm and Suicides in Cotton Bel'. *Economic and Political Weekly* 35(38), September.

Business Line (Newspaper) 2002. India, December 9.

Chaturvedi, Sachin. 2002. 'Agricultural Biotechnology and New Trends in IPR Regime: Challenges before developing Countries'. *Economic and Political Weekly* 37(13), March.

Dong, Hezhong, Li Weijiang, Tang Wei, and Zhang Dongmei. 2004. 'Development of Hybrid Bt Cotton in China—A Successful Integration of Transgenic Technology and Conventional Techniques'. *Current Science* 88(6): 778–82.

Ghosh, P.K. 2000. 'Genetically Modified Crops in India with Special Reference to Cotton'. Indian Society of Cotton Improvement, Central Institute of Research on Cotton Technology, Mumbai.

Indian Express. 2003. 'Government Defends Bt Cotton'. Indian Express Bureau, New Delhi.

Iyengar, S. and N. Lalitha. 2002. 'Bt Cotton in India: Controversy Visited'. *Indian Journal of Agricultural Economics* 57(3): 459–66.

James, C. 2002. 'Global review of commercialized transgenic crops: 2001' (ISAAABrief No. 26-2001).

Joseph, Reji K. 2005. 'Is GM Technology Desirble?'. *Economic and Political Weekly*, 40(49), December.

Mayee, C.D. 2002. 'Cotton Scenario in India. Challenges and Approaches for Increasing Productivity'. In Workshop, TNAU, Coimbatore.

Mayee, C.D. and M.R.K. Rao. 2002. 'Likely impact of Bt Cotton Cultivation on Production and Utilization in India'. National Seminar on Bt Cotton in India at UAS, Dharward.

Naik, Gopal et al. 2005. 'Bt Cotton Controversy: Some Paradoxes Explained'. EPW. April.IIMA _ INDIA.

Pray, Carl E., et al. 2002. 'Five Years of Bt Cotton in China–the Benefits continue'. *The Plant Journal* 31(4).

Pray, et al. 2001. 'Impact of Bt Cotton in China'. *World Development* 29(5).

Qayum, Abdul and Kiran Sakkhari. 2003. 'Did Bt Cotton Save Farmers in warrangal?'. Deccan Development Society, Hyderabad.

Shai, Suman and S. Rahman. 2003. 'Performance of Bt Cotton: Data from First Commercial Crop'. *Economic and Political Weekly* 38(30).

Shiva, Vandana, et al. 1999. 'Globalization and Threat to Seed Security: Case of Transgenic Cotton Trials in India'. *Economic and Political Weekly* 34(10 and 11).

Venkateshwarlu, K. *The Hindu*, India, December 30, 2002.

12

Potential and Prospects of Rabi Crops Cultivation in Assam

Moromi Gogoi and Rupam Kr. Bordoloi

Agriculture is considered as the mainstay of the economy of Assam. Agriculture and allied activities in Assam have overriding importance as sources of livelihood to its people. It still contributes more than one fourth (26.19 per cent) to the Net State Domestic Product (NSDP) and supports about 70 per cent of its population. The net cropped area of the State is 27.53 lakh hectares against gross cropped area (GCA) of 39.57 lakh hectares. The State of Assam experiences plenty of rainfall and possesses fertile land which is extremely advantageous for growing variety of crops. The soil, topography, rainfall, and climate of the State are quite congenial for producing different crops in different seasons. But, agriculture in the State is characterized by low level of productivity due to recurring natural calamities, low level of mechanization, inadequate availability of quality inputs, poor soil health, low level of assured irrigation, and inadequate marketing infrastructure.

POTENTIAL OF RABI CROPS CULTIVATION IN ASSAM

Although, agriculture continues to occupy a pre-eminent place in the economy of Assam and the farmers living in rural areas constitute the backbone of Assam agriculture, the State's agriculture is exposed to vagaries of monsoon. During monsoon, heavy rainfall causes extensive damage to summer and kharif crops. Therefore thrust now is being shifted to production of rabi crops which can be grown in flood free season. The emergence of rabi crops is not only an opportunity to enhanced production, but also reduces the burden of production loss due to floods. In view of increasing demand for food crops for fast growing population, the state government has come up with a host of new schemes/activities to bring more areas under rabi crops and also to upscale the production to meet the increasing requirements. The development programmes undertaken by the government are expected to increase the production and it is certainly going to enhance the prospects of rabi crops in the State if taken up in right earnest.

In Assam, farmers grow crops mainly in two seasons that is, kharif and rabi season. The major kharif crops are autumn rice, winter rice, maize, pulses, kharif oilseeds like sesamum, soybean, groundnut, kharif vegetables, etc. The non-food crops such as jute, mesta, cotton, etc., are also grown to some extent by the farmers in the kharif season. On the other hand, major rabi crops cultivated are summer rice, wheat, grams, rapeseed, and mustard, various rabi oilseeds, rabi vegetables, potato, etc.

Considering the importance of rabi crops cultivation in Assam, the study was undertaken to examine the potential and prospects of summer rice, rape, and mustard and pulses cultivation which are the major food crops of Assam.

SUMMER RICE CULTIVATION IN ASSAM

Excessive and untimely rain intercepted during kharif and autumn seasons and the havoc created by river flood have adversely affected the winter and autumn rice cultivation in the state. The farmers in the flood-prone areas are in search of some alternative cropping pattern for producing staple grains and cash crops for income generation. Thus, summer rice has emerged as an important cereal crop in the state. During the past decade, the crop has taken a swing in the saucer-shaped marshy areas as a traditional rainfed crop and also in the non-traditional areas as an irrigated crop. Massive efforts have been made by the Government for production of summer rice in the State through installation of Shallow Tube Wells (STWs) in the non-traditional areas and its cultivation in those areas has been identified as a capital intensive, high cost but highly productive crop, against the low capital low-cost and high-return rainfed summer rice in the traditional areas. Summer rice which triggers green revolution in the state during 2000–01 with the aid of STWs installed under different developmental programmes resulted in 1.3 per cent increase in production in 2008–09 over 2002–03. But the area coverage under the crop decreased due to increase in the production cost mainly because of hike in price of diesel used in STWs.

Although rice is the major food crop in the state, its vulnerability to natural disasters like flood, submergence and even drought affect the production system drastically. However, notwithstanding the dismal performance, the farmers have shown dynamism in favour of summer rice (boro). The area under rice dominated the cropping system sharing about 64.70 per cent of total cropped area in 2008–09. The major rice is winter rice which occupied 1,805 thousand hectares in 1991 declined to 1,773 thousand hectares in 2008–09 while the area under boro rice increased from 128 thousand hectares to 360 thousand hectares during the same period. The productivity of boro rice is more than that of winter rice with 30–40 per cent yield premium. It is also relatively a safer option as it is cultivated in the flood free summer season. The emergence of this newer crop provided not only an opportunity to enhance production, but also reduces the burden of production loss due to floods. In recent years, with the expansion of new farm technology, particularly in the flood affected areas, the farmers have shown interest in growing summer rice. The area under summer paddy has increased by almost 60 per cent during the period 1991–92 to 2008–09 with a CGR of 6.19 per cent per annum. Area under autumn rice declined slightly showing a CGR of −3.20 per cent during the same period. Winter rice has been still maintaining its predominant position due to higher demand notwithstanding its lower yield compared to summer rice. The slight decline in the area under winter rice can be attributed mainly to floods in some regions of the State, but the area under this crop has again increased in the subsequent flood free year.

OILSEED PRODUCTION IN ASSAM

Rape and mustard are the principal oilseeds grown in the State which occupy 8.00 per cent of the total cropped area. The area of this crop was 2.13 lakh hectares in 1980–81 which steadily increased to 2.94 lakh hectares in 1999–91, then declined to 2.45 lakh hectares in 2004–05, and further to 2.26 lakh hectares in 2008–09. Total production increased from 102.4 thousand metric tonnes (MT) in 1980–81 to 157.91 thousand MT in 1990–91, then it declined to 129 thousand MT in 2004–05, and 123 thousand MT in 2008–09. It indicates that area and productivity of rape and mustard had not increased to the desired level. It is significantly lower than the agriculturally advanced states like Punjab, Gujarat, and Haryana. This lower productivity is mainly due to non-adoption of improved method of cultivation, use of traditional seed varieties, low use of' chemical fertilizer and other soil nutrients. Moreover, method of cultivation remains traditional and no technological breakthrough has been achieved and hence productivity of this crop has not increased to the desired extent.

PULSES PRODUCTION IN ASSAM

Pulses are considered as an indispensable constituent of Indian diet. The major pulses grown in Assam are black gram, green gram, lentil, peas, beans, etc. Assam's share in India's pulses production is very negligible. The area coverage was only 2.97 per cent of the gross cropped area in the year 2008–09. The cropping pattern of the State reveals that the farmers usually pay top priority in growing food crops and pulses and other crops are grown in small patches. In Assam, pulses are generally grown in rabi season. However, some varieties are grown in kharif season in certain districts, but the area under such varieties is very small. Nearly 94 per cent of the total cropped area of pulses comes under rabi pulses.

In order to assess the potential and prospects of rabi crops cultivation in Assam, the present study was undertaken with the following objectives:

1. to analyse the area, production and productivity of rabi crops in the district of Nagaon vis-à-vis Assam;

2. to study the economics of production of rabi crops in the sample farmers;
3. to study the factors affecting productivity of rabi crops; and
4. to find out the constraints of production of rabi crops and suggest policy measures.

METHODOLOGY AND DATA

The study was undertaken in Nagaon district of Assam. The district was purposively selected as most of the rabi crops are grown extensively in the district. Among the rabi crops; summer rice, pulses, and oilseeds were covered under the study. From the district, two community development blocks that is, Kathiatoli and Batadrawa were selected for the study in consultation with the district agriculture office. From each block, three villages were selected for field investigation where rabi crops were extensively grown by the farmers. Accordingly, from Kathiatoli block, Chang Chaki, Uttar Chang Chaki, and Charia Hagi villages and from Batadrawa block; Solouguri, Athgaon Chapori, and Kandhulimari villages were selected for data collection. Then from each selected village, 20 farmers of different farm size groups were selected by following random sampling method. Thus a total of 120 farmer households were covered by the study.

Reference Period

This study relates to the year 2009–10. The study was based on both primary and secondary level data. The relevant primary data were collected with the help of a set of specially designed schedules and questionnaire. The data were collected from the sample households through personal interview method.

The required secondary-level time series data of area, production, and productivity of rice, pulses, and oilseeds were collected from the publications of the Directorate of Economics and Statistics, Khanapara, Guwahati. Moreover; relevant secondary level data of various published and unpublished sources were also used in the study.

RESULTS AND DISCUSSIONS

Analysis of primary level data, generated from the study area yielded the following major findings:

The sample households comprised a total population of 813 persons with 422 males and 391 females. The average family size was 6.77 persons, which was comparatively higher than the State average of 4.87 according to 2011 Census. The sex ratio of population was worked out at 926 females per thousand males which was slightly lower than the State average of 954 in 2011 Census.

In the sample, 43.17 per cent population were literate up to Primary level, 25.09 per cent up to high school level, 10.21 per cent H.S.L.C. passed, 4.18 per cent H.S. passed, 1.85 per cent Graduate, 0.49 per cent had education up to Post Graduate, and above level and remaining 15.01 per cent population had Technical Education.

Of the total population, 66.79 per cent were earner or worker, 12.42 per cent helper, and the remaining 20.79 per cent were dependent or non-worker. Of the total working population, 21.16 per cent were primarily engaged in cultivation comprising of 70.35 per cent males and 29.65 per cent females.

Again, of the total 120 households, there were 40.83 per cent marginal, 10.83 per cent small, 20.00 per cent medium, and 8.34 per cent large farmers. The total operational holding of the sample farmers was 224.05 ha, comprising 93.75 per cent own land, 2.95 per cent leased in land and only 1.59 per cent mortgaged land. Of the total operational holdings, 73.10 per cent were found to be irrigated and only 26.90 per cent were unirrigated. The average size of operational holding was found at 1.90 hectares which was considerably higher than the state average of 1.15 ha recorded in the year 2005–06.

LAND UTILIZATION PATTERN

Out of total cultivated land, 95.39 per cent of land holdings were allocated to field crops and only 4.16 per cent were under horticulture crop cultivation. Of the total uncultivable holding of 23.77 ha, 46.57 per cent were under homestead, 11.44 per cent fallow land, and remaining 41.99 per cent were allocated under miscellaneous tree crops and grooves. Total land possessed by the sample farmers in the size group of below 1.00 hectare was 9.47 per cent, size group of 1.00 to 2.00 ha was 30.10 per cent, size group between 2.00 and 4.00 ha was 32.45 per cent, and 27.97 per cent land was possessed by the size group of 4.00 and above ha.

CROPPING PATTERN

It was seen that rice was the dominant crop in the cropping system which occupied more than 55.50 per cent of the total cropped area. In the sample district of Nagaon, sali paddy covered more than 69.77 per cent and summer paddy occupied 20.67 per cent of total area under rice. Total pulses and rapeseed and mustard, occupied 2.63 per cent and 5.21 per cent of the gross cropped area, respectively. Assured irrigation through STW had encouraged the farmers to grow boro paddy and vegetables during pre-kharif and summer seasons.

With the introduction of HYV seeds, the farmers' attitudes towards acceptance of new technology package steadily started changing. The area under HYV rice stood at 1,16,803 hectares which was about 33.21 per cent of the total cropped area. Although the district was traditionally a mono cropped district, with the adoption of modern technology as well as with the provision of irrigation facilities in limited areas, the farmers had started growing at least two crops in a year which helped in increasing the cropping intensity to 150 per cent in the year 2007–08.

PRODUCTIVITY OF CROPS

Under irrigated conditions, the productivity of HYV autumn paddy varied from 3,369 kg/ha to 3,483 kg/ha with an overall average yield of 3,430 kg/ha. The productivity of local autumn paddy under irrigated condition varied from 3,252 kg/ha to 3,258 kg/ha and the overall average yield was recorded at 3,253 kg/ha. The yield rate of winter paddy, both HYV and local, under rainfed condition was comparatively lower.

The yield of HYV sali paddy in irrigated condition varied between 3,309 kg/ha to 3,658 kg/ha and the overall average was worked out at 3,460 kg/ha. The overall yield of local sali paddy in irrigated condition was 3,271 kg/ha and in unirrigated situation, it was 3,102 kg/ha. The sample farmers reported to have used irrigation water occasionally, only when there were drought spell in monsoon season.

The summer paddy recorded a remarkable production in irrigated holdings. The productivity of HYV boro paddy in irrigated condition varied from 5,976 kg//ha to 6,145 kg/ha. The yield of local summer paddy under irrigated condition was also quite high; the aggregate yield rate was found at 4,548 kg/ha.

The yield rates of all crops were found to be higher in irrigated conditions than the un-irrigated conditions. So far as yield of the other crops grown by the sample farmers were concerned, mustard, vegetables, potato, pulses, etc., were found to be quite encouraging.

CROPPING INTENSITY

The cropping intensity in the sample irrigated holdings varied between 174 per cent to 187 per cent and in unirrigated holdings it varied from 153 per cent to 159 per cent among the different farm size groups. The overall cropping intensity was worked out at 178 per cent for irrigated holdings and 155 per cent in unirrigated holdings. Cropping intensity under both the conditions were highest in marginal group followed by medium holdings.

INCOME FROM CROP PRODUCTION

The total income from the rabi crops (summer rice, pulses and rape and mustard) cultivated by the sample farmers were worked out at Rs 5,890,147.50. Taking the value of the main products at farm harvest prices (as per prevailing market rate), the per hectare income from crop production of the sample farmers found to have varied from Rs 30,303.48 to Rs 40,350.33 and overall per hectare income was worked out at Rs 37,180.58. Per hectare income from summer paddy was higher than that of pulses and rapeseed and mustard. The per household income from rabi crops cultivation was found at Rs 8,384.92 against marginal farmers, Rs 45,126.35 against small farmers, Rs 84,898.10 against the medium farmers and Rs 1,77,380.10 against the large farm size groups with an average household income of Rs 49,084.56.

THE BENEFIT–COST ANALYSIS

The benefit–cost ratios (BCRs) were worked out by taking in to account the value of main product at the prevailing market price for different varieties of crops and the total operational costs.

The per hectare BCR of the local summer paddy cultivation varied between 1.47:1 and 1.34:1 with an overall average of 1.40:1 and in HYV summer paddy cultivation, it varied from 1.64:1 to 1.54:1 and the overall BCR were found at 1.57:1.

Analysis of BCR in pulses cultivation under irrigated condition showed that the farmers in all the farm size groups enjoyed positive BCR which varied between 1.46:1 and 1.23:1 and the overall average was 1.30:1. The per-hectare BCR for the sample farms in pulses cultivation in rainfed condition in aggregate was found at 1.3:1.

The BCR of rapeseed and mustard cultivation under irrigated and unirrigated condition were worked out and in aggregate it was found at 1.38:1 for both the condition. Table 12.1 shows the per-hectare BCR of the sample farmers in rabi crops cultivation. In the study area, it was observed that a major section of the farmers had opted for HYV even under rainfed condition and applied chemical fertilizers and adopted pest and disease control measures as and when required.

Table 12.1 Benefit–Cost Ratio of the Sample farmers in Summer Paddy, Pulses, and Rape and Mustard Cultivation

Farm Size	Local Summer Rice (Irrigated)	HYV Summer Rice (UnIrrigated)	Pulse (Irrigated)	Pulse (UnIrrigated)	Rape and Mustard (Irrigated)	Rape and Mustard (UnIrrigated)
Below 1.00 ha	1.47:1	1.64:1	1.46:1	1:39:1	1.57:1	1.48:1
1.00 – 2.00 ha	1.43:1	1.60:1	1.36:1	1.28:1	1.45:1	1.36:1
2.00 – 4.00 ha	1.38:1	1.55:1	1.29:1	0	1.39:1	1.32:1
4 ha and above	1.34:1	1.54:1	1.23:1	0	1.27:1	1.26:1
Overall	1.40:1	1.57:1	1.30:1	1.33:1	1.38:1	1.38:1

FACTORS AFFECTING PRODUCTIVITY OF RABI CROPS

The new agricultural technology such as use of machinery, application of required doses of fertilizer, HYV seed, plant protection measures, water management, and intercultural operations, etc., are the factors which effect the production of crops.

It was tried to examine how different inputs affect the productivity of summer paddy, rapeseed and mustard, and pulses in the sample households (Table 12.2). Accordingly, for summer paddy cultivation, no inputs were found to have significant effect on production. For pulses cultivation, fertilizer and irrigation were found to be significant at 5 per cent and 1 per cent level, respectively. Other variable inputs such as bullock labour, machine labour, and plant protection had a positive but insignificant effect on productivity. In mustard cultivation, fertilizer and irrigation were found significant at 5 per cent and 1 per cent level, respectively. The other variables like human labour, bullock labour, and machine labour had positive but insignificant impact on productivity.

POLICY IMPLICATIONS

Based on the findings of the study and the problems identified at the grassroots level, following policy measures are suggested for increased production and productivity of rabi crops:

1. **Consolidation of Holding:** Due to scattered holdings, full potential of irrigation facility could not be utilized by the sample farmers. The consolidation of holdings would go a long way in increasing the efficiency of' production.
2. **Supply of Institutional Credit:** Lack of institutional credit for investment on modern inputs is one of the important problems faced by the farmers for adoption of scientific package of practices.

Table 12.2 Factors Affecting Productivity of Rabi Crops

Variables	Summer Paddy		Pulses		Rapeseed and Mustard	
	Marginal Product	t value	Marginal Product	t value	Marginal Product	t value
Constant	34877	4.0901	13641	12.51993	15924	8.4688
Human Lab.	−27.64631	−0.6843	−5.34651	−0.42674	−28.64479	−1.6624
Bullock Lab.	0.17687	0.2417	0.11111	0.30019	−0.32421	−0.5296
Machine Lab.	3.19287	1.2228	1.15151	1.30163	−0.74729	−0.9116
Seed	2.65833	0.9280	−0.51007	−0.39476	2.39348	0.2262
Fertiliser	1.93294	0.7146	−3.45217	−2.41857**	−3.34432	−2.2975**
Plant protection	4.67057	1.2173	3.77144	1.36769	2.52691	0.9125
Irrigation	0.78531	0.3400	7.68663	11.06305*	7.59369	10.1154*
R^2	0.03440		0.56016		0.52745	
F-Value	0.56997		20.37668		17.85916	
No. of observation	120		120		120	

Note: ** indicates 5 per cent and * indicates 1 per cent level of significance.

Institutional credit in the form of crop loans should be provided in easy terms so that the farmers are not compelled to take loans from the private traders. The government may enforce the banking institutions with required regulations in this regard.

3. **Strengthening of Agricultural Extension and Farmer—Linked Support Services:** Due to poor and inadequate extension services and farmer-linked support services, the farmers were not fully aware of the government's programmes and policies for agricultural development. Therefore, frequent training and visit programme of extension officials particularly, the subject matter specialist can help the farmers in increasing the production and income, thereby improving their economic condition.
4. **Creation of Single Window Input Delivery System:** Inputs like certified seed, fertilizer, manure, plant protection chemicals, planting materials, etc., should be made available to the farmers at their doorstep at right point of time. The single window system of input delivery may help the farmers to get their inputs within a short time at a reasonable price.
5. **Supply of HYV Seeds within the Easy Reach of the Farmers:** Good quality certified seeds should be made available to the farmers at reasonable prices at the time of need. The Department of Agriculture should popularize short duration crop varieties of all the crops together with judicious application of fertilizers particularly, phosphatic fertilizers in pulses cultivation.
6. **Soil Testing Services:** Soil testing services should be made available in the study area. The numbers of soil testing laboratories should be increased to cater the needs of the farmers so that every farmer can test their soil for better harvest.
7. **Improvement of Marketing System:** There is need to strengthen the regulated market systems which can eliminate unhealthy practices and ensure fair price to the producer. The state government must take effective measures for regulating the marketing of agricultural produce in general and oilseeds and pulses in particular.
8. **Improvement of Post-harvest Technology:** The state government should take initiatives for development of rural godowns, storage facilities at market places, establishment of quality processing mills and for creation of infrastructural facilities at strategic locations.
9. **Development of Rural Roads and Transport Facilities:** It is of utmost necessity to develop the road communication system in order to facilitate the transportation of marketable produce to the places of assembling and marketing.
10. **Field Demonstrations and Adaptive Trials at Farmer's Field:** Field demonstration and application of scientific methodologies at farmer's field on improved technologies are considered essential to provide technological support to the farmers. Practical field trial and demonstration on intensive cropping with high yield variety (HYV) seed, application of soil nutrients, prophylactic measures, water management, etc., would encourage the farmers to take up crop cultivation more seriously. For introduction and popularization of new crop varieties, result oriented field demonstration may be encouraged.

CONCLUSION

The study highlighted that rabi crops have enormous potential in the study area despite a number of constraints being faced by the farmers. The officials in the State Agriculture Department and the scientists of' Agricultural Universities are to work in tandem to ensure that the fruits of technology reach the farmers at the grassroots.

In order to achieve the desired level of productivity of rabi crops, the Government of Assam must come up in an effective way in creating basic infrastructural facilities and in coordinating with related departments. A selective 'area approach' has been considered more effective to consolidate the situation and to boost up the production of rabi crops. In chronically flood affected areas, special programmes should be taken up for oilseed, pulses, and summer rice cultivation in rabi seasons. In view of the situation, it is necessary for the State Government to make concerted effort to bring all the potential areas under rabi cultivation with suitable adjustment of cropping sequence to attain self-sufficiency in foodgrain production. And in doing so, it is desirable to involve the farmers in the decision making for successful implementation of agriculture development programmes.

REFERENCES

Agro-Economic Research Centre for N.E. India. 1988. 'Prospects of Changing Cropping pattern in Favour of Oilseeds and Pulses Production in Assam'.

Agro-Economic Research Centre for N.E. India. 1998. 'Economics of Pulses Production and Identification of Constraints in Raising Their Production in Assam'.

Basavaraja, H. Mahajanashetti, and P. Sivanagaraju. 2008. 'Technological Change in Paddy Production: A Comparative Analysis of Traditional and SRI Methods of Cultivation'. *Indian Journal of Agricultural Economics* 63(4): 629–40.

Borah, B.C. 2007. 'Strategies for Agricultural Development in North-East India; Challenges and Emerging Opportunities'. *Indian Journal of Agricultural Economics* 62(1):13–31. Indian Society of Agricultural Economics, Mumbai.

Despande, S.D. and G. Singh 2001. 'Long Term Storage Structures in Pulses National Symposium on Pulses for Sustainable Agriculture and Nutritional Security'. Indian Institute of Pulses Research, New Delhi.

Gogoi, M. 2008. 'Impact of Irrigation on Crop Production in Assam with Special Reference to Shallow Tube Well Irrigation'. Unpublished PhD Thesis.

Government of India. 2000. 'National Agriculture Policy'. Ministry of Agriculture, p. 1.

Government of Assam. 2009–10, 2010–11. 'Economic Survey of Assam' Directorate of Economics and Statistics, Government of Assam, Guwahati.

Government of Assam. 2010, 2011, 2012. 'Handbook of Agriculture'. Directorate of Economics and Statistics, Government of Assam, Guwahati.

Gupta, S.K. 2008. 'Pulse Production in India'. S.M. Jharwal (ed.), *Glimpses of Indian Agriculture*, Vol. 1, pp. 207–54.

Gurjur, Madan Lal and K.A. Varghese. 2005. 'Structural Changes Over Time in Cost of Cultivation of Major Rabi Crops in Rajasthan'. *Indian Journal of Agricultural Economics* 60(2): 249–63.

Mellor, Jhon W. 2002. *The Economics of Agricultural Development*. New York: Cornell University Press, 1967, p. 141.

Reddy, Amarender. 2005. 'Pulse Crops Production Technology, Problems and Opportunities'. Agricultural Situation in India, Directorate of Economics and Statistics, Department of Agriculture and Co-operation, Govt. of India, pp. 779–88

Sengupta, K. and N. Roy. 2003. *Economic Reforms and Agricultural Development in North-East India*. New Delhi: Mittal Publication.

59th Round Survey of NSSO. 2003. Ministry of Statistics and Programme Implementation (MOSPI), Government of India. 'State of Indian Farmers—A Millennium Study', Volume-V. Ministry of Agriculture, Government of India, 2004.

Saikia, T.N. 1989. 'Prospects of Oilseed cultivation in Assam.' P.C. Goswami (ed.) *Agriculture Development in Assam*, p. 111.

Talukdar, K.C. and B.B. Deka. 2005. 'Cultivation of Summer Rice in the Flood Plain of Assam–An Assessment of Economic Potential on Marginal and Small Farms'. *Agricultural Economics Research Review* 18: 21–38. Agricultural Economics Research Association, Pusa, New Delhi.

13

National Horticulture Mission Scheme Impact Evaluation

Parmod Kumar

Diversification in consumption pattern is observed in India. The economy is moving from being a supply-driven economy to a demand-driven economy. With the change in consumption pattern, diversification of the production basket is also evident. Total area and production of fruits and vegetables in India have gained considerable momentum in the last two decades. The area under fruits increased from 2.8 million hectares in 1991–92 to 4 million hectares in 2001–02 that further increased to more than 6 million hectares by the end of the 2009–10. Area under vegetables increased from 6.2 million hectares in 2001–02 to 8 million hectares in 2009–10. Area under floriculture and aromatic plants has also experienced increase although area under such crops before the beginning of 2000s was almost negligible. Area under floriculture increased from 1 lakh hectares in 2001–02 to 1.8 lakh hectares in 2009–10. Similarly, area under aromatic and medicinal plants increased from 1.3 lakh hectares in 2004–05 to 5 lakh hectares in 2009–10. At the aggregate, area under horticultural crops increased from 12.8 million hectares in 1991–92 to 16.6 million hectares in 2001–02 and further to 20.9 million hectares in 2009–10. Production of horticultural crops increased from 96.6 million tonnes in 1991–92 to 146 million tonnes in 2001–02 and further to 223 million tonnes in 2009–10.

In the mid-1980s, the horticulture sector was identified by the Government of India as a promising emerging sector for agricultural diversification to enhance profitability through efficient land use, optimum utilization of natural resources and creating employment for rural masses. In the period 1948 to 1980, the main focus of the country was on cereals. During 1980 to 1992 there was consolidation of institutional support and a beginning was made for the development of horticulture sector. In the post-1993 period a focused attention was given to horticulture development through an enhancement of plan allocation and knowledge-based technology. This decade had been called a 'golden revolution' in horticultural production. The productivity of horticultural crops increased marginally from 7.6 tonnes per hectare in 1991–92 to 8.8 tonnes per hectare in 2001–02 while area under horticultural crops increased from 12.8 million hectares in 1991–92 to 16.6 million hectares in 2001–02.

The Indian horticulture sector is facing severe constraints like low crop productivity, limited irrigation facilities, and underdeveloped infrastructure support like cold storages, markets, roads, transportation facilities, etc., add to the problem. Heavy post-harvest and handling losses also result in low productivity per unit area and high cost of production. On the other hand, India's extended growing season, diverse soil, and climatic conditions provide sufficient opportunity to grow a variety of horticulture crops. Thus, efforts are needed in the direction to capitalize on strengths and remove constraints to meet the goal of moving towards rapid horticultural growth in India. The foreign trade policy in 2004–09 emphasized the need to boost agricultural exports, growth, and promotion of exports of horticultural products. Horticulture contributes nearly 28 per cent of GDP in agriculture and 54 per cent of export share in agriculture.

During the Tenth Five Year Plan, several schemes were launched to promote the horticulture sector in the country. The National Horticulture Mission (NHM) was initiated in 2005–06 by the Government of India, covering 340 districts and the Technology Mission for Integrated Development of Horticulture in North Eastern states including Sikkim, Jammu & Kashmir, Himachal Pradesh, and Uttarakhand. The main objective of this mission is to promote integrated development in horticulture, to help in coordinating, stimulating and sustaining the production and processing of fruits and vegetables and to establish a sound infrastructure in the field of production, processing, and marketing with a focus on post-harvest management to reduce losses.

The NHM was a Centrally sponsored scheme in which Government of India provided 100 per cent assistance to the State Mission during the Tenth Plan. During the Eleventh Plan, the assistance from Government of India (GoI) was 85 per cent with 15 percent contribution by the state governments. The main objective of the Scheme was to develop horticulture to the maximum potential available in the states and to augment production of all horticultural products including fruits, vegetables, flowers, plantation crops, spices, and medicinal and aromatic plants. The Scheme has been approved 'in principle' for implementation up to the end of Eleventh Five Year Plan. For implementation of the NHM programme in 18 states, an amount of Rs 630 crore was provided during 2005–06, an amount of Rs 1000 crore during 2006–07, Rs 1150 crore during 2007–08, while the outlay for the scheme during 2008–09 was Rs 1100 crore. Presently, the Scheme is being implemented in 18 States and 2 UTs covering 344 districts of the country for the development of potential crops. The pattern of assistance was 100 per cent to the state governments during the Tenth Plan. With effect from the Eleventh Plan (2007–08), the state governments are contributing 15 per cent of the share.

The activities that can be covered by co-operatives are post-harvest management; cold storages; pack houses; refrigerator vans and mobile processing units. The Mission provides support for processing activities and marketing activities, like wholesale market; rural markets/Apni-mandis/direct markets; functional infrastructure for collection, grading, etc; extension, quality awareness and market led extension activities for fresh and processed products. The scheme was in operation effectively for about four years up to 2009. It was felt necessary to analyse the impact of the programme vis-à-vis objectives of the NHM scheme especially for the major focused crops in terms of area expansion, increase in production and productivity. Therefore, this study was carried out for impact evaluation in different NHM states through the AERCs under the overall leadership of the Agricultural Development and Rural Transformation Centre (ADRTC) Bangalore and the consolidation of the report was done by the ADRTC. The study focuses on the following objectives:

i. assess the impact in terms of increase in area, production, and productivity of identified horticultural crops covered under NHM, keeping 2004–05 as the base year in the state in general and for the identified crops/districts in particular;
ii. assess the extent to which the scheme has helped in creating employment opportunities and enhancement of income of the farmers; and
iii. suggest measures in improving the implementation strategy.

METHODOLOGY AND DATA

The study is based on both primary and secondary data. The secondary information has been collected from the National Horticulture Board (NHB) and the Horticultural Department in the states as well as Directorate of Economics and Statistics, GoI. The primary data was collected through intensive sample survey of the selected farmers through structured questionnaire by visits and personal interviews. Main reliance of the study has been on primary data. Following the directions from the Ministry of Agriculture and Farmers Welfare, two crops namely mango and sweet orange were selected in Andhra Pradesh; mango and litchi were selected in Bihar; kinnow, guava, anola, and garlic were selected in Haryana; grapes, pomegranates, flowers, and aromatic/medicinal crops were selected in Karnataka; anola, papaya, coriander, and mango were selected in Rajasthan; banana, mango, and flowers were selected in Tamil Nadu and pineapple and mandarin oranges were selected in West Bengal. For these selected crops, 200 households each were surveyed in Andhra Pradesh and Rajasthan; 150 households were surveyed in Haryana and Tamil Nadu each; 100 households were surveyed in West Bengal and Bihar each; and 212 households were surveyed in Karnataka.

For the selection of districts, villages, and households the following procedure was adopted.

Specified districts as instructed by the Ministry of Agriculture and Farmers Welfare, GoI, were selected for the detailed primary survey in each state. From each districts, two villages were selected, keeping into account the cropping pattern in each of these districts taking one village near the periphery of district headquarters or accessible mandi/ market and one village from a distant place to realize the effect of distance factor in the findings. From each selected village, 25 beneficiary farmers were selected using random sampling method for detailed household survey. In this way, a total number of 50 beneficiary farmers were surveyed in details in each selected district. While selecting the sample care was taken to represent all the section of the society such as small and marginal farmers, SC/ST farmers, and women folk, so that outreach of the programme to these sections is also reflected in the study. Members of Growers Associations, Gram Pradhan, Block, and District level concerned functionaries were also contacted or interviewed. The reference period for the selected sample was 2008–09 while survey was carried out during the year 2009–10. The study covers the implementation of the NHM programme from 2005–06 to 2008–09.

RESULTS AND DISCUSSION

Growth in Area, Production, and Productivity: Based on Secondary Data

This section presents the details of selected crops in seven selected states during the historical period and their growth rate for the historical period and pre-NHM (2000–01 to 2004–05) and post-NHM (2004–05 to 2008–09) period.[1] In Karnataka horticultural crops area expanded by significant rate of around 3 per cent per annum during the 1980s and slightly less than 3 per cent per annum in the 1990s. However, against the impressive growth in area under horticultural crops in Karnataka in the decades of 1980s and 1990s, the yield growth rate of the same was either insignificant or it was even negative in most of the crops. Overall, yield of horticultural crops increased by less than one per cent in the 1980s but no increase was realized in the 1990s. In the 2000s, there was some reversal in area and yield growth rates. Area under horticultural crops in Karnataka grew at an annual growth rate of 1.6 per cent in the 2000s while yield increased at 1.9 per cent whereby yield growth was not significant. At the aggregate, NHM seems to have overall positive effect on the horticultural crops as area growth increased from 0.8 per cent per annum during the before NHM period (2000–04) to 2.7 per cent per annum during the post-NHM period (2005–08). Similarly, yield growth rate that was negative during the before NHM period (–4.1 per cent per annum), observed a very high growth rate of 6.5 per cent per annum during the post-NHM period. Among the selected crops, although effect of NHM was visible on the area expansion of selected horticultural crops, however, its effect on the yield rate was yet to be seen.

In Andhra Pradesh, area under horticultural crops underwent a significant negative rate (–2.5) per cent per annum, while the yield recorded a positive growth (14.6) per cent per annum during the decade of 1980s. During the 1990s, significant positive growth was reported in area and yield of fruits, vegetable crops, and the total horticultural crops. In the 2000s, fruits, vegetables, and commercial flowers reported a positive and significant growth in area and yield. At the overall, area under all horticultural crops increased at a rate of 3.7 per cent per annum in the pre-NHM period that increased to 4.3 per cent per annum in the post-NHM period while yield growth rate increased from 1.7 per annum in the pre-NHM to 7.4 per cent per annum in post-NHM period. In the case of selected crops, both mango and sweet orange observed rapid growth in area during both pre and post-NHM periods, while area growth was slightly less in the post-NHM period compared to pre-NHM period. On the other hand, the yield growth rate was negative in the pre-NHM period for both the crops that turned positive for mango while remained negative for the sweet orange in the post-NHM period. Nonetheless, the NHM did not appear to have helped the selected crops in expanding its area or increasing the yield level of these two selected crops in Andhra Pradesh.

A detailed analysis of percentage growth in area and yield of different horticultural crops in Bihar revealed that during the 1990s fruits area increased by less than 2 per cent per annum while vegetables area decelerated significantly. The growth in yield was negligible in fruits and negative in vegetables in Bihar. The expansion in area and the rise in the yield levels of horticultural crops in the 2000s may be indicative of the positive contribution of NHM in Bihar. The fruits area declined in the post-NHM period but its yield rate increased at a rapid rate. In all other crops, namely, vegetables, spices, and flowers, not only area expanded in the post-NHM period but also

[1] For working out time growth rate following semi-logarithm trend equation was used:

Logn Y = a + b × Time; where b represents trend growth rate and its significance was checked using 't' test. For the pre and post-NHM growth rate comparisons, given the less degree of freedom instead year on year growth rate was used.

their yield increased indicating the positive effect of NHM programme in Bihar. In the case of selected crops of mango and litchi in Bihar, during the pre-NHM period there was almost no growth in the area as well as yield of mango crop while its yield rate increased by almost 13.5 per cent per annum during the post-NHM period. In the case of litchi, there was 7 per cent per annum growth in area in the pre-NHM period while yield growth rate was negative during the same period. In the post-NHM period, its area as well as yield grew by around 1.5 to 2 per cent per annum. Thus, it is difficult to conclude that NHM really helped in expansion of area of the selected crops while there is some conclusive evidence of rise in the yield rate of the selected crops during the post-NHM period.

In Rajasthan area under horticultural crops grew at an annual rate of 4.5 per cent per annum while yield rate increased by 11 per cent per annum during the 1980s. Growth in area further increased to 4.63 per cent per annum during the 1990s while yield growth decelerated to almost zero during the same time period. The growth rate of area decreased to 0.64 per cent per annum whereas yield rate increased by 7.5 per cent per annum during the 2000s. The area under horticultural crops was almost stagnant in the pre-NHM period whereby it grew by only 0.5 per cent per annum, the area increased at a rapid pace of above 5 per cent per annum in the post-NHM period. However, yield rate of horticultural crops increased by around 5 per cent per annum before NHM period (2000 to 2004) while it decelerated to 1 per cent per annum in the post-NHM period. The growth in both area and yield of selected horticultural crops was not found to be steady. There were fluctuations in both area and yield of selected crops. Dividing the period between pre and post, area under amla decelerated from 23 per cent growth rate in the pre to 15 per cent in the post-harvest period. Area under papaya accelerated from negative growth of –2 per cent per annum to positive growth of 4 per cent per annum. Area under coriander increased from almost zero percent to more than 10 per cent per annum. Area growth under mango remained negative during both the periods. Yield rate improved for two crops namely amla and mango while papaya yield decelerated and coriander yield growth remained negative during both the period. In conclusion, impact of NHM on selected crops in Rajasthan was inconclusive both in area as well as yield level.

The growth rate of area and yield of all horticultural crops in Tamil Nadu during the period of 1980s worked out to 2.5 per cent and 3.0 per cent, respectively, whereas there was a fall in the growth rate of area (1.9 per cent) and yield (1.8 per cent) during the period of 1990s. Again a significant growth rate of 2.5 per cent in area and 4.8 per cent in yield was observed during the 2000s. The growth rate of area and yield of all horticultural crops during the post-NHM period worked out to 3.47 per cent and 1.52 per cent, respectively. The growth rate, in area however, was negative in the pre-NHM period while growth in yield was positive 4 per cent during that period. So, in the latter period, there was substantial increase in area growth but there was decline in yield growth in the aggregate horticultural crops implying that the effect of NHM at the overall was inconclusive. Among selected crops, mango registered slight deceleration in the area as well as yield growth rate during the above two periods. On the contrary, acceleration in the growth rate of area as well as yield was observed in the case of flowers. Thus, in the selected crops in Tamil Nadu, visibly there was some positive impact of NHM, on both area and yield rate in the case of flowers but not in mango crop.

In West Bengal, the area and yield of aggregate horticultural crops grew at an impressive growth rate of above 12 per cent per annum during the 1980s. During the next 1990s, while yield growth rate was sustained at the same rate of 12 per cent per annum, area growth came down to less than 5 per cent per annum. In the next decade, while the trend rate of growth of area stood at 6.6 per cent per annum (2000–01 to 2009–10), the yield growth was once again sustained at 11.1 per cent per annum. At the aggregate, area under horticultural crops grew at a rate of 3 percent per annum in the pre-NHM period that accelerated to 10 per cent per annum in the post-NHM period. Similarly, yield growth of all horticultural crops jumped from 3 per cent per annum in the pre-NHM period to 20 per cent per annum in the post-NHM period. It comes out that horticulture in West Bengal has exhibited an impressive growth in both area as well as yield. A detailed analysis of growth rate in area and yield rate of pineapple and mandarin oranges in the state showed that area under pineapple grew at around at 5 per cent per annum before NHM period that declined to –6 per cent per annum in the post-NHM period. Its yield increased from almost zero growth rate in the pre-NHM to 2 per cent per annum in the post-NHM period. Similarly, yield and area under mandarin orange also increased from zero per cent in the pre-NHM period to around 2 per cent per annum in the post-NHM period. This confirms that the growth in area and production of pineapple and mandarin orange lagged far behind the growth observed for aggregate horticulture, especially of fruits in West Bengal.

In Haryana, horticultural crops occupied only minor importance as the share in all India area of fruits and vegetables taken together was only 1.28 per cent during 1996–97 which reached to 1.98 per cent during 2004–05.

The area expansion appeared to be commendable by indicating a growth rate of 9.5 per cent per year during the above-mentioned period. Further details of area of horticultural crops in Haryana were not available.

Household Characteristics, Cropping Pattern and Production Structure of Sample Households

After glancing through the growth rates based on secondary data, in this section we analyse the primary survey data collected though household survey. Among our selected sample in all the seven states, the household size varied between 4.2 members to 7.5 members while number of earners varied between 1.6 and 3.2. On the average, two-thirds of the members of a family were in the working age and rest, were children below 16 and senior citizens (above 60 years). More than one third to one-half of the household members were illiterate or literate up to the primary level, while around 1/3rd were educated up to secondary level. Very few members were educated up to graduate level or above. More than 80 per cent occupation of the selected households belonged to farming alone. The size of operational holdings averaged at around 7 acres among the selected farmers. Average cropping intensity was 152 and it was highest 189 in Haryana and lowest 112 in West Bengal. Among the sources of irrigation of selected households, electric tube well was the main source of irrigation, followed by canal and tanks. There were some instances of leasing-in and leasing-out land among the selected farmers. The terms of tenancy were mostly fixed rent in cash in Rajasthan, Tamil Nadu, West Bengal, and Haryana and rent in cash and kind both were found in Karnataka and Bihar. On average, credit per acre varied from Rs 810 in Bihar and Rs 23 thouand in Karnataka. Among different sources of credit, institutional credit constituted the major amount, around 75 per cent and non-institutional had a share of around 25 per cent. The purpose of credit was generally for agriculture and animal husbandry uses. Overall, per household assets were measured at Rs 2 lakh and their value ranged from Rs 0.37 lakh in Bihar, to Rs 3.69 lakh in Rajasthan.

The selected farmers were growing, paddy, wheat, coarse cereals, pulses, oilseeds sugarcane, cotton, and horticultural crops. The horticultural crops included fruits, vegetables, plantation, flowers, medicinal, and aromatic crops. Looking at the proportion of horticultural crops grown by the selected farmers, the highest percentage share of horticultural crops in the gross cropped area was in West Bengal (60 per cent), followed by Andhra Pradesh (42 per cent), Tamil Nadu (41 per cent), Karnataka (38 per cent), Haryana (30 per cent), Bihar (24 per cent), and Rajasthan (only 13 per cent). Share of large farmers in horticultural crops like fruits, vegetables, floriculture, medicinal, and aromatic crops, which were mainly more labour intensive crops, was less than that of small and marginal farmers who had comparatively higher share among these crops in comparison to traditional crops. Net household income from farm activities after subtracting the cost of production was measured at Rs 159 thousand in Tamil Nadu, followed by Rs 109 thousand in Haryana, Rs 93 thousand in Karnataka, Rs 87 thousand in Rajasthan, Rs 53 thousand in Andhra Pradesh, Rs 26 thousand in West Bengal, and Rs 20 thousand in Bihar. Adding farm and non-farm income, the aggregate household income was measured at between Rs 1.8 to 2 lakh in Rajasthan, Tamil Nadu, and Haryana, Rs 1.2 lakh in Karnataka, above Rs 50 thousand in Andhra Pradesh, and less than Rs 50 thousand in West Bengal and Bihar.

The Production Structure and Resource Use under Horticultural Crops

Per acre net profits (over total cost) of grapes was measured at Rs 26 thousand and net profit from pomegranate was measured at Rs 13.5 thousand per acre in Karnataka. The realized profit of floriculture crops in Karnataka averaged at Rs 15.7 thousand whereby only large and marginal farmers had positive profit amounting to Rs 27 thousand and Rs 17.7 thousand, respectively while small and medium farmers incurred losses amounting to Rs 5 thousand and Rs 3 thousand, respectively. The net profit over variable cost of medicinal and aromatic crops averaged at Rs 11 thousand and net profit over total cost averaged at Rs 4.6 thousand. In Andhra Pradesh both the crops namely mango and sweet orange were in the gestation period and therefore there was no revenue or income earned in the cultivation of these two crops. In the case of mango in Bihar, net returns (after subtracting fixed and variable cost) were measured at Rs 23247 per acre. In contrast to mango cultivation, the net return per acre from litchi cultivation turned out to Rs 9999. In Rajasthan, on average, about Rs 5277 was the net loss per acre of amla mainly due to crop being in gestation period as majority of our sample farmers had undertaken new plantation during the last three to four years. The per-acre net return from coriander was, on an average, Rs 5434. Like the case of amla, mango also did not reach fruit bearing stage, and therefore, the output and net returns were negligible. In Tamil Nadu, our analysis was restricted to three crops namely banana, mango, and flowers. The average revenue earned by a farmer from the cultivation of an acre of banana was to the tune of Rs 155833. The net returns earned by a sample farmer by cultivating mango were to the tune of Rs 2740 per acre. Net returns obtained by a sample household from the cultivation of flower crop in Tamil Nadu was Rs 81218 per acre. In the case of economics of pineapple production in West Bengal, net returns turned out to Rs 25406. Net return per acre from mandarin orange cultivation turned out to Rs 8189. The net returns from kinnow cultivation in Haryana were Rs 14327 per acre at the aggregate. Farmers reaped Rs 28339 per acre from guava during 2008–09. In the case of garlic, net return per acre was computed as Rs 40612 at the aggregate level.

One of the objectives of NHM was to create opportunities for employment generation for skilled and unskilled persons, especially unemployed youth in the villages. In Karnataka, horticultural crops had advantage over traditional food crops not only in terms of higher productivity, value addition, and net profitability but they also provided better employment to the households in comparison to all other crops grown in the field. There was possibility of further value addition and more employment generation in case farmers preferred to undertake some processing of the horticultural crops at the field level. In Andhra Pradesh, among horticultural crops vegetable crop namely chilies employed more labour than tomato and onion crop. In the case of selected crops in the state, mango and sweet orange reported the same amount of labour employed for cultivation. In Bihar, the requirement of human labour in mango and litchi crops were 64 and 59, that were comparatively two to three times more than that in maize, wheat, lentil, and gram crops. It was 1.5 times that of paddy. In Rajasthan, horticultural crops were more labour intensive compared to non-horticultural crops as average man-days required for an acre of horticultural crops was higher than that of non-horticultural crops. In Tamil Nadu, the man days used for all the kharif crops by the sample households on an average worked out to 20.76 per acre. For horticultural crops, the sample farmers used 27.57 man days on average per acre. In West Bengal, total man days employed varied between 74 in paddy to 80 in millet and maize, 65 in chilies, 54 in ginger, 48 in potato, 89 in turmeric, and 139 in pineapple. In Haryana among traditional crops, cotton, flowers, and moong generated around 20 days of employment while average use of labour in horticultural crops was around 29 man days per acre.

After harvesting and packing where do farmers market their produce is of interest. In the case of grapes in Karnataka, whole sale market, local market in the nearby town and intermediaries at the farm gate level were used by the selected farmers to dispose of their produce. Similar to grapes, marketed surplus of pomegranates was sold through government agency and through some sort of cooperative marketing and pre-arranged contract with the merchants. Half of the produce of flowers was sold though wholesale market and rest of the half was sold in local market, through government agency and through merchants and pre-arranged contracts. Aromatic and medicinal crops were sold through merchants, intermediaries at the farm gate level and through wholesale markets.

There was absence of regulated markets for fruits and vegetables in the selected districts in Andhra Pradesh. Sample farmers marketed vegetable crops namely onion, tomato, and chilies through local municipal marketing authorities, Rythu Bazars and wholesale regulated markets. In Bihar, mango output was marketed through intermediaries at the farm gate, merchant/traders through pre-arranged contracts and directly to the consumers at village and wholesale markets. Litchi was also mostly sold to merchant/traders on pre-arranged contracts. In Rajasthan, amla, papaya, coriander, and mango were sold through wholesale market, pre-arranged contracts and through intermediaries. The selected horticultural crops namely mango, banana, and flowers in Tamil Nadu were marketed through wholesale markets, local market and intermediaries at farm gate. The pineapple and orange produced by the sample farmers in West Bengal were marketed through intermediaries at the farm gate, wholesale market, and local market. In Haryana, the horticultural commodities were yet in the gestation period. Horticultural crops were mostly sold without indulging into any processing except the case of grapes in Bangalore where few farmers were found undertaking some processing at the field level.

IMPACT OF NHM ON THE EXPANSION OF HORTICULTURAL CROPS

The objective of NHM was to ensure an end-to-end holistic approach covering production, post-harvest management, processing, and marketing to assure appropriate returns to growers. In Karnataka, out of 1.03 acres per household sown area of grapes during 2008, around 0.37 acres obtained input quantity through NHM. In pomegranate, out of 0.60 acres per household, 0.16 acres were provided input support. Similarly, in flowers, out of 0.3 acre per household cultivated, less than 0.1 acres obtained input support. In the case of aromatic and medicinal plants, out of 0.3 acres per household planted, around 0.2 acres per household were provided input support. All the resources procured were provided by Department of Horticulture (DOH), Karnataka Government, advanced contract with the buyers and private nursery. Total area rejuvenated among all the selected households was 21 acres of grapes, 17.5 acres of pomegranate, and only 1.5 acres of flowers.

The few households who rejuvenated their plants indicated significant increase in productivity after rejuvenation. The sample farmers indicated that the best help provided by the NHM was in terms of providing them planting material like nursery or seedling. Among all activities, the highest numbers of households received subsidy for nursery and planting material followed by drip irrigation, plant protection, and maintenance activities. Although majority of the farmers were happy about the subsidy provision under the NHM but the amount of subsidy was a cause of concern for many of them. The infrastructure building especially that of post-harvest management as well as capacity building under NHM was found lacking although some attempts were made in that direction.

All the area under selected crops in Andhra Pradesh was in the gestation period and thereby no rejuvenation activity was observed among the selected farmers in the state. Regarding resource procurement, majority of respondents received planting material. Under the subsidy programme, 98 per cent sample farmers were benefitted under area expansion followed by 75 per cent Integrated Pest Management (IPM)/Integrated Nutrient Management (INM), 66 per cent under mechanization, and 4 per cent under organic farming units. Under the capacity building, training sessions were organized at village and nearby villages and highest number of training programmes were organized by non-governmental organizations. As large farmers received more assistance from NHM scheme, thereby a need was felt for strong government intervention to help marginal and small farmers opting horticultural crops.

In Bihar also no case of rejuvenation was found in the sample crops. As regards the NHM reaching to the households with resource provision, it was found that the majority of sample farmers were benefitted through various promotional activities undertaken under NHM. About 27 per cent farmers obtained good quality planting material like nursery through NHM. Nearly 26 per cent farmers indicated having obtained help in INM/IMP and around 25 per cent benefited through capacity building via training programme under NHM. There was complete lack of facilities such as tissue culture, mother stock block maintenance, fogging and sprinkler irrigation, and post-harvest management. The aggregate amount of subsidy was Rs 24345 per household. The percentage of subsidy to total investment was around 61 per cent.

No cases of rejuvenation was found in amla, papaya, and coriander in Rajasthan. Four farmers were found in the case of mango who were involved in rejuvenation activities through NHM. Only about 8 per cent sample farmers cultivating mango were supported for rejuvenation/protection. The average increase in productivity as a result of rejuvenation was 22 quintals per acre of mango. On resource procurement it was found that around 98 per cent farmers obtained good quality planting material like nursery through NHM. About 50 per cent farmers were found to use poly-house with ventilation, insect proof netting, fogging, and sprinkler irrigation. Not a single farmer was found to use the modernized post-harvest management system such as pack house, storage unit, mobile processing unit, etc. The planting material, fertilizer, pesticides, and other inputs and drip/sprinkler were the major items for which subsidy was provided to the beneficiary farmers. On capacity building, it was found that the training was provided to the sample farmers through various sources on an average of 1.62 times per household per year. Around half of the selected farmers suggested that processing facilities should be provided and necessary infrastructures should be developed in their villages or nearby villages to accelerate the adoption of horticultural crops.

The resource provision under NHM and rejuvenation activities in Tamil Nadu were observed among the selected crops. Majority of the sample respondents availed the promotional activities such as availability of good quality planting materials like nursery, rejuvenation with improved cultivators, integrated nutrient management or integrated pest management and so on. With regards to subsidy, it was provided for planting materials, fertilizers, pesticides, other inputs drip/sprinkler irrigation, vermi compost, and modern nursery. Training was imparted to the farmers under NHM. In West Bengal, positive impact of the NHM could be noticed in the case of area expansion through rejuvenation and protection resulting into increase in production and productivity of the selected crops. Nonetheless, for procurement of resources for pineapple and mandarin orange cultivation, informal sources like private nurseries and fellow farmers continued to play an important role. However, the survey revealed that the subsidy provision only formed a negligible amount of the total investment carried out by the sample farmers. Furthermore, very little was done in dissemination of technology through training and capacity building activities. At the same time, there was a complete absence of post-harvest management facilities like pack house, storage units, and mobile processing units in the study regions in West Bengal.

In Haryana farmers did not opt for rejuvenation due to low level of subsidy. Further, perceptions of farmers about employment generation and increase in household income through cultivation of kinnow, guava, amla, and garlic were positive. Subsidy provision was listed as the most important positive factor by 95 per cent farmers. However, response of the farmers' regarding infrastructure and capacity building was found poor.

POLICY SUGGESTIONS

For growth of horticultural crops in general and selected crops in particular, the following policy suggestions may help strengthening the implementation of NHM in the selected states.

The main objective of NHM is to increase area and yield rate of horticultural crops. In our analysis of historical data before and after implementation of NHM as well household survey data, a large degree of spatial and temporal variations was observed especially in the yield rate of different horticultural crops. One of the foremost reasons for yield fluctuations was erratic rainfall and periodic occurrence of droughts. Although, among our selected households, horticultural crops were grown more on the irrigated area, however, quality of irrigation was a big question as the latter was subject to fluctuating power supply in the villages. Thus, there is a clear need to increase irrigation facilities and erratic power supply for agriculture need to be corrected. Many of the farmers expressed need for more subsidy provision for putting up infrastructure for drip irrigation. Additional financial assistance is needed from the Central government to raise subsidy on drip irrigation that can prove much helpful not only increasing area under horticultural crops but also increasing productivity and reducing fluctuations in the yield rate of horticultural crops.

A majority of our respondents indicated that NHM has helped increasing the employment opportunities for farmers through expansion of area under horticultural crops. There was however, no special subsidy for farmers having shifted their area from the field crops to horticultural crops. Introducing specified subsidy for the area shifted from field to horticultural crops may lead to not only expansion in area under horticultural crops but also enhancement in employment opportunities as man-days employment is higher on horticultural crops compared to foodgrains and other commercial crops.

In our study area more than 90 per cent of the household beneficiaries of horticultural crops were not happy with infrastructure and capacity building activities under the NHM programme. Farmers need training for proper use of various inputs and performing best farm practices such as planting and pruning, INM, IPM, organic farming, use and maintenance of drip/sprinkler irrigation system, etc. Imparting training is needed for plant protection, rejuvenation and so on. Lack of proper capacity building at the farm level reflected in poor participation of the farmers in some of the activities underlined in NHM like upgrading existing tissue culture units, raising root stock seedling under net house conditions, soil sterilization, etc. The awareness camps for various components of NHM and procedures and norms for availing such subsidies by the farmers under NHM need to be arranged frequently and at the door steps of the farmers. Extensive publicity of NHM programme is required at the Gram Panchayat level.

There were large numbers of farmers who expressed their dissatisfaction regarding marketing facilities. For marketing their produce farmers depended in many cases on merchants and intermediaries who were exploitative in nature. Suitable wholesale and terminal markets with in-built cold chain and where-house facility for the sale of horticultural crops in general and flowers and aromatic crops in particular shall be opened in big cities and towns in the horticulture production belt. There is also a need for creation of chain of collection centres of farmers' produce in rural areas to feed the terminal/wholesale markets mentioned above.

Post harvest management was found completely lacking in the selected areas. Requirement of pack house and cold storage was expressed by most of the farmers given the perishable nature of horticultural crops. The selected horticultural crops were mostly sold by the farmers without indulging into any processing except the case of grapes where few farmers were found undertaking some processing at the field level. Lack of post-harvest handling and storage also lead to post harvest losses. Management of post-harvest losses, processing, transportation, storage, and marketing needs considerable attention. There is need for advancement of good post-harvest management practices like usage of cold chain, ripening chambers, and processing activities for more value addition in the horticultural produce. Creation of additional post-harvest management infrastructure under NHM is required. Value addition in floriculture, for example, cut flower and aromatic crops at the farm level need to be popularized to provide more employment to unemployed youth in the rural areas.

In addition to the above, other measures required for the expansion of horticultural area includes the following steps: more area needs to be brought under drip irrigation, rain water harvesting, and expansion of other micro irrigation systems need to be emphasized; establishment of community seed banks with identification of genotypes for specific agro climatic regions; INM and IPM practices and rejuvenation and replacement of senile plants need to be popularized whereby appropriate subsidy and capacity building should be provided to the farmers; vermi compost/bio-digester units need to be established for the promotion of organic farming; contact farming in horticultural crops need promotion as there is a lot of scope for the agri-business and corporate sector to enter in horticulture on a big way and therefore the relevant Act needs amendment favouring written and legal contract between the corporate and small farmers.

REFERENCES

Government of India. 2007. *Report of Working Group on Horticulture, Plantation Crops, and Organic Farming for the XI Five Year Plan (2007–12)*. New Delhi: Planning Commission, Government of India.

Government of India. 2010. 'NHM Operational Guidelines 2010', Ministry of Agriculture, New Delhi.

Government of Rajasthan. 2009. 'NHM Annual Action Plan 2009–10'. Rajasthan Horticulture Development Society, Directorate of Horticulture, Pant Krishi Bhawan, Jaipur.

Kumar, Parmod and Anil Sharma. 2006. 'Perennial Crop Supply Response Functions: The case of Indian Rubber, Tea and Coffee', *Indian Journal of Agricultural Economics* 61(4): 630–76.

ANNEXURE 13A

Table 13A The Details of Agro-Economic Research Centres and the Lead Person Involved in the State Report

State	AERC/Unit	Authors
Andhra Pradesh	AERC Vishakhapatnam	G. Gangadhara Rao
Bihar	AERC Bhagalpur	Basanta Kumar Jha
Gujarat	AERC Vallabh Vidyanagar	Mrutyunjay Swain
West Bengal	AERC Shantiniketan	Kali Sankar Chattopadhayay, Debashis Sarkar
Haryana	AERC Delhi	Usha Tuteja
Tamil Nadu	AERC Chennai	K. Jothi Sivagnanam
Rajasthan	AERC Vallabh Vidyanagar	Mrutyunjay Swain and R.H. Patel
Karnataka	ADRTC Bangalore	Parmod Kumar

14

Growth and Trends of Horticulture Crops in the North-East and the Himalayan States

Komol Singha

The horticulture sector has been one of the driving forces of overall agricultural development in India (Dev 2012; Mehta 2009). The sector covers a wide range of crops, namely, fruit crops, vegetable crops, tuber crops, ornamental crops, medicinal, aromatic crops, spices, and plantation crops, etc. (Government of India [GoI] 2001). With the growth of technology, modernization and changes in food habits of the people, the sector has undergone a major shift in the recent past, got a tremendous potential to push overall agriculture growth above the country's targeted 4 per cent level (Bahadur 2010). Also, the sector can directly or indirectly address poverty and food security issues in both urban and rural areas of the developing world (Abou-Hadid 2005). At present, the sector has, perhaps, become the most profitable venture of all farming activities, as it provides ample employment opportunities and scope to raise the income of the farming community in the country (Choudhary, Singha, and Vishnu 2013).

The area under the horticultural crops in the country has increased significantly from 12.77 million ha in 1991–92 to 21.83 million ha in 2010–11, accounting for a compound annual growth rate (CAGR) of 2.7 per cent in the last 20 years. During the same period, its production has increased from 96.6 million metric tonnes (MT) to 240.5 million MT, accounting for a CAGR of 4.7 per cent, and the yield rate has also increased from 7.55 MT/ha in 1991–92 to 11.03 MT/ha in 2010–11 (National Horticulture Board [NHB] 2011). In totality, at present, India produces 257.2 million tonnes of horticulture products from the area of 23 million ha (Directorate of Economics and Statistics [DES] 2012). Despite the sector's impressive growth, unlike other traditional crops, it has received very little attention from the policy makers, planners, and investors, till very recently, not only in the country, but also in the world (Singha, Choudhary, and Vishnu 2014).

Realizing these potentials, the GoI has started putting a thrust on the development of horticultural sector, after the 1960s, in order to exploit the country's vast potential horticulture resources and to generate the much needed value addition in the sector. Further, greater emphasis has been given on the sector since the Seventh Five Year Plan. As a result of which, presently, the country has registered itself as the second largest producer of fruits and vegetables in the world, next to China. Of the fruits, the country is the largest producer of mango, banana, coconut, cashew, papaya, and pomegranate, and also, the largest exporter of spices. Besides, the country ranks first in the productivity of grapes, banana, cassava, peas, and papaya (Singha, Choudhary, and Vishnu 2014).

With the growth of income and employment, the agriculture sector of the country has also diversified towards horticulture crops, to a great extent, in the recent past. According to Mittal (2007), the rationale behind the shift from cereals to fruits and vegetable has been the economic opportunity. Some scholars (Weinberger and Lumpkin 2007) found that the shift towards horticulture was due to institutional supports including infrastructure, while, others (Chand 1996) opined that environmental advantages enthused the farmers to diversify from traditional crops to high value cash crops of horticulture. The potential for diversification through high value cash crops like fruit, vegetables, spices, plantation crops, and medical plants, etc., seems to be more favourable in the hilly regions of North-eastern Region (NER) and Himalayan States (HS) of India where traditional crops have little growth potential. Also, there are number of indigenous horticulture crops in these States which have not been estimated and surveyed by any of the agencies in the country. To a great extent, these crops provide livelihood and sustainability to the local communities. Nevertheless, the estimation and survey of horticulture crops, unlike any other traditional crops, are not so simple especially in the NER and HS. It requires special attention and efforts. Therefore, before taking up any development initiatives for the sector, a thorough understanding of growth and structure of the same is very much needed and baseline survey of the horticulture crops can give clearer picture of the sector's development in the country.

The very arbitrary data[1] on horticulture crops in NER, available in the public domain, pose a serious problem in understanding sector's contribution to overall development of the larger agriculture sector. Besides, there is no systematic data on some of the marginal and minor horticulture crops of these states. To fill this gap, it is necessary to devise a methodology that can be followed while collecting horticulture data. It is also necessary to identify the problems that are being faced while collecting data of horticultural crops by various agencies and take some remedial measures to make data on horticulture sector more scientific and factual. Therefore, the present study is the modest attempt in this very direction with a special reference to NER and HS. The specific objectives of the study are:

1. to compare and contrast the growth and pattern of major agriculture and horticulture crops in NER and HS vis-à-vis the all-India level;
2. to identify the horticultural crops which can provide livelihood to the local people to a great extent but compilation or survey of such crops has not been initiated by any of the organization; and
3. to identity the problems that are being encountered by the grassroots-level officials while collecting the horticultural data.

DATA SOURCES AND METHODOLOGY OF THE STUDY

The present study mainly relies on the primary field data. Four states have been selected from the two regions—NER and HS. Based on the agro-climatic conditions, Assam and Sikkim have been selected from the NER, Shimla and Uttarakhand from the HS. To supplement the primary data, secondary data were also employed, and they were collected from the Department of Horticulture (Manual on Horticultural Statistics), NHB, Planning Commission, and DES of the respective States. Growth rate of area, production, and productivity of horticulture crops were estimated by employing log-log regression model and further to test acceleration or deceleration of growth trends of area, production, and yield of horticulture crops (structural break), a modified exponential growth model was used. Additive and multiplicative models were also applied to identify the major factor(s) that enhanced the growth of horticulture sector in the selected states. Using descriptive statistics like, percentages and decadal growth rates, cropping pattern, and importance of horticulture crops were analysed.

As of the primary data, altogether 1872 households were surveyed from the four States through a stratified sampling technique. Pre-tested questionnaire schedule was adopted for collection of baseline data on area, production, and yield of horticulture crops. However, all the horticulture crops categorized by the NHB was not included in this present study. The crops included in the present survey were—vegetables, fruits, flowers, spices, and plantation crops. The selection of villages under the block and districts of the four States were made on the basis of highest area of the respective crop selected and complete enumeration of the village was done for the purpose. The state-level secondary information (four states) were collected from the four AERCs of the country: (1) AERC for North-East India, Assam Agricultural University, Jorhat, (Assam) surveyed Assam; (2) AERC Visva-Bharati, Santiniketan, Kolkata (West Bengal) surveyed Sikkim; (3) AERC Himachal Pradesh University, Shimla (Himachal Pradesh) surveyed Himachal Pradesh; and (4) AERC Delhi University, Delhi surveyed Uttarakhand. Secondary data were used for

[1] The data of Assam, often, represents other sister states in NER.

corroboration of the primary survey. The year 2012–13 was adopted as the year of reference for the present study. This helped us to understand the differences (if there is any) between the secondary and the present base-line survey. This also helped us to develop a comprehensive method for data collection and formulation of horticultural policies in the NER and HS where estimation of primary data was quite difficult.

RESULTS AND DISCUSSION

Area, Production, and Yield of Horticulture

Understandably, growth of any crop can be enhanced by expanding area, keeping the yield rate constant. Horticulture Mission for North East and Himalayan States (HMNEH) has been the backbone of horticulture development mission in these States with a substantial enhancement in funds. The aim of the HMNEH was to achieve horizontal and vertical integration of horticultural programmes to ensure adequate, appropriate, timely, and concurrent attention to all the links in production, post-harvest management and consumption chain to maximize economic, ecological, and social benefits. Over the years, there has been a significant improvement in the area, production, and productivity of various horticulture crops. Table 14.1 depicts growth rate of area, production and yield of horticulture crop in four States vis-à-vis all India level from 2002–03 to 2012–13.

From Table 14.1 we can see that the area under the horticulture crop was found to be increasing over the years. Of the four states, Assam being the largest geographical area and Sikkim the smallest, the area covered by horticulture crops in these States were also found to be the highest and the lowest respectively. The same holds true for the production as well. Larger states produced larger output and vice versa. But, the yield rate was found to be very different, not aligned with trends of area and production. The area of horticulture cultivation in the states is not comparable with the country and so the unit of production is also not comparable. Yield can be used as a comparable unit. Unfortunately, the yield level of these states turned out to be very low compared to the national level. Also, among the States, Assam was relatively a better performer compared to other States taken for the study, and Sikkim performed the least. This is, probably, due to either, lack of improved technology or small cultivable land area (limited plain area) for vegetables and other field crops in the three states of Himachal Pradesh, Sikkim, and Uttarakhand.

Of the selected states, Uttarakhand registered the highest productivity with 8.63 million tonnes per hectare Metric tonnes (MT)/ha) of horticulture in 2003–04, higher than that of national level (8.12 MT/ha). At the bottom, Sikkim registered at 1.94 MT/ha in 2003–04. Despite its concerted effort, over the years, the states could not become at par with national level. Nevertheless, the state increased the yield of horticulture crops from 1.94 MT/ha in 2003–04 to 3.28 MT/ha in 2012–13, compared to 11.04 MT/ha of national level, while Uttarakhand could register yield at to 6.43 MT/ha. This implies that the pace of growth and development of horticulture crops in NER and HS is still very slow and lagging behind the national level.

In terms of the growth trends of area, production, and yield of horticulture crops, Sikkim was found to be statistically significant for all series (that is original, 3 MA, and 5 MA), higher than that of the other three states (Table 14.2). Although slightly lower than in Sikkim, the growth rates of area, production, and yield of horticulture crops in Himachal Pradesh were all found to be statistically significant in all the three series, barring the yield growth rate for original series. In Uttarakhand, the growth trends of area and production were found to be statistically significant for all the series, but the yield was not found to be significant. In totality, the growth of area and production of horticulture crops in all the four states were found to be increasing and statistically significant. But, the same was not true in case of yield during the study period. What makes this growth of area and production of horticulture significant needs further verification. Why is the yield trailing?

To understand the growth trend of horticulture production, we decomposed the production into three effects: (1) Yield Effect, (2) Area Effect, and (3) Interaction Effect. This can help us in identifying the sources of growth through which the production of horticulture crops has enhanced in NER and HS during 2003–04 to 2012–13. The decomposition equation can be written as:

$$(P_n - P_0) = A_0 (Y_n - Y_0) + Y_0 (A_n - A_0) + (A_n - A_0)(Y_n - Y_0)$$

Or, it can be written as below:

$$\Delta P = \text{Yield Effect} + \text{Area Effect} + \text{Interaction Effect}$$

where, $(P_n - P_0)$ implies change in production (ΔP); A, P and Y imply Area, Production and Yield respectively; Subscript n and 0 imply Current Year and Previous Year respectively. Table 14.3 shows the decomposition of horticulture output growth.

Conventional economics of crop production depends mainly on land, labour, and technology or technical know-how. From Table 14.3 we can see that the area effect has been the driving force for the growth of horticulture crops in Assam and Uttarakhand but yield effect

Table 14.1 Area, Production, and Yield of Horticulture

A/P/Y	Year	Assam	Sikkim	Himachal Pradesh	Uttarakhand	All India
Area	2003–04	415.40	58.20	271.80	130.90	19449.00
	2004–05	410.10	57.40	277.20	285.50	20198.90
	2005–06	536.95	26.56	256.96	255.18	18666.63
	2006–07	582.10	61.00	279.31	267.50	19392.80
	2007–08	583.10	63.48	284.10	274.60	20213.80
	2008–09	479.59	58.72	284.11	274.92	20661.60
	2009–10	488.71	67.64	304.37	298.91	20875.70
	2010–11	575.50	65.90	302.50	273.00	21824.10
	2011–12	596.32	63.01	305.88	298.16	23241.97
	2012–13	624.22	72.56	305.83	296.39	23548.08
Production	2003–04	3158.20	112.70	1470.30	1130.00	157834.90
	2004–05	3189.60	118.30	1739.70	1778.50	169828.80
	2005–06	5799.25	92.36	1765.90	1604.38	184849.40
	2006–07	6041.58	136.12	1555.37	1711.10	191813.30
	2007–08	6077.80	152.21	1899.10	1773.90	211250.50
	2008–09	4679.31	155.44	1919.57	1832.79	214715.90
	2009–10	6327.63	207.88	1803.35	1755.69	223089.00
	2010–11	5074.80	199.10	2526.20	1790.80	240426.00
	2011–12	5427.05	230.48	1988.89	1909.40	257277.15
	2012–13	6009.73	238.21	2143.48	1905.83	260062.68
Yield	2003–04	7.60	1.94	5.41	8.63	8.12
	2004–05	7.78	2.06	6.28	6.23	8.41
	2005–06	10.80	3.48	6.87	6.29	9.90
	2006–07	10.38	2.23	5.57	6.40	9.89
	2007–08	10.42	2.40	6.68	6.46	10.45
	2008–09	9.76	2.65	6.76	6.67	10.39
	2009–10	12.95	3.07	5.92	5.87	10.69
	2010–11	8.82	3.02	8.35	6.56	11.02
	2011–12	9.10	3.66	6.50	6.40	11.07
	2012–13	9.63	3.28	7.01	6.43	11.04

Note: Horticulture crops include–Fruits, Vegetables, Plantation Crops, Medicinal and Aromatic Plants, and Spices.
Area in '000 ha; Production in '000 MT; Yield MT/ha.
Source: Compiled from NHB (2006; 2008; 2009; 2010; 2011).

was the main driver of horticulture crop development in Sikkim and Himachal Pradesh. In this present study, the impact of interaction effect was found insignificant. As defined in basic statistics, an interaction may arise when considering the relationship among three or more variables, and describes a situation in which the simultaneous influence of two variables on a third is not an additive. Area effect and yield effect have explained the change in production of horticulture crops in the four states. Further, using quadratic equation we can predict growth trend of area, production, and yield of horticulture crops in three series—original, 3 years' moving average, and 5 years' moving average.

KEY FINDINGS

Food security, nutritional security, sustainability, and profitability are the main focus of the future agricultural development. The high value agricultural outputs particularly the horticultural crops are considered as catalysts for the next wave of growth in the farm sector, not only in the larger parts of the country but also in the hilly regions of

Table 14.2 Growth Rate of Area, Production and Yield (2003–04 to 2012–13)

	Area			Production			Yield		
	b	Adj. R^2	Gr. Rate	B	Adj. R^2	Gr. Rate	b	Adj. R^2	Gr. Rate
				lnY = a + bt					
				Assam					
Original	0.035	0.453	3.648	0.052	0.300	5.427	0.017	–0.005	1.719
	(2.91)*			(2.21)*			0.98		
3 MA	0.023	0.393	2.403	0.028	0.224	2.906	0.008	–0.076	0.897
	(2.35)*			1.74			0.71		
5 MA	0.017	0.926	1.758	0.023	0.302	2.328	0.011	–0.006	1.191
	(8.02)*			1.78			0.98		
				Sikkim					
Original	0.044	0.136	4.595	0.100	0.856	10.541	0.055	0.492	5.669
	1.55			(7.41)*			(3.12)*		
3 MA	0.056	0.861	5.783	0.112	0.986	11.895	0.041	0.715	4.221
	(6.66)*			(22.27)*			(4.32)*		
5 MA	0.048	0.865	4.962	0.109	0.990	11.619	0.048	0.894	4.943
	(5.76)*			(23.47)*			(6.60)*		
				Himachal					
Original	0.017	0.734	1.726	0.039	0.559	4.065	0.022	0.213	2.294
	(5.09)*			(3.53)*			1.85		
3 MA	0.020	0.964	2.086	0.044	0.949	4.544	0.023	0.852	2.371
	(13.88)*			(11.57)*			(6.44)*		
5 MA	0.020	0.974	2.033	0.043	0.958	4.438	0.022	0.909	2.318
	(13.88)*			(10.77)*			(7.15)*		
				Uttarakhand					
Original	0.050	0.309	5.178	0.035	0.432	3.615	–0.015	0.097	–1.490
	(2.24)*			(2.80)*			–1.40		
3 MA	0.028	0.631	2.873	0.024	0.748	2.453	–0.007	0.147	–0.791
	(3.61)*			(4.68)*			–1.49		
5 MA	0.028	0.719	2.917	0.024	0.799	2.430	–0.009	0.268	–0.897
	(3.72)*			(4.57)*			–1.68		

Note: 3 MA implies 3 years moving average; 5 MA means 5 years moving average; r means growth rate; Figures in parentheses are t values.
* indicates statistically significant at 5 per cent
Source: Author's computations.

Table 14.3 Decomposition of Horticulture Output Growth (2003–04 to 2012–13)

States	Area Effect	Yield Effect	Interaction Effect	Total
Assam	0.70	0.30	0.00	1.00
Sikkim	0.70	1.00	–0.70	1.00
Himachal	0.35	0.74	–0.09	1.00
Uttarakhand	1.80	–0.30	–0.50	1.00

Source: Author's estimation from the Table 14.1.

NER and HS. In the light of discussion made in the previous sections, it has been exemplified by the sector's growth rate in the last few decades in these two regions. Fruits, flowers, vegetables, spices, plantation, and medicinal crops were grown mainly by the marginal farmers in these hilly States. Despite natural and man-made difficulties, the cultivation of cash crops is turned out to be beneficial for the farmers, as it fetches them with higher income and employment throughout the year (Singha, Choudhary, and Vishnu 2014).

As depicted by the secondary data, the growth of area and production of horticulture crops in all the states taken for the present study were found to be increasing constantly and statistically significant. But, the same does not hold true in the case of yield during the same study period. Of the crops, fruits, flowers, plantation, and perennial crops were found to be suitable in these states. Motivating factors for cultivation of horticulture crop is concerned; remunerative price has been at the forefront. As revealed by the farmers, especially in Sikkim and Uttarakhand, the support of government in the sector's development was very important and influential for taking up horticulture cultivation. Despite commonality of the states' land-locked geographical condition, the major horticulture crops grown across these States were found to be different. Therefore, findings and recommendations should be analysed separately for all the four states taken for the study.

Assam

Assam is endowed with unique agro-climatic condition, which is suitable for growing a wide range of horticultural crops. Horticultural crops cover an area of 5.75 lakh hectares, out of the total cultivable area of 40.99 lakh hectares in the state. In the percentage term, it is around 14.04 per cent of the total cultivable area of the state. The growth of area and production of the crop in the state during the last two decades has been very significant. Within the sector, the growth of area and production of fruit crops has been 19.12 per cent and 22.70 per cent during the period of 2004–05 to 2011–12 respectively. However, fruit crops productivity growth was very negligible, estimated at 4.40 per cent during the same period. In the case of vegetables, the growth rate of area increased by 16.17 per cent and the production by 20.76 per cent during the same period. Similar to fruit, the growth rate of vegetable was found to be increasing as low as 5.54 per cent from 2004–05 to 2011–12. While, the spices growth rate of area, production, and productivity was found to be 15.31 per cent, 16.94 per cent and 2.09 per cent during the period from 2004–05 to 2011–12 respectively.

Out of the total 1119 sample horticulture households collected from Assam, 52.99 per cent were marginal farmer, 26.27 per cent small, 12.42 per cent medium, and 8.31 per cent large farmers. The highest percentage (27.79 per cent) of sample farmers who were cultivating horticulture crops were enthused primarily by the better price, followed by the factor of close to the market, estimated at around 22.70 per cent of the total farmers. Suitability of the region and climate (easy to grow) was also another important factor for growing horticulture crops by the sample farmers, accounted for around 18 per cent of the total. Quality of the soil or suitability of the soil was also another factor for cultivating this crop, accounted for 13.76 per cent of the total sample farmers. The government support also motivated the farmers to cultivate horticulture crops, accounted for 6.52 per cent of the sample farmers. A total of 5.81 per cent and 5.36 per cent of horticulture sample farmers were motivated by easy availability of seed and low cultivation cost respectively. Hardly, 0.18 per cent of horticulture farmers were found to be motivated by their neighbour farmers to grow horticulture crops.

As of the problems faced by the farmers, majority of the sample farmers in Assam were not in a position to adopt modern technology mainly due to high cost of inputs and materials, lack of proper knowledge, non-availability of inputs, and poor extension support. In addition to that, at macro level, road connectivity to reach market was also one of the major problems faced by 84.63 per cent of the sample farmers. Limited of storage facility and shortage of quality seed facility were reported as major problem by 42.90 per cent and 23.50 per cent of the sample respondents respectively. Also, the other constraints in growing horticultural crops as mentioned by the sample farmers were the shortage of labour (8.85 per cent), packaging of products (7.69 per cent) and lack of potential market (7.95 per cent). On the side of agencies involved in horticulture data collection, the main difficulties faced by them are—the lack of updated records, problems in applying uniform statistical methodologies, and lack of cooperation from farmers.

Sikkim

Sikkim is declared as an 'Organic State'. The main motivating factor for horticulture crop cultivation in the State was the encouragement of government, accounted for 35 per cent of the total sample farmers. It was followed by the factors like, good price, accounted for 27.9 per cent, easy availability of inputs with 21.1 per cent, and proximity to market with 10.5 per cent and so on. In the case of technological adoption, it was observed from the survey that around 25 per cent of respondents adopted poly-house cultivation in their horticulture farms, 18.53 per cent adopted greenhouse cultivation and Integrated Nutrient Management (INM)/Integrated Pest Management (IPM) was used by 27.45 per cent of the sample farmers.

As of the yield status, it is estimated at 938.24 kg per hectare for leafy vegetables, 3059.86 kg per hectare for the mandarin orange and 4500 kg per hectare for ginger in kharif season. In rabi season, the vegetables like cabbage, cauliflower, and beans were cultivated under both irrigated and unirrigated condition. Most of the cabbage and cardamom areas, estimated at 62.22 per cent, were covered by irrigation, leaving 37.78 per cent as unirrigated area. The highest productivity of crop was registered by the tomato in south Sikkim with 5094.49 kg per hectare, followed by cabbage with 5091.38 kg per hectare, cauliflower with 4843.69 kg per hectare, carrot with 3307.88 kg per hectare, broccoli with 307.63 kg per hectare, and bean with 1409.57 kg per hectare. The farmers of west Sikkim have got both irrigated and unirrigated area for growing tomato, lady's finger and chilli. Understandably, the productivity of lady's finger was higher in irrigated land with 1150 kg/ha, while it was 777.78 kg/ha in unirrigated land. As of the ginger and turmeric, they recorded a productivity of 5629.94 kg per hectare and 3353.61 kg per hectare respectively. As of the annual crops, the banana, cymbidium orchids, and papaya were found to be cultivated in Sikkim and the productivity of banana was higher in the irrigated condition. The flower, cymbidium orchid gave on an average of yield 62784 (in number) sticks per hectare.

As of the mixed cropping system, most of the horticulture areas in Sikkim were found to be cultivated mixed cropping. A remunerative profit was earned from banana cultivation. Around 89.09 per cent of the sample areas in the State were found to have cultivated spice crop, especially cardamom. Like cabbage, cardamom gave an attractive income to its growers in Sikkim. Cultivation of papaya and banana gave the highest and second highest income respectively to the horticulture farmers in the State. Tomato was another attractive profit earning vegetable crop in the state. Despite these opportunities, the State faced certain drawbacks while cultivating horticulture crops. It was observed that labour availability and problem related to road infrastructure were found to be quite serious. Besides, lack of good packaging materials and marketing facility, lack of storage also posed serious problem for further expansion of the sector. At the macro level, for the policy makers, main difficulty was to assess appropriate data on area, production, and productivity of horticultural crops across the state. For further development of the sector, the opinion from the maximum respondents, estimated at 34.05 per cent of the total sample farmers, went in favour of providing better storage facility. Around 18.4 per cent of the sample farmers suggested for providing transport facility and packaging. Marketing facility was sought by 14.42 per cent of the sample farmers.

Himachal Pradesh

Horticulture sector plays a significant role in development of the state's economy in Himachal Pradesh (HP). The area under horticulture crops in HP has increased significantly, especially the fruits and flowers from 67.531 thousand hectares in 2007–08 to 79.024 thousand hectares and 0.583 thousand hectares in 2007–08 to 0.682 thousand hectares in 2009–10 respectively, while, the area under spices was almost the same during this same period. The area of vegetables has slightly decreased from 35.764 thousand hectares to 35.672 thousand hectares during the same period.

The yield of spices, vegetables, and flowers have increased significantly, whereas, the yield of fruits decreased during the study period, despite growth in its area and production. It is also understood that the horticulture sector in the state has reached to the marginal lands; consequently, a negative impact on productivity and profitability of the same was realized in the recent past. Therefore, it is advocated that policy should increasingly cater to productivity enhancement rather than on increasing area. In term of technological adoption, all the fruits and flower growers adopted the technology of INM/IPM in the area of 42.77 ha and 9.48 ha respectively. While, none of the vegetable and spice growers were found to be adopted any kind of technology. In the case of fruit growers, almost all the sample farmers received the benefit from the government under the National Horticulture Technology Mission (NHTM) in the form of plants and equipment. Barring vegetables and spices, about 96 per cent flower growing households received benefits from the government in the form of subsidy.

As of the problems, poor quality of planting material including seed and root stocks; poor layout of orchards; lack of appropriate poly-materials for the orchard, lack of proper training, and pruning of the fruit trees; inadequate plant nutrition and organic matters; lack of adequate use of plant protection materials; poor overall management of orchards were the major problems for horticultural crops in the state. From the present study, we can find that 61 per cent of horticulture growers belonged to marginal category. The proportion of land under irrigation was relatively higher in the flower growing farmers as compared to other crops. As of the motivating factors, horticulture has been the major crops for more than 10 years, for almost all the sample farmers of the surveyed areas. It fetches good price, higher income, and employment for most of the sample farmers. Some of the administrative and institutional problems were also encountered by this sector. To mention a few, lack of trained staff, their unwillingness to participate and coordination among themselves to arrange the data systematically after its collection and lack of motivation, were some of the important ones. In the context of financial difficulty, it includes—lack of sufficient funds available for collection and compilation of data, no specific allocation of funds for different activities were involved in data development and its management. Technical problems include—deficiencies in the use of methodology to estimate the production by the trained staff, ineffective application of methods and procedures, and as of the infrastructural difficulties, non-availability of good transport facilities to visit the fields and also non-availability of equipment for compilation of data.

Uttarakhand

Like any other developing economies, Uttarakhand is also shifting from the agricultural based to the secondary and tertiary sectors. The horticulture sector has become one of the most important sub-sectors of agriculture and been playing an important role in the State's economy for long. Within the sector, specifically among the fruit crops, the mango, apple, and citrus occupy larger shares; while the peas, potato, tomato, French beans, etc., are the major contributors of vegetable crops. The contribution of Uttarakhand in the total area in India under fruits and vegetables was 2.81 and 1.07 per cent respectively during 2011–12. But, share of the State's production for horticultural crops was much lower than that of the area, due to low rate of productivity.

In kharif season, marigold was grown on 5.12 hectares in Kangri village of Haridwar district. A marginal area was devoted to marigold in Prateetpur village of Dehradun district and the yield of marigold was around 325 quintals per hectare. Also, the sample respondents were found to be growing a large variety of vegetable crops in Dehradun district, where Okra, peas, French beans, and cucumber were the major crops. Also, the crops like, tomato, onion, radish, bitter gourd, and potato were also found to be grown. On an average, the yield of tomato was estimated at 9375 kg per hectare. In the case of radish, it was 13125 kg per hectare and potato was estimated at 10544 kg per hectare. The yield of these (radish and potato) vegetables was also found to be slightly higher than that of any other vegetable crops grown in the State. During the summer season, respondents of Tehri Garhwal district were mainly found to be cultivated ginger, potato, peas, and French beans. A miniscule area of the sample farmers was devoted to cauliflower, tomato, chilli, onion, and turmeric. The productivity of ginger, potato, peas, and French beans was found to be around 174 qtls/ha; 105 qtls/ha; 70 qtls/ha, and 54 qtls/ha respectively. Among annual crops, mango and litchi were found to be cultivated. The yield of these fruits was estimated at 66 qtls per hectare and 30 qtls per hectare respectively. It is encouraging that the yield of mango in Dehradun district was much higher than that of the state level, probably due to the availability of improved varieties, which yielded higher output. Also, the fruit cultivation was further contributed by favourable climatic conditions in the state. The major sources of irrigation for horticultural crops in selected villages were canal, tube wells, and natural spring.

CONCLUDING REMARKS AND POLICY RECOMMENDATIONS

In nutshell, horticulture is one of the major sectors in the economy of the hilly states. It has shown a much needed opportunity for diversification and increased employment. The sector has a few advantages for faster growth of production and productivity in the near future, due to its market and climatic condition of the states. However, besides weak physical infrastructures, majority of the farmers belong to marginal holdings and as a result of which speedy development initiative is slowed down. The area, production, and yield of horticultural crops have improved during the past two decades in these hill States. But, because of non-availability of comprehensive official data on basic parameters at the grassroots level puts a serious limitation in designing and planning for improved productivity through the extension, input supply, and efficient marketing logistics. It is understood that the data collection in hill areas is a serious problem due to its small, fragmented, and scattered land holdings. This also leads to low production and productivity since adoption of technology in these holdings is very low compared to that of the national average. Farmers often under-estimate their area and production, exaggerate the cost with the expectation of receiving government subsidies. This affects accuracy of data collected by the mobile teams, government officials. Appropriate data on horticulture crop of these States are not found in the public domain.

This study faced some of the major difficulties like lack of uniform and accurate baseline data, appropriate and uniform methodology for data collection, technological loopholes, shortage of skilled manpower, and updated data on area, production, and productivity of horticulture crops. To overcome these problems, some of the possible recommendations have been made.

1. Consolidated horticulture data cannot be collected unless the categorization of crops within the larger categories (for example, fruits, vegetables, tuber, annual crops, etc.), is identified. As the individual crops included in the broader horticulture sector (fruits, vegetables, etc.), of a region/state is different from other states. For instances, master leaf is one of the important leafy vegetables in North-east India, while it is not so in the HS. Therefore,

baseline data on horticulture crops should be made for individual crops, not for broader crop category as fruit, vegetables, spices, and so on.

2. Most of the horticulture farmers belonged to marginal and small category. This limits in adoption of large and modern technological implements under horticulture cultivation. As a result of which, export volume cannot be enhanced. Therefore, some initiatives should be taken up for the agglomeration of the marginal lands and adopt large scale horticulture cultivation with new technologies.
3. Estimation of area, production, and yield of horticulture crops under the mixed cropping system that have been grown with other non-horticulture crops should also be made. Besides, roof top gardening/ courtyard gardening and field bund plantation should also be estimated with special care while collecting horticulture data.

REFERENCES

Abou-Hadid, A.F. 2005. 'High Value Products for Smallholder Markets in West Asia and North Africa: Trends, Opportunities and Research Priorities', The Global Forum on Agricultural Research, 191. Roma, Italy: Food and Agriculture Organization of the United Nations.

Bahadur, S. 2010. 'Horticulture–Key to India's Agriculture Growth', available at: http://www.commodityonline.com/news/horticulture-key-to-indias-agriculture-growth-34627-3-34628.html (accessed on 30 July 2013).

Chand, R. 1996. 'Ecological and Economic Impact of Horticultural Development in the Himalayas: Evidence from Himachal Pradesh'. *Economic and Political Weekly*, 31(26): A 93–A 99.

Choudhary, R. Singha, K., and Vishnu, K. 2013. 'Growth of Horticultural Sector: An Indicator of Urbanisation and Overall Development', Paper presented at National Seminar on Urban and Peri-Urban Agriculture, at Institute for Social and Economic Change, Bangalore, India.

Dev, S.M. 2012. 'Small farmers in India: Challenges and opportunities'. Working Paper No. 14. Mumbai, India: Indira Gandhi Institute of Development Research.

Directorate of Economics and Statistics (DES). 2012. *Agricultural Statistics at a Glance*. New Delhi: Ministry of Agriculture and Farmers Welfare, Government of India.

Government of India (GoI). 2001. *Report of the Working Group on Horticulture Development for the Tenth Five Year Plan*, Report No.14/2001. New Delhi: Planning Commission, Government of India.

Mehta, P.K. 2009. 'Diversification and Horticultural Crops: A Case of Himachal Pradesh', Unpublished doctoral thesis, Institute for Social and Economic Change, University of Mysore, Bangalore.

Mittal, S. 2007. 'Can Horticulture be a Success Story for India?', Working Paper No. 197, Indian Council for Research on International Economic Relations New Delhi.

National Horticulture Board (NHB). 2006–11. 'Indian Horticulture Database', Ministry of Agriculture and Farmers Welfare, Government of India, New Delhi.

Singha, K., R. Choudhary, and K. Vishnu. 2014. 'Growth and Diversification of Horticulture Crops in Karnataka: An Inter-District Analysis', SageOpen, p. 1–7: DOI: 10.1177/2158244014548018.

Weinberger, K. and T.A. Lumpkin. 2007. 'Diversification into Horticulture and Poverty Reduction: A Research Agenda', *World Development*, 35: 1464–80.

ANNEXURE 14A

Table 14A The Details of Agro-Economic Research Centres and the Lead Person Involved in the State Report

State	AERC Centre/Unit	Authors
Assam	AERC Jorhat	Moromi Gogoi and Debajit Borah
Uttrakhand	AERC Delhi	Usha Tuteja
Himachal Pradesh	AERC Shimla	Ranveer Singh
Sikkim	AERC Shantiniketan	Debashis Sarkar, Debanshu Majumder , Ranjan Kumar Biswas

15

Potential of Horticultural Crops in Assam and Meghalaya

Gautam Kakaty

India's varied agro-climatic conditions allow it to produce a wide variety of horticultural crops such as fruits, vegetables, tropical, and tuber crops, along with ornamental crops like coconut, cashew nut, cocoa, etc. Commercial importance of fruits in recent time has increased manifold all over the World. Besides their nutritional and social importance, fruits contribute significantly a country's economy. According to the *Handbook of Horticulture Statistics, 2014*, vegetables, fruits, plantation crops, spices, and flowers, and aromatic plants contributed to the extent of 60.33, 30.23, 6.32, 2.14, and 0.98 per cent of the total horticultural production, respectively in 2013–14. India is now second largest producer of fruit in the World next to Brazil. The country ranks first in the production of mango, banana, sapota, and acid lime and, in recent years, recorded the highest productivity in grapes.

Assam is endowed with unique agro-climatic condition, which permits growing of wide range of horticultural crops. It accommodates various fruits, vegetables, flowers, spices, medicinal and aromatic plants, nut crops, tuber crops, and also plantation crops. Meghalaya has immense potential for the development of horticulture. The variation of altitude, soil, and climatic conditions and temperature regime of the state provide ample scope for growing of different types of horticultural crops including fruits, vegetables, spices, plantation crops, medicinal, and aromatic plants of high economic values.

TECHNOLOGY MISSION FOR INTEGRATED DEVELOPMENT OF HORTICULTURE (TM-IDH)

Considering the potential of horticulture in the north-eastern states, the Government of India (GoI) introduced the TM-IDH in North-Eastern Region (NER) including Sikkim. An outlay of Rs 229.38 crore was approved for Ninth Five Year Plan and the programme was launched in the year 2001–02.

The broad objectives of the Mission are:

1. to establish convergence and synergy among numerous ongoing governmental programmes in the field of horticulture development to achieve horizontal and vertical integration of these programmes;
2. to ensure adequate, appropriate, timely, and concurrent attention to all the links in the production, post harvest, and consumption chain;
3. to maximize economic, ecological, and social benefits from the existing investment and infrastructure created for horticulture;

4. to promote ecologically sustainable intensification, economically desirable diversification, and skilled employment; and
5. to generate value addition, promote the development and dissemination of eco-technologies based on the blending of the traditional wisdom and technology with frontier knowledge such as bio-technology, information technology, and space technology; and to provide the mission links in ongoing horticulture development projects.

The TM-IDH is being implemented through four mini missions. These four components are: (i) Mini Mission Mode I: Research is coordinated and implemented by the ICAR in N.E. Region/Assam Agricultural University. (ii) Mini Mission Mode II: Production and productivity improvement activities are coordinated by the Department of Agriculture & Co-operation and implemented by the Agriculture/Horticulture Department of the States. (iii) Mini Mission Mode III: Post harvest management, marketing, and export are coordinated by National Horticulture Board. (iv) Mini Mission Mode IV: Processing is coordinated and implemented by the Ministry of Food Processing Industries. The present study has been undertaken with the following objectives:

1. to study the potentialities of horticultural crops in the study area;
2. to study the programmes under Technology Mission in the sample States;
3. to study the market accessibility of horticultural crops in the study area;
4. to study the problems of horticultural crops cultivation; and
5. to suggest policy implications.

METHODOLOGY AND DATA

Keeping in view the objectives of the study, it was decided to undertake the study in two states of NER—Assam and Meghalaya. Following multi-stage stratified random sampling technique; two districts dominated by horticultural crops, where TM-IDH is in progress, were selected in consultation with the Directorate of Horticulture, Assam. Thus, Kamrup and Nagaon districts in Assam were selected at the first stage. In the next stage two orange growing blocks namely, Boko and Bangaon from Kamrup district were selected. In the third stage, five villages were selected and from each village eight beneficiary households were selected randomly. Similarly, two banana growing blocks namely, Bajaya and Laokhowa were selected in consultation with the district agriculture officer, Nagaon. From each block, 5 villages were selected. From each village, eight beneficiary households were selected at random. Thus, a total number of 80 beneficiaries were selected from the two districts of Assam.

In the State of Meghalaya too, following the same methodology, two districts under TM-IDH with larger area under horticultural crops were selected at the first stage. They were East Khasi Hills and Ri-Bhoi districts. In the next stage, two orange growing blocks namely, Mylliem and Mawphlang from East Khasi Hills district were selected. In the third stage, five villages were selected and eight beneficiary households were selected at random from each village. Similarly, two pineapple growing blocks namely, Umling and Umsning were selected in consultation with the District Agriculture Officer, Ri-Bhoi. From each block, five villages were selected and from each village, eight beneficiary households were selected at random. Thus, a total of 80 beneficiaries were selected from the two districts of Meghalaya.

RESULTS AND DISCUSSION

The total population in the sample beneficiary households in Assam was 496 (52.22 per cent males and 47.78 per cent females) and in Meghalaya, it was 424 (54.01 per cent males and 45.99 per cent females). It was found that in Kamrup and Nagaon district, the percentage of literate person to the total population was 85.83 per cent and 83.13 per cent, respectively. In East Khasi Hills and Ri-Bhoi district, the literacy rate was 82.95 per cent and 79.23 per cent, respectively.

In Kamrup district, out of the total population, 41.30 per cent were workers, 38.06 per cent were helpers, and 20.65 per cent were non-workers. In Nagaon district, of the total population; 31.73 per cent were workers, 44.18 per cent were helpers, and 24.09 per cent were non-workers. The economic status of sample households of East Khasi Hills showed that out of the total population, 40.55 per cent were workers, 37.33 per cent were earning dependents, and 22.12 per cent were non-workers. In Ri-Bhoi district, 30.92 per cent were worker, 43.00 per cent were helper, and 26.08 per cent were non-worker.

Out of the total working population in the sample district of Kamrup, 33.33 per cent were primarily engaged as cultivators, 10.78 per cent were as agricultural labourers, 8.82 per cent were as non-agricultural labourers, and 8.82 per cent were engaged in the livestock, forestry, and fishery sector. Under non-agricultural sector, 16.67 per cent of the workers were engaged in household cottage industries, 7.84 per cent in weaving, 7.84 per cent in services, and

5.88 per cent were engaged in petty trade, commerce, and transport.

Industrial category-wise classification of population in the sample of Nagaon district showed that out of the total working population, 49.37 per cent were primarily engaged in agriculture, 8.86 per cent were agricultural labourers, 6.33 per cent were non-agricultural labourers and 12.66 per cent were engaged in the livestock, forestry, and fishery sector. In the non-agricultural sector, 8.86 per cent were engaged in household cottage industries, 4.81 per cent in weaving, 3.80 per cent in services and professions, and 6.33 per cent in petty trade, commerce, and transport.

The occupational distribution of population in the sample of East Khasi Hills district showed that out of the total working population, 35.23 per cent were primarily engaged in agriculture, 12.50 per cent as agricultural labourers, 9.09 per cent as non-agricultural labourers, and 11.36 per cent were engaged in livestock, forestry, and fishery sector. In the non-agricultural sector, 12.50 per cent of the workers were engaged in household cottage industries, 10.23 per cent in weaving, 4.55 per cent in services and professions, and 4.55 per cent were engaged in petty trade, commerce, and transport. Out of the total working population in the sample District of Ri-Bhoi, 39.06 per cent were primarily engaged as cultivators, 20.31 per cent were as agricultural labourers, 10.00 per cent were as non-agricultural labourers and 7.81 per cent were engaged in the livestock, forestry, and fishery sector. Under the non-agricultural sector, 12.50 per cent of the workers were engaged in household cottage industries, 4.69 per cent in weaving, 3.13 per cent in services and professions, and 4.69 per cent were engaged in petty trade, commerce, and transport.

In case of operational holdings in Kamrup district, 16.15 per cent were under field crops and 83.85 per cent of the cultivated land were under horticultural crops. In Nagaon district, 15.38 per cent were field crops, and 84.62 per cent of the cultivated area was under horticultural crops.

Out of the total operational holdings in East Khasi Hills district, 8.42 per cent was under terrace land, 71.61 per cent were under orange orchards, 14.22 per cent were under horticultural crops, 2.91 per cent were under jhum land, and 2.84 per cent were under tea land. The average size of operational holding per family varied from 0.61 hectares to 4.24 hectares with an overall average of 1.74 hectares. Out of the total operational holdings in Ri-Bhoi district, 8.42 per cent were under terrace land, 76.32 per cent were under pineapple orchards, 3.17 per cent were under other horticultural crops, 3.73 per cent were under jhum land, and 8.37 per cent were under tea land. The average size of operational holding per family varied from 0.37 ha to 4.12 ha with an overall average of 1.95 hectares. It was observed that in both the districts of Meghalaya, jhum land was community land. Allotment of such jhum land to a particular family is done by the village community.

In Kamrup district, the average yield of irrigated paddy was found at 3,553 kg/ha and the average yield of unirrigated paddy was found at 2,987 kg/ha. The sample beneficiary farmers used 54.50 hectares of land for orange cultivation in the reference year. The yield of orange varied from 10,630 kg/ha to 11,750 kg/ha with an overall average of 10,951 kg/ha. The sample farmers used 2.30 hectares of land for vegetable cultivation. The productivity of vegetables varied from 13,948 kg/ha to 15,560 kg/ha with an overall average of 14,587 kg/ha.

The average yield of irrigated paddy in Nagaon district was found at 3,714 kg/ha. The average yield of un-irrigated paddy per hectare was found to be 3,188 kg/ha only. The sample beneficiary farmers found to have paid high priorities in horticultural crops because of the suitability of their soil, climate and marketing potential and assistance provided under the TM-IDH. The average productivity of banana was found at 14,413 kg/ha and that of Assam lemon was found at 6,582 kg/ha. The sample farmers also cultivated vegetable crops in 2.29 hectares of land. The average productivity of vegetables was 14,276 kg/ha, while the average yield of potato was found at 11,636 kg/ha.

The sample beneficiary households in East Khasi Hills raised 5.87 hectares of land for terrace cultivation. The average yield of terrace paddy in the hilly areas was 1,144 kg/ha. Though, jhum cultivation is the main traditional practice of crop production in hill areas, some major changes were found in the present study. All the sample orange beneficiary (40) households raised orange cultivation in 49.95 hectares of land. The average productivity of orange was 5,639 kg/ha. These farmers considered orange gardening as the main source of income for their families. The productivity of orange was highest in case of the marginal farmers (5,953 kg/ha) followed by the small farmers (5,842 kg/ha) and the semi-medium farmers (5,716 kg/ha). It was observed that the medium group of beneficiaries could produce only 5,639 kg/ha. It was because of the fact that the farmers in the medium farm size group brought some new area under orange cultivation under the area expansion programme of TM-IDH and were yet to attain the fruit bearing stage. The productivity of pineapple cultivation varied from 7,435 kg/ha to 8,788 kg/ha. The sample farmers had only 2.03 hectares of land under jhum cultivation in the reference year. The average yield of jhum cultivation was found at 1,046 kg/ha. Tea cultivation became a lucrative enterprise in Meghalaya too. Some of the sample beneficiary households already started mini tea gardens in 1.98 hectares of land indicating a kind of transition in the hill farming. These new tea gardens had not yet attained the optimum productivity stage. The average yield rate from these tea gardens varied from 11,926 kg/ha to 12,011 kg/ha with pooled average of 11,978 kg/ha.

The sample pineapple beneficiary households in Ri-Bhoi district of Meghalaya raised 6.57 hectares of land for terrace cultivation. The average yield of terrace paddy in the sample varied from 1,088 kg/ha to 1,297 kg/ha with pooled average of 1,166 kg/ha. The productivity of pineapple varied from 11,448 kg/ha to 12,420 kg/ha with pooled average of 11,740 kg/ha. The sample farmers raised 2.47 hectares of land for orange cultivation. The average yield of orange was found at 4,722 kg/ha. The productivity of jhum cultivation in Ri-Bhoi district varied from 997 kg/ha to 1,088 kg/ha. The productivity of recently introduced tea cultivation varied from 11,125 kg/ha to 13,438 kg/ha with an overall average of 12,138 kg/ha.

The total production of orange was 5,968.03 quintals among sample households in Kamrup district that valued at Rs 28,82,631.10. The total marketable surplus of orange for 40 sample beneficiary farmers of Kamrup district of Assam was 5,731.60 quintals for example, 96.04 per cent of the total production. The total gross return from sale of the marketable surplus of orange was Rs 27,65,437.00 and per household return was Rs 96,136.00. The marketable surplus of banana produced by the sample beneficiary farmers of Nagaon district of Assam was 96.01 per cent of the total production. Total sale proceeds from banana for the sample beneficiary farmers were Rs 40,83,096.23 and the average per household income from the sale of banana was Rs 1,02,077.00.

The sample beneficiary farmers of East Khasi Hills district in Meghalaya produced 2,816.73 quintals of orange and marketed surplus was 95.85 per cent. The total sale proceeds from the marketed surplus were found at Rs 12,65,510.00 and the average per household receipt from sale proceeds was Rs 31,638.00.

The total marketed surplus of pineapple of the sample beneficiary farmers of Ri-Bhoi district in Meghalaya was 6,700.05 quintals accounting for 95.84 per cent of the total production. The total receipt (Rs 95,433.00) from the sale proceeds of pineapple was higher in medium farms. The average per household receipt from the sale proceeds of pineapple was Rs 49,650.00.

The identified major marketing channels of orange, pineapple, and banana in Assam and Meghalaya were:

1. Producer–Retailer–Consumer
2. Producer–Commission Agent–Retailer–Consumer
3. Producer–Commission Agent–Wholesaler–Retailer–Consumer

Price spread of orange in Channel 1 was worked out for Shillong, Guwahati, and Boko markets. The producer's share of consumer's Rupee was 48.75 per cent in Shillong market, 50.00 per cent in Guwahati, and 52.78 per cent in Boko market. The retailer's net margin was found at 46.87 per cent in Shillong market, 45.75 per cent in Guwahati, and 43.06 per cent at Boko market. The analysis of price spread for orange in Channel 2 shows that the grower's share of consumer's Rupee was 41.87 per cent in Shillong market, 39.00 per cent in Guwahati market, and 38.50 per cent in Boko market. The commission agent's share of consumer's Rupee was 22.12 per cent in Shillong market, 21.80 per cent in Guwahati market, and 22.00 per cent in Boko market. Excluding the transportation, sorting, and market charges the retailers share of consumer's Rupee was 32.25 per cent in Shillong market, 34.45 per cent in Guwahati market and 34.50 per cent in Boko market.

The price spread of orange in Channel 3 showed that producer's share of consumer's Rupee was 37.50 per cent in Shillong market, 35.50 per cent in Guwahati market, and 34.72 in Boko market. The commission agent's share was 15.56 per cent in Shillong market, 14.35 per cent in Guwahati market, and 15.22 at Boko market. The wholesaler's net margin was found at 12.50 per cent in Shillong market, 12.65 per cent in Guwahati market, and 12.06 per cent in Boko market. The retailer's margin was recorded at 29.43 per cent in Shillong market, 31.75 per cent in Guwahati market, and 31.88 per cent in Boko market. Handling, transportation, grading, storage charges, market charges/market fees, etc., also varied from market to market depending upon the local rates.

The price spread of pineapple in Channel l, for Nongpoh and Guwahati markets showed that the grower's net share of consumer's rupee was highest in Nongpoh market (50.00) followed by Guwahati market (46.67). The growers viewed that in this channel they were getting better prices than selling at farm site on contract. The retailers were making handsome margin, the highest being 40.40 per cent in Guwahati and 40.00 per cent in Nongpoh market. The price spread of pineapple in Channel 2 shows that the grower's share of consumer's rupee was 46.43 per cent in Nongpoh market and 43.33 per cent in Guwahati market. The commission agent's share of consumer's Rupee was 25.00 per cent in Nongpoh market and 26.93 per cent in Guwahati market. Excluding the transportation, sorting, and market charges, the retailer's share of consumer's Rupee was 20.00 per cent in Nongpoh and 16.80 per cent in Guwahati market.

The price-spread analysis of pineapple in Channel 3 shows that the producer's share of consumer's rupee was 42.86 per cent in Nongpoh and 40.00 per cent in

Guwahati market. The commission agent's share was 24.86 per cent in Nongpoh market and 26.00 per cent in Guwahati market. The wholesaler's net margin was found at 10.14 per cent in Nongpoh market and 10.13 per cent in Guwahati market. In channel 3, retailer's margin varied from 12.14 per cent in Nongpoh market to 10.26 per cent in Guwahati market.

The price spread of banana in Channel 1 showed that the grower's net share of consumers Rupee was highest in Nagaon market (49.20 per cent) followed by Sonitpur market (48.08 per cent). The retailers were found to earn more margins; it was 43.80 per cent in Nagaon market and 43.83 per cent in Sonitpur market. The price-spread analysis of banana in Channel–2 showed that the grower's share of consumer's Rupee was 47.30 per cent in Nagaon market and 44.17 per cent in Sonitpur market. The commission agent's share of consumer's Rupee was 25.20 per cent in Nagaon market and 27.16 per cent in Sonitpur market. Excluding the transportation, sorting, and market charges, the retailer's share of consumer's Rupee was 21.00 per cent in Nagaon market and 20.58 per cent in Sonitpur market. The price-spread of orange in Channel 3 for Nagaon and Sonitpur markets shows that the producer's share of consumer's Rupee was 40.90 per cent in Nagaon market and 38.33 per cent in Sonitpur market. The commission agent's share was 22.40 per cent in Nagaon market and 23.50 per cent in Sonitpur market. The wholesaler's net margin was found at 10.00 per cent in Nagaon market and 11.00 per cent in Sonitpur market. In channel-3, retailer's margin varied from 12.14 per cent in Nagaon market to 10.26 per cent in Guwahati market.

PROBLEMS OF COMMERCIAL CULTIVATION OF HORTICULTURAL CROPS

The marked improvement of horticultural sector in the study area was due to the implementation of TM-IDH which contributed immensely towards increased fruit production. However, horticultural crops cultivators encountered a number of problems which include:

1. Lack of adequate horticultural research & development and extension/services
2. Inadequate technology and unscientific method of cultivation
3. Undulating topography and land ownership pattern
4. Inadequate road, transport, and communication facilities
5. Lack of storage facilities
6. Small-scale production and perishability
7. Inadequate fruit processing units
8. Lack of proper packaging, handling, and refrigerated transport
9. Lack of grading and standardization
10. Inadequate marketing and distribution network

CONCLUSION AND POLICY IMPLICATIONS

Based on the findings of the study and the observations made, the following policy implications have been suggested for development of horticulture in the States of Assam and Meghalaya:

1. **Improved Production Possibilities and Research:** Research and development activities on pineapple, banana, and orange should be strengthened to increase production and productivity with improvement in quality to increase profit per unit of area. This will also increase the export potential of these commodities.
2. **Strengthening Input Supply and Extension Services:** Extension services in the states should be strengthened with field demonstration on scientific cultivation of pineapple, banana, and orange.
3. **Development of Grower's Marketing Cooperative:** Establishment of a multipurpose grower's marketing cooperative society with cold storage and transport facilities can go a long way in eliminating the monopolistic trade practices of market functionaries. This will ensure remunerative prices for the fruits/products produced by the society members.
4. **Establishment of Cold Storage Facilities:** There is an urgent need to create and maintain cold storage facilities at the assembling market places. The cold storage facilities may be developed either in private and public sectors or even in PPP-mode.
5. **Development of Rural Roads and Transport Facilities:** It is utmost necessary to develop the road communication system to facilitate the transportation of marketable surplus to the assembling and marketing centres.
6. **Grading, Standardization, and Packaging:** The fruit growers may be encouraged to adopt some post-harvest technology including grading and standardization of pineapple according to size, shape, and degree of ripeness.
7. **Setting up of Regulated Markets:** The regulated market(s) should be developed in strategic location with all facilities, for which a comprehensive plan is a must.

8. **Setting up of Fruit Canning and Processing Units:** There is an urgent necessity of establishing processing units in the vicinity of fruit growing areas for canning and processing of surplus fruits.
9. **Market Information and News Services:** The market information and news may be linked with agricultural extension services, adult literacy centres, and village panchayats to educate the farmers on the prevailing market prices, market arrivals, etc., which would be helpful to enhance the bargaining power of the growers.
10. **Use of Standard Weights and Measures:** As per the Standard Weights Acts, the government introduced metric system of weights and measures since 1958 for various items. It is, therefore, desirable that uniform system of weights and measures be introduced and implemented for the advantage of both producers and consumers of horticultural produce.
11. **Pricing Policies and Provisions of Support Price:** The government policy of announcing support price helped the growers of certain important cereal and cash crops only. But there is no provision of fixation of prices for commercial horticultural crops like pineapple, banana, and orange. It is therefore suggested that the concerned State Governments should be empowered to enact pricing policy to fix minimum prices for the principal horticultural crops in the state.
12. **Provisions of Institutional Credit:** The provision of institutional credit particularly from cooperative and institutional sources should be strengthened, so that the crop growers can tide over the difficulties, whenever necessary.
13. **Market Inspection and Survey:** Considering the complex problems in agricultural marketing, the State Governments should conduct regular market survey and market inspection to study the various problems and situations.
14. **Scope of Export Marketing:** To achieve the untapped potential of export, systematic market survey should be conducted. Some of the private exporters may be encouraged for setting up of modern fruit processing industries in the region for promotion of export. With steady increase of exports there is scope for increasing producer's return from fruit crops.
15. **Strengthening the State Agricultural Marketing Boards:** It is desirable that the commercial horticultural crops, like pineapple, orange, and banana should also be within the purview of the State Agricultural Marketing Board. Various methods of control over sales can be used by the Marketing Boards to improve the farmer's return through market control devices.

The chapter has highlighted that the prospect of horticultural crops in Assam and Meghalaya is bright provided the marketing facilities and the needed infrastructural supports are ensured. It has amply demonstrated that the establishment of fruit processing industry and improvement of marketing network may go a long way in commercialization of horticultural crops in Assam and Meghalaya.

REFERENCE

Government of India. 2014. *Handbook of Horticulture Statistics*, Ministry of Agriculture, New Delhi.

16

Government Intervention in Horticulture Development in Maharashtra

A Case Study of Ratnagiri District

Sangeeta Shroff

The agricultural sector continues to dominate Maharashtra's economy with respect to employment though the share of this sector to state domestic product is declining rapidly. This clearly indicates low productivity of resources invested in agriculture. The main cause of non-remunerative returns is due to agriculture in the state being mainly rainfed and this problem is aggravated as one-third of the area in the state falls under rain-shadow region where rainfall is scanty and erratic. Barely 16 per cent of the gross cropped area is under irrigation. These factors largely explain the dominance of low value coarse cereals such as jowar in the cropping pattern of the state and also the low cropping intensity. Keeping these factors in mind, the Government of Maharashtra (GoM) has been making concerted efforts to improve the productivity of land in the state. One way of doing this is by promoting horticulture in a big way. This is possible because there exist in the state, a number of factors which can promote the cultivation of horticultural crops. The state has diverse soil, topography, and climatic conditions, which are conducive for the cultivation of less water intensive horticultural crops. The cultivable waste land can also be used for the promotion of fruit crops and various fruits can be cultivated in different parts of the state. The state also has the advantage of four state agricultural universities, four colleges of horticulture, and three national research centres, which can help to make rapid progress in developing new varieties of crops, supplying genuine quality planting material, improving productivity, profitability, and sustainability of farming methods. The state also has infrastructure to support horticulture. Keeping in mind the potential of horticulture, the GoM introduced the Horticulture Developed Programme linked with Employment Guarantee Scheme (EGS) in 1990–91.

This scheme was introduced in 1990–91 for 25 fruit crops with the main objective of utilizing 2.9 million hectares of cultivable waste area, generating employment opportunities, and controlling soil erosion. The scheme also aimed at converting land from low value agriculture to high value agriculture which would improve the socio-economic condition of farmers. The scheme provides subsidy as per norms fixed by the government for each fruit crop for a period of three years.

As a result of the implementation of the above mentioned scheme, the area under horticulture rapidly increased in Maharashtra. The area which was 0.24 million hectares in 1990–91 increased by 1.25 million hectares by

2006–07. Presently area under fruits is 1.5 million hectares making Maharashtra the highest producer of fruits among all states in the country. Maharashtra also has highest area in the country (25 per cent) under fruits and produces 17.4 per cent of fruits in the country.

Thus the Horticulture Development Programme linked to EGS helped the state to promote horticulture. Konkan Divison benefitted maximum from this scheme and out of total area under this scheme, 25 per cent was in Konkan division. The maximum area under EGS linked Horticulture was under mango which comprised 36.44 per cent of the area. Cashew nut, orange, and pomegranate were also important horticultural crops under this scheme.

About 18.05 lakh beneficiaries have availed this scheme and the maximum number were in Konkan Division which accounted for 24 per cent of total beneficiaries, followed by Pune division which accounted for 18.5 per cent of beneficiaries.

In the light of the above, this chapter attempts to evaluate the scheme in terms of benefits and problems faced by beneficiaries from Ratnagiri district in Konkan division.

METHODOLOGY AND DATA

This district was selected because it has maximum area under this scheme, not only in Konkan division but in the entire state. Further, area under mango was highest out of all fruit crops and occupies 36.44 per cent of area under this scheme. Thus mango crop in Ratnagiri district was selected to study the benefits that accrued to the beneficiaries under this scheme.

Ratnagiri district is divided into nine talukas and has 1418 villages. Over 85 per cent of the land surface in Ratnagiri is hilly. The predominant soil in the district is red or lateritic soil. Irrigation facilities in the district are very restricted and crop production is mainly concentrated in the Kharif season. About 61 per cent of main workers are in the agricultural sector and with respect to marginal workers it can be observed that 81.2 per cent workers are engaged in agriculture. Thus agriculture is the predominant activity in the district. With respect to land holding pattern, about 53 per cent of the number of holdings belong to small and marginal category but occupy only 26 per cent of the area. In contrast, the number of large and very large holdings is 28.6 per cent but they occupy 57 per cent of the area. This land is however mostly on hill slopes and less fertile.

In Ratnagiri district about 1.35 lakh beneficiaries have availed subsidy under EGS linked Horticulture Programme. As noted above, maximum area under this scheme was under mango. Accordingly, in Table 16.1, the taluka-wise area under mango under EGS linked Horticulture Programme in Ratnagiri district is presented. Table 16.1 shows that out of the nine talukas in Ratnagiri district, the maximum area under mango was in Ratnagiri taluka which constituted 25.5 per cent of the total area in the district. Rajapur taluka had 14.07 per cent area under mango while in case of other talukas this share ranged between 7 and 9 per cent.

Table 16.1 Taluka-Wise Area (hectares) under Mango under EGS Linked Horticulture Programme in Ratnagiri District (1990–91 to 2005–06)

Taluka	Area (hectares)	Percentage to Total Area
Mandangad	3281	9.03
Dapoli	3219	8.87
Khed	3279	9.03
Chiplun	3313	9.12
Guhagar	2572	7.08
Devrukh	3335	9.20
Ratnagiri	9254	25.50
Lanja	2940	8.10
Rajapur	5110	14.07
Total	36303	100.00

Source: GoM, unpublished data.

In Table 16.2, the number of villages covered under mango crop under this scheme, the number of beneficiaries who availed this scheme, subsidy disbursed and man-days generated in Ratnagiri district for mango crop is presented.

It can be observed from Table 16.2 that Ratnagiri Taluka had maximum beneficiaries for mango crop which constituted 25 per cent or one-fourth of total beneficiaries.

Table 16.2 Beneficiaries, Subsidy, and Man-Days Generated under EGS linked Horticulture Programme for Mango Crop in Ratnagiri District (1990–91 to 2006–07)

Taluka	No. of Villages	No. of Beneficiaries	Subsidy (Rs lakhs)	Man-Days Generated (lakhs)
Mandangad	179	4048	531.18	6.78
Dapoli	178	3662	519.75	6.63
Khed	192	3593	531.82	6.78
Chiplun	125	3364	522.77	6.67
Guhagar	122	2702	417.32	5.32
Devrukh	185	3554	547.06	6.98
Ratnagiri	170	10004	1523.36	19.45
Lanja	118	3093	471.44	6.01
Rajapur	220	5384	823.06	10.50
Total	1489	39404	5887.06	75.12

Source: GoM, Ratnagiri District, unpublished data.

Accordingly the subsidy disbursed and man-days generated were also highest in Ratnagiri taluka. Thus, from Tables 16.1 and 16.2, it is clear that Ratnagiri taluka had highest area under mango crop and maximum beneficiaries for this crop were also in this taluka. In view of this, we have selected Ratnagiri taluka for our sample survey on EGS linked Horticulture Programme. The crop selected was mango and up to 2007–08 the area under fruit crops under the EGS linked Horticulture Programme in Ratnagiri Taluka was 14107 hectares. Out of this area, 69 per cent was under mango crop while 28 per cent was under cashewnut.

Thus, from the secondary data it is clear that mango was the most important crop under EGS linked Horticulture Scheme in Ratnagiri Taluka. Accordingly, a field survey was conducted in Ratnagiri Taluka to study the implementation of the scheme and the returns accruing to the beneficiaries. The entire area under mango was under alphonso variety and hence data for this variety was collected.

With respect to mango crop (variety alphonso), in order to study the economic returns from this crop, data was collected from a sample of 30 farmers (9 small, 9 medium, and 12 large) from Ratnagiri Taluka in Ratnagiri District. It may be mentioned here that small farmers were those having land holdings below 2 hectares, medium farmers had land holdings between 2 to 4 hectares, and large farmers were those with land holdings above 4 hectares. These farmers had availed subsidy under the EGS linked Horticulture Programme in 2001–02, which is therefore the reference year of the study. Information was collected from beneficiaries who availed the subsidy in the same year so as to maintain uniformity in the analysis as all these beneficiaries would have the same gestation period. Data on initial year costs, recurring costs, area under the fruit crop, number of trees planted, number of trees survived, output per tree, price per dozen was collected. Thus, by calculating cost and returns, the profitability of the horticulture venture was observed.

RESULTS AND DISCUSSIONS

Cost of Cultivation of Mangoes (Alphonso) in Ratnagiri Taluka

The cost of cultivation of mangoes (Alphonso) is presented from Table 16.3 to Table 16.5. The mango graft had a gestation period of 6 years and it was in the sixth year that the grafted mango plant began to yield fruit. The yield in the fifth year was negligible. In Table 16.3, the initial year or the cost in the first year, that is, 2002–03, are presented. In Table 16.4 the costs from the second to the sixth year

Table 16.3 Cost of Cultivation (Rs per tree) in Ratnagiri Taluka for Mango Graft under EGS Linked Horticulture Programme (First Year cost)

	1st Year Cost			
	Small Farmer	Medium Farmer	Large farmer	All
Area under EGS linked Horticulture Programme (Ha)	0.58	1.16	3.34	1.86
Trees planted (Number)	58.30	116.67	334.20	186.20
No. of trees survived	58.30	116.67	334.20	186.20
Survival rate (percentage)	100	100	100	100
Cost of cultivation				
Land preparation	32.67	31.36	25.78	27.50
Digging of pits & filling	28.11	36.20	193.14	87.50
Fencing	108.58	76.77	89.05	88.62
Cost of Sapling	28.50	29.31	30.85	30.32
Fertilizer (organic + chemical)	20.32	17.05	17.44	17.65
Pesticide	7.60	6.50	19.60	6.20
Water charges	34.7	23.93	14.9	18.63
Labour Charges	66.89	53.34	31.66	39.4
Miscellaneous Charges	8.72	4.48	4.81	5.13
Total Cost (per tree)	336.09	278.94	427.23	320.95
Cost of cultivation (per hectare)	33609	27894	42723	32095

Source: Field Survey.

Table 16.4 Cost of Cultivation (Rs per tree) in Ratnagiri Taluka for Mango Graft under EGS Linked Horticulture Programme (Second to Sixth Year cost)

	2nd to 6th Year Cost			
	Small Farmer	Medium Farmer	Large Farmer	All
Fertilizer (organic + chemical)	110	95	90	98
Pesticide	40	35	57	57
Water charges	175	120	75	123
Labour Charges	340	275	160	258
Total (per tree)	665	525	382	536
Cost of cultivation (per hectare)	66500	52500	38200	53600

Source: Field Survey.

Table 16.5 Cost of Cultivation (Rs per tree) in Ratnagiri Taluka for Mango Graft under EGS Linked Horticulture Programme (Total Cost from First to Sixth Year)

	Total Cost			
	Small Farmer	Medium Farmer	Large Farmer	All
Area under EGS linked Horticulture Programme (Ha)	0.58	1.16	3.34	1.86
Trees planted	58.30	116.67	334.2	186.20
No. of trees survived	58.30	116.67	334.2	186.20
Survival rate (percentage)	100	100	100	100
Cost of cultivation				
Land preparation	32.67	31.36	25.78	27.5
Digging of pits & filling	28.11	36.2	193.14	87.50
Fencing	108.58	76.77	89.05	88.62
Cost of Sapling	28.50	29.31	30.85	30.32
Fertilizer(organic + chemical)	130.32	112.05	107.44	115.6
Pestcide	47.60	41.50	76.60	63.20
Water charges	209.7	143.93	89.9	141.63
Labour Charges	406.89	328.34	191.66	297.4
Miscellaneous Charges	8.72	4.48	4.81	5.13
Total cost (per tree)	1001.09	803.94	809.23	856.90
Cost of cultivation (per hectare)	100109	80394	80923	85690

Source: Field Survey.

are presented and in Table 16.5 the total cost for 6 years is presented.

It can be observed from Table 16.3, that on an average across all farmers the initial year cost was Rs 320.95 per tree or Rs 32095 per hectare as 100 trees are planted in one hectare. Out of this total cost in the initial year, 73 per cent was the fixed cost while 27 per cent was the variable cost. In case of small farmers the fixed cost was 59 per cent whereas in case of large farmers it was 79 per cent. The digging of pits was 57 per cent of the fixed costs in case of large farmers as the sample had two farmers who had rock land. In case of rock land, blasting was required and hence this increased the cost of digging pits.

In Table 16.4, the recurring costs incurred from the second to the sixth year are indicated. After the initial year expenses, the recurring expenses are in the form of fertilizers, pesticides, and water charges. Across all farmers these expenses were Rs 536 per tree and were Rs 382 per tree for large farmers and Rs 665 for small farmers.

Table 16.5 gives a consolidated picture of the total cost incurred for all size group of farmers over the entire gestation period, that is, from the first year when the sapling is planted till the sixth year when the tree begins to yield fruit. From Table 16.5 the following observations can be noted:

1. Across all size group of farmers, on an average 1.86 ha was under mango graft under the EGS linked Horticulture Programme and 186.2 trees were planted. In case of small farmers, the average area under the scheme for mango graft was 0.58 ha and in case of large farmers the corresponding figure was 3.34 ha. Thus, small farmers on an average planted 58.3 trees while in case of large farmers it was 334.2.
2. The survival rates of trees planted on an average across all size groups was 100 per cent. The survival rate was 100 per cent because in case of mortality of plants, they were normally replaced the next year. According to the scheme, if plants are well managed but still do not survive, they are replaced next year. The sample was uniformly chosen so that all farmers had availed the subsidy in 2001–02 and all trees were in the fruit bearing stage in 2006–07.
3. The total cost has been calculated for the initial 6 years because after incurring costs for 6 years, the mango graft begins to yield fruit. The respondents stated that while the trees do sometimes give fruit in the fifth year, the yield is barely 25 fruits or a little more per tree. The total cumulative cost for 6 years, that is, the gestation period was highest at Rs 100109 per hectare for small farmers and on an average across all size groups was Rs 85690 per hectare.

Gross Returns from Mango Cultivation (alphonso) under EGS Linked Horticulture Programme in Ratnagiri Taluka

The gross returns from EGS linked Horticulture Programme in Ratnagiri Taluka are indicated in Table 16.6.

From Table 16.6 the following observations can be noted:

1. The yield per hectare across all size groups of farmers was 11581 fruits or 965 dozens. Small farmers had the highest yield of 1058 dozens per hectare while in case of large farmers it was lowest at 944 dozens per hectare.

Table 16.6 Gross Returns for Mango Cultivation under EGS Linked Horticulture Programme (Rs) in Ratnagiri Taluka in the First Year of Fruit Bearing Stage

	Small Farmer	Medium Farmer	Large Farmer	All
Total fruits per hectare (number)	12695	12008	11324	11581
Total fruits (in dozen) per hectare	1058	1001	944	965
Sale price per dozen(Rs)	127.7	116.2	136	131
Total Gross Returns per hectare (Rs)	135107	116316	128384	126415
Gross return per tree (Rs)	1351	1163	1283	1264

Source: Field Survey.

2. Although the large farmers had lowest yield they obtained a higher price of Rs 136 per dozen as compared to small and medium farmers. This is because of the larger size of the fruit and perhaps better bargaining power. Many of the farmers in this sample were graduates and had better access to information on prices prevailing in regulated markets.
3. The gross returns per hectare across all size group of farmers was Rs 126415 per hectare and was highest in case of small farmers amounting to Rs 135107 per hectare.

Net Returns from Mango Cultivation in Ratnagiri Taluka

The important point to note however is that it is not gross returns per hectare but net returns per hectare that determine the viability of the horticulture venture. Accordingly in Table 16.7, the net returns per hectare have been calculated. In order to calculate the net returns the total cost per tree has been deducted from the gross returns per tree. As there are 100 mango grafts in a hectare the net returns per hectare are also indicated.

From Table 16.7, it can be observed that:

1. The gross return across all size groups of farmers was Rs 1264 per tree and the cumulative cost per tree was Rs 856.90. This indicates in the sixth year itself, net revenue of Rs 407 per tree was obtained which translates to Rs 40710 per hectare.
2. The maximum net returns were earned by large farmers which amounted to Rs 47377 per hectare.

Table 16.7 Net Returns (Rs) from Mango Cultivation under EGS Linked Horticulture Programme in Ratnagiri Taluka

	Small Farmer	Medium Farmer	Large farmer	All
Gross return per tree (Rs)	1351	1163	1283	1264
Cost per tree (Rs)				
1st year cost (per tree)	336.09	278.94	427.23	320.95
2nd to 6th year cost (per tree)	665	525	382	536
Total cost (per tree) (1st to 6th year)	1001.09	803.94	809.23	856.90
Net Return per tree	349.91	359.06	473.77	407.10
Net Return per hectare	34991	35906	47377	40710
Subsidy per hectare				
1st year subsidy	19137	15434	15434	16668
2nd year subsidy	7628	5955	5955	6513
3rd year subsidy	7280	5635	5635	6183
Total subsidy	34045	27024	27024	29634
Net Return + subsidy (Rs) per hectare	69036	62930	74401	70344
Subsidy as a percentage of total cost of initial 3 years from year of plantation	56	55	46.6	55
Subsidy as a percentage of total cost up to fruit bearing stage i.e. 1st to 6th year	34	33.61	33.4	34.58

Source: Field Survey.

Although small farmers had the highest gross returns, the net returns were more in case of large farmers as they had lower costs than small farmers and also obtained a better price.

3. The above-mentioned net returns were earned by farmers without considering the subsidy disbursed to them by the GoM. If the subsidy amount per hectare is added to the net return, then across all size groups, the net return plus subsidy is Rs 70344 per hectare. As per this scheme small farmers are entitled to a higher subsidy than medium and large farmers. This improved their returns to Rs 69036 per hectare.
4. In the first year of fruit bearing stage itself, the beneficiaries earned a positive return which was augmented with the subsidy component. The yield of the mango graft increases over the years as branches in the tree increase. Further, recurring

costs are only in the form of fertilizers, pesticides, and water charges. Thus, returns in the first harvest of the fruit bearing stage can be considered to be the lowest returns (unless crop is destroyed due to weather or other conditions in successive harvests).

As mentioned earlier, mango crop (Alphonso variety) is popular in the Konkan region of the state. Therefore the productivity of the crop in the districts in the Konkan division is presented in Table 16.8.

It can be observed from Table 16.8 that the productivity of mango crop ranged between 1.08 tonnes per hectare to 2.90 tonnes per hectare during the period 2000–01 to 2004–05. There are approximately 5 mango fruits in 1 kg which means that the per hectare yield ranges from 454 dozens to 1208 dozens. Taking a very conservative price of Rs 125 per dozen (after interviewing several mango growers) this translates into gross returns ranging between Rs 56750 to Rs 151000 per hectare. After incurring costs during the gestation period (which is more or less recovered in the first year of fruit bearing stage), the variable costs per year range between Rs 10000 to Rs 15000 per hectare. This means that the mango grower can earn between Rs 40000 to Rs 135000 per hectare, which is a conservative estimate.

With respect to sale of mangoes, the farmers explained that often the trees were given to pre-harvest contractors who accordingly harvested the fruit and graded the fruits according to size. The fruit reached the final consumer through several agencies such as wholesalers or commission agents. The field visit to Ratnagiri taluka also revealed that very recently online trading for sale of mangoes had started. Hence potential buyers and sellers could settle their contract using this facility. Of course, as mentioned this facility had just started and farmers were yet to get conversant with this method of sale.

Table 16.8 District Wise Productivity (tonnes per hectare) of Mango Crop (Alphanso variety) in Konkan Division

Year/District	2000–01	2001–02	2002–03	2003–04	2004–05
Thane	2.69	2.90	1.08	1.20	2.40
Raigad	1.98	2.29	1.60	1.40	1.25
Ratnagiri	2.50	1.19	1.86	1.56	2.29
Sindhudurg	1.63	1.09	2.68	3.24	1.50
Konkan Division	2.17	1.63	1.88	1.84	1.98

Source: GoM, unpublished data from Dept. of Horticulture.

CONCLUSIONS AND POLICY IMPLICATIONS

Government intervention in horticulture has brought about a substantial increase in area under horticulture in the state. A large number of trees are still in the fruit bearing stage. As production of fruits begins to further increase after the gestation period is over, horticultural growth may come under pressure unless supported by markets and other complementary infrastructure. The production of fruits in Maharashtra which was 35 lakh tonnes in 1991–92 has increased to 98 lakh tonnes in 2006, 2012–13. Horticultural produce is highly perishable and in order to transport them from production centres, often in interior villages to consumption points located in urban areas, it is necessary to increase investment in public infrastructure (roads, communication, electricity). Cooperatives and organized retail can help to mop up supplies rather than small fragmented markets.

In India, horticultural produce is mainly consumed as a table fruit. However, due to the perishable nature of the commodity, this sector is marked by a very high wastage and value loss. It is therefore necessary to invest in post-harvest technology, storing, processing, cold storage, etc., so as to increase the shelf life of the produce.

With the opening up of the economy and economic liberalization reforms, there now exist tremendous opportunities to tap export markets. Exports are required to comply by the stringent export quality standards failing which would lead to barriers in trade. In order to meet the challenges of WTO, the Sanitary and Phyto-Sanitary standards imposed by the developed world must be applied so that they do not serve as an impediment to trade. However, the government is not complacent to the problems faced by horticulture sector and is attempting to address these issues through National Horticulture Mission, amendment to Agricultural Produce Market Committee Act, etc., to pave way for direct marketing, contract farming, etc. Efforts are also being made by private sector, but there lies a long road ahead. If the potential of this sector is realized, the agricultural sector is bound to show better performance.

REFERENCES

Government of Maharashtra. 2006–07. *District Socio-Economic Abstract of Maharashtra.*

———. *Economic Survey of Maharashtra*, various issues. Directorate of Economics & Statistics.

———. Department of Horticulture (unpublished data). Pune and Ratnagiri.

17

Economics of Mushroom Farming in Himachal Pradesh

C.S. Vaidya, Pratap Singh, and Ranveer Singh

Mushrooms are the fruiting bodies of some members of lower group of plants, called fungi. Due to this reason the mushrooms are also called fleshy fungi. The fungus and hence mushrooms are characterized by the absence of chlorophyll which is responsible for imparting green colour to plants. Due to absence of chlorophyll, mushrooms are not able to synthesize their own food and have to depend upon outside sources for their nutritional requirements. It is because of this that mushrooms grow saprophytically on dead organic matter or parasitically with other living matter. The mushrooms are fruit bodies or reproductive structures emanating from mycelium, which under natural conditions remain buried under the soil. There is about 100 countries all over the world where mushrooms are cultivated which together produce about 50 lakh tonnes of mushrooms. Of the total mushroom production, 50 per cent is accounted by Europe, 27 per cent North America and about 14 per cent by East Asian countries. Presently, the production of mushrooms is increasing at a rate of 7 per cent the world over whereas in India this growth rate is 30 to 40 per cent per annum. It is expected that the world production of mushroom would increase to 70 lakh tonnes per annum by the year 2010 and to 110 lakh tonnes by the year 2021. According to estimates of National Research Centre for Mushrooms the production in India was 40,000 MT during 1996–97 which was expected to increase to 6 lakh MT by the year 2025. The exports from India, during 1993, were insignificant but presently, it is reported that, India has pushed back Taiwan to gain position of top exporter of whole white button mushroom in the world. India has also gained the second position in the export of cut mushrooms. During 1997–98 total export of fresh and dried mushrooms touched 57 crore rupees. Haryana, Himachal Pradesh, Uttar Pradesh, Punjab, and Tamil Nadu are the main mushroom-producing states in India.

The increasing population has put tremendous pressure on scarce and fixed land resource with a consequence that about 80 per cent of the holdings have become marginal or small. As a result, the income generation from farms is continuously going down. The limited availability of land has made extension of farm limits almost impossible. The only viable alternative is the introduction of non-land-based activities having good income generation capacity. Himachal Pradesh has wide variations in agro-climatic conditions and provides ideal situation for cultivation of mushrooms. Any region which is about 2000 ft above MSL, temperature varies between 10°C to 30°C and has humidity of 75–85 per cent has good potential for cultivation of mushrooms. Many places in the state such as Chail, Solan, Shimla, Mandi, and Dalhousie, as well as the regions around them fulfil these criteria and it is possible to have four harvests per annum at these places. The importance of mushroom cultivation also stems out from facts: Mushroom cultivation generates direct and indirect employment; it requires very little land as it is cultivated in closed rooms; the used compost can be reused as good manure in other field crops; it has capacity of being exported and earning foreign exchange; it provides rich diet to vegetarians; being rich in proteins and low in carbohydrates and fat, it is very good for the patients of heart, diabetes and obesity, etc.

The study has been conducted with the objectives:

1. to study the socio-economic profile of mushroom cultivators;
2. to work out economics of mushroom cultivation on different sizes of farms;
3. to examine the different marketing channels, margins and costs; and
4. to study the socio-economic constraints and problems in production and marketing of mushrooms.

METHODOLOGY AND DATA

The study was conducted in the state of Himachal Pradesh where there were 876 registered growers of mushrooms by the end of December 2009. However, only 112 were actually engaged in mushroom production. Out of 112 registered growers who actually were growing mushrooms 49 were located in district Shimla, 55 in Solan, six in Sirmour, and only two in Bilaspur. Thus, districts Solan and Shimla were purposely selected for the detailed study. It was decided to draw a sample of 40 growers from each district. For this purpose in each of the districts, five locations were identified where maximum number of growers was located. The requisite sample of 40 producers was selected randomly in district Solan but it was found that in Shimla, only 30 producers could be contacted due to very thin spread of activity. Thus, the study has been based on 70 mushroom cultivators located in two districts and ten locations. Further sampled producers were divided into three size classes on the basis of scale of the operation. The producers having less than 100 trays have been categorized as small, 101 to 250 trays as medium and those having more than 251 trays were categorized as large mushroom producers. During data collection it was found that almost all of the producers were using polythene bags instead of wooden trays. In terms of input use and output, it was determined that four polythene bags are equivalent to one wooden tray. Accordingly, the polythene bags were converted to wooden trays for the purpose of determination of their size class.

RESULTS AND DISCUSSION

Socio-Economic Profile of Mushroom Farmers

Family size: The family size forms the basis for determination of working force available for farm activities. Though it is affected by factors such as occupational structure and age composition, it still remains the starting point. The average family size of the sampled farmers was 5.24 persons per family at overall level. The family size was highest among the small farmers (5.92) followed by marginal and large farmers.

Educational level: The literacy levels and more importantly the formal education plays a great role in opening the minds of people to venture into the new fields. About 86 per cent of the persons at overall level were literate and of them about 1 per cent had obtained some formal qualifications. About 3 per cent of the sampled persons were illiterate. And rest 11 per cent were non-school going children. This clearly indicates the very low percentage of persons with formal education which is so important for such unconventional and highly technical vocation like mushroom cultivation.

Off-farm income: The mushroom farmers have been deriving off-farm income from two main sources, namely government job and trade/business. It was found that each family at overall level, derived an annual income of about Rs 43,373. The off-farm income increased with the increase in farm size.

Land resources: Each household at overall level owned 1.36 ha of land of which 0.63 ha land was cultivated. The land resources of large farmers were higher than the small farmers but the categorization of the farmers is not on the basis of land holding but on the basis of scale operation in mushroom cultivation and hence the variation is justified. The extent of irrigated land was almost insignificant which might have increased the risk of production in the field crops owing to dependence on rains which are becoming increasingly erratic and scarce. This scenario might have motivated the farmers to opt for mushroom cultivation which does not depend on rains and the area is agro-climatically suitable for this venture.

Livestock profile: Overall, each farmer had on an average 1.79 cows, 0.09 buffaloes, and 0.65 heads of other livestock. This gave them an income of Rs 31,749 per year.

Economics of Mushroom Cultivation

Cost of Cultivation

The cost of cultivation and production was worked out separately for each category of the mushroom cultivators. The analysis has been carried out by working out the labour costs involved on various items such as application of compost, insecticides and pesticides, cleaning, sorting, and harvesting. It was found that the mushroom cultivators were purchasing the readymade bags for growing mushrooms and thus avoided purchasing separately the inputs such as polythene bags, compost, wheat straw, chicken manure, gypsum, urea, and casing oil. Hence, the costs of these items have not been taken separately (Table 17.1).

Table 17.1 Cost of Production of Mushrooms on Sampled Farms
(Rs/Farm)

Cost Components	Marginal	Small	Large	All farms
A. Variable Cost				
1. **Labour**	23,300.90	69,934	18,4130	43,731.86
	(25.96)	(25.95)	(20.45)	(22.71)
– Compost	713.20	2,946	6,870	1,756.62
– Insecticides/Pesticides	2,629.24	5,712	4,530	3,672.28
– Cleaning			–	
– Sorting			–	
– Watch and Ward	430.18	12,184	8,300	1,141.68
– Harvesting	3,809.43	10,008	25,200	6,965.06
– Washing	1,791.50	6,942	26,100	4,807.22
– Packing	13,927.35	32,142	1,13,130	25,389
2. **Material Cost**	53,184.84	165,056.28	600.000	120,502
	(59.27)	(61.25)	(66.65)	(62.59)
3. **Other Costs**	8,240.15	26,507	92,982.50	8,730.25
	(9.18)	(9.83)	(10.32)	(4.53)
– Electricity charges	514.71	2,700	13200	1,937.10
– Water charges	23.18	35	45	160.00
– Interest on working capital	7,366.68	23,772	79,737.50	16,635.15
– Total variable cost	84,725.94	261,497.28	877,112.50	182,964
	(94.41)	(97.04)	(97.41)	(95.03)
B. Fixed Cost				
Depreciation on:				
– Building	1,366.88	238.59	4,868.42	2,102.06
– Racks	1,800.00	3,650.00	7,300.00	4,250.00
– Implement and machinery	53.38	107.36	364.22	100.55
– Fans	195.97	39.21	1,060.00	94.37
– Refrigerator	–	–	–	–
– Air ducts	1,141.32	3208	7,600.00	2,146.00
Interest on fixed capital	455.75	724.31	2,119.26	869.29
Total fixed cost	5,013.30	7,967.47	23,311.90	9,562.27
	(5.59)	(2.96)	(2.58)	(4.97)
C. Total Cost	89,739.24	269,464.75	900,424.40	192,526.94
	(100.00)	(100.00)	(100.00)	(100.00)
Total Production	20.27	55.76	208.00	42.27
Cost per bag	106.70	88.99	75.03	88.64
Cost per kg.	44.27	48.32	43.29	45.54

Note: The figures in parentheses are percentages to total.

Marginal Farms: The cost of cultivation of mushrooms has been worked for an average farm and per bag basis the details of which have been presented in Table 17.1. It may be seen from the Table 17.1 that total cost of cultivation on marginal farms was Rs 89,739.24 per farm. The highest component of this was of material cost, 59.27 per cent followed by labour cost amounting to 25.96 per cent. Other costs like water and electricity charges and interest on working capital accounted for 9.18 per cent of the total cost. The fixed cost was 5.59 per cent of the total cost which was calculated by taking in to account the depreciation on fixed items such as building, refrigerator, and other equipment. The marginal farmers by spending this amount were able to produce an average of 20.27 quintals of mushrooms on each farm. This amounted to cost of Rs 106.70 per bag and the cost of production on marginal farms was Rs 44.27 per kg.

Small Farms: The cost of cultivation of mushrooms for small category of farmers was Rs 2,69,464.75 per farm. The highest component of this was of material cost, 61.25 per cent followed by labour cost amounting to 25.95 per cent. Other costs like water and electricity charges and interest on working capital accounted for 9.83 per cent of the total cost. The fixed cost was 2.96 per cent of the total cost. The small farmers by spending this amount were able to produce an average of 55.76 quintals of mushrooms on each farm. As a result, the cost of mushroom cultivation was Rs 88.99 per bag and the cost of production on small farms was Rs 48.32 per kg (Table 17.1).

Large Farms: The cost of cultivation of mushrooms for large category was Rs 900424.40 per farm. The highest component of this was of material cost, 66.65 per cent followed by labour cost amounting to 20.45 per cent. Other costs like water and electricity charges and interest on working capital accounted for 10.32 per cent of the total cost. The fixed cost was 2.58 per cent of the total cost. The large farmers by spending this amount were able to produce an average of 208 quintals of mushrooms on each farm. The resultant cost of mushroom cultivation was Rs 75.03 per bag and the cost of production on large farms was Rs 43.29 per kg (Table 17.1).

All Farms: Total cost of cultivation for all farms averaged at Rs1,92,526.94 per farm. The highest component of this was of material cost constituting 62.59 per cent of total cost followed by labour cost amounting to 22.71 per cent. Other costs like water and electricity charges and interest on working capital were 4.53 per cent of the total cost. The fixed cost was 4.97 per cent of the total cost. The average production was 42.27 quintals of mushrooms on each farm. This amounted to total cost of Rs 88.64 per bag and the cost of production on all farms was Rs 45.54 per kg.

Production Pattern

The production pattern of mushrooms depends upon number of crops grown in a year and number of harvests. The farmers were generally taking two crops a year. The number of harvests or pickings depends upon the maturity of individual fruits and market demand. It was observed that during the tourist season in Shimla, the farmer tend to increase the number of pickings as they did not want to wait for taking advantage of good prices. The numbers of harvests per year were 68 for marginal, 72 for small and 73 for large farmers. At overall this figure was 71 harvests. The average production per harvest was higher (284.93 kg) for large category as compared with only 77.44 kg per harvest for small and 29.81 kgs for marginal category. At overall level per harvest production was 59.54 kg per farm. The average production per farm was 2,027 kgs for marginal, 5,576 kgs for small and 20,800 kgs for large category. At overall level the per-farm production was 4,226 kgs per farm.

Marketing of Mushrooms

The mushroom marketing channels starts with a farmer and ends with the ultimate consumer involving a number of intermediaries in between. The involvements of these marketing intermediaries increase the cost of marketing. It was observed that the farmers did not sell mushroom directly to the consumers. This study analysed the marketing channels, producer's share, marketing margins and marketing efficiency in the state.

Marketing Channels: The various marketing channels used by the sampled mushroom cultivators are following:

I. Producer–Retailer–Consumer
II. Producer–Wholesaler-Retailer–Consumer
III. Producer–Transporter–Consumer

The marketing margins have been studied for these marketing channels and it was found (Table 17.2) that the channel I is most efficient from both producer and consumer point of views. The producers are getting the mushrooms at the least cost of Rs 110 per kg as compared with Rs 115 in channel II and Rs 112 per kg in channel III. On the other hand, the producer gets highest percentage of about 77 per cent in consumer rupee in this channel as compared with about 71 per cent in channel II and only 66 per cent on channel III. The marketing cost is highest in channel II and after deducting the marketing costs from the gross returns, the producers' net margin is highest in channel I (71 per cent) as compared to channel II (63 per cent) and channel III (60.2 per cent). However, the

Table 17.2 Marketing Costs and Margins through Different Marketing Channels

Particulars	Channel –I	Channel –II	Channel –III
Price received by growers	84.68 (76.98)	81.44 (70.71)	73.58 (65.70)
1. Marketing costs incurred by growers	6.52	8.70	6.16
– Packing	6.26	6.86	5.88
– Transportation	–	1.60	–
– Others	0.26	0.24	0.28
Net margin of grower	78.16 (71.05)	72.74 (63.25)	67.42 (60.20)
2. Marketing Costs of Trader		–	–
– Packing	–	–	–
– Wastage & spoilage	–	–	–
– Marketing fee	–	–	–
– Handling	–	–	–
– Rent for shop	–	1.10	–
– Wages for labour	–	–	–
– Retailer purchase price	84.68 (76.98)	81.44 (70.71)	73.58 (65.70)
3. Expenses by retailer			–
– Carriage	–	0.30	–
– Losses	–	4.00	–
– Handling		0.20	–
– Market fee	–	0.50	–
Retailer margin-Market fee	25.32 (23.02)	28.56 (24.83)	38.42 (34.30)
Consumer price	110.00 (100.00)	115.00 (100.00)	112.00 (100.00)

Note: The figures in parenthesis are percentages to total

retailers' purchase price was least in channel III being about 66 per cent of the consumers' rupee, it was highest in case of channel I, about 77 per cent of the consumers' rupee. The retailers' margin was highest in channel III, about 34 per cent of the consumers' rupee as compared with about 25 per cent in channel II and only 23 per cent in channel I.

Problems Faced by Mushroom Farmers

During the course of investigation it was observed that many of the mushroom farmers had either temporarily or permanently given up the vocation. Such a scenario could emerge only because the vocation is riddled with many problems. Such problems were envisaged on two stages namely production stage and marketing stage. Thus, the problems related with these aspects were listed and analyzed with the help of multiple response analysis.

Production Stage Problems: It was heartening to note that none of the farmers reported lack of up to date knowledge of production techniques and they revealed that any doubts can be cleared by visiting National Research Centre for Mushrooms, Solan or consulting the fellow farmers. The lack of capital is a big hurdle in undertaking the activity and further increasing the scale of operation. In fact, this was most important problem, about 71 per cent farmers reporting it at overall level. The marginal farmers were managing the problem more effectively with about 55 per cent reporting it. But all the small and large farmers were facing this problem acutely due to larger scale of operation. The cultivation of mushrooms being highly scientific endeavour requires specialized labour for the purpose. About 65 per cent farmers reported that it was very difficult to find required labour force conversant with the intricacies of the activity. The problem was observed to be more acute for small farmers; about 72 per cent small farmers faced the problem. This problem was being faced by about 60 per cent of marginal and large farmers. The inputs required for mushroom cultivation like spawn have to be prepared by employing proper scientific techniques or has to be purchased from National Research Centre for Mushrooms, Solan, and private dealers. A few sources of critical inputs, many times, create problems of their availability, 52 per cent farmers reporting this problem. However, the problem was not as acute for marginal and large farmers, 40 per cent farmers facing it. But for small farmer this problem was acute as 80 per cent encountered the unavailability of inputs. The mushroom cultivation being highly capital intensive venture makes provision of credit an important issue. Although the banks have a provision of credit for this activity and NABARD has refinance scheme for commercial banks for loans granted for this activity, the farmers usually face many difficulties in obtaining credit for either taking up this activity or increasing the scale of operation. The long procedures and difficult requirements are reported to be the main cause. About 61 per cent farmers at overall level faced problems in obtaining credit for mushroom cultivation. The percentage of such farmers was 55 in case of marginal and 100 in case of large farmers. The inputs for production of mushrooms were reported to be very costly. This problem was reported by about 58 per cent farmers at overall level. The percentage of farmers reporting this problem was about 57 per cent in case of marginal farmers, 64 per cent in case of small and 40 per cent in case of large farmers.

Marketing stage problems: Majority of the mushroom farmers were small and as a result had low volume of production. The marketing of small quantities increases the unit cost of marketing resulting in lower profits.

This problem was reported by about 63 per cent farmers overall. The category wise analysis indicates that all the marginal and none of the small and large farmers faced this problem. It was observed that this problem was not acute for marginal farmers as they marketed the produce in nearby markets due to low volume of production. About 47 per cent of the marginal farmers reported this problem. But small and large farmers faced this problem more acutely due larger volume of production which required larger markets for disposal. Consequently 80 per cent of the small and all the large farmers faced this problem. Overall, about 60 per cent of sample reported this problem of markets being located far away; increasing the cost of marketing and the problem was compounded by highly perishable nature of mushroom. The local demand here means the demand by the households located within the village itself where the farm is situated or in the near vicinity. About 69 per cent farmers at overall level thought that it would have been ideal if their entire produce was consumed within the village and they would be saved from marketing problems and its costs. In the process they were ready to forego higher profits and contend with lower returns. The percentage of farmers reporting this problem was about 66, 72, and 80 in case of marginal, small, and large farmers, respectively. The highly perishable nature of mushrooms poses problems during marketing as the time available for disposal was quite low. This is especially true in absence of processing which increases the shelf life. This problem was reported by about 69 per cent farmers and was more pronounced in case of marginal farmers. Many farmers complained of low prices, about 73 per cent farmers at overall level felt that prices were low in comparison with other fresh vegetables and did not match the trend of prices of other vegetables. About 58 per cent farmers at overall level felt that marketing cost of mushrooms was quite high. The problem was more acute for small farmers, 64 per cent farmers complaining on this account as compared with 57 per cent marginal farmers and 40 per cent large farmers. Those farmers who either were marketing the produce in distant markets or were planning to do so complained that no market information for these markets was available. In absence of this it was impossible to tap full potential and comparative advantage of these markets. Overall, 77 per cent farmers had complained in this respect.

POLICY SUGGESTIONS

The private units for production of compost and spawn are encouraged as these are the inputs whose availability is critical for the adoption and spread of the activity. The benefit of training should be extended to larger number of people. More persons can be motivated for attending these training by increasing the daily allowance and other benefits can be included as a package, for example, some quantity of free compost or spawn or other inputs like polythene bags etc. The importance of the activity and the training programme schedules and importance should be widely advertised in local Hindi newspapers, read in rural areas. Transportation subsidy should be provided on the produce for bringing it to the market. The extension services should be geared up for providing technical advice on the doorstep of the farmers. The department of horticulture should provide the compost not only to registered growers but also to anyone who grows mushrooms, whether registered or not. The farmers should be advised to reduce the use of labour for mushroom cultivation. The working capital per bag needs reduction. None of the mushroom growers was observed to be processing the mushrooms. The importance of this should be told to them. The extent of the activity can be increased many times without having any fear of market demand.

18

Evaluation of Five Decades of Livestock Development in Maharashtra and Threats and Opportunities in WTO Regime

Deepak Shah

Encompassing a wide geographical area and reflecting different political system, differing levels of economic development, social systems and changes in tastes, preferences and traditions, the approach to livestock development has varied widely from region to region in India, especially with respect to consumption of milk and milk products. Viewing our livestock spectrum in the light of such variability, it is pertinent to ask whether the future of our livestock will remain as bright as in the past. It is perceived that free world trade regime ushered in by the WTO not only poses many threats to India's livestock industry but also opens up many opportunities for the industry. This is certainly a point that needs to be investigated in the today's WTO regime, which has been marked with sustained dumping of cheap imports of dairy products on to the developing countries and which needs to be checked by taking stringent measures to revive tariff rates and quotas.

In view of the strategic importance of livestock sector in the agricultural development of our economy, its systematic and well conceived development becomes the pre-condition for successful agricultural policy. Technological changes in agriculture associated with the green revolution have brought about significant changes not only in the structure of milk production but also in the size and composition of animal draught power in several areas of the country. The state of Maharashtra is not an exception to this phenomenon. The issues of increase in milk production and demand for and supply of inputs have acquired new dimension in the state of Maharashtra. Undoubtedly, therefore, for designing appropriate policies of livestock development and thereby giving a further boost to their contribution, it is extremely essential to focus on the nature and significance of changes taking place in the animal husbandry sector across various regions of the state. This study, thus, comprehensively evaluates various issues relating to livestock sector of the state, especially in respect of changing structure in livestock production, changes in size and composition of availability of draught animal power, impact of mechanization on draught power availability and changes in policies governing development of livestock sector in Maharashtra in terms population and production dynamics and infrastructure development, aside from assessing prospects of developing livestock sector in WTO regime.

OBJECTIVES OF THE STUDY

The major objectives of the study are: (i) to assess the changes in size, composition, and availability of draught animal power in relation to mechanical power with a view to assess to extent of mechanization in the state of Maharashtra, (ii) to examine the changes in the composition of bovine population over time and also to analyse growth of breedable female population vis-à-vis total stocks of bovines in the State, (iii) to evaluate the structural changes in milk production and also to identify factors responsible for imbalances in milk production across different regions of the State, and (iv) to examine opportunities as well as threats to livestock sector in WTO regime.

METHODOLOGY AND DATA

The reference period of study for the changing dynamics of livestock population for Maharashtra and India is from 1951 to 2003. However, the pattern of livestock production with respect to Maharashtra state is evaluated encompassing the period between 1985–86 and 2005–06, as before 1985–86 the estimates for various livestock products, including milk, are not available consistently for various districts of the state. Efforts have been made to collect information on other important aspects of livestock development encompassing last 4–5 decades. Data used for this study were chiefly collected from various secondary sources. Data on livestock population for different districts of Maharashtra and India were collected from various livestock census reports of Maharashtra and India. The analytical techniques in this study include estimation of growth rates, instability in growth, estimation of coefficient of variations and rank correlation coefficients between two periods. Rank correlation coefficients were calculated to examine the changes in ranking of districts over time in terms of production of various products of livestock origin. In this study, exponential trend equations have been fitted to the time series data obtained for various parameters from various sources in order to compute compound rates of growth that were also tested for their significance by the student 't' statistics. With a view to understand growth performance of various parameters better and in order to capture year to year fluctuation in the same over the given period of time, an index of instability was also incorporated in the analysis, which appeared to have taken care of the trend component in the time series data. The instability was also estimated with the help of computation of Coefficient of Variation.

RESULTS AND DISCUSSION

The major findings mainly revolve around various estimates relating to changes in livestock population over time in Maharashtra vis-à-vis India, variations in livestock population across regions of Maharashtra, changes in size and composition of livestock, growth in breedable population vis-à-vis total stocks of bovines, availability of draught animal and mechanical power over time in the state, the extent of tractorization and mechanization of irrigation, the impact of farm mechanization on draught power availability, and assessment of factors responsible for the adoption of advanced technology. Besides providing estimates relating to livestock population dynamics, the findings include estimates with respect to the livestock production scenario of Maharashtra over the last two decades with major foci of attention on structural changes in milk, meat, wool, and egg production across different districts and regions of the State, growth and instability in livestock production, productivity variations of milch animals, especially bovines, identifying factors responsible for regional imbalances in livestock production, changes in ranking of districts over time in terms of livestock production, etc. The findings also encompass a broader insight into various livestock development policies, programmes and schemes that were initiated in the state of Maharashtra for improving the overall livestock resource base of the State, besides examining the likely impact of trade liberalization under WTO regime on the domestic market in general and the livelihood of farming community of India in particular.

AGRICULTURAL MECHANIZATION IN MAHARASHTRA

It has been noticed that in certain parts of the country, the level of mechanization has reached such a stage where it has gradually started displacing work animals. A section of researchers have also put forward the argument that the growth in mechanization has met the additional draught power requirement of the green revolution in agriculture by supplementing the work animal population (Mishra and Sharma 1990; Nair and Dhas 1990). The issue of demand for draught animals arising out of technological and institutional changes in agriculture coupled with their supply has received very scantly attention in the past. However, the works of Binswenger (1978), Sharma (1981), and Vaidyanathan, Nair, and Hariss (1982) provide good insight into this important aspect of the livestock economy and its linkages with agriculture.

The bovine economy of Maharashtra witnessed a number of changes in terms of its size, composition and productivity. The size of bovine herd increased from about 17 million in the mid-1950s to 21.7 million in the late 1980s. While the sex composition of bovine has shifted in favour of females, its breed composition has shifted considerably in favour of crossbreeds. The draught animal population of Maharashtra has shown a slight increase over time. The issue of economic viability along with socio-economic acceptability of species and breeds will become more pertinent under the changed situation in the State.

Draught Animal Population

Slow growth in the stock of draught animals was observed in the state of Maharashtra. During the period between 1951 and 2003, the draught animal population in Maharashtra increased only by 0.31 per cent per annum. The Pune region showed a declining trend in the total stock of draught animals due to fall in its draught bovine population after 1966 (Table 18.1). The other regions of Maharashtra also showed very slow growth in draft animal population after 1966.

The decline in draught animal population in Pune region and its slow growth in other regions of the state after 1966 might be due to interaction of a number of factors, some of which are decline in the average size of cultivated holdings, shift in cropping pattern and increase in the cost of rearing work animals. As human population pressure on land increases, the size of land holding tends to decline. Because of the indivisibility of work animals, the density of work animal population per unit of cultivated area tends to increase. However, beyond a point when the average size of cultivated holding falls below the critical minimum needed to maintain a pair of work animals, there will be a tendency to do away with work animals and to cling to milch animals.

Mechanization of Irrigation

The growth of mechanization of irrigation has contributed significantly to the increase in the availability of draught power in agriculture. A rapid increase in the number of electric pumpsets and oil engines was seen in Maharashtra during the last five decades. The number of electric pumpsets and oil engines in Maharashtra, which stood at only 0.2 lakhs during 1951, increased to some 5.9 lakhs in 1982 and 10.3 lakhs by 2003 (Table 18.2). The bulk of the increase was contributed by electric pumpsets.

In Maharashtra, out of the total net irrigated area, 56 per cent was well irrigated in 1972–73, which marginally rose to 58 per cent in 1982–83 and sharply to 68 per cent in 2001–02. The rising importance of well irrigation in the state implies that the requirement of various sources of draught power for lifting water has been increasing. Notably, the installation of more electric pumpsets and oil engines has not had any significant influence on the changes in the draught animals stock. Given the utilization pattern of bullock, an increase in the intensity of mechanized irrigation would not have resulted in a reduction in the draught animal stock. Though it might contribute to the displacement of bullock labour from irrigation, this must have been more than compensated by the increase in the cropping intensity consequent to the increase in the intensity of mechanization. It follows that if mechanization of irrigation is followed by mechanization of land preparation and other cultivation operations

Table 18.1 Region-Wise Draught Animal Population in Maharashtra

Region	Year	Draught Animal Population (10^6)			CGR	NSA (10^6 Ha)	DDA
		Cattle	Buffalo	Total			
Maharashtra	1951	5.24	0.31	5.55	–	17.77	0.31
	1966	6.44	0.32	6.76	1.46	18.01	0.38
	1982	6.57	0.34	6.91	0.12	17.96	0.38
	2003	6.32	0.21	6.53	−0.20	17.50	0.37
India	1951	60.73	6.54	67.27	–	134.98	0.50
	1966	71.43	7.59	79.02	1.14	137.72	0.57
	1982	70.07	5.97	76.04	−0.12	142.51	0.53
	2003	54.32	5.83	60.15	−1.26	138.27	0.44

Notes: CGR – Compound Growth Rate; NSA – Net Sown Area; DDA – Density of Draught Animal. Net sown area figures used are three-year average centered on the respective years in the table.
Source: All India Livestock & Farm Equipment Census, Maharashtra State (various years), Directorate of Animal Husbandry, Maharashtra State, Pune.

Table 18.2 Region-Wise Number of Electric Pumpsets and Oil Engines in Maharashtra

Regions	Year	Electric Pumpsets and Oil Engines (10^4)						Density of Electric Pumpsets and Oil Engines per 100 Ha of NSA
		Oil Engines		Pumpsets		Total		
		Number	CGR	Number	CGR	Number	CGR	
Maharashtra	1951	1.76	–	0.10	–	1.86	–	0.10
	1966	14.68	16.03	3.80	28.61	18.48	17.43	1.03
	1982	13.92	−0.32	44.86	15.97	58.78	7.19	3.27
	2003	11.45	−0.97	91.14	3.61	102.59	2.82	5.86
India	1951	8.25	–	2.62	–	10.87	–	0.08
	1966	47.10	12.97	41.46	21.33	88.56	15.82	0.64
	1982	329.60	12.38	358.10	13.81	687.70	13.09	4.83
	2003	7237.40	16.70	8448.30	17.12	15685.70	16.92	113.44

like harvesting and threshing, it would result in a reduction in the work animal stock.

Farm Power Availability

A steady growth in pumpsets, oil engines and tractors has resulted in a significant increase in the availability of total farm power in Maharashtra. The estimated gross availability of farm power from animal and mechanized sources in Maharashtra was about 2.3 million horse power (HP) units in 1951 and this increased to 3.7 million HP in 1966 and 6.2 million HP in 1982. By 2003, it is estimated to have increased to 10.24 million HP units in Maharashtra (Table 18.3). Bulk of the increase in total farm power availability over time was contributed by regions like Nashik, Pune, and Aurangabad. The composition of farm power in Maharashtra has also undergone a marked change over time. While the share of power from mechanized sources in total farm power availability has shown a rising trend in the last five decades, a declining trend in this respect was noticed in the case of draught animal power, though in absolute terms, the draught animal power marginally increased from 1966 to 1982 with decline in the same thereafter.

In Maharashtra, the share of animal power declined to 26 per cent by 2003, which was about 73 per cent in 1966 and as much as 95 per cent in 1951. The study also revealed that in Maharashtra the share of irrigation equipment in total farm power availability increased from 4 per cent to 50 per cent and that for tractors from 1 per cent to

Table 18.3 Availability of Mechanical and Draught Animal Power (HP) in Maharashtra

Region	Year	Mechanical Power (MP) (10^4 HP)				DAP (10^4 HP)	TFP (10^4 HP)	Share of MP in TFP (%)	Share of DAP in TFP (%)	Share of Pumpsets and Oil Engines in TFP (%)
		Oil Engines	Pumpsets	Tractors	Total					
Maharashtra	1951	8.78	0.52	1.79	11.09	222.11	233.19	4.75	95.25	3.99
	1966	73.39	18.99	8.06	100.44	270.41	370.85	27.08	72.92	24.91
	1982	69.60	224.32	51.69	345.60	276.54	622.14	55.55	44.45	47.24
	2003	57.26	455.69	249.34	762.29	261.33	1023.62	74.47	25.53	50.11
India	1951	41.24	13.09	21.59	75.91	2690.62	2766.53	2.74	97.26	1.96
	1966	235.48	207.31	135.03	577.82	3160.92	3738.74	15.45	84.55	11.84
	1982	1648.00	1790.50	1296.25	4734.75	3041.44	7776.19	60.89	39.11	44.22
	2003	36187.00	42241.50	5903.00	84331.50	2406.12	86737.62	97.23	2.77	90.42

Note: DAP – Draught Animal Population; TFP – Total Farm Power; HP – Horse Power. It is assumed that one animal is equivalent to 0.4 HP, Oil engines/pumpsets 5 HP, and tractor to 25 HP

Source: Computations are based on figures compiled from various livestock census reports of Maharashtra and India.

24 per cent between 1951 and 2003. Around 67 per cent of the mechanized power in Maharashtra's agriculture was estimated to be derived from oil engines and electric pumpsets and the latter has been increasing at a faster rate in recent years. Consequently, the consumption of electric power per hectare of cultivated land has increased rapidly.

Tractorization

In the state of Maharashtra, tractorization has taken place at a faster rate from the mid-1960s onwards. The annual growth in tractor population of Maharashtra hovered at around 12 per cent between 1951 and 1982 and declined to 8 per cent between 1982 and 2003. In 1951, there were only 714 tractors in the state, which increased to 3224 in 1966 and further to 20674 in 1982. By 2003, the number of tractors in the state increased to as many as 99,735, showing thereby nearly five folds rise in the same between 1982 and 2003 and 30-folds between 1966 and 2003. As for mechanization, one of the major observations of this study is that though the intensity and spread of mechanization has progressed rapidly in Maharashtra, even now it is operating at a low level. Therefore, its impact on displacement of work animals is likely to be low. The incidence of displacement is likely to be more in the large and medium farms where the intensity of tractorization is higher. Further, mechanization of irrigation has taken place at much faster rate than mechanization of tillage in Maharashtra. Thus, the combined effect of mechanization on displacement of work animals is likely to be low in this State.

BOVINE POPULATION TRENDS IN MAHARASHTRA

The bovine population in the State has been showing rising trend over time. It increased by about 42 per cent during the five decades between 1951 and 2003. A notable feature was that while the adult male bovine population remained stable over the period of five decades, the adult female bovine population has shown significant growth. Consequently, the sex ratio of adult bovine has shifted rapidly in favour of females. Another important observation was the declining proportion of cattle population in relation to total stock of bovines. It has also been noticed that about 96 per cent of the total adult male bovine population in Maharashtra was used for draught purposes. However, the growth of adult male bovine population over time has been very slow. This was perhaps due to the increasing farm mechanization in the State.

Growth Rates in Bovines

The growth in buffalo population was much faster than cattle population in Maharashtra state. Though Aurangabad region possessed the largest bovine population, it ranked 4th in terms of overall growth (1951–2003) in buffalo population and 3rd in cattle population. The Nashik region had the second largest stocks of bovines in Maharashtra, which ranked 3rd in terms of overall growth in buffalo and 1st in cattle population. In terms of stocks of bovines, Pune was the third most important region of Maharashtra, though it ranked 2nd with respect to overall growth in buffalo and 5th in cattle population between 1951 and 2003. Both these regions showed positive growth in cattle and buffalo population during the entire period between 1951 and 2003. In fact, increase in buffalo population was mainly contributed by increase in stocks of breedable buffaloes. Its population increased in all the regions in all the periods thereby giving positive and substantially higher growth rate for the whole State.

Structural Changes in Milk Production

An overall analysis revealed that during the period between TE 1987–88 and TE 2005–06 the increases in milk production figures were much sharper for Nashik, Pune and Aurangabad regions compared to other regions of the State. However, as for the total contribution, Pune region showed the highest contribution to the State's total milk production. The contribution of Pune region to the State's total milk production stood at about 35 per cent during the period between triennium ending (TE) 1987–88 and TE 2005–06 (Table 18.4). The other major contributors to the State's total milk production were Nashik and Aurangabad regions—contributing 26 per cent and 18 per cent to the State's total milk production during TE 2005–06.

As regards milk production expansions, while Aurangabad region showed the maximum expansion in milk production of crossbred cows during the period between TE 1987–88 and TE 2005–06, the increase during the same period for indigenous cow, buffalo and goat milk production was found to be the highest in the case of Nashik region. In general, the past two decades showed 160 per cent increase in total milk production for the state of Maharashtra, which has been due mainly to the production increases of crossbred cow milk as the increases in the case of indigenous cow, buffalo, and goat milk production have been very slow during this period.

Table 18.4 Changing Milk Production Pattern in Maharashtra: (1985–86 to 2005–06)
(Milk Production in lakh kgs)

Regions	Indigenous Cow			Crossbred Cow			Buffalo			Goat			Total Production		
	P-I	P-II	% Chg	P-I	P-II	% Chg	P-I	P-II	% Chg	P-I	P-II	% Chg	P-I	P-II	% Chg
Konkan	572	894	56	179	440	146	2096	4178	99	46	110	139	2892	5622	94
Nashik	1671	3348	100	870	7352	745	1878	5387	187	361	760	111	4781	16846	252
Pune	1293	1740	35	1828	8927	388	4872	11699	140	472	837	77	8466	23202	174
Aurangabad	2037	2741	35	320	3461	982	1916	4628	142	371	693	87	4644	11523	148
Amravati	1076	1212	13	148	844	470	1264	2145	70	169	269	59	2656	4470	68
Nagpur	736	939	28	130	1264	872	842	1641	95	115	203	77	1823	4046	122
Maharashtra State	7385	10874	47	3475	22288	541	12868	29678	131	1534	2872	87	25262	65711	160

Notes: Period-I: 1985–86 to 1987–88 (Triennium Average); Period-II = 2003–04 to 2005–06 (Triennium Average); Chg: Change.
Source: Estimates are based on figures compiled from various reports on 'Milk, Eggs, Wool and Meat Production and Livestock and Poultry Keeping Practices in Maharashtra, Directorate of Animal Husbandry, Maharashtra State, Pune.

Regional Imbalances in Growth of Milk Production

Although Pune region showed very high share in State's total milk production through crossbred cow, buffalo and goat, in respect of production expansions of indigenous cow milk this region lagged far behind Nashik and Aurangabad regions. The higher share of Pune region in total milk production of Maharashtra was mainly due to its significant contribution to State's total milk production of crossbred cow, buffalo and goat milk production. In the case of crossbred cow, Nashik region showed significant expansion in terms of its share in total crossbred cow milk production of the State, which increased from 25 per cent during 1985–86 to 1994–95 to 33 per cent during 1995–96 to 2005–06, whereas this share for Pune region declined from 53 per cent to 40 per cent during the same period. In fact, Nashik, Pune and Aurangabad regions also showed very high share in total goat milk production of the State as there three regions accounted for almost 70 per cent of the total goat milk production of the State. The Nagpur region accounted for the lowest share in terms of all the breeds of milch animals, followed by Amravati and Konkan regions.

Structural Changes in Egg Production

The egg production in Maharashtra through both improved and *deshi* breeds of poultry was estimated at 2,597 million during TE 1996–97, which increased to 3,440 million during TE 2005–06, showing thereby 32 per cent rise in the same over the last one decade. Among various regions, Pune region was found to account for substantial share in total egg production of Maharashtra, though its share in total egg production of the State declined from 57 per cent during TE 1996–97 to 47 per cent during TE 2005–06. The next important region in terms of egg production in Maharashtra was found to be Nashik, which accounted for 16 per cent share in total egg production of the State during TE 1996–97 and 26 per cent during TE 2005–06. Thus, a decline in share of Pune region in total egg production of the State was compensated by an increase in share in this respect by Pune region during the same period. The share of other regions in total egg production of the State remained by and large constant over the last one decade, and hovered at around 3–4 per cent for Amravati region, 6–7 per cent for Aurangabad and Nagpur regions.

Structural Changes in Wool Production

Although the state of Maharashtra accounts for a reasonable share in total wool production of India, there has not been much increase in wool production in the State, which has grown hardly 0.63 per cent annually during the period between 1994–95 and 2005–06. The wool production in Maharashtra was estimated at 1,550 metric tonnes during TE 1996–97 and this increased to 1,645 metric tonnes during TE 2005–06, showing hardly 6 per cent rise in the same during this period. However, some of the regions of Maharashtra like Pune, Nashik, and Aurangabad were noticed to account for substantially very high share in total wool production of the State. Pune region accounted for as much as 45 per cent share in total wool production of Maharashtra over the last one decade. The region of Nashik was the second in terms of total wool production of the State, which accounted for 30 per cent share in the State in wool production during TE 1996–97 and 34 per cent share in this respect during TE 2005–06. The Aurangabad region showed a share of 15–16 per cent in total wool production of Maharashtra over the last one decade. This amply demonstrates the fact that the regions of Pune, Nashik and Aurangabad cornered more than 90 per cent share in total wool production of the State mainly because of their significantly high share of sheep population as compared to other regions of the State. As for wool production, the estimated rank correlation coefficient in respect of cross-section of districts was positive as well as very high, implying not much of a change in the ranking of districts in respect of their contribution to State's total wool production during the period between 1994–95 and 2005–06.

Structural Changes in Meat Production in Maharashtra

The meat production in Maharashtra was estimated at 1.58 lakh tonnes during TE 1987–88, which increased to 1.72 lakh tonnes during TE 1996–97 and by the TE 2005–06, it had grown to 2.31 lakh tonnes, showing thereby nearly 50 per cent increase over the last two decades. It is to be noted that though cattle accounted for the major share in total meat production in Maharashtra during eighties and even nineties period, the scenario changed thereafter and at present buffaloes account for major share in total meat production of the State. Another species of livestock contributing significantly to the total meat production of Maharashtra was goat, which accounted for 28 per cent share in total meat production of the State during TE 1987–88 and 23 per cent during TE 2005–06. The goat meat production in Maharashtra grew at the rate of 5.32 per cent a year between 1995–96 and 2005–06, though it recorded a negative annual growth in the same at 1.92 per cent between 1985–86 and 1994–95. Similarly, sheep meat production in the State grew at rate of 5.76 a year between 1995–96 and 2005–06. This is an indication of the fact that meat production through sheeps and goats in Maharashtra expanded tremendously only in more recent times.

TRADE LIBERALIZATION AND LIVESTOCK SECTOR

Though over the past two decades India has been net exporter of meat and meat products with negligible dependence on import trade of these products, the scenario obtaining in terms of export trade of milk and milk products during this period is not very encouraging. Despite the fact that India's dependence on import trade of butter, ghee from cow milk, cheese and curd animal fats, etc. has come down sharply over the past two decades in the face of rise in export trade in the same, the trade balance of India in these products remains negative due to higher value associated with imports as against export (Table 18.5). In the era of WTO regime India faces significant threat in the case of import trade of some of the dairy products like butter, ghee, cheese and curd, animal fats and some other livestock based products like hides and skins.

Of late the distortions in global livestock trade are taking place due to subsidized production of livestock products in EU and USA. These subsidized livestock products are exported in the world markets much below their true cost of production. This coupled with trade barriers, restrictive trade policies and stringent health and sanitary standards restrict many producers in developing world to enter in higher priced international markets. In the dairy sector, the subsidized exports of EU have adversely affected the dairy industry in India, Brazil and Jamaica.

A very recent study comes out with several interesting observations insofar as the impact of trade liberalization on domestic producers and consumers of milk and milk products is concerned. In order to evaluate the effect of trade liberalization on domestic market of milk and milk products, the study considers three alternative scenarios of world prices (fob) of milk with Rs 647 (US$1500) per quintal as marked with lower range of world price, Rs 1140 (US$ 2650) per quintal as higher range of world price, and Rs 884 (US$ 2050) per quintal as intermediate range of world price (Table 18.6). The study observes a steep fall in the domestic prices of milk following the decision to import milk products at a low international price (Rs 640 per quintal).

Since supply of milk is highly price elastic, this situation will lead to adverse affect on milk production in states like Haryana, Maharashtra, Tamil Nadu, West Bengal, and Uttar Pradesh that account for more than 40 per cent share in total milk production of India. The decline in milk supply will further translate into negative producer's surplus in all the states of India. However, the magnitude of negative surplus is likely to vary across states. In this sequel, Uttar Pradesh will show the highest negative surplus, followed by Maharashtra, Tamil Nadu, Haryana, and West Bengal.

In the WTO regime, surging imports have not only affected farm incomes but also employment in many

Table 18.5 Export and Import Trade of India in Livestock Products
(in '000' US $)

Exports/ Imports	Milk Equivalent	Milk Condensed, Dry & Fresh	Butter	Ghee from Cow Milk	Cheese & Curd	Animal Fats	Hides & Skins	Hen Egg	Meat@ Products (Buffalo)
Export									
TE 1983	1525	333	1186	1170	19	–	374	2526	37333
TE 1993	3910	2545	1270	1251	8	5	653	4197	72244
TE 2001	45361	20470	5127	4713	304	628	730	12477	224270
CGR (%)									
1981–90	0.90	7.06	–3.08	–4.57	2.38	–60.34*	–10.93	–14.02	7.89*
1991–2001	29.92*	20.09	16.85*	15.93*	56.12*	50.32*	–4.52	18.14*	18.77*
1981–2001	18.35*	20.73*	5.96*	5.65*	12.12**	–11.42	7.27	19.84*	14.17*
Imports									
TE 1983	132054	74009	57973	47975	72	49610	1552	11	–
TE 1993	11701	8555	2976	2976	71	295	24930	–	–
TE 2001	20265	9487	9557	9543	958	893	64299	64	–
CGR (%)									
1981–90	–25.02**	–21.67**	–48.23*	–48.11*	–0.94	–52.49*	56.64*	1.14	–
1991–2001	1.37	–7.53	16.32	16.29	35.14*	14.61*	12.15*	11.40	–
1981–2001	–13.63*	–18.31*	–9.96	–8.34	10.82*	–10.76	27.50*	1.51	–
Net Exports/ Trade Balance									
TE 1983	–130529	–73676	–56787	–46805	–53	–49610	–1178	2515	37333
TE 1993	–7791	–6010	–1706	–1725	–63	–290	–24277	4197	72244
TE 2001	25096	10983	–4430	–4830	–654	–265	–63569	12413	224270

Note: *: significance of growth rate at 1 per cent level of probability.
**: significance of growth rate at 5 per cent level of probability.
@: Buffalo meat accounts for nearly 90 per cent in total meat and meat products exports of India.
Source: Computations are based on figures obtained from Occasional Paper, NABARD and FAO Trade Yearbook.

developing countries. Consequent upon cheap imports and absence of adequate protection measures, safeguarding income and livelihood of poor farmers have emerged issues that need to be addressed by policy makers.

POLICY IMPLICATIONS AND CONCLUSIONS

An analysis into farm mechanization in Maharashtra reveals that technological changes in agriculture associated with the green revolution has brought about significant changes in the size and composition of animal draught power as there is relatively increasing dependence on mechanized sources of farm power as against animal draught power. The reduction in the demand for work animals has two major consequences: a proportionate release of animal feed; and a change in the composition of livestock population in favour of milch stock. Although the rising importance of well irrigation in the state implies that the requirement of various sources of draught power for lifting water has been increasing, the rise in the intensity of mechanization did not appear to have affected the draught animal population in the state. Further, given the utilization pattern of bullock, an increase in the intensity of mechanized irrigation would not have resulted in a reduction in the draught animal stock. Though it might contribute to the displacement of bullock labour from irrigation, this must have been more than compensated by the increase in the cropping intensity consequent to the increase in the intensity of mechanization. In fact, the intensity and spread of mechanization is operating at low level in Maharashtra, its impact on displacement of work animals is likely to be low. Interestingly, while the adult male bovine population remained stable over the last five decades in Maharashtra, the adult female bovine population increased significantly, implying a rapid shift in the sex ratio of bovines in favour of females.

Table 18.6 Impact of Free Import of Milk on Producer, Consumer and Net Social Welfare in Selected States of India with Different Range of World Prices in 1999

Particulars	Low Range of World Price (Rs 640 per quintal)					Intermediate (Rs 850/quintal)	
	Haryana	Maharashtra	TN	WB	UP	Maharashtra	Other States
Production in million quintals	46.8	57.1	45.7	34.7	141.5	57.1	
Producer price – existing	1030	1090	995	1020	960	1090	
Producer price – free trade	824	750	734	759	788	1073	
Supply elasticity with price	0.68	0.74	0.61	0.52	0.58	0.74	
Supply – existing	46.8	57.1	45.7	34.7	141.5	57.1	
Supply under free trade	40.4	43.9	38.4	29.5	126.8	56.5	No change
Change in producers surplus	−8985	−17155	−10981	−8388	−23019	−965.10	
Unit change in producers surplus	−213	−300	−240	−231	−166	−16.9	
Aggregate demand	42.1	57.1	45.7	36.3	138.7	57.1	
Wholesale price – existing	1225	1300	1200	1250	1175	1300	
Wholesale price under free trade	980	895	885	930	965	1280	
Price elasticity of demand	−0.45	−0.47	−0.44	−0.36	−0.36	−0.47	
Existing demand	42.1	57.1	45.7	36.3	138.7	57.1	
Demand under free trade	45.8	65.4	50.9	39.6	147.6	57.4	No change
Change in consumer's surplus	10779	24819	15227	12151	30064	1146	
Unit change in consumer's surplus	256	435	333	335	217	20.1	
Change in total surplus/welfare	1793	7663	4246	3763	7045	181	
Employment (change in million man-days)	−35.7	−76.2	−42.1	−26.5	−84.8	−3.5	
Forex (change in million US $)	−56.8	−125.4	−79.2	−50.2	−133.9	−8.3	

Note: WB – West Bengal; TN – Tamil Nadu; UP – Uttar Pradesh
Source: Jha (2004).

Now, India has emerged as an exporter of milk products in contrast to its import in the earlier years. With the increasing number of milch animals and milk production in rural settings, it may be possible for India to emerge in future as the largest producer and exporter of milk products in the world. Thirdly, since the use of animals for draught purposes is showing no significant growth, farmers should resort to modern techniques of tilling as substitute to animal power in the face of growing farming activities. This may in turn bring opportunities for improvement in land productivity and a still higher income generation from the farmers' scarce land and other resources. Finally, the increasing use of mechanical equipment for cultivation in lieu of additional power will progressively increase the rural demand for energy, that is, for electricity, diesel, etc. Therefore, there must be adequate investment in power sector in the future to accommodate this increasing rural demand for power.

As for the prospects of livestock sector in the era of WTO regime, one of the recent studies observes a steep fall in the domestic prices of milk following the decision to import milk products at a low international price (Rs 640 per quintal). Since supply of milk is highly price elastic, this situation will lead to adverse affect on milk production in states like Haryana, Maharashtra, Tamil Nadu, West Bengal, and Uttar Pradesh that account for more than 40 per cent share in total milk production of India. The study further finds that at the intermediate range of world price (Rs 850 per quintal), the derived duty paid (DDP) price at one of the Indian port will be Rs 1204 per quintal. On the other hand, price differences in the wholesale market suggest that import would take place only in Mumbai, which would have limited effect on the producers and consumers of milk. The producers and consumers in other states would by and large remain unaffected. However, the study shows concern for the protection of livestock sector in India in view of loss of employment and the wide ramifications this has for the rural economy. It emphasizes upon the fact that a high import at the low range of the world price would cause enormous loss of employment in the country and, therefore, on this account the sector requires protection from low world price of milk. Since as of now protection in the form of moderate tariff (35–40 per cent) and tariff rate quota appears to be sufficient, any argument for further reduction of tariff must be resisted.

As for scope for the expansion of Indian dairy industry in new liberalized trade regime is concerned, it has been observed that, in general, the Indian dairy sector would be competitive only if the export subsidies on dairy products are abolished. In more relaxed market environment, the real challenge before Indian livestock sector would be in terms of Sanitary and Phytosanitary Measures (SPS), Agreement on Technical Barriers to Trade (TBT) and animal welfare related issues. With a view to meet these requirements—both domestically and in the world markets—modernization of supply chain encompassing producer as well as consumer is the need of the hour.

Undoubtedly, India is already price competitive in the world market and when subsidies from competitive producers like USA and EU countries are removed, the situation will make India more price competitive. In case India is not able to capture the world market in the event of removal of subsidies from the modern bloc countries, the other competitors like Australia and New Zealand would capture this market and enter in a big way to flood markets with their dairy products, making us losing our competitiveness and a great opportunity in the new trade regime.

REFERENCES

Binswanger, Hans. 1978. "The Economics of Tractors in South Asia: An Analytical Review," Research Report. Published by Agricultural Development Council, New York, and International Crop Research Institute for Semi Arid Tropics, Hyderabad, India, pp. 1–96.

Jha, Brajesh. 2004, 'Implications of Trade Liberalization for the Livestock Sector', *Indian Journal of Agricultural Economics*, 59(3): 566–77.

Mishra, S.N. and Rishi K. Sharma. 1990. *Livestock Development in India: An Appraisal* New Delhi: Vikas Publishing.

Nair, K.N. and A.C. Dhas. 1990. 'Cattle Breeding Technology and Draught Power Availability: An Unresolved Contradiction', in Martin Doombos and K.N. Nair (Eds.) *Resources, Institutions and Strategies: Operation Flood and Indian Dairying*. New Delhi: Sage Publications.

Sharma, R.K. 1981. 'Draught Power Planning in Indian Agriculture', PhD Thesis, Delhi School of Economics, University of Delhi (Unpublished).

Vaidyanathan, A., K.N. Nair and Marvin Harris. 1982. 'Bovine Sex and Special Ratio in India', *Current Anthropology*, Vol. 23, No. 4: 365–73.

19

Evaluation of Integrated Dairy Development Project in North-Eastern Region

T.N. Saikia and Gautam Kakaty

Animal husbandry in India is an integral part of agriculture sector and plays an important role in providing employment and income to the rural people. The livestock sector has been receiving significant priority in India in the last couple of decades. The dairy farming is an important enterprise that provides employment, income and nutritive food to the people and also supplies cow dung as organic manure to enrich soil fertility and thus help in increasing crop production.

The growth of dairy sector has been very impressive in the sense that the bovine population during last few decades has increased by about 34.0 per cent. The production of milk has been doubled during the last two decades mainly due to the Government policy adopted for the development of dairy sector. In order to increase milk production, Intensive Cattle Development Project (ICDP), Key Village Scheme, Operation Flood (OF) programme and later on, Integrated Dairy Development Project (IDDP) were launched for the development of the dairy sector. In recent years, India became the second largest milk producers in the World, next to USA. The dairy sector in India derives its strength from 288 million cattle and buffaloes which is about 52 per cent of Asia's bovine population. The livestock sector contributed 25.5 per cent of national agricultural GDP and 5.6 per cent of total national GDP in 2001–02. The share of livestock sector in terms of gross value of agricultural output has increased from 8.6 per cent in 1971–72 to 35.5 per cent in 2001–02 and 38.14 per cent in 2010–11. Milk production in India has increased from 31.6 million tonnes in 1980–81 to 60.8 million tonnes in 1993–94, 84.6 million tonnes in 2001–02 and further to 132.43 million tonnes in 2012–13.

The IDDP, a Central Sector Scheme with 100.00 per cent grant-in-aid, has been implemented in non-OF, Hilly and Backward Areas by the concerned state governments since 8th Five Year Plan and continued during 9th and 10th Plan periods.

DAIRY SECTOR IN THE NORTH-EASTERN (NE) REGION

Livestock is an important subsidiary occupation of the rural households in the NE states. A large percentage of animals in this part of the country is indigenous, less productive, and poorly managed. The local cattle and buffaloes are poor milk yielders; yet, it is one of the major subsidiary sectors in rural areas. The production of milk in the region was 1237 thousand tonnes in 2012–13. The per capita availability of milk is abysmally low, that is, 84 grams per head per day whereas the national average is 226 grams.

In the NE India, crop cultivation and livestock rearing are interlinked and play an important role in the State's economy. The agriculture and allied sector contributed nearly 31.0 per cent of the state income at current prices in 2002–03. The analysis of allocation of resources revealed that of the total annual allocation to animal husbandry and veterinary sector, the share of dairy sector has been very low which varied from 10.0 per cent to 18.0 per cent only. With the increase in demand for milk and milk products the cross-breed milch animals have been developed in the region only during the last couple of decades. The concentrations of milch animals especially of indigenous cows are still in dominance and cross-breed cows are limited.

IMPLEMENTATION OF IDDP

In view of the importance of dairy sector in the non-OF, hilly and backward regions, the Central Sector Scheme, IDDP has been implemented in four NE States. The basic objectives of the scheme are: (i) Development of milch cattle through cross breeding, (ii) Increase of milk production by providing technical guidance; training and supply of input services, (iii) Procurement, processing and marketing of milk in a cost-effective manner, (iv) Ensuring remunerative prices to milk producer, (v) Generation of additional employment and income, and (vi) Improvement of social, economic and nutritional status of people living in the disadvantaged hilly and backward areas.

The present study has been undertaken:

1. to assess the impact of IDDP in generation of additional employment and income to the different categories of beneficiaries;
2. to assess the impact of IDDP in terms of genetic improvement of cattle through selective breeding/cross breeding and in making availability of feed, fodder and other essential items for the development of dairy sector;
3. to assess the impact of IDDP in milk production and in development of marketing and processing infrastructure in the project area;
4. to assess whether the implementing agencies followed the guidelines in selection of beneficiaries and imparted training through dairy extension services amongst the farmers;
5. to study the problems faced by the implementing agencies in execution of the project as per guidelines laid down by the Department of Animal Husbandry, Dairying and Fisheries (DADF); and
6. to suggest policy implications.

METHODOLOGY AND DATA

As desired by the DADF, Ministry of Agriculture and Farmers Welfare, Government of India, the evaluation of IDDP study covered four NE states, namely Meghalaya, Arunachal Pradesh, Mizoram, and Sikkim. Out of three IDDP covered districts in Meghalaya, two districts were selected for the study. In Arunachal Pradesh, out of two IDDP-covered districts, one district was selected; in Mizoram, out of three IDDP-covered districts, two districts were taken into account; and in Sikkim as IDDP covered only one district, the same district was included in the study.

A multi-stage stratified random sampling technique was adopted for selection of society and beneficiary farmers (see Figure 19.1). From the control group farmers belonging to society but non-beneficiary and non-member but owner of milch animals were selected. Keeping in view the manpower resources and time constraint, a sample of 316 beneficiary households were selected from four States of NE region. It was decided that from each district at least three milk co-operative societies were selected randomly and within each society a cluster of 3–5 villages were selected to draw beneficiary households. From each society 16 to 17 beneficiary members were selected at random with probability proportional to the number of members in each society. However, at least 50 beneficiary farmers were selected from each district except Sikkim. From each district, 15 non-beneficiary but members of the societies were selected to study the reasons of not becoming the member of the society under the IDDP scheme. This category of universe however was not covered in Sikkim. Information from the non-member but the owner of milch animals was collected by case study method to assess whether the IDDP has any spin-off effect in the minds of the non-beneficiaries. Thus, in all 19 milk producer's co-operative society, 316 beneficiary member households, 50 co-operative society members but not the beneficiary of IDDP scheme and 46 non-member but owner of milch animals constituted the sample size of the study. The data for the study were collected from both the primary as well as secondary level sources. The reference year of the study is 2005–06.

RESULTS AND DISCUSSION

Looking at the distribution of respondents in Meghalaya 99.00 per cent were literate with educational qualification up to High School Leaving Certificate (HSLC) and above levels. In Arunachal Pradesh, of the 80 sample beneficiary respondents, 7.50 per cent was illiterate and the rest 92.50 per cent were literate. In Mizoram only 1.00 per cent

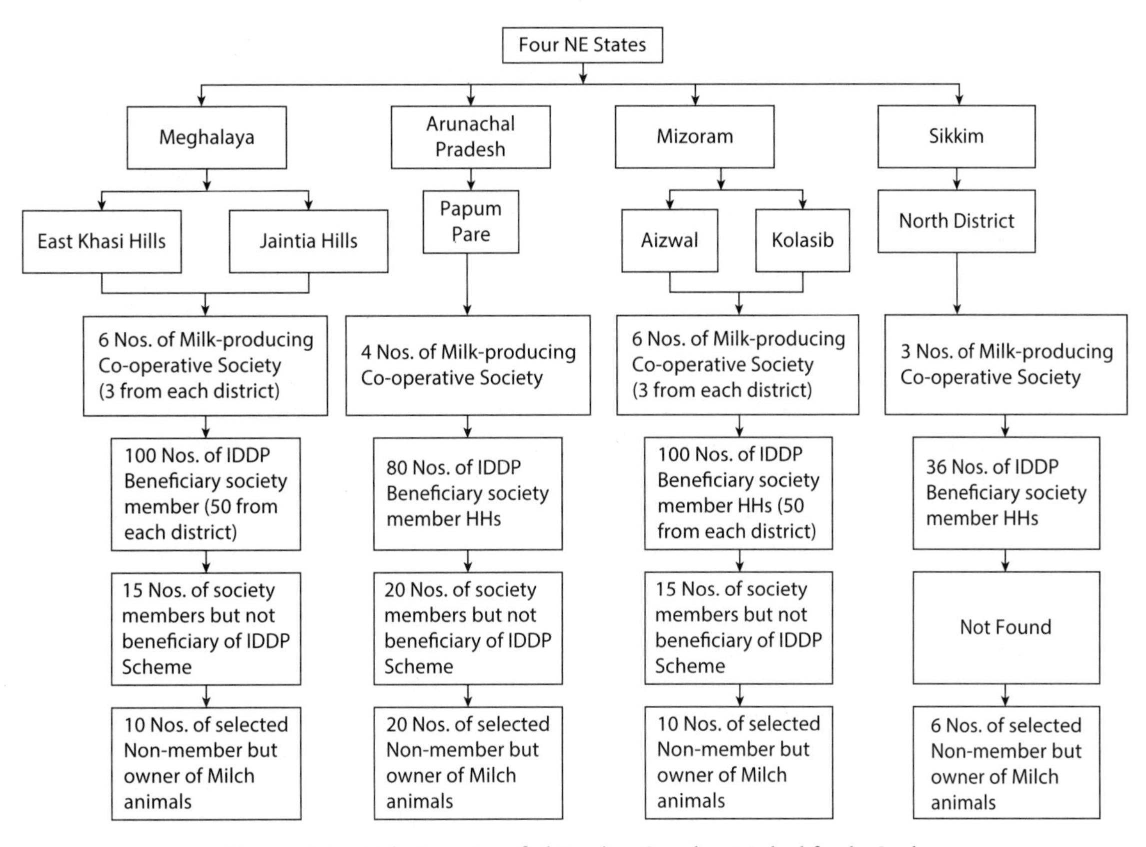

Figure 19.1 Multi-Stage Stratified Random Sampling Method for the Study

Source: Authors.

respondent beneficiary was illiterate and the rest 99.00 per cent literate. In Sikkim, of the total 36 sample beneficiary respondents, 19.44 per cent was illiterate and the rest 80.56 per cent literate. The overall educational status in the sample was 4.75 per cent illiterate and 95.25 per cent literate.

The average size of operational holding in Meghalaya was 1.47 hectares, 1.43 hectares in Arunachal Pradesh, 1.51 hectares in Mizoram and 1.10 hectares in Sikkim. So far as member non-beneficiary and non-member non-beneficiary farmers were concerned, the educational status, economic status and occupational did not show significant differences.

The sample dairy farmers of Meghalaya possessed 689 milch cows of which 67.49 per cent were in milk and 32.51 per cent dry at the time of field study. In Arunachal Pradesh, the sample beneficiary farmers possessed 589 numbers of cows of which 57.38 per cent were in milk and remaining 42.62 per cent dry. In Mizoram, the sample farmers possessed 985 numbers of cross-breed cows. Of the total milch cows in the sample, 59.90 per cent were in-milk cows and 40.10 per cent cows in dry period. In Sikkim, 36 sample farmers possessed only 75 milch animals of which 74.67 per cent were in milk and 25.33 per cent dry.

The investment patterns in dairy farms were comprised of fixed capital assets such as milch animals, cattle-shed/stores, and feeding equipment. In 100 samples of Meghalaya, the overall investment in dairy enterprise was Rs 61,83,405. The overall investment in dairy enterprise in Arunachal Pradesh sample was worked out at Rs 17,17,300. In the 100 samples of Mizoram state, the overall investment in dairy enterprise was found at Rs 1,51,22,580. The overall investment in dairy enterprise in Sikkim was recorded at Rs 5,43,605.

The study revealed that, of the total cows in milk in Meghalaya, 44 numbers of indigenous cow produced 19,468 litres per annum, that is, per cow per day production was 1.80 litres and on the other hand 421 numbers of cross-breed cows produced 8,72,507 litres per cow and per day production was worked out at 7.21 litres. The sample beneficiaries of Meghalaya possessed buffaloes also, and 3 numbers of she-buffaloes were in milk and produced 5652 litres of milk and per day per buffalo production was worked out at 5.78 litres. In Arunachal Pradesh, the IDDP beneficiaries possessed 265 numbers of indigenous cow in milk and produced l 1,03,377 litres, that is, per cow/day production was 1.07 litres. The sample beneficiaries also possessed 73 numbers of cross-breed cow and produced

1.49,475 litres, that is, per day/cow production of milk was found at 7.13 litres. This indicated that per day yield of cross-breed cow was higher by 151 per cent than the indigenous cows.

The beneficiary farmers in Mizoram possessed 590 numbers of cross-breed cows of which 69.90 per cent were in milk and produced 12.12,761 litres of milk per annum, that is, per cow/day production was found at 7.28 litres. In case of 36 beneficiary farmers in Sikkim, the sample beneficiaries possessed 37 numbers of indigenous cows in milk and produced only 39,468 litres of milk, that is, per cow per day production was worked out at 2.92 litres. The sample beneficiaries possessed 19 cross-breed cow in milk and produced 66,936 litres of milk per annum and per cow/day production was worked out at 9.65 litres. This showed that per day yield of milk in case of cross-breed cow was higher by 230.48 per cent than the indigenous cows in Sikkim.

The overall annual expenditure on dairy farms of sample member beneficiary farmers in Meghalaya, Arunachal Pradesh, Mizoram, and Sikkim revealed that under the head, variable costs, the expenditure incurred on fodder and feed concentrate was 48.86 per cent followed by expenditure on human labour (24.79 per cent), then came the expenses on veterinary charges (8.78 per cent). Taking all the variable costs together, it came to 87.10 per cent (Rs 30,175,496.09). So far as fixed costs were concerned, it was worked out at 12.90 per cent (Rs 4,467,722.35). Of the total fixed costs, depreciation on animal was found at 10.21 per cent followed by interests on different items of fixed capital comprising of cattle shed, storage feeding pan, bucket and other dairy equipment etc. (2.69 per cent).

The economic analysis of dairy enterprise in Meghalaya showed that, of the total income of Rs 14,404,276 about 89.97 per cent was derived from milk and 10.03 per cent was the estimated value of young stock in the year under study. In Arunachal Pradesh, of the total income of Rs 4,451,342, about 91.48 per cent income was obtained from production of milk and 8.52 per cent was the estimated value of young stock. In Mizoram, of the total income of Rs 23,204,022 about 95.24 per cent income was derived from milk and 4.76 per cent was from the estimated value of the young stock. Similarly, in Sikkim, of the total income from dairy farms, 93.94 per cent income was derived from production of milk and 6.06 per cent was from the estimated value of the young stock.

The analysis of four NE States indicates that, of the total income Rs 42,929,470 about 93.05 per cent of income was derived from production of milk and 6.95 per cent was from the estimated value of young stock. The profitability in dairy enterprise is, by and large depends on the breeds of milch animals maintained by the dairy farmers and lactation period of cows in milk.

The breed-wise analysis of cost of production of milk in the sample states showed that in case of local cows the average cost of production of a litre of milk was estimated at Rs 15.42 and for cross-breed cows at Rs 13.93 with an overall average of Rs 14.03/litre.

In order to substantiate the finding of income generation over investment, the Benefit–Cost Ratio (BCR) analysis on the profitability of dairy farmers was undertaken. The estimated BCR in Meghalaya for local cow was only 1.05:1 while for cross-breed cows, it was found at 1.14:1 and the BCR for buffaloes was worked out at 1.17:1. The overall BCR was worked out at 1.14:1. In Arunachal Pradesh, the BCR for local cows was only 1.07:1 while for cross-breed cows it was found at 1.34:1 and the overall BCR was worked out at 1.21:1. In Mizoram, the BCR for cross-breed cows was found at 1.32:1. In Sikkim, the BCR for local cows was 0.94:1 while for cross-breed cows, it was found at 1.41:1 and the overall BCR was worked out at 1.14:1.

The analysis of overall BCR in four different sample States revealed that for the local cow, BCR was only 1.05:1 and for cross-breed cows, it was found at 1.25:1. The BCR for the buffaloes was worked out at 1.17:1. The overall BCR was found at 1.24:1. This indicated that the dairying is by and large an economically viable enterprise. The BCR analysis indicated that there is much potential to make the livestock farming more remunerative by way of cross-breeding through Artificial Insemination and by improving the nutritional status from locally available feed and fodder resources. The dairy farming is expected to make a real break by transforming the dairying enterprise into a commercially viable proposition.

GENERATION OF EMPLOYMENT

Table 19.1 gives the details of generation of employment (in man days) of the respondent beneficiaries in the sample States by size group of milch animals.

So far as generation of employment in terms of man days was concerned, of the total man days involved in all activities in Meghalaya, the dairy sector provided employment opportunity to family members to the extent of 60.41 per cent to 93.82 per cent, with an overall average of 85.90 per cent for all farms. In Arunachal Pradesh, dairy sector provided employment opportunity from 49.26 per cent to 86.69 per cent of the total man days involved in all activities with an overall average of 69.28 per cent for all farms. In Mizoram, dairy sector provided employment opportunity from 59.45 per cent to 96.41 per cent of the

Table 19.1 Generation of Employment (Man days) of the Respondent Beneficiaries by Size Group of Milch Animals in the Sample States

Size Group (in No.)	Agriculture Proper	Agriculture and Other Sources	Dairy	Total	% of Dairy Sector to Total
Meghalaya					
Below 5	1540	2220	10795	14555	74.17
5 to 10	2720	2025	33359	38104	87.55
10 to 20	720	1150	28374	30244	93.82
20 to 30	860	625	2266	3751	60.41
30 to 40	640	325	3362	4327	77.70
Total	**6480**	**6345**	**78156**	**90981**	**85.90**
Arunachal Pradesh					
Below 5	1201	890	2030	4121	49.26
5 to 10	3975	3050	12271	19296	63.59
10 to 20	1601	610	8903	11114	80.11
20 to 30	329	280	3967	4576	86.69
Total	**7106**	**4830**	**27171**	**39107**	**69.48**
Mizoram					
Below 5	530	260	1158	1948	59.45
5 to 10	4955	1906	26802	33663	79.62
10 to 20	3894	1240	39621	44755	88.53
20 to 30	170	0	4567	4737	96.41
Total	**9549**	**3406**	**72148**	**85103**	**84.78**
Sikkim					
Below 5	9400	6310	19985	35695	55.99
5 to 10	820	640	3725	5185	71.84
Total	**10220**	**6950**	**23710**	**40880**	**58.00**
Overall					
Below 5	12671	9680	33968	56319	60.31
5 to 10	12470	7621	76157	96248	79.13
10 to 20	6215	3000	76898	86113	89.30
20 to 30	1359	905	10800	13064	82.67
30 to 40	640	8561	102681	111882	91.78
Total	**33355**	**21531**	**201185**	**256071**	**78.57**

Source: Field Survey Data.

total man days of work involving all farm activities. The average for Mizoram was worked out at 84.78 per cent for all farms. In Sikkim, dairy sector provided employment opportunity from 55.99 per cent to 71.84 per cent of the total man days in all the activities, with an overall average of 58.00 per cent.

GENERATION OF INCOME

The annual income of the beneficiaries have been worked out by the farm size of operational holdings for all sample states and are presented in Table 19.2. So far as generation of income by the sample dairy farms was concerned, in the samples of Meghalaya, the proportion of income of dairy units to the total family income was found at 88.26 per cent. The farm-size-group-wise share of income from dairy enterprise varied from 51.52 per cent to 92.21 per cent. In Arunachal Pradesh, the proportion of income of dairy units to total family income was found at 74. 10 per cent and farm-size-group-wise share of income from dairy enterprise varied from 72.20 per cent to 81.08 per cent. In Mizoram, the proportion of income of dairy units to total family income was 93.68 per cent and farm-size-group-wise share of income from dairy enterprise varied from 63.62 per cent

Table 19.2 Generation of Income of the Beneficiaries by Farm Size Group of Operational Holdings in the Sample State

Size Group (in Ha)	Agriculture Proper	Other Allied Activities		Dairy	Total	% of Dairy Sector to Total
		Farm	Non-farm			
Meghalaya						
Below 1 Ha.	184,220	5,400	368,400	6,608,403	7,166,423	92.21
1 to 2 Ha	307,957	21,600	216,400	5,697,446	6,243,403	91.26
2 to 4 Ha	167,776	28,800	55,000	1,214,891	1,466,467	82.84
4 to 10 Ha	129,444	0	94,600	526,501	750,545	70.15
10 Ha & above	252,003	0	84,000	357,035	693,038	51.52
Total	**1,041,400**	**55,800**	**818,400**	**14,404,276**	**16,319,876**	**88.26**
Arunachal Pradesh						
Below 1 Ha.	29,924	0	259,200	847,932	1,137,056	74.57
1 to 2	260,822	0	396,000	1,709,856	2,366,678	72.25
2 to 4 Ha	365,883	0	61,200	1,109,188	1,536,271	72.20
4 to 10 Ha	183,054	0	0	784,366	967,420	81.08
10 Ha & above	0	0	0	0	0	0.00
Total	**839,683**	**0**	**716,400**	**4,451,342**	**6,007,425**	**74.10**
Mizoram						
Below 1 Ha	97,124	40,200	74,400	370,288	582,012	63.62
1 to 2 Ha.	259,744	0	159,600	8,771,226	9,190,570	95.44
2 to 4 Ha.	555,903	0	208,800	12,772,957	13,537,660	94.35
4 to 10 Ha.	170,533	0	0	1,289,551	1,460,084	88.32
10 Ha & above	0	0	0	0	0	0.00
Total	**1,083,304**	**40,200**	**442,800**	**23,204,022**	**24,770,326**	**93.68**
Sikkim						
Below 1 Ha	157,6000	0	342,000	479,850	978,850	49.02
1 to 2 Ha	180,000	0	0	243,166	423,166	57.46
2 to 4 Ha	116,200	0	0	101,273	217,473	46.57
4 to 10 Ha	25,000	0	0	45,541	70,541	64.56
10 Ha & above	0	0	0	0	0	0.00
Total	**478,200**	**0**	**342,000**	**869,830**	**1,690,030**	**51.47**
Overall						
Below 1 Ha	468,268	45,600	1,044,000	8,306,473	9,864,341	84.21
1 to 2 Ha	1,008,523	21,600	772,000	16,421,694	18,223,817	90.11
2 to 4 Ha	1,205,762	28,800	325,000	15,198,309	16,757,871	90.69
4 to 10 Ha	508,031	0	94,600	2,645,959	3,248,590	81.45
10 Ha & above	252,003	0	84,000	357,035	693,038	51.52
Total	**3,442,587**	**96,000**	**2,319,600**	**42,929,470**	**48,787,657**	**87.99**

Source: Field Survey Data.

to 95.44 per cent. In Sikkim, the proportion of income of dairy units to total family income was 51.47 per cent and farm-size-group-wise share of income varied from 46.57 per cent to 64.56 per cent.

For assessing the impact of IDDP on non-beneficiary, 50 cooperative members but not the beneficiary of IDDP were selected randomly from three sample states, namely Meghalaya, Arunachal Pradesh, and Mizoram, for comparative analysis. Due to nonavailability of data on member non-beneficiary of IDDP in Sikkim State, the consolidation work could not be done for the state. In Meghalaya, 15 member non-beneficiary farmers possessed 49 milch cows, 20 member non-beneficiary farmers of Arunachal Pradesh possessed 105 milch cows and 15 member non-beneficiary farmers of Mizoram possessed 86 milch cows. Altogether 50 member non-beneficiary farmers in the sample possessed 240 numbers of milch cows.

The overall annual expenditure of the four states of the sample member non-beneficiary farmers under IDDP stood at Rs 2300596.84. Of this total expenditure, the variable cost was found at Rs 1946806.99 (84.62 per cent) while the total fixed cost stood at Rs 353789.85 (15.38 per cent). Of the total variable cost, the highest, 43.53 per cent expenditure was found against fodder and of the total fixed cost, the highest (11.36 per cent) percentage was attributed to depreciation on animals.

In Meghalaya, the BCR for the indigenous milch cows was found at 1.04:1 and in case of cross-breed cows, the BCR was found at 1.15:1. In Arunachal Pradesh, the BCR for the indigenous milch cows was recorded at 1.07:1 and for cross-breed cows at 1.18:1. In Mizoram, the BCR for the indigenous cows was found at 1.10:1 and in case of cross-breed, it was at 1.23:1. The overall estimated BCR for all the sample States were worked out. For the indigenous cows, the BCR was found at 1.07:1 and for cross-breed cows, it was 1.20:1. The overall BCR was found at 1.18:1.

Altogether, 46 non-member non-beneficiary of IDDP, but owners of milch animals were selected. It was found that majority of the non-member non-beneficiary sample farmers possessed both indigenous and cross-breed cows reared mostly to meet the domestic requirement of milk, only the surpluses were marketed.

In Meghalaya, the annual expenditure on dairy enterprise for non-member non-beneficiary of IDDP was 87.63 per cent of the variable cost, of which feed costs accounted for 42.82 per cent. The total fixed costs were worked out at 12.37 per cent. In Arunachal Pradesh, of the total annual expenditure on dairy enterprise, 88.53 per cent were variable cost, of which feed costs were estimated at 46.45 per cent. The total fixed costs were worked out at 11.47 per cent. In Mizoram, of the total annual expenditure, 79.45 per cent were variable cost of which feed costs accounted for 41.67 per cent. The total fixed costs were worked out at 20.55 per cent. Of the total annual expenditure on dairy units of non-member non-beneficiary farmers in Sikkim, 88.76 per cent were on variable cost, of which feed cost accounted for 54.46 per cent followed by expenditure on human labour (22.81 per cent). The total fixed cost was estimated at 11.24 per cent.

The overall annual expenditure on dairy farms of sample non-member nonbeneficiary farmers in Meghalaya, Arunachal Pradesh, Mizoram, and Sikkim were worked out. Of the total annual expenditure on dairy enterprise, 84.44 per cent were variable cost, of which feed costs accounted for 45.41 per cent. The total fixed cost was worked out at 15.56 per cent. The analysis showed that the per unit annual costs of maintenance of indigenous cows possessed by the IDDP beneficiaries, member non-beneficiaries and non-member non-beneficiaries did not have much variations.

The estimated BCRs for non-member non-beneficiary farmers in the sample States were worked out. In Meghalaya, BCR for local cow was found at 1.03:1. In Arunachal Pradesh, the BCR for local cows was found at 1.06:1 while for cross-breed cows it was at 1.13:1. In Mizoram, the BCR for local cow was 1.10:1 while for cross-breed cows, it was found at 1.22:1. Similarly, in Sikkim also, the BCR for local cow was only 0.78:1 while for cross-breed cows, it was found at 1.18:1. The overall BCR in the four sample states for local cow was found at 1.00:1 while for cross-breed cows, it was at 1.20:1 and the overall BCR was found at 1.12:1.

The above analysis sufficiently has established that dairy farming provided ample employment opportunity and income to the farmers living in the remote hilly and backward areas.

CONSTRAINTS FOR DAIRY DEVELOPMENT IN HILLY AND BACKWARD AREAS

The major constraints hindering the development of animal husbandry and dairy sector in the study area were identified and recorded as follows:

1. Lack of policy on animal breeding and livestock development in the context of dairy farming in the state plan in proper perspective was considered as a major hindrance in the States under the study.
2 Technology intervention on Artificial Insemination has not been fully put into gear in all the areas of the States under the study.
3. Shortage of green fodder and feed concentrate was one of the root causes of poor performance of dairy sector in general, as the genetic milk production potential of cross breed animals could not be exploited fully in the absence of proper nutrition.
4. Stall feeding of animals is not an acceptable proposition. There was no provision for development of fodder cultivation under the IDDP.
5. Lack of perception on the part of the farmers, inadequate research focus on animal husbandry and poor linkages among different stakeholders and associated agencies including the Government were quite rampart.

6. Non-availability of compound feed manufacturer in the sample States was yet another major snags for which feed concentrates and balanced feed were not available within easy reach of the dairy farmers.
7. Payment for supply of milk to the society was not only lower but also irregular in payment which caused great hardship to the members of' the society and to the member non beneficiary who supplied milk in the 'Milk Booth' of the society.
8. Lack of surveillance and monitoring of infectious and contagious diseases and inadequate facility of Artificial Insemination and pregnancy diagnosis were some other problems as reported by the sample dairy farmers.
9. Unorganized and fragmented market for all livestock products and milk in particular involved a chain of middlemen who reap the lion's share of actual benefit, depriving the producers from their due share.

CONCLUSIONS AND POLICY IMPLICATIONS

On the basis of the field survey, careful observations and discussions held with the respondents, the following suggestions are offered for the improvement of dairy farming in non-OF, hilly, and backward areas of NE region.

1. The concerned state government should prioritize the strategies of dairy development in the State Plan to make a real breakthrough in the dairy sector.
2. There is need to evolve a comprehensive dairy development policy in the States through genetic improvement of indigenous milch animals.
3. The state government agencies dealing with dairy sector and the IDDP must work together in order to work out a feasible arrangement to provide green and dry fodder in adequate quantities and at a reasonable rate to the dairy farmers.
4. In order to overcome the fodder deficit, the Animal Husbandry and Veterinary Department of the concerned states should take up the programme for enhancement of fodder production through increase in the area under fodder crop.
5. There is necessity to expand the net work of the village-level milk cooperatives to all villages falling under the jurisdiction of the IDDP. In fact, it should be extended to other potential areas of the region as well.
6. There is need to educate and assist the milk producers in respect of breeding, feeding, animal management technique and marketing of milk and milk products in a cost effective manner.
7. The village level dairy co-operative society should revise the milk procurement price at par with the prevailing market rate and adequate measures should be taken to pay the prices of milk on the spot.
8. Infrastructural development like road communication and transport is needed for transportation of fodder, feed concentrates, veterinary medicines and transportation of milk to the consuming centres round the year.
9. With the improvement of livestock through crossbreeding, the susceptibility of the livestock to various diseases may increase. In order to reduce the mortality of livestock, efforts should be made to control the animal diseases though health care and disease control measures. It requires timely prophylactic measures and emergency services for treatment of livestock.
10. Intensive epidemiological studies of diseases, particularly infectious diseases should be undertaken for control and eradication.
11. Utilization of straw of cereals and other food crops with proper treatment can bridge the requirement of feed for the dairy animals.
12. Suitable plan and strategies for cultivation of green fodder in the fallow land should be evolved to meet the crisis of green fodder.
13. Facility for Artificial Insemination and Pregnancy Test at the doorstep of the dairy farmers needs to be created.
14. There is necessity of establishment of organized network of market so that the dairy farmers get due share for their produce.

Thus, the findings of the study amply demonstrated that the present composition of livestock population with adequate number of crossbreed animals can boost up milk production. Keeping in view the hill agro-ecosystem, the farmers in the study area were found to have using their wisdom to exploit the resources in a sustainable manner. The introduction of IDDP coupled with adoption of co-operative system of dairy farming with distribution of inputs and marketing of milk through milk booths has opened up a new vista of milk marketing and processing in the backward hilly areas. The sustainable development of dairy farming in the hilly areas through optimum utilization of natural resources, adequate health-care facilities for livestock, improvement of breeding programme through artificial insemination and timely vaccination can go a long way in the field of animal husbandry in general and dairy development in particular.

REFERENCES

Alvares, Claud. 1983. 'Operation Flood—The White Lie', The Illustrated Weekly of India, 8–13.

Bansil, P.C. 2002. 'Agricultural Statistics in India', CBS Publications and Distributors, New Delhi. Central Statistical Organization. 2003. 'National Accounts Statistics', Department of Statistics, Ministry of Planning, Government of India, New Delhi.

Chand, Romesh. 1995. 'Livestock in Himachal Pradesh: Factors Affecting Growth Composition and Intensity', *Indian Journal of Agricultural Economics*, 50(3): 299–310.

CMIE. 1994. 'Basic Statistics Relates to Indian Economy', CME Publications, Bombay.

Dhaka, J.P., D.K. Jain, V.K. Kesavan, and Lotan Singh. 1998. 'A Study of Production and Marketing of Surplus Functions for Milk in India', National Dairy Research Institute, (ICAR), Haryana.

FAO. 2004. 'FAO Statistics', Database, August.

Government of Arunachal Pradesh. 2003a. 'State Report on 17th Quinquennial Livestock Census, 2003', Report Department of Animal Husbandry and Veterinary, Nirjuli, Vol. 1 & 2.

Government of Arunachal Pradesh 2003b. 'Arunachal Pradesh at a Glance', Directorate of Economics and Statistics, Itanagar.

Government of Arunachal Pradesh. 2004. 'Economic Review of Arunachal Pradesh', Directorate of Economics & Statistics, Itanagar.

Government of Arunachal Pradesh. 2004. 'Statistical Abstract of Arunachal Pradesh', Report, Directorate of Economics & Statistics, Itanagar.

Government of India. 2002. 'Basic Statistics of North-Eastern Region', NEC, Shillong.

Government of India. 2003. 'Livestock and Poultry Results (Provisional)," Report 17th Livestock Census, Department of Animal Husbandry & Dairying, Ministry of Agriculture, New Delhi.

Government of Meghalaya. 2004. *Pocket Statistical Handbook, Meghalaya, 2003*. Mizoram: Directorate of Economics and Statistics.

Government of Mizoram. 2004. *Statistical Handbook—Mizoram*. Report Mizoram: Directorate of Economics and Statistics.

Grewal, S.S., and P.S. Rangi. 1983. 'Economics and Employment of Dairying in Punjab', *Indian Journal of Agricultural Economics*, 35(4): 120–5.

Huria, V.K. and K.T. Acharya. 1980. 'Dairy Development in India—Some Critical Issues', *Economic and Political Weekly*, 15(45 and 46): 1931–42.

Jain, D.K., A.K. Sharma, and V.K. Kesavan. 1998. 'Demand Analysis of Milk and Milk Products in India', National Dairy Research Institute (ICAR) Karnal, Haryana.

Kahlon, A.S. 2002. 'Dairy Financing in Kurukshetra and Kaithal Districts of Haryana—An Ex-Post Evaluation Study', NABARD Regional Office, Chandigarh.

Kanwar, P.N. 1975. 'Economics of Cross-breed Cows', *Indian Journal of Agricultural Economics*, 15(3): 151–2.

NABARD. 2002. 'An Ex-Post Evaluation Study of Dairy Development in Kollam District of Kerala', NABARD Regional Office, Triruvananthapuram.

Patel, R.K. 1993. 'Present Status and Promise of Dairying in India', *Indian Journal of Agricultural Economics*, 48(1), January–March: 1–33.

Prabaharan, R. 2002. 'Livestock Development in India—Some Constraints,' Agricultural Research Review (Conference Proceedings).

Rao, V.M. 2005. 'Cooperatives and Dairy Development—Changing Destiny of Rural Women', Mittal Publications, New Delhi,

Singh, A.J., P.K. Dhillon, and Kiran Sethi. 1996. 'Status and Prospects for Dairying in India with Special Reference to Punjab State', Agricultural Situation in India, May

Singh, Surendar. 1996. 'Agro-Economics of Dairy Development in India', Criterion Publications, New Delhi,

Sharma, Usha and S.K. Sharma. 2005. 'Discovery of North-East India, Mizoram', Vol. 8, Mittal Publications, New Delhi.

Sharma, Vijay Paul. 2004. 'Livestock Economy of India: Current Status, Emerging Issues and Long-Term Prospect', *Indian Journal of Agricultural Economics* Vol. 59. No. 3 July–September: 512–54.

Sharma, V.P. and Ashok Gulatti. 2003. 'Trade Liberalization, Market Reforms and Competitiveness of the Indian Dairy Sector', MTID Discussion Paper, 61, Washington, DC, USA, April.

Tripathi, R.S. 1995. 'Cow Milk Production in H.P. Hills—An Economic Approach', *Indian Journal of Dairy Science* Vol. 48(2): 98–102.

20

Problems and Prospects of Fish Farming in Bihar and Jharkhand

Ranjan Kumar Sinha and Rosline Kusum Marandi

Fish is foods of majority of people. It provides proteins and contains fat, inorganic substances and vitamins. It is more valuable for human, especially for a population whose staple food is rice. Besides, it helps in generating employment and revenue. Its production in India has increased from 752 thousand MT in 1950–51 to 9040 thousand MT in 2012–13, indicating more than twelvefold increase during the last six decades. Out of total production nearly 36.74 per cent is contributed by marine and remaining by inland fisheries in the country. It is the source of livelihood to over 14.48 million people largely belonging to socially and economically backward groups. It contributed 0.8 per cent in 2012–13 to the country's GDP at 2004–05 prices (Government of India [GoI] 2014). During the last 10 Five Year Plans, the GoI has substantially increased the outlays for fisheries development. During the Fourth Plan (1969–74), on the recommendation of a Technical Committee set up by Government of India, a pilot scheme of Fish Farmers' Development Agencies (FFDAs) was launched for development of fisheries and delivery of sustainable aquaculture throughout the country, which was renamed as Development of Inland Fisheries and Aquaculture in the 10th Plan, and it continued in the 11th Plan also.

Fishery is an ancient activity of mankind. It has developed throughout the world from centuries till today. Almost all countries and world institutions have fishery development programmes. In 1981, FAO asserted a resolution that occurred in the potential of fisheries to contribute to a new international order, its intention to take a lead by helping the developing countries to secure their rightful place in world fisheries. Its gradual development has opened up new dimension of research, particularly relevant to the policy makers and other stakeholders. In India, the process of transformation of the fishery sector from subsistence to commercial status and subsequently the growing scope of linking with the global market have opened up interests. Fishery is a key allied sector of agriculture providing income, employment and the much-needed nutritional security. Since natural fishing in coastal waters has reached maximum sustainable yield further growth in fishery has to come through commercial aquaculture. Technological progress in commercial aquaculture has substantially diminished the level of production risk, compared to traditional fishery (Kokate and Upare 2005). Recent research relating to socio-economic nature revealed that the income, price and supply elasticities vary substantially across fish species and it is wrong to group them together in any policy analysis (Kumar 2004). Impressive growth in inland fish production in West Bengal is attributed to higher profitability (Rs 22227/ha) by Kar and Kumar (2004).

Mishra (1997) analysed fish production and marketing structure in community ponds of Chhattisgarh and found that the yield per ha was 1538/kg for medium farm size, which was the highest and sold at a price of Rs 23.8/kg. In Punjab producer's share in consumer rupee varied between 38.00 to 45.00 per cent of fresh fish (Godara et al. 2006). Singh and Pandey (2004) analyse marketing efficiency of fish in Uttar Pradesh and observed that the producer's share ranged from 28.00 to 38.00 per cent. Kant et al. (2000) in a study of Azamgarh district (UP), found that the CB ratio was strong positive indicating 1:3.14 in production of fish per acre and thus concluded that fishery enterprise is most profitable proposition. An Evaluative Study of NABARD (2000) in Punjab on Inland fisheries development indicate that net income per acre of fish pond was Rs 26141 as compared to Rs 10100 from the competing crops. Singh and Singh (2004) in their study on 'Stocking Density and Species Mix in Composite Fish Culture in North Bihar: A Techno Economic Analysis' found that the stocking rate in fish production is much higher in North Bihar.

The present chapter is based on a study entitled 'Problems and Prospects of Fish', conducted by the Agro-Economic Research Centre (AERC) for Bihar and Jharkhand, TM Bhagalpur University, Bhagalpur, Bihar during the work plan year 2007–08 with following objectives:

1. to estimate the cost of cultivation and production of fish;
2. to identify the various channels and system of fish marketing;
3. to identify the existing constraints of fish farming in the area;
4. to examine the future prospects of fish farming in the area; and
5. to suggest policy measures for the development of fish farming in the area.

METHODOLOGY AND DATA

To pursue the objectives of the study the data were collected from primary and secondary sources. The primary data was collected through duly structured fish farmers' schedule. The selection of the respondent was made through a multi-stage stratified sampling method. At the first stage, the selection of one district was made from each of the agro-climatic sub-zones of both the states on the basis of highest number of total ponds (government *jalkars* (tank) and private ponds) in the district among the districts of respective sub-zones. Accordingly, Madhubani, Purnea and Bhagalpur districts were selected from North Bihar Plains, North East Plains and South Bihar Plains respectively in Bihar and Dumka and West Singhbhum districts from Chhotanagpur North-Eastern Hills and Plateau and Chhotanagpur South Hills Plateau respectively in Jharkhand. Accordingly one *anchal* (Teshil for Revenue collection) from each of the sample districts were selected. Similarly Benipatti, Dagarua and Sahkund anchals were selected from Madhubani, Purnea and Bhagalpur districts respectively in Bihar and Saraiyahat and Jagarnathpur anchals from Dumka and West Singhbhum districts respectively in Jharkhand. Subsequently, on the basis of the lists of jalkars of the sample anchals, along with the names of the lessee of those jalkars obtained from the offices of District Fisheries Officer (DFO) of the respective sample districts and classified the fish farming households into three popular categories, namely, small (up to 0.5 ha), medium (0.5 to 2 ha) and large (above 2 ha). A total of 90 fish farming households from Bihar and 60 fish farming households from Jharkhand were randomly selected for in-depth investigation. The secondary data was collected from different published and unpublished sources. The reference year of the primary data collection is 2007–08.

RESULTS AND DISCUSSIONS

Fisheries in Bihar and Jharkhand

Bihar is one of the few states with large inland fisheries and aquaculture resources. Till 1970, Bihar used to supply fresh fish in neighbouring states, but around the year 1990 the inflow of fishes from other states, particularly Andhra Pradesh, started gravitating the fish markets in the state. At present, the annual consumption of fish in the state is nearly 6.00 lakh MT against the annual production of around 4.00 lakh MT. The state has larger number of non-vegetarian population, but has lower rate of per capita consumption than all India averages on both accounts. The state has a stretch of 3200 km rivers and canals, 0.60 lakh ha reservoirs, 0.95 lakh ha tanks and ponds, 0.05 lakh ha flood plain lakes and derelict water and 1.60 lakh water bodies constituting 1.60 per cent, 2.06 per cent, 3.93 per cent, 0.62 per cent and 2.17 per cent, respectively of the total of all India's inland fishery resources. The state has the largest fishermen population (49.60 lakh), which accounts for 34.23 per cent of country's fishermen population. It has 532 primary fishermen co-operative societies with membership of nearly 40000 fishermen at the primary level. The major portion of fish production is realized from ponds and tanks, which are over 40000 in numbers covering total areas of 68 lakh ha. Out of three agro-climatic sub-zones in the state, northBihar plains is the most potential region in terms of total number of ponds (55.00 per cent) and the water spread area (39.20 per cent) followed by south Bihar plains (26.42 per cent and 36.60 per cent) and North-East plains (18.33 per cent and 24.20 per cent), respectively.

Jharkhand has advantage of having a sizeable number of medium and large reservoirs and substantial number of ponds and tanks of different sizes. But, these resources are largely untapped and thus, the state depends on the supply line of Andhra Pradesh and West Bengal, which usually meet its annual fish demand. The annual consumption of fish in the state is nearly 83 thousand MT against the present annual production of nearly 62 thousand MT, having a shortfall of 21 thousand MT (25.30 per cent) of the total annually. The state has a stretch of 4,298 km rivers and its tributaries, 0.94 lakh ha reservoirs, 0.29 lakh ha tanks and ponds and 1.23 lakh ha water bodies accounting for 2.15 per cent, 3.23 per cent, 1.20 per cent and 1.67 per cent of the total of all India's inland fishery resources respectively. In regard to rivers and tributaries, these are seasonal in nature. The state has the second largest fishermen population (19.30 lakh), next to Bihar accounts for 13.32 per cent of India's fishermen's population. The state has 66 primary fisherman cooperative societies with membership of 9,150. The major portion of fish production comes from tanks and reservoirs, which are spread over 94,000 ha and 29,000 ha, respectively. Amongst two agro-climatic sub-zones in the state, Chhotanagpur North Eastern Hills and Plateau leads, which accounts for nearly 69.00 per cent of the area under tanks and 52.00 per cent under ponds.

Study Area and the Sample Respondents

In Bihar, out of three sample districts, Madhubani district covers an area of 3,501 sq. km, constituting 3.72 per cent of state's total area (94,163 sq. km). Its total population is 4.48 million (2011 census). Purnea district covers an area of 3,229 sq. km constituting 3.43 per cent of state's total area. Its total population is 3.26 million (2011 census). Bhagalpur district covers an area of 2,569 sq. km, constituting 2.73 per cent of state's total area. The population of the district is 3.03 million (2011 census). There are 3,555 jalkars distributed across 21 anchals in 1853.21 ha of water spread area in Madhubani district; the largest in number and water area in the state. Purnea district has 691 jalkars distributed across 13 anchals in 948.42 ha of water spread area whereas that of in Bhagalpur, there are 781 jalkars distributed across 12 anchals in 871.68 ha of water spread area. These ponds/jalkars are leased out for short and long periods.

Out of the total sample respondents, 51.11 per cent of the fish farmers belonged to the age group of 46 to 60 years followed by 31 to 45 years (36.67 per cent), 18 to 30 years (6.67 per cent) and 61 years and above (5.55 per cent). Of the total 96.67 per cent were married and 95.00 per cent belonged to Hindu religion. Majority of the respondents have attained the secondary level of education (51.11 per cent) followed by primary (36.67 per cent), Graduation and above (8.89 per cent) and intermediate level (3.33 per cent). Among the caste groups 93.33 per cent dominated with intermediate castes (particularly gorhi, nishad, etc.). Nearly 87.78 per cent reported that fishery was their main occupation and remaining 12.22 per cent under took agriculture as main occupation. The most important subsidiary occupation was agriculture (63.33 per cent). The total population was 1254, with an average of 13.93 members per family. The sample fish farming households have an owned area of 98.61 hectare. In addition they leased in 5.60 hectare. There were no leased out area. The total cultivated/operated area was 104.21 hectare. Of the total operated area irrigated area was 91.61 hectare, giving the percentage of irrigated area at 87.91 per cent. The total cropped area of the sample households was higher in Purnea (82.12 ha) followed by Madhubani (45.85 ha) and Bhagalpur (29.04 ha). At the overall level it was 157.01 hectare. Paddy remained the most prominent crop accounting for 42.16 per cent of the GCA followed by wheat (26.57 per cent), maize (9.71 per cent), jute (8.17 per cent), lentil (3.44 per cent), gram (2.64 per cent) and mustard (2.44 per cent). The data revealed that taking together the area of paddy, wheat and maize; came to 78.44 per cent of the GCA, which showed the concentration of cereal crops in the region.

In Jharkhand, out of the two sample districts, Dumka (Santhal Pargana) covered an area of 5518.20 sq. km constituting 7.28 per cent of state's total area (75,834.29 sq. km). Its total population was 1.32 million (2011 census). West Singhbhum covered an area of 8,012 sq. km constituting 10.57 per cent of the state's total area. Its total population was 1.50 million (2011 census). There were 658 jalkars/ponds distributed across 10 anchals in 380.16 ha of water spread area in Dumka district. Similarly, there were 500 jalkars/ponds distributed across 15 anchals in 688.66 ha in West Singhbhum district. These ponds/tanks were leased-out for short and long periods.

Out of the total, 51.66 per cent of the sample households belonged to the age group of 46 to 60 years followed by 36.67 per cent in 31 to 45 years group, 6.67 per cent in 18 to 30 years group and 5.00 per cent in 61 years group. All of them are married and belonged to Hindu religion. Nearly 48.33 per cent have attained the education up to primary level, 45.00 per cent secondary level and 6.67 per cent intermediate level. Among the social groups, 73.34 per cent are from intermediate castes (fisherman community), 23.33 per cent scheduled tribes and only 3.33 per cent scheduled castes. Of the total 75.00 per cent opted fishery as primary occupation and remaining 25.00 per cent mainly on agriculture. Fishery is also leading secondary occupation for 43.33 per cent sample fish farming households. The total population of the 60 fish farming households is 743 comprising 12.38 members per family. The sample households owned 31.15 ha. In addition, they had leased in 0.50 ha and leased out 1.95 ha. Overall they possessed 29.70 ha and out of it 43.80 per cent is irrigated and 56.20 per cent unirrigated. The total cropped area was 23.27 ha in West Singbhum and 13.19 ha in Dumka district. Paddy remained the most important crop, accounting for 60.97 per cent of the Gross Cropped Area (GCA) followed by wheat (17.69 per cent), mustard (7.41 per cent), maize (6.31 per cent), etc. It revealed that kharif crops are mainly grown in the state. In fact, due to undulated topography of the state, agriculture is mainly dependent on monsoon for irrigational purposes.

Economics of Fish Farming

In Bihar, of 90 fish farming households, 30 (33.33 per cent) had small ponds (up to 0.5 ha), 37 (47.11 per cent) medium ponds (0.5 to 2 ha) and 23 (25.54 per cent) large size ponds (above 2 ha) and operating altogether 107 ponds comprising 95 (88.78 per cent) government jalkars and 12 (11.22 per cent) private ponds. The sample households had 1.307 ha of ponds per household. However, it was higher in Purnea (1.644 ha/household) followed by Bhagalpur (1.378 ha/household) and Madhubani (0.898 ha/household). Ponds were either owned or leased in. Leasing of ponds is made for short term (up to 3 years) and long term (up to 10 years). Among the selected fish farming households only 8 (8.89 per cent) had own ponds and remaining 82 households had leased in ponds.

The analysis of cost included both fixed cost and variable costs. Fixed cost includes rent paid for the leased-in ponds. In case of private ponds the rental value of land has been taken into consideration. Variable costs included all cash and kind expenses incurred for production. The details of per hectare cost and return are depicted in Table 20.1. On the selected total farms the total cost came to Rs 51410.35/ha. Out of it the share of variable costs was Rs 50177.14/ha, that is, 97.62 per cent and fixed cost Rs 1233.21/ha, that is, 2.40 per cent. In case of private ponds the rental value of land/ponds area was Rs 2052/ha. Of the variable costs the value of fingerlings/seeds was highest constituting 61.35 per cent (Rs 31539.10/ha). The next important item was labour constituting 8.70 per cent hired labour and 3.90 per cent imputed value of family labour; followed by feeds Rs 2750.96/ha (5.35 per cent), watch and guard Rs 2303/ha (4.48 per cent), interest on purchase of all inputs Rs 2224.71/ha (4.32 per cent), harvesting Rs 1949.64/ha (3.79 per cent), etc. The total return was estimated at Rs 93088.36/ha and the net return (total return minus total cost) came to Rs 41,678.01/ha on overall farms. Per quintal cost of production was calculated at Rs 2,481.19 and the yield of fish was 20.32 qtls/ha on overall farms. The average price received per quintal was Rs 4,581.12. The Cost–Benefit Ratio (CBR) was 1:1.81. It is almost similar in all the three sample districts and there is very little or no relationship between the CBR and the size of fish farms.

Total production of fish on total farms was 2390.79 qtls. It was found that the output per hectare was lower (18.40 qtls) in Purnea whereas it was 21.10 qtls in Bhagalpur and 22.65 qtl the highest in Madhubani district. Out of total production 2328.18 qtls (97.38 per cent) were marketed and 62.61 qtls (2.62 per cent) were used in home consumption. It revealed that a very high percentage of produce is marketed. A total quantity of 2328.18 qtls of the produce was marketed through three identified marketing channels. These are (i) Producer–Consumer; (ii) Producer–Retailer–Consumer and; (iii) Producer–Wholesaler–Retailer–Consumer. Out of the total marketed quantity, 1135.15 qtls was marketed through the channel No. II, that is, one level (48.76 per cent) followed by 876.58 qtls by channel no. III, that is, two level (37.65 per cent) and 316.45 qtls by channel no. I (13.59 per cent), that is, zero level. In Madhubani and Purnea districts out of the total marketed quantity, channel no. III, prominently figured at 50.63 per cent and 48.31 per cent followed by channel no. II at 35.61 per cent and 40.73 per cent and channel no. I at 13.76 per cent and 10.96 per cent respectively. In case of Purnea district, the largest quantity was sold through channel no. II (64.97 per cent), followed by channel no. III (19.08 per cent) and channel no. I (15.95 per cent). Out of the total selected farmers, only 28.89 per cent took loans from different sources. The average amount of borrowing was Rs 41250. Out of the total borrowers, six have repaid their loan amount, 19 have repaid partially and one has not yet started repayment of the amount. Thus, the average amount repayment was estimated at Rs 18379.63 and the outstanding amount of Rs 22870.37.

In Jharkhand, of the 60 fish farming households, 25 (41.67 per cent) constitute from small ponds (up to 0.5 ha), 25 (41.67 per cent) medium farms (0.5 to 2 ha) and 10 (16.67 per cent) large size ponds (above 2 ha). On an average the sample households had 1.08 ha of ponds per household. It was 1.48 ha/household in Dumka whereas 0.76 ha in West Singhbhum. It clearly revealed that there are almost small and medium size fisheries in the state. Government ponds/tanks are leased in to the fisherman

Table 20.1 Per Hectare Cost and Return of Fish Farming in Bihar

Costs	Madhubani	Purnea	Bhagalpur	Overall*
A. Fixed Cost (in Rs)				
Surakshit Jama or	1484.81	845.25	1369.56	1233.21 (2.40)
Rental Value of Land	3725.00	1790.00	1015.00	2052.00
B. Variable Cost (in Rs)				
Lime	816.12	522.94	944.34	760.91 (1.48)
Manure	1082.61	1783.34	974.16	1280.04 (2.49)
Fertilizer	229.17	992.65	793.31	641.38 (1.25)
Fingerlings/Seeds	32299.80	32395.84	30921.66	31539.10 (61.35)
Feeds	2038.00	2676.40	3538.46	2750.96 (5.35)
Medicines & Other Chemicals	366.67	230.84	256.66	251.39 (0.49)
Hired Labour	4245.34	4267.08	4906.66	4473.03 (8.70)
Harvesting	2140.91	1194.67	2513.33	1949.64 (3.79)
Family Labour (Imputed)	1918.28	2572.00	1518.66	2002.98 (3.90)
Watch/Guard	2409.00	2086.67	2413.33	2303.00 (4.48)
Interest on Variable Cost	1618.48	2788.07	2267.56	2224.71 (4.32)
Total	44761.85	51510.50	51048.13	50177.14
Grand Total (A + B)	**46246.66**	**52355.75**	**52417.69**	**51410.35 (100.00)**
Gross Return (in Rs)	87406.18	89754.00	98193.06	93088.36
Net Return (in Rs)	41159.52	37398.25	45775.37	41678.01
Cost of Production/qtl	2041.79	2845.42	2484.25	2481.19
Yield of Fish (qtl)	22.65	18.40	21.10	20.32
Cost Benefit Ratio	**1:1.89**	**1:1.71**	**1:1.88**	**1:1.81**

Note: * values in parentheses represent percentage
Source: Primary survey.

societies against which lessee have to pay rent, commonly known as Jamabandi or Reserve deposits. There are two periods of leasing, namely, short period (up to 3 years) and long term (for 10 years). Most of the ponds were found leased in for short period. Among the selected fish farming households, only 7 had owned ponds and remaining were operating on leased in ponds. The overall rent was fund at Rs 2244.55 per ha per annum.

The details of per hectare cost and return are shown in Table 20.2. On the selected total farms, total cost was estimated at Rs 26785.25/ha. Out of it, the share of fixed cost was Rs 1845.27, which accounts for 6.89 per cent of the total cost. Total variable cost was calculated at Rs 24939.98, accounts for 93.11 per cent of the total costs. Of the total cost, the cost of fingerlings/seeds was highest constituting 56.45 per cent (Rs 15119.50/ha) followed by labour (11.12 per cent) comprising 7.03 per cent (Rs 1882.97/ha) for hired labour and 4.09 per cent (Rs 1097.67/ha) for family labour, feeds (6.34 per cent) etc. The total return was calculated at Rs 40640.95 and the net return came to Rs 13855.70/ha. The average price realized out of the state was R. 3412.33/ qtl, indicating the CBR of 1:1.52. The costs and returns trend were almost similar in both the sample districts. The district wise and farm wise analysis revealed that there is no significant relationship between the CBRs and farm sizes. The total production of fish on total farms was estimated at 774.64 qtl. The overall per hectare yield rate was 11.91 qtl. It was 13.73 qtl/ha in Dumka district whereas that of in West Singhbhum 8.54 qtl/ha.

In regard to disposal of the produce, the data revealed that of the total 50 qtls (6.45 per cent) is consumed at home and 724.64 qtls (93.55 per cent) marketed, indicating marketing of quite higher percentage of the total quantum of produce. A total quality of 724.64 qtls of the produce was marketed. For marketing the produce, three marketing channels were identified, namely, (i) Producer–Consumer; (ii) Producer–Retailer–Consumer, and; (iii) Producer–Wholesaler–Retailer–Consumer. Of the total marketed quantity, 554.76 qtls (76.56 per cent) was marketed by the channel no. I, followed by channel II (16.85 per cent) and channel III (47.76 qtls). In Dumka

Table 20.2 Per Hectare Cost and Return of Fish Farming on Total Farms in Jharkhand

Costs	Dumka	West Singhbhum	Overall*
A. Fixed Cost (in Rs)			
Surakshit Jama or	2166.67	1523.86	1845.27 (6.89)
Rental Value of land	–	–	–
B. Variable Cost (in Rs)			
Lime	416.95	561.03	489.00 (1.83)
Manure	751.67	1936.67	1344.77 (5.02)
Fertilizer	326.17	–	326.17 (1.22)
Fingerlings/Seeds	15696.67	14542.33	15119.50 (56.45)
Feeds	1717.05	1677.00	1697.03 (6.34)
Medicines & Other Chemicals	336.67	445.00	390.84 (1.46)
Water	160.03	427.00	293.52 (1.09)
Hired Labour	2621.67	1144.27	1882.97 (7.03)
Harvesting	606.67	697.20	651.94 (2.43)
Family Labour (Imputed)	1776.67	404.67	1097.67 (4.09)
Watch/Guard	66.67	901.05	483.86 (1.81)
Interest on Variable Cost	1259.21	1067.40	1163.31 (4.34)
Total	25736.10	23803.62	24939.98
Grand Total (A + B)	**27902.77**	**25327.48**	**26785..25 (100.00)**
Gross Return (in Rs)	44345.89	36939.07	40640.95
Net Return (in Rs)	16443.12	11611.59	13855.70
Cost of Production/qtl	2032.25	2965.75	2248.97
Yield of Fish (qtl)	13.73	8.54	11.91
Cost Benefit Ratio	**1:1.59**	**1:1.46**	**1:1.52**

Note: * values in parentheses represent percentages.
Source: Primary survey.

district, out of the total marketed quantity channel no. I (75.61 per cent) was prominently used followed by channel nos. II (19.51 per cent) and III (4.88 per cent). In case of West Singhbhum district, the largest quantity was sold through channel no. I (79.41) followed by III (11.77 per cent) and II (8.82 per cent). Among the sample households 11 households (18.33 per cent) could avail credit and out of them only 7 households have received it from the formal sources and remaining from informal sources. The data revealed that average amount of borrowing was Rs 15700.96 on total farms. Among the borrowers, one has repaid the full amount, 3 have paid partially and 7 are yet to start repayment. The average amount of repayment was Rs 6187.36 on total farms and the outstanding was found to be Rs 9513.60.

Problems of Fish Farming

In Bihar, the constraints as perceived by the sample household are siltation of ponds/tanks (47.77 per cent), 42.22 per cent difficulties of capital/credit; 36.67 per cent lack of technical guidance and same for fish diseases 28.88 per cent lack of quality fingerlings; 23.33 per cent had difficulties of fish theft/insecurity of ponds from anti-social elements; 22.20 per cent reported about lack of proper transportation and marketing facilities; 22.22 per cent said about the fishery department, which is mainly involved in leasing out of the jalkars and collection of revenue rather than facilitating the prospective fish farmers; 20.00 per cent complained about lack of proper boundary around the ponds, which sometimes creates social tension; 12.22 per cent were of the view that Jalkar Management Act, 2006, is no doubt a welcome step of the Government but it did not promote professionalism in fisheries rather it has socially empowered to the fishermen community and 10.00 per cent reported about the ill wills of dominant people of the area in regard to grabbing of ponds (Table 20.3).

In Jharkhand, the constraints as perceived by the sample farmers are lack of capital/credit (41.67 per cent), poor socio-economic status of fisherman (40.00 per cent), shortage of water in the ponds (40.00 per cent), high mortality of fingerlings (30.00 per cent), lack of technical and extension backup (18.33 per cent), lack of cooperation of Matsya Mitra (16.67 per cent), forcible use of ponds water by the strong people for irrigating the fields adjoining to the ponds (15.00 per cent), lack of infrastructural facilities such as nets and vans (15.00 per cent), and theft of fish (15.00 per cent) (Table 20.4).

Prospects of Fish Farming

In spite of various constraints faced by the sample households in Bihar, the state is blessed with vast and varied fisheries and aquaculture resources. The current situation of disappointing fisheries development can be mainly attributed to poor institutional setup, almost non-existence of extension services, lack of adequate resources and infrastructural

Table 20.3 Constraints Faced by the Sample Households in Bihar (in %)

S.N.	Problems	Madhubani (N = 30)	Purnea (N = 30)	Bhagalpur (N = 30)	Overall (N = 90)
1.	Jalkars Management Act, 2006 don't encourage professionalism in Fisheries	10.00	20.00	6.67	12.22
2.	Fishery Department mainly involved in collection of Revenue & settlement of Jalkars	23.33	13.33	30.00	22.22
3.	Lack of Capital/Credit	40.00	30.00	56.67	42.22
4.	Lack of Quality Fingerlings (Seeds)	23.33	36.67	26.67	28.88
5.	Insecurity of the Ponds/theft of Fish	13.33	20.00	36.67	23.33
6.	Fish Diseases	46.67	33.33	30.00	36.67
7.	Silted Ponds/Tanks	56.67	40.00	46.67	47.77
8.	Lack of proper boundary/area of ponds	26.67	20.00	13.33	20.00
9.	Grabbing of Ponds by the dominant people of the area	6.67	10.00	13.33	10.00
10.	Lack of Transportation & Marketing Facilities	20.00	16.67	30.00	22.22
11.	Lack of Technical Guidance	23.33	46.67	40.00	36.67

Source: Primary survey.

Table 20.4 Constraints Faced by the Sample Households in Jharkhand (in %)

S.N.	Problems	Dumka (N = 30)	West Singhbhum (N = 30)	Overall (N = 60)
1.	High mortality rate of fingerlings	33.33	26.67	30.00
2.	Due to undulated topography water stays in the ponds for very short period	36.67	43.33	40.00
3.	Ponds' water is forcibly used for irrigating field crops	20.00	10.00	15.00
4.	Lack of capital/credit	40.00	43.33	41.67
5.	Due to naxalism and poverty pre-mature harvesting is commonly done	10.00	30.00	20.00
6.	Lack of infrastructure for fishing and marketing	6.67	23.33	15.00
7.	Theft of Fish/lack of Security	20.00	10.00	15.00
8.	Poor socio-economic conditions of the fisherman (like illiteracy, poverty, etc.)	36.67	43.33	40.00
9.	Lack of Technical and Extension backup	23.33	13.33	18.33
10.	Lack of Co-operation of Matsya Mitra	10.00	23.33	16.67

Source: Primary survey.

facilities, devoid of conducive policy environment, defunct fisheries cooperatives, lack of professionalism among fisheries personnel, fragmented social set up, poverty and illiteracy among the primary producers, etc. In view of the vast potentiality present abundantly in the state a road map for fisheries has been prepared by the government for 11th Plan period, which aimed at implementing activities like: conservation of water bodies like ponds and tanks, intensive and semi-intensive fish culture in ponds, construction of inlet and outlet for easier passage in *maun* for culture-based fisheries, culture up to an optimism size, raising annual production of fry up to 65 crore from the present level of 35 crore, and; developing the market system to support farmers for different price. There is much scope for developing culture based fisheries in mauns (Ox-bow lakes) and bringing ponds into intensive and semi-intensive culture to attain the desired level of 4.56 lakh tonnes of annual fish production. In addition, it will create employment to fishermen community, which is at present 23 lakh in the state, constituting 50.00 per cent of the total fishermen population.

Jharkhand has rich inland fishery resources in the form of rivers and its tributaries (42.98 kilometres), reservoirs (94000 ha) and tanks (29900 ha). It has 16 fish farmers' development agencies (FFDAs) and 66 Fisherman Cooperative Societies (FCSs). The average fish production in ponds under FFDA is 9.5 qtl/ha/year. The state produces 62000 MT against the demand of 1 lakh MT. There are various constraints for realizing higher production levels such as access to inputs including seed and feed in production areas, low stocking of seasonal reservoirs, and lack of market connectivity. The contribution from reservoir of the state in total fish production is very low, having average productivity level of 5 to 6 kg/ha. Based on the nutrient status of these reservoirs vis-à-vis scientific technologies available in the country the production levels of 30 to 35 kg/ha in large, 50 to 60 kg/ha in medium and 250 to 750 kg/ha in small reservoirs could be easily achieved by judicious and systematic efforts. There are few intricate issues, particularly managerial and financial. Sick Fisheries Co-operative Societies (FCSs), lack of adequate harvest, post harvest and market infrastructure facilities should be circumvented for raising the production level from these water bodies from abysmally very low levels at present. Some other measures that should be put in place are strict enforcement of management rules, observation of closed season, providing training and fishing tools to fishermen, intensive extension practices and observing ethics of responsible fisheries.

POLICY IMPLICATIONS

In the light of the emerging scenario and empirical inputs obtained from both the states following suggestions are emerged for policy implications, which are presented in Tables 20.5 and 20.6 for the states of Bihar and Jharkhand respectively.

Table 20.5 Suggestions Given by the Sample Households in Bihar (In %)

S.N.	Suggestions	Madhubani (N = 30)	Purnea (N = 30)	Bhagalpur (N = 30)	Overall (N = 90)
1.	Availability of Fingerlings be ensured	40.00	30.00	36.67	35.55
2.	Professionalism in Fisheries be encouraged	13.33	10.00	20.00	14.44
3.	Extension backup should be strengthened	53.33	36.67	43.33	44.44
4.	Renovation of ponds be made	50.00	40.00	30.00	40.00
5.	Availability of Credit facility be made	40.00	30.00	20.00	30.00
6.	Availability of Quality feeds be made	16.67	23.33	26.67	22.22
7.	Measurement of Ponds' area should be made	10.00	13.33	23.33	15.56
8.	Training and Follow-up of Training should be made	26.67	13.33	6.67	15.56
9.	Fish Festival be Celebrated	6.67	3.33	13.33	7.78
10.	Fish Diagnostic Centres (FDCs) be established	10.00	16.67	6.67	11.11
11.	Social Security measures for fisherman (like Insurance, Pension, Housing, etc.) be taken up	26.67	36.67	13.33	25.25
12.	Transportation and Marketing facilities be extended	26.67	26.67	10.00	21.11

Source: Primary survey.

Table 20.6 Suggestions Given by the Sample Households in Jharkhand (In %)

S.N.	Suggestions	Dumka (N = 30)	West Singhbhum (N = 30)	Overall (N = 60)
1.	Availability of Fingerlings be ensured	30.00	43.33	36.67
2.	Financial assistance to the fishermen be given to stop the pre-mature harvesting	23.33	20.00	21.67
3.	Availability of credit facility be made	30.00	40.00	35.00
4.	Renovation of ponds be made	23.33	30.00	26.67
5.	Matsya Mitra should be incentivized to propagate the new techniques	36.67	26.67	31.67
6.	Strengthening of Extension back-up	43.33	30.00	36.67
7.	Rearing of fingerlings be promoted	13.33	16.67	15.00
8.	Fish calendar be maintained	6.67	16.67	11.67
9.	Fish festival be arranged	10.00	20.00	15.00
10.	Social Security measures for fishermen be taken	36.67	30.00	33.33

Source: Primary survey.

REFERENCES

Godara, A.S., Ram Singh, and Satya Pal Sharma. 2004. Marketing Pattern of Fisheries in Haryana (Summary), *Indian Journal of Agricultural Economics* Vol. 59, No. 3, July–September, 497.

Government of India. 2014. *Agriculture Statistics at a Glance, 2014*. New Delhi: Ministry of Agriculture.

Kant, R. S.K. Singh, B.B Singh, and R.P. Singh. 2000. 'Production and Marketing of Inland Fish, Bihar', *Journal of Agricultural Marketing* Vol. VIII, No. 4, October–December: 460–5.

Kar, A. and S. Kumar. 2004. Opportunities and Constraints of Fish Production—A Case Study of West Bengal (Summary), *Indian Journal of Agricultural Economics*. Vol. 59, No. 3, July–September.

Kokate, K.D. and S.M. Upare. 2005. Role of Fisheries in Rural Development, Kurukshetra, Vol. 53, No. 9, July issue, 17.

Kumar, P. 2004. 'Fish Demand and Supply Projections in India', Research Report, Division of Agricultural Economics, IARI, New Delhi and World Fish Centre. Penang, Malaysia.

Mishra, A.M. 1997. *Evaluation of Fish Farmers' Development Agencies in Madhya Pradesh* Study No. 73. Jabalpur: AERC, JNKVV.

NABARD. 2000. A *Report on Inland Fisheries in Patiala and Bhathinda Districts of Punjab, An Ex-post Evaluation Study* Chandigarh: NABARD.

Singh, J.P. and Y.K. Pandey. 2004. 'Price Spread of Fish in Different Marketing Channels: Constraints and Policy Implications of Fish Farming in Moradabad District of Western UP', *Indian Journal of Agricultural Economics* Vol. 59, No. 3, July–September, 497.

Singh, R.K.P. and T.T. Singh. 2004. Stocking Density and Species Mix in Composite Fish Culture in North Bihar: A Techno Economic Analysis, *Indian Journal of Agricultural Economics* Vol. 59, No. 3, July–September, 500.

21

Economic Analysis of Fish Ponds in Himachal Pradesh

Ranveer Singh, Meenakshi, S.P. Saraswat, and Partap Singh

Fisheries play an important role in India's economy in augmenting food supply, raising nutritional levels and earning foreign exchange. Pisciculture is becoming more and more alluring due to its low capital investment, short gestation period and generation of high profit. Its importance from social and economic point goes to augmentation of nutritional level, employment generation, earning foreign exchange. It is suitable proposition for rural development and to improve the economic conditions of the rural people (Biswas 2006). In Himachal Pradesh, where the average size of holding is about one hectare, pisciculture in ponds is becoming more and more popular as a supplementary enterprise with agriculture in the rural areas to increase the income of the farmers to a substantial extent. Fish can be raised in small ponds as well as big ponds, *kuhls* (gravity water channel), and channels which are generally found in village and by constructing either ponds in agriculture farms or by the construction of exclusive fish farms. A rational development of pond fisheries is based on culturing those valuable species and varieties of fish for food, which over a short period provide a high quality product (Martyshev 1983: 11). The estimated network of the state fisheries water resources is about 3,000 km out of which 600 km have been classified as trout water's which can be judiciously trapped for trout culture. At present, trout is a highly priced fish in the country. The state has taken a major leap in production of indigenous schizothoracids and exotic salmonids such as rainbow and brown trout. The main trout fish producing districts in the state are Kullu, Mandi, Kinnaur, and Chamba. Himachal Pradesh has also become the first state in the country to introduce trout farming in the private sector besides emerging as a number one producer of this species of fish. With the technical support of the state government large number of trout farms has been set up in the private sector in the state.

With this background, it was considered pertinent to examine the economic feasibility of the fish ponds/raceways with specific objectives:

1. to examine the costs and returns of fish pond; and
2. to examine the problems faced by the fish farmers.

METHODOLOGY AND DATA

The required primary data and information for this chapter was obtained through multi-stage stratified random sampling technique. At the first stage Kangra district was selected purposively on the basis of having largest area under fish ponds. Secondly, from selected district, six administrative blocks were selected on the basis of highest area under pond fisheries. A list of pond owners in these blocks was obtained, out of which one third, that is, 22 fish ponds were selected randomly. They were classified into four categories according to the gross area of fish ponds (i) Marginal: having pond size below 100 sq m; (ii) Small–having pond size 100 to 200 sq m; (iii) Medium: having pond size 200 to 300 sq m; and (iv) Large: having pond size above 300 sq m. Among the sampled fish ponds, there were 5 each in marginal and small category and 6 each in medium and large category. The same technique was used in the selection of trout fish farms. At the first stage Kullu, Mandi, Shimla, and Kinnaur districts were selected purposively on the basis of having largest area under trout fish raceways. Secondly, a list of trout fish farms was obtained from the district fishery office, and then 20 trout fish units were selected randomly for the detailed study. They were classified into three categories according to the size of raceways: (i) Small: trout farms having water area below 100 cu. m; (ii) Medium: having water area 100 to 200 cu m; and (iii) Large: having water area above 200 cu m. Thus in the selected sampled trout fish farms there were 10 small, 6 medium, and 4 large sized trout farms. The data used in this chapter pertains to the agricultural year 2006–07. The standard cost components have been used to work out the economics of trout fish units. Prorated establishment cost was computed by using the following formula:

$$P_c = P\,(I + i/100)^{-n}$$

P_c = Prorated Establishment cost (Rs/pond/raceway)
P = Initial capital investment (Rs/pond/raceway)
i = Rate of interest (12%)
n = Life span of pond (10 years for pond fisheries and 15 years for trout fisheries)

Here the prorated establishment cost is the sum that amounts to the initial cost in 10 years for pond fish farms and 15 years for trout fish farms at the rate of 12 per cent per annum compounded yearly.

RESULTS AND DISCUSSION

Pond Pisciculture

A viable pisciculture practice primarily depends on the selection of a suitable site, which in turn depends upon water retentive quality of the soil and availability of adequate water supply during the culture period (Santhanam, Sukumaren, and Nutrajan 1990: 27). After having selected a proper site, fish cultivation begins with the construction of fish pond and its liming. The pond is kept dry for 15 days after the application of lime and then it is filled with good quality of water of required volume. After filling the pond with water, ponds are manured with organic manure. For sustained production of commercial fish, unwanted fauna and flora and harmful insects are eradicated by means of some suitable chemical control measures or it is also done manually. Generally, in the sampled fish farms, composite fish farming is prevailing which is quite profitable. It involves a mixed farming of fish species. Fish fingerlings (approximately 5–6 thousand per hectare) are added to the pond and the fish species which are generally mixed to add in the pond are silver carp, grass carp, and common carp. Other than manure, fish are also fed with supplementary feeding of mustard oil cakes and wheat/rice bran. Vegetable leaves and kitchen refuse is also used as a supplementary feed for pisciculture.

The average size of sampled fish ponds was 187.59 sq. m. Most of the fish ponds (95 per cent) were on agricultural land. The main source of water for these ponds was kuhls (86 per cent). As far as the investment is concerned, majority of the fish farmers (45 per cent) constructed their ponds by using their own resources. Twenty seven per cent of the fish farmers availed the finance from state fisheries department fully and 22 per cent partially.

Costs and Returns from Sampled Pond Fish Farms

The average cost of construction of fish pond was Rs 7143 whereas the average prorated establishment cost of these ponds was Rs 2300. Total fixed and variable costs constituted 23 and 77 per cent respectively of the total cost incurred by all the sampled pond fish farmers. The main cost components were feed, prorated establishment cost of ponds, and labour charges which accounted for 53, 21, and 13 per cent of the total cost, respectively. Cake and bran were the major components whose cost constituted about 42 per cent of the total cost. Total cost incurred per farm by all the sampled pond fish farmers was observed to be Rs 11160. The net income per farm increased with the size of fish farms from Rs 5090 in marginal category to Rs 17942 in the large category whereas it averaged at Rs 11585 for all farms (Table 21.1). The output per unit of input was highest in medium category (1:2.59) followed by marginal (1:2.11), small (1:2.01) and large category (1:1.80). The output per rupee of input was Rs 2.05 on all the sampled fish farms. The output–input ratios indicate that the medium category farmers were operating efficiently as compared to others. The time spent per farm on the various activities of feeding of fish, maintenance of pond, fish catching and watch and ward by the farmers was 439 hours per annum, whereas the share of income

Table 21.1 Costs and Returns in Production of Pond Fish on Sampled Farms

Cost Components	Marginal Farms (Below 100 M^2)	Small Farms (100–200 M^2)	Medium Farms (200–300 M^2)	Large Farms (above 300 M^2)	All Farms
A. Variable cost (Percentage to total cost)					
1. Value of fingerlings	4.99	4.95	6.13	7.97	6.87
2. Value of feed	43.03	42.57	49.66	58.09	52.71
(i) Dung @ Rs 50/Qtl.	1.38	2.16	0.99	2.09	1.78
(ii) Lime @ Rs 10/Kg.	0.33	0.20	0.21	0.46	0.36
(iii) Grass @ Rs 50/Qtls.	8.96	9.14	14.05	5.14	8.05
(iv) Cake and ran	32.36	31.07	34.41	49.28	41.91
(v) Vegetable waste	–	–	–	1.12	0.61
3. Value of human labour					
(i) Family labour	16.00	12.25	12.83	8.34	10.59
(ii) Hired labour	–	–	–	5.17	2.83
4. Interest on working capital	3.19	2.99	3.43	3.97	3.65
Total variable cost (A)	67.21	62.76	72.05	83.54	76.65
B. Fixed cost (Percentage to total cost)					
1. Prorated pond cost	28.35	32.37	24.08	14.98	20.61
2. Interest on implements and tools	1.90	2.05	1.61	0.61	1.15
3. Depreciation of implements and tools	2.54	2.82	2.26	0.87	1.59
Total fixed cost (B)	32.79	37.24	27.95	16.46	23.35
Total cost (A + B) (Rs/farm)	4574	6565	9250	22385	11160
Total production (Qtls/farm)	1.48	2.40	3.58	8.17	4.10
Value of total production (Rs/farm)	9664	13262	23965	40327	22745
Net returns (Rs/farm)	5090	6697	14715	17942	11585

from fisheries (out of total gross income) was 12 per cent which was reasonably good considering the time spent on the various activities of fisheries.

Marketing of Pond Fish

Out of total per farm production of 4.10 qtl of fish, 3.54 qtl was marketed after keeping part of the produce for home consumption and given away in the form of gifts by all the sampled pond fish farmers. Contractors were the main functionaries involved in the marketing of pond fish in large category, as 58 per cent of the produce was sold through contractors. In the case of other categories, most of the produce directly went to consumers around the producing centres. This trend shows that most of the produce was used by local population.

Trout Pisciculture

Keeping in view the vast potential of trout in the perennial rivers, Himachal Pradesh has become the first state in the country to introduce trout farming in the private sector. The total production of trout in Himachal Pradesh was 0.54 tonnes in the year of 1996–97 which increased to 25 tonnes in 2005–06 (Government of Himachal Pradesh 2007), thereby showing an increasing rate of growth of 23 per cent per year.

Raceway Construction

As far as the construction and dimensioning of raceways are concerned these may be of three types: (i) mud ponds, (ii) made of RCC or cement mortar and (iii) made of stone pitched sloping sides with cement mortar pointing. The most of the raceways in the study area were of second type. The raceways vary in size from 15–30 m in length, 2–3 m in width, and 0.8 to 1.3 m in depth. The average size of raceway was 152.70 m^3 for all the sampled trout fish farms. Most of the raceways (90 per cent) were on agricultural land. The source of water for these raceways was only the kuhls. Majority of the fish farmers (45 per cent) constructed their ponds by using their own resources.

Twenty-seven per cent of the fish farmers availed the finance from state fisheries department fully and 22 per cent partially.

Fingerlings and Feed for Trout Fish

Fingerlings of trout fish are generally obtained from State's hatcheries. There are in all six hatcheries at Patlikuhl, Nagani (Kullu), Barot (Mandi), Holi (Chamba), Dhamwari (Shimla) and Sangla (Kinnaur). After obtaining fingerlings from the hatcheries, these are stocked in raceways/circular ponds with 30–50 fingerlings/m^2. The basic requirement during the rearing period is the regular flow of abundant and silt free cold water in the raceways or circular ponds. Feeding also warrants full attention of the culturist. Artificial dry feed is getting popular in view of their better conversion, easy handling, storage and transportation. The main ingredients used in feed formulation are fish meal, soybean, bone meal, whole wheat, yeast, linseed oil, methionine and sodium alginate. The crude protein value of such a feed is kept at 40–50 per cent level. The feed given to fish is kept fresh and of high quality. Low quality feed causes disease and mortality. Wide varieties of diseases are known to afflict the trout at different stages of its life. Most of these diseases are attributable to viral, bacterial, fungal infections or nutritional deficiencies. The feed was processed by the state government trout fish farm at Patlikuhl with an annual production of 64.3 tonnes during the year 2006–07, which was not enough to meet the requirement of trout farmers of the State.

Costs and Returns from Sampled Trout Fish Farms

The average cost of construction of a raceway was Rs 210215 whereas the average prorated establishment cost of these raceways was Rs 38469. Total fixed and variable costs constituted 15 and 85 per cent, respectively of the total cost incurred by all the sampled trout fish farmers. The main cost components were feed, prorated establishment cost of raceway, value of fingerlings, and labour charges which accounted for 64, 15, 10, and 6 per cent of the total cost, respectively (Table 21.2). Total cost incurred by all the sampled trout fish farmers was observed to be Rs 265214 per farm, whereas category wise, the same varied from Rs 136255 for small category to Rs 508541 for large category indicating positive relationship with the increase

Table 21.2 Costs and Returns in Production of Trout Fish on Sampled Farms

Cost Components	Small Farms (Below 100 M^3)	Medium Farms (100–200 M^3)	Large Farms (Above 200 M^3)	All Farms
A. Variable Cost (Percentage to total cost)				
1. Value of fingerlings	9.98	10.04	8.70	9.59
2. Value of feed	66.61	63.88	63.90	64.33
3. Value of salt & medicine	0.29	0.49	1.42	0.80
4. Value of human labour	7.56	5.84	5.98	6.39
– Family labour	2.79	0.44	0.01	0.90
– Hired labour	4.77	5.40	5.93	5.49
5. Interest on working capital	4.22	4.01	4.00	4.05
Total variable cost (A)	88.66	84.26	84.00	85.17
B. Fixed Cost (Percentage to total cost)				
1. Prorated raceway cost	10.91	15.45	15.69	14.50
2. Interest on value of implements and tools	0.08	1.00	0.12	0.10
3. Interest on covering net	0.09	0.01	–	0.03
4. Depreciation of implements & tools	0.18	0.16	0.19	0.17
5. Depreciation on covering net	0.08	0.01	–	0.03
Total fixed cost (B)	11.34	15.74	16.00	14.83
Total cost A + B (Rs/farm)	136255	325070	508541	265214
Total production (Qtls/farm)	8.99	25.54	34.25	19.01
Value of total production (Rs/farm)	186893	610661	843132	445556
Net returns (Rs/farm)	50638	285591	334591	180342

in the size of trout fish farms. The net income per farm increased with the size of trout fish farms from Rs 50638 for small category to Rs 334591 in case of large category whereas the net income of all the sampled trout fish farms was Rs 180342/farm during the study period.

Employment and Income from Trout Fish Farming

Human labour used in various activities in trout fish farming include maintenance of tank, cleaning of tank, feeding of fish, watch and ward, fish catching, and grading of fish. The use of hired labour was higher as compared to family labour on all the sampled trout fish farms. The proportion of both family and hired male labour was more as compared to female labour in all the categories of trout fish farms. Out of the total time spent on these activities, by all the sampled trout fish farmers, maximum time went to the activity of watch and ward followed by fish catching, feeding, grading, and cleaning of tank and maintenance of tank. The time spent per farm on various activities annually, by the trout fish farmers, was 213.25 days, whereas the share of income from trout fisheries (out of gross income) was 41 per cent, that is, Rs 445241 per farm, which is very good considering the time consumed by various activities of trout rearing.

Marketing System for Trout Fish

Trout fish marketing involves activities such as catching, dressing, packing, and transportation. Before packing, fish are dressed and the intestine and other unwanted parts are removed by hand. After dressing, fish are packed in thermocol box. The ice is also put in the box for protecting the fish by maintaining proper temperature. Trout fish are sold throughout the year however main supply season of fish is September to April. Demand for fish in the consuming markets is higher in winter season. The producers supply the fish to hotels as per demand. The major quantity of marketed surplus, that is, 57 per cent was sold to hotels at Delhi and in the state. Nearly 22 per cent was sold to local contractors, 19 per cent directly to consumers and only 2 per cent to wholesalers at Delhi.

Problems Faced by Fish Farmers

The non-availability of required breed of fingerlings, inadequate financial assistance from the government and shortage of water in summer and winter were the main problems faced by pond fish farmers. The required hydrological conditions for the construction of raceways limit the number of sites where a farm could be constructed. Therefore, location of trout farms is usually away from the house as reported by majority of the farmers. High cost of construction of raceways becomes a major bottleneck due to lack of finance. Majority of the trout fish farmers also reported that fingerlings and feed were not available at desired places. Most of the farmers were of the view that there was no proper market for trout fish in the area and the market was away from the producing area. Due to far away markets, the problem of transportation was also faced by them.

POLICY SUGGESTIONS

It can be concluded from the above analysis that fish farming is quite a profitable proposition. Pond fisheries enterprise requires less time, having low investment and has no marketing problem whereas high investment is required for trout farms and the markets of trout fish are far away from producing areas. The pond fish farms are located mainly in warm districts whereas the trout farming is practiced in the areas where cold water is available. The trout pisciculture has a very high potential but this venture has not developed as fast as it could due to various constraints and problems faced by this sector. For success of trout farming still there is a need for proper input supply with appropriate extension services. Feed is the most important component for the proper development of pisciculture, but still most of the farmers have little knowledge about fish feed. Training related to management of farms, appropriate fish feed and marketing of fish should be imparted to the farmers for the proper growth of trout fisheries. There is a need to improve the production of feed and ensure its supply through establishing feed distribution centres for fish farmers as well as for trout farmers. Proper supply of fingerlings of required breed at appropriate places and time can also improve the production. To ensure the timely and required supply of fingerlings, the production of fingerlings at hatcheries and fisheries department should be increased.

There is also a need to strengthen and promote institutions such as cooperatives, producer's organizations and contract farming that link producers to markets and reduce marketing and transaction costs. Adequate financial assistance should also be provided to fish farmers for construction of new raceways and rejuvenating old raceways. For the success of trout farming as a small industry, Himachal Pradesh has some grey areas related to the key factors such as quantity and quality of water supply, feeding and feed management, level of hygiene maintenance schedule for healthcare and disease investigation (Sandhu 2007). For the more success of pond fisheries, farmers should be encouraged to construct new ponds and to renovate old ones. For this, adequate financial assistance at reasonable rate of interest should be provided to them. To ensure adequate supply of water throughout the year, hand pumps/tube wells should be installed by providing financial assistance to the farmers.

REFERENCES

Biswas, K.P. 2006. *Economics in Commercial Fisheries*. New Delhi: Daya Publishing.

Government of Himachal Pradesh. 2007. Directorate of Fisheries, Bilaspur, available at Website: www.hpfisheries.nic.in

Martyshev, F.G. 1983. *Pond Fisheries*. New Delhi: Amerind Publishing.

Sandhu, K. 2007. 'Lackluster Rainbow', Himachal Plus, *The Tribune*, 26 September.

Santhanam, R., N. Sukumaren, and P. Nutrajan. 1990. *A Manual of Fresh-Water Aquaculture*. New Delhi: Oxford and IBH Publishing.

Section II
Technology, Inputs, and Services for Agricultural Development

22

Biotechnology in Agriculture

Potential, Performance, and Concerns*

Vasant P. Gandhi, Dinesh Jain, and Aashish Argade

Technological advances are urgently needed in Indian agriculture to raise yields and in this context, biotechnology offers a huge new potential and perhaps even a new green revolution. Remarkable scientific advances in the recent decades have made it possible to identify genes, know their functions, and also transfer them from one organism to another. These advances in biotechnology are offering numerous possibilities such as the development of Bt cotton.

Bt cotton gets its name from a bacterium called *Bacillus thuringiensis*, which is an aerobic bacterium, a natural enemy of bollworms, characterized by its ability to produce crystalline inclusions during sporulation. This bacteria was first discovered by a Japanese bacteriologist in 1901 and subsequently in 1915 a German scientist isolated crystal toxin in Thuringen region of Germany. Bacillus thuringiensis was registered as a microbial pest control agent in 1961 under the federal Insecticide and Rodenticide Act in the US. In India Bt formulations have been registered under Pesticides Act 1968. With the advent of biotechnology, the bacterial gene was introduced genetically into the cotton genome. The worms feeding on the leaves of a Bt cotton plant become lethargic and sleepy, and are finally eliminated. Bt cotton was first developed by Monsanto and it is currently one of the most widely grown transgenic crops in numerous countries including United States, China, India, Australia, Argentina, South Africa, and Indonesia.

Cotton is the most important cash crop in India and the country ranks first in cotton area and second in cotton production in the world. About 15 million farmers in the country across 10 states are engaged in cotton production. However, cotton yields in India are one of the lowest in the world, a major reason being susceptibility to severe pest attacks. As a consequence, there was a heavy usage of pesticide on cotton crop, to the tune of 55 per cent of total pesticide consumption, leading to environmental and human health hazards. The introduction of Bt cotton in such a scenario was a pertinent breakthrough. After its approval in 2002 in India with hesitation, Bt cotton has spread rapidly across the country, helping to raise production and incomes, and bring a second green revolution in some states (Table 22.1).

The analysis from several years of Indian trial data had demonstrated the superiority of Bt technology in terms of reduced pesticides application and increase in effective

* The authors acknowledge with thanks the contributions and assistance of Darshan Ajudia, Varsha Khandker, Yogendra Rajyguru, Narinder Arora, Pravin Gagadani, P.V. Sethumadhavan, CMA, and the Ministry of Agriculture and Farmers Welfare.

Table 22.1 Growth in Production, Area and Yield of Cotton—All India

Year	Production (in Lakh Bales)	Area (in Lakh Ha)	Yield (in Kg per Ha)
1950–51	33	58.8	95
1960–61	57	76.1	127
1970–71	54	76.1	120
1980–81	78	78.2	170
1990–91	117	73.9	269
2000–01	140	85.8	278
2001–02	158	87.3	308
2002–03	136	76.7	302
2003–04	177	77.9	387
2004–05	213	89.7	404
2005–06	185	86.8	362
2006–07	226	91.4	421
2007–08	259	94.1	467
2008–09	229	94.1	413
2009–10	240	101.3	403
2010–11	330	112.4	499
2011–12**	361	119.9	512
Annual Growth Rates			
1981–82 to 2001–02	2.657	1.090	1.566
1991–92 to 2001–02	−0.422	2.015	−2.442
2001–02 to 2011–12	13.148	3.577	9.578
2005–06 to 2011–12	9.683	5.138	4.558

Notes: **First Advance Estimates released on 14 September 2011.
Source: Directorate of Economics and Statistics, Department of Agriculture and Cooperation.

yields. The impact assessment commissioned by Mahyco–Monsanto Biotech claimed sizable benefits for Bt adopters (Nielson 2004). The major advantages claimed for Bt cotton include reduction in the use of insecticides by almost 50 per cent, reduction in the harmful effect on the environment, good quality of cotton fibre at par with that of non-Bt cotton, better yield per unit of input use, and lesser residue of pesticides in the fibre resulting in reduced harmful effects such as allergic reactions.

However anti-biotechnology activists declared the technology as a complete failure (for example, Shiva and Jafri 2003). The voices against Bt cotton indicate that the gene may spread and its impact on the eco-system is not known, the Bt cotton seed would be very expensive compared to non-Bt seed for the farmers, some companies may have a monopoly on Bt seed, the Bt cotton farmers may still need to use insecticides, the Bt cotton seed cake will cause harm to the animals, Bt may enter in the human food chain and cause harm, transgenic varieties will lead to disappearance of native varieties and biodiversity in the country, and insects will soon become resistant to Bt cotton making the pest control even more difficult in the near future.

Even though the performance of Bt cotton has been known to be satisfactory in government circles, there is discontent in other quarters with Bt cotton. Strong views both for and against Bt technology have surfaced. Thus biotechnology has continued to be controversial, and new biotechnology innovations for agriculture have faced much resistance, and have yet to be officially approved.

OBJECTIVES

In view of the continued concerns and diverse views on Bt cotton and considering the importance of cotton in Indian agriculture, it seemed important to undertake a comprehensive and systematic review to study the economic returns and other related aspects of the cultivation of Bt cotton as opposed to non-Bt cotton in major cotton producing states in the country. The Centre for Management in Agriculture (CMA) at IIM Ahmedabad conducted this study at the request of the Ministry of Agriculture and Farmers Welfare to make an assessment of the benefits and concerns of agri-biotechnology in India with a focus on Bt cotton.

In this context the following are some of the important questions:

1. What is the promise and potential of biotechnology for agriculture and are these promises important for India?
2. What is the performance of agri-biotechnology vs its promise. Does bio-technology make economic sense for India?
3. What are the concerns and the nature of the risk-perception and resistance faced by biotechnology, and the reality regarding the possible harm/risks?
4. What are the challenges, and what should be the approach and policy of the country given the findings?

The research first surveys extant literature on the promise, performance, and concerns of agri-biotechnology, also using secondary data for India. It examines the issues of performance and concerns through primary data collected from four major cotton growing states of Andhra Pradesh, Gujarat, Maharashtra, and Punjab. Wherever possible, statistical and econometric techniques are used for in-depth analysis. The study derives conclusions and possible implications for policies and path of action for biotechnology in India.

PROMISE OF BIOTECHNOLOGY AND ITS IMPORTANCE FOR INDIA

One of the most important benefits that biotechnology can bring to many crops in India is the resistance to pests and diseases. Biotechnology offers a major advantage of reducing pesticide use and therefore the environmental harm that they are causing and also avoid crop loss due to pests. In crops like rice, traits such as resistance to a host of pests and diseases can be transferred from wild species to cultivated species using biotechnology. Gaur and Choudhary (2010) have shown that before the advent of Bt cotton in India in 1998, cotton accounted for 30 per cent of the total pesticide market and 42 per cent of insecticide market in India. In 2006, the share of cotton in the Indian pesticide and insecticide market fell to 18 per cent and 28 per cent, respectively. The number of sprays reduced by 36 to 50 per cent in case of Bt cotton compared to non-Bt cotton, with comparable reduction in the cost of pesticides (Gandhi and Namboodiri 2006).

The technology can offer increase in yields and through this increase in production, exports and incomes; and even reduce the area required for production, thereby reducing environmental harm. Edgerton (2009), demonstrates that maize yields in US can be increased by agronomic practices to enhance yields at historical rates but enhanced yields are possible due to biotechnology traits, which can potentially double the productivity.

Another major possibility that biotechnology offers is improvement in the output quality like nutritional enrichment, reduction in fats or harmful fats in the food, and reducing allergens. Biotechnology was used to increase the content of iron and beta-Carotene in rice by incorporating genes from different sources like beans, daffodils and soybeans (Khush 2002).

Biotechnology can incorporate salt tolerance in plants which would be a boon for large areas affected by salinity. Gill and Tuteja (2010) suggested use of biotechnology to develop traits in rice, tobacco, tomato, apple, and pears, for tolerance to salinity, temperature extremes, and drought. It can also help to reduce fertilizer use and runoff by improving the nutrient availability and absorption efficiency of plants in the soil.

Biotechnology can find applications in enhancement of the shelf life of food, making food products last longer and, thereby reducing wastage. India loses almost 35–40 per cent of fruits and vegetable production due to excessive softening. Softening aggravates the condition of the produce during transportation, handling and eventually adversely impacts consumer preference and taste. Meli et al. (2010) demonstrated that transgenic tomato had firmer fruits and shelf life was enhanced by around 30 days as compared to non-transgenic tomato.

PERFORMANCE OF BIOTECHNOLOGY VERSUS ITS PROMISE IN INDIA

To answer this section, we first use literature survey on performance of Bt cotton. Several cross sectional and longitudinal studies have been studied here to establish that the direct benefits of Bt cotton are increased yields and profits, particularly in India which is dominated by resource-constrained farmers. The most prominent advantages of Bt cotton include increase in productivity and reduction in use of insecticides.

After the advent of Bt cotton in India in 2002 and its steady and steep adoption at the macro level, the average yield of cotton in India increased from 308 kg per hectare in 2001–02 to 526 kg/ha in 2008–09. Consequently, cotton production in India rose from 15.8 million bales in 2001–02 to 31.5 million bales in 2007–08. The country became a prominent exporter by 2007–08 when it exported 8.8 million bales (Chaudhary and Gaur 2010).

A cross-sectional survey by Gandhi and Namboodiri (2006) of 694 cotton farmers in Gujarat, Maharashtra, and Andhra Pradesh found higher yields (30.7 per cent), lower pesticide consumption and higher profit (87.6 per cent) associated with the cultivation of Bt cotton compared to non-Bt cotton. A similar study across the states of Andhra Pradesh, Maharashtra, Karnataka, and Tamil Nadu showed insecticide use reduction by 50 per cent, 34 per cent higher yields, and higher profits despite higher seed cost (Qaim et al. 2006). Stone's (2011) longitudinal study conducted between 2003 and 2007 in Warangal district of Andhra Pradesh indicates that cotton yields increased by 18 per cent and mean sprayings dropped by 54.7 per cent in the sample villages. A panel study by Krishna and Qaim (2012) indicates that the decline in pesticide usage in Bt cotton has been sustainable.

Zilberman, Ameden, and Qaim (2007) analysed several earlier studies and found that transgenic varieties resulted in higher yield increases in countries where pesticide usage was low and pest infestation was high, as in case of India. Farmers in low-income countries are risk averse and are willing to pay a premium to reduce risk; hence they cultivated Bt cotton not only for profitability, but also as an insurance against pest attack. Scale neutrality of Bt proved all the more useful for small farmers.

The first impact of any increase in income on a sustained basis is a change in consumption pattern, which is typically expected to increase. Kathage and Qaim (2012) found that, consumption expenditure of households increased only after the profit increase was sustainable. As such, between 2002 and 2004, though the adoption rate of Bt and profit increased among farming households, there was no significant change in the consumption behaviour of the households adopting Bt cotton. However, in the period between 2006 and 2008, the annual consumption of Bt-adopting households increased by an average of Rs 15841. This was 18 per cent higher compared to non-adopters of Bt cotton. Subramanian and Qaim (2009), through micro Social Accounting Matrix (SAM) model, found that Bt cotton led to increase in aggregate labour returns by 42 per cent, and 55 per cent for hired female labourers, while incomes of poor and 'vulnerable' were 134 per cent higher compared to returns from non-Bt cotton.

DATA AND METHODOLOGY

In order to address the topic and objectives of the research, a primary sample survey of cotton farmers and consumers/urban people was planned. Multi-stage stratified and random sampling was done. Based on area, production and yield of cotton, the highest four cotton producing states of Gujarat, Maharashtra, Andhra Pradesh, and Punjab were selected in the study sample. Together, these four sample states account for 75 per cent of the cotton area and 76 per cent of the cotton production in the country. Given the limitations of time and resources, it was decided to limit to one sample district in each state. The highest cotton producing district was selected in each state. Thus, Guntur in Andhra Pradesh, Rajkot in Gujarat, Jalgaon in Maharashtra, and Bhatinda in Punjab were selected.

Villages within each district were randomly selected to cover cotton growing areas including Bt cotton, and provide a diversity of agro-ecological settings. This was done in discussion with district officials and/or seed dealers at the district headquarters who were knowledgeable about the district. Farmers were selected in each village through a random process and effort was made as far as possible to have both Bt and non-Bt farmers in the sample, and if not, to have Bt farmers who could reflect on their non-Bt growing experience. Effort was also made to cover both irrigated and unirrigated farms, as well as small, medium and large farms. In each state, a sample of 100 was planned consisting of 80 farmers and 20 consumers. Overall 98 farmers were small, 185 were medium, and 43 were large. 207 farmers had irrigation whereas 119 did not have irrigation (Table 22.2).

Table 22.2 Sample Distribution

State	District	Farmers		Consumers	
		Sample Size	(%)	Sample Size	(%)
Andhra Pradesh	Guntur	82	25.2	33	28.7
Gujarat	Rajkot	81	24.8	22	19.1
Maharashtra	Jalgaon	82	25.2	32	27.8
Punjab	Bhatinda	81	24.8	28	24.4
Overall		326	100.0	115	100.0

Source: Own field survey.

A very detailed questionnaire was developed for the study based on the objectives, research questions and the behavioural framework presented earlier. The farmer questionnaire covered a large number of aspects covering, but not limiting to profile of respondent farmer, experience with cotton/Bt cotton, comparison of cotton varieties grown in last few years, comparative cost of cultivation of cotton, pest incidence observed on the cotton crops, details of pesticide on cotton, perception of farmers on various aspects of Bt cotton and its cultivation; advantages of Bt cotton vis-à-vis non-Bt cotton, direct or indirect impact of Bt cotton technology, secondary pest resurgence, overall judgement, etc.

RESULTS AND DISCUSSION

Pest Incidence

Substantial resistance/substantially lower incidence in the case of boll worms including American, pink, and spotted bollworms, particularly pink bollworm on Bt cotton compared to non-Bt cotton has been reported. Bt cotton also shows resistance towards foliage feeding pests such as leaf rollers and caterpillars. However, there is greater incidence of sucking pests like mealy bugs, aphids, jassids, and white fly; and also shows a higher incidence of the disease of Alternaria leaf spot on Bt cotton.

Factors Influencing Adoption of Bt Technology

Agronomic potential: Most farmers indicate that Bt cotton has shown good pest resistance and is responsive to fertilisers and irrigation. Almost all farmers indicate that Bt cotton yields more than non-Bt cotton. However, there is little difference in the by-product yield and Bt cotton is not as tolerant to drought and salinity as non-Bt cotton. On the whole the agronomic potential of Bt cotton appears to be strong except for the issues of drought and salinity tolerance.

Agro-economic potential: There is good demand for Bt cotton even though the price may not be very different compared to non-Bt. Government procurement and price support does not exist for most farmers and neither does contract farming. A huge majority of farmers find Bt cotton quality to be better than non-Bt cotton, even as seed, fertilizer, water, and labour costs for Bt cotton are higher. Thus, Bt cotton entails higher cost of production per hectare, but almost all farmers indicate that it is substantially more profitable than non-Bt cotton, indicating a strong agro-economic potential, based on the open market and not government support.

Effective demand: The findings indicate that the cotton farmers are enterprising and willing to take risks and be opinion leaders for other farmers. Almost all of them are aware of the benefits of Bt cotton and the package of practices to follow. Lack of awareness about right varieties and brands and insufficient access to credit were reported by some farmers. The villages of most farmers are well-connected with markets. Cotton is extremely important for family income and livelihoods of the farmers who grow it, and this would strongly drive demand for good technology.

Aggregate supply and distribution: Farmers indicate that a large number of companies supply Bt cotton seeds and numerous varieties are available in sufficient quantity when needed. There is no barrier to access Bt technology. On the distribution front, farmers indicate that large numbers of dealers nearby are ready to sell Bt cotton seeds, though they charge a high price for Bt cotton seeds and often do not provide credit. Dealers provide guidance on the kind of seeds to use and most farmers are satisfied with the quality of the seeds.

SOURCES OF INFORMATION AND ADVICE

It is important of know the source of information and advice as it indicates who influences farmers' decision-making. Seed dealers are found to be the most common and most important source of information, followed by fellow farmers. The seed company and other input dealers also play a small role; mass media like newspapers and television play a limited role; government sources such as extension workers and call centres do not play much of a role as far as information on Bt cotton is concerned. The main advantages conveyed by the information sources include yield advantage, pest resistance and profitability.

Very little negative information is conveyed, and this mainly relates to the risk and high seed cost, rather than harm to human beings and environment. The source of such information is fellow farmers and not newspapers or NGOs.

COSTS, YIELDS, AND PROFITABILITY OF BT COTTON

The findings indicate that there is a substantial difference in the seed, fertlizer, harvesting, and marketing costs. As a result the total cost increases by 72 per cent. However, the yield increases by 33 per cent and the revenue by 79 per cent with the adoption of Bt cotton. As a result there is a substantial increase of 83 per cent in the profits. The findings also indicate that farmers perceive a substantial increase in the yields and particularly in profits with the adoption of Bt cotton. This economic advantage explains the rapid adoption of Bt cotton and its huge popularity with the farmers. It may be noted that non-Bt cotton findings are based on a limited set of responses and usually based on recall by farmers since hardly any currently grow non-Bt cotton.

There is a considerable increase in the seed cost and varies from over 300 per cent in Andhra Pradesh to just 57 per cent in Maharashtra. Fertilizer costs, including farmyard manure, show a huge increase in most states. Pesticide costs do not show much change, except in Punjab where they show a decrease. Harvesting and marketing costs also show a change. On an all India average, total cost shows a 71 per cent increase, ranging from as high as 205 per cent in Andhra Pradesh to just 50 per cent in Gujarat. The highest yield increase is seen in Andhra Pradesh followed by Maharashtra. Revenue increase is most substantial in Gujarat at over 300 per cent and the least in Maharashtra at just 20 per cent. The profit increase is the greatest in Gujarat at 140 per cent and lowest in Punjab at 34 per cent.

ECONOMETRIC ANALYSIS OF PERFORMANCE OF BT COTTON

Regression analysis is used here with the yield and other relevant variables examined as the dependent variables, and a Bt cotton dummy variable as the independent variable to compare the impact of Bt varieties as against other varieties. The results would be identical to that of obtained through analysis of variance (Green and Carroll 1978). Note that these results may not match fully match since the sample numbers and responses vary. The findings indicate that Bt cotton is statistically significant in increasing the yields, and on an average gives an impact of 35 per cent increase in the yields (Table 22.3). The impact on the value of output is also statistically significant and is found to be 93 per cent. However, the total cost increase is also large and significant and is of 111 per cent. This derives from increases in pesticide cost and seed cost of 42 per cent and 184 per cent, respectively and are statistically significant. The findings indicate that there is also a 54 per cent increase in price (but this may be partly related to historical cotton prices). Despite the cost increases, the yield, value of output, and price advantages of Bt cotton lead to a statistically significant 75 per cent increase in the profits, which is very substantial.

PERCEIVED PERFORMANCE AND SATISFACTION WITH BT COTTON

A majority of the farmers see advantage of Bt cotton in the quality and availability of seeds, reduction of the pest incidence and problem, and the need to use pesticides. They also see advantage in the boll size, staple length, fibre colour and cotton price. Strong advantage is seen in yield and profit. It is also seen as suitable for early sowing. Disadvantages are seeing in seed cost and fertilizer need. No difference is seen in machinery need, irrigation and harvesting cost as well as in marketing and byproduct output.

Table 22.3 Regression Results on the Impact of Bt Cotton on Various Cost and Performance Variables—All Sample States–India

Dependent Variable		Variables		N=652
		Constant	Bt	Per cent Impact of Bt
Yield	Coefficient	1977.48	691.48	34.97
	t-stat	20.97	5.84	
	Significant	***	***	
Value of Output	Coefficient	58491.94	54592.66	93.33
	t-stat	15.36	11.69	
	Significant	***	***	
Total Cost	Coefficient	17016.12	18918.24	111.18
	t-Stat	5.59	5.59	
	Significant	***	***	
Pesticide Cost	Coefficient	4601.92	1920.19	41.73
	t-Stat	3.87	1.50	
	Significant	***		
Seed Cost	Coefficient	1301.82	2393.32	183.84
	t-Stat	2.31	3.86	
	Significant	**	***	
Price	Coefficient	2739.83	1493.40	54.51
	t-Stat	49.38	21.45	
	Significant	***	***	
Profit	Coefficient	44050.28	33016.56	74.95
	t-Stat	9.43	5.90	
	Significant	***	***	

Note: *** = significant at 99 per cent, ** = significant at 95 per cent, * = significant at 90 per cent.
Source: Own field survey

It is interesting to see that farmers disagree with most of the concerns that are being raised. Majority of the farmers disagree that Bt cotton cultivation is a risky business and not suitable for small farmers. More than 80 per cent disagree that farmers are compelled to grow Bt cotton since local varieties are not available and that Bt cotton has led to poverty and distress among the farmers. Over 85 per cent of the farmers disagree that there has been an increase in the suicides among farmers due to Bt cotton, or that there had been cases of pesticide poisoning among Bt farmers. Over 90 per cent disagree that cattle have died after eating Bt cotton plants, and the majority do not believe that the seeds, oil or oil cake are unsafe for human consumption. 90 per cent or more farmers do not find the government, non-government agencies, seed dealers or pesticide and other input dealers against the cultivation of Bt cotton. The majority in fact indicate that the current policies are not in favour of Bt cotton and that improvements are required, many also indicating that the market linkages are weak and need improvement.

Farmers indicate that Bt cotton has had a substantial, widespread positive impact on their village including all social groups, small farmers, women, tribals, traders and the poor. To some extent large farmers, upper caste villagers, and farmers with irrigation have benefited more, but substantial beneficiaries include labour-wage earners too. Majority of the respondents indicate that there has been no impact on humans, land, water, air, other crops, and beneficiary insects. While some indicate negative impact on humans, land and animals, what exactly is meant is not known.

In terms of future expectations, a large majority of the farmers suggest more resistance towards bollworms and other emerging pests, herbicide tolerance and drought tolerance of cotton. Higher yields are most strongly suggested, and a large number of request for field demonstrations and a lowering of seed cost.

Overall, farmers almost all strongly agree that Bt cotton has a strong yield advantage over non-BT cotton and requires less pesticide, though the seed cost and cultivation cost are high. Almost all indicate very high profitability of Bt cotton and that the technology has improved their household economic status. Nearly 80 per cent indicate that they are completely satisfied with Bt cotton, and over

90 per cent indicate that they would definitely grow Bt cotton in the future.

CONSUMER AWARENESS, PERCEPTIONS, AND AWARENESS ON BIOTECHNOLOGY

To understand the awareness, opinions and risk perception of consumers/common people regarding GM foods and crops a survey was conducted of a sample of urban residents/consumers in the four sample states. It includes people living in the largest city/town, but not metropolitan cities, of the sample districts in each of the four states covered in the study. The respondents belong to the middle-age group of 31 to 50, the average age being 37 years.

Findings show that the familiarity with genetically modified (GM) is far from universal and over 50 per cent indicate that the public does not have enough information about GM food/crops, and over 70 per cent think that the available information is unauthentic. Of those who had heard, only 9.7 per cent were very familiar and 47.8 per cent were not very familiar. Newspapers, TV, and Radio are the dominant sources of information. Over 50 per cent agree that the uncertainty generates fear; almost everyone agreed that the media fills the void of uncertainty, and nearly 90 per cent attribute the fear of GM due to media scare. For most people, public opinion becomes more important than the opinion of the experts.

Most of the people are aware about the success of Bt cotton in India, and most people are fine with consuming cottonseed oil made of Bt cotton, and wearing cotton clothes made of Bt cotton and most have never come across issues of skin allergies due to Bt cotton clothes. Over 80 per cent think that the resistance to GM is due to poor awareness and information. Findings indicate that much can be changed on this front by strong communication and awareness building especially by the government and experts.

People are aware that GM foods are not of natural origin, but most don't think they are harmful to environment or human health. There is a fear though that GM technology will make people dependent on MNCs and seed companies. However, they don't consider GM technology against religion. Findings indicate that about one-third of people may refuse to accept GM foods or crops in the present state of awareness, but the rest would accept them. Most respondents favour the right to know how the food is produced and request mandatory labelling of foods so that people can decide. Almost all people are in favour of thorough testing of GM foods and most indicate willingness to accept GM foods provided they are found safe in other countries and if the government accepts them. Most agree that the technology is required so that there is enough food and starvation is avoided.

CONCLUSIONS AND POLICY IMPLICATIONS

Biotechnology has tremendous promise and for a country like India that is heavily dependent on agriculture for food security and farmer well-being, biotechnology seems to offer the much needed technological innovation. Bt cotton is a significant example of the impact of biotechnology on crop performance, farmer incomes and profitability.

After the unauthorized appearance of Bt cotton in Gujarat in 2001, in March 2002 the Genetic Engineering Approval Committee (GEAC), the regulatory authority of the Government of India for transgenic crops approved the commercial cultivation of three Bt cotton varieties. It was realized soon that three hybrids were too less and was a major limiting factor for a country of the size of India. By 2009, 522 Bt cotton hybrids of 35 companies secured approval for commercial cultivation.

Since the introduction of Bt cotton in 2002, the production growth rate shot up to 13.14 per cent and yield growth rate to 9.57 per cent. Even the area has grown at 3.17 per cent, going up to 5.13 per cent in the latter part of the period under study.

Results indicate that Bt cotton appears to tackle the problems of boll worms and leaf feeding insects which are major pests, but it shows a higher incidence for sucking pests and Alternaria leaf spot. Almost all farmers indicate that Bt cotton yields more than non-Bt cotton. Factors of agronomic potential indicate that Bt cotton has good pest resistance and is responsive to fertilizers and irrigation. Almost all farmers indicate that Bt cotton is substantially more profitable than non-Bt cotton, indicating a strong agro-economic potential based on the open market and not government support. The results on factors affecting creation of effective demand indicate that the cotton farmers are willing to take risks and be opinion leaders for other farmers. The findings on information sources indicate that seed dealers are the most common and important source of information. Although the total costs for Bt cotton cultivation increased by 72 per cent, the yield increased by 33 per cent, revenue by 79 per cent and profits by 83 per cent. Econometric analysis results indicate that Bt cotton is statistically significant in increasing the yields, and on an average has an impact of 35 per cent increase in the yields. The impact on the value of output is also statistically significant and is found to be 93 per cent. Despite the cost increases, the yield, value of output, and price advantages of Bt cotton led to a statistically significant 75 per cent increase in the profits, which is very substantial.

Farmers perceive benefits of Bt cotton in the quality and availability of seeds, reduction of the pest incidence, the need to use pesticides; in boll size, staple length, fibre colour and cotton price. Strong advantage is seen in yield and profit. It is also seen as suitable for early sowing. Disadvantages are perceived in seed cost and fertilizer need. No difference is seen in machinery need and irrigation and harvesting cost as well as in marketing and byproduct output.

A survey of consumers was also conducted to assess awareness, opinion and risk perception of consumers. The awareness about GM foods/crops was in general very low. Newspapers and other mass media constitute the major source of information. Respondents feel that fear of such products is due to media scare. Government and expert intervention to create awareness, thorough testing of GM foods, adequate labelling were some of the requirements of the consumers that came out of the survey.

REFERENCES

Chaudhary, B., and K. Gaur. 2010. 'Bt Cotton in India: A Country Profile', *ISAAA Series of Biotech Crop Profiles.* Ithaca, New York: ISAAA.

Edgerton, M. 2009. 'Increasing Crop Productivity to Meet Global Needs for Feed, Food, Fuel', *Plant Physiology,* 149(1): 7–13.

Gandhi Vasant P., and N.V. Namboodiri. 2006. 'The Adoption and Economics of Bt Cotton in India: Preliminary Results from a Study', IIMA Working Paper No. 2006-09-04, Indian Institute of Management, Ahmedabad.

Gill, S., and N. Tuteja. 2010. 'Polyamines and Abiotic Stress Tolerance in Plants', *Plant Signaling and Behaviour,* 5(1): 26–33.

Kathage, J., and M. Qaim. 2012. 'Economic Impacts and Impact Dynamics of Bt (Bacillus thuriengiensis) Cotton in India', *Proceedings of the National Academy of Sciences of the United States of America,* 109(29): 11652–56.

Khush, G.S. 2002. 'The Promise of Biotechnology in Addressing Current Nutritional Problems in Developing Countries', *Food and Nutrition Bulletin,* 23(4): 354–7.

Knight, J., and A. Paradkar. 2008. 'Acceptance of Genetically Modified Food in India: Perspectives of Gatekeepers', *British Food Journal,* 110(10): 1019–33.

Krishna, V.V., and M. Qaim. 2008. 'Consumer Attitudes Towards GM Food and Pesticide Residues in India', *Review of Agricultural Economics,* 30(2): 233–51.

Krishna, V., and Qaim, M. 2012. 'Bt cotton and sustainability of pesticide reductions in India'. *Agricultural Systems,* 107: 47–55.

Li, G.P., K.M. Wu, F. Gould, J.K. Wang, J. Miao, X.W. Gao, and Y.Y. Guo. 2007. 'Increasing Tolerance to Cry1ac Cotton from Cotton Bollworm, Helicoverpa Armigera, was Confirmed in Bt Cotton Farming Area of China', *Ecological Entomology,* 32, 366–75.

M. Lattin, James, J. Douglas Carroll, Paul E. Green. 2003. 'Analyzing Multivariate Data'. English, Book edition.

Meli, V., S. Ghosh, T. Prabha, N. Chakraborty, S. Chakraborty, and A. Datta. 2010. 'Enhancement of Fruit Shelf Life by Suppressing n-glycan Processing Enzymes', *Proceedings of the National Academy of Sciences of the United States of America,* 107(6): 2413–18.

Ministry of Agriculture. Various years. *Agricultural Statistics at a Glance* New Delhi: Government of India.

Nielsen, AC. 2004. 'Nationwide survey by AC Nielsen ORG-MARG Underscores Benefits of Bollgard™ cotton'. Press release, Mumbai, India.

Qaim M., A. Subramanian, G. Naik, and D. Zilberman. 2006. 'Adoption of Bt Cotton and Impact Variability: Insights from India', *Review of Agricultural Economics,* 28(1): 48–58.

Rao, Chandrasekhara N. 2013. 'Bt Cotton Yields and Performance: Data and Methodological Issues', *Economic & Political Weekly,* 48(33): 17 August, 66–9.

Sharma, D.C. 2010. 'Bt cotton has Failed Admits Monsanto'. *India Today,* 6 March.

Shiva Vandana, and Afsar H. Jafri. 2003. Failure of the GMOs in India, Research Foundation for Science, Technology and Ecology.

Stone, Davis Glenn. 2011. 'Field Versus Farm in Warangal: Bt Cotton, Higher Yields, and Larger Questions', *World Development,* 39(3): 387–98.

———. 2012. 'Constructing Facts: Bt Cotton Narratives in India', *Economic & Political Weekly,* 47(38): 62–70.

Subramanian, A., and M. Qaim. 2009. 'Village-wide Effects of Agricultural Biotechnology: The Case of Bt Cotton in India', *World Development,* 37(1): 256–67.

Zilberman, D., H. Ameden., and M. Qaim. 2007. 'The Impact of Agricultural Biotechnology on Yields, Risks, and Biodiversity in Low-income Countries', *The Journal of Development Studies,* 43(1): 63–78.

23

Evaluation of Participatory Irrigation Management in India

Study of Andhra Pradesh, Gujarat, and Maharashtra*

Vasant P. Gandhi and N.V. Namboodiri

Participatory Irrigation Management (PIM) has assumed great importance in India in the last few decades due to the growing difficulties faced in water resource management, and the realization that stakeholder involvement and participatory management leads to substantial improvements. This research examines the performance and results of PIM across three major states of India, namely, Andhra Pradesh, Gujarat, and Maharashtra. It was undertaken in cooperation with Agro-Economic Research Centres (AERCs) in these three states, and with the support of the Ministry of Agriculture and Farmers Welfare, Government of India.

THE PROBLEM

There is a growing crisis in water resource management in India and this is becoming increasingly serious as development accelerates (Gandhi and Namboodiri 2002). Scarcities of water are becoming common and frequent and the quality of water is suffering as well. The management of water distribution across the vast areas of the country, and amongst millions of users, in a sustainable manner is becoming a major challenge. There is crisis in the management of surface water because of the huge investment requirements, project implementation delays, problems of maintenance, institutional difficulties, and environmental concerns. There is crisis in the management of ground water because of excessive exploitation against inadequate recharge resulting in receding water tables in many areas. The crucial role of irrigation in food production as well as livelihoods needs no emphasis. The technical and economic solutions to these problems are typically known and often simple, but their institutional management in a participative political economy framework is becoming very difficult and posing a serious challenge (Gandhi 1998).

The role of governments in the construction and management of irrigation systems has existed for a long time (Randhawa 1980). However, since independence, irrigation development has become part of a positive

* The authors gratefully acknowledge the contributions of B. Chinna Rao of AERC Vishakaptnam, Andhra Pradesh; A. Narayanamoorthy and S.S. Kalamkar of AERC Pune, Maharashtra; and H.F. Patel and V.J. Dave of AERC Vallabh Vidyanagar, Gujarat, to this research, as well as the support of all the AERC heads, state governments, and the Ministry of Agriculture and Farmers Welfare, Government of India. Review and comments from IEG Delhi are also appreciated.

government strategy of development, and canal irrigation development has expanded substantially. In many states, governments have also installed tube wells under public management for irrigation to make available ground water for the farmers. However, government controlled irrigation systems frequently show low water use efficiency, poor maintenance, weak financial sustainability and excessive dependence on subsidies. The efficiency of irrigation systems in various operations from water accumulation to extraction, diversion to its actual use through various stages has been poor (Majumdar 2000). The poor utilization of the irrigation potential created over the planning bears this out. The analysis of the shortcomings of the conventional irrigation management points substantially to the lack of meaningful involvement of the farmers in decision-making and in various physical activities (Sivamohan and Scott 1994). This realization has led to a growing emphasis on participative irrigation management.

NEED FOR PARTICIPATORY IRRIGATION MANAGEMENT

The experience over the last two decades shows that if farmers actively participate in irrigation management there is marked improvement in water utilization efficiency (Gandhi and Namboodiri 2002). Uphoff (1986) has highlighted some of the important benefits, drawing upon international studies. With participation, there is increase in the area under irrigation and also in the number of farmers who gain access to irrigation. In Pochapad, the irrigated area increased by 25 to 30 per cent after Warabandi and the formation of pipe committees. Similar findings have come from the Mula Command in Maharashtra and the water users cooperatives in Gujarat. Cooperation between farmers was found to increase and due to this, many water related disputes get sorted out. The agency was able to supply water with great control and economy. In Mula, for example water logging had perceptibly declined after the formation of Pani Panchayats (Singh 1991).

A change/transfer in irrigation management whereby farmers take over the management of operation and maintenance while government agencies mainly focus on developing and improving the management of water at the main system level has been supported by many researchers including Vaidyanathan (1999); Subramanian et al. (1997); and Meinzen-Dick and Mendoza (1996). Such ideas have led to the promotion of PIM. Several states have modified the old irrigation acts to accommodate group management by farmers. Some are in the process of enabling farmers to form water cooperatives and charging for water by volume as against the usual crop acre rate.

The PIM broadly refers to the formation of groups of water users/farmers in a formal body for the purpose of managing parts or whole of an irrigation system. The bodies are often called Water Users' Associations (WUA) but may also go by other names such as irrigation cooperatives or partnerships. PIM implies the involvement of water users in different aspects and levels the management of the water including planning, design, construction, maintenance, and distribution as well as financing. The primary objective of PIM is typically to achieve better availability and utilization of the water through a participatory process that gives farmers a significant role in the management decisions of water in their hydraulic units (Salman 1997).

This system of user management is preferred since it is felt that the users have a stronger incentive to manage water more productively, and can respond more quickly to management problems in the system, particularly at the farm level (Brewer et al. 1999; Grocenfeldt and Svendsen 1997; Subramanian et al. 1997). Moreover, transferring responsibilities has also come to be seen as a way to reduce pressures on thinly stretched government finances, while at the same time improving irrigated agricultural production and ensuring the long term sustainability of irrigation systems (Geijer, Svendsen, and Vermillion 1996; Vermillion 1991; Mitra 1992). The intention is also to encourage efforts by individuals to take responsibility for the management of the resource, in the belief that individuals have greater stake and better information for making efficient resource allocations (Brewer et al. 1997).

Participatory irrigation management or the user participation in the management of irrigation systems in India typically seeks to address the following aspects:

1. Improve efficiency of irrigation systems
2. Ensure sustainability of irrigation systems
3. Improve performance of irrigated agriculture
4. Reduce pressures on government finances
5. Permit farmers to play a greater role, which is a major shift away from conventional government policy

THE PIM POLICY IN INDIA

The PIM policy of the Government of India covers the management of diverse water resources through a participatory approach. According to the policy, this is to be done by involving users, other stakeholders and various governmental agencies in the decision-making. This must cover various aspects including planning, design, development and management of the water resources. Necessary legal and institutional changes should be made at various levels for this purpose.

The proposed major objectives of the government's PIM policy (India, Ministry of Water Resources) are:

1. to create a sense of ownership of water resources and the irrigation system among the users, so as to promote economy in water use and preservation of the system;
2. to improve service delivery through better operation and maintenance;
3. to achieve optimum utilization of available resources through better/sophisticated methods, accurately as per crop needs;
4. to achieve equity in water distribution;
5. to increase production per unit of water, where water is scarce and to increase production per unit of land where water is adequate;
6. to make best use of natural precipitation and ground water in conjunction with canal irrigation for increasing irrigation and cropping intensity;
7. to encourage better use of water through better choice of crops, cropping sequence, timing of water supply, period of supply and frequency of supply, depending on soils, climate and other infrastructure facilities available in the commands such and roads, markets, cold storage etc. so as to maximize the income and returns;
8. to encourage collective and community responsibility of the farmers for collecting water charges and making payments to irrigation agency; and
9. to create a healthy atmosphere between the irrigation agency personnel and the users.

OBJECTIVES

The major objectives of the research are as under:

1. to examine the evolution of PIM in Andhra Pradesh, Gujarat, and Maharastra;
2. to examine if the devolution of power to the WUAs has taken place in the selected states;
3. to examine if the WUAs have contributed towards regular water supply, efficiency in water use, collection of water charges, and operation and maintenance of the water delivery systems;
4. to examine if there is any change in the performance/pattern of agriculture and well-being seen by the beneficiaries;
5. to identify constraints in the effective implementation of PIM, problems in the coordination between the WUAs and the irrigation agencies, and ways to remove the constraints including training needs and proper organization structure; and
6. to suggest ways by which successful PIM models can be replicated in other states and possible actions for the effective implementation of PIM in all states and Union Territories.

The research was conducted in the states of Andhra Pradesh, Maharashtra, and Gujarat through the involvement of the respective AERCs. The survey instruments were prepared and finalized in consultation with the AERCs. Efforts were made to maintain uniformity across states through meetings, communications and visits. Data was collected during the 2004–05 season and effort was made to cover information of 2003–04 were ever possible based on recall. This research consolidates the findings of the research conducted by the different state AERCs. Findings from Andhra Pradesh, Gujarat and Maharashtra were consolidated.

RESEARCH METHODOLOGY: PROFILE OF SAMPLE WATER USERS' ASSOCIATIONS AND FARM HOUSEHOLDS

Data was collected from a broad sample of Water Users' Associations (WUAs) under various irrigation systems in the three states. Within these, a sample farm households were selected using stratified random sampling with the intent of covering small (small and marginal – less than 2 ha), medium (2–4 ha) and large (above 4 ha) farmers. The coverage of WUAs and the sample number of beneficiary farmer households selected from each of these WUAs is outlined in the Table 23.1 below. The total sample size was 435. The composition of number of small, medium, and large farm households are respectively 222, 124, and 89.

RESULTS AND DISCUSSION

Participation, Involvement, and Activity Levels of Different Functionaries and Groups in the WUAs

The involvement, participation, and activity levels of different functionaries and groups in functioning of the WUAs may vary but this is a very important indicator of the desired outcome and success of the WUAs. What is the level of participation and activity level of different functionaries and groups in the working of the WUAs of different kinds?

It was observed the chairman and/or secretary in particular and the managing committee in general are actively

Table 23.1 Household Sample Distribution Across States, Irrigation Systems and Farm Sizes

Irrigation System	State	Abbreviation Used	No. of Institutions	Small (Below 2 ha)	Medium (2–4 ha)	Large (Above 4 ha)	Total
Canal: Major	Andhra Pradesh	CMAP	2	12	10	8	30
Canal: Medium	Andhra Pradesh	CMEAP	3	18	15	12	45
Canal System	Maharashtra	CM	5	30	25	20	75
Canal System	Gujarat	CG	3	36	6	3	45
Tank System	Andhra Pradesh	TAP	4	30	20	10	60
Tube wells	Gujarat	TWG	4	48	8	4	60
River Lift System	Maharashtra	RLM	3	18	15	12	45
Pani Panchayat	Maharashtra	PPM	2	12	10	8	30
Check Dams	Gujarat	CDG	3	18	15	12	45
Total			29	222	124	89	435

involved in the affairs of almost all the WUAs studied here except PPM where they play only a marginal role. While the role of government officials was greater under the canal systems WUAs but the local institutions such as village panchayat have played only a passive role. The farmer member households irrespective of their farm size have been actively involved in WUAs. The landless labourers play an active role mainly under the CDG system compared to other irrigation systems. This could be due to the fact that the check dam construction was a village-wide activity involving the entire village population and the benefit accrued to the landless households as well by way of higher water table leading to increased farming activity. Under the canal system both the head and tail reach farmers have been actively participating or rendering help to carry out various functions of the WUAs. One of the noticeable features was the interest showed by the members of these WUAs irrespective of their socio-economic background. On the whole the role played by various socio-economic group in various activities of the WUAs studied here substantially indicate the active participation of people with the respective WUAs across economic and social divisions. The active role of the chairman, secretary and general body are noticeable. The involvement of local institutions such as Panchayat is however very limited in this activity.

Devolution of Powers and Decision-Making

One of the most important aims of PIM is empowering farmers and giving them the decision-making and responsibility for managing the irrigation systems. The devolution of powers to the WUAs for the management of the irrigation system is a major aim of PIM and is considered very important for improving the water use efficiency. This devolution of powers can be examined by observing the devolution of decision-making related to planning, implementation, revenue, conflict resolution, and meeting the equity and efficiency considerations. High degree of devolution of power would mean less burden of external agencies in various activities of water resource management. To what extent the devolution of powers to the WUAs has taken place has been examined for different irrigation systems studied using a set of questions on who makes different important decisions of water resource management.

With respect to the canal system water institutions, the findings on the devolution of powers indicate the following. The control of government continues to be high in respect of assessment of water availability, release of water, water pricing, and collection of water dues from the farmers. However, the devolution of powers to the WUA is high in the distribution of water, maintenance of irrigation structures, and equitable distribution of water. As far as the capital investments are concerned, the powers are by and large jointly held by the WUA and the irrigation authorities.

For the WUAs studied under the tank irrigation system in Andhra Pradesh, the devolution of power to the WUA has taken place very significantly only in terms of its maintenance and repairs of the irrigation structures, and the choice of deciding the cropping pattern. To some extent the WUA has the powers in planning for release of water and in taking punitive action against members in case of misuse of water. Under the TWG system all powers are rested with the WUA. This is because the TWG studied here have been either handed over the structure and equipment or own them entirely, and if at all they depended on the government only for availing of the government capital subsidies at the initial stage and for power supply.

Under RLM except for pricing of water the WUA has the sole power with respect to carrying out various functions as and when required. Under the PPM system, except in pricing and collection of due from the users, the devolution of powers to the WUA is nearly complete. Finally under the check dam system the powers rested with the government was only in terms of release of the investment subsidy to the WUA. Thus as far as the devolution of powers to the WUAs are concerned, the government agencies continues to have greater powers under the canal systems in terms of pricing of water, collection of dues from the farmers and release of water to the canals. For the rest, the devolution is substantial and the WUAs studied here have significant powers over the management of the water resource.

The Impact of PIM on the Agricultural Economy

A major objective of PIM and the establishment of WUAs are to improve agricultural productivity, production and incomes through better utilization and efficiency in water resource use. What has been the impact of PIM and the WUAs on agriculture such as in terms of increase in area irrigated, shift in cropping pattern, change in input use including use of improved and high yielding varieties, and changes in productivity? This has been studied under the different irrigation systems. With the availability of water for irrigation, farmers may opt for water intensive but more remunerative crops and the availability of irrigation may also have impact on the use of various inputs. Whether the PIM has resulted in such shifts towards more irrigated high value crops, use of modern inputs and if there are any significant change in the levels of productivity, has been examined here.

Findings are based on farmer responses on the change in cropped area and the level of irrigation as of now compared to that at the time of the formation of the WUAs. Note that the findings would be influenced by the conditions prevailing in the survey year, and the farmer recall of the position in the pre-PIM time. Under CMAP, although there was a marginal increase in irrigated area during the kharif and summer seasons, a decline in the irrigated area during the rabi season during the reference year appeared as an overall decline in irrigated area. On the contrary under CMEAP there was an increase in irrigated area during all seasons. Under the CM, the irrigated area during the kharif and summer seasons went up by almost 50 per cent on an average. The expansion in irrigated area under canal system in Gujarat (CG) has also been very dramatic since the WUA took over the management of the system. In fact the irrigated area registered a fivefold increase.

The performance of the tank system in Andhra Pradesh (TAP) during the reference year in terms of area irrigated was very poor and this could be attributed to inadequate rainfall and no water in the tanks. Under the TWG, the sample households have been cultivating crops without irrigation until the WUAs came into existence. Since then almost two-thirds of the cropped area has received irrigation. Under RLM except during the summer there was only a small addition to the irrigated area since the establishment of the WUA. The most dramatic increase in irrigated area was under CD where there was a seven fold increase in irrigated area after the check dams have been constructed.

The change in the cropping pattern across the command area of the selected WUAs showed large variation. Under CMAP there was no major change in the cropping pattern since the formation of the WUA. The only notable change was in terms of the cultivation of some fodder crops during the summer under irrigated conditions. However, under CMEAP the irrigated area under fruits and vegetables during the kharif season and the cultivation of pulses under irrigation during the rabi season showed an increase. Under CM there was a very significant increase in irrigated area under vegetables and oilseeds during the kharif and rabi seasons. Under CG, both the cropped area and irrigated area under cash crops like cotton and castor and area under irrigated wheat have registered a significant increase. But no major shift in cropping pattern was observed under TAP. Under TWG, the cropped area under tobacco and wheat as well as the area irrigated under them have increased significantly. Under RLM the cultivation of vegetables under irrigation increased during the kharif and rabi seasons, and of oilseeds during the rabi season since the establishment of the WUA. Under PPM the cultivation of foodgrains under irrigated conditions became more common among the sample households. A marked shift in cropping pattern in favour of high value cash crops like cotton away from bajra and jowar and cultivation of fodder crops during the rabi and summer have been a major change noticed under CD.

Another aspect examined here was the change in the input use since PIM and establishment of WUAs. The change in the use of agricultural inputs have been measured on a five-point scale ranging from large increase to large decrease and the finding based on the response of sample households on these aspect. It is found that irrespective of the irrigation system, there have been a decline in the use of local varieties of seeds and the use of bullock labour. However, the use of improved and high yielding varieties of seeds, other modern inputs such as fertilizers and pesticides, and the use of farm machinery have show significant increases.

Whether PIM has had an impact on increasing agricultural productivity has been examined based on the response of the sample households. Under CMAP there was noticeable increase in yields of major crops such as paddy, banana, and pulses since the devolution of power to the WUA. While 43.3 per cent of the sample households reported large increase in paddy yield, the rest 56.7 per cent also reported an increase in yield of paddy. However under CMEAP, none of the households reported large increase in yield, but a majority of them reported increase in yield. Since no change in the yield of unirrigated crops were reported by the sample households, the results indicate a positive impact on yields after the introduction of WUAs. This was also true for other irrigation systems studied here.

The Impact of PIM on Improving the Performance of Water Resource Management

What has been the impact of PIM on the performance of irrigation systems in improving water resource management? The expected impacts may include the performance on addressing water scarcity and use efficiency, improving the empowerment and equity, reducing adverse environmental impact, and improving financial viability. This has been examined in the study. In order to measure the efficiency of the WUA in managing the irrigation system we have considered six broad indicators, namely, timely and adequate water availability, increase in irrigated area, change in cropping pattern, better maintenance of the irrigation structure and finally reduction in cost of maintenance. The equity related issues probed here are equitable distribution of water, empowerment of farmers, volumetric pricing and all landholders taking membership in the WUA. The equity and empowerment were also assessed in terms of beginning a sense of ownership, unification of diverse groups, freedom to raise resources, resolution of disputes, and active involvement of all classes of farmers.

The responses were obtained from the farmers on a five-point scale, namely, highly positive, positive and no impact, negative, and highly negative. Sometime this reduced de facto to a 3-point scale of highly positive, positive, and no impact. Under CMAP the major positive factors stands out under the efficiency parameters are timely water availability, and better maintenance; under the equity considerations, three factors that stands out are more equitable distribution of water, empowerment of farmers and all landowners becoming member of the WUA. Beginning of a sense of ownership and active involvement of all member farmers stood positive and highly positive. Other factors that are positive to highly positive are deciding the quantum of water to be used, and transfer of power to the WUAs. More or less a similarly views were expressed by the sample farmer households under CMEAP with the exception that the active involvement of all member farmers was not as strong as it was under CMAP.

Positive impacts of the WUA for the sample under CM are: adequate water availability, better maintenance of the irrigation system, equitable distribution of water, empowerment, freedom to raise resources, more farm employment and diversified economic activities. The farmers of CM reported positive impacts with respect to all the factors considered here except adequate water availability, reduction in cost of maintenance, volumetric pricing and transfer of power to the WUA in deciding water charges. Under TAP, except some marginal positive impacts on diversified economic activities no other major positive impact was reported by the sample households under. Under TWG, the WUA could not make much dent on empowerment of farmers to manage the irrigation system, all land owners becoming members, year round availability of water and choice of deciding irrigation timings. The overall impact of RLM was positive except those related to diversified economic activities. Whereas under PPM and CDG all sample farmer households reported a strong and positive impact on most indicators of equity, efficiency and social empowerment.

Difficulties Faced by the WUAs in the Operation of PIM

What are the difficulties faced by the WUAs in achieving effective operation of the PIM? What are the problems that need to be addressed to make PIM more effective? The farmers were asked about a range of possible problems associated with supply, management and distribution of water including financing and investment. The responses have been obtained on a five point scale ranging from very major to none. The study probed a number of problems that may be faced. We report here mainly those problems that are reported as very major to major.

Under CMAP the major problems that are stated are inadequate field channels, lack of start-up financial support from the government, lack of consensus on deciding the cropping pattern and the lack of freedom to decide on the water rates. However, the major problems faced by the WUAs under CMEAP are very different and they include non-availability of water, conflict among members about timing of water, complaint from tail-end farmers and lack of start-up financial support from the government. The farm households under CM reported few very major to major problems but these included inadequate maintenance, high cost of maintenance, inadequate field channels, lack of government support and little raining to staff members. The farmer households under CG reported only light to occasional problems namely inadequate maintenance, high cost of maintenance, non availability of water at the canal, inadequate field channels and complaints from the tail-end farmers, particularly when there is acute scarcity of water in the canal. Some of the major problems reported by the farmers under TAP are non-availability and poor quality of water in the tank, high cost of maintenance, lack of member cooperation, and complaints from tail end farmers on non-availability of water. Under TWG three major problems have been stated by the farmers, namely, high cost of electricity, high cost of repairing, fast receding water table in the wells. The major problems faced by the members of WUA under RLM are high cost of maintenance, high cost of electricity, lack of government support, while under PPM the major problems faced by the farmers were lack of financial support from the government and high cost of electricity. Under CDG problems such as receding water table, lack of mechanisms to control water use, and lack of training to members were the prominent problems reported.

The Impact of PIM on the Village Economy and on Different Groups in the Village Society

What is the impact of PIM on the village economy? What is its impact on different social and economic groups in the village? Does PIM and the creation of WUAs help improve the general economy of the area? The preceding section showed the effects on aspects such as cropping patterns, input use and yields and water management and this could have implications for the economy, wage rates, employment and subsidiary occupations like dairying. The study sought to examine the impact of PIM on the village economy and its various socio-economic groups.

Findings indicate that CMAP, WUAs have had a positive impact on the village as a whole. The benefits accrued have not been confined to a particular class, caste, religious or social group nor to those belong to the head or tail end of the canal. A very much similar pattern emerges from the water user association under CMEAP. One major difference under this WUA was that the impact of WUA on the village as a whole was positive in a smaller percentage of cases as compared to CMAP. The WUAs under the canal system in Maharashtra (CM) reported an altogether different picture. The impact on various socio-economic groups ranged from positive to negative to no impact. The negative or non impact responses were more relevant for the people belong to tribals, lower caste, scheduled caste and those who do not have any cropping activity. The study reported that even after the establishment of the WUAs, the benefits are mainly to the farmers and not much to the non-farming groups. The impact of the WUAs on general economy reported by the CG was either substantially positive or positive indicating that it has a favourable impact on all social and economic classes of people in its command. The positive impact of the WUA under the tank system in Andhra Pradesh (TAP) was not broad based and was confined mainly to the upper income groups in the command. The responses from the WUAs under the RLM and PPM in Maharashtra showed little broad based impact on the general economy of the area. On the other hand the responses from WUAs under TWG and CDG in Gujarat showed a strong and positive impact on various socio-economic classes in the irrigation commands.

Synopsis of the Results on the Performance and Impact of PIM

This section seeks to provide a synopsis of the disaggregate findings reported above on the performance and impact of PIM. It does this by providing simple aggregates or averages of selected findings as well as providing a comparative picture through figures and tables. Note that these are based on the reports of the state studies and the analysis presented above. Individual household survey observations were not available from the AERCs for the analysis. Broad overall assessment of the WUAs by the farmers of the performance and the financial viability is also covered here.

Participation by Members in the Activities/Decision Making of WUAs

Taking all the types of WUAs into account, the average rate of participation (active to very active) by members in the WUAs was found to be quite high at almost 80 per cent. This shows that the participation by the members in the WUAs is quite high on an average. However, there is considerable variation. Under CDG and TWG the participation by members was almost 100 per cent and it was also high under the canal system in Gujarat and the RLM system in Maharashtra. However a low participation rate was observed under tank irrigation system in Andhra Pradesh and Pani Panchayat system in Maharashtra.

Involvement/Participation of Different Functionaries and Socio-economic Groups

The aggregate picture shown in the Table 23.2 indicates that the most active participation (active to very active)

Table 23.2 Active Role Played by Various Functionaries–Percentage Reporting

Institutions	General Body	Chairman	Secretary	Village Panchayat	Small/Marginal Farmers
1 CMAP	66	100	10	6	94
2 CMEAP	68	94	14	2	66
3 CM	46	90	76	0	48
4 CG	100	100	100	100	100
5 TAP	64	74	8	2	50
6 TWG	100	100	100	100	100
7 RLM	80	98	84	0	96
8 PPM	4	24	24	0	0
9 CDG	100	100	100	100	100
All Institutions	60	86	58	34	72

was that of the Chairman at 86 per cent, followed by the General Body at 60 per cent and the Secretary at 58 per cent. Whereas 72 per cent of the small/marginal farmers on an average were actively involved, only 34 per cent of the Panchayats showed active involvement. A look across WUA types showed that the Chairman of the WUA played an active role across the WUAs except in Pani Panchayats in Maharashtra. The involvement of Secretary in various functions and activities of the institutions found to be relatively low under CMAP, CMEAP, TAP, and PPM. The village Panchayats have played a passive role except for most institutions in Gujarat studied.

Devolution of Powers and Decision-Making

A key objective of PIM is the devolutions of powers and decision-making to the WUAs. This has been examined through questions on who now makes the decisions regarding several important matters of water resource management. The results based on a simple average across decisions and WUA types indicates that in 78.3 per cent of them the decisions are made by the WUAs or Jointly, and only in 21.7 per cent of them are the decisions made by the government. This indicates a good degree of devolution of powers. However, there is some variation. The devolution of powers to the WUA was nearly complete in terms of monitoring the distribution of water, maintenance of irrigation structures, monitoring use of water and freedom of choosing the cropping sequence. However the government agencies continues to have greater power in terms of assessment of water availability, pricing of water, collection of dues from the farmers and release of water to the canals. Across WUA types, the devolution is less in TAP and CMEAP, and the greatest in TWG and RLM.

Impact of PIM on the Agricultural Economy

The impact of the WUAs on farm economy was examined in terms of bringing more area under irrigation, increased use of inputs and better yields. The average results indicate that the cropped area increases by 8.28 per cent and the irrigated area increases by 31.43 per cent (Table 23.3). This indicates a substantial impact of PIM activity. The results indicate a substantial increase in irrigated area in summer season, followed by increase in the kharif season and only a marginal increase in the rabi season. PIM also has a substantial positive impact on increasing the use of yield

Table 23.3 Average Cropped Area per Sample Household (ha)

Crop Season	At the Time of Establishment of WUA		At Present		Percent Change	
	Total Area	Irrigated Area	Total Area	Irrigated Area	Total Area	Irrigated Area
Kharif	2.11	1.29	2.10	1.52	−0.32	17.89
Rabi	0.92	0.97	1.02	0.98	10.71	0.69
Summer	0.31	0.31	0.50	0.49	59.86	59.50
Total	3.34	2.27	3.62	2.99	8.28	31.43

Table 23.4 Average Increase in Various Inputs for Crop Production

Inputs	Per cent Increase
1. Seed Local	33.1
2. Seed HYV	61.5
3. Seed Improved	63.9
4. Fertilizer	83.1
5. Pesticides	65.1
6. FYM	37.8
7. Bullock Labour	13.3
8. Machine Labour	89.4
9. Family Labour	38.2
10. Hired Labour	64.2
11. Irigation Cost	77.0
12. Other Costs	50.0
13. Others	5.6
All Inputs	52.4

increasing inputs such as HYV and improved seeds and fertilizers. The input use increase is the highest for CDG. The average crop yield has increased by more than 50 per cent except under TAP and CMEAP (Table 23.4). The yield increase was the highest for CG and TWG.

Impact of WUA on Village Economy, Water Use Efficiency, Equity, and Empowerment of Users

A positive to highly positive contribution of PIM/WUAs to the village economy as a whole is reported by 78 of the sample households on an average across irrigation system types. It ranges from below 60 per cent for PPM and CMEAP to 100 per cent for CG, TWG, RLM, and CDG. The benefits have been reported positive by equal numbers for both large/medium and small/marginal farmers on an average.

With respect to the impact of PIM on water use efficiency, equity and empowerment of users, a positive to highly positive impact is reported on an average by 78 per cent for adequate availability of water, 72 per cent for timely availability of water, 76 per cent for better maintenance, 85 per cent for more equitable distribution, 71 per cent for empowerment. However, TAP reports no positive impact on any of these, and the impact is relatively low in CMAP on timely availability of water, CM on better maintenance and empowerment, and TWG and PPM on empowerment was reported by more than 75 per cent of the sample except those from Andhra Pradesh.

Problems and Difficulties Faced by the WUAs in the Operation of PIM

On an average across the WUA types, the most important problems and difficulties reported are inadequate field channels, lack of government support or funding, lack of training to members, difficulties in handling extreme water scarcity, high cost of maintenance, high cost of electricity, non-availability of water in the canal, complaints from tail-reach farmers, and lack of freedom in determining water rules. However, these problems vary across the institutions. For example, inadequate field channels was a major problem for WUAs under canal irrigation in Andhra Pradesh and Maharashtra. Inadequate support/funding was a major problem for CMAP, CM, TAP, and RLM. Lack of training staff/members are largely reported by members of canal and tank systems in Andhra Pradesh. Conflict among members is a major problem in CMEAP, non-payment of water charges in CM and PPM, and leadership in TAP and PPM.

One of the issues that has not been covered in this study is that of accountability to the users, and some of the problems highlighted here stem from a lack of proper accountability. There is great need for good financial audit as well as social audit of these institutions in order to improve the reliability and confidence of the users. One concern in this context is the observation that WUA presidents in AP frequently play a major role in the execution of works, like contractors. This could result in poor accountability, political interference, antisocial activities, and low member participation. Another problem that has been only partially examined is that of financial viability of these institutions. Inadequate government support/funding is a problem for a large number of these institutions, and the financial health of most of these institutions is either just satisfactory or poor. This is indicative of a financial sustainability problem and the need for local or alternate resource mobilization to overcome the problem.

For broad overall assessment of performance, the WUAs were rated by the beneficiary farmers on a five point scale. As per the response of the sample farmers, CMAP, CDG, and RLM were rated as the most successful WUAs followed by CG, TWG, CM, and CMEAP. The TAP and PPM had a poor rating by a large number of beneficiaries. The rating on the financial health of the selected WUAs showed that CG and CDG were rated as the most financially viable institutions followed by CM, RLM, PPM, and TWG. The CMEAP, TAP, and to some extent CMAP were rated as having relatively poor financial viability.

CONCLUSIONS AND POLICY IMPLICATIONS

Research on the shortcomings of the conventional irrigation management in the country indicates a lack of meaningful involvement of the farmers in decision making, planning and various activities. A change in irrigation management whereby farmers are involved and even take over part of the operation and maintenance while government agencies mainly focus on developing and improving the management of water at the main system level has been proposed in PIM. PIM implies the involvement of irrigation users in different aspects and levels of management of the water resource including planning, design, construction, maintenance, and financing, but particularly in distribution. It generally involves a grouping of farmers into bodies or institutions, often called Water Users' Associations (WUAs) for the purpose of managing a part or more of the irrigation system. The primary objective of PIM is to achieve better utilization of available water through a participatory process that endows farmers with a major role in the management decisions over water in their hydraulic units. This research was conducted in cooperation with the AERCs in the states of Andhra Pradesh, Maharashtra, and Gujarat.

Has PIM worked? To what extent has PIM resulted in benefits such as better availability of water for irrigation, greater efficiency in water use, better recovery of water charges, and better operation and maintenance of the irrigation structures? The study begins with examination of the evolution of PIM in states of Andhra Pradesh, Gujarat and Maharashtra. It then examines the role and functions of WUAs, devolution of powers to the WUA, impact of WUA on farm economy, impact of WUA on the village economy, equity, efficiency and social justice, and the major problems faced by the farming households with respect to the functioning of WUA.

For broad overall assessment of performance, the WUAs were rated by the beneficiary farmers on a five point scale. As per the response of the sample farmers, CMAP, CDG, and RLM were rated as the most successful WUAs followed by CG, TWG, CM, and CMEAP. The TAP and PPM had a poor rating by a large number of beneficiaries. The rating on the financial health of the selected WUAs showed that CG and CDG were rated as the most financially viable institutions followed by CM, RLM, PPM, and TWG. The CMEAP, TAP, and to some extent CMAP were rated as having relatively poor financial viability.

The study indicates that there has been considerable progress in bringing participation and devolution of powers in irrigation management in the three states but substantial further efforts are required and will help improve performance. It is found that increased participation commonly brings significant benefits to performance in water resource management but some kinds of WUAs have performed better than others. Many of these institutions require not just setting-up but also inputs in institution design, institution building, and training in order to make them strong and sustainable. Greater accountability also needs to be incorporated through proper financial audit, performance evaluation, and social audit, and the financial viability and sustainability of these institutions needs to be enhanced through local resource mobilization as well as external development support.

REFERENCES

Brewer, J., S. Kolavalli, A.H. Kalro, G. Naik, S. Ramnarayan, K.V. Raju, and R. Sakthivadivel. 1999. *Irrigation Management Transfer in India: Policies, Processes and Performance.* New Delhi: Oxford & IBH.

Gandhi, Vasant P. 1998. 'Rapporteur's Report on Institutional Framework for Agricultural Development', *Indian Journal of Agricultural Economics*, 67(3): 487.

Gandhi Vasant P, and N.V. Namboodiri. 2002. 'Investments and Institutions for Water Management in India's Agriculture: Policies and Behaviour', in Donna Brennan (ed.) *Water Policy Reform: Lessons from Asia and Australia.* Canberra, Australia: Australian Centre for International Agricultural Research.

Geijer, J.C.M.A., M. Svendsen, and D.L. Vermillion. 1996. *Transferring Irrigation Management Responsibility in Asia: Results of a Workshop*, Short Report Series on Locally Managed Irrigation No. 13, International Irrigation Management Institute, Colombo.

Grocenfeld, David, and Mark Svendsen. 2000. *Case Studies in Participatory Irrigation Management*, (ed.), World Bank Institute, Washington DC.

Majumdar, D.K. 2000. *Irrigation Water Management: Principles and Practice* New Delhi: Prentice-Hall of India.

Meinzen-Dick, Ruth and M. Mendoza. 1996. 'Alternative Water Allocation Mechanisms: Indian and International Experiences', *Economic and Political Weekly*, Vol. XXXI (26): A75–A82.

Mitra, Ashok K. 1992. 'Joint Management of Irrigation Systems in India: Relevance of Japanese Experience', *Economic and Political Weekly* June 27.

Randhawa, M.S. 1980. *A History of Agriculture in India*, Vol. I, New Delhi: ICAR.

Salman, M.A.S. 1997. "The Legal Framework for Water Users Associations: A Comparative Study", The World Bank, Washington DC. Available at http://documents.worldbank.org/curated/en/272041467980487313/pdf/multi-page.pdf.

Singh, K.K. 1991. *Farmers in the Management of Irrigation Systems*. New Delhi: Sterling Publishers.

Sivamohan, M.V.K., and C.A. Scott. 1994. 'Moving towards the concept of "partnerships"', in 'Irrigation Management in India'. In M.V.K. Sivamohan, C.A. Scott (eds). *India: Irrigation Management Partnerships*. Hyderabad: Booklinks, pp. 1–13.

Subramanian A., N.V. Jagannathan, and R. Meinzen-Dick. 1997. 'User Organizations for Sustainable Water Services'. Technical Paper 354, The World Bank, Washington, DC.

Uphoff, Norman. 1986. *Improving International Irrigation Management with Farmers' Participation*, London: West View.

Vaidyanathan, A. 1999. *Water Resource Management: Institutions and Irrigation Development in India*. New Delhi: Oxford University Press.

Vermillion, D.L. 1991. *The Turnover and Self Management of Irrigation Institutions in Developing Countries*. Colombo: International Irrigation Management Institute.

ANNEXURE 23A

Table 23A The Details of Agro-Economic Research Centres and the Lead Person Involved in the State Report

State	AER Centre/Unit	Authors
Andhra Pradesh	AERC Vishakhapatnam	B. Chinna Rao
Maharashtra	AERC Pune	A. Narayanamoorthy, S.S. Kalamkar
Gujarat	AERC Vallabh Vidyanagar	H.F. Patel, V.J. Dave

24

Status and Perception about Resource Conservation Technology

A Case Study of Drip Irrigation

Vaibhav Bhamoriya

Agriculture today suffers from multiple crises and one of the worst crisis is that pertaining irrigation and water resources. It is a crisis of poor management of water resources that is leading to deeper crisis of availability and quality of water. Enhancing water availability and amenability for use and managing the distribution are difficult challenges due to dynamic nature of the resource and unequal distribution. Such challenges make it difficult for agriculture to remain sustainable. The search for solutions to these crises led to experimentation with new technologies such as micro-irrigation.

Micro-irrigation techniques including drip and sprinkler irrigation were introduced as water saving technology. Their potential to solve many issues facing modern agriculture and farmer is well known. Some researchers and practitioners believe they can transform agriculture into a very profitable and low risk occupation. It is also expected that micro-irrigation will help improve the availability of water and therefore the soil indirectly. It is often showcased as a technology that will save Indian agriculture from certain doom. However, a technology cannot exist without limitations and these have to be kept in mind. Its' impact on water resource conservation is being debated by some researchers arguing that field-level gains due to micro-irrigation result in basin level losses. Molle and Turral (2004) argue that water savings are notional and the farmer is likely to use the 'saved' water for irrigating another crop. The government tool upon itself to promote the technology and launched various schemes to promote micro-irrigation in the country. Initially the National Committee on use of Plasticulture in Agriculture took up various schemes for the promotion of micro-irrigation systems followed by NABARD financing drip irrigation systems from 1985. Maharashtra was the first state to introduce subsidies in 1986–87. Subsidies ever since have been a regular phenomenon of promotional efforts for drip-irrigation. Three states of Tamil Nadu (TNAHODA), Andhra Pradesh (APMIP) and Gujarat (GGRCL) have created special purpose vehicles for promotion of micro-irrigation and management of subsidies.

The history of micro-irrigation in India now spans a little over four decades dominated by government push and subsidy support. The subsidy burden makes the study of this technology also more important from the perspective of public policy choices.

Table 24.1 Country Wise Coverage of Drip and Sprinkler Irrigation (Mha in 2010)

Country	Total Area Equipped for Irrigation	Sprinkler Irrigation	Drip Irrigation	Total Micro-Irrigation	% of Total Irrigated Area	Year of Reporting
USA	24.7	12.3	1.64	13.99	56.6	2009
India	60.9	3.04	1.90	4.94	8.1	2010
China	59.3	2.93	1.67	4.60	7.8	2009
Russia	4.5	3.50	0.02	3.52	78.2	2008
Brazil	4.45	2.41	0.32	2.74	61.6	2006
Total	211.8918	35.07	10.08	45.15	21.3	

Source: Working Group on Farm Irrigation Systems, ICID.

Data collected by ICID shows that an area of 1.89 million ha (8.1 per cent of total irrigated area) in 2010. This is compared to the top five countries in the world in the Table 24.1.

In India, Maharashtra (0.48 million ha), Andhra Pradesh (0.36 million ha) and Karnataka (0.17 million ha) account for more than 70 per cent of the total area under drip irrigation. However most of the potential area for micro-irrigation has not been brought under adoption thus far as shown by Figure 24.1 (Raman 2009). Till very recently, Maharashtra, Andhra Pradesh, and Tamil Nadu accounted for more than two thirds of the area under drip irrigation. Drip irrigation is widely believed and found to be more suitable to widely spaced horticultural crops but finding increasing use and popularity in vegetables, field cash crops and also at times in cereal crop cultivation (Narayanmoorthy 2003).

Micro-irrigation has made its mark as an agri-input that enhances productivity and enables cash crop cultivation. Various field experiments have shown this technique to increase water use efficiency up to 80–90 per cent depending on the crop and soil type (Indian National Committee on Irrigation and Drainage [INCID] 1994; Sivanappan 1994). Drip Irrigation, a type of micro-irrigation, in particular is one of the most efficient methods of irrigation (Keller and Bilsner 1990). The benefits of micro-irrigation and drip irrigation are not restricted to water saving. Various researchers have established other benefits of the technology as listed below (see Andal 2010; Basker et al (2011); Mitra 2011):

1. It increases the productivity and yields of crops thus increasing farm incomes.
2. It reduces weeds and soil erosion. It may also reduce pests.
3. It also reduces water logging, salinity and ground water pollution.
4. It reduces the cost of cultivation mainly due to savings in labour costs and energy savings. According to some estimates, the system can save electricity of 278 kWhr/ha for wide spaced orchard crops and 100 kWhr/ha for closely grown crops. (Raman 2009)
5. It results in better crop output quality.
6. It enables balanced use of nutrients and better fertilizer use efficiency (Narayanamoorthy 2010). The use of water soluble fertilizers (WSF) recommended with drip irrigation systems are very suitable for fertigation ensuring nutrient-supply to root zones, causing marginal or no loss of nutrients. Thus fertilizer use efficiency can be increased up to 95 per cent (as per KRIBHCO).
7. It is well suited to all soil types and undulating terrains as the water flow rate can be controlled (INCID 1994).
8. It can lead to social empowerment especially for women in villages Verma et al (2006).

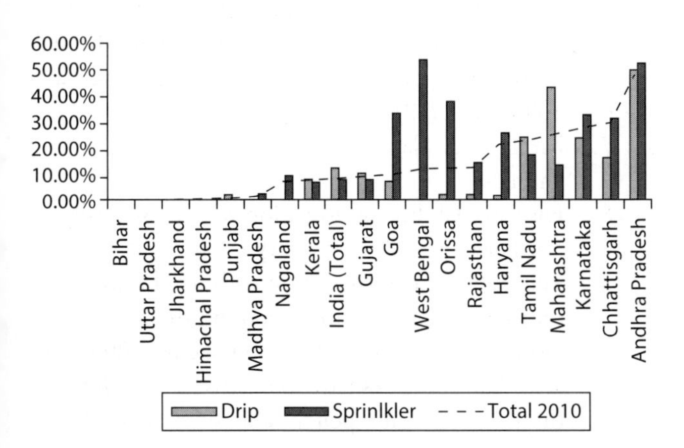

Figure 24.1 Actual vs Potential Area for Micro-Irrigation Across States

Source: Raman 2011.

The spread of micro-irrigation has been restricted to a few pockets in a few states of India in areas where groundwater withdrawal is also high. There is an urgent need to study the technology and its experience and impact to get insights that can inform public policy and choice. There is huge unutilized potential for drip irrigation in India (Raman 2011).

The lukewarm response has been attributed to several causes including lack of access to groundwater, lack of cash, crop specificity and lack of know-how, poor product quality and absence of credit facilities (Narayanamoorthy 1996 and IWMI Policy briefing 23).

Some technical issues were reported as well. A disadvantage of drip irrigation as stated in literature is the accumulation of salt especially in the salinity-prone regions (Hanson and May 2011). Drip irrigation also gives benefits in growth of crops but there are other factors which influence whether this advantage can be transformed into actual yield and economic advantage (Myburgh 2012).

Different researchers have looked at different methods to measure the effectiveness of drip irrigation systems but they have generally looked at two main advantages only—water saving and productivity enhancement. Apoorva Oza (India Infrastructure Report 2007) also highlights the issue by mentioning that investments in drip technologies can result in an estimated annual water savings of 2.2 million cubic metres of water in India. Most studies have used perceptional data as well as experimental field data to show the positive impacts of drip irrigation at the farm level. The results are only complicated further by the lack of uniform spread signalling some part of the story stays undiscovered.

Thus some researchers and also practitioners believe that drip irrigation may have an adverse impact on water resource conservation. The counterargument to the benefits of drip irrigation therefore is that by enabling irrigation with lesser water availability there is increased usage of even the marginal quantities of water and sources rather than their conservation.

Micro-irrigation systems may also promote an increase in cropping intensity or a shift to high value water-intensive crops which in turn increase the pressure on ground water resources (IWMI 2006). Empirical studies highlighting this aspect of the technology may give us more insight into the complex nature of the problem. A model shows that it can affect the aquifer recharge cycle and the environment suffers a net loss.

An evaluation of the Centrally Sponsored Scheme on micro-irrigation found that while a farmer appreciated the water savings effect of drip irrigation, it wasn't a sufficient condition for adoption. Caution is advised that shifting from rainfed irrigation to drip- irrigation may result in loss of water saving benefits because of a net expansion in irrigated area (Oza 2007).

The main factors responsible for the limited spread of the technology have been documented by quite a few researchers and are enlisted below:

1. High initial costs of up to Rs 1,25,000 per hectare make the technology unfeasible for small and marginal farmers. Such a huge investment requires advance crop planning on the part of the farmers and an assured income for the produce which may be true only for high value crops.
2. High emitter clogging rates due to dust and salinity. The system requires proper filtration so that dust and other particles do not block the small emitter holes.
3. Unsuitable cropping patterns. Drip irrigation has been used for irrigating only a few selected crops in India. It is adopted mostly for coconut (19 per cent penetration), banana (11 per cent), grapes (10 per cent), mango (9.4 per cent), citrus fruits (7.9 per cent) and pomegranate (6.2 per cent) [Task Force Report 2004]. It may not be suitable for closely planted crops like cereal grains which are grown across large areas in the country.
4. It requires a lot of technical and management skills for setting up and upkeep. Lack of technical support and follow up by the government, private companies and NGOs may be a hindrance for adoption.
5. The process of applying and being approved for the subsidy however is complex and involves numerous agencies. As a result, farmers are dependent on manufacturers and middlemen to facilitate the process.
6. Only selected, pre-approved drip kits qualify for the subsidy which stifles creative marketing strategies on the part of manufacturers as well as efforts to bring down the cost of drip systems through innovative technology or product designs (AgWater Solutions 2012).
7. Mechanical damage by farm labour, birds and animals.
8. Easy availability of irrigation water especially in northern parts of the country.
9. The shift to micro-irrigation without the attendant measures is expected to compound the over exploitation of the aquifer (Molle and Wester 2009). The attendant measures have to be driven by policy and this usually where a policy paralysis exists as of now.

There is a need to go from the farm to the basin level in order to understand the water savings on scale. However, it is difficult to find a basin In India with a relatively high adoption rate of drip irrigation to conduct basin level studies. This calls for the challenge of using some other method to study the impact due to scale in terms of both space and time.

Most studies have focused on adoption and performance in a limited geography or crop types only. The scale effects have not been studied. The role of different actors in different stages of adoption and use remains to be explored deeper. One very important aspect of perception of users and non-users about the impact of the technology on various aspects of agriculture, performance and resource conservation also needs to be documented.

This study focuses on the potential of micro-irrigation and more specifically drip irrigation as a resource conservation technology for conserving the water resources, which are being plundered for irrigation purposes. The study also sought to study the other aspects of economics of irrigation and agriculture as the adoption decisions are taken at the farm level but the water conservations effects will be visible only at a larger scale.

Earlier studies have failed to capture the adopters' viewpoint and have not gone in depth about the adoption process and its various stages and are therefore are unable to evolve useful insights into the adoption process and managing it better to promote the technology and its impacts. They have concentrated in analysing experimental data on economics than studying viewpoints and experiences of adopters and non-adopters. The perception of adopters and non-adopters is more important from the pre-adoption consideration of whether to adopt or not.

There is a dearth of literature that talks about the role of various actors in the adoption process. Not enough studies have dealt with the various aspects associated with sales like after sales service and hassles of repairing the equipment.

Existing studies have covered a limited geography only and the impact of technology on resource conservation itself over a larger area than a farm and over increasing adoption rates is yet to be studied. The impact of technology in existing studies has been studied mostly across crop, geography and irrigation specificity on agriculture.

OBJECTIVES

The survey of literature and identification of the research gaps combined with the main points from the discussion with experts in the domain led to the framing of the research questions for the study. The research questions for the study were:

1. to study the impact of micro-irrigation on water resource availability, use and conservation from a water conservation approach;
2. to study the impact of micro-irrigation on economics of agriculture in the regions of adoption under study;
3. to study the economics of irrigation in the regions of micro-irrigation adoption by combining the water resource conservation and agriculture economics; and
4. to suggest probable implications of adoption of micro-irrigation on sustainability of agriculture.

DATA AND METHODOLOGY

Since some of the research issues being studied had not been studied in detail before, it was expected that there would be aspects not identified or detailed and it was important to explore such dimensions before an analysis be set up. Therefore a mixed methodology was applied and case studies were conducted first to determine the variation and new issues missing in earlier in studies about drip irrigation. Case study locations were chosen to get sufficient variation of crops, agronomic zones and administrative setups. It was also considered important to focus on locations of relatively higher adoption for exploring scale effects. The study design required studying pockets of relatively higher adoption over a significant area and for a significant number of years to be able to gauge the perception of the farmers (adopters and non-adopters) about the impacts of drip irrigation technology when adopted at scale. Also the need to survey across geographies was evident to make the results generalizable. Before choosing such pockets it was required to make a choice of the states to be studied to enable generalization within and across geographies.

The sample states were chosen for their higher adoption percentages in terms of area and spread of crops. Two zones were selected—the southern and western zone, where drip irrigation has performed better, of the country. Two states were selected from each zone to allow variability and triangulation of results within a zone. From both the zones the top two states in terms of drip irrigation adoption were selected—Andhra Pradesh and Tamil Nadu were selected in the south zone and Gujarat and Maharashtra in west zone. Rajasthan, which has a higher area under sprinkler irrigation, was not chosen as the study focused on drip irrigation and not sprinkler irrigation and some of the research issues are significantly different with regards to adoption of the two technologies.

All four states have a different administrative mechanism for the promotion of drip irrigation. Andhra Pradesh was the first state to pioneer with a special purpose vehicle APMIC (Andhra Pradesh Micro-Irrigation Corporation) was created as a result of the APMIP (Andhra Pradesh

Table 24.2 Sample Spread Across States

	Andhra Pradesh	Gujarat	Maharashtra	Tamil Nadu	Sub-total
Adopters	121 (75.6%)	76 (70.4)	91 (79.8)	82 (69.5)	370
Non-adopters	39 (24.4.%)	32 (24.5)	23 (20.27)	35 (30.5)	129
Total	160	108	114	117	499

Source: Own field survey.

Micro-Irrigation Project) and it was entrusted with the task of promoting drip irrigation and channelizing subsidies for better results. Tamil Nadu created a specialized agency for implementation of horticulture projects. Drip irrigation is one such major initiative taken up by the TANHODA (Tamil Nadu Horticulture Development Agency). Tamil Nadu has also seen substantial rise in area under drip irrigation since TANHODA was entrusted the task. Gujarat Created GGRCL (Gujarat Green Revolution Company Limited) a 100 per cent government owned autonomous entity as a limited company to administer subsidies and promote drip irrigation ensuring the objectives of the government are fulfilled. With a dedicated subsidy funding it has seen a dramatic rise in area under drip irrigation. Maharashtra has seen the growth in certain pockets mainly driven by state subsidies and promotion by private companies. Thus all the four chosen states had different administrative set ups for promoting and managing the spread of drip irrigation.

The case study findings along with the variables identified from literature and the research framework were used to design a survey instrument (questionnaire). The survey instrument was then pretested and modifications were made based on the pretesting for finalizing it.

For conducting the survey two relatively high adoption pockets were chosen within each state with differences in crops wherein drip was applied. This allowed collecting perceptions on scale impacts and to study the crop specificity of the technology by choosing areas with different topographies. In Gujarat the chosen pockets were in two different districts—Sabrakantha and Amreli. In Maharashtra they were in two different talukas of the district of Nashik but separated by more than 50 km and natural phenomenon and different topography. In Tamil Nadu the pockets chosen were in Coimbatore district spread across different Talukas in different directions from the district headquarters with different crops and entirely different cropping patterns and agriculture techniques. At the same time in Andhra Pradesh the pockets chosen were in Rangareddy and Medak districts and again for variation in the application crops and topography and soil type.

Trained personnel administered the survey over 1 to 1.5 hours for each respondent. The total sample size was 499 respondents (370 adopters and 129 non-adopters). The analysis involved tabulations across states and different responses and finally a regression analysis correlating the various impacts of the technology on performance parameters with the resource conservation impact.

The survey spanned 4 states and 16 identified pockets and a total of 499 respondents were administered the survey instrument and their responses collected. The spread of the sample across the various states and adopters–non-adopters is given in Table 24.2.

A Total of number of 499 responses were collected from 370 adopters and 129 non-adopters from four states, viz., Andhra Pradesh, Tamil Nadu, Maharashtra and Gujarat. The numbers of adopters surveyed from Maharashtra were 91 and non adopters 23, thus a total number of 114 respondents were surveyed from Maharashtra.

RESULTS AND DISCUSSIONS

More than two quarters of the respondents were from other backward classes and other minorities categories (Table 24.3). 63.2 per cent of respondents in Maharashtra were from the general category, while 94.5 per cent of the farmer respondents in Tamil Nadu were from other backward classes. Respondents from other minorities constituted half the sample in Andhra Pradesh and about two

Table 24.3 Caste-Wise Break-up of Respondents (State Wise and Combined)

Caste	Andhra Pradesh (%)	Gujarat (%)	Maharashtra (%)	Tamil Nadu (%)	Overall (%)
Scheduled castes	10.0	0.9	2.6	2.7	4.05
Scheduled tribes	0.6	0	4.4	0	1.25
Other backward classes	27.5	8.3	23.7	94.5	38.5
Other minority	58.1	67.6	6.1	2.7	33.65
General category	3.8	23.1	63.2	0	22.55
Total	100.0	100.0	100.0	100.0	100

Source: Own field survey.

Table 24.4 Water Availability Benefit of Drip Irrigation

DI Is Beneficial for Water Availability	Andhra Pradesh (%)	Gujarat (%)	Maharashtra (%)	Tamil Nadu (%)	Overall (%)
Very high	11.3	25.0	1.8	79.3	28.0
High	45.6	38.0	38.1	14.7	35.0
Medium	35.0	33.3	53.1	2.6	31.2
Low	6.9	0.9	7.1	2.6	4.6
Very low	1.3	2.8	0.0	0.9	1.2

Source: Own field survey.

Table 24.5 Positive Impact on Water Table with Drip Irrigation

Water Table Increase	Andhra Pradesh (%)	Gujarat (%)	Maharashtra (%)	Tamil Nadu (%)	Overall (%)
Highly positive	14.6	12.0	1.8	3.5	24.7
Positive	50.0	37.0	43.4	47.0	41.7
No impact	33.5	48.1	54.0	48.7	32.4
Negative	1.9	2.8	0.9	0.9	1.2
Highly negative	0.0	0.0	0.0	0.0	0.0

Source: Own field survey.

thirds of the sample in Gujarat. This contradicts the popular notion that only 'higher caste farmers' can afford drip irrigation. The low participation of the scheduled tribes as respondents is also noticeable in the sample despite special incentives provided by most state governments.

The common sources of irrigation for the respondents were tube well, open well, canal as reported by 78.5 per cent, 45 per cent, and 12.4 per cent respondents, respectively. Majority of the farmers had more than one source for irrigation water.

In Andhra Pradesh, Gujarat, and Tamil Nadu, the major source of irrigation water was tube well (95.0 per cent, 58.30 per cent, and 78.40 per cent) followed by open well (3.10 per cent, 43.50 per cent, and 52.60 per cent). In Maharashtra 97.4 per cent of farmers reported open well as the major source of irrigation, 74.6 per cent and 10.5 per cent of respondents source their irrigation water requirement from tube wells and lift irrigation from canal. Drip irrigation appears to be popular across varied sources of irrigation.

It was found that 4 out of every 5 adopters rate a high or very high level of satisfaction with the experience and results of the technology. About 60 per cent of overall respondents perceived it as beneficial for water availability. In Tamil Nadu almost 80 per cent, of the respondents perceive a very high impact and 45.6 per cent in Andhra Pradesh and 38 per cent in Gujarat. 53 per cent respondents in Maharashtra perceived only medium impact.

It is expected that drip irrigation as a water saving technology, has a positive impact on the water table and the overall water situation in the village (Table 24.5). More than 57 per cent of the adopters and also 43 per cent non-adopters perceive the water table to have increased or improved. Two thirds of the overall sample reported a high or very high positive impact on the water table. The perception of non-adopters is very similar to adopters confirming the positive impact of drip irrigation on water tables.

Almost 55 per cent of adopters perceive an improvement in overall water situation in the village, a longer term measure (Table 24.6). About 70 per cent of adopters and 38 per cent of non-adopters reported improvement in water availability since drip became popular in the cluster of villages. Only Gujarat had less than half respondents reporting a positive impact on overall water situation in the village with adoption of drip irrigation.

Overall, 86.6 per cent adopters and 99.23 per cent non-adopters believe that drip irrigation results in water savings at the farm level. The perception that drip irrigation saves water is very strong amongst both adopters as well as non-adopters alike (Table 24.7). Forty four per cent respondents agreed that drip irrigation had a high or

Table 24.6 Overall Water Situation in the Village with Drip Irrigation

Overall Water Situation in the Village	Andhra Pradesh (%)	Gujarat (%)	Maharashtra (%)	Tamil Nadu (%)	Overall (%)
Highly positive	5.1	4.6	0.9	3.6	3.7
Positive	53.2	33.3	53.1	75.0	53.8
No impact	41.1	54.6	44.2	11.6	38.1
Negative	0.6	7.4	1.8	9.8	4.5
Highly negative	0.0	0.0	0.0	0.0	0.0

Source: Own field survey.

Table 24.7 Reduction in Water Quantity Used for Irrigation

Drip Irrigation Saves Water	Andhra Pradesh (%)	Gujarat (%)	Maharashtra (%)	Tamil Nadu (%)	Overall (%)
Very high	8.6	24.1	5.3	8.6	10.9
High	26.7	41.7	31.9	26.7	34.6
Medium	44.8	31.5	54.0	44.8	43.3
Low	19.0	2.8	8.8	19.0	11.1
Very low	0.9	0.0	0.0	0.9	0.2

Source: Own field survey.

Table 24.8 What Happens to 'Saved' Water?

What Happens to the Saved Water?	Andhra Pradesh (%)	Gujarat (%)	Maharashtra (%)	Tamil Nadu (%)	Overall (%)
Used for irrigating more crops	56.3	55.0	31.3	30.0	67.4
Used for expanding area under agriculture	70.0	36.9	64.4	23.1	60.6
Used for more irrigation to the same crops	28.1	21.9	6.9	47.5	37.0
Used for other agriculture and related purposes	53.8	10.6	45.0	9.4	25.8
Used for other non-agricultural purposes	41.9	6.9	8.1	.6	16.8
Water table rises	**11.3**	**18.1**	**1.3**	**1.3**	**10.8**
Used to share/sell to other farmers in need of irrigation water	2.5	4.4	2.5	0.0	2.2
Don't know/Can't say	2.5	4.4	1.9	.6	2.4

Source: Own field survey.

very high impact on reduction in water quantity used for irrigating a farm.

The more important findings are about the use of 'saved' water. In Maharashtra and Gujarat it is used for irrigating more crops while in Andhra Pradesh for expanding cropped area. In Tamil Nadu the major use is for more irrigation to the existing crops enabling farmers to overcome deficit irrigation. The 'saved' water is also used for other agriculture related purposes and other non-agriculture purposes in Andhra Pradesh and Gujarat. The technology has the potential to impact intra as well as inter-sectorial allocations and consumption of water when implemented in the cluster mode in Gujarat, Andhra Pradesh, Tamil Nadu water savings have enabled a few famers to sell or share saved water with another farmer.

10.8 per cent or one out of every nine farmers report that the water was not used up and gets recharged and contributes to raising the water table in the region (Table 24.8). This has not been reported earlier from farm level studies and is a very promising result and might be partly due to the scale effects of drip irrigation with significant adoption in certain pockets. A possibility therefore exists that when drip irrigation is adopted at a scale within a region or area there are possibilities of real 'savings' of water as well both at the farm level and at the area or region level as well. Theoretically real 'savings' are possible at the basin level as well.

Drip irrigation is shown to impact the economics of agriculture positively through various impacts such as increase the cropping intensity, increase yield from existing crops. 48.2 per cent respondents observed a positive impact while another 6.9 per cent observed a very positive impact of drip irrigation on reduction of farming costs. Another 42.1 per cent observed no impact on the total costs of farming.

The impact of adoption is positive on reduction of total labour used on the farm. The labour does not lose out, as they are able to move to a better lifestyle with a loss of occupation due to shift to higher labour intensive cash crops.

The technology also has a direct impact on the prosperity of farmers as majority of adopters confirmed a positive impact on the income. Drip irrigation is reported to help increase incomes and all respondents agreed with this without fail. Drip irrigation also helps to influence the adoption of high value and less water intensive crops. The technology also helps to make agriculture more sustainable and profitable by having a positive impact on better market prices as well as better market power.

It is found that drip irrigation has emerged as one of the main coping mechanisms to protect the farmer and agriculture from the various problems that plague modern agriculture such as shortage of power, labour and also water. However drip irrigation needs a better financial model to ease adoption at the initial stage as only 16 per cent adopters and 12 per cent non-adopters agreed to drip irrigation as a financial proposition without subsidy.

The positive impact on soil quality also impacts the sustainability of agriculture favourably. At the same time the ability of technology to provide assured and timely irrigation only enhances the sustainability, as does its potential to save water at the farm level and also at a larger scale. Drip is therefore found to have a positive impact on sustainability of agriculture.

The awareness levels of farmers is very low for trainings concerning the use of drip irrigation and almost one third of the adopters felt the need for special training to make it more profitable for farmers.

The realization of the water conservation benefits at the aggregate level is possible only when there are concentrated clusters of relatively high proportion of adoption such that the scale effects of micro-irrigation technology appear and these effects are primarily the resource conservation type.

CONCLUSIONS FOR POLICY IMPLICATIONS

The following policy recommendations are arrived at and suggested based on the results obtained from analyzing the data collected through the survey.

1. Drip irrigation technology has the potential to show conservation effects but these effects are visible only when the adoption is large scale at a cluster level. The government and other promoting agencies need to focus on the cluster approach and on geographic pockets based on hydrology and crop economics (farm level) to get the best conservation impact. This should be supported with policies that help farmers to regularly use the technology over a sustained period of time.
2. While framing policies related to drop irrigation and its promotion it must be kept in mind that drip irrigation impacts not only irrigation and its economics but also agriculture as a whole and its economics positively as well. Multiple benefits ought to be pooled to increase the adoption rates. Funding from various schemes can be dovetailed to increase the subsidy allocations for drip irrigation.
3. Drip irrigation aids more and assured income to the farmers and this increases the possibility of using low cost drip irrigation technology as a tool to reduce vulnerability and also for poverty alleviation for the vulnerable masses. This will need a radical shift and political will on part of the government as so far it has only dealt with the ISI certified high quality and high cost drip irrigation technologies.
4. There is an urgent need to create better market linkages in order to ensure better market prices for adopters of drip irrigation. The technology provides little price advantage or protection against price risks until the adoption is at a fairly large scale so as to stabilize the commodity market of a particular drip amenable crop. There is a need for government participation for fulfillment of pre-conditions for an efficient market to function. The SIMI project in Nepal (Smallholder Irrigation and Marketing Initiative) executed by the International Development Enterprises (IDE) is an initiative from which learnings can be drawn.
5. Drip irrigation suffers from some shortcomings such as costly after sales service. Thus there is a need to innovate business models. Private entrepreneurs can be motivated to take up the challenge of evolving the solutions for these challenges rather than control the business so tightly.
6. Dissemination and communication should make use of crop specificity of drip irrigation and capitalize with a cluster approach in a crop specific cluster. This will ensure that farmers get their farm level benefits whereas the conservation impacts are realized at the cluster level.
7. Drip irrigation is not easy to master. Trainings are important to enable farmers to benefit as much as potential from the adoption and maximize return from application as well. The trainings will need to focus on not only the technical aspects but also on the managerial aspects like marketing and value chain fundamentals and institution formation to enable the farmers the best economic deal possible.
8. A better understanding of the needs of the farmers in the various stages of adoption and use and the role of various actors in each of these stages can help to enable more targeted and focused efforts at promoting drip irrigation.
9. Formal institutions and their officers need incentives to play the roles as exemplified by the special purpose vehicles in the states of Gujarat, Tamil Nadu and Andhra Pradesh.
10. The subsidy procedure is viewed as largely opaque, complicated and cumbersome as well as time consuming by many farmers and this calls for reforming policies and maybe also rules and regulations to overhaul the subsidy procedure to make it more convenient, clear, faster and much more fair towards all sections of the society.

REFERENCES

Andal, G. 2010. *Assessment of micro irrigation technology on yield water use salinity nitrate contamination in ground water in Rangareddy district of Andhra Pradesh*, Mimeo, Jawaharlal Nehru Technological University, Hyderabad.

Bhamoriya, Vaibhav and Susan Mathew. 2014. '*An Analysis of Resource Conservation Technology: A Case of Micro-Irrigation System (Drip Irrigation)*'. A Mimeo, Centre for Management in Agriculture, Indian Institute of Management, Ahemdabad.

Bhaskar, K.S., M.R.K. Rao, P.N. Mendhe, and M.R. Suryavanshi. 2011. 'Micro Irrigation Management in Cotton', *CICR Technical Bulletin* (31).

Hanson, B. and D. May. 2011. 'Drip Irrigation Salinity Management for Row Crops', *University of California Mimeo,* University of California Agriculture and Natural Resources Publication (8447). June.

Keller, Jack and R.D. Bilesner. 1990. Sprinkler and Trickle Irrigation, Chapman & Hall, New York.

Mitra, L. 2011. *'Achieving Targeted Growth through Micro Irrigation'*, Proceedings of 21st Congress on Irrigation and Drainage. Abstract of 8th ICID International Micro-Irrigation Congress. IRNCID Publication.

Molle, F. and H. Turral. 2004. 'Demand management in a basin perspective: Is the potential for water saving overestimated?' presented at International Water Demand Management Conference, Dead Sea, Jordan.

Molle, F. and P. Wester. 2009. *River Basin Trajectories: Societies, Environments and Developments*, IWMI Comprehensive Assessment of Water Series, Vol. 8. CABI Publishing.

Mybrugh, P.A. 2012. 'Comparing Irrigation Systems and Strategies for Table Grapes in the Weathered Granite-gneiss Soils of the Lower Orange River Region', *South African Journal for Enology and Viticulture*, 33(2): 257–63.

Narayanamoorthy, A. 1996. 'Evaluation of Drip Irrigation System in Maharashtra', Mimeograph Series No. 42, AERC, Pune.

———. 2003. 'Averting Water Crisis by Drip Method of Irrigation: A Study of Two Water-Intensive Crops', *Indian Journal of Agricultural Economics*, 58(3): 427–37.

———. 2010. 'Can Drip Irrigation be Used to Achieve the Macro Objectives of Conservation Agriculture?', *Indian Journal of Agricultural Economics*, 65(3), July–September.

Oza, Apoorva. 2007. 'Irrigation and Water Resources', *India Infrastructure Report,* Rural Infrastructure, 3inetwork, by Prem Kalra and Anupam Rastogi, Oxford University Press, New Delhi.

Palanisami, K., K. Mohan, K.R. Kakumanu, and S. Raman. 2011. 'Spread and Economics of Micro-irrigation in India: Evidence from Nine States', *Economic & Political Weekly*, 46(26 & 27).

Raman S. 2009. *Micro Irrigation for Electricity Saving in Gujarat—A Potentiality Assessment*, Mimeo, GGRCL.

Sivanappan, R.K. 1994. 'Prospects of micro-irrigation in India: https://link.springer.com/journal/10795Irrigation and Drainage Systems', 8 (https://link.springer.com/journal/10795/8/1/page/1 1): 49–58.

Task Force. 2004. *Report of Task force on Micro irrigation*, Delhi. Department of Agriculture and cooperation, GOI.

Varma, S., S. Verma and R.E. Namara. 2006. *IWMI Water Policy Briefing*, Vol. 23.

25

Irrigated Agriculture in Tamil Nadu

Pattern and Scope for Diversification*

K. Jothi Sivagnanam

This study is on irrigation and crop diversification in the State of Tamil Nadu. The State has been fighting for its share of Cauvery water. Cauvery is traversing through four states namely, Kerala, Karnataka, Tamil Nadu, and Puducherry. As Cauvery being the largest river system running across Tamil Nadu, more than a third of the population's livelihood chances are linked with its waters. Since the mid-1970s, however, the river has become dry and the state has been left to fend for itself to meet its irrigation requirements in the Cauvery basin. The alternate sources of irrigation like canal, tank, and other systems of irrigation are also limited due to varying reasons like scarcity and poor quality. Hence, the farmers have turned to shore up the available ground water at a considerable cost and have taken resource alternative means for meeting the scarcity. Crop or agricultural diversification has been found to be found to be one of the means of managing the water scarcity. Crop diversification has also been an effective strategy for achieving food and nutrition security, income growth, poverty alleviation, employment generation, judicious use of land and water resources, sustainable agricultural development and environmental improvement.

Crop diversification is a strategy for overcoming many problems, most important of which are food and nutrition insecurity and also for poverty alleviation, and income and employment generation and maximizing crop production. Variations in the annual rainfall received; coupled with the changing seasonality, put enormous pressure on the farmers as monsoons largely determine the success of agriculture. But the monsoon more often fails causing drought, often in consecutive years.

Water scarcity for agriculture has resulted in the state adopting a two-pronged approach:

1. Making a shift to an alternative cropping pattern, which is less centered around the traditional hydrophilic crops such as paddy, banana and sugarcane; and
2. Encouraging water conservation measures through a variety of schemes: for example, watershed development schemes.

The alternative strategy of crop diversification takes into account the water requirements of various crops and their

* This chapter is based on the report 'Irrigated Agriculture in Tamil Nadu: Pattern and Scope for Diversification' submitted earlier by R. Rajkumar, A. Abdul Salam, and T. Vasanthakumaran of AERC, Chennai. Acknowledgements are due to them. I also thank Dr R. Srinivasan and Dr V. Loganathan for helpful comments and suggestions.

yield per unit of water. Rice for example requires 1,250 mm of water. With 3162 kg/ha of yield in the state, its productivity is 0.25 kg/m^3 of water and, in monetary terms, return is Rs 1.3 per m^3 of water. Banana yields 31656 kg/ha with a requirement of water of 2000 mm, productivity of 1.88 kg/m^3 of water and a return of Rs 6.5 of water.

The specific objectives of the study are:

1. to examine the nature of irrigation and cropping pattern in the three select areas of study, namely Thiruvarur (canal), Pudukottai (tank), and Villupuram (well) districts;
2. to assess alternative cropping patterns, or crop diversification, in response to land and water use issues, particularly in relation to water scarcities and cropping response in the select canal, tank, and well commands; and
3. to suggest strategies for overcoming the problems emerging from irrigation and cropping systems operation in the select study areas, in respect of crop diversification as a response to water scarcities.

METHODOLOGY AND DATA

Three districts have been chosen to examine the nature and status of crop diversification in the context of different systems of irrigation. For canal system, Thiruvarur district has been chosen; for tank system, Pudukkottai district has been chosen; and for well irrigation system, Villuppuram district has been chosen. Since the purpose is a quick assessment of crop diversification in the three (canal, tank, and well systems) commands, one specific area in each district has been selected for the field survey. The areas so chosen are: Saliperi canal command of Thiruvarur district, Kannangudi tank command of Pudukottai, and Kanai block well commands (in Kedar, Siruvazhai, and Ezhusempon villages) of Villuppuram district.

The methodology adopted here is simple, and it is primarily the questionnaire based survey. The sample of farmers chosen for interviews is 144, including 48 from Thiruvarur district, 45 from Pudukottai district, and 51 from Villuppuram district. The survey was created using the custom-designed and pilot-tested questionnaire.

In making the selection of commands, two main aspects have been kept in view: the predominance of a particular system of irrigation and sustained efforts over the last few years in improving the systems and the cropping patterns in the area. Saliperi canal irrigation command has been focused on, and developed by, the efforts of the irrigation management Training Institute at Thuvakudi. Saliperi has become a model canal system for participatory irrigation management. Kannangudi tank irrigation command has been part of a large, European Union-Ford Foundation tank modernization programme and thus has seen considerable improvement, although some years later, now, the tank command is rather neglected by the farmers and the Water Users' Association. Kanai block also has a tank system in Kedar, but the concern for us is the well irrigation at the village. The two other villages chosen from Villuppuram for well irrigation and cropping pattern study are Siruvazhai and Ezhusempon.

The questionnaire has been designed in such a way that by using the data collected, from the farmers of the three different systems of irrigation, would help us in fulfilling the three objectives of the study and in assuring the research questions posed in the study.

RESULTS AND DISCUSSION: IRRIGATION, WATER SCARCITIES, AND CROP DIVERSIFICATION

This section analyses the data collected from the field survey and interprets them in regard to irrigation, water scarcity, and crop diversification in the three districts, taking nine select villages, three from each district.

Irrigation in Tamil Nadu

Tamil Nadu has a history and culture of canal, tank and well irrigation. Thiruvarur district is in the Cauvery delta and hence it is irrigated largely by the canal, even though tanks and wells are also used for irrigation.

Canal Irrigation

About 30 per cent of the farmers interviewed have indicated the names of the rivers from which they received water for irrigation in the canals: obviously, they are all from Thiruvarur district and the rivers are: Puthur river (18.1 per cent), Valappar river (5.6 per cent), Kizhkudi, and Veerakan rivers (2.1 per cent), Cauvery (1.4 per cent), and Sukkar river (0.7 per cent). They are all rivers that feed the canals from which the farmers of Thiruvarur draw water for their irrigation. The ayacuts (or commands) of the canals, as reported by the farmers, have a range of 40 acres to 300 acres: there are ayacuts of varying sizes that are irrigated by the canals: 100 acres (6.3 per cent), 120 acres (6.3 per cent), 140 acres (3.5 per cent), 250 acres (2.1 per cent) and 300 acres (6.9 per cent).

Being the delta, starved of water for several years with the exception of the monsoon and cyclonic weather benevolently providing the much needed water, the season of surplus for all of them depending on the canals is the cyclonic weather period: October–December. Some farmers (4.2 per cent) however report of surplus waters in the first month of the period while most (20.1 per cent) report surplus in the end of the season. The season of deficit is always the period before the cyclonic weather (June–September). Nearly all of them (25 per cent) indicated June–August as deficit months while 4.2 per cent of them indicated August as the most deficit month. This happens because the Mettur dam which is opened usually on June 12 gets delayed for shortage of water in the dam, or pre-monsoon showers fail the farmers miserably. In the recent 30 years, there have been several years of rainfall falling short of the normal or requirement.

Tank Irrigation

One third of the farmers interviewed are from Pudukkottai district and they are chosen for undertaking tank irrigation related crop diversification. Only about 30 per cent of the farmers interviewed indicated the tanks from which they receive their irrigation waters. Periyakulam has been reported as the source by 18 per cent of them, Kulathur tank by 5 per cent, Kedar by 3 per cent and Veerankulam by 2 per cent. About 29 per cent of the farmers use tanks for irrigation whereas others use other sources.

The tanks irrigating lands have commands ranging from 25 acres to about 3,000 acres. For 4.2 per cent of the farmers, their tank commands are less than 100 acres; for 9.4 per cent of them, the commands are anywhere between 100 and 200 acres; for 11.8 per cent of them, the commands are between 300 and 500 acres; and for 3.5 per cent of them their commands are larger than 500 acres.

The total area under canal and tank irrigation, for individual farms, is between less than 1 acre to more than 35: while 22 per cent of the farmers irrigate 1–2 acres, 33.4 per cent irrigated 3–4 acres, 14.6 per cent 5–6 acres, 9.1 per cent 7–8 acres, 6.3 per cent 9–10 acres and 3.5 per cent more than 10 acres. It must be pointed out that 65.5 per cent of the farmers interviewed have reported these as total under canal and tank irrigation. Mean area under canal and tank irrigation is 4.8 acres.

There are two major crops, namely, paddy and sugarcane, which are grown by 77 per cent and 13.5 per cent of the farmers, respectively, as the first most important crops. Of course, there are several other crops (like groundnut, switch wood) but they are grown by small proportion of them. The total area irrigated in respect of the first most important crops grown by the individual farmers of the three districts range anywhere from less than 1 acre to as much as 15 acres. Similarly, area under second most important crops irrigated has more or less the same pattern.

Well Irrigation

Area under well irrigation has a range of 1 acre to about 20 acres. About 45 per cent of the farmers have well irrigation which means that some of the farmers in canal and tank commands also use wells as substitute. It is indeed one of the strategies adopted by farmers during water deficits periods.

Well irrigation costs, particularly where it is a substitute for canal and tanks, is an additional cost to the farmers. As much as 17.4 per cent of the farmers interviewed have reported costs on digging bore wells ranges from Rs 5,000 to Rs 1,50,000. From our survey, we gather that 5 per cent of the farmers interviewed have spent between Rs 5,000 and Rs 10,000, 4 per cent between Rs 10,000 and Rs 20,000 and 2 per cent between Rs 20,000 and Rs 30,000. Those who spent between Rs 30,000 and Rs 50,000 account for 4.2 per cent, between Rs 50,000 and Rs 1,00,000 account for 3.5 per cent and Rs 1,50,000 account for about 1 per cent.

Water Scarcities

About 97 per cent of the farmers have reported that during the last two decades, 14 years had witnessed water deficit due to various reasons. The year 2004 was considered the largest deficit year by about 28 per cent of the farming people, whereas 2009 was considered so by about 11 per cent of them. The year 2007 (11.1 per cent) was the next most deficit year, followed by 2002 (9.7 per cent), 2003 (7.6 per cent), 2010 (5.6 per cent), 2005 and 2008 (4.2 per cent each) and 2006 (92.1 per cent). Thus, overall 97 per cent of the farmers perceived one or more years as having been deficit and causing problems for agriculture.

It does not mean that the other 6 years had surplus water. There indeed were surplus years: in fact, farmers have reported 17 surplus years in the last three decades. There are also some years which have been reported as both deficit and surplus years: that is deficit for some and surplus for some others. Excess or heavy rainfall has been reported by 96 per cent as the reason for surplus years. Decline in rainfall and failure of monsoons particularly lack of northeast monsoons and not opening the Mettur dam in time are the major reasons for water deficits in study areas.

Strategies for Meeting Water Deficits

The farmers who face water deficit adopt several strategies for meeting it. Most farmers (17.4 per cent), according to the survey, leave their land fallow during the deficit period. A considerable proportion of them (11.1 per cent), on the other hand, use well irrigation during the deficit periods. About 8 percent of them borrow water from other farmers at the time of crop stress. Some farmers (1.4 per cent) use oil-driven pump sets to irrigate their fields by pumping water from their wells or from their neighbours (0.7 per cent). Some farmers (1.4 per cent) reduce their cropped area during deficits.

Other farmers choose alternative crops for cultivation during water deficits: cereals 3.5 per cent, groundnut 0.7 per cent, maize 0.7 per cent, pulses 1.4 per cent and oil seeds 0.7 per cent. So only a small proportion of the farmers (7 per cent) grow different crops for meeting the deficits. That is, crop diversification is favoured only by a small proportion of them, if any. It is significant to note that the crops so grown are also the recommended alternatives during the deficits periods or drought period by the government agencies under agriculture diversification.

Crop Diversification

Overall, the one noteworthy conclusion the study leads us to is that most farmers of the three districts follow the traditional patterns of cropping, using the wisdom of their parents and grandparents at the times of water deficits. Everywhere the team went, the one voice heard is the voice of the traditionalists, who believed in 'what worked for my father/grandfather works well for me'. Crop diversification, in the last three decades, has been ignored by the farmers and there is little evidence for wider adoption of the crop diversification or agriculture diversification in the canal, tank and well command area we have studied.

In real terms, only a few crops dominate the crops among the first most important crops indicated by the farmers interviewed. As mentioned earlier, paddy (77.1 per cent) and sugarcane (15.3 per cent) are the most grown by the largest proportion of the farmers. Groundnut is grown by 2.1 per cent of them, cereals by 0.7 per cent of them and mango by a similar negligible proportion. Thus, the staple (paddy) and the cash crop (sugarcane) dominate the first most important crops.

Things are not very different with the second most important crop either, although there are more diverse crops among them. Paddy still dominated in the case of 60.4 per cent of the farmers. Groundnut is the favourite of 12.5 per cent of the farmers, a crop that has taken the place of sugarcane even though it is a short term, (4 months) crop. Pulses are preferred by 5.6 per cent of the farmers while cereals are favoured by 4.2 per cent of them. Cotton (1.4 per cent), coconut, flowers, sugarcane, switch wood and water melon are all crops grown 0.7 per cent the farmers each type of crops.

Most important crop by season is May–September (southwest monsoon period when water is available from rivers and canals) for 50.7 per cent of the farmers and annual crops are the preference of 15.3 per cent of the farmers (mostly sugarcane). Perennial crops such as the coconut and banana are preferred by a small proportion (0.7 per cent). Season beginning in April has been reported as the season for 29.9 per cent of the farmers.

SUGGESTIONS FOR IMPROVEMENT

The farmers interviewed have been asked three most important suggestions they consider to resolve the problems they face in irrigation and crop diversification.

Natural fertilizers use dominates the suggestion in the first list with 47.2 per cent of the farmers backing it up. Resolving labour shortages is the second best with 14.2 per cent of the farmers wanting it done. Subsidy for bore well (6.3 per cent) and bore well facilities (3.5 per cent) are sought by the farmers as bore wells are useful in tapping groundwater to meet the water deficit periods. Introduction of machinery has been suggested as one of the strategies for overcoming problems by 3.5 per cent of the farmers. Drip irrigation, subsidized seeds, and deepening tanks are suggested by 1.4 per cent of the farmers for each suggestion as solutions.

As many as 16 different suggestions have been given by various farmers for resolving the problems faced in the three districts of our study. Resolving shortage is the pick of 16 per cent of the farmers as the second best suggestion. Stopping power shedding and natural fertilizers use have been suggested by 9 per cent of the farmers each. Deepening tanks (6.3 per cent), subsidy for bore wells (4.2 per cent), cleaning canal (4.2 per cent) and sprinkler technologies (3.5 per cent) have been suggested by sizable proportions of the farmers. All the other suggestions have been preferred by very small proportion of the farmers.

Looking at the suggestions, most appear to be suggestions for resolving their most common problems rather than problems relating to irrigation, water deficit and crop diversification. The wisdom is: when the common problems are out of the way, the pestering problems can be dealt with easily.

SUMMARY AND CONCLUSIONS

Having interpreted the results of the data analysis pertaining to irrigation and crop diversification in three districts of Tamil Nadu, using nine select villages, the summary of the findings and conclusions of the study are given below.

Canal, tank and well irrigation commands in the three districts of Thiruvarur, Pudukkottai, and Villupuram, in that order, do have water surpluses and water deficits during different times: surpluses at the end of the cyclone rains during November–December and deficits during August–September.

Cropping is therefore affected in the different commands and the farmers take corrective measures, either in the form of substitutes for irrigation, especially from bore wells and often from neighbours.

Some of them, resort to growing different crops which could face the water deficit situation, while a large number of them leave their lands fallow.

There is also the undercurrent that the farmers are generally aware of the problems of water shortages, especially since the Cauvery dispute began that they have taken to cultivating crops following the traditional wisdom of their parents and grandparents and thus there is very little crop diversification in the canal, tank and well irrigated areas.

It follows from the above that the farmers have taken an easy way out of the problems emerging from irrigation and crop diversification in the state of Tamil Nadu and their activities are a direct corollary of the water scarce situation prevailing. From this, the following conclusions have emerged.

The ayacuts (or commands) of the canals, as reported by the farmers, have a range of 40 acres to 300 acres: there are ayacuts of varying sizes that are irrigated by the canals: 100 acres (6.9 per cent), 120 acres (6.3 per cent), 140 acres (3.5 per cent), 250 acres (2.1 per cent), and 300 acres (6.9 per cent).

A little more than a third of them (10.5 per cent) report of 1–2 acres of land under irrigation while 7 per cent of them report of 2–4 acres, 6.3 per cent of them 4–6 acres, and 5.6 per cent of them 6 plus acres.

Being in the delta, starved of water for several years with the exception of the monsoon and cyclonic weather benevolently providing the much needed water, the season of surplus for all of them depending on the canals is the cyclonic weather period: October–December.

Nearly all of the (25 per cent) indicate June–August as deficit months while 4.2 per cent of them indicate August as the most deficit month. This happens because the Mettur dam which is opened usually on June 12 gets delayed for shortage of water in the dam, or pre-monsoon showers fail the farmers miserably.

The tanks irrigating the lands have commands ranging from 25 acres to about 3000 acres. For 4.2 per cent of the farmers, their tank commands are less than 100 acres; for 9.4 per cent of them the commands are anywhere between 100 and 200 acres; for 11.8 per cent of them, the commands are between 300 and 500 acres; and for 3.5 per cent of them, their commands are larger than 500 acres.

Nearly half the farmer-owned lands under irrigation are either small holdings (17.7 per cent) or medium holdings (32 per cent). Large and very large holdings under irrigation account for 27.5 per cent but there are area differences between the holdings: 5–6 acres account for 16.7 per cent, 7.8 acres for 7 per cent and 9–10 acres for 3.5 per cent. Ten plus acre holdings account for 2.8 per cent.

There are two major crops, namely, paddy and sugarcane, which are grown by 77 per cent and 13.5 per cent of the farmers, respectively, as the first most important crops. Of course, there are several other crops but they are grown by small proportions of them (groundnut and switchwood, for example). The total area irrigated in respect of the first most important crops grown by the individual farmers of the three districts under study range anywhere from less than 1 acre to as much as 15 acres.

The total area under canal and tank irrigation, for individual farms, is between less than 1 acre to more than 35 acres: while 22 per cent of the farmers irrigate 1–2 acres, 33.4 per cent 3–4 acres, 14.6 per cent 5–6 acres, 9.1 per cent 7–8 acres, 6.3 per cent 9-10 acres, and 3.5 per cent more than 10 acres.

It has also been learnt that the large farmers also usurp the land of the poor and marginal farmers for the loans given to them and thus generally have large farms.

Area under well irrigated has a range of 1 acre to about 20 acres. The fact that 44.6 per cent of the farmers report of land under well irrigation means that some of the farmers in canal and tank commands also substitute their irrigation from canal and tanks, as the case may be, from wells.

As much as 17.4 per cent of the farmers interviewed have reported costs on digging bore wells, from Rs 5000 to Rs 150000. That is, small and marginal farmers cannot afford bore wells and if they decide to have one, they go for bore well of cheaper cost.

Every year in the last decade (2001–10) and some years in the previous decades (1979, 1984, 1986, 1995) have been indicated by the farmers as having been deficient in water for various reasons and by a sizable proportion of the farmers (96.5 per cent, overall) interviewed.

There indeed were surplus years: in fact, by the farmers' indication, 17 surplus years in the last three decades: 1980, 1987, 1990–94, 2001 (all indicated by 0.7 per cent of the farmers each), 2003 (1.4 per cent), 2004 (4.9 per cent), 2009 (6.9 per cent), 2006 (9 per cent), 2007 (11.1 per cent), 2000 and 2002 (13.2 per cent each), 2008 (14.6 per cent), and 2005 (16.7 per cent).

Excess or heavy rainfall (95.9 per cent) has been reported as the reason for surplus years. Decline in rainfall

(1.4 per cent), failure of monsoon (86.9 per cent) and, particularly, lack of northeast monsoon (71.6 per cent) and the Mettur dam not having been opened in time (17.4 per cent) have all been cited as the major reasons behind the deficits in water for irrigation in the districts of our study.

The duration of surplus for 6.3 per cent of the farmers lasts for about two months, usually November and December, during the cyclonic rains. The duration of deficit also lasts for two months, during August and September, prior to the cyclonic period, for 6.3 per cent of the farmers. It is often the same set of farmers who face the prospect of surplus as well as deficit.

As much as 14.6 per cent of the farmers face water deficits during the season of first most important crops. When considered in the context of canal irrigation, 2.8 per cent of the farmers in Thiruvarur district face the deficits during the second cropping season. Between 3 acres and 5 acres of lands under canal irrigation of the small proportion of the farmers get affected. Even more farmers (7.6 per cent) of the canal commands face water deficits during the third cropping season and the area of crops affected range from 1 acre to 10 acres: 5.6 per cent of the farmers have landholdings under stress between 2 acres and 5 acres.

The farmers with water deficits adopt several strategies for meeting them. Most farmers (17.4 per cent), according to the survey, leave their land fallow during the deficit period. A considerable proportion of them (11.1 per cent) on the other hand use well irrigation during deficits. About 8 per cent of them report using bore wells for overcoming the situation, and 4.9 per cent of them borrow water from other farmers at the time of crop stress.

Natural fertilizers use dominates the suggestions in the first list with 47.2 per cent of the farmers backing it up. Resolving labour shortage is the second best with 14.2 per cent of the farmers wanting it done. Subsidy for bore well (6.3 per cent) and bore well facilities (3.5 per cent) are sought by the farmers as bore wells are useful in tapping ground water to meet the water deficit periods. Introduction of machinery has been suggested as one of the strategies for overcoming the problems by 3.5 per cent of the farmers.

It must be understood that the non-availability of water or the scarcity of water alone cannot determine an alternative cropping pattern. The alternative cropping system needs to be worked out for the state, keeping in view the agro-climatic zone as the base and keeping in mind the nature of the soils, traditional cropping patterns and the availability of water resources. State level committees could be constituted with a team of different experts to arrive at a suitable, alternative cropping system. The success of crop diversification as a strategy for overcoming the water scarcity depends on the factors like economic returns to the farmer, availability of credit and easy access to markets.

REFERENCES

Arputharaj, C. 1982. 'Problems and Prospects of Tank Irrigation in Tamil Nadu', Workshop of Modernization of Tank Irrigation, Problems and Issues, Centre for Water Resources (mimeo), Centre for Water Resources, Madras, India.

Dhawan, B.D. 1988. *Irrigation in India's Agricultural Development: Productivity, Stability, Equity*. New Delhi: Sage Publications, p. 265.

International Bank for Reconstruction and Development. 1970. *India, Cauvery Delta Irrigation and Drainage Rehabilitation Project*, Appraisal Report. Washington DC: International Bank for Reconstruction and Development.

Oswald, Odile. 1992. 'An Expert System for the Diagnosis of Tank Irrigated Systems', International Irrigation Management Institute, Colombo.

Prasad, Kamta. 2000. *Water Resources and Sustainable Development: Challenges of 21st Century*, New Delhi: Shipra Publications.

26

Study of Tanks in Watershed Development Area in Karnataka

P. Thippaiah

Tank irrigation is one of the cheapest sources of surface irrigation. This was sustained until the 1960s in various states and the country as a whole. After this period, there has been a sluggish growth in irrigated area under tanks in the country more particularly in the state of Karnataka, which is having less proportion of irrigated area as compared to national average and southern states.

One of the interesting developments in the case of Karnataka relates to the stagnation of agricultural production and productivity during the 1980s. The expert committee, constituted in 1993, to examine the reasons for stagnation indicated that the decline in irrigated area was one of the reasons for the stagnation in agricultural production. This committee had suggested Watershed Management in the Catchment areas of the tanks, desilting of tanks, modernization of distribution systems under tanks and initiating conjunctive use of tank and groundwater (Government of Karnataka 1993). Perhaps these recommendations were responsible for the initiation of various programmes by the state government, either with its own resources or external funds for restoration/rejuvenation of tanks with a view to sustaining agriculture, ensuring food security and alleviating rural poverty especially in high variable rainfall areas.

Two schemes are very important in this respect namely, watershed development programmes including the World Bank funded Sujala Watershed Development Programme and Karnataka Community Based Tank Management Project (KCBTMP) funded by the World Bank and implemented by the Jala Samvaradhane Yojana Sangha (JSYS). These programmes are now relatively old. Therefore, this is the right time to examine how these programmes are being implemented and how they have minimized the various factors that had rendered the tanks useless. In this study, an attempt has been made to look at the coordination between these schemes in terms restoring the tanks and also to examine the pattern of investment, community participation in tank rehabilitation and impact it has made in increasing the efficiency of tanks and increasing the productivity and income of the farmers. It has now become necessary to take stock of the situation and identify the constraints, if any, before replicating these in other proposed areas. With this background in view, the present study outlines the following objectives:

1. to study the background of the selected tanks;
2. to study the socio economic conditions of command area farmer households of the rehabilitated tanks;
3. to present the earlier initiatives of tank rehabilitation programmes;

4. to study the investment pattern in various components of tank rehabilitation;
5. to study the impact of tank rehabilitation programme;
6. to study the impact of watershed development programmes on tanks; and
7. to suggest policy recommendations.

METHODOLOGY AND DATA SOURCES

The study selected two tanks each in Kolar and Chickballapur districts where more number of tanks had been rehabilitated and more area treated under watershed development programmes. These tanks were rehabilitated under KCBTMP in watershed and non-watershed development programmes areas. The idea behind this sample framework was to examine three important aspects: first, to examine the impact of tanks rehabilitation programme in terms of making the tanks more efficient and providing benefits to farmers and also to examine if there were any shortcomings leading to failure of programmes. Second, the study intended to examine the impact of watershed development programmes on the status of tanks. Third, it was also examine the nature of coordination between the watershed development programmes implemented by the Watershed Department and Tank Rehabilitation programme under KCBTMP, implemented by the JSYS.

The study has used both secondary and primary data for analysis. The secondary date were collected from the Water Resources Department (Minor Irrigation), Zilla Panchayat Engineering Department about salient features of selected tanks and Watershed Development Department for about structures created in the catchment areas of tanks, JSYS for understanding about the costs of tank rejuvenation, tank filling, and Directorate of Economics and Statistics for the number of tanks and irrigated area. In consultation with the watershed department, four tanks were selected. The information on watershed areas and non watershed areas was also obtained from the watershed department at the district level, and with their help, the rehabilitated tanks were randomly identified, as the JSYS, which implemented the KCBTMP, had not cooperated well in providing the required information.

The data required for the proposed study were gathered from a sample of 20 farmers/households from each tank, benefiting from tank rehabilitation. Altogether, a total of 80 sample beneficiaries (households) from four tanks (two in watershed development area and two in non-watershed area) were selected for an in-depth analysis (see Table 26.1). A structured questionnaire was used for collecting information from the tank beneficiaries. Both qualitative and quantitative information was gathered through these questionnaires. Useful discussions were also held with site officials of Watershed Development Department, ZP Engineers, NGOs and, tank user groups for understanding the tank rehabilitation process and impact assessment. The reference period for the study is 2007–08.

MAJOR FINDINGS AND DISCUSSION

Socio Economic Conditions of the Command Farmer Households

A majority of households in the command area belonged to Vokkaliga community (63.75 per cent) and their holdings are largely small in nature (81.25 per cent). The landholdings size is still less in the case of command area households (0.28 ha). The proportion of people speaking Kannada as a mother tongue is found to be high (58 per cent) among the sample households. This finding runs against the general belief that the people living in the adjoining sample areas speak Telugu as they are close to Andhra Pradesh, where the mother tongue is Telugu. Even then, this finding is valid as a majority of them claim Kannada as their mother tongue as they have studied Kannada in schools.

The total population of the 80 sample households stands at 468, which works out to 5.85 members per family. However, the family size is found to be higher with 6.23 persons per family in non-watershed area households. The literacy levels of the household members accounted for 64 per cent. This is less than state literacy rate of 66.60 per cent. The annual average household income works out to Rs 97,120. About 70.28 per cent of the household income in all villages is accounted by agriculture followed by dairy activities (9.59 per cent) and jobs (9.49 per cent). According to the poverty criteria, the

Table 26.1 Sample Areas of the Study

Name of the Tank & Village	District	Taluk	Programme	Area
Belaganahalli (Emal Kere)	Kolar	Kolar	JSYS	Watershed
Yerrakote (Marap kere)	Chickkaballapur	Chintamani	JSYS	Watershed
Jambapura (Thippareddy Kere)	Kolar	Kolar	JSYS	Non-Watershed
Gadadasanahalli (Hosakere)	Chickkaballapur	Chintamani	JSYS	Non-Watershed

proportion of households below poverty line constitutes 6.25 per cent. However, the incidence of poverty is very high in the villages according to BPL and Anthyodaya cards (58 per cent). These background characteristics are almost similar across the villages as well as watershed and non-watershed areas.

Past and Present Initiatives of Tank Rehabilitation

In Puducherry, Tamil Nadu, Rajasthan, and Gujarat, the rehabilitation of tanks has been taken up in the past with the assistance of external agencies such as European Economic Community and the World Bank. These rehabilitation programmes largely involved NGOs and the village communities in the implementation process. The experience of these shows that the desired results were not achieved. The tank user groups created in these areas for maintaining the tanks after rehabilitation had failed in this respect.

In Karnataka, under the Drought Prone Area Programme, Integrated Western Ghats Development Programme, and the Desert Development Programme, desiltation of tanks had been taken up along with other components. However, this component, that is, desiltation of tanks was found ineffective, unlike other components.

The World Bank Food Programme was implemented in Karnataka in 1978. Under this, there was a provision made for Rs 1,412 lakh in 5 districts for the improvement of 353 tanks (Government of Karnataka 1977). The project was aimed at improving tank irrigation, promoting employment generation and increasing fish and foodgrains production. An examination of these aspects by using the secondary data shows that there was negative growth before and after tank rehabilitation. However, the negative growth rate gap was found reduced.

Under Herehalla Watershed Development Project in Shimoga district of Karnataka, both Zilla Panchayat tanks (with less than 40 ha of command area) and Minor Irrigation Department tanks (with more than 40 ha of command area) were covered for improvement. The financial as well as physical progress accounted for more than 90 per cent. However, the farmers were not happy with the desiltation work as well as the lining of main canals (Raju and Paramesha 2002).

One more major scheme was implemented in the state in November 2001 called Raitha Kayaka Kere Programmes with a view to rehabilitating 745 tanks with an estimated cost of Rs 195 crore. The test check of the completed tanks by the Comptroller and Auditor General showed various gaps in the implementation of this scheme by pointing out that the intended objective of holistic rejuvenation of tanks had not been fulfilled. In addition to these, some NGOs had also rehabilitated a few tanks with the financial assistance from National Bank for Agriculture and Rural Development (NABARD) and Council for Advancement of People's Action and Rural Technology (CAPART). Unfortunately, the data related to the impacts of these attempts are not available for making value judgement.

Investment in Karnataka Community Based Tank Management Project

The KCBTMP was funded by the World Bank and implemented by the JSYS. Under this programme, a total of 440.43 crore had been earmarked for rehabilitation of 2005 tanks from 2001–02 to 2006–07. As against this, only 55.73 per cent was found spent, showing that the financial performance was not satisfactory. Similarly, the tanks rehabilitated numbered 932 as against the target of 2005, indicating not completing the programme in time. Similar trend could be noticed in the selected districts. As against the target of 1,021 tanks in the Kolar and Chickkaballapur districts, only 942 tanks had been taken up for rehabilitation. Out of these we found only 577 tanks completed and entrusted to the local Tank User Groups (TUGs) for future maintenance. A sum of Rs 50.11 crore had been allocated for these tanks covering both civil and non civil works (Table 26.2). As against this, 42.52 crore were spent accounting for 84.86 per cent of the allocations.

One of the significant observations is that the allotment per tank was found to be high in the case of Minor Irrigation Tanks as compared to Zilla Panchayat Tanks. As per the guidelines of the Project Implementation Plan (PIP), a sum of Rs 54300/ha of the command area was to be earmarked for civil works related to tanks with a command area of less than 40 ha, and a sum of Rs 37900/ha of command area in the case of tanks with a command area of more than 40 ha. However, a total amount of Rs 52003 and Rs 20864 was found allocated, respectively. All the components of tanks were found executed by NGOs with the involvement of local communities.

In respect of 4 sample tanks, the allotment was 19.86 lakhs accounting for Rs 4.97 lakh per tank as per the information displayed in villages and low as compared to Rs 7.42 lakh per tank in the case of Zilla Panchayat tanks in Kolar district and Rs 7.63 lakh in the case of Chickkaballapur district. A major proportion of this was allocated to tank area and bund improvement accounting for 80.55 per cent. Only 3.63 per cent on catchment area treatment and 7.04 per cent for command area development were allocated. Besides this, 5.24 per cent was

Table 26.2 Estimated Cost and Expenditure on Handed Over ZIlla Panchayat and Minor Irrigation Department Tanks in Kolar (May-2008) and Chickkaballapur (March-2008)

Rs in Lakhs

District	No. of Tanks	Estimated Cost			Expenditure			Expenditure as % of Cost		
		Civil Works	Non-Civil Works	Total	Civil Works	Non-Civil Works	Total	Civil Works	Non-Civil Works	Total
Kolar	93	847.38	95.68	**943.07**	642.85	90.66	**733.51**	75.86	94.75	**77.78**
Malur	68	377.50	58.75	**436.26**	292.33	51.64	**343.97**	77.44	87.88	**78.85**
Mulbagal	185	1,443.76	157.38	**1,601.15**	1,306.30	146.11	**1,452.72**	90.50	92.84	**90.73**
Total	**346**	**2,668.65**	**311.82**	**2,980.47**	**2,241.78**	**288.41**	**2,530.20**	**84.00**	**92.49**	**84.89**
Bagepalli	54	378.93	52.67	**431.60**	332.84	49.23	**382.07**	87.84	93.47	**88.52**
Chinthamani	112	782.00	124.62	**906.62**	668.25	115.78	**784.03**	85.45	92.91	**86.48**
Gudibanda	16	109.20	14.90	**124.10**	98.02	14.06	**112.08**	89.76	94.40	**90.32**
Sidlaghatta	49	500.42	67.78	**568.20**	385.31	58.81	**444.12**	77.00	86.77	**78.16**
Total	**231**	**1,770.55**	**259.96**	**2,030.52**	**1,484.42**	**237.88**	**1,722.31**	**83.84**	**91.51**	**84.82**
G Total	**577**	**4,439.20**	**571.78**	**5,010.99**	**3,726.21**	**526.30**	**4,252.50**	**83.94**	**92.04**	**84.86**

Source: Computed from raw data provided by DPUs (JSYS) of Kolar and Chintamani.

Table 26.3 Total Estimated Cost of Civil and Non-civil Works

(Rs Lakhs)

Village/Tank	Major Items	Catchment Area	Tank Area	Tank Bund	Sluice	Waste Weir	Command Area	Social	Administrative	Environmental	Total
Yerrakote	**Shramadhan (6% Community Contribution)**	0.05	0.03	0.10	0.00	0.00	0.03	0.00	0.00	0.00	0.20
	JSYS Amount	0.20	1.32	1.48	0.04	0.02	0.25	0.40	0.20	0.00	3.90
	Total	0.25	1.35	1.57	0.04	0.02	0.27	0.40	0.20	0.00	4.10
Belaganahalli	**Shramadhan**	0.03	0.05	0.10	0.00	0.00	0.01	0.00	0.00	0.00	0.18
	JSYS Amount	0.02	1.09	1.68	0.02	0.00	0.01	0.64	0.20	0.00	3.66
	Total	0.05	1.14	1.77	0.02	0.00	0.02	0.64	0.20	0.00	3.84
Jambapura	**Shramadhan**	0.01	0.04	0.12	0.01	0.00	0.02	0.00	0.00	0.00	0.19
	JSYS Amount	0.00	1.54	1.15	0.03	0.00	0.30	0.00	0.00	0.00	3.02
	Total	0.01	1.59	1.27	0.04	0.00	0.32	0.00	0.00	0.00	3.22
Gadadasanahalli	**Shramadhan**	0.03	0.30	0.11	0.02	0.01	0.05	0.00	0.00	0.00	0.50
	JSYS Amount	0.39	4.73	1.69	0.26	0.08	0.75	0.00	0.00	0.30	8.21
	Total	0.42	5.04	1.80	0.27	0.09	0.80	0.00	0.00	0.30	8.71
Total of Four Tanks	**Shramadhan**	0.11	0.43	0.41	0.02	0.01	0.10	0.00	0.00	0.00	1.08
	JSYS Amount	0.61	8.69	6.00	0.35	0.10	1.31	1.04	0.40	0.30	18.79
	Total	0.72	9.11	6.41	0.37	0.11	1.40	1.04	0.40	0.30	19.86

Source: Computed from the information displayed in the villages.

allotted the upliftment of weaker sections through Income Generating Activity (IGA) and the rest was used for crop demonstration and environmental purposes (Table 26.3). Across tanks, 89.77 per cent was allotted to tank beds and bunds in Jambapura. In the respect of other tanks, it was around 76 per cent. Among the tank area investments, a major proportion was spent on desiltation (45.87 per cent) as against the state average of 37.32 per cent. The allotment for sluice and waste weir repairs constituted just 1 per cent of the total civil works costs.

Impact of Karnataka Community Based Tank Management Project

As per the objectives of the project, it was expected that the irrigated area by tanks would improve after rehabilitation. However, the area did not increase. In fact, the cultivated area under tanks during the Kharif season was found declined from 19.30 ha to 14.24 ha after rehabilitation with a major proportion of this area being under ragi crop at 64.46 per cent before rehabilitation and declining to 45.51 per cent after rehabilitation. This decline was due to a decline in the total cultivated area under tanks. The other crops included vegetable crops as they were better suited to agro climatic conditions in the region and bore well irrigation. Rabi crop cultivation was not found significant (5.82 ha). During this season, mulberry and vegetables are grown under rain fed conditions although tank irrigation showed an improvement from 25.52 per cent to 34.47 per cent after tank rehabilitation, in absolute terms, it had declined. The low area under tank irrigation was attributed to the failure of rainfall.

It was expected that the rehabilitation of tanks would increases yields in the command area. However, yield levels under tanks in respect of ragi was 29–30 qtl/ha followed by 350 quintals in the case of potato and tomato and 30 quintals in the case of paddy. These yield levels had remained more or less same for a long time, though income per ha after tank rehabilitation did show some improvement. This increase was mainly due to rise in prices of the produce rather than increase in yield levels.

Under the tank rehabilitation programme, it was expected that all the households in the villages would be members of Tank User Groups. The field data shows that among the sample households/farmers nearly 30 per cent had enrolled as members, thus contrasting, the claim made by TUGs that all the households had enrolled as members. Many household/farmers reported that they had been enrolled as members as per their knowledge so as to fulfil the project guidelines. It was also expected that all the sample households/farmers would share the cost of tank rehabilitation in the form of labour and cash. The data shows just 35 per cent of the farmers contributed towards the total cost.

Under the tank rehabilitation programme, it was expected that the farmers would lift the silt that was excavated. However, out of 80 sample farmers, 21 farmers were found lifted silt. The average silt lifted came to 47 tractor loads. This silt was used for enriching dry and bore well irrigated lands. Many farmers revealed that they had not lifted silt as they found the transport cost very high. Those who lifted silt for enriching their fields revealed the yield levels had increased after using silt to the extent of 1 qtl/ha to 3 qtl/ha. They further added that the yield levels would increase further after a few years once the silt adjusted to the original soil system.

Under the programme, it was expected that tank fillings would improve due to restoration of feeder channels and improvement in tank structures. But, according to 75 per cent of the sample farmers, the tanks did not fill after rehabilitation. They further indicated that tanks got filled once in 5 years. This finding contrasts against the norm adopted for tank rehabilitation that the selected tanks would get surplus 5 times in 10 years. Most farmers attributed the failure of rainfall and catchment disturbances and construction of several structures by the watershed development department in the catchment area as responsible for non-filling of tanks in the downstream.

Encroachment was one of the major problems facing tank rehabilitation. According to the survey of the Revenue Department, on an average, 1.45 ha of water spread area of tanks had been encroached. It is said, all these encroachments took place while rehabilitation tanks had been going on. The field data and observation reveal that no encroacher had surrendered encroached land.

Bore wells are predominant in all the sample villages. About 50 per cent of the sample farmers found with owning bore wells. Altogether, a total of 69 bore wells were found in the sample villages. Out of these, 21.42 per cent were located in the command area and 37.14 per cent around the tanks. This concentration in the vicinity of the tanks led to loss of tanks water as they are sucking tank water, affecting efficiency of the tanks. It was expected that the number of bore wells would increase after a rehabilitation of tanks as it improves the ground water levels. But only two bore wells were found to have come up after rehabilitation.

The structural improvements effected across tanks were of sub-standard quality not sustainable. The bunds strengthened with the silt as well as gravel excavated from foreshore area of the same tanks were found already eroded due to improper treatment. The main canals linings were damaged in many places. Further, we found the canal stone slabs already fallen down and there were gaps in between the slabs. Partial desiltation was undertaken in all tanks as against the farmer's wish for complete desiltation. Only, 20 per cent of the accumulated silt had been removed.

The institutions Tank User Groups (TUGs) created at the time of commencement of tank rehabilitation programmes were supposed to take care of tank management for future operation and maintenance. The experience across selected tanks shows that the TUGs were not at all functioning nor generating income from tank based usufructs for the maintenance of tanks.

Fodder seeds sown to strengthen bunds and supply fodder to the villages turned out to be ineffective. The plantation in the foreshore area under the tank rehabilitation programme failed to survive. The survival rate was 20 per cent as compared to 42 per cent survival rate at the state level.

There were conflicts between the forest and watershed departments, JSYS and TUGs over the ownership of surviving plants as they had been planted by the forest and watershed departments before tank rehabilitation. There were also conflicts between watershed departments and the JSYS during planting when both the schemes were operating in the same area.

The crop demonstrations conducted by the Agriculture University at the time of tank rehabilitation for promotion of water efficiency and higher yields did not materialize, as the farmers failed to practised them.

The conservation and maintenance of water were undertaken in the same area under different schemes such as watershed development programme, Jala Nirmala Yojana Scheme and Karnataka Community Based Tank Management Project. Due to lack of coordination, there was overlapping of schemes leading to negative effects. While executing tank rehabilitation, the services of line departments such as Zilla Panchayat Engineering Division and Minor Irrigation Department were not sought. Corrupt practices indulged in by NGOs and the District Planning Units of JSYS while executing the works were reported in Gadadasanahalli and Jambapura villages.

The only positive development of tank rehabilitation was the creation of dead storage which facilitated water to stay long as to provide water for small ruminants and cattle during summer. To some extent, the dead storages helped fish rearing and enriching of ground water.

Watershed Development Programmes and Their Impact on Tanks

In the country as well as in the state, several watershed development programmes have been implemented. Under these programmes, treatment of catchment areas of tanks was taken up as one of the components with a view to preventing soil erosion and silt deposit in the downstream tanks and arresting rain water run off for enriching ground water. Under this component, a variety of structures such as form ponds, check dams, nala bunds and percolation tanks, contour bunding, trenching, and boulder checks had been taken up both in arable lands and non arable lands and across drainages. Some of the structures, particularly, the check dams, farm ponds, percolation tanks had the capacity of storing several cubic metres of rain water. According to farmers, these are not allowing runoff from the catchments leading to non filling of tanks. The available data for five year from 2001–02 to 2005–06 shows that in all 59,504 structures in the state and 5,447 structures in the two sample districts had been built. Similarly, in the catchment areas of two watershed area sample tanks, an average of 26.60 to 31.16 structures had been built.

Lack of coordination between tank rehabilitation programmes of the JSYS and watershed Development Departments while going for the structures in the catchment areas of tanks was affecting the storages of tanks. Several studies have highlighted this fact. Lack of coordination and independent nature of works by each department promoted neither increase in the tank irrigated area nor substantial increase in agricultural production despite the combined efforts of these programmes in the state.

Suggestions for Tank Rehabilitation Programme

For the improvements of the tank rehabilitation programme and tank efficiency, the respondents and focus groups have suggested some measures. They have been presented below.

Suggestions for the Improvement of Catchment Area of Tanks

Various structures have been built across the drainages of tanks for conserving rain water under the watershed development programmes. This type of conservation of water in the catchment areas has been preventing rain water flow towards the tanks. Therefore, steps are needed to construct the structures in the catchment areas without affecting the run off. In the catchment areas of many tanks, the drainages/feeder channels have been found blocked/encroached or diverted affecting the run-off water into the tanks. A survey of catchment areas of tanks should be conducted and measures taken to clear these obstacles.

The silt removed during tank rehabilitated was not lifted many farmers due to high transport cost and lack of bullock carts and bullocks. The government has to lift this silt and dumped on the farmer's lands particularly, on the fields of small and marginal farmers who could not offered to hire the tractors. Silt removed in the tanks was a drop in the ocean. It has only created dead storage rather than facilitation the flow irrigation. Complete desiltation is necessary to restore original capacity and provide irrigation to the entire command area.

Suggestions for the Improvement of Command and Tanks Area

The main canals have been lined to prevent the transmission loses. However, in many places the slabs have not been properly aligned. This is causing transmission losses. In some places, they had already fallen down. The persons executed these works have to be made liable for this and they have to be properly repaired.

The rehabilitation has focused only on physical works rather than addressing food security. The tank rehabilitation programme has failed to convince the farmers to cultivate crops under command areas of many tanks rain fed conditions instead of leaving it fallow, in the event of non filling of tanks. This shows that the programme had not addressed food security. This aspect has to be considered in view of food insecurity problem in the country. The tank rehabilitation programme authorities claim that the encroachers of foreshore areas of tanks have been evicted. However, the filed observations show the encroached lands still under the possession of encroachers and they are cultivating these lands. Strict action and permanent boundaries around the tanks only can prevent encroachments.

The desiltation and repairs of tanks have been carried out without having proper data on water yield levels from catchment areas, flows from drainages, storage, discharges, command developments, encroachment of water spread area, conditions of feeder channels, main canals, sluice, distribution systems and improvements made to the tanks under the earlier schemes. In order to maintain data on these, register for each tank should be maintained by the authorities concerned for the benefit of future rehabilitation of tanks.

There has been an enormous growth of bore wells in the vicinity of tank resulting in the depletion of water storage in tanks. They should be regulated/controlled to avoid excess water sucking. Sand extraction and processing in the tanks has caused heavy water percolation at the cost of tanks. To overcome this problem, the state should not only regulate sand mining and but also ensure that this activity would not affect the performance of tanks.

Suggestions for the Watershed Development Programme Improvements

The watershed development programme, instead of looking at it in relation to tank rejuvenation programme has been independently operating its work. This is a major problem coming in the way of the non-success of tank rejuvenation and increasing the efficiency of tanks in terms of filling and irrigating the command area. Therefore, the tanks in the watershed area have to be rehabilitated along with watershed programmes keeping in mind the interest of tanks in the downstream.

Many water storage structures built in upstream of the tanks by watershed development programmes (most of the micro watersheds are named after tanks). Therefore, there is need for directing the watershed development department to prohibit the construction of certain kinds and size of structures in the catchment areas of tanks as the state has invested on many tanks. There is also need to specify more appropriate technical strategies for the development of structures in watershed areas.

Suggestions for the Institutional Improvements

The TUGs have been formed for each tank so that these institutions take care of maintenance of tanks and distribution of water. However, it is noticed that they were tend to be active when the money is flowing and once the work gets over they become inactive or sick. The TUGs have to be made accountable to this for any laps of the above. The government has given powers to Tank User Groups to raise funds trough selling tank fish, silt, trees and grass after the execution of tank rehabilitation programme so that they can utilize these funds for further development of tanks. However, many of the TUGs have not generated any revenue so far. In this kind of a situation, the government has to collect some revenue from the farmers for maintaining the tanks. A majority of the farmers have not contributed their share or participated in the tank rehabilitation programme, as they are found with other sources of irrigation (bore wells) and incomes. They have to be educated about the long term benefits of tank irrigation and motivated to take care of tanks.

People in the villages have indicated that the NGOs involved in tank rehabilitation have acted as contractors and executed the woks with hiring untrained staff. This is one of the reasons for carrying out substandard works. Instead of involving inexperienced NGOs, the line departments such as Water Resources and Public Works Departments (PWDs) should be involved in carrying out quality works. Although the Zilla Panchayats Engineering and Minor Irrigation Departments are custodians of besides being technical departments and specialized in irrigation systems, they have been kept out of the tank rehabilitation process. While rehabilitating tanks, they have to be involved at all stages.

Policies for the Proper Selection of Tanks for Rehabilitation

Tanks with less command area have been taken up for rehabilitation. In future, tanks with less water spread area and more command area should to be given priority over other tanks. The views of the local farmers on tank fillings, dependability of tanks for irrigation and drinking purposes, and the feasibility and effects of desiltation of tanks are found to have ignored. This is one of the major reasons for many tanks non filling even after rehabilitation. Therefore, the departments concerned have to collect this information every year, so as to give priority to those tanks with a good storage capacity every year. In many villages, the community wanted more silt to be removed but this was rejected by the bureaucracy. In other words, the village communities have not prioritized the works. Majority of the tanks are found to have been selected by the government and carried out under the directions of the government and NGOs. This has to be minimized.

Suggestions for the Improvement of Tank Structures

Due to inclusion of many tanks related activities under the tank rehabilitation programme less amounts were available for the improvement of tanks. Some of the components could have been left to the government to implement under other programmes, so that the entire budget on tanks would have gone to the improvement of tanks. All physical structures have been repaired at high cost without a proper assessment of their strengths. According to the needs of tanks, the repairs have to be prioritized. The tanks already rehabilitated under other programmes were again taken up for restoration. Despite, there is no significant improvement of tanks except duplication of works and waste of resources without tangible benefits to the community. This type of wastage should be avoided by getting information on already rehabilitated tanks from the departments concerned.

Various demonstrations and training programmes across various income generating activities such has fisheries and animal husbandry and artisan work have been conducted for the villagers under tank rehabilitation programme. These components are unnecessary as the Departments of Industry and Agriculture is carrying out these activities either through the state schemes or central and centrally sponsored programmes.

Other Suggestions

Many well off farmers have their private irrigation source (bore wells). They have not shown any interest in the tanks nor participated in the development activities of village tanks. Such farmers should be denied tank water, so that small farmers could benefit from the available water. In some of the districts, the district administration has issued ordered preventing the use of tank water with a view to improving ground water levels so as to make available drinking water for both human beings and livestock without considering the agricultural production. Therefore, the district administration has to review the policy of closure of tank sluices in the light of JSYS tank rehabilitation programme to help several small and marginal farmers under the tanks who are heavily dependent on tanks for their livelihood and enhance agricultural production. A major proportion of the command area of several tanks has not been cultivated for several years as the tanks have not filled. There is a need to educate them to undertake cultivation of crops in the command area as under rain fed conditions without leaving them fallow, so that one can produce some foodgrains under dry land agriculture.

Under several programmes such as employment generation activities and area development programmes, importance was given to works which were not sustainable. These programmes should focus first on tanks rather than other works. The implementers of the tank rehabilitation programme claims that the programme has made good progress benefiting several groups /stakeholders in the villages. In fact, this is far from truth. There is some kind of exaggeration by the NGOs and the JSYS. Added to this, there has been given a wide publicity of tank rehabilitation by way of putting name boards and write-ups on the walls of public buildings in the villages. Instead of this approach, focus should be made on achieving the tangible results.

REFERENCES

Government of Karnataka. 1977. 'Project for Desiltation of Tanks for Improved Irrigation and Development of Fisheries with the Assistance of World Food Programme', Project Formulation Division Bangalore, Planning Department, December.

———. 1993. 'Stagnation of Agricultural Productivity in Karnataka during 1980s', Report of the Expert Committee, Government of Karnataka, Bangalore.

Raju K.V., and J.H. Paramesha. 2002. 'Herehalla Watershed Project in Shimoga: A Concurrent Evaluation'. Bangalore: ISEC, Mimeo.

27

Impact Evaluation of Xth Plan

National Watershed Development Project for Rainfed Areas: Impact Evaluation of Rajasthan*

V.D. Shah

Rainfed areas constitute about 63 per cent of 142 million hectare arable land in India, which contributes only 45 per cent of total foodgrains output. The green revolution considered as cornerstone of India's agricultural growth bypassed the development of rainfed areas and remained confined primarily to the irrigated tracts. This neglect of rainfed areas created unintended agricultural, ecological and socio-economical imbalance between irrigated and rainfed areas. After 1980s, government realized these disparities and felt that the target of 4 per cent agricultural growth as envisaged in the National Agriculture Policy (NAP) is not possible to achieve without integrated development of rainfed areas on a sustainable basis. The NAP seeks to improve rural livelihood and reinforce farming system/productivity of vast rainfed areas through adoption of holistic participatory watershed development approach. Watershed approach is a low-cost location specific technology. It is a vehicle for enhancing the productivity and simultaneously preserving the natural resources of rainfed areas. Watershed comprises three features—arable land, non-arable land and network of natural drainage lines. Technically, watershed is an area with a common drainage point implying that rainwater falling within watershed flows through one or more natural courses and converges at a common point. Watershed development refers to conservation, regeneration and judicious use of all natural resources like land, water, plants, animal and human resources within watershed areas.

For the overall development of rainfed areas on a sustainable basis, a large number of projects based on watershed approach are being implemented by the state/central government, NGOs, externally aided projects and privately by local communities. Among these projects, the largest project in terms of scope is 'The National Watershed Development Project for Rainfed Areas (NWDPRA)', being implemented by the Ministry of Agriculture and Farmers Welfare (MoAFW), Government of India (GoI). The NWDPRA was launched in 1990–91 in 25 states and 2 UTs and continues to be implemented during IXth plan. On the basis of lessons learnt and experience gained, NWDPRA was restructured in the year 2001. For bringing uniformity in implementation approach of watershed programmes among various agencies,

* This is a part of research project undertaken and report submitted to the Ministry of Agriculture, Government of India in 2010.

'WARASA Jan Sahbhagita (WJ)' guidelines were formulated and adopted jointly by the MoA and the Ministry of Rural Development (MoRD), GoI. During the Tenth Five Year Plan period (2002–07), the revised NWDPRA was implemented as per WJ in 6315 blocks spread over 28 states including Rajasthan and 2 UTs covering an area of 23.30 lakh hectares by spending about Rs 1148 crore. NWDPRA is now planned, implemented, monitored and maintained by watershed communities themselves. In WJ, 'bottom-up management approach' is adopted for organizing NWDPRA.

All the community development blocks having less than 30 per cent cultivated land under assured means of irrigation qualify for selection under Xth Plan NWDPRA project. The maximum permissible unit cost in NWDPRA was Rs 4500 per ha for area with less than 8 per cent slope and Rs 6000 per ha for area with slope greater than 8 per cent. Of the total project fund, 22.5 per cent allocated was for Management Component, 50 per cent for Natural Resource Management (NRM), 20 per cent for Farm Production System (FPS) and 7.5 per cent for Livelihood Support System (LSS).

In Rajasthan, in view of many adversities such as scanty and erratic rainfall, limited irrigation, desert/semi-desert conditions in large areas, frequent drought conditions etc., soil and water conservation, efficient use of water resources and harvesting rain-water through adoption of NWDPRA assumes vital importance. In 2007, under different programmes such as CDP, DDP, DPAP, IWDP, TAD, NWDPRA, etc., total 8642 watershed projects were under implementation in Rajasthan. Under Xth Plan NWDPRA, total 1138 watersheds covering 545496 ha of 201 blocks spread over 31 districts were taken up by spending about Rs 202 crores. It was implemented with 90 per cent of the central and 10 per cent share of the state. Up to May 2004, Xth Plan NWDPRA was implemented in Rajasthan by the Directorate of Watershed Development and Soil Conservation (WDSC) and thereafter by Ministry of Rural Development and Panchayati Raj (MoRD & PR). As revised NWDPRA (Xth Plan) was completed by the end of 2006–07, it became imperative to examine and to document effectiveness, relevance, sustainability, and socio-agro-economic impact of NWDPRA.

OBJECTIVES

This study was conducted with specific objectives as under:

1. to evaluate the overall impact on agro-economic, socio-economic parameters due to intervention of the Xth Plan NWDPRA programme;
2. to assess the qualitative performance of the NWDPRA programme;
3. to locate the problems/constraints faced at different stages of the programme and to suggest remedial strategies to eliminate it;
4. to have suitable policy implications, if need be.

DATA AND METHODOLOGY

Selection of Watersheds and Sample Households

Four sample watersheds falling in four districts and agro-climatic zones were selected purposively. From each selected watershed, 40 beneficiary households (B) and 40 non-beneficiary households (NB) comprising marginal (MF, 1 < ha), small (SF, 1–2 ha), medium (MF, 2–4 ha), big/large (BF, >4 ha) and Landless farmers (LL) were selected randomly (Table 27.1). Through well structured schedules, the field data were collected by recall from sample households for pre-project year 2001–02 and project ending year 2006. The secondary data were collected from concerned government offices and websites.

Table 27.1 Details of Sample Watersheds and Sample Households (HH)

Watershed (District)	Selected Block	Selected Villages	B/NB	Sample HHs. (Nos.)					
				MF	SF	MD	BF	LL	T
Kirap (Ajmer)	Masuda	Kirap	B	8	8	8	8	8	40
			NB	8	8	8	8	8	40
Sakariya (Chittorgarh)	Chhoti Sadari	Sakariya	B	8	8	8	8	8	40
		Sandikheda	NB	8	8	8	8	8	40
Modak-VI (Kota)	Kherabad	Dhuniya	B	8	8	8	8	8	40
		Nimana	NB	8	8	8	8	8	40
Dhar (Udaipur)	Badgaon	Dhar	B	8	16	8	–	8	40
		Badanga	NB	8	16	8	–	8	40

The difference between post-project and pre-project parameters reveals mixed impact of NWDPRA plus non-NWDPRA activities. The change in parameters for non- beneficiary households shows impact of non-NWDPRA activities. In order to ascertain realistic impact of NWDPRA, changes observed for beneficiary households are compared with changes observed for non-beneficiary households.

RESULTS AND DISCUSSION

Profile of the Selected Watersheds

In all the four selected watersheds, rainfall is variable, scant, erratic, and recharge level of wells, sown area, and crop-productivity is dependent on the pattern of rainfall. Soil erosion occurs through wind and water. Soils have low level of productivity. It is poorly drained with poor recharging capacity. Soil slope is below 8 per cent in all watersheds. Soil type is clay loam or sandy loam. The primary occupation of majority of the households was agriculture and secondary occupation was dairy and wage labour.

Socio-Economic Characteristics of Sample Households

From the total 320 sample households, more than 82 per cent are either ST or SC. The average family size for beneficiary households ranged between 5.68 in Kirap and 7.58 persons in Dhar. For non-beneficiary, it was below 6 persons. In selected watersheds, about 42.19 per cent family members of beneficiary as well as non-beneficiary households were illiterate. Only 0.78 per cent of total members of beneficiary households had education up to higher secondary and above. In selected watersheds, majority of sample households had agriculture as principal and dairy/livestock rearing as subsidiary occupation.

Agro Economic Impact of NWDPRA

Operational Landholdings

The average size of operational landholdings of beneficiary households in 2006–07 worked out to 2.42 ha in Kirap, 2.29 ha in Sakariya, 2.55 ha in Modak-VI and only 1.45 ha in Dhar watershed. For NB households, it worked out 2.08 ha in Kirap, 2.45 ha in Sakariya, 3.26 ha in Modak-VI and 1.45 ha in Dhar. In each watershed, category-wise average size of landholding in pre-project year 2001–02 and in 2006–07 remained almost the same.

Net Cropped Area (NCA)

In Kirap and Modak-VI watersheds, proportion of NCA to size of landholding was 97 per cent or more for both beneficiary and non-beneficiary households. In Dhar, on account of sloppy and hilly soil, nearly 30 per cent operational land of beneficiary and 40 per cent of NB households turned as permanent fallow. As a result, net cropped area reduced significantly for both beneficiary and non-beneficiary households.

Crop Pattern

(a) **Kirap watershed:** Compared to base year 2001–02, beneficiary as well as non-beneficiary households recorded marginal difference in respect of area under kharif crops. The area under rabi crops increased by 3.08 ha for beneficiary as against 3.56 ha for non-beneficiary households. The beneficiary households increased the irrigated area by 2.59 ha as against 3.43 ha by non-beneficiary. This clearly indicates negligible role of NWDPRA in expansion of irrigated area.
(b) **Sakariya watershed:** Compared to base year, beneficiary households increased the area allocation to more remunerative and higher moisture/water demanding crops such as soybean and groundnut in 2006–07, whereas, in case of non-beneficiary, it remained stable for soybean and declined for groundnut. In 2006–07, beneficiary households increased area under rabi crops and GCA by about 9 per cent. For non-beneficiary households, it was meager. The non-beneficiary households were able to put additional area under irrigated wheat and rapeseed in 2006–07. This indicates that NWDPRA intervention impacted positively and affected crop-diversification and shift in crop- pattern.
(c) **Modak-VI watershed:** As compared to 2001–02, beneficiary households increased area under rabi crops and GCA by 13.60 ha and 18.60 ha, respectively. For NB households, it was only 3.24 ha for rabi crops and 3.56 ha for GCA. The beneficiary households recorded 91 per cent increase in area under irrigation as against only 22.04 per cent by NB households. These observations clearly demonstrate that NWDPRA activities helped beneficiary households in enhancing irrigation potential and subsequently crop area under irrigation.
(d) **Dhar watershed:** As compared to pre-project year, beneficiary households increased area under rabi crops by 5.26 ha in 2006–07 as against 1.51 ha by

non-beneficiary households. Almost similar trend was observed for GCA. In 2006–07, 20.13 per cent area under kharif was irrigated by beneficiary households as against only 3.03 per cent by non beneficiary households. This illustrates that NWDPRA intervention impacted positively irrigated and cropped area.

Cropping Intensity

In selected watersheds, compared to 2001–02, cropping intensity recorded notable increase in 2006–07 for beneficiary as well as non-beneficiary households. However, this increase in percentage and absolute terms was much higher for beneficiary households. The intervention of NWD-PRA improved the water availability for irrigation and subsequently it helped beneficiary households to increase cropped area and area under supplementary irrigation. Thus, NWDPRA impacted favourably cropping intensity.

Cost of Cultivation

As compared to the year 2001–02, average cost of cultivation per hectare in 2006–07 for beneficiary households moved up by 58.80 per cent in Kirap, 43.56 per cent in Sakariya, 48.29 per cent in Modak-VI and 81.97 per cent in Dhar watershed. For non-beneficiary, it ranged between 43.25 per cent in Kirap and 86.10 per cent in Dhar. The substantial increase in cost of cultivation was primarily due to higher use of costly inputs such as HYV seeds, fertilizer, higher rate of application of inputs, and increase in input and labour rates. Thus, watershed treatments brought changes in pattern of input use and also enhanced the cost of cultivation.

Yield of Crops

In all the four watersheds, compared to base year both, beneficiary and non-beneficiary farmers achieved higher yield for all crops (barring few cases) in 2006–07. In Sakariya, incremental yields achieved by beneficiary farmers varied from 35.96 per cent for gram to 188.46 per cent for isabgul. For non-beneficiary, it varied from 3.98 per cent for gram to 100 per cent for isabgul. In Kirap, for beneficiary farmers, it varied from 23.07 per cent for bajra to 58.18 per cent for udad. For non-beneficiary, it varied from −22.50 per cent for gram to 38.74 per cent for jowar. In Modak-VI, yield increment for beneficiary households varied from 15.01 per cent for soybean to 90.02 per cent for jowar. In Dhar also, increment in yield (except gram) obtained by beneficiary households was far superior for each crop as compared to the same for NB households. In all 4 watersheds, NWDPRA had noticeable positive impact on crop-yield. However, scale of impact varied across watersheds due to variation in soil-climatic conditions, soil-moisture level, terrain, rainfall, input use, etc.

Gross Value of Farm Production, Net Farm Income, and Output–Input Ratio

In all four selected watersheds, as compared to base year, value of gross farm produce per hectare of cropped area shot up sharply for both, beneficiary and non-beneficiary households. For beneficiary farmers, it moved up by 73.45 per cent in Kirap, 111.21 per cent in Sakariya, 175.62 per cent in Modak-VI and 63.92 per cent in Dhar watershed. For non-beneficiary households, it ranged between 51.92 per cent in Kirap and 117.76 per cent in Modak-VI. The significant upsurge in the value of gross produce for beneficiary households was basically due to higher farm harvest prices and higher yield. In all four sample watersheds, net farm income per hectare of GCA and output–input ratio (except Dhar) for beneficiary and non-beneficiary households in 2006–07 were found much higher than those in 2001–02. Further, net farm income and output input ratio for beneficiary households was found substantially higher than those for non-beneficiary households. This suggests that intervention of NWDPRA contributed in generating higher net returns from farm enterprises.

Improvement in Water Level

As compared to non-beneficiary, net increase in water table in case of beneficiary households was more than 4.43 feet in kharif, 1.88 feet in rabi and 0.62 feet in summer season. This clearly illustrates that water conservation technology adopted under NWDPRA is most effective in recharging of rainwater. This improvement in water table eases the problem of drinking water in watershed community.

Livestock, Cattle Grazing and Milk Production

As expected, in selected watersheds number of milch animals and livestock increased moderately in 2006–07. In each watershed, the percentage of beneficiary households reporting increase in milk production and productivity in 2006–07 has been found moderately higher than non-beneficiary households. This indicates positive contribution of NWDPRA in enhancing fodder availability and milk yield.

Farm Employment Generation and Outmigration

In selected watersheds, requirement of human labour for farming sector recorded noticeable upsurge in 2006–07. Compared to 2001–02, beneficiary households in 2006–07 generated per ha/annum additional farm employment of 42 man days in Kirap and Sakariya, 36 man days in Modak-VI and 56 man days in Dhar watershed. Additional farm employment generation was observed relatively low for non-beneficiary households. This indicates that NWDPRA is capable in generating additional farm employment. In selected watersheds, average period of out-migration in 2006–07 reduced by 36 per cent for beneficiary households as against 24 per cent for NB. This shows that NWDPRA impacted favourably and reduced time period of outmigration.

Economic Appraisal

Economic appraisal of the project was carried out using discounted cash flow method. Using 10 per cent discount rate, Benefit–Cost Ratio (BCR), Internal Rate of Return (IRR) and Net Present Value (NPV) have been worked out for 10 and 20 years time horizon and presented in Table 27.2. It is evident that BCRs, IRRs and NPVs worked out for 20 years horizon are higher than 10 years time horizon. For each selected watershed, IRR is greater than opportunity cost of capital and BCR is greater than one. This clearly indicates that investment in NWDPRA is economically attractive and viable. A positive and high NPV for each sample watershed implies positive worth of the project in generating returns in excess to costs.

Implementation Constraints and Suggestions

The nature of constraints observed and suggestions for elimination of the same are given below:

Table 27.2 BCR, IRR, and NPV of Selected Watersheds (Discount rate: 10 per cent)

Item	Time Horizon	Kirap	Sakariya	Modak-VI	Dhar
BCR	10 Yr.	3.50	3.82	9.02	1.17
	20 Yr.	5.31	5.68	13.14	1.68
NPV (Rs in lakh)	10 Yr.	51.78	60.05	83.11	16.17
	20 Yr.	82.64	94.60	135.70	23.28
% IRR	10 Yr.	53	62	144	16
	20 Yr.	59	64	144	23

S. No.	Problem/Constraint	Problem and Suggestion
1	Per hectare unit cost of Rs 4500 (<8% slope) and Rs 6000 (>8% slope) is very low	During Xth plan period, minimum wage rate, material cost and hiring cost of skilled human resources, etc. recorded abnormal increase. Due to heavy increase in input cost, activities undertaken in the mid/later phase of the project, failed to meet target. Several works completed at this stage were found inadequate in size and scale and of substandard quality. This ultimately reduced the effectiveness of such works. In this situation, it is desirable to review cost norm and to fix it at a realistic level (around Rs 12000–14000/ha). Also, make provision to review cost norm regularly at the interval of every two years or so.
2	Delay in fund transfer at grassroots level	It may be observed that after release of funds by MoAFW, GoI, it takes 2–4 months to reach at WC/WDT. This delays timely implementation of the project. Therefore, it is desirable to evolve a system which ensures timely transfer of funds to WC/PIA/WDT.
3	No provision for project survey	It is necessary to conduct survey at the initial stage of the project for deciding an appropriate location and type of treatments required to be given for generating maximum benefits for the watershed community. Conducting survey and preparation of such project report requires some expenses. Hence, there is a need for adequate provision of funds for survey and project work.
4	Internal head-wise reallocation of funds	In WJ guidelines, there is a provision for reallocation of the budget to the extent of 10 per cent from one subcomponent to another. In Rajasthan, NRM is very important. Majority of beneficiaries of NWDPRA are demanding for NRM works. There are some sub-heads where there is no scope of work and have surplus funds. Hence, it would be appropriate to increase fund reallocation limit from present 10% to 25%.
5	Contribution under Plantation is high	In NWDPRA, planting of Agro forestry/horticulture trees envisage for 50 per cent contribution by participants which they viewed as too high. Under similar Haryali project, the same is 10 per cent for general and 5 per cent for SC/ST farmers. Therefore, it should be like Haryali.
6	Low Fund Allocation for NRM works	In WJ guidelines, there is a provision of 50 per cent fund for NRM works. The requirement of fund for NRM works depends upon the degraded condition, slope and type of land. Looking at the high proportion of degraded land in many watersheds of Rajasthan, higher funds are needed for NRM works. Therefore, there is a need of upward revision of fund allocation for NRM works.

(cont'd)

(cont'd)

S. No.	Problem/Constraint	Problem and Suggestion
7	Inadequate staff at implementation level	The total technical staff available with State Nodal Agency WDSC is about 700. In addition to 1,128 NWDPRA watersheds, they are also looking after other 5000–5500 watersheds of DDP, DPAP, CDP, IWDP, etc. Hence, unfavourable staff ratio at implementation level is adversely impacting timely and effective implementation of the project. Therefore, it is suggested to have separate department with adequate staff strength for better implementation.
8	Extension of capacity building phase in 2nd year	The Nodal Agency is facing a problem to complete the capacity building work of different personnel categories in first year project. Therefore, provision in WJ for tending capacity building phase in 2nd year is suggested.
9	Poor monitoring at state/district level	State Level Watershed Committee (SLWC) and District Level Watershed Committee (DLWC) were formed as per WJ guidelines. SLWC is supposed to meet regularly at least once in six months for periodic review and overall monitoring of NWDPRA in the state. Instead of having at least 10 meetings of SLWC, during entire project period, only 3 meetings were held. This in turn affected the qualitative level and time schedule of the project activities. The same situation was observed in case of district/block level monitoring. Hence, there is a need to strengthen the project monitoring system at state, district, block and watershed level.

CONCLUSIONS AND POLICY IMPLICATIONS

It can be inferred that NWDPRA holds the key to the development of country's vast rainfed areas. It is observed that the programme improved the groundwater aquifers as well as in situ soil moisture. Further, the study reveals that NWDPRA impacted positively irrigation, cropping intensity, crop-pattern, farm employment, fodder and bio-mass, out-migration, etc. It improved the status of land less households and boosted village economy. The NWDPRA is beneficial to watershed community but it lacks certainty regarding its sustainability in future. Though, it is very difficult to identify a single factor, upsurge in crop yields and farm income owing to improvement in irrigation and in situ moisture seem to be the key driving forces for the noticeable performance of NWDPRA.

The study further reveals that quantum of benefits derived were notable but below the desired level. Effecting necessary corrections to eliminate constraints discussed in forgoing analysis, benefit level of the programme can be raised further. The people's participation was found lower at all the stages of the programme. The awareness level about project activities also found low to moderate. This calls for higher efforts to enhance the people's participation at all the stages of the programme, decision making process and particularly in activities related to development of common property resources. Further, additional efforts are needed to raise the awareness level and building capacity of the stakeholders/beneficiaries. Regular arrangement of meetings of WC/WA will bring more transparency. The regular interaction between PIA/WDT/WC and beneficiaries will be helpful in identifying problems and evolving solutions in a participatory mode. Though NWDPRA has an essential component of institutional building, functioning of most of the created institutions (SSGs, UGs) was found weak. FPS, LSS and capacity building activities, WDT/PIA were had paid little attention. Hence, there is a need for WDT, PIA, and WC to provide attention to these aspects. The inclusion/support of local NGOs in the programme will be helpful in reducing implementation problems. Some beneficiaries expressed an apprehension that inadequate attention towards maintenance and repair of created structures is likely to affect impact and sustainability of the programme in future. Therefore, there is an urgent need of evolving effective arrangement on a permanent basis.

REFERENCES

Badal, P.S., Pramod Kumar, and Geeta Basaria. 2006. 'Dimensions & Determinants of Peoples Participation in Watershed Development Programmes in Rajasthan', *Agricultural Economics Research Review*, 19(1): 57–69.

Deshpande R.S., and N Rajasekaran. 1995. 'Impact of NWDPRA in Maharashtra', Gokhale Institute of Politics and Economics GIPE, Pune.

Deshpande, R.S., and A. Narayanamoorthy. 1999. 'Impact of National Watershed Development for Rainfed Areas (NWDPRA): An Analysis across States', AERC, Pune.

Government of Rajasthan. 2005. *Vital Agriculture Statistics 2005*. Jaipur: Department of Agriculture, Government of Rajasthan.

Govind Babu, R.K. Singh, and B. Singh. 2004. 'Socio-Economic Impact of Watershed Development in Kanpur', *Agricultural Economics Research Review*, Vol. 17, October: 23–45.

Majumdar, D.K. and P.P. Saikia. 2006. 'People Participation and Constraints in Watershed Development Programme', Agricultural Situation in India, MoAFW, GoI.

Narayanamoorthy, A., and K.G. Kshirsagar. 2000. 'Rapid Impact Evaluation of NWDPRA in Maharashtra', GIPE, Pune.

Raju, K.V., A. Narayanamoorthy, Govinda Gopakumar, and H.K. Amarnath. 2004. 'State of Indian Farmers—A Millennium Study, Vol. 3, Department of Agriculture and Cooperation, Government of India, New Delhi.

Rao, C.H. Hanumantha. 2000. 'Watershed Development in India—Recent Experience and Emerging Issues', *Economic and Political Weekly*, Vol. 35, No. 45, 4 November: 3943–7.

Shah, Amita. 2000. 'Watershed Programme—A Long Way to Go', *Economic & Political Weekly*, pp. 3155–64.

———. 2005. 'Economic Rational, Subsidy and Cost Sharing in Watershed Project', *Economic and Political Weekly*, vol. XI, NO. 26 25 June–1 July: 2663.

Shah, V.D. 1999. 'Impact of National Watershed Development Project for Rainfed Areas (NWDPRA) in Gujarat', Report No. 112, AERC, Vallabh Vidyanagar.

Shah, V.D., and V.G. Patel. 1996. 'Impact of National Watershed Development Project for Rainfed Areas (NWDPRA) in Gujarat', AERC, Vallabh Vidyanagar.

The Energy & Resources Institute. 2007. 'Mid-Term Evaluation of Tenth Plan Watershed Implemented under NWDPRA in Rajasthan', The Energy & Resources Institute, New Delhi.

28

Impact Evaluation of Watershed Development for Rainfed Areas

A Case of WARASA Jan Sahbhagita

Kali Sankar Chattopadhyay, Debajit Roy, and Ranjan Kumar Biswas

Development, promotion, and management of appropriate watershed technologies in dry land regions have been viewed as major priorities to ameliorate the problem of natural resource degradation. This results in multiple benefits such as ensuring food security, enhancing viability of farming and restoring ecological balance. The present strategy of watershed development programme is to protect and sustain the livelihoods of resource poor farmers who are experiencing production constraints in addition to problems created by soil erosion and moisture stress. Watershed development is to ensure the availability of drinking water, fuel wood, fodder and helps in raising income and employment for farmers and landless labourers through improvement in agricultural productivity and production.

In view of the above, this study has been undertaken to assess the long-term economic impact on agriculture productivity, land use and cover, groundwater recharge watershed system and sustenance of watershed technologies/practices of different states in India. The aspects that have been covered in this study are:

1. community organization and institutional aspects,
2. planning and implementation aspects,
3. environmental, social aspects, economic and institutional aspects,
4. indirect benefit, and
5. overall impacts and sustainability and people's responses.

METHODOLOGY AND DATA

This study covers four states, namely West Bengal, Bihar, Maharashtra, and Rajasthan and all primary data collected across states pertain to the reference year 2006–07. From each of these four states, four numbers of watershed areas/projects were selected purposively belonging to different agro-climatic conditions. From each of the states, a total of 320 households (80 from each watershed) were selected. Thus the total number of sample households for the study works out to be 1,280.

RESULTS AND DISCUSSION

The major results of the study have been discussed here for the all four states as follows.

West Bengal

In West Bengal, it is evident that there is no uniformity in family size in between the selected watershed areas (WP). The literacy rate is higher among males (82.29 per cent) than females (64.47 per cent). In non-watershed (NWP) area literacy rate is lower for both male and female at 71.41 per cent and 55.38 per cent, respectively. The size of land holding is 1.02 hectares and 0.77 hectares in WP and NWP, respectively. It has been found that the farmers in NWP are somehow well equipped with tractor and sprayer than WP.

The average size of holdings in WP is 1.02 hectares comprising of cultivated (operational), cultivable fallow, permanent fallow, home stead, irrigated and non-irrigated area. In NWP, the average size of holding is 0.77 hectares. It indicates that the size of holdings is lower in WP than NWP. Total cultivated area of the sample farms in watershed area is 100.96 hectares, out of which 22.14 per cent is under pond irrigation followed by 1.88 per cent under canal irrigation, 8.40 per cent under STW, 1.23 per cent under other wells and 3.41 per cent under other sources. The non-irrigated area in WP is 62.95 per cent. In NWP, the total cultivated area is 87.42 hectares of which 26.66 per cent of area is irrigated under different irrigational sources followed by 73.34 per cent under non-irrigation. It indicates that the WP area is well irrigated in comparison to NWP area. This could be attributed to impact of watershed on groundwater augmentation in watershed area.

The contribution of watershed as reflected in gross returns from rainfed crops was considered as the dependent variables, since the watershed impact is direct and implicit. Accordingly, gross returns from rainfed field crops in 2007 was regressed on dry land cropped area in hectares (X_1), human labour (X_2), bullock labour (X_3), seeds in Rs (X_4) and fertiliser in Rs (X_5). The adjusted R^2 for the watershed and non-watershed area was 87 per cent and 94 per cent which indicate adequacy of fit of the model (see Table 28.1).

The regression coefficients are the estimates of the elasticity of production with respect to the independent variables. In WP, elasticity coefficient for human labour, bullock labour and fertilizer are 0.02, –0.01, and –0.03, respectively, and are statistically significant at 5 per cent. For land, the elasticity coefficient is 1.01 and significant at 5 per cent. The coefficient for seed is –0.03 and is not significant.

In NWP, variables land and seed are significant and their elasticities are 0.93 and 0.07. For human labour, bullock labour and fertiliser, the elasticity coefficients are 0.06, –0.03, and 0.01, respectively and significant at 5 per cent. The returns to scale are 1.01 and 1.04 in WP and NWP areas, implying constant returns to scale. This shows that the production technology used in watershed and non-watershed is scale neutral.

Table 28.1 Factors Contributing to Gross Returns From Rainfed Field Crops on Sample Farms in Selected Watershed in West Bengal, 2007

S. No.	Variables	Beneficiary		Non-beneficiary	
		Coefficient	t stat	Coefficient	t stat
1.	Log of intercept	4.31	15.55	4.14	20.11
2.	Log of land (acres)	1.01	10.98	0.93	13.05
3.	Log of human labour (Rs)	0.02	0.37	0.06	1.30
4.	Log of bullock labour (Rs)	−0.01	−0.16	−0.03	−1.20
5.	Log of seed (Rs)	−0.03	−0.47	0.07	1.36
6.	Log of fertiliser (Rs)	0.02	0.41	0.01	−0.09
7.	R^2	.93	–	.95	–
9.	Returns to Scale	1.01	–	1.04	–

Source: Primary Data.

The geometric mean levels of gross returns for WP and NWP sample farms are Rs 11500.83 and Rs 11764.65, respectively. The geometric level of inputs land, human labour and bullock, seed, fertilizers are computed both watershed and non-watershed sample farms as 0.49, Rs 2300.87, Rs 413.75, Rs 172.43, Rs 612.60, and 0.48, Rs 2302.69, Rs 418.49, Rs 163.07, and Rs 617.26, respectively in that order. The net returns per farm has been observed to be Rs 189.68, Rs 518.48, and Rs 1057.91 for marginal, small, and medium farms, respectively.

A large number of farmers in WP are rearing livestock on a small scale after the WDP. Farmers expressed during the discussion that due to availability of fodder on farm and common lands, the number of bullocks, cows, buffaloes, sheep, and goat has increased. The net return from livestock per farm and per acre are Rs 24.12 and Rs 38.22, respectively in WP area and Rs 21.42 and Rs 5.15 in NWP area.

The equity in the distribution of income among different categories of farmers due to WDP has been analysed using Gini coefficients. Gini coefficients are computed for marginal, small and medium farms. Gini coefficients for WP and NWP areas are 0.44 and 0.41 for all farms, respectively. This indicates a fairly equitable distribution of income in WP area than that of NWP area.

Rajasthan

In Rajasthan during 2006–07, compared to base year 2001–02, beneficiary as well as non-beneficiary households recorded marginal increase in respect of area under kharif crops and area allocation to different crops in Kirap watershed. The area under rabi crops has also increased. Similarly, beneficiary and non-beneficiary have also registered increase in GCA. The beneficiary households increased the irrigation area by 2.59 hectares as against 3.43 hectares by non-beneficiary households. This gives clear indication of no role of NWDPRA in expanding irrigation area in this watershed.

Compared to base year, beneficiary households increased the area allocation to more remunerative and higher moisture/water demanding crops such as soybean and groundnut in 2006–07 in Sakariya watershed. Whereas, in case of non-beneficiary, it remained nearly stable for soybean and declined to a few extent for groundnut. In 2006–07, beneficiary households increased area under rabi crops and GCA by about 9 per cent. The increase in rabi area and GCA for non-beneficiary households was meagre. Beneficiary households were able to put additional area under irrigated wheat and rapeseed in 2006–07. This clearly indicates that NWDPRA intervention impacted positively on shifting of crop pattern and crop-diversification.

In crop-pattern, soybean and maize among kharif crops and coriander and wheat among rabi crops occupied the dominant position in Modak-VI watershed. As compared to 2001–02, for beneficiary households, increase in area under rabi crops and GCA was by 13.60 hectare and 18.60 hectare, respectively. Whereas for non-beneficiary households, it was only 3.24 hectare for rabi crops and 3.56 hectare for GCA. The beneficiary households recorded 91 per cent increase in area under irrigation, whereas, it was only 22.04 per cent for non-beneficiary households. Compared to non-beneficiary households, higher quantum of incremental area under irrigation and GCA for beneficiaries clearly demonstrates positive impact of NWDPR activities on irrigation and crop-pattern.

As compared to pre-project year, beneficiary households increased area under rabi crops by 5.26 hectare in 2006–07 as against 1.51 hectare by non-beneficiary households in Dhar watershed. A similar trend was witnessed in respect of GCA. In 2006–07, 20.13 per cent of kharif crop area was irrigated by beneficiary households as against only 3.03 per cent by non-beneficiary households. This indicates positive impact of NWDPRA intervention on irrigation and cropped area.

In all 4 watersheds, compared to base year 2001–02, cropping intensity recorded notable increase in 2006–07 for beneficiary as well as non-beneficiary households. However, this increase in percentage and absolute term was much higher for beneficiary households. The NWDPRA intervention improved the ground water aquifers and soil-moisture which subsequently helped beneficiary households to increase double cropped areas and supplemental irrigation. This helped beneficiary households in enhancing cropping intensity.

As compared to 2001–02, the overall average cost of cultivation per hectare in 2006–07 for beneficiary shows an increase of 58.80 per cent in Kirap, 43.56 per cent in Sakariya, 48.29 per cent in Modak-VI and 81.97 per cent in Dhar watershed. For non-beneficiary, it ranged between 43.25 per cent for Kirap and 86.10 per cent for Dhar. The increase in cost of cultivation was mainly due to higher use of costly inputs such as HYV seeds, fertilizers, higher rate of application of inputs and increase in input prices. Thus, watershed treatments brought changes in use pattern of inputs and also enhanced cost of cultivation. In total cost of cultivation, most important items were human labour, bullock labour and machine labour.

In all the 4 watersheds, compared to base year, beneficiary and non-beneficiary farmers achieved higher yield for all crops (barring few cases) in 2006–07. At the same time, value of gross produce per hectare of cropped area shoot up sharply for both, beneficiary and non-beneficiary households. Overall, for beneficiary farmers, it went up by 73.45 per cent in Kirap, 111.21 per cent in Sakariya, 175.62 per cent in Modak-VI and 63.92 per cent in Dhar watershed. For non-beneficiary households, it ranged from 51.92 per cent in Kirap to 117.76 per cent in Modak-VI. The significant upsurge in the value of gross produce was mainly due to higher farm harvest prices and higher yield achievement. Also, net farm income per hectare of GCA and output–input ratio (except Dhar) for beneficiary and non-beneficiary households in 2006–07 were found much higher than those in 2001–02. Further, net farm income and output input ratio for beneficiary households was found substantially higher than those for non-beneficiary households. This suggests quite positive impact of NWDPRA on net return from farm enterprise.

In selected watersheds, as compared to 2001–02, the average annual net income per household from various sources recorded impressive upsurge in 2006–07, for both, beneficiary and non-beneficiary households. For beneficiary, increase was Rs 25427 in Kirap, Rs 16068 in Sakariya, Rs 37270 in Modak-VI and Rs 13819 in Dhar. The corresponding numbers for non-beneficiary were Rs 14489, Rs 11144, Rs 25745 and Rs 10196, respectively. The sharp increase in the net annual income per beneficiary households shows positive impact of NWDPRA on livelihood security of different stakeholders of the watersheds.

As compared to base year 2001–02, the average rise in water level in wells during Kharif 2006–07 recorded by beneficiary households ranged from 7.03 feet in Dhar watershed to 8.55 feet in Kirap watershed. During summer, it ranged from 1.88 feet in Dhar to 2.66 feet in Sakariya watershed. As compared to non-beneficiary, net increase in water table for beneficiary households was more than 4.43 feet in Kharif, 1.88 feet in rabi and 0.62 feet in summer season. This clearly indicates that water conservation technology adopted under NWDPRA is effective. This improvement in water table situation eased the drinking water problems of watershed community to some extent.

In all 4 selected watersheds, as compared to base year, the proportion of beneficiaries as well as non-beneficiaries who adopted various improved farming practices is found higher in 2006–07. As compared to non-beneficiary households, the adoption rate was found moderately higher for beneficiary households which indicates positive impact of NWDPRA on adoption of improved farm technology.

As expected, in all selected watersheds, number of milch animals and total number of livestock increased moderately in 2006–07.

In selected watersheds, requirement of human labour for farming sector shows noticeable upsurge in 2006–07. Compared to 2001–02, beneficiary households in 2006–07 generated per hectare per annum additional farm employment of 42 man days in Kirap and Sakariya, 36 man days in Modak-VI and 56 man days in Dhar watershed. Additional farm employment generation was observed relatively very low for non-beneficiary households.

The perceptions of beneficiaries indicate that most of the indicators determining the quality of life are showing positive changes in all the selected watersheds. Beneficiaries reported moderate improvement in transportation, communication, educational facilities. They also reported moderate to high positive changes in respect of farming aspects, irrigation and household income. The impact has been found positive but somewhat below the expectation in respect of out-migration, availability of drinking water, etc.

Using 10 per cent discount rate, BCR, IRR, and NPV have been worked out for 10 and 20 years time horizon. For 10 years horizon, Benefit–Cost Ratio (BCR) was 3.50 for Kirap, 3.82 for Sakariya, 9.02 for Modak-VI and 1.17 for Dhar watershed. And the Net Present Value (NPV) was Rs 51.78 lakhs for Kirap, 60.05 lakhs for Sakariya, 83.11 lakhs for Modak-VI and 16.17 lakhs for Dhar watershed. The Internal Rate of Return (IRR) was 9 per cent for Kirap, 62 per cent for Sakariya, 144 per cent for Modak-VI and 23 per cent for Dhar. BCR, IRR, and NPV worked out for 20 years horizon are higher than 10 years time horizon. For each selected watershed, IRR are greater than opportunity cost of capital and BCR are greater than one which clearly indicates that investment on NWDPRA is economically very attractive and viable. A positive and high NPV for each sample watershed implies positive worth of project in generating returns in excess of all costs.

Bihar

In Bihar, the work activities commenced in 2002–03 and completed in 2006–07. Land and water resource development activities constitute the primary areas of intervention. The expenditure on management constitutes about 18.38 per cent whereas 81.62 per cent incurred on development components, which includes resource management (51.64 per cent), farm production system for land owning families (20.58 per cent) and livelihood support system for landless families (9.10 per cent). The impact of the project on various items may be briefly seen as follows.

In WS-I, the area under private wasteland decreased by 16.67 per cent indicating development of waste lands by way of plantation, etc. the benefits from which would also be available to the non-landholders. Similarly in WS-II, the area under govt. wasteland and private wasteland decreased by 15.00 per cent and 22.00 per cent, respectively, which reveals that community as well as private waste land by 21.92 per cent and 21.43 per cent and 31.44 per cent, respectively have been found, clearly indicating increase in community and private plantations.

The change in irrigational status of agricultural land in 2006–07 over 2001–02 of the watershed indicate marginal increase in irrigated area in all the selected watersheds and almost in all the crop seasons, which may be due to increase in number of water harvesting structures (tanks, check dams, ponds, etc.). The increase was mainly found to big farms, which showed that perceived benefits are concentrated on large farms. Of course it is not a new concern. In fact, it needs group owned water harvesting structures in real sense rather jointly owned by own relatives/neighbours or raiyets. The approach to sharing the benefits of water harvesting structure among the resource poor farmers is to develop well, which has been found important sources of irrigation.

The land development and creation of new water harvesting structures in all the watershed areas have not much effectively brought some additional areas under the important crops both in kharif and rabi. The data indicate that there is increase in the area under paddy crops from 0.64 per cent to 4.37 per cent, maize 0.65 per cent to 3.37 per cent, pulses 0.99 per cent to 2.08 per cent and oilseeds up to 1.85 per cent. Of course, there is increase in area of important crops but it is not much appreciable. It is worth to mention here that almost similar increase has been indicated by the non-beneficiary respondents.

In regard to production, it increased from 1.11 per cent to 4.87 per cent in case of paddy, 1.25 per cent to 6.97 per cent in case of wheat, 2.28 per cent to 6.61 per cent in case of maize, 1.24 per cent to 3.97 per cent in case of pulses and oilseeds witnessed negative growth. The findings indicate that the production increase is higher in rabi season for wheat, pulses and oilseeds across all the watersheds and this indicates the overall effectiveness of the watershed activities. Similarly change was also indicated in case of non-beneficiary respondents, which related that benefits were not centered on the beneficiaries rather shared with non-beneficiaries also.

It is generally presumed that if the facilities are extended to farmers, the cost of the production of the crops will come down provided the prices of the inputs are constant. But things are different. Neither the cost fallen nor is the prices of any inputs constant. Among the beneficiary farmers, it rose at the overall level to 8.16 per cent in WS-I, 5.54 per cent in WS-II, 4.38 per cent in WS-III and 13.08 per cent in WS-IV. Among the non-beneficiary farmers, it increased to 8.53 per cent in WS-I, 12.36 per cent in WS-II, 12.39 per cent in WS-III and 5.16 per cent in WS-IV. The reason for increase in cost of cultivation is mainly due to increase in prices of the inputs like fertilizer, irrigation, seeds, etc. The watershed development programme could not slash to the cost of production. The reason is obvious lesser the impact of the programme.

The disposal for all the crops level in WS-I is lower among the beneficiary households. However it is a bit higher among the non-beneficiary households. The reason behind low disposal may be lower production. Among the beneficiary households, the percentage of disposal is comparatively higher across all the three watersheds, for namely, 34.47 per cent in WS-II, 18.82 per cent in WS-III and 19.86 per cent in WS-IV. It is by 0.39 per cent in WS-I, 6.46 per cent in WS-II, 17.15 in WS-III and 21.93 per cent in WS-IV among the non-beneficiaries households. It revealed that the volume of disposal has increased, which may be due to distribution of benefits amongst the households or villagers.

The total average income of beneficiary group has increased in all the sample watersheds but it recorded higher in WS-III (25.24 per cent) followed by WS-II (19.22 per cent), WS-IV (11.30 per cent) and WS-I (0.31 per cent). Almost similar is the case of non-beneficiary group. It increased by 23.18 per cent in WS-IV followed by 14.72 per cent in WS-I, 5.13 per cent in WS-II and 2.56 per cent in WS-III.

The data suggest in all watersheds milk and meat generating animals/birds are kept by a large number of families to supplement their food items and cash resources, while cows and buffaloes are kept for sourcing domestic milk consumption of children and course for generating income. In all the selected watersheds the total number of livestock increased. It increased as much as 73.00 per cent in WS-I, 30.74 per cent in WS-IV, 21.32 per cent in WS-III and 10.78 per cent in WS-II. It reveals that the project has facilitated in keeping larger number of livestock. But in absence of clear and agreed livestock holding and grazing practices there cannot be favourable long term impact on conservation of common land resources.

The perception of beneficiary farmers indicate that positive changes have taken place in recharging of groundwater level and qualitative aspects of livelihoods by about 15.00 to 20.00 per cent across the watersheds. Irrigation, afforestation and availability of irrigation have changed positively to the tune of 17.50 per cent, absorption of women in various activities (7.50 to 15.00 per cent), production (10.00 to 15.00 per cent), cropping intensity (7.50 to 10.00 per cent) etc. Non-beneficiary farmers also indicated positive change of the programme on improvement in groundwater conditions (7.50 to 15.00 per cent), qualitative aspect of livelihood (5.00 to 12.50 per cent), production (2.50 to 7.50), availability of irrigation (5.00 to 15.00 per cent). The analysis reveals that there is a general improvement in quality of life but in overall sense, the impact of the programme in these watersheds has been somewhat lower.

In the initial years of the programme no UGs/SHGs could be formed in any of the sample districts, which may be due to delay in launching of the programme. These could be formed after 2003–04. SHGs formed by landless and women particularly of SCs received sewing machines, she-goats, leaf plate making machine, dhankutti machine, etc. for undertaking non-farm group activities. 3 to 4 training programmes relating to know-how of the programme and land management practices are organized across all the watersheds. But due to poor knowledge, skill and now level of maintenance of the assets substantial support to the livelihood has not been found.

The overall approaches of all the PIAs have been to implement the plan/activities within the prescribed budget limit with almost no planning for user groups. The WDT is not effective in the area of community organization. However, they all have performed well in terms of level of achievements of physical (93% and above in number and 83% and above in overage) and financial (98% and above).

In fact, there is no single indicator of successful watershed development, so the most feasible approach is to compare the performance of a variety of indicators, which also reflect the diversity of project objectives. It is noteworthy that the cost per hectare is helpful in assessing their cost effectiveness. It is calculated at Rs 8213/hectare in WS-I, Rs 8144/hectare in WS-II, Rs 7103/hectare in WS-IV and Rs 6561/hectare in WS-III. The programme has significant positive impact on creation of employment opportunities. It has been created about 7142 man days in Ws-I to the highest of 8915 of man days in WS-III. The internal rate of return calculated on the basis of the additional income over and above the pre-project income from agriculture, micro-enterprises, wages etc. within the village, varies from 187.00 per cent to 202.00 per cent (average of 4th and 5th year) across the sample watersheds. The cost and benefit ratio also varies from 1:1.87 to 1:202. The average employment generation per hectare works out to 12.75 man days in WS-I, 14.80 man days in WS-IV, 16.31 man days in WS-II and 17.58 man days in WS-III. The quantitative impact on productivity of the crops indicates that expect pulses (–2.55 per cent) in WS-III, the productivity of major crops have noted positive change but in case of cereals, pulses (–)2.55 per cent to 10.44 per cent, oilseeds from 0.59 per cent to 6.78 per cent and vegetables and others form 0.19 per cent to 2.40 per cent across the watersheds. The cropping intensity has fallen by 4.72 per cent in WS-I. No change has been found in WS-IV. As regards the income benefit it has increased from 8.22 per cent to 13.28 per cent per hectare per annum. Similarly annual per hectare family income has also increased from 5.45 per cent to 10.49 per cent across the sample watersheds. However, its equity depends on the magnitude of the households of the area. Positive change has also been found in case level of groundwater and coverage of green/biomass in the villages.

Maharashtra

In Maharashtra, watershed changed the status of the rain fed agricultural land into irrigated land and thus, paved the way for enhanced agricultural productivity, employment and income of the farmers in the villages covered the selected watershed. Enhanced irrigation potentiality has been created due to watershed and visible increase in the area of cultivation has taken place in all the watersheds. Watershed has positive impact in the beneficiary villages as it ensures assured sources of drinking water facilities to the stakeholders.

Among the four selected watershed, watershed-I (Kolhapur) manifest a remarkable progress do far as various live stock position is covered during the period 2001–02 to 2006–07, increase of cow calf is by 94.84 per cent followed by Buffalo (74.43 per cent), Goat (71.67 per cent) and Sheep (70.83 per cent). In the watershed-II (Nagpur) the increase of Goat in 138.23 per cent followed by Buffalo calf (115.62 per cent). In watershed-III (Raigarh) during the period 2001–02 to 2006–07, the increase of cow calf is by 100 per cent followed by buffalo calf (50 per cent). Similarly, in watershed –IV (Nanded) the number of cows has increased by 33.33 per cent followed by bullock (25 per cent).

Though the basic facilities of medical services and post offices are found in all most all beneficiary villages but it is deplorable that expect the watershed-I (Kolhapur), we find that in all most all other watersheds there is conspicuous absence of latrines facilities.

It reveals from the observations that the watershed beneficiary villages have recorded impressive growth in terms of crop production recorded impressive growth in terms of cost of cultivation. In the watershed beneficiary villages the marginal farmers have impressive growth of marketable surplus during 2001–02 to 2006–07.

With regard to percentage change in the annual income in the 'before' the operation of watershed and 'after' its operation, it is revelled that the highest percentage of (146.92 per cent) increased in the annual income has occurred during the period 2001–02 to 2006–07 in the watershed-IV (Nanded) followed by the watershed-II (Nagpur) with 139.48 per cent. the watershed-III (Raigarh) demonstrates a record increase of 192.06 per cent in the annual income during the period 2001–02 to 2006–07, followed by the watershed-II (Nagpur) with 67.24 per cent.

As per the performance indicator of the selected watershed in Maharashtra, it reveals that the highest area has been developed in the watershed-II (Nagpur) (91.01 per cent), followed by the watershed-IV (Nanded) (77.44 per cent). In all the watersheds there has been encouraging number of man days employment generated, the highest position in occupies by the watershed-I (Kolhapur) with 46765 man days, followed by the watershed-IV (Nanded) with 36907 man days. The additional area brought under cultivation also indicates a growing trend the highest position occupied by the watershed-IV (Nanded) with 65 hectare, followed by the watershed-III (Raigarh) with 49 hectare. There are also positive performance indicates with regard to additional area brought under supplemental irrigation. The watershed-I (Kolhapur) has 142.50 hectare, the watershed–III (Raigarh) has 64 hectare and the watershed-IV (Nanded) has 34 hectare of additional area brought under supplemental irrigation. On the contrary, due to lack benefits accruing to the non-beneficiary big farmers, the productivity in agriculture, crop intensity, irrigation, quality of land, recharging of water, availability of irrigation, absorption of women in various activities, change forestry, literacy level and quality aspects of livelihood all remained standstill.

With regard to crops like cereals, pulses and oil seeds there has been positive co relationship so far as irrigated land and its productivity of these crops are concerned ('x' denote quantity of irrigated land (in hectare) cultivated, 'y' is the production in quintals). The crop-wise co relation shows positive correlation. Since fruits and sugar cane are in the category of cash crops, we have subtracted the figures and also found a positive correlation.

The foregone analysis in assessing the impact of NWD-PRA on the rural agricultural economy of Maharashtra has concluded that watershed developments have greater potential to generate employment opportunities to the rural people. This is due to the increased availability of water resources, diversified cropping pattern including cultivation of labour-intensive vegetable crops and other horticultural crops. This additional employment generation from a watershed program varies across regions depending on the cropping intensity, and the labour-intensity crops grown in that region. This additional employment generation in the villages led to minimizing migration of landless and other labour. Thus, watershed programmes also contributed towards checking migration of rural people to the urban areas. This migration has greater concern for planning and devising rural development strategies. Water shed approach has captured development as a strategy for raising agricultural productivity has been indispensable particularly in dry land areas- one that integrates sectors and provides the foundation for subsequent development.

CONCLUSIONS AND POLICY IMPLICATIONS

In view of the findings of the study, certain concluding observations can be made with respective policy suggestions pertaining to selected states as follows.

In West Bengal, Watershed development programme intervention in natural resource conservation resulted in diversified land use and cover. Therefore, for sustainability of the programme other incentive augmenting rural development programmes could be linked in watershed development programme in phased manner. In the aggregate, the watershed development programme can be considered as an appropriate rural development strategy by implementing all land based rural development programmes under the concept of watershed development programme. Hence, higher budgetary allocation in watershed development programme could be given to dry land horticulture development to maintain the environmental economic goal of maximized net farm income of marginal and small farmers together conserving the ecosystem. Promotion of local institutions through training and education of members for maintenance of water harvesting structures is crucial for sustainability of the watershed development programme.

In Rajasthan, it can be inferred from primary survey data that NWDPRA programme brought very positive changes in respect of irrigation, cropping intensity, crop-pattern, farm employment, fodder and bio-mass, out-migration, status of land less households etc. Though it is very difficult to identify a single key factor, improvement in water availability for irrigation and in situ moisture lead to rise in crop-yields and farm income seem to be the driving force behind the noticeable performance of NWDPRA.

The study in Rajasthan further reveals that quantum of benefits derived were below the expected level. By effecting necessary corrections to eliminate constraints discussed in forgoing analysis, benefit level of programme can be raised further. The participation of beneficiaries was low at the stages of planning, implementation and in village meetings. The awareness level about project activities was also low to moderate. This call for higher efforts to increase the people's participation at all the stages of programme, decision making process and particularly activities related to common property resources. Further, additional efforts are needed to raise the awareness level and building capacity of the stakeholders/beneficiaries.

The NWDPRA is economically very attractive and viable and has succeeded in boosting people's empowerment. The goals of upliftment of farming communities of rainfed areas, equity, employment and food-security would not look distant, if NWDPRA is pursued in earnest. In the years to come, the NWDPRA deserves higher financial allocation and large scale replication in untreated rainfed areas of Rajasthan.

In Bihar, people's participation in watershed activities is poor except in case of wage earners/subsidy beneficiaries. Most of the farmers expressed that improved, certified and guaranteed seeds in addition to enlarging water potential and providing market would usher agriculture in rainfed agro-eco-regions. In fact, people's participation is expected only when provisions of direct benefits to the farmers are made. So watershed activities should be taken up in such a way (PRA and action research) that majority of villagers could be encouraged/ incentivized to participate. It has been also found that although rainfed and water scarce areas have been chosen for the programme, the land areas developed are essentially private croplands. The community land development activities do not get much attention. As the target of PIA is to develop a total area of 500 hectare, with no minimum expenditure or area earmarked for community land. PIAs usually opt for the easier course of developing only the flatter terrain of cropland areas, where quick participation of land owning households is also possible. In such a situation land beneficiaries are deprived of any direct benefits. In order to avoid such problem and conflict between beneficiaries and non-beneficiaries, development of community land resources and introduction of income generating activities for the landless and other weaker sections should be considered.

There should be a Detailed Project Report (DPR) of the selected micro watershed area in the initial year of project and get it known to all by displaying the list of activities to be undertaken during the project period. It should be prepared by a team of technical experts on the basis of felt needs of local people. The effectiveness of community organisation and sustaining watershed activities largely depend on the training and awareness of the members of WA, WC, and WDT. The roles and responsibilities of these groups are defined but not in practice, which need to be activated by regular reviewing and monitoring of the programme.

In Maharashtra, Watershed development needs to integrated into the main stream strategy for agricultural growth, if a large part of it is going to realised from the hitherto rainfed areas. Regular training at watershed committee, PAI/block and district level should continue all along the year. Training on innovative activities, local skills, improved technology, etc. should be given priority. In fact, a training and community organisation activities calendar should be prepared and accordingly the programme be organized. Nursery is a vital need in all the watersheds. Provision of saplings of fuel and fodder plantation, fruit bearing trees, vegetable cultivation should be ensured either through individual nursery or from central nursery at every watershed area. Establishment of a medicinal/herbal plantation garden is felt essential in the watershed. Community based grain banks and seed banks should be established in the watershed and government support should be ensured at the beginning for food and seed security. Since the climate of Maharashtra is conducive for the cultivation of flowers and it has a high market value in the neighbouring state of Andhra Pradesh, floriculture should be promoted for the economic upliftment of the rural poor. In all the watershed projects, it is necessary to fix target and allocate fund for other activities like soil and moisture conservation, development of non-arable land, drainage line treatment etc. are indispensable for the all round development of the watershed project.

On the whole, it may be concluded that there should be a holistic approach to rural agriculture development through watershed programme, primarily aiming at integration of several development activities such as soil conservation, land and water management, agriculture, afforestation and animal husbandry with special emphasis to relate these actions with human issues and to develop the capability of the target population at the micro level befitting to the local conditions. The impact evaluation study has demonstrated that watershed development programme to large extent is able to regenerate natural resources including land, forest, and water and play a crucial role in augmenting agricultural growth, productivity, cropping intensity, and cropping pattern.

REFERENCES

Bhatnagar, Pradip. 1996. "Growth Potential of Rural Employment Progarmmes: The Watershed Approach," *Journal of Rural Development*, National Institute of Rural. Development Rajendranagar, Hyderabad, vol. 5, no. 2 April-June: 169–78.

Majmudar, D.K., and P.P. Saikia. 2006. "Peoples Participation and Constraints in Watershed Development Programmes," *Agricultural Situation in India* LXIII(4), July.

Rao, C.H. Hanumantha. 2000. "Watershed Development in India: Recent Experience and Emerging Issues," *Economic and Political Weekly* 35(45), November: 3943–7.

Reddy, G. Sastry, and H.P. Singh. 2001. "Watershed Approach for Rural Development," *Agriculture Situation in India*, Oct.

Reddy, Y.V. Shriniwas Das, S.K. Sharma, Vishnu Murthy, T. and B.V. Ramanarao. 1991. "Watershed Programme Approach for Dryland Development in India," *Agricultural Situation in India*, December.

Shah Amita. 2000. "Watershed Programme: A Long Way to Go," *Economic & Political Weekly*, vol. 35: 3155–64.

——— 2005. "Economic Rational, Subsidy and Cost Sharing in Watershed Project," *Economic and Political Weekly*, vol. XL, no. 26, June, 25 July, 1.

29

Economic Policy Reforms and Indian Fertilizer Sector

Vijay Paul Sharma and Hrima Thaker

Chemical fertilizers are a key element of modern technology and have played an important role in the success of Indian agriculture. Some argue that fertilizer was as important as seed in the Green Revolution (Tomich, Kilby, and Johnson 1995), contributing as much as 50 per cent of the yield growth in Asia (Hopper 1993; Food and Agriculture Organisation 1998). Others have found that one-third of the world cereal production is due to the use of fertilizer and related factors of production (Bumb 1995).

For over last four decades, India has relied on increased crop yields to meet an ever increasing demand for food. According to Ministry of Agriculture data, total foodgrains production rose from about 102 million tonnes in triennium ending (TE) 1973–74 to over 260 million tonnes in TE 2013–14, more than 155 per cent increase (GoI 2015). Meanwhile, the total area under foodgrains, which accounted for about 75 per cent of total cropped area in TE 1971–72, declined to about 63 per cent in TE 2012–13. This dramatic increase in foodgrains production was the result of over 155 per cent increase in crop yields, from 833 kg per ha to 2,103 kg per ha between TE 1973–74 and TE 2013–14. With shrinking arable land resources, and burden of increasing future population, development of new technologies and efficient use of available technologies and inputs will continue to play an important role in sustaining food security in India. Therefore, the only way to increase agricultural production is to increase crop yields through scientific use of fertilizers and other inputs like high yielding variety seeds, irrigation, etc. using the limited arable land, with an emphasis on protecting the environment.

The government has been consistently pursuing policies conductive to increased availability and consumption of fertilizers in the country. Over the last four decades, production and consumption of fertilizers have increased significantly. The country had achieved near self-sufficiency in urea and DAP, with the result that India could manage its requirement of these fertilizers from indigenous industry and imports of all fertilizers except MOP were nominal till early 1990s. However, during the last decade there has been a significant increase in imports of fertilizers because there has not been any major domestic capacity addition due to the uncertain policy environment.

The significance of fertilizer industry and its related policy in the country arises from the fact that agriculture still employs about half of country's workforce and more importantly, it supports nearly two-third of the population. Therefore, fertilizer policy in India has been mainly driven by the socio-political objectives of making fertilizer available to farmers at affordable prices and increasing fertilizer consumption. Given the socio-political importance of fertilizer pricing on one hand and ever increasing subsidies, on the other hand, the need for streamlining the sector has been felt for a long time. However, fertilizer has become the most contentious issue in reforming Indian economy exposing deep contradictions between economics and politics in the democratic set-up. In view of the importance of fertilizers in agricultural growth and the changing policy environment, it is important to examine changing structure of fertilizer markets, policy environment and the role of various factors influencing fertilizer consumption. The present chapter attempts a comprehensive and in-depth analysis of the Indian fertilizer sector under the new economic policy regime.

The specific objectives of the study are:

1. to analyse the trends and patterns in consumption of fertilizers in the country;
2. to analyse extent of fertilizers subsidies and issues related to inter-crop, and inter-farm size equity in distribution of fertilizers subsidy; and
3. to study the impact of important factors on fertilizer consumption.

DATA AND METHODOLOGY

The study is based on secondary data related to fertilizer production, consumption, and imports along with fertilizer prices, output prices, the area under irrigation, high yielding varieties, rainfall, subsidies, etc. In order to examine trends and pattern of growth of fertilizers, compound annual growth rates were computed.

The fertilizer demand model was estimated using annual time series data, from 1976–77 to 2009–10 using ordinary least squares (OLS) method. The empirical model is specified as follows:

$$F_t = b_0 + b_1 \, HYV_t + b_2 \, GIA_t + b_2 \, CI_t + b_3 \, Pfert_t + b_4 \, Pr{+}w_t + b_5 \, Credit_t + U_t$$

where, F_t is fertilizer consumption in thousand tonnes; t denotes year.

The following independent variables were hypothesized to influence the consumption positively (+), negatively (–), or either negatively or positively.

HYV = Percentage of area under HYV to gross cropped area (+).
GIA = Percentage of gross irrigated area to gross cropped area (+).
CI = Cropping intensity (%) (+).
Pfert = Prices of fertilizers are represented by price of N through Urea, average price of P_2O_5 through DAP and SSP, price of K through MOP and N+P+K price is the price of N, P, and K and weighted by their consumption shares (–).
Pr+w = Output price is represented by procurement price of rice and wheat (main users of fertilizers) and weighted by the share of their production (+).
Credit = Short-term production credit per hectare of gross cropped area (Rs) (+).

Two forms of functions, namely, linear and Cobb–Douglas were tried in this analysis. The results of the linear regression equation were used for interpretation as it was found better when compared with Cobb–Douglas production function.

RESULTS AND DISCUSSIONS

Trends in Fertilizer Use

India is the second largest consumer of fertilizers in the world with estimated consumption of 25.8 million tonnes in TE 2013–14, after China. At the time of onset of green revolution in the mid-1960s, the consumption of fertilizers was less than 1 million tonnes and increased to about 2.84 million tonnes by the end of Fourth Five-Year Plan, which further increased to 11.57 million tonnes by the end of seventh Plan (Table 29.1). The rapid expansion of

Table 29.1 Consumption of Fertilizer Nutrients (N, P_2O_5 and K_2O) during Plan Periods

Year	Consumption ('000 tonnes)			
	N	P_2O_5	K_2O	Total
Plan I End (1955–56)	107.3	13.0	10.3	130.6
Plan II End (1960–61)	211.7	53.1	29.0	293.8
Plan III End (1965–66)	574.8	132.5	77.3	784.6
Plan IV End (1973–74)	1829.0	649.7	359.8	2838.5
Plan V End (1978–79)	3419.5	1106.0	591.5	5117.0
Plan VI End (1984–85)	5486.1	1886.4	838.5	8211.0
Plan VII End (1989–90)	7385.9	3014.2	1168.0	11568.1
Plan VIII End (1996–97)	10301.8	2976.8	1029.6	14308.2
Plan IX End (2001–02)	11310.2	4382.4	1667.1	17359.7
Plan X End (2006–07)	13772.9	5543.3	2334.8	21651.0
2010–11	16558.2	8049.7	3514.3	28122.2
2011–12	17300.3	7914.3	2575.5	27790.0
2012–13	16820.9	6653.4	2061.8	25536.2
2013–14	17018.8	5645.4	2057.0	24721.2

Source: Fertilizer Association of India (FAI) (2015).

irrigation, spread of HYV seeds, introduction of Retention Price Scheme (RPS), distribution of fertilizers to farmers at affordable prices, expansion of dealer's network, improvement in fertilizer availability and virtually no change in farm gate fertilizer prices during the 1980s were major reasons for increase in fertilizer consumption during 1971 to 1990. During the 1990s, total fertilizer consumption fluctuated between 12.15 and 16.8 million tonnes with the exception in 1999–2000 when fertilizer consumption was over 18 million tonnes. In the last decade, fertilizer consumption increased at a faster rate and total fertilizer consumption reached a record level of 28.1 million tonnes during 2010–11 and then declined for three consecutive years and reached 24.72 million tonnes in 2013–14.

Urea is the largest straight nitrogenous fertilizer in terms of total nutrient production and accounted for 84.2 per cent of total production while share of other straight nitrogenous fertilizers such as ammonium sulphate, calcium ammonium nitrate, and ammonium chloride was about 1.2 per cent and complex fertilizers contributed 18.5 per cent in 2013–14. The share of public sector in N capacity has declined over time while the share of the private and cooperative sector has increased. However, an important issue confronting the N sector is with respect to the feedstock because natural gas, which is the main feedstock of nitrogenous fertilizers (accounts for 78.3% of installed capacity), is available in limited quantities and the industry competes with the power sector for its share. With the Government policy favouring conversion to gas based units, the demand for gas is expected to go up in the future, which may in turn lead to further shortages.

In case of phosphatic fertilizers, DAP constitutes about 42.1 per cent of total production and share of SSP is about 16.8 per cent and the rest are constituted by other NP/NPK complexes. Over the years, public sector has lost its share to private and cooperative sectors and today about two-thirds of the phosphatic fertilizer capacity is in the private sector. Due to limited availability of phosphatic raw materials/intermediates such as phosphoric acid and rock phosphate in the country, domestic units are highly dependent on imports. The high dependence on imports of raw materials exposes the Indian phosphatic industry to highly volatile markets.

Intensity of Fertilizer Use

On per-hectare basis, fertilizer consumption was less than 2 kg during the 1950s and increased to about 5 kg in 1965–66. However, after introduction of green revolution in 1966–67, per hectare fertilizer consumption more than doubled in the next five years from about 7 kg in 1966–67 to about 16 kg in 1971–72, which further increased and reached a level of 50 kg in mid-1980s. Average fertilizer consumption crossed 100 kg per ha in 2005–06 and reached a record level of 146.3 kg in 2010–11 and then declined to 126.6 kg in 2013–14, mainly due to steep increase in prices of phosphatic and potassic fertilizer after introduction of Nutrient Based Subsidy (NBS) scheme in 2010–11. However, fertilizer consumption in India is highly skewed, with wide inter-regional, inter-state, inter-district and inter-crop variations. Intensity is generally higher in southern (173 kg/ha) and northern region (167.4 kg/ha) and lower in the eastern (112.3 kg/ha) and western region (93.1 kg/ha) (FAI 2015).

The average intensity of fertilizer use in India at the national level is still much lower than in other developing countries, but there are many disparities in fertilizer consumption patterns both between and within regions of India. During the TE 1986–87, only three districts were using more than 200 kg/ha of fertilizer and another 12 districts were consuming between 100 to 150 kg/ha of fertilizer (Table 29.2). In contrast about 60 per cent of the districts were using less than 50 kg fertilizer (N+P+K) per hectare. However, the number of districts in the high-fertilizer use category (>200 kg/ha) increased significantly during the second-half of the 1990s and 2000s. In the TE 1999–2000, out of 470 districts, 31 districts (6.6%) were using more than 200 kg/ha while about one-third of

Table 29.2 Classification of Districts According to Ranges of Fertilizer Consumption (N + P + K)

Consumption (kg/ha)	TE 1986–87	TE 1999–2000	TE 2002–03	TE 2011–12
Above 200	3 (0.9)	31 (6.6)	36 (7.5)	135 (25.4)
150–200	12 (3.4)	45 (9.6)	47 (9.7)	77 (14.5)
100–150	32 (9.2)	94 (20.0)	92 (19.0)	115 (21.6)
75–100	34 (9.7)	62 (13.2)	61 (12.6)	57 (10.6)
50–75	55 (15.8)	78 (16.6)	79 (16.4)	59 (11.1)
25–50	92 (26.4)	80 (17.1)	97 (20.1)	55 (10.3)
<25	121 (34.7)	79 (16.8)	71 (14.7)	35 (6.5)

Note: Figures in parentheses show per cent to a total number of districts.
Source: FAI (2014a).

the districts were consuming less than 50 kg. Between the TE 2002–03 and TE 2011–12, the number of districts consuming more than 200 kg/ha more than tripled from 36 to 135. Increasing number of districts consuming consistently higher amounts of fertilizer (>200 kg/ha) is a cause of concern as it might lead to environmental degradation particularly land and water resources. On the other hand, still about 170 per cent of the districts use less than 50 kg/ha of fertilizers. Therefore, there is a need have two-pronged strategy, one to monitor districts with high intensity of consumption and take corrective actions to reduce adverse effects on environmental resources and two, promote fertilizer consumption in low-use districts to improve crop productivity.

One of the major constraints to fertilizer use efficiency in India is the imbalanced use of nutrients. Nitrogen (N) applications tend to be higher in comparison to potassium (K) and phosphate (P). This is partly the result of a difference in price of different nutrients, and partly due to the lack of knowledge among farmers about the need for balanced fertilizer applications. The N:P:K ratio was little skewed towards N in the mid-1970s but started improving in the late 1970s and 1980s and reached a level of 5.9:2.4:1 in 1991–92 (Table 29.3). This improvement was due to tight controls by the government on fertilizer prices and sales and distribution during the decade of the 1980s. The fertilizer prices remained unchanged during the 1980s. However, in August 1992, prices, distribution and movement of phosphatic and potassic fertilizers were decontrolled while urea remained under statutory price control. The subsidy on phosphatic and potassic fertilizers was withdrawn resulting in a sudden increase in retail prices of these fertilizers. The NPK use ratio got distorted significantly from 5.9:2.4:1 during 1991–92 to 9.5:3.2:1.0 in 1992–93. The share of N, P, and K in total fertilizer consumption which was 63.2, 26.1, and 10.7 per cent, respectively in 1991–92, N share increased to about 71 per cent in 1993–94 while share of P declined to 21.6 per cent and that of K to 7.3 per cent. In order to correct the imbalance in use of N, P, and K fertilizers, Government of India implemented a scheme of ad-hoc concession on sale of decontrolled fertilizers to the farmers from 1 October 1992 but still there was significant disparity in prices of N, P, and K fertilizers which led to more use of N and less use of P and K fertilizers leading to more imbalance in use of fertilizers (NPK ratio reached a level of 10.0:2.9:1.0 in 1996–97). Concerned with this deteriorating NPK ratio, Government of India announced a substantial increase in concessions on phosphatic and potassic fertilizers in subsequent years, which led to improvement in NPK ratio and reached a level of 4.3:2.0:1.0 in 2009–10.

Table 29.3 Consumption Ratio of N and P_2O_5 in Relation to K_2O and N in Relation to P_2O_5 in India: 1975–76 to 2013–14

Year	$N:P_2O_5:K_2O$			$N:P_2O_5:K_2O$	
	N	P_2O_5	K_2O	N	P_2O_5
1975–76	7.7	1.7	1	4.6	1
1981–82	6.0	1.9	1	3.1	1
1991–92	5.9	2.4	1	2.4	1
1992–93	**9.5**	**3.2**	**1**	**3**	**1**
1993–94	**9.7**	**2.9**	**1**	**3.3**	**1**
1994–95	8.5	2.6	1	3.2	1
1996–97	**10.0**	**2.9**	**1**	**3.5**	**1**
2002–03	6.5	2.5	1	2.6	1
2005–06	5.3	2.2	1	2.4	1
2006–07	5.9	2.4	1	2.5	1
2007–08	5.5	2.1	1	2.6	1
2008–09	4.6	2.0	1	2.3	1
2009–10	4.3	2.0	1	2.1	1
2010–11	4.7	2.3	1	2.1	1
2011–12	**6.7**	**3.1**	**1**	**2.2**	**1**
2012–13	**8.2**	**3.2**	**1**	**2.6**	**1**
2013–14	**8.3**	**2.7**	**1**	**3.1**	**1**

Source: FAI (2014 and 2014a).

To ensure balanced use of fertilizers, the government intended to move towards a Nutrient based Subsidy (NBS) regime instead of product pricing regime and introduced NBS policy for P and K fertilizers from 1 April 2010 and market prices of all fertilizers (except urea) were to be determined by markets forces based on demand-supply situation. However, the steep increase in prices and declining NBS rates of P and K fertilizers have led to the imbalanced application of nitrogen vis-à-vis phosphatic and potassic fertilizers. These developments have led to worsening of NPK ratio and it reached a level of 6.7:3.1:1 in 2011–12 (post-NBS period) and became even worse (8.7:3.4:1) in 2012–13 and 8.3:2.7:1.0 in 2013–14.

The NPK ratio also shows wide inter-regional and inter-state disparities. While existing variation from the ideal ratio (4:2:1) is nominal in the South (4.9:1.9:1) and the East region (4.6:1.3:1), it is very wide in the North (36.0:8.5:1) and western region (8.1:3.4:1). State-wise consumption ratio of N and P in relation to K shows that greatest degree of N:P:K imbalance was seen in case of Rajasthan (180.3:54.6:1), Haryana (64.0:12.8:1), and Punjab (56.8:13.5:1) in 2013–14. It is also worth noting that NPK ratio has deteriorated in almost all States in the post-NBS period, which is a cause of concern.

Fertilizer Imports

The fertilizer consumption in India has generally exceeded the domestic production except for few years. During the 1950s and 1960s, about two-third of the domestic requirement of N fertilizers was met through imports. With the introduction of the high-yielding varieties of wheat and rice in the mid-1960s, the demand for fertilizers increased and was mainly met through. However, the introduction of the Retention Price Scheme (RPS) in the late 1970s led to the development of a large domestic industry and helped in achieving near self-sufficiency. However, the scheme attracted criticism for not providing sufficient incentives to encourage efficiency leading to high-cost fertilizers and fertilizer subsidy. India was nearly self-sufficient in fertilizers in the early 1990s, but during the last decade due to low/no addition in domestic capacity coupled with a rise in demand for fertilizers, imports have increased significantly. India imported about 13 million tonnes of NPK fertilizer nutrients in 2012–13 as against about 2 million tonnes in early 2000s. The growth of imports was rather slow in the 80's and 90's and accelerated in 2000s. The fertilizer imports increased significantly in 2005–06, and the trend continued thereafter (Figure 29.1). The share of imports in total fertilizer consumption declined from over 60 per cent in 1960s to 43.8 per cent in 1970s, further to about 25.3 per cent in 1980s and reached a level of 21.6 per cent in 1990s. However, imports increased significantly during the last decade, and import share in total consumption increased to about 47 per cent in 2011–12.

Fertilizer Use by Crops and Farm Size

It is generally perceived that major benefit of fertilizers goes to the areas having better access to technology, irrigation facilities, infrastructure and growing fertilizer-intensive crops like rice, wheat, sugarcane, fruits and vegetables. Trends in usage of fertilizers in India show that rice is the largest user of fertilizer (about 1/3rd of total consumption), followed by wheat (24.2%) in 2006–07. Rice and wheat accounted for over 60 per cent of total fertilizer consumption in the country in 1995–96 and the share declined to 56.8 per cent in 2006–07. Fruits, vegetables, and sugarcane combined represent another 11 per cent of fertilizer use. Cotton accounts for about 5.6 per cent of total use. In all the years, rice was the dominant crop fertilized. Fruits and vegetables appear to be increasing in importance. Fertilizer intensity measured as average kg per hectare does not follow exactly the same pattern across crops; intensity tends to be higher on sugarcane (234.9 kg/ha), vegetables (253.8 kg/ha), cotton (183 kg/ha) and fruits (158.6 kg/ha) and lower on cereals (rice, 129.2 kg/ha and wheat 162.6 kg/ha) and pulses. It is evident that farmers growing input-intensive crops are the main beneficiary of fertilizer use (Sharma and Thaker 2011).

Table 29.4 shows farm size wise consumption of fertilizers in India in 1991–92, 1996–97, 2001–02, and 2006–07. It is evident from the table that average use of fertilizers was higher on small and marginal farmers compared to medium and large farmers. The average fertilizer consumption per hectare of gross cropped area was the highest (139.74 kg) on marginal farms and the lowest on large farms (67.64 kg) in 2006–07. A similar trend was observed during 1991–92, 1995–96, and 2001–02. Moreover, there has been a significant increase in fertilizer intensity per hectare of gross cropped area on all farm size holdings during the period 1991–92 and 2006–07. However, the increase was the largest (95.9%) on small farms, followed by marginal holdings (93.5%) and the lowest (47%) on large farms. The average fertilizer subsidy per hectare of cropped area is significantly higher in case of small and marginal farmers compared with large farmers because average fertilizer consumption is also higher on small and marginal farms. Sharma and Thaker (2010)

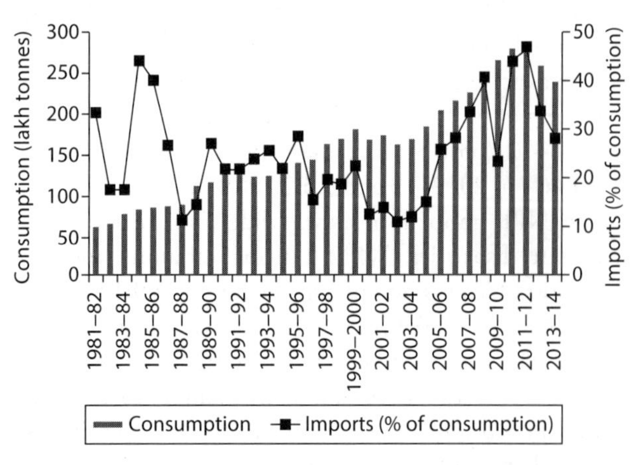

Figure 29.1 Trends in Consumption and Imports of Fertilizer Nutrients in India: 1981–82 to 2013–14

Source: FAI (2014 and 2014a).

Table 29.4 Pattern of Fertilizer Use Intensity (kg/ha of Gross Cropped Area) by Farm Size

Year	Marginal	Small	Semi-medium	Medium	Large	All Households
1991–92	72.2	65.5	61.7	56.3	46.0	60.7
1996–97	103.8	82.6	75.3	68.1	51.1	77.1
2001–02	126.2	100.6	88.8	75.8	55.9	92.6
2006–07	139.7	128.3	108.3	95.1	67.6	112.8

Source: GoI (2007), GoI (2008), and GoI (2012).

have reported that the benefits of fertilizer subsidy are not restricted to only resource-rich states but have spread to other states also. It is worth mentioning that the benefits of fertilizer subsidy have spread to unirrigated areas as the share of the area treated with fertilizers has increased on unirrigated lands. Likewise, the share of unirrigated areas in total fertilizer use has also increased during the last decade, indicating that fertilizer subsidies have befitted farmers in rainfed areas.

FERTILIZER SUBSIDY

The estimates of fertilizer subsidy over the years in the post-reforms era show that fertilizer subsidy has increased significantly (Table 29.5). Fertilizer subsidy has increased from Rs 5185 crore in 1991–92 to Rs 70967 crore in 2014–15 representing an increase of over 13 times. The fertilizer subsidy in India as a percentage of the GDP varied from 0.47 in 2002–03 to 1.9 per cent in 2008–09 and declined to about 0.8 per cent in 2011–12. However, a steep increase in the cost of inputs to fertilizer production, high import prices of fertilizers and constant farm gate prices have led to substantial increase in fertilizer subsidy in the recent period. The fertilizer subsidy has increased by over 5.5 times between TE 2003–04 and TE 2014–15, from Rs 11853 crore to over Rs 67973 crores. The share of fertilizer subsidy in total subsidies varied from about 25 per cent in 2002–03 to about 59 per cent in 2008–09. The fertilizer subsidy reached a peak of Rs 99495 crore in 2008–09 and then witnessed a declining trend. After two consecutive annual decreases in 2009–10 and 2010–11, fertilizer subsidy increased during 2011–12 mainly due to rise in world prices of fertilizers and fertilizer prices in 2011 averaged 43 per cent higher than 2010. However, after the introduction of NBS scheme, fertilizer subsidy recorded a decline during 2012–13 but increased in subsequent years and is budgeted to be 72969 crores during 2015–16.

Table 29.5 Trends in Fertilizer Subsidy (Rs crore) in India: 1991–92 to 2011–12

Period	Concession on Decontrolled Fertilizers		Subsidy on Urea		Total Fertilizer Subsidy	Share (%) in Total Subsidies[1]
	Indigenous P&K	Imported P&K	Indigenous	Imported		
1991–92	–	–	3500	1300	5185[2]	42.3
1992–93	–	–	4800	996	5796	48.3
1995–96	500[3]		4300	1935	6735	53.2
1998–99	3790	–	7433	333	11596	49.2
2001–02	3760	744	8044	148	12695	40.4
2002–03	2488	737	7799	1.2	11016	25.3
2003–04	2606	4720	8521	0.8	11848	26.7
2004–05	3977	1165	10243	742	16128	34.6
2005–06	4499	2097	10653	2141	19390	38.8
2006–07	6648	3650	12650	5071	28019	42.0
2007–08[4]	10334	32598	1640	9935	43319	43.7
2008–09	32957	32598	20969	12971	99495	59.1
2009–10	16000	23452	17580	7000	64033	43.3
2010–11	40766		15081	6454	62301	35.9
2011–12	36089		20208	13716	70013	33.1
2012–13	30480		20000	15133	65613	25.5
2013–14	29300		26500	11538	67339	26.4
2014–15 *(RE)*	20667		38200	12100	70967	26.6
2015-16 (BE)	22469		38200	12300	72969	29.9

Source: GoI (2015a).

[1] Share is computed from subsidy figures given in various issues of Expenditure Budget Vol. I, Ministry of Finance, Government of India.

[2] Includes Rs 385 crore fertilizer subsidy given to small and marginal farmers.

[3] Total subsidy on imported and indigenous P&K fertilizers.

[4] Subsidy figures for 2007–08 and 2008–09 include both cash and bonds for both urea and decontrolled fertilizers.

FACTORS AFFECTING FERTILIZER DEMAND

The results of regression analysis for the total fertilizer consumption equation are reported in Table 29.6. The high R^2 value (0.99) indicates that explanatory variables in the model have accounted for over 99 per cent variation in fertilizer use and the model best fits when predicting fertilizer demand. The model was significant at 1 per cent level. All explanatory variables used in the model were statistically significant and had theoretically expected signs. The price of fertilizers was negatively related with fertilizers demand while the area under high yielding varieties, irrigation, cropping intensity, the price of output, and credit had a positive relationship with fertilizer demand.

The results show that non-price factors were more important determinants of fertilizer use. Among the non-price factors, irrigation was the most important factor influencing fertilizer demand, followed by cropping intensity. The price of fertilizers was the third important determinant of fertilizer use in the country. The price of output is less important compared with input price. The results clearly indicate that increase in area under irrigation and cropping intensity will accelerate fertilizer consumption in the country. In case of pricing policy instruments, an increase in prices of fertilizers would lead to a reduction in fertilizer use while output price had a positive impact on fertilizer consumption but was less powerful than input prices. Therefore, it is necessary to prioritize input price policy mechanism over higher output prices.

Table 29.6 Estimated Regression Equation for Total Fertilizer (N + P + K) Use in India

	Coefficient	Standard Error	't' Value	Rank@
Constant	−59,461.75	9,260.82	−6.421	–
HYVs	56.792	38.521	1.474	5
GIA	437.355***	105.735	3.812	1
CI	426.189***	86.385	4.934	2
P_{fert}	−603.725***	140.827	−4.287	3
P_{r+w}	5.420*	2.843	1.906	4
Credit	0.029**	0.012	2.393	6
Adj. R Square	0.994	–	–	–
F	947.3181***	–	–	–
D-W statistics	1.683	–	–	–

Note: @ Based on standardized coefficients (ignoring signs) given coefficients x (s.d. of X_i/s.d of Y_i), where s.d. is standard deviation, Xi is i^{th} explanatory variable and Y is dependent variable.
*** Significant at one per cent; ** Significant at 5 per cent;
* Significant at 10 per cent.

CONCLUSIONS AND POLICY IMPLICATIONS

There is undisputable need for continuous rapid growth in fertilizer use especially in less-consuming regions in the country in the coming years to increase agricultural production and productivity at the desired rate. In order to meet the additional demand, there is a need to increase fertilizer supplies and generate effective demand. Sustained growth in fertilizer demand mainly depends on an increase in supplies, creation of adequate and efficient distribution network and increase in effective demand for fertilizers at the farm level. Therefore, there is a need to enlarge domestic capacity and production.

For sustained agricultural growth and to promote the balanced nutrient application, it is imperative that an appropriate fertilizer pricing policy is put in place. Current pricing and subsidy schemes generally do not promote the balanced use of nutrients. Therefore, there is a need to keep parity between N, P, and K prices and also address the issue of the deficiency of secondary and micronutrients. A long-term fertilizer pricing policy that promotes fertilizer use as well as production is needed.

The burden of fertilizer subsidies has increased substantially during the last few years, but these subsidies are justified on several grounds. Although there is a high degree of inequity in the distribution of fertilizers across states and crops, there is a fair degree of equity in the distribution of these subsidies across different farms sizes. Small and marginal farmers are key beneficiaries of fertilizer subsidies. Moreover, benefits of fertilizers subsidies are not restricted to only irrigated areas but have spread to rainfed areas. A reduction in fertilizer subsidy is, therefore, likely to have an adverse impact on the income of marginal and small farmers. The increase in fertilizer prices would lead to a reduction in fertilizer use on these farms and consequently lower production and productivity. An increase in prices of fertilizers is also likely to have an adverse impact on agricultural production in low-fertilizer using regions growing mainly coarse cereals, pulses, and oilseeds. The targeting of fertilizer subsidies (geographical targeting between regions, states, and districts, and farm size targeting between different categories of households) is a critical and sensitive issue. Since it is practically not feasible to develop an effective targeting system that reaches poorer households/regions, comprehensive coverage of all farm households is a better alternative than ineffective targeting. However, efforts are required to contain subsidies through periodic revisions of farm-gate prices of fertilizers and reducing costs of production of fertilizers. If there is a significant reduction/withdrawal of fertilizer subsidy, it would have serious adverse effect on agricultural production and consequently threaten the national food security. On the other hand, no change in prices of fertilizers over a period of time and disparity in prices of different nutrients also lead to an adverse impact on fertilizer production and land productivity.

REFERENCES

Bumb, B. 1995. 'Global Fertilizer Perspective, 1980–2000: The Challenges in Structural Transformation', Technical Bulletin T-42. Muscle Shoals, AL: International Fertilizer Development Center.

Fertilizer Association of India (FAI). 2014. 'Annual Review of Fertiliser Production and Consumption 2013–14: Highlights', *Indian Journal of Fertilisers*, 10(4): 117–68.

——— 2014a. *Fertilizer Statistics*, 2012–13 and Earlier Issues. New Delhi: The Fertilizer Association of India.

FAI. 2015. 'Fertilizer Consumption in India: Fertilizer Association of India'; http://www.faidelhi.org/general/Consumption%20of%20Fertilisers.pdf

FAO. 1998. '*Guide to Efficient Plant Nutrition Management*', FAO/AGL Publication, FAO, Rome.

Government of India (GoI). 2007. *All India Report on Input Survey 1996–97*. New Delhi: Agricultural Census Division, Ministry of Agriculture, Government of India.

———. 2008. *All India Report on Input Survey 2001–02*. New Delhi: Agricultural Census Division, Ministry of Agriculture, Government of India.

———. 2012. *All India Report on Input Survey 2006–07*. New Delhi: Agricultural Census Division, Ministry of Agriculture, Government of India.

———. 2015. 'Third Advance Estimates of Production of Foodgrains for 2014–15', Agricultural Statistics Division, Ministry of Agriculture, Govt. of India, New Delhi, May.

———. 2015a. *Union Budget*, Various Issues from 1991–92 to 2015–16. New Delhi: Ministry of Finance, Government of India.

Hopper, W. 1993. 'Indian Agriculture and Fertilizer: An Outsider's Observations', Keynote address to the FAI Seminar on Emerging Scenario in Fertilizer and Agriculture: Global Dimensions, The Fertilizer Association of India, New Delhi.

Sharma, Vijay Paul, and Hrima Thaker. 2010. 'Fertilizer Subsidy in India: Who are the Beneficiaries?' *Economic and Political Weekly*, 15(12): 68–76.

———. 2011. *Economic Policy Reforms and Indian Fertilizer Sector*. New Delhi: Allied Publishers.

Tomich, T., P. Kilby, and B. Johnson. 1995. *Transforming Agrarian Economies: Opportunities Seized, Opportunities Missed*, Ithaca, NY: Cornell University Press.

30

Fertilizer Consumption in Tamil Nadu*

K. Jothi Sivagnanam

Additional growth in agricultural production is needed for meeting the basic necessaries of a larger size of the population. It has also to generate additional surpluses required for further economic development with an emphasis on employment and equity. The bulk of the growth in agricultural production has to come from increase in agricultural productivity and productivity of land can be increased by removing constraints in the soil fertility and by promoting technological advancement.

There is a widespread deficiency of nitrogen in the Indian soil. The availability of phosphorous and potash is also very low. Deficiency in sulphur and micronutrients is widespread in Indian soils. Experience world over suggests that chemical fertilizers are important to remove the deficiency of soil nutrients, fertility constraints to raise land productivity through facilitating technological change. Even China, which is popular for the use of organic source of plant nutrients, is no exception as its use of chemical fertilizers is enormous.

Indiscriminate fertilizer and pesticide usage coupled with mono-cropping and non-application of base manure like cow dung and green manure (dead plants) have taken a sever toll on the organic content in the soil and thereby soil fertility in Tamil Nadu got reduced by half in 30 years, according to the Draft Organic Farming Policy of the Government of Tamil Nadu (*The Times of India*, 17 June 2015).

The availability of some fertilizer at subsidized rates, have induced the farmers to over-utilize them and this, in turn, has caused environmental hazards. So there is a suggestion by environmental economists that subsidies for such types of fertilizers should be either reduced or discontinued altogether. Organic matter in the soil dropped from 1.2 per cent in 1971 to 0.68 per cent in 2002. Organic matter further declined to 0.5 per cent in many districts in the State in 2014. This is against a desirable level of 0.8 per cent to 1.3 per cent of organic carbon (Department of Soil Science and Agriculture Chemistry of Tamil Nadu Agricultural University [TNAU] 2014).

The need for further use of chemical fertilizers is obvious from the experience of HYVS on irrigated land. Even in unirrigated land, the complementarities of chemicals fertilizers and high yielding varieties are proved. It is a fact that nearly half of the cultivated land does not come under fertilizer use and it points out the scope for raising the fertilizer consumption. But raising the fertilizer use is not an end in itself and the end is the growth in agricultural productivity and increase in production. We have to keep in mind that the productivity has also to be sustained

* This chapter is based on the report 'Factors Affecting Fertilizer Consumption in India with special reference to Tamil Nadu State' submitted earlier by R. Swaminathan and M. Muthuraj of AERC, Chennai. The author thanks them. The author also wishes to thank Dr V. Loganathan for going through the draft.

for ecological balance. Unfortunately, there is stagnation in the production of foodgrains. There is also stagnation in the area under high yielding varieties (HYVs) and the area under irrigation too is stagnated in recent years. This resulted in the stagnation of the application of chemical fertilizers which mainly depends upon the climatic conditions. If the trend continues, it may be difficult to achieve required growth rate in the coming years. There is not much scope for increasing the area under cultivation as we have already exploited more land. The only way to increase production is to increase the per hectare productivity of crops. That is where the fertilizer plays a critical role in improving the agricultural production as well as the productivity.

The benefits of fertilizer use in improving the yield and productivity are well documented (Baanante, Bumb, and Thompson 1989). In modern agriculture, special emphasis is placed on improved techniques of crop production. Improved varieties play a prominent role in increasing per hectare yield. However, full potential of the improved varieties can be realized only if essential inputs, particularly fertilizers are applied both in requisite quantities and in appropriate time.

Optimum fertilizer application plays a key role in improving the productivity of various crops. It is a critical component of the strategy for self-sufficiency in foodgrains to feed a rapidly growing population. Concerted efforts in this direction have resulted in a substantial increase in agricultural production and productivity. From a modest level of 52 million tons in 1951–52, the foodgrains production has increased to the level of 257 million tonnes during 2014–15 as per *Economic Survey 2014–15*. The Survey also pointed out that to improve resilience of the agricultural sector and bolster food security–including availability and affordable access the strategy for agriculture has to focus on improving yield and productivity. The increase in fertilizer consumption has contributed significantly to such improvement in yield and productivity on more or less a sustained manner in the country.

The present level of fertilizer use is relatively low with reference to the objective of accelerating the growth in the agriculture sector, as well as the consumption level prevailing in other countries, including some of the developing countries in Asia. Moreover, the consumption of chemical fertilizers in the country is unevenly distributed, being much higher in regions with assured irrigation. In view of the limited scope for increasing the land area under cultivation, further increases in agricultural production can be achieved only through better water management, expansion of the area under irrigation, improved farming practices, research and development in the use of scientific inputs and seeds, and last but not the least, more extensive and balanced use of fertilizers through fertilizer education. Hence, the critical importance of the fertilizer sector in the Indian economy, especially in increasing the production and productivity of all the crops.

Chemical fertilizers have played a vital role in the success of India's green revolution and consequent self-reliance in foodgrain production. The increase in fertilizer consumption has contributed significantly to sustainable production of foodgrains in the country. The Government of India has been consistently pursuing policies conductive to increased availability and consumption of fertilizers in the country.

There are different types of fertilizer like inorganic fertilizer, mixed fertilizer, biological fertilizer, etc. available in solid, liquid and gaseous forms. Out of 13 elements required by the crops, the three major nutrients are nitrogen, phosphorous, and potassium (NPK) followed by calcium, magnesium and sulphur. A look at the fertilizer consumption in India shows that there is a major imbalance in use. All-India ratio of NPK was 8.2:3.2:1 in 2012–13 whereas the recommended ratio is 4:2:1. Punjab has the highest fertilizer consumption in India at 27.8:17.3:1. The per-hectare consumption of total NPK was highest in Puducherry with 509 kg and the least in Arunachal Pradesh at 2 kg. The all India average was 128 kg.

Fertilizer application should depend on (i) nutrient status of the soil and (ii) nutrient requirement of the crop. Efficient use of fertilizer should be advocated through different approaches including Integrated Nutrient Management (INM), site specific nutrient management, precision farming, fertigation, etc. There should be synergy between the use of chemical fertilizer and organic manure, green manure, biofertilizer, composted waste, crop residues, industrial waste, vermi compost and other sources.

The production of nitrogen (N) and phosphorus (P) fertilizer together has increased from mere 0.3 lakh MT in 1950–51 to about 224 lakh MT in 2013–14. Since there are no commercially viable sources of potash (K) in the country, its requirement is met through imports. The per-hectare consumption of NPK has increased from less than a kg in 1951–52 to 125.39 kg in 2013–14.

There is a positive correlation between the application of balanced fertilizers (NPK) and the yield rate. Soil fertility (capacity of a soil to provide crops with essential nutrients) is to be increased to raise crop productivity given the limitation of extensive cultivation. Realizing the importance of application of fertilizer, the government extends subsidy to the farming community.

Further, during the study period (1980–2000) the net area sown in Tamil Nadu declined from 46 per cent to

43 per cent; but this was accompanied by increase in the production of foodgrains which was mostly through increased productivity. There may be varied sources for the increase in productivity; but the present study aims to understand the relative significance of fertilizers in improving the production and productivity.

It is against this backdrop that the present study aims to analyze the fertilizer consumption in Tamil Nadu for different categories of farmers and for major crops with a view to understand the trend, pattern determinants and the efficiency level of its use.

The specific objectives of the study are:

1. to examine the trends and pattern of fertilizer use in major crops grown by different categories of farmers in Tamil Nadu;
2. to analyse the factors affecting the trends and pattern of fertilizer use in Tamil Nadu;
3. to estimate the marginal returns and costs of fertilizer use in major crops in Tamil Nadu; and
4. to suggest measures to boost the fertilizer consumption.

METHODOLOGY AND DATA

The study was suggested by the Department of Fertilizers, Ministry of Chemicals and Fertilizers, Government of India (GoI) to find out reasons for the stagnation in fertilizer consumption in recent times and to suggest measures to increase fertilizer consumption to achieve the target of foodgrains production. The study was conducted by the AERCs of New Delhi, Chennai, Jorhat, Ludhiana, and Vishva-Bharati, and AERC, Ludhiana was the coordinating centre. The secondary data for the respective state were collected by the respective centres. For the collection of primary data, all the districts in the state were grouped into three categories—low, medium, and high—with regard to fertilizer consumption. The district with consumption of more than 50 thousand tonnes is treated as high-level application district. The districts with application between 25 thousand tonnes and 50 thousand tonnes are treated as middle-level application districts and the districts with consumption of less than 25 thousand tonnes were treated as low level application districts. Among the groups, Villupuram was selected for the highest application, Vellore was selected for middle level application and Ramanathapuram was selected for low-level application. From each district, 50 farmers were selected randomly from two villages by selecting one block in the district. The distribution of the selected categories of farmers in the three districts is given in Table 30.1. The reference period of the study is 2004–05.

Table 30.1 The Sample Farmers from the Three Districts

Farmers	Sample Selected from		
	High Application Villupuram District	Middle Application Vellore District	Low Application Ramanathapuram District
Marginal	25	28	30
Small	12	11	10
Medium	10	9	8
Large	3	2	2
All	50	50	50

Source: Own field survey.

Various statistical tools like simple averages, coefficient of variation (CV), constant growth rates, and log linear multiple regression analysis were used to analyse the data. In order to identify the determinants of per-acre value of consumption of fertilizer for marginal, small, medium, and large farmers producing cotton and paddy crops, the following form of log-linear multiple regression model was estimated by the method of least squares.

$$Log\ Y = \beta_0 + \beta_1 logX_1 + \beta_2 logX_2 + \beta_3 logX_3 + \beta_4 logX_4 + \beta_5 logX_5 + \beta_6 logX_6 + \beta_7 logX_7 + u \ldots$$

where, Y = Per-acre consumption of fertilizer in tonnes,
X_1 = Availability of irrigation,

X_2 = Soil fertility $\quad$ X_3 = Prices of fertilizer

X_4 = Availability of seed $\quad$ X_5 = Application of new technologies

X_6 = Household income $\quad$ X_7 = Availability of transport

u = Disturbance term and $\beta_0, \beta_1 \ldots \beta_7$ are the parameters to be estimated system.

RESULTS AND DISCUSSION

Trends and Pattern of Fertilizer Consumption in Tamil Nadu

A steady increase in the quantum of consumption of fertilizers over the years is being noticed in the state of Tamil Nadu. During nineties, the fertilizer consumption in the state increased at an annual rate of 2.89 per cent as against the national level growth rate of 4.11 per cent per annum. The growth rate of fertilizer consumption has been decelerating both at the state and national levels. However, in absolute terms, the average fertilizer consumption per hectare had increased from 40 kgs/ha in 1970–71 to 149 ha in 1999–2000. Area under HYV and increase in irrigation facilities have contributed to the increase in consumption levels of fertilizers in the state. Intensive fertilizer promotion programme has also been implemented in selected districts where the fertilizer consumption has been on the low side. On the extension side, farmers have been educated to select the right type of fertilizers for the different varieties of crops and to adopt the proper methods of application of fertilizer to the crops. The marking mechanism has also been functioning smoothly.

Based on the figures of fertilizer consumption for the years 1970–71 to 1985–86, growth rates have been worked out for total consumption as well per hectare consumption of the fertilizer nutrients. In the case of actual consumption, the growth rates for 'P' and 'K' are found to be higher than that for 'N'. The consumption of 'N' had been increasing at 5.75 per cent, whereas 'P' and 'K' had figures of 6.31 per cent 6.12 per cent, respectively. This is due to the fact that in the earlier years, consumption of 'P' and 'K' were at very low levels.

The trend and the compound growth in the consumption of fertilizers (NPK) in Tamil Nadu have been computed. The trend co-efficient was found to be statistically significant at 5 per cent level. The results show that overall compound growth rate for the entire study period however, is around 4 per cent. Consumption of fertilizers per hectare has also increased significantly during the study period.

The application of fertilizers in the three sample districts, Villuppuram, Vellore, and Ramanathapuram, has been examined by collecting primary data from the farmers. The important crops in all the three sample districts are paddy, sugarcane, cotton and pulses. Most of the farmers apply the recommended dosages of fertilizers with a few exceptions where they apply more than recommended dosages. But there is no application less than the recommended dosage. The sample results also show that almost all the farmers have used more urea than that of super and potash. The major determinants of the consumption of fertilizers in the sample districts are the fertilizer price and the levels of soil moisture. Of these two, almost all the farmers pointed out that the price of the fertilizer is the major determinant of its application. The average yield of paddy in the three districts is 1,450 kgs/acre (Villuppuram), 1,475 kgs/acre (Vellore) and 1210 kgs/acre (Ramanathapuram).

DETERMINANTS OF FERTILIZER CONSUMPTION

Theoretically, there are many possible factors that affect the level of fertilizer consumption. The present study took seven factors namely irrigation, soil fertility, fertilizer prices and application of new technologies, household income and availability of transportation by running log linear regression across different categories of farmers cultivating two crops (paddy and cotton) to estimate the influence of all these variables in determining the level of fertilizer consumption in the three sample districts.

The seven variables were responsible for 76 per cent of the variations in the per acre fertilizer consumption. The coefficients of irrigation availability, fertilizer prices, application of new technologies and household income are statistically significant at 5 per cent level. Irrigation availability and household income have greater influence on per acre fertilizer consumption determination. Thus, it may be concluded from the above results that all the seven variables have some influence on per acre fertilizer consumption of marginal, medium, small and large farmers producing cotton. But irrigation availability has a greater influence on per acre fertilizer consumption than other factors.

Similar regression analysis has been carried out for the paddy farmers also. In the case of paddy, all the variables, except transport facilities and availability of seeds emerged statistically significant at 5 per cent level. It indicates that an additional percentage increase in availability of irrigation, soil fertility, fertilizer prices and application of new technologies could increase per acre fertilizer consumption by 0.1654, 0.1321, 0.1104, and 0.1741 per cent, respectively. However, after complete review, the study has narrowed down to the following simple regression model.

$$FC = \beta_0 + \beta_1 logX_1 + \beta_2 logX_2 + \beta_3 logX_3 + \beta_4 logX_4 + u$$

where, FC = Fertilizer consumption per acre, X_1 = Price of Urea, X_2 = Rainfall, X_3 = Yield, X_4 = area.

The results for paddy cultivation are:

$$FC = 142.183 + .017 + .095 + .045 - .690$$

$$(-2.251) + (1.876) + (4.235) + (2.626) - (-.209)$$

And the results for cotton cultivation are:

$$FC = -172.257 + .0290 + .071 + .382 + 11.235$$

$$(-1.779) + (1.986) + (3.226) + (3.439) + (.667)$$

The values in parenthesis are 't' values. The R square value is 0.952 for paddy and it fits well as it explains most of the variations in the fertilizer consumption. The results indicate that the fertilizer consumption per acre for the cultivation of paddy mostly depends on the rainfall, the yield and the price of urea. Similarly, the results for cotton also indicate that all the four independent variables could have positive influence in determining the fertilizer consumption per acre. These four independent variables have explained more than 90 per cent variations in fertilizer consumption per acre as indicated by the R square value of 0.915 for cotton cultivation.

RESOURCE USE EFFICIENCY

In order to examine the resource use efficiency, fertilizer consumption of cotton and paddy crops of marginal, medium, small and large farmers, marginal value productivity (MVP) of different inputs are compared with their respective marginal factor cost (MFC). The MVP of a factor input is defined as the change in value of fertilizer consumption resulting from a change of one factor while keeping all other factors constant. The basic condition to be satisfied is to obtain efficient resource use in the equality of MVP to MFC. In order to compare MVP with respective MFC, MVP has been converted into monetary terms by multiplying it with price of per unit fertilizer consumption.

Resource-use efficiency is examined by estimating the ratio of value of MVP of different factor inputs namely irrigation availability, soil fertility, fertilizer prices, availability of seeds, application of new technologies, household income and transport facilities to the respective factor price. Equality of MVP and factor price (MVP/MFC = 1) indicates optimum resource-use efficiency of a particular input. Inequality of MVP and factor cost (that is MVP/MFC ≠ 1) indicates the degree of resource use efficiency. If the ratio is more than one and the regression coefficient is significant, the resource is said to be under-used. Similarly where the ratio is less than one and the regression coefficient is significant, the resource is said to be over-used.

The ratio of MVP to MFC for inputs irrigation, soil fertility, fertilizer prices and application of new technologies for marginal farmers producing cotton was 0.11, 0.84, 0.47, and 0.63, respectively. This indicates that for every additional rupee spent on these variables, gross revenue of cotton increased by Rs 0.11, Rs 0.84, Rs 0.47, and Rs 0.63, respectively. Among the significant variables, soil fertility was found to be the most important factor input in the production of cotton. The ratios of all other variables namely availability of seed, household income and transport facilities were more than unity but they were found to be non-significant.

The MVP to MFC ratio for all the variables was more than unity for medium farmers producing cotton. It shows that there was scope to increase these resource-inputs to maximize their return. Among the seven variables, four variables namely, availability of irrigation, fertilizer prices, application of new technology and household income were found to have significant influence. The ratio of MVP to MFC for inputs namely irrigation, soil fertility, fertilizer prices and application of new technologies for small farmers producing cotton was 0.11, 0.85, 0.48, and 0.67, respectively.

Thus, it may be inferred from the analysis that the inputs namely availability of irrigation, fertilizer prices and application of new technologies were under-utilized in cotton production in the study area for all the groups of farmers. In the case of marginal and small farmers, household income was underutilized whereas in the case of medium and large farmers, soil fertility was found to be underutilized. It could be concluded that there is a scope for increasing the use of resource inputs in cotton for all the groups of farmers to maximize the return.

Similar ratios have also been computed for marginal, medium, small and large farmers producing paddy. The MVP/MFC ratio for inputs namely irrigation, fertilizer prices and application of new technologies for marginal farmers producing paddy was 0.18, 0.54, and 0.48, respectively. Among the significant variables, fertilizer prices were found to be the most important factor input in the production of paddy. The MVP/MFC ratio for irrigation, fertilizer prices and application of new technologies for small farmers producing paddy was 0.19, 0.55, and 0.48, respectively. The ratios of all other variables namely soil fertility, availability of seeds and transport facilities were more than unity. The MVP/MFC ratio for all the variables was more than unity for large farmers producing paddy. It shows that there was scope to increases these resource-inputs to maximize their return. Among the seven variables, four variables namely, availability of irrigation, soil fertility, fertilizer prices and application of new technology were found to have significant influence.

Thus, it may be inferred from the analysis that the inputs namely availability of irrigation, fertilizer prices and application of new technologies were under-utilized in paddy production in the study area for all the groups of farmers. In the case of marginal and small farmers, household income was underutilized whereas in the case of medium and large farmers, soil fertility was found to be underutilized. It could be concluded that there is a scope for increasing the use of resource inputs in paddy for all the groups of farmers to maximize the return. Thus, it may be inferred from the analysis that the inputs like irrigation, fertilizer prices and application of new technologies were underutilized in paddy production in the study area in the case of all the groups of farmers.

SUMMARY AND CONCLUSIONS

The continuing growth in population calls for an increase in foodgrain production from its current level of 257 MT to even higher levels in the coming years. In the face of serious limitations on increasing land area under cultivation, which is saturated or declining as in the case of Tamil Nadu, the only option is raising farm productivity and fertilizers are an essential input for increasing productivity of foodgrains and other agricultural crops.

Application of chemical fertilizers depends basically upon soil fertility and previous crop grown and amount of organic manure applied. Soil fertility status varies in different agro-climatic zones to a considerable extent. Therefore, common fertilizer dose cannot be recommended for all regions. Due to variations in soil fertility, rainfall and climatic conditions, a common dose of fertilizer cannot be recommended for all regions.

The present study has made an earnest attempt to analyse the trend, pattern, and determinants of fertilizer consumption among different categories of farmers of major crops and suggest the following measures to increase the fertilizer use to achieve the target of agriculture production. The main findings are:

1. The average size of sample operational holding was 1.5 acres for marginal farmers, 3.5 acres for small farmers, 6.5 acres for medium farmers, and 13.0 acres for large farmers, in Villuppuram district. In Vellore districts, it was 2.25 acres, 4.00 acres, 8.00 acres, and 15.00 acres, respectively and in Ramanathapuram district it was 1.5 acres, 4.00 acres, 10.00 acres, and 16 acres, respectively.
2. In the sample areas, farmers usually apply recommended dosages of fertilizer. Some farmers are over-ambitious and they applied more than the recommended level. Generally in the irrigated areas, usually the recommended dosages of fertilizer were applied by all the categories of farmers in the sample areas.
3. There were no major differences in the adoption of fertilizer use technology in the sample villages among the different categories of farmers. The major difference is the presence of personal labour participation the farming operations. Medium and large farmers hire more labourers and the other categories use mostly household labour.
4. Farmers in the dry land agriculture said that the rainfall is not sufficient and timely and hence it had affected the consumption of fertilizers.
5. In Tamil Nadu, green manuring is applied if there is waterlogging. This is applied only for paddy cultivation.
6. All the farmers strongly pointed out that the price of fertilizers is the single most important determinant of the application of fertilizer. They also pointed out that the reduction in the application will reduce the yield of crops.
7. Fertilizer prices have a greater influence on the determination per acre consumption of fertilizer. It implies that 1 per cent increase in the price of the fertilizer will bring down its consumption by 0.1789 per cent.

To increase productivity and sustain soil health, it is essential to adopt improved agronomic practices. This calls for timely sowing, appropriate spacing depending on the variety of hybrids, adequate irrigation during the critical stage, and the balanced use of fertilizers according to soil requirements. Organic manures such as farmyard manure, green manure, crop residue, and recyclable waste are good sources of nutrients.

Selection of HYVs or hybrid seeds based on soil and agro-climatic conditions and seed treatment will also give the desired yield (Saini 1979).

An analysis of the secondary data on the application of chemical fertilizers reveals that the growth rates of fertilizers are higher than agricultural growth rates. The application of N fertilizers is high in the sample areas and the application of P and K are also to the recommended level. The marginal returns compared to cost are also positive.

Among the various determinants of fertilizer consumption, the study found that price of urea, rainfall, and yield have turned out to be the most important variables. The analysis of the sample data also have revealed that the farmers are using more of urea than that of super and potash. Hence, any increase in its price must certainly have implications on the consumption. Therefore the most important policy implication of this study is that the system of fertilizer subsidy needs to be strengthened and its scope may be expanded. Farmers are aware of the comparative advantage of fertilizers and they are willing to apply the recommended dosages and the yield levels respond positively. Minimum support prices for the crops and fertilizer subsidy are to be extended to get the maximum benefit from higher application of chemical fertilizers.

REFERENCES

Baanante, C.A., Bumb, and T.P. Thompson. 1989. 'The Benefits of Fertilizer Use in Developing Countries', IFDC Paper series: P-8, Muscle Shoals, Alabama.

G.R. Saini. 1979. *Farm Size, Resource Use Efficiency and Income Distribution*. New Delhi: Allied Publication.

Times of India. 2015. 'Organic farming: Tamil Nadu dug deep'. *Times of India*. New Delhi.

TNAU. 2014. 'Soil and water test based advisories in Agriculture'. Department of Soil Science and Agriculture Chemistry, Tamil Nadu Agricultural University.

31

Impact Study of Soil-Testing Analysis in Madhya Pradesh

Hari Om Sharma and Deepak Rathi

Soil testing is a chemical process by virtue of which requirement of nutrients for plant can be analysed so as to sustain the soil fertility. The farmers find it extremely difficult to know the proper dose and type of fertilizer, which is suitable for his soil. While, using a fertilizer one must take into account the requirement of the crops and the characteristics of the soil.

The basic objective of the soil-testing programme is to provide a service to farmers for better and more economic use of fertilizers and better soil management practices for increasing agricultural production in their farm. Higher production from high yielding varieties (HYVs) cannot be obtained without applying proper dose of fertilizers to overcome existing deficiencies of soils. Efficient use of fertilizers is a major factor in any programme designed to economically bring an increase in agricultural production.

There are more than 514 soil-testing laboratories in India with a capacity of about 6.5 million samples per annum (2008). In order to provide soil-testing facilities to all 106 million farm holdings in a reasonable period of time, the existing analyzing capacity of the soil-testing programme needs to be augmented almost 15–20 times. Madhya Pradesh is having presently 70 soil testing laboratories and four mobile laboratories to analyze approximately more than 4 lakh samples per annum (2008). The main objective of soil-testing laboratory is to maintain the soil health by analyzing nutrient status of the soils and to give suggestions on the quantities of major nutrients like nitrogen, phosphorus and potassium to be applied to the soils. Micronutrient analysis is also important to know the status of manganese, boron, zinc, and iron, etc. present in the soil and accordingly suggest supplemental application for better plant growth.

Success or failure of soil-testing programmes largely depends on rapidity of providing correct information to farmers, ability of the programme to provide service to a large group of farmers in a particular area, proper analysis and interpretation of results and recommendations that when followed will be profitable to the farmers for the effective utilization of this service to improve local agricultural production Time and quality consciousness in the service is a real challenge for the analysts in the new millennium. This compels the laboratories to adopt rapid, reliable, time saving procedures and methods to meet future requirements. The farmer's confidence in the programme can be established only by demonstrating that it actually provides a means of improving his profit. Looking to the importance of the soil testing in farmers' field, the present study had been taken to evaluate the adequacy, usefulness, effectiveness and contribution of these soil-testing laboratories to the development of agriculture with the following objectives:

1. to assess the soil-testing infrastructure available across different agro-climatic regions/districts of Madhya Pradesh;
2. to analyse the gap between target and achievement of soil test samples and recommendation adopted by the farmers;
3. to evaluate the cost effectiveness of the soil-testing analysis; and
4. to identify constraints in adoption of soil-testing technology by the farmers and to suggest ways and means for proper utilization of these soil-testing laboratories.

METHODOLOGY AND DATA

Soil-testing laboratories of Sagar and Dhar have been selected purposively for the study. The soil-testing laboratory of Sagar district covers farmers of Sagar and Damoh districts and soil-testing laboratory situated at Dhar covers only Dhar district.

Both primary and secondary data were collected for the study. The primary data were collected from respondents with the help of pretested interview schedule related to the year 2009–10 and 2010–11. The secondary data for the period of 10 years, that is, 2001–02 to 2010–11 were collected from the office of Joint Directorate Soil Testing Department of Agriculture, Vindhyachal Bhawan, Bhopal and from respective soil-testing laboratories of Sagar and Dhar (MP) from their published and unpublished records.

RESULTS AND DISCUSSION

Soil-Testing Infrastructure

There were 70 soil-testing laboratories existing in the state covering all the districts. The maximum number of laboratories was found to exist in Malwa Plateau (13), followed by Kymore Plateau and Satpura Hills (11), and Vindhya Plateau (10). The other agro-climatic zones also had more than one soil-testing laboratory in their area. On an average the coverage or catchments per laboratory was found to be 0.63 lakh farmers and 0.47 lakh hectare land or cultivable land. As for as agro-climatic regions are concerned the largest number of farmers covered by laboratories were found in Central Narmada Valley (1.15 lakh), followed by Vindhya Plateau (1.06 lakh), Chhattisgarh Plains (0.70 lakh), and Kymore Plateau and Satpura Hills (0.67 lakh). The coverage of area under each laboratory revealed that laboratory situated in Chhattisgarh plain (Bhalaghat district) covered 0.72 lakh hectares, followed by Central Narmada Valley (0.65 lakh hectares), Northern Hills of Chhattisgarh (0.60 lakh hectares) and Kymore Plateau and Satpura Hills (0.51 lakh hectares). Laboratories situated in other Agro-Climatic Regions covered significant area and provide service to needy farmers. It is also observed from the data that laboratories situated in Satpura Plateau (0.34 lakh hectares) covered the lowest area. This indicates that infrastructure available per lakh hectare was appreciable in Satpura Plateau. In Madhya Pradesh each soil-testing laboratory covered 0.66 lakh farmers and 0.51 lakh hectares.

Gap between Target and Achievement of Soil Test Samples

The gap of 19.95 and 21.18 per cent was noted between target and achievement of soil test samples in Sagar and Dhar districts, respectively. The target of Sagar soil-testing laboratory was found to be same in base year (2001–02) and the current year (2010–11) and the target was found to have decreased in current year compared to base year in Dhar district of Madhya Pradesh. The target was found to be same for both the districts, that is, 10000 soil samples per year. The average of gap of 40 per cent was found between target and achievement. The achievement was found to be increase from 2,197 (base year) to 9,615 (current year) showing growth of 10.87 per cent per annum with the rate of 657.21 sample per year.

Adoption of Soil Testing Recommendation

About 71 per cent farmers received soil-testing report from the respective laboratories of their district. Only 49 (69.01 per cent) farmers adopted the recommendations and applied the fertilizer or other chemical for improvement of their crops, while remaining 22 (30.99 per cent) did not follow these recommendations due to several constraints (Table 31.1).

Table 31.1 Distribution of Sample Respondents (Numbers)

Particulars	Sagar	Dhar	Overall
Total Respondents	50	50	100
Who Received Report	36	35	71
	(72.00)	(70.00)	(71.00)
Who Adopted Recommendation	26	23	49
	(72.22)	(65.71)	(49.00)

Note: Figures in parentheses show percentages to total.
Source: Own field survey.

Table 31.2 Incremental Return after Adoption of Soil-testing Recommendation by the Farmers in Different Crops

Particulars	Soybean		Wheat		Gram		Potato		Garlic	
	Before	After	Before	After	Before	After	Before	After	Before	After
Yield Physical Unit (qtl/ha)										
Main product	15.23	19.76 (29.74)	38.61	46.88 (21.42)	12.35	17.29 (40.00)	123.50	172.90 (40.00)	22.23	29.64 (33.33)
By-product	22.84	27.78 (21.63)	19.30	22.44 (16.27)	7.41	10.37 (39.95)	0.00	0.00 (0.00)	0.00	0.00 (0.00)
Returns (Rs/ha)										
Main product	44171.83	57304.00 (29.73)	46343.29	53860.57 (16.22)	27170.00	38038.00 (40.00)	86450.00	121030.00 (40.00)	66690.00	82992.00 (24.44)
By-product	2284.75	2778.75 (21.62)	1930.97	2244.19 (16.22)	592.80	829.92 (40.00)	0.00	0.00 (0.00)	0.00	0.00 (0.00)
Gross returns	46456.58	60082.75 (29.33)	48274.26	56104.76 (16.22)	27762.80	38867.92 (40.00)	86450.00	121030.00 (40.00)	66690.00	82992.00 (24.44)
Net Income (Rs/ha)										
At variable cost	30685.00	48273.08 (57.32)	34676.69	39881.50 (15.01)	17062.83	27067.07 (58.63)	31534.57	64131.47 (103.37)	37430.30	52524.57 (40.33)
At total cost	22674.50	37915.62 (67.22)	26352.85	30209.12 (14.63)	12271.40	20363.16 (65.94)	16636.23	47446.56 (185.20)	25934.96	38221.76 (47.38)
Return Per Rupee Investment (Rs/ha)										
At variable cost	2.95	5.09	3.55	3.46	2.59	3.29	1.57	2.13	2.28	2.72
At total cost	1.95	2.71	2.20	2.17	1.79	2.10	1.24	1.64	1.64	1.85

Note: Figures in parentheses show percentage change.
Source: Own field survey.

The per-hectare expenditure on seed, fertilizer and plant protection measures increased for all crops after adopting soil-testing analysis recommendation. The per-hectare expenditure on labour was also found to be higher in all crops except soybean. The cost of cultivation and cost of production of all the crops reduced drastically, while return per rupee investment was found to be increased after adaption of recommendation of soil testing (see Table 31.2).

Constraints in Adoption of Soil-Testing Recommendations

The lack of knowledge about soil-testing technology (70 per cent), non-receipt of soil-testing report (62 per cent), less cooperation from officers of agriculture department (46 per cent) and complicated method of testing soil sample (30 per cent) were found to be the main constraints in adoption of soil-testing recommendations.

CONCLUSIONS AND POLICY IMPLICATIONS

The present infrastructure of soil-testing facility is found to be insufficient in different agro-climatic regions of Madhya Pradesh. Whatever infrastructure is available is not functioning properly. Therefore, targets and achievements need to be increased by employing skilled and trained staff in these laboratories. The quantity and quality of soil samples tested also need to be increased in the state.

There is ample scope to improve the analysing capacity as well as dissemination ability of the soil-testing laboratories. Coupled with professional management and proper linkage with farms, all this can bring radical changes in the soil-testing service in the state to meet the farmers' satisfaction to the maximum extent.

Special care may be taken for collection of representative soil samples. Validity of sample has to be ensured at all levels-starting from collection stage to storage in laboratory even after analysis. Since the reports, when sent through usual postal system, are often not received in time by the farmers, a system of online communication of reports may be started by which the soil-testing laboratory may send the report to the Block Development Officer (BDO) at least to reduce postal delays. The farmers often visit BDO's office for various other activities and may be able to collect the reports from there. This however also presupposes that all the soil-testing laboratories are provided with computer facilities. Keeping the cost in mind, the system of on-line communication reports may be started in the selected laboratories initially and then all the laboratories will have to be covered simultaneously.

The laboratories may be kept informed on the outcome of the recommendations made by them on fertilizer use at least on representative and typical case by case basis, for example, where the recommendation has given as expected or better than expected results and where results fell short of expectations.

The Department of Agriculture should be assigned to ensure effective and live linkages between the field and the laboratory. It is to be appreciable if each laboratory may adopt at least one nearby village from where sample may be collected by the laboratory staff and recommendations are also communicated or handed over directly by the laboratory staff to the farmers and to follow the outcome of the programme. Each laboratory can take up one village as a mission to see the utility of the programme by itself and find out shortcomings so that the whole programme can be improved based on such direct observation and intense study. Presently, the laboratories are literally cut off from the field and they work in isolation from the whole programme.

The state government in Madhya Pradesh is already charging a fee of Rs 5.00 per sample but it is too less. A sufficient fee will bring accountability on the part of the laboratory to make a sound recommendation because farmers will participate in sample collection or at least will know that a sample has been collected and will be expected to appreciate the value of the report received based on the part of the cost borne by them. They will start asking the question if report is not received in time or is not found to be useful when the recommendation is followed as advised by the laboratory. Charging the fee will also help the states to supplement the requirement of funds by the laboratories. A minimum fee of Rs 20 per sample analysis may be suggested to be charged. Estimated cost of analysis of a sample is approximately Rs 80 for physical parameters plus NPK analysis while with tests for micronutrients the cost would be about Rs 100 (Only chemicals and 20 per cent of glass breakages are considered as part of the cost for this purpose).

Soil analysis and fertilizer recommendation is only a part of the soil-testing service. To a good measure, the efficiency of the service depends upon the care and efforts put forth by extension workers and the farmers in the collection and dispatch of the samples to the laboratories and in obtaining reports timely. Its effectiveness also depends upon the proper follow up in conveying the recommendations to the farmers, including the actual use of fertilizer according to the recommendations. The role of extension service, soil chemists and the agronomists in the field is important. The service is suffering both from technological shortfalls and due to inadequate and untrained manpower involved. Weakness of the programme in its various aspects as discussed above needs improvement.

The soil health card issued to the farmers may be periodically updated so that the farmers can be made aware about the changing fertility status of their land. This card may also be useful to the farmers in getting loans for agriculture purposes where agricultural value of the land may be one of the factors.

If the fertilizer industry will venture to produce and promote the products based on requirement created by specific soil nutrient deficiency, the industry will have to get into the soil-testing programme in a big way and generate such information as a measure supplementary to soil-testing programme basically being run by the government. The fertilizer industry may adopt at least one district in a state and monitor how the fertilizer in the adopted district is used based on the basis of plant nutrient deficiency record determined through accurate soil-testing.

The awareness about soil-testing facility, its need and significance at the farmers' level is important for the success of the programme. Extension activities will have to build up awareness among farmers that the adoption of recommendations of soil testing reduces their cost of production and increases returns leading to the popularization of the fact that soil testing benefits farmers. Sufficient field staff with trained personal quartered at the village level to take up new methods as well as local demonstrations of technologies will help in the dissemination.

32

Agricultural Machinery Industry in India

A Study of Growth, Structure, Marketing and Buyer Behaviour

Sukhpal Singh

Agricultural machinery and implements are an important factor in agricultural production and productivity enhancement. There are direct as well as indirect effects of agricultural machinery and implements on productivity through better use of other inputs, more efficient and timely completion of agricultural operations and increase in cropping intensity (Venugopal 2004). But, the adoption of machines is the result of many factors at the farm level such as size of landholding, irrigation, labour, credit and risk orientation, and socio-economic profile of the farmer. Still, level of mechanization of agriculture in India remains low.

Agricultural machinery and equipment industry comprises of a large number of segments even in the organized sector. Tractor industry is one of the most capital-intensive industries in agricultural machinery industry with more than half a dozen major players in the market at present. In combine harvesters which got manufactured in India in 1970 for the first time, Punjab occupies a dominant place with more than 87 per cent of the units located in the state. They were producing two types of combines—self-propelled and tractor operated. The other major parts of the industry are electric motors, diesel engines, pump sets, power tillers, drip and sprinkler systems, and tractor-driven implements. Many of these industries are characterized by subcontracting and ancillary systems where small units work for the larger parent units supplying components or performing specific tasks in manufacturing process.

Though tractor industry is well established and there have been studies of tractor purchase and use behaviour during the past, there have been no recent studies on the growth, structure, and competition dynamics and the firm behaviour and strategies in the light of the changing nature of the tractor market in terms of its location shift and saturation in the traditional pockets of it symbolized by the emergence of second hand tractor markets and entry of many new homegrown players in the market.

On the other hand, not much is known about the structure and organization of the combine harvester and micro irrigation industry. At least, there is little documented evidence on these industries in India. Also, many of the combine harvester makers have diversified into tractor manufacturing and marketing more recently. There are some of the homegrown tractor producers in the country producing either both tractors and combine harvesters or only tractors. They have carved significant niche for themselves in the recent past. Therefore, it is interesting and important to examine this phenomenon of indigenous growth and diversification in an increasingly competitive market for tractors. Further, there are no studies on micro irrigation equipment industry which is growing in importance due to changing cropping pattern and need for and focus on water saving technologies.

OBJECTIVES

In the light of the above, the chapter examines the nature of growth, and market structure in the tractor, thresher/reaper/combine harvester and micro irrigation equipment industries; explores the changing nature of demand in these industries especially tractor and harvesting industry; analyses the marketing and business strategies of the various types of players in this market such as bundling of inputs, diversification, consolidation, etc., and examines the tractor, combine harvester and micro irrigation equipment market with focus on buying aspects and underutilization including the emergence and functioning of the second-hand tractor markets. The next section details out the methodology followed. The third section discusses major findings followed by conclusions and policy implications in the last section.

RESEARCH METHODOLOGY

Three industries within the agricultural machinery sector—tractor, combine harvester and micro irrigation have been studied with focus on major locations of the industry especially that of the tractor and the combine harvester industry. The study begins with secondary data based analysis and goes on to case studies of firms in the three industries with a view to understand their marketing and business strategies. It also analyses dealer and farmer level data to understand distribution-level issues and strategies. Given the concentration of combine harvesting industry in a few towns in Punjab (Nabha and Bhadson in Patiala District) which accounts for 87 per cent of the total units in India, and the emergence of some of these players as tractor manufacturers and marketers, besides the location of more than half a dozen tractor players in that region (Punjab, Haryana, and Himachal Pradesh) (International Tractors, Preet Tractors, Standard Tractors, Indo Farm Tractors, Punjab Tractors, and the HMT), most of the case studies are conducted in these areas and focus on players who are into both combine and tractor business or were earlier into threshers business. Bhadson in Patiala is one of the natural modern small-scale-industry-based high export potential horizontally coordinated agricultural implements cluster which has high scope for technological upgrade. Similarly, Moga district is horizontal natural cluster for wheat threshers with low export potential and high scope for technological upgrade. Further, Punjab's tractor density (179 tractors per 1000 farm households) compares extremely with Gujarat (31) and Maharashtra (13) as the two regions are at almost opposite ends of the scale (NSSO Report No. 497).

On the other hand, some of the micro irrigation equipment firms, most of whom are based in western India, have been studied to understand their growth and business strategies (Table 32.1). Dealer-level analysis is carried out to understand the distribution dynamics in tractor and micro irrigation equipment sectors. There was no practice of dealership in combine harvester industry. Farmer-level analysis is carried out to understand the purchase and use practices in all the three industries.

RESULTS AND DISCUSSIONS

Tractors

The landholdings of tractor owners were by and large medium or large with average owned holding being 11.15 acres and average operational holding being 12.38 acres. Small and very large farmers did not lease in land. It was medium and large farmers who leased in some land to augment their own holdings. Almost 83 per cent of the sample farmers had purchased new tractors and the rest second hand tractors. Requirement for farming was the only reason

Table 32.1 A Profile of Units, Dealers and Farmers Studied Across Industries

Industry Sector	Location/s (State/s)	Companies/Units Studied	Dealers Interviewed	Farmers Interviewed
Tractors	Punjab and Gujarat	Two new large and two small tractor manufacturers	36	23
Combine harvesters	Punjab, Gujarat and Maharastra	3 large and 16 small-scale	–	42
MI equipment (drip/sprinkler)	Gujarat	4	14	34
All		26	50	99

Source: Own field survey.

for buying tractor. Major consideration in purchase of the tractor were oil efficiency, horse power, strength, price and life of the tractor in that order across both brands, that is, Sonalika and Standard. Most of the farmers (78 per cent) had bought 50 hp tractors with the next highest (13 per cent) being 60 hp showing a clear preference for higher hp tractors. Most of the new tractors were bought from dealer and about 17 per cent from other farmers, second hand market or directly from the company as the company was locally located. About 2/3rd of the farmers had bought tractors on cash basis with the rest buying on credit from banks.

Most common use of tractors was 401–600 hours per annum with 30 per cent farmers reporting that (Table 32.2). The next major group was the one which used it between 1001–1500 hours per annum which is more than the minimum use of tractor for viability. Further, kharif had more tractor usage than rabi largely due to paddy cultivation. But, still, on an average, the tractor was used only for 751 hours which is much below the NABARD norm for viable use of the machine. It was surprising that about 9 per cent farmers used it for less than 200 hours and another 13 per cent for less than 400 hours altogether during the year. 70 per cent of the owners were using it for less than 1,000 hours—a minimum prescribed by NABARD for viability.

Though about 40 per cent farmers did not report any major problem in purchase and usage of tractors, more frequent problem for many others was poor quality of spare parts followed by other problems which included main seal leak, high price of spare parts, leaking in zian (leakage happening in the instrument), high noise, slow lift, low flexibility, starting trouble, not powerful on load, and slip problem.

Second Hand Markets for Tractors in Punjab

There are daily, weekly, and fortnightly markets for the sale and purchase of second hand as well as new tractors in various market towns of Punjab where thousands of farmers participate and sell or buy tractors (primary survey). In a state which today has more than 4.6 lakh tractors accounting for one-fourth of the total population of tractors in the country with just 2.5 per cent of cultivated area, this phenomenon is both encouraging as well as disturbing. With only about 11.17 lakh operational holdings in the state, it means that every third holding in the state is equipped with a tractor. In some villages, there is a tractor for every five acres of land. Added to this is the fact that more than 70 per cent of the farms are below 10 acres each (Government of Punjab [GoP] 2005). Moga and Barnala (both district towns) are the largest markets in terms of number of farmers visiting and the number of tractors brought for sale with Talwandi Sabo (in Bathinda district) emerging as the third largest.

There are more than a dozen tractor markets mainly in the cotton belt of the state, which facilitate buyers and sellers of second hand tractors in quick transactions. There are different reasons for the sale of old tractors which relate to the larger agricultural economy of the state. The non-viability of the tractors due to the small size of the holdings,

Table 32.2 Use Pattern of Tractors

Particulars	Uses in Hours							
	≤100	101–200	201–300	301–400	401–500	501–600	601–650	Total
Kharif								
No. of farmers	4	3	4	1	6	4	1	23 (355.39)
% farmers	17.39	13.04	17.39	4.35	26.09	17.39	4.35	100.00
Rabi								
No. of farmers	5	4	6	3	2	3	–	23 (295.00)
% farmers	21.74	17.39	26.09	13.04	8.70	13.04	–	100.00
Other Uses								
No. of farmers	9	2	1	1	1	–	–	14 (165.71)
% farmers	64.29	14.29	7.14	7.14	7.14	–	–	100.00
Total								
Category of use hours	**≤200**	**201–400**	**401–600**	**601–800**	**801–1000**	**1001–1200**	**1201–1500**	**Total**
No. of farmers	2	3	7	2	2	3	4	23 (751.26)
% farmers	8.70	13.04	30.43	8.70	8.70	13.04	17.39	100.00

Note: Figures in brackets indicate average hours of tractor use in respective season.
Source: Own field survey.

Table 32.3 State and Company-Wise Distribution of Dealers by Monthly Sales of Tractors/Outlet

Sale (No. of Tractors)	Company			State		All
	Indofarm	Sonalika	Standard	Gujarat	Punjab	
	No. of Dealers					
≤5	3 (75.00)	12 (66.67)	13 (92.86)	8 (57.14)	20 (90.91)	28 (77.78)
6–10	1 (25.00)	5 (27.78)	1 (7.14)	5 (35.71)	2 (9.09)	7 (19.44)
11–15	–	–	–	–	–	–
16–20	–	1 (5.56)	–	1 (7.14)	–	1 (2.78)
Total	4 (100)	18 (100)	14 (100)	14 (100)	22 (100)	36 (100)

Note: Figures in brackets show percentage of total sample dealers.
Source: Own field survey.

domestic financial crisis, repayment of bank or other loans, purchase of land, change of model/brand/horse power of the tractor, lack of business for tractors due to their over-population and competition, sale and purchase of tractors as a business proposition, and change of occupation are the major reasons for sale of tractors in these markets (Singh and Kolar 1998). The buyers are satisfied with the operations of these markets as they find a ready market and can liquidate the tractors and other equipment in a short time to meet the exigencies. However, only a small proportion of those who sold tractors, bought another one from the tractor market (primary survey).

The operations of these markets are in the hands of a number of informal groups known as 'mandis' in each market. They are nothing but collectivities of a few individuals (5–15) who operate as commission agents inside as well as outside the market. These groups generally lease in land for the market, outside the town on an annual basis and share this cost among them. This is the major cost. Other costs are working costs for facilities like tents, chairs, etc. About a few hundred tractors in a small market and a few thousand tractors in bigger markets are brought every time (weekly or fortnightly) a market is held, and a few (20–50) are sold every such time. The tractor model and price are displayed on the tractor to facilitate buyer–seller interaction. The commission agents also provide other facilities like space for parking the tractors, and chairs and tents for the farmers. On every transaction, the agents charge Rs 300–500 up to a transaction of rupees one lakh and Rs 600–1000 on a transaction of above rupees one lakh each from the buyer as well as the seller. The payments are made either on the spot or within a week after paying security. The commission agents are responsible for ensuring the payments (Table 32.4).

Besides the facilitation of business among farmers at one place which lowers search and transaction costs, these markets also generate employment for those who cater to the needs of the people assembled in the market. Further, since these markets supply second-hand tractors, small and medium farmers are able to mobilize money to buy these tractors which are low cost and easily available besides being relevant for these classes of farmers. Recently, even the tractor sales agencies have realized the value of these markets and have started displaying their new models of tractors in these markets so that farmers are made aware of their features and the new product is publicized among the potential buyers.

These markets are totally unregulated by any government agency, but they function fairly efficiently so far as farmers are concerned. Only in one of the markets, the District Collector has allowed the union of these mandis to issue licenses (identity cards as agents for tractor sales and purchase) to the members of the mandis. Also, the buyers and sellers are made to pay a Red Cross fee of Rs 100 per transaction each (primary survey).

Combine Harvesters

The combine owners were large landholders in Gujarat with average size of operations holding being 23 acres and owned holding being 19 acres compared with 14 and 9 acres in Maharashtra and 15 and 13 acres in Punjab. In Gujarat, no combine owner was smallholder or even medium holder (<10 acres) while in Maharashtra and Punjab there were some small, medium landholder and even landless persons in Maharashtra had combines (Table 32.5).

In Maharashtra, both types of combines were equally preferred [TD:SP = 55:45]. Punjab and Gujarat had huge majority of tractor driven combines over self-propelled. Reasons for preferring self-propelled combine included high work efficiency, followed by multi crop use, low maintenance and long life. This version was mostly preferred by large farmers for whom efficiency of farm operations was essential to reap economies of scale. On the other hand, tractor driven was driven by tractor's use in other

Table 32.4 Profile of Second-Hand Tractor Markets in Punjab

Name of Market Parameter	Talwandi (Bathinda)	Jhunir (Mansa)	Mour (Mansa)
Frequency of market	Weekly	Weekly	Weekly
Day of market	Wednesday	Thursday	Friday
No. of facilitators	60	7	12
Major buyers	Farmer, scrapper, tractor agent	Farmer and tractor agent	Farmer and tractor agent
No. of tractors in market for sale	20–30	20–40	15–20
Staff at each facilitation centre	6–7	6–7	5–6
Major brand for resale	Eicher, Farmtrack	Eicher (90%)	Eicher (90%)
Place of origin of second-hand tractors	Punjab, Haryana, Rajasthan	Malwa region only	Malwa region only
Rent of plot used as market place (annual)	Rs 30000 to 60000	–	–
No. of tractors sold per facilitator	6–7	3–4	1–3
Share of second hand tractors in market	79%	99%	100%
Facilitators' cost per mandi day	Rs 800 to 1000	Rs 800 to 1000	Rs 800 to 1000
Facilitator's Commission	1% of sale price	Rs 300 each from both the parties	Depending on competition
Margin for each facilitator (annual)	Rs 60000 to 100000	No response	No response
Regulation by committee	Yes	No	No
Other fees	Rs 100/tractor given to Red Cross	Rs 100/tractor given to Red Cross	Rs 100/tractor given to Red Cross
Other businesses of facilitators	Agriculture, labour	Agriculture, labour	Agriculture, labour

Source: Primary Survey.

Table 32.5 Distribution of Combine Owners by Type of Combine Owned—State and Farm Category-Wise

LHC (ha) Particulars	Landless		<2.00		2.00–4.00		4.01–10.00		10.01–25.00		>25		Total	
	No.	%	No.	%	No.	%	No.	%	No.	%	No.	%	No.	%
Gujarat														
Tractor Driven	0		0		0		1		3		2		6	
Self Propelled	0		0		0		0		0		1		1	
Total	0		0		0		1	14.3	3.0	42.9	3.0	42.9	7	100.0
Maharashtra														
Tractor Driven	0		0		0		2		2.0		1.0		5	
Self Propelled	1		1		0		0		1		1		4	
Total	1	11.1	1	11.1	0		2	22.2	3	33.3	2	22.2	9	100.0
Punjab														
Tractor Driven	0		0		4		9.0		8.0		3.0		24	
Self Propelled	0		0		0		0.0		2.0				2	
Total	0		0		4	15.4	9	34.6	10	38.5	3	11.5	26	100.0
All														
Tractor Driven	0		0		4		12.0		13.0		6.0		35	
Self Propelled	1		1		0		0.0		3.0		2.0		7	
Total	1	2.4	1	2.4	4	9.5	12.0	28.6	16.0	38.1	8.0	19.0	42	100.0

Source: Own field survey.

field operations. It was, therefore, logically preferred by farmers with low landholdings. Other important factors were low price, low fuel and labour cost, and suitability for less land area required for harvesting. Seventy per cent of the farmers in Gujarat and 80 per cent of the farmers in Punjab used combines of a capacity of 50–60 HP. This ratio does not hold for Maharashtra where almost 45 per cent of the farmers use combines of capacity 105–110 HP.

The capacity of combines was higher in Maharashtra due to predominant use of these machines for custom hiring there.

In Punjab, popular brand image played a very important role in the purchase decision with 100 per cent farmers factoring this in their purchase decision. Goodwill, experience of other farmers and resale value came next with almost the same proportion of farmers giving importance to each of these factors (around 20 per cent). On the other hand, in Gujarat, dealer advice (57.2 per cent) and popular brand image (42.9 per cent) influenced the purchase decision of a majority of farmers. Maharashtra farmers went by popular brand image (33.3 per cent) and good experience of other farmers (44.4 per cent), though other factors also had an influence in case of 77.7 per cent farmers. Other factors included affordable price, clearance, low maintenance, power steering, high work efficiency, local dealer, easy availability of spare parts, relationships with company, easy for use/driving, fine cutter, pressure of relatives, and dealing.

Apart from Maharashtra, where economic consideration was the major reason to buy a combine, the prime reason was farming in Gujarat and Punjab. This might be due to the high land and labour costs in Maharashtra as compared to Gujarat and Punjab. Brand name and low repair and maintenance cost formed the major chunk of decision making in Gujarat and Maharashtra. However, in Punjab, the prime considerations were cleaning and the material of the combine. Some other factors considered were work efficiency, easy to operate, power steering, engineering and technology, low fuel consumption, multi crop use, low man power requirements, no crack, long life, shafts, cross, facilities, combine barma (mechanical instrument), roller, quality of parts, belt, bearing, should be all material of one company, heavy, and resale value.

Annually, combine was used for 41–55 days in general by one-third of owners, 26–40 hours by 16 per cent farmers and 56–70 hours by 19 per cent farmers. It was used for higher number of days (>70 days by 88 per cent owners) in Maharashtra due to custom hiring practice and lesser number of combines in the region so far. Hiring of combines was prevalent in Gujarat and Maharashtra. Around 45 per cent farmers in Gujarat and 80 per cent farmers of Maharashtra hired out combines for more than 90 per cent of their usage.

MICRO IRRIGATION

On the whole, 65 per cent of the farmers had part of their land under micro irrigation scheme (MIS) and another 35 per cent their entire land under MIS and they mostly used electric motors (99 per cent) for extraction of groundwater. More than 50 per cent farmers has one electric motor each and another 24 per cent two motors each with rest owning more than two motors each. This was so given the depth of groundwater in the regions. The major crops under MIS were groundnut, cotton, potato, mango and papaya and other horticultural crops. 95 per cent the MIS area was on landholdings of above 10 hectares with only 5 per cent being on holdings of below 10 ha. So far as source of information for MIS technology was concerned, a large chunk each of the farmers had come to know of the technology from neighbouring farmers, company dealers and from both of the channels each (26 per cent and 17 per cent each).

Water saving, energy saving and solving labour problem were major reasons behind MIS purchase. It was followed by economical, effective and careful use of fertilizers and pesticides for buying MIS. Farmers selected sprinkler system in water rich and sandy soil areas and preferred drip in water scare areas. Major brands were Jain, Netafim, Prixit, and Captain and Nandan in terms of proportion of farmers and Netafim, EPC, Captain, and Nandan and Jain in that order in terms of share of area under MIS (Table 32.6). Four brands—Netafim, Jain, Prixit, and Nandan were the major players in terms of area covered. The major reasons for preferring a particular brand were good product quality and/or ISI standards, good experience of other farmers, company approaching farmers and the brand image besides relationship with dealer, good services and low cost of the

Table 32.6 Distribution of MIS Dealers by Annual Sales Turnover (Rs in crore)

Company	Total Sale	MIS Sale in Total Sale	Average Business Sale/Dealer	Average MIS Sale/Dealer	Share of MIS Sale in Business Sale (%)	Share of Main Outlets in Business Sale (%)
Jain	6.30	3.85	1.58	0.96	61.11	96.43
Netafim	15.92	15.92	5.31	5.31	100.00	100.00
Prixit	2.71	2.59	0.90	0.86	95.39	47.97
Plastro	12.24	11.84	3.06	2.96	96.73	75.69
Total	37.17	34.20	2.66	2.44	92.00	87.60

Source: Own field survey.

Table 32.7 Distribution of Area Irrigated by Crops under MIS (Area in ha)

Category of Farmers Crops	<2 ha	2–4 ha	4–10 ha	10–25 ha	>25 ha	Total	% of Seasonal Area	% of Gross MIS Area
Kharif Crops								
G'nut	0.00	0.80	12.00	26.00	12.00	50.80	43.90	17.26
Cotton	0.00	0.40	25.43	12.48	4.00	42.31	36.57	14.37
Castor	0.00	0.00	5.68	4.00	0.00	9.68	8.37	3.29
Fennel	0.00	0.00	3.12	2.40	0.00	5.52	4.77	1.88
Others	0.00	1.40	0.00	6.00	0.00	7.40	6.40	2.51
Season Total	**0.00**	**2.60**	**46.23**	**50.88**	**16.00**	**115.71**	**100.00**	**39.31**
Rabi Crops								
Potato	0.00	0.00	11.52	39.28	37.60	88.40	92.70	30.03
Wheat	0.00	0.80	0.24	4.40	0.00	5.44	5.70	1.85
Others	0.00	0.00	0.72	0.80	0.00	1.52	1.59	0.52
Season Total	**0.00**	**0.80**	**12.48**	**44.48**	**37.60**	**95.36**	**100.00**	**32.39**
Summer Crops								
Bajra	0.00	0.00	2.40	9.60	0.00	12.00	59.17	4.08
Mung	0.00	0.00	0.00	5.20	0.00	5.20	25.64	1.77
Others	0.00	0.00	1.08	2.00	0.00	3.08	15.19	1.05
Season Total	**0.00**	**0.00**	**3.48**	**16.80**	**0.00**	**20.28**	**100.00**	**6.89**
Perennial Crops								
Mango	0.00	2.40	4.40	14.40	0.00	21.20	33.63	7.20
Papaya	0.00	0.00	0.00	0.00	14.72	14.72	23.35	5.00
Vegetables	2.60	3.20	0.00	0.00	0.00	5.80	9.20	1.97
Others	0.40	5.52	4.60	8.80	2.00	21.32	33.82	7.24
Season Total	**3.00**	**11.12**	**9.00**	**23.20**	**16.72**	**63.04**	**100.00**	**21.41**
All Crops	**3.00**	**14.52**	**71.19**	**135.36**	**70.32**	**294.39**		**100.00**
% of Grossed MIS Area	**1.02**	**4.93**	**24.18**	**45.98**	**23.89**	**100.00**		

Source: Own field survey.

equipment and documentation responsibility being taken by the dealer.

The cost of MIS varies from crop to crop. It is higher for narrow spacing crops such as potato and lower for wider spacing crops such as castor and cotton. However, the return on investment is higher in commercial crops and lower in traditional crops or subsistence crops. Nearly 83 per cent farmers did not face any problem at the time of purchasing MIS, 12 per cent farmers got late delivery and 6 per cent farmers could not get the system in time due to delay in processing of documents by the MIS companies. Farmers faced major problem with Jain dealers, especially in Bayad taluka of S.K. district. Surprisingly, 37 per cent farmers did not receive any after sales service, another 20 per cent only need based and another 11 per cent agronomical, technical and acidification services. Only large players like Jain, Netafim, and Prixit provided some after sales services, not the local companies. No farmer was getting after sale services before Gujarat green revolution company Ltd. (GGRCL) intervention.

Most of the farmers (80 per cent) bought the equipment from dealers followed by sub-dealers (11 per cent). One farmer each bought it from manufacturer, local shop, and farmer each (3 per cent each). The place of purchase was district or block headquarter in 60 per cent and 35 per cent case respectively. Two-thirds of farmers had bought it on credit and 1/3rd on cash payment basis. 71 per cent farmers had knowledge about GGRCL and 69 per cent farmers had taken benefit of its subsidy. 51 per cent farmers had a good experience, 23 per cent did not have direct contact with GGRCL and equal percentage of farmers did not have any experience of dealing with GGRCL.

Interestingly, many dealers across companies had appointed sub-dealers or agents ranging from three to as many as 13 with the average number of such sub-dealers being seven. The level of competition measured in terms of number of dealers in the area ranged from minimum of two in Bayed to as high as 11 in Bhuj district. Each dealer on an average covered 350 ha in a year ranging from minimum of 145 ha to as much as 570 ha. Further, three-fourths of the sales were to loanee farmers and only one-fourth to non-loanee farmers. About 86 per cent dealers faced different types of business problems. Short supply of material is a major problem for about 43 per cent dealers. About 21 per cent dealers faced problems such as lack of support from company in providing after sales services, following up bank finance procedure, and delay in releasing dealer's commission.

Farmers are now becoming more aware about the benefits of micro irrigation system. They know that in the increasingly water-scarce environment, only micro irrigation could give them sustainable means for agriculture. The increasing sale of drip system in the reduced subsidy regime is also an indicator of farmer's reducing dependence on subsidy and they are now ready to pay higher amount for improved technology. But considering the economic condition of average Indian farmer, the following reasons would be helpful for irrigation companies and the government at various levels, in promoting micro irrigation (see Table 32.8).

To the average farmer, finance is a major problem. So, companies should come up with some solutions like tie-up with banks, financial institutions, etc., to provide easy loans to the farmers. The limit of Rs 50000 on getting loans without collateral should be increased to rupees one lakh or the redeemable value of the micro irrigation equipment and lease financing for micro irrigation by the manufacturing firms to provide credit support, like in the case of a car, should be promoted. Preference should be given in the matter of bank loans for digging wells and electricity connection to those opting for drip irrigation.

Table 32.8 Distribution of Farmers by Reasons for Buying MIS (multiple responses)

Reasons	No of Farmers	% of Farmers
Water saving	31	88.57
Energy saving	23	65.71
Labour saving/solving labour problem	19	54.29
Increase crop productivity	18	51.43
Reduce fertilizer cost	9	25.71
Soil improvement	3	8.57
Increase area under irrigation	3	8.57
Solving weeding problem	3	8.57
Commercial crop cultivation	2	5.71
Water quality problem	1	2.86
Good soil aeration	1	2.86
Saving land leveling cost	1	2.86
Reduce production cost	1	2.86

Source: Own field survey.

After sales service to the farmers is a must to enable them to derive the maximum benefits that drip irrigation technology can offer to them. Surprisingly, 37 per cent farmers did not receive any after sales service, another 20 per cent only need based and another 11 per cent agronomical, technical and acidification services. Only large players like Jain, Netafim, and Prixit provided some after sales services, not the local companies. No farmer was getting after sale services before the GGRCL intervention. GGRCL has included the after sale services, especially agronomical and technical (repair and maintenance) in its MIS package being offered to farmers who are availing subsidy. Companies were very poor in providing agronomical services but good in providing technical services. This was mainly because local dealers were responsible for technical services while agronomical services were provided by the respective company. Companies appointed agronomists on workload basis. These agronomists had to cover larger area under their jurisdiction and hence sometimes they could not visit all client farmers' farms but provide telephonic services to the farmers. They prepared and followed their own schedule of visit as a part of their routine job responsibilities. Many companies lack of adequate number of agronomists.

Pepsee systems are not complete substitutes for highly sophisticated drip technologies. Though returns offered by micro-tubes and drip kits are higher than those offered by Pepsee. Pepsee systems are viewed as a 'stepping stone' to adoption of a higher degree of sophistication and higher cost technologies and if these technologies are designed in such a way that the transition is made simple and modular, the results can be very positive. Thus, as the farmers are convinced about the results, become familiar with the technology and possibly also improve their financial status in the process, they will shift to the more efficient technologies being marketed today.

CONCLUSION AND POLICY IMPLICATIONS

Given the small size of holdings, it is necessary to encourage custom hiring of agricultural machinery especially tractors and combines to raise productivity while maintaining costs of cultivation. This can be done by encouraging panchayats, PACS and agricultural machinery cooperatives/companies, clinics or even private entrepreneurs with loans, training, and subsidies. Easier and wider availability of credit for second-hand machines can also come in handy to deepen agricultural mechanization. Since every village now owns 15–20 tractors in northern region, it is necessary to establish agricultural machinery service centers at the village level or the cluster of village's level. Also, contract farming can encourage mechanization as the crops grown under this arrangement are new, more amenable to mechanization and come with good agricultural practices which cut costs or raise yields and productivity.

Combine harvester is a costly machine and requires good and careful usage to attain viability. As seen in average annual use figures, in Punjab and Gujarat, it is not being used enough (only around 50 days in a year) compared with Maharashtra where it was primarily bought for custom hiring and was used for as much as 90 days. In order to ensure its viability there is need to exercise caution in its funding and ownership. Banks should insist on definite business plans before approving loans for farmers so that second hand combine markets like tractors do not emerge in India. There is need to provide for collective purchase of combine harvesters by farmers' groups or PACS like tractors so that local availability during times of need especially during peaks of harvesting seasons can be ensured.

Further, due to the labour displacing nature of combine harvesting, there is a need to exercise caution in promotion of combine harvesters. It not only ends up displacing labour but also causes shortage of fodder. But, despite this obvious cost and time advantage from harvester combines, the 2008–09 seasons witnessed something unusual: Manual harvesting is back and combine are out of favour suddenly. This time, not more than 40 per cent of the wheat has been combine harvested. The main reason given by them is better straw recovery. One acre yields 20 qtls of wheat and an equal quantity of straw. Through manual harvesting, one can recover almost this entire straw, whereas the combine-reaper would salvage only 10–12 qtls. This is because the combine operates 30–40 cm above the ground and the left-over stalk gives less straw. In normal years, straw yields do not matter much. But this time, with straw prices ruling at Rs 400–500 per quintal (against last year's Rs 200–300), farmers have found it worthwhile to invest extra time and money in manual harvesting, instead of combines. This can hit the dairy industry hard which is an important allied sector crucial for local livelihoods in many parts of India.

In the drip irrigation technique, labour saving on land preparation is a major benefit which farmers have not yet realized. Irrigation companies should highlight this benefit to attract customers for their products. The companies should ensure quality components like drippers, emitting pipes, filters, etc., so that farmers face no major technical problems like system clogging, which affect their motivation on drip irrigation. The decreasing subsidy indicates that low cost would be key factor for survival in the future. Therefore, companies should come up with more cost effective products and business strategies.

There is need for educating the farmer that the plant can do with less water than provided to the plant in conventional furrow irrigation. Farmers need to be educated through extension services and publicity on the effectiveness of drip irrigation, especially, for narrow spaced crops like sugarcane, cotton, etc. Micro irrigation equipment manufacturing companies should be involved intensively in promoting the method through frequent field demonstrations at the farms. Word of mouth and demonstration are the biggest promotional strategies in the industry. What is needed is a promotion at the grassroots to change perceptions and educate farmers on how the drip can benefit them apart from water saving.

Rather than advertisement, personal selling is a more effective way of MIS marketing. Farmers should be educated on concomitant use of liquid fertilizers through pipe network and their reservations on system clogging should be dispelled through frequent demonstrations. Further, importance of liquid fertilizer in increasing input efficiency and bringing down the cost of cultivation should be clearly brought home by effective extension.

REFERENCES

Government of India. 2005. 'Situation Assessment Survey of Farmers—Income, Expenditure and Productive Assets of Farmer Households'. NSS report 497, National Sample Survey Organization, Ministry of Statistics and Program Implementation, New Delhi.

Government of Punjab (GoP). 2005. *Statistical Abstract of Punjab, 2004*, Publication No. 905. Chandigarh: Economic Adviser to Govt. of Punjab, Economic and Statistical Organization (ESO).

Singh, Joginder and J.S. Kolar. 1998. 'Why Farmers are Forced to Sell Tractors?', *The Tribune*, Chandigarh, 6 June.

Venugopal, P. 2004. *Input Management*, volume 8 of series: *State of the Indian Farmer–A Millennium Study*. New Delhi: Academic Foundation.

33

Farm Mechanization in India with Special Focus on Eastern Region

C.S.C. Sekhar and Yogesh Bhatt

The share of agriculture in aggregate gross domestic product (GDP) has come down rapidly over the past decade. However, the proportion of the population still dependent on the sector makes it crucial in development policy making. Although farm mechanization played a major complementary role to input usage in the initial stages of the green revolution in India, it has not progressed as desired (Rao 1975). Mechanization is important for the progress of modernized agriculture. Tractors, seed drills or tube wells (pump sets) enable a farmer to save time and grow an extra crop, or devote more area to existing crops. Tube wells help the farmer exercise better control over the quantum and the timing of irrigation, thereby helping him realize higher yields. Although presently India is the top producer of four-wheeled tractors with growing exports to markets like USA (Rajdou 2009), Indian agriculture is far less mechanized than that of other South Asian countries, such as Bangladesh and Sri Lanka. While India has about 22 per cent of area under mechanized tillage, Bangladesh and Sri Lanka have about 80 per cent of their agricultural area mechanized (Biggs, Justice, and Lewis 2011). Even within India, the extent of mechanization is extremely varied, and disparities between regions are large. Punjab and Haryana have the highest levels of mechanization while eastern states like West Bengal, Bihar, and Odisha have the lowest.

Recently, the central government and various state governments have taken several measures. Two central sector schemes—'Promotion and Strengthening of Agricultural Mechanization through Training & Testing' and 'Demonstration and Post Harvest Technology and Management'—were launched during the 11th Five Year Plan. At the same time, mechanization is also being promoted through other programmes, for namely, Macro Management of Agriculture (MMA), Rashtriya Krishi Vikas Yojana (RKVY), National Horticulture Mission (NHM) and National Food Security Mission (NFSM), etc. Eastern region of the country is receiving renewed focus in view of the potential of this region to bring about the next green revolution.

Against this backdrop, the present chapter intends to assess the effect of this increased focus on mechanization on agricultural growth in the eastern region, where the second green revolution is being planned. The specific objectives of the study are:

1. to assess the pattern of mechanization at the crop and operation level;
2. to analyse the comparative economics of labour and machinery in the region; and
3. to estimate impact of recent mechanization on agricultural growth, if any.

The study is organized into four sections. A brief introduction is presented in the first section along with the conceptual framework and an overview of mechanization in India and the study states. The second section describes the methodology followed and the database used. The third section presents the results of the study. The final section summarizes the key findings, draws concluding observations based on the findings and attempts to outline the policy implications accordingly.

CONCEPTUAL FRAMEWORK AND BRIEF OVERVIEW OF LITERATURE

Mechanization in the developed economies is greatly facilitated by technical progress and industrial development, leading to substitution of labour with capital due to reduction in cost of machines and fuel. Also, because of the rapid economic growth in these economies, demand for labour in non-agricultural sectors increases more rapidly, which leads to shortfall in labour supply to agriculture. This, in turn, leads to rise in agricultural productivity, thereby increasing per capita incomes in agriculture. The increase in agricultural incomes results in lesser preference for manual labour and increases preference for mechanization (Heady 1960; Rayner and Keith 1968).

The context in which farm mechanization started in India in the mid-1960s is different from that of developed countries. The population was growing at a significant rate, which resulted in an increasing supply of labour for the agricultural sector. The growth in per capita incomes was also negligible during this period. Yet there was increase in mechanization during the period. This was contrary to the pattern observed in developed countries, and could not be easily explained by neo-classical theory which posits mechanization as a response to changes in relative factor prices or factor prices relative to output. Rao (1975) argued that Ricardo's framework is more appropriate in the Indian context. Rapid increases in population increase supply of labour but at the same time increase demand for food and other agricultural commodities. Such increase in food prices result in corresponding rises in agricultural wages (cash and kind) leading to mechanization, particularly in the absence of public investment in other inputs such as irrigation. Mechanization impacts economic growth in general and agricultural growth in particular (Steckel and White 2012; Olmstead and Rhode 2001). In India, mechanization and related issues such as farm size and productivity were at the centre of agricultural debates during the 1960s and the 1970s. However, with the shift in focus to technological inputs into agriculture, such as high-yielding varieties (HYVs) and chemical fertilizers, the focus shifted away from farm mechanization. Many important studies on farm mechanization in India appeared in the 1960s and the 1970s. These studies explored the broad themes of (i) effect of mechanization on employment, (ii) effect of mechanization on output and (iii) cost–benefit analysis of mechanization.[1]

MECHANIZATION TRENDS IN INDIA AND EASTERN REGION

Total farm power availability in the country increased from 0.30 kw/ha in 1971–2 to 1.66 kw/ha in 2009–10. The share of draught power came down by 36 percentage points, from 45 per cent to 9 per cent, during this period. The share of agricultural workers also came down from 15 per cent to 5 per cent (a decline of 10). On the other hand, the share of tractors increased from 7 per cent to 42 per cent (+35 per cent). There is a large increase in share of tractors in farm power availability during the three decadal periods, that is, 13 per cent in 1981–82 compared to 1971–72, 10 per cent and 12 per cent, respectively during the next two decades. However, there is stagnation (0 per cent increase in share) from 2001–2 to 2009–10, which is a matter of concern.

The percentage of mechanization in India is 40 per cent, while the percentage of population engaged in agriculture is 55 per cent, leading to a mechanization intensity of 0.73. As compared to this low level of mechanization intensity, developed countries such as US and western European countries have an intensity of 40 and 24, respectively. Even countries like Brazil and Argentina have corresponding figures of 5 and 8, respectively. This gives an indication of the low level of mechanization in India. All the three states in the eastern study region are below the national average of farm power availability (Table 33.1).

Table 33.1 State-wise Farm Power Availability (kw/ha)

State	1997–98	2001
Bihar	0.7	0.8
Odisha	0.5	0.6
West Bengal	1.1	1.3
India	1.2	1.4

Source: Srivastava 2006.

[1] For links between mechanization and employment see Foster and Rosenzweig (2011); Rao (1972); Rudra, Majid, and Talib (1969, 1969a, 1970, 1971); Sarkar and Prahladachar (1966); Sharma (1972); Singh (1968). For effect of mechanization on output see Rudra (1971); Rao (1972); Sapre (1969) and Sharma (1972); Vashishtha (1972).

Table 33.2 State-wise Annual Sale of Tractors–Compound Annual Growth Rate

States	CAGR			Average % Share		
	1997–98 to 2004–05	2005–06 to 2012–13	1997–98 to 2012–13	1997–98 to 2004–05	2005–06 to 2012–13	1997–98 to 2012–13
Bihar	2	19	6	6	5	5
Odisha	32	8	13	3	2	2
West Bengal	−2	19	13	1	2	2
India	−4	12	7	100	100	100

Source: Agricultural Research Data Book 2013, IASRI.

Odisha and Bihar are way below the national average, while West Bengal is closer.

State-level data on tractor sales shows that the total share of the three eastern study states during this period is less than 10 per cent of the total national sales, showing the poor state of tractorization in these states (Table 33.2).

METHODOLOGY AND DATA

Primary Data Surveys

There is a general lack of consensus on the impact of mechanization on agricultural growth and employment (for details on this, see the references in footnote 1). As regards the objectives listed in the first section to assess the extent and pattern of mechanization primary data surveys in two states—Bihar and West Bengal have been used. The reference years for the survey were 2008–09, 2009–10 and 2010–11. For assessing the effect of mechanization on growth, a longer time-series data were required, data from Cost of Cultivation Studies, which provide operation-wise labour use details were used. The relevant data were also extracted from various issues of *Agricultural Statistics at a Glance* and such other sources.

For primary survey, multi-stage sampling was adopted. At the first stage, two districts in each state were selected. One covered under the mechanization promotion programmes or alternately highly mechanized in the state and the second district not covered by any of the programmes (or with low density of mechanization). At the second stage, one village from each of the two districts was randomly chosen. At the third stage, a complete listing of all the households using machinery for farm operations was prepared and further 50 households were selected randomly from each village, thus totalling 100 households in each state. The number of households from each size-group in the sample was in proportion to the total number of households of that size-group in the population. In West Bengal, the districts of Hooghly (highest density of tractors) and Purulia (lowest density of tractors) were taken for the study. District Bhagalpur was selected in Bihar and further two villages/cluster of villages were chosen in two different blocks of the district. Tabular analysis, supplemented with econometric analysis was the broad methodology followed.

Econometric Analysis

To estimate the effect of mechanization on agricultural growth at the national and at the state level for eastern region separately, the methodology was different at the national and state level because of data availability and therefore, the results might not be strictly comparable.

Model at the National Level

Two channels through which mechanization could impact agricultural growth were through the increase in cropping intensity (or GCA) or/and through improvements in yield. At the All-India level, estimated two behavioural equations—one each to capture the effect of mechanization on GCA and foodgrain yield (output per ha), respectively. In both the equations, all other important effects such as irrigation, fertilizer consumption, technology, etc. were controlled. The time period of analysis was 1960 to 2009.

At the national level, the following econometric model has been estimated.

$$GCA_t = f\,(GCA_{t-1}, GIA_t, Fert_t, \text{Farm Power}_t, e_t)$$

$$FGYld_t = f\,(FGYld_{t-1}, GIA/GCA_t, Fert/ha_t, \text{Farm Power}/GCA_t, e_t)$$

where, GCA_t = GCA at time t; $FGYld_t$ = Foodgrain yield at time t; GIA_t = Gross irrigated area in time period t; Fert = Fertilizer consumption in time period t; TRPWR/MECPWR t = Tractor/mechanical power available in time t from Singh, Singh, and Singh (2010); DRAUGHT POWER t = Draft power available in time t from Singh, Singh, and Singh (2010); et = Error term.

Irrigation and fertilizer consumption in addition to a measure of mechanization to isolate the effect of machinery on area and yield were used. Two measures of mechanization from Singh, Singh, and Singh (2010)—TRPWR and MECPWR were used. The first measure gives the per hectare farm power available from tractors alone whereas the second measure gives the farm power available per hectare from all mechanical sources, including diesel engines and electric motors, which are mainly used for irrigation.

Model for the Eastern States

Unlike at the national level, long time-series data on mechanical power was not available at the state level. Therefore, data on the cost of cultivation from 1997 to 2009 were used for estimation at the state level. The effect of expenditure on various inputs on value of production (VoP) was analysed. The expenditure approach was adopted because two of the important inputs, irrigation and machinery, were only available in value terms in CoC data. There were no measures of the quantity of these inputs available. The effect of machinery cost on VoP was estimated after controlling for draught power, seeds, irrigation, and fertilizer. The following econometric model has been estimated for states of Odisha, Bihar and West Bengal:

$$VOMP_t = f\,(SEED_t,\ FERT_t,\ IRRGN_t,\ ALV_t,\ HLV_t,\ MLV_t\ e_t)$$

where, VOMP = value of main product (Rs/ha); SEED = Expenditure on seed (Rs/ha); FERT = Expenditure on fertilizer (Rs/ha); IRRGN = Expenditure on irrigation (Rs/ha); ALV = Expenditure on animal labour (Rs/ha); HLV = Expenditure on human labour (Rs/ha); MLV = Expenditure on machinery (Rs/ha); e_t = Error term.

RESULTS AND DISCUSSION

Pattern of Mechanization

The major crops in the sample region of Bihar and West Bengal are paddy, wheat and potato. Maximum area is devoted to these crops and these crops are also sown for longer duration compared to other crops. The crop duration index (CDI)[2] is less than 50 per cent in all three study years in both the study states, indicating that more than 50 per cent of the land is uncultivated. In Bihar machines are mainly used for only three operations—ploughing, irrigation, and marketing. The percentage of farmers using machines is higher for these operations but percentage of expenditure is lower. This indicates that the cost of mechanization appears lower in Bihar. In West Bengal also the operations for which machines are used are ploughing, irrigation, and marketing. The percentage of farmers using machines for ploughing and marketing is about 50 per cent each. Tractors allow quicker operations—time use of tractors is less than 10 per cent of total time use in these operations. However, cost of machinery is disproportionately high probably because of lack of custom hiring facilities. Overall, in the two states, machines are only used for ploughing, irrigation, and marketing. Percentage of farmers using machines is higher for these operations but percentage of expenditure is lower for ploughing while for marketing it is higher. Therefore there is a need to encourage more use of tractors and power tillers in ploughing through custom hiring centres.

There is a discrepancy between the preference for tools and machines and their actual use for many operations—particularly ploughing, irrigation, harvesting, threshing, and marketing (Table 33.3). For ploughing, many farmers expressed preference for tractor rotavator and disc harrow whereas currently most of the ploughing operations are done using the plough. In case of irrigation, there is a strong preference for electric pumps while all the irrigation at present is through diesel pumps. For threshing there is a strong preference for power threshers whereas currently manual/animal threshers are being used. For marketing, there is a strong willingness to move away from animal cart to tractor trolley (Table 33.4). Therefore, efforts should be made to provide the appropriate and preferred tools and machinery to farmers through development of custom hiring centres and other appropriate facilitating mechanisms.

Costs of Mechanization

In this section, the costs of mechanization vis-à-vis total input costs, value of production and respective growth rates have been presented. Tables 33.3 and 33.4 are based on analysis of the secondary data in the study states. Results from primary data are presented in Table 33.5. In Bihar, the share of machinery costs in total input costs is much higher than in West Bengal (Table 33.5). This share is higher than 10 per cent for most crops except potato (3%). The share of machinery costs in value of production is also higher at more than 5 per cent for all the crops except potato. As for the growth trends in these costs the growth rate of cost of machinery is lower than that of human labour for some crops, particularly pulses (Table 33.6). This is mainly because of faster growth in the wage rates of hired labour. These growth trends in the costs of machinery and labour indicate some scope for substitution of labour with machinery in the state for pulse.

[2] For details of computation of this index see Sekhar and Bhatt (2014).

Table 33.3 Pattern of Utilization of Machinery/Tools in the Study States

S. No.	Operation	Type of Machine	% Distribution of Farmers using the Machinery/Implement	% of Distribution Total Time Use of Machinery/Implements	% of Distribution Total Cost of Machinery/Implements Use
			Bihar		
1	Ploughing	Animal plough	10	83	64
		Tractor plough	90	17	36
2	Irrigation	Pump	100	100	100
3	Transportation and Marketing	Cart	21	69	60
		Tractor trolley	79	31	40
			West Bengal		
1	Ploughing	Animal plough	51	95	59
		Tractor plough	49	5	41
2	Irrigation	Pump	100	100	100
3	Transportation and Marketing	Cart	50	92	16
		Tractor trolley	50	8	84
			Combined		
1	Ploughing	Animal plough	31	92	61
		Tractor plough	69	8	39
2	Irrigation	Pump	100	100	100
3	Transportation and Marketing	Cart	35	88	26
		Tractor trolley	65	12	74

Table 33.4 Usage of Machines–Actually Used and Preferred

State	Actual Use	Preferred
	Ploughing	
West Bengal	Animal plough (51%), tractor plough (49%)	Animal plough (50%), tractor plough (18%), **tractor rotavator (32%)**
Bihar	Animal plough (10%), tractor plough (90%)	Animal plough (15%), tractor plough (60%), **disc harrow (10%), others (15%)**
Overall	Animal plough (31%), tractor plough (69%)	Animal plough (32%), tractor plough (39%), **tractor rotavator (16%), disc harrow (5%), others (8%)**
	Sowing	
West Bengal	No machinery (100%)	No machinery (100%)
Bihar	Manual seed drill (94%), tractor seed drill (6%)	Manual seed drill (82%), tractor seed drill (18%)
Overall	Manual seed drill (94%), tractor seed drill (6%)	Manual seed drill (82%), tractor seed drill (18%)
	Irrigation	
West Bengal	Diesel pump (100%)	Diesel pump (46%), **electric pump (54%)**
Bihar	Diesel pump (100%)	Diesel pump (100%)
Overall	Diesel pump (100%)	Diesel pump (73%), **electric pump (27%)**
	Plant Protection	
West Bengal	Manual machinery (100%)	Manual machinery (100%)
Bihar	Manual machinery (100%)	Manual machinery (100%)
Overall	Manual machinery (100%)	Manual machinery (100%)
	Harvesting	
West Bengal	Sickle (71%), animal potato digger (29%)	Sickle (77%), animal potato digger (23%)

(*Cont'd*)

Table 33.4 (*Cont'd*)

State	Actual Use	Preferred
Bihar	Sickle (100%)	Sickle (100%)
Overall	Sickle (86%), animal potato digger(14%)	Sickle (69%), animal potato digger (25%), **reaper (6%)**
		Threshing
West Bengal	Animal/manual paddy thresher (100%)	Animal/manual paddy thresher (100%)
Bihar	Animal/manual paddy thresher (100%)	Animal/manual paddy thresher (50%), **power thresher (50%)**
Overall	Animal/manual paddy thresher (100%)	Animal/manual paddy thresher (75%), **power thresher (25%)**
		Marketing
West Bengal	Cart (50%), tractor trolley (50%)	**Tractor trolley (100%)**
Bihar	Cart (21%), tractor trolley (79%)	Cart (20%), tractor trolley (80%)
Overall	Cart (35%), tractor trolley (65%)	Cart (10%), **tractor trolley (90%)**

Table 33.5 Share of Human Labour and Machinery Costs (Secondary Data)

Crops	HL (Rs/ha)	MC (Rs/ha)	HL as % of OC	MC as % of OC	HL as % of VoP	MC as % of VoP
			Bihar			
Gram	2168	1279	33	19	10	6
Lentil	2187	1047	41	19	11	5
Maize	4397	1456	38	12	17	6
Paddy	5935	1233	56	12	36	8
Potato	8800	1027	27	3	14	2
Wheat	3123	2460	28	22	15	12
			West Bengal			
Jute	14399	519	70	3	46	2
Paddy	11406	894	58	5	43	3
Potato	13686	2018	26	4	22	3
Mustard	6320	489	49	4	31	2
Wheat	7938	693	42	4	35	3
			Combined			
Gram	2168	1279	33	19	10	6
Jute	14399	519	70	3	46	2
Lentil	2187	1047	41	19	11	5
Maize	4397	1456	38	12	17	6
Paddy	17341	2127	57	7	40	5
Potato	22486	3045	27	4	18	2
Mustard	6320	489	49	4	31	2
Wheat	11061	3152	37	10	25	7

Notes: 1. Data Source: Cost of Cultivation Data Average of 2002–03 to 2009–10.
2. HL: Human Labour in Rs/ha; MC: Machinery Cost in Rs/ha; OC: Operating costs in Rs/ha; VoP: Value of Production in Rs/ha.

Costs of Mechanization in the Study States: Primary Data

In Bihar paddy and wheat have a higher share of machinery costs compared to other crops. Share of machinery cost in VoP is quite low, ranging from 3 per cent to 14 per cent across different crops. This is true even for crops with marketed surplus exceeding 50 per cent, that is, paddy and maize. This implies that either the use of machinery was low or the price of machinery was low, indicating scope for increasing

Table 33.6 Growth Rates of Costs and VoP

Crops	HL	MC	Production
	Wage Rate	Total	Value of Production
		Bihar	
Gram	4	2	11
Lentil	7	–4	8
Maize	9	7	14
Paddy	8	12	10
Potato	10	20	11
Wheat	9	7	13
		West Bengal	
Jute	7	28	13
Paddy	8	9	11
Potato	8	13	5
Mustard	9	13	5
Wheat	14	39	12
		Combined	
Gram	4	2	11
Jute	7	28	13
Lentil	7	–4	8
Maize	9	7	14
Paddy	8	11	10
Potato	9	16	8
Mustard	9	13	5
Wheat	11	13	12

Notes: 1. Growth rates are for the period 1996–97 to 2009–10.
2. HL: wage rate in Rs per person; MC: Machinery Cost in Rs/ha; OC: Operating costs in Rs/ha; VoP–Value of Production in Rs/ha.
Source: Cost of Cultivation Data.

Table 33.7 Share of Human Labour and Machinery Costs (Primary Data)

Crop	Machine Costs (MC)	% of MC to VoP	% of MC to Input Costs	% of MS to VoP
		Bihar		
Paddy	5376	9	19	61
Wheat	9309	14	22	32
Maize	3791	10	13	57
Gram	800	3	12	40
Lentil	1000	7	21	42
		West Bengal		
Kharif Paddy	6683	18	26	59
Boro Paddy	3103	7	11	84
Potato	8540	11	10	87
Mustard	5746	33	27	30
Wheat	1754	10	16	70
		Combined		
Paddy	5376	9	19	61
Kharif Paddy	6683	18	26	59
Boro Paddy	3103	7	11	84
Wheat	11063	13	21	40
Maize	3791	10	16	57
Gram	800	3	12	40
Lentil	1000	7	21	42
Mustard	5746	33	27	30
Potato	8540	11	10	87

Note: MC: Machinery Cost in Rs/ha; OC: Operating costs in Rs/ha; VoP: Value of Production in Rs/ha.
Source: Cost of Cultivation Data Average of 2002–03 to 2009–10.

use of machinery. In West Bengal share of machinery costs in total cost was large for kharif paddy and mustard but quite low for wheat and potato. As a proportion of VoP also, the machinery costs were higher in case of mustard and paddy (Table 33.7). Overall, costs of machinery were generally higher in absolute terms for crops that have higher MS (to production) such as paddy and potato. Wheat and mustard showed a reverse trend because of higher VoP.

ECONOMETRIC ANALYSIS—ALL-INDIA RESULTS

Mechanization showed an insignificant effect (Table 33.8). The variables that appeared to have significant effect on GCA were irrigation (GIA) and fertilizer use (NPK). In the foodgrain yield equation again, mechanization did not have any significant effect on the dependent variable. This is in contrast to some of the recent studies such as Singh (2001), MoA (2013). The Singh et al. study did not control for the effects of any other major inputs such as irrigation, fertilizer, seeds, etc. This was a classic omitted variable problem that could lead to highly biased results. In such cases, the included variable tends to pick up all the effect due to other variables as well. As the MoA (2013) study was basically a reproduction of the Singh (2001) results, both studies suffer from the same methodological shortcoming. The analysis attempted to address this limitation by including other relevant explanatory variables like irrigation and fertilizer consumption, in addition to a measure of mechanization, to isolate the effect of machinery on area and yield (Table 33.8). The study used two measures of mechanization from Singh et al. (2010)—TRPWR and

Table 33.8 Econometric Results (All India)

Dependent Variable	Explanatory Variables				R^2	DW
	GIA	FERT	DRAUGHT POWER	TRPWR/MECPWR		
GCA	0.29***	0.03**		−0.04	0.97	2.56
FGYLD (with TRPWR)	0.63*		−0.91**	−0.02	0.99	1.72
FGYLD (with MECPWR)	0.48		−0.97**	0.01	0.99	1.52

*, **, *** indicate significant at 1%, 5%, and 10%.
Source: Data from *Agricultural Statistics at a Glance* and Singh, Singh, and Singh (2010).

MECPWR. The first measure gives the per hectare farm power available from tractors alone whereas the second measure gives the farm power available per hectare from all mechanical sources, including diesel engines and electric motors, which are mainly used for irrigation. When these variables are included, mechanization turns insignificant. The results in Table 33.9 indicate that the TRPWR and MECPWR had a statistically insignificant effect, and that the use of machinery as a negligible effect on production—once the effect of other relevant variables was accounted for. The input that significantly affected yield turned out to be the percentage of irrigated area. The results therefore show that when a proper account was made of other important explanatory variables, results could be vastly different.

Results in the Eastern Region

The results indicate that in Bihar, the expenditure on machinery showed a significant effect only for potato (Table 33.9). For paddy, gram and maize it was insignificant. The expenditure on seed for gram, fertilizer for maize and potato and irrigation for paddy (negative) and potato showed significant effect on the dependent variable. The expenditure on human and animal labour generally shows a significant negative effect, perhaps indicating scale economies. In Odisha, both the major crops, paddy and moong, showed a significant positive effect for seed (Table 33.9). Machinery expenditure showed significant negative effect for moong possibly because of the indivisibility effect. In West Bengal, seed, fertilizer, irrigation and human labour showed significant positive effect on VoP (Table 33.9). Machinery showed significant positive effect in case of jute and potato but in the case of paddy it showed significant negative effect. Overall, in all the three states, seed appeared to be the major determinant of production followed by irrigation and fertilizer. Machinery was mostly insignificant in its effect.

Farmer's Perceptions: Farmers' feedback is invaluable in assessing the prospects and problems of any programme.

Table 33.9 Econometric Results–Study States (Based on Cost of Cultivation Data)

	Dependent Variable: Value of the Main Product (Rs/Ha) All the Variables are in Expenditure Terms							
State/Crop	Explanatory Variables						R^2	DW
	SEED	FERT	IRRGN	HLV	ALV	MLV		
Bihar								
1. Paddy		0.36	−0.04*	−0.9***		0.02	0.90	2.26
2. Gram	0.87***	−0.03				0.11	0.91	2.71
3. Maize		0.49***			−0.24**	−0.14	0.87	1.92
4. Potato		0.38**	0.45***		0.05	−0.15*	0.95	2.11
Odisha								
1. Paddy	1.25**		−0.15			−0.11	0.91	2.19
2. Moong	0.88**					−0.09**	0.92	1.81
West Bengal								
1. Paddy	1.01**		0.56***			−0.81*	0.91	2.09
2. Potato		0.67**		0.40***		0.54***	0.82	2.14
3. Jute	0.37***					0.91*	0.88	2.47

Notes: *, **, *** indicate significant at 1%, 5%, and 10%.
Source: *Cost of Cultivation*, GoI data and authors' computations.

Table 33.10 Farmer's Perception about Machinery Use (Percentage)

	West Bengal	Bihar	Overall
Reasons for Using Machine			
Quicker Operations	70	60	65
Higher yield	19	5	12
Economical	11	35	23
Total	100	100	100
Major Uses of Machines			
Better land utilization	19	29	27
Reduced drudgery	6	27	17
Higher income		12	6
Higher yield	25	11	18
Higher social esteem		10	5
Other		11	3
Total	50	100	76
Awareness and Assistance under Govt. Programmes			
Aware	0	60	30
Found programmes useful	NA	40	20
Assistance Received	NA	18	9
Type of Assistance			
Subsidy on consumables	NA	8	4
Subsidy on purchase of machine	NA	4	3
Demonstration	NA	4	2
Training	NA	2	0

Notes: The figures indicate % distribution of farmers.
Source: Primary Data Surveys.
NA denotes Not Applicable.

Therefore, farmers' views have been elicited on their reasons for using machinery, benefits with machinery, awareness and assistance availed under the government programmes. Reasons for using Machinery: Quicker operation was cited as the major reason for using machinery in both states (Table 33.10). However, it is notable that much larger percentage of farmers in Bihar found using machinery economical as compared to their counterparts in West Bengal. This indicates that the transaction costs of using machinery were much higher in West Bengal.

Usefulness of the Machinery: In Bihar farmers reported better land utilization (29 per cent) as the main benefit of use of machines (Table 33.10), followed by reduced drudgery (27 per cent), higher income (12 per cent), higher yield (11 per cent) and higher social esteem (10 per cent). In West Bengal only 50 per cent of the farmers found usage of machinery useful. Higher yield (25 per cent), better land utilization (19 per cent) and reduced drudgery (6 per cent) have been reported as some of the uses. Overall, of all the farmers in two states, 75 per cent found using machines useful. Better land utilization (24 per cent), higher yield (18 per cent) and reduced drudgery (17 per cent) are some of the major uses reported.

Awareness and Assistance under Government Programmes: In Bihar about 60 per cent of farmers were aware of the government programmes, and 18 per cent received some form of assistance under the programmes (Table 33.10), such as subsidy on consumables (8 per cent), subsidy on purchase of a machine (4 per cent) and demonstration (4 per cent) each and training (2 per cent). In West Bengal none of the respondents was aware of the government mechanization programmes. None had received any assistance. Overall, in the two states, only 30 per cent of farmers were aware of government mechanization programmes, and only 9 per cent had received some kind of assistance under such programmes. As for the type of assistance, subsidy on consumables (4 per cent) ranks first. Of the 30 per cent of the farmers aware of the programmes, only 20 per cent found them useful.

POLICY SUGGESTIONS

The level of mechanization in the three study states of the eastern region was very low, and below the national average. Odisha and Bihar were way below the national average, while West Bengal was closer. The cost of machinery was disproportionately high in West Bengal, probably because of the lack of custom hiring facilities. The cost of mechanization appeared lower in Bihar. Machines were mainly used for only three operations—ploughing, irrigation, and marketing. The percentage of farmers using machines for these operations was higher. But the proportion of expenditure was lower for ploughing. Therefore, it might be necessary to encourage more use of tractors and power tillers in ploughing through custom hiring centres. Also, there was a discrepancy between the preference for tools and machines and their actual use for many operations, particularly for ploughing, irrigation, harvesting, threshing, and marketing. Efforts should be made to provide the appropriate tools and machinery to the farmers.

As per CoC, in Bihar, the share of machinery costs in total input costs was much higher than in West Bengal. The share of machine cost relative to hired human labour was also higher, and the growth rate of cost of machinery was lower than that of human labour for some crops, particularly pulses. This was mainly because of faster growth in the wage rates of hired labour. These growth trends in the costs of machinery and labour indicated some scope for substitution of labour with machinery in Bihar for pulses.

In West Bengal, the share of machine labour in total input costs was quite low, ranging from 3 per cent in jute to a maximum of 5 per cent in paddy. The total cost of production was higher than the VoP for two major crops—paddy and potato—but the operational costs were lower. This indicated that in West Bengal, paddy and potato farmers operated at mere subsistence level with gross revenue merely covering the operational costs.

The primary survey results broadly support the findings from the CoC data. Share of expenditure on machinery was far lower than that of human labour, which was about 50 per cent for major crops. Costs on machinery were generally higher in absolute terms for crops that have higher marketed surplus relative to production such as paddy and potato. In Bihar, the share of machinery costs in VoP was quite low even for crops with marketed surplus exceeding 50 per cent, that is, paddy and maize. This implies that, in Bihar, either the use of machinery was low or its price, indicating the scope for increasing the use of machinery.

Results of econometric analysis at All-India level showed that use of machinery so far played only a minor role in increasing production either through area or the yield effect. The major determinants of agricultural production were irrigation, seed and fertilizer. Irrigation (GIA) and fertilizer use (NPK) were significant in explaining GCA. The variable significantly affecting yield was irrigated area. These results are also supported at the state level in the study region.

Given the very low level of mechanization in the two states and the extent of land remaining underutilized (low CDI), mechanization needs to be promoted through appropriate policies supplemented with suitable input and output policies. Custom hiring centres need to be established as a large number of farmers expressed lack of such facility as a major problem. Provision of other inputs, particularly irrigation, seed and fertilizer needs to be improved as our econometric results show that these inputs affect the area and yield levels more than the machinery use.

REFERENCES

Biggs, Stephen, Scott Justice, and David Lewis. 2011. 'Patterns of Rural Mechanization, Energy and Employment in South Asia: Reopening the Debate', *Economic and Political Weekly* Vol. XLVI, No. 9, 26 February, pp. 78–82.

Foster, Andrew D., and Mark R. Rosenzweig. 2011. 'Are Indian Farms Too Small? Mechanization, Agency Costs, and Farm Efficiency' Mimeo, Brown University, Providence, RI, USA, June 2011.

Heady, Earl O. 1960. 'Extent and Conditions of Agricultural Mechanization in the United States', in J.L. Meij (ed). Mechanization in Agriculture, North-Holland Publishing Company, Amsterdam.

IARI. 2013. Agricultural Research Data Book Indian Agriculture Research Institute, New Delhi.

IASRI. 2006. Study Relating to Formulating Long-term Mechanization Strategy for each Agro-climatic Zone/State in India. Indian Agricultural Statistics Research Institute, New Delhi, India.

MoA. 2013. Farm Mechanization in India, Presentation by Mechanization and Technology Division, Dept. of Agriculture and Cooperation, Ministry of Agriculture.

Olmstead, Alan L., and Paul W. Rhode. 2001. 'Reshaping the Landscape: The Impact and Diffusion of the Tractor in American Agriculture, 1910–1960', *The Journal of Economic History* 61(3): 663–98.

Rao, C.H. Hanumantha. 1972. 'Employment Implications of Green Revolution and Mechanization in Agriculture in Developing Countries: A Case Study of India', Mimeo, Institute of Economic Growth, Delhi.

———. 1975. 'Technological Change and Distribution of Gains in Indian Agriculture', Institute of Economic Growth, Delhi and The Macmillan Company of India Ltd, Delhi.

Rayner A.J., and Keith Cowling. 1968. 'Demand for Farm Tractors in the United States and the UK', *American Journal of Agricultural Economics*, Vol. 50(4): 896–912.

Ricardo, David. 1966. On the Principles of Political Economy and Taxation, edited by Piero Sraffa with the collaboration of M.H. Dobb, Cambridge University Press.

Rudra, A. 1970. 'Employment Patterns in Large Farms of Punjab', *Economic and Political Weekly*, 6(26): A89–A94.

———. 1971. 'Use of Shadow Prices in Project Evaluation', *Indian Economic Review* vol. 7(1): 1–15.

Rudra, A., Majid A., and Talib B.D. 1969. 'Big Farmers of Punjab', *Economic and Political Weekly*, Review of Agriculture September 1969, pp. A-143 to A-146.

———. 1969a. 'Big Farmers of Punjab: Second Installment of Results', *Economic and Political Weekly* Vol. 4, No. 52 (December 27, 1969), pp. A213–A219.

Sapre, S.G. 1969. A Study of Tractor Cultivation in Shahada, Gokhale Institute of Politics and Economics, Poona.

Sarkar K.K. and Prahladachar M. 1966. 'Economics of Tractor Cultivation: A Case Study', *Indian Journal of Agricultural Economics* Vol. 21, pp. 171–82.

Sekhar, C.S.C., and Yogesh Bhatt. 2014. 'Effect of Farm Mechanization on Agricultural Growth and Comparative Economics of Labour and Machinery in India', Report submitted to the Ministry of Agriculture, Government of India, May 2014.

Sharma, R.K. 1972. 'Economics of Tractor versus Bullock Cultivation', Agricultural Economics Research Centre, Delhi.

Singh, B. 1968. 'Economics of Tractor Cultivation: A Case Study', *Indian Journal of Agricultural Economics* Vol. 23, pp. 83–88.

Singh, Gajendra (2001). 'Relationship between Mechanization and Agricultural Productivity in Various Parts of India', *AMA* 32(2): 68–76.

Singh, Surendra, R.S. Singh, and S.P. Singh. 2010. 'Farm Power Availability and Agriculture Production Scenario in India', *Agricultural Engineering Today* 34(1): 9–20.

Srivastava, N.S.L. 2006. 'Farm Power Sources, their Availability and Future Requirements to Sustain Agricultural Production' in IASRI, Study relating to Formulating Long-term Mechanization Strategy for each Agro-climatic Zone/State in India. Indian Agricultural Statistics Research Institute, New Delhi, India.

Steckel Richard H., and William J. White. 2012. 'Engines of Growth: Farm Tractors and Twentieth-Century US Economic Welfare', Working Paper 17879, National Bureau of Economic Research.

Vashishtha, P.S. 1972. 'Impact of Farm Mechanization on Employment and Output: Some Preliminary Results', Mimeo, presented at the 12th Indian Econometric Conference, Kanpur.

ANNEXURE 33A

Table 33A The Details of Agro-Economic Research Centres and the Lead Person Involved in the State Report

State	AERC/Unit	Authors
Bihar	AERC, Department of Economic and Sociology, Punjab Agricultural University, Ludhiana	Basant Kumar Jha, Rajiv Kumar Sinha, Rosline Kusum Marandi
West Bengal	AERC, Visva-Bharati, Santiniketan	Debashis Sarkar, Debajit Roy, Kali Shankar Chattopadhyay

Section III
Agricultural Marketing, Post-Harvest Management, and Value Addition

34

Assessment of Marketed and Marketable Surplus of Major Foodgrains in India

Vijay Paul Sharma and Harsh Wardhan

The contribution of the agricultural sector to national gross domestic product (GDP) has witnessed a secular decline during the last few decades with the consequent increase in shares of other sectors, particularly services. Agriculture contributes about 12 per cent of national GDP but is the largest employer and main source of livelihood to majority of the rural population (Central Statistical Office [CSO] 2014). Improving performance of agriculture is, therefore, crucial for achieving food security, rural development and poverty reduction. Indian agriculture, which witnessed a visible deceleration during the 9th and 10th Five Year Plans, recorded a robust growth during the 11th Plan. The foodgrains production touched a new peak of 265 million tonnes in 2013–14, an addition of about 55 million tonnes between triennium ending (TE) 2005–06 and TE 2013–14 (Government of India [GoI] 2015).

Indian agriculture has also witnessed structural changes with the composition of agricultural output shifting from traditional foodgrains to high-value products. Agriculture is increasingly being driven by expanding demand for livestock products, fish and other high-value crops like fresh fruits and vegetables, processed foods and beverages. At the All-India level, the share of high-value commodities/products (fruits and vegetables, livestock products, fisheries) has increased from about one-third in TE 1983–84 to over 50 per cent in TE 2011–12 (Sharma 2011). The composition of export trade has also changed, away from traditional products towards horticulture, meat and meat products, and processed products. The Indian food consumption basket has become increasingly diversified and expenditure on fruits, vegetables, milk, eggs, meat and fish, beverages, and processed food is rising rapidly, leading to changes in cropping pattern in the country.

Indian agriculture has also become increasingly market-oriented and monetized. The proportion of agricultural production that is marketed by the farmers has increased significantly over the last few decades. In the early 1950s, about 30–35 per cent of foodgrains output was marketed, which has increased to more than 70 per cent. The marketed surplus is relatively higher in case of commercial crops than subsistence crops. As Indian agriculture has undergone significant transformation, and no reliable estimates of marketed and marketable surplus are available, understanding of marketing behaviour of producers and reliable estimates of marketed surplus, as well as factors affecting it, can be of significant help in designing appropriate production, procurement, storage, distribution and pricing policies. The present study was undertaken to estimate marketed and marketable surplus of major foodgrains in leading producing states and examine important factors which determine the level of marketed surplus for various categories of farms. It is expected that the results of this study would be useful in designing effective food procurement, distribution and price policy.

OBJECTIVES

The main objectives of the study are:

1. to estimate marketable and marketed surplus of selected cereals (rice, wheat, maize, and bajra) and pulses (gram and tur) in selected states; and
2. to examine the impact of various socio-economic, technological, institutional, infrastructure, and price factors on marketed surplus of selected crops.

RESEARCH METHODOLOGY

As the main focus of the study was on the estimation of the marketed and marketable surplus of foodgrains and response of marketed surplus to price and other exogenous variables, the study uses both primary and secondary data pertaining to selected foodgrains. In order to understand the emerging trends and patterns in production and yield performance, secondary data on area, production, and productivity of selected crops was collected from different published sources.

In order to estimate marketed surplus of major foodgrains on different categories of farms, primary data was collected from 918 households selected from nine districts in four major rice-producing states, that is, Haryana, Punjab, Uttar Pradesh, and West Bengal; 1,193 wheat farmers from 15 districts of Rajasthan, Madhya Pradesh, Uttar Pradesh, Haryana, and Punjab; 358 maize growers from Rajasthan, Maharashtra, and Karnataka; 500 bajra farmers from seven districts of Haryana, Rajasthan, and Uttar Pradesh; 553 farmers from major gram-growing states, namely Rajasthan, Maharashtra, Karnataka, and Madhya Pradesh; and 441 households cultivating tur from seven districts of Uttar Pradesh, Madhya Pradesh, Maharashtra, and Karnataka (see Table 34.1). The household survey was conducted by participating AERCs/Units. The reference period for the study is 2011–12.

Conceptual Framework and Theoretical Model for the Study

The concept of marketed surplus has been used in a variety of ways. In some of the earlier studies on foodgrains marketing in the developing countries, three concepts of marketed surplus have been generally used; gross marketed surplus, net marketed surplus, and marketable surplus (Bhargava and Rustogi 1972; Farruk 1970; Harriss 1982; Hussein and Rajbanshi 1985; Krishnan 1965; Nadkarni 1980; Narain 1961; Rahman 1980; Raquibuzzaman 1966; Sharma and Gupta 1970). For the purpose of this study, the marketable surplus has been estimated by subtracting total retention from total production. The retention consists of quantity kept for self-consumption, for seed purpose, for feed, and payments in kind to labourers, gifts, and others. Gross marketed surplus is calculated by estimating the total quantity of produce sold in the market without considering whether there is any buy back by those sellers later on. Net marketed surplus, on the other hand, excludes the amount of produce which is bought back.

Determinants of Marketed Surplus

Many studies have observed that marketed surplus of a crop depends on various price and non-price factors. Empirical studies of marketed surplus have found that farmers respond positively to price changes, and this is consistent with economic theory. In addition to price, a number of other socio-economic, institutional, technological, and infrastructure factors influence marketed surplus. Among these are farm size and production, family size, wealth/income, risks, access to modern technology, markets, market information, etc. A number of studies have reported that in most cases there exists a strong linear, and in some cases a strong non-linear relationship between the quantity sold and variables like farm size, quantity produced, family size, prices, and socio-economic and institutional variables for different categories of farmers. The linear relation may be written as:

$$MS = \alpha + \beta_i X_i$$

where, MS denotes the marketed surplus and X_i ($i = 1, 2, \ldots, n$) represents the independent variables influencing marketed surplus. The dependent variable, marketed surplus, is defined as sales as a share of total output per household. The independent variables include farm size, family size, awareness of minimum support price (MSP), access to regulated market, distance of farm from main market, per household production of the crop, source of off-farm income, access to institutional credit, access to roads, access to market and market information, and price received for the produce. We hypothesize that with the increase in farm size and production, higher income and output price and better access to various institutional and technological

Table 34.1 List of Selected Crops, States, and Farm Category-wise Sample Size

States	Marginal	Small	Semi-Medium	Medium	Large	Total
	<1 ha	1–2 ha	2–4 ha	4–10 ha	>10 ha	
			Rice			
Haryana	58	79	34	23	6	200
Punjab	36	60	96	84	24	300
Uttar Pradesh	61	21	11	7	0	100
West Bengal	124	97	65	32	0	318
Total	279	257	206	146	30	918
			Wheat			
Rajasthan	21	100	70	79	23	293
Madhya Pradesh	42	16	21	19	2	100
Uttar Pradesh	126	41	22	11	0	200
Haryana	86	110	59	36	9	300
Punjab	36	60	96	84	24	300
Total	311	327	268	229	58	1193
			Maize			
Rajasthan	9	38	33	29	9	118
Maharashtra	37	37	20	6	0	100
Karnataka	40	43	39	14	4	140
Total	86	118	92	49	13	358
			Bajra			
Haryana	21	31	27	18	3	100
Rajasthan	18	80	69	100	33	300
Uttar Pradesh	65	20	11	4	0	100
Total	104	131	107	122	36	500
			Gram			
Rajasthan	11	28	46	95	32	212
Maharashtra	36	35	19	10	0	100
Karnataka	27	34	26	36	18	141
Madhya Pradesh	24	23	17	20	16	100
Total	98	120	108	161	66	553
			Tur			
Uttar Pradesh	52	24	12	12	0	100
Madhya Pradesh	9	13	28	34	16	100
Maharashtra	33	42	20	5	-	100
Karnataka	19	39	33	36	14	141
Total	113	118	93	87	30	441

Source: Field Survey.

factors, marketed surplus should increase. Family size, distance from market, and poor access to infrastructure, on the other hand, are expected to have a negative effect on the marketed surplus. In this chapter, we used multiple linear regression analysis to examine the impact of various factors on marketed surplus of selected crops.

RESULTS AND DISCUSSIONS

Rice

Rice is the most important crop in India occupying about 43.2 million ha of the total cultivated area and having a total production of over 102 million tonnes (TE 2012–13).

Rice had the highest contribution (14.5 per cent) to the total value of output from agriculture and allied activities in TE 2012–13 and also emerged as India's top agricultural export commodity with about 15.2 per cent of the total agricultural export value in TE 2013–14 (GoI 2015). Rice production in the country increased at an annual compound growth rate of 2.35 per cent during the period 1971–2012, of which yield accounted for nearly 84 per cent and area, 16 per cent of the production growth rate (Table 34.2). Rice production has continued to increase during the last four decades; however, rice production (4.2 per cent) and yield (3.58 per cent) recorded the highest growth rate during the 1980s and the lowest (1.86 per cent in production and 1.07 per cent in yield) during the 1990s. However, growth rate picked up during the last decade.

Table 34.2 Trends in Compound Annual Growth Rates (%) in Area, Production, and Yield of Selected Crops in India: 1971–72 to 2012–13

Area/ Production/ Yield	1970s	1980s	1990s	2000s	All Period
			Rice		
Area	0.92***	0.60	0.78***	0.08	0.36***
Production	2.58*	4.20***	1.86***	2.10***	2.35***
Yield	1.65	3.58***	1.07**	2.03***	1.96***
			Wheat		
Area	2.34***	0.36	1.40***	1.53***	0.99***
Production	4.91***	3.39***	3.11***	3.13***	3.25***
Yield	2.51***	3.02***	1.69***	1.58***	2.24***
			Maize		
Area	0.03	0.07	1.17***	2.62***	0.96***
Production	1.40	2.60	3.74***	5.88***	3.28***
Yield	1.36	2.52	2.55***	3.17***	2.34***
			Bajra		
Area	−1.20	−0.93	−1.06*	−1.02	−0.83***
Production	−0.08	1.35	1.58	2.16	1.71***
Yield	1.12	2.30	2.67	3.22*	2.49***
			Gram		
Area	−0.81	−1.42	0.24	3.35***	0.02
Production	−0.64	−0.51	1.19	5.51***	1.05***
Yield	0.17	0.92	0.95	2.10***	1.03***
			Tur		
Area	1.59***	2.22***	−0.22	1.63***	1.04***
Production	1.30	1.70	0.73	2.22**	0.94***
Yield	−0.29	−0.51	0.95	0.57	−0.11

Note: ***, **, and *: Significant at 1, 5, and 10 per cent level.
Source: Authors' computation using GoI (2015a) data.

Rice yields, which were low (about 1393 kg/ha) during the early 1980s, witnessed a steady increase during the last three decades and reached a level of 2,175 kg/ha in the recent period (2006–11). However, rice yield in the country is lower compared to other major rice-producing countries such as China (6.74 t/ha), Indonesia (5.14 t/ha), and Vietnam (5.63 t/ha) as well as the world average (4.39 t/ha). At the state level, Punjab has the highest yield (3,949 kg/ha), followed by Andhra Pradesh (3,134 kg/ha) and Haryana (3,024 kg/ha), while Madhya Pradesh (933 kg/ha), has the lowest yield. Rice yields are relatively lower in eastern states of Assam, Jharkhand, Chhattisgarh, and Odisha.

Due to effective government procurement policy, rice procurement increased significantly from about 21 million tonnes in 2000–01 to 35 million tonnes in 2011–12, with a slight decline to 34 million tonnes in 2012–13 and 31.3 million tonnes in 2013–14. Procurement as percentage of production has also increased during these years from about 24 per cent in 2000–01 to about 33.7 per cent in 2011–12, declined in the next three years, and reached 29.4 per cent in 2013–14. It is estimated that government procures about 40 per cent of marketed surplus at national level and it varies from less than 5 per cent in Karnataka and Assam to over 90 per cent in Chhattisgarh, Punjab (76 per cent), Andhra Pradesh (68 per cent), and Odisha (66 per cent). The procurement of rice, which was highly concentrated in few states like Punjab, Haryana and Andhra Pradesh up to the late 1990s, has become more diversified. Punjab is still the largest contributor (24.1 per cent) to national procurement, and Andhra Pradesh ranks number two (22.9 per cent), but both states have lost their shares between TE 2002–03 and TE 2012–13 (Sharma and Wardhan 2014). While states like Chhattisgarh, Odisha, West Bengal, and Bihar have increased their share in rice procurement. The share of decentralized procurement states, namely Andhra Pradesh, Chhattisgarh, Karnataka, Kerala, Madhya Pradesh, Odisha, Tamil Nadu, Uttarakhand, West Bengal, and Bihar, has increased significantly and crossed 50 per cent share in TE 2013–14.

The pattern of marketed surplus of rice, based on the household data collected from rice farmers in selected states showed that gross marketed surplus (sales as a proportion of production) was marginally lower than marketable surplus (Table 34.3). Medium farms had the highest rate of marketed surplus (83.2 per cent), followed by semi-medium (80.9 per cent) and the lowest on marginal farms (62.8 per cent). In the case of selected states, Punjab and Haryana farmers sold more than 95 per cent of their rice output in the market. West Bengal farmers, on the other hand, sold about 61 per cent of the total output. Since rice is a

Table 34.3 Marketable and Marketed Surplus of Rice on Sample Households

	Marketable Surplus (% of Production)	Gross Marketed Surplus (% of Production)	Net Marketed Surplus (% of Production)
State			
Haryana	95.5	95.5	95.1
Punjab	99.4	99.4	99.4
Uttar Pradesh	77.0	75.8	75.8
West Bengal	62.6	40.5	37.9
State Total	85.5	78.0	77.1
Farm Size			
Marginal	64.7	63.0	59.6
Small	73.6	74.0	72.2
Semi-medium	81.0	80.9	79.9
Medium	90.3	83.2	82.7
Large	99.1	72.1	72.1
All farms	85.5	78.0	77.1

Source: Field Survey, 2011–12.

Table 34.4 Factors Influencing Marketed Surplus of Selected Crops in India

Factors	Crop			
	Rice	Wheat	Maize	Bajra
Constant	−11.8679*** (3.3728)	−9.4746*** (2.3290)	40.1851 (5.3899)	57.3603*** (4.3660)
Farm Size	0.5615*** (0.1262)	1.5778*** (0.1398)	0.74451*** (0.3093)	0.8218*** (0.3036)
Family Size	−0.1990 (0.1736)	−0.7677*** (0.1232)	−0.8819*** (0.2691)	−0.3047 (0.2612)
Price Received	0.0827*** (0.0029)	0.0502*** (0.0018)	0.0449*** (0.0049)	–
Awareness about MSP	3.5946* (2.0850)	13.2723*** (1.1372)	–	12.7090*** (2.3069)
Access to Regulated Market	9.8557*** (1.3326)	7.6013*** (0.9754)	–	7.1415*** (2.7322)
Distance to Market	0.4571*** (0.1277)	−0.0440 (0.0665)	.56498*** (0.1361	0.4340*** (0.1484)
R^2	0.57	0.62	0.32	0.18

Note: ***, **, and *: Significant at 1, 5, and 10 per cent levels.
Source: Field Survey, 2011–12.

staple crop in eastern and southern regions, a significant proportion of crop output was kept for self-consumption. The average farm retention (self-consumption, seed, and other purposes) on sample households was 14.5 per cent but varied from less than one per cent on large farms to 35.3 per cent on marginal farms. In the case of states, average retention was lowest (less than one per cent) in Punjab and the highest (37.4 per cent) in West Bengal. Some farmers purchased for self-consumption, even after they have sold their produce in the market. Since farmers need cash for the next crop and for other requirements, they (particularly small and marginal farmers) are forced to sell part of the grains after harvest and buy at a later date at a higher price. Farmers' market participation was quite high in all the states and varied from 94.7 per cent in West Bengal to 100 per cent in Punjab and Haryana.

The results of regression analysis to examine the factors affecting marketed surplus revealed that output price, farm size, and market access have significant positive impact on the marketed surplus of rice (Table 34.4). Family size matters too on marginal and small farms and has a negative impact on marketed surplus. Household's awareness of MSP has a positive and significant impact on marketed surplus, and so has access to regulated markets. The relative importance of factors influencing marketed surplus as measured by standardized regression coefficients indicated that the price received by farmers was the most important factor, followed by access to regulated markets, farm size and awareness of MSP. The family size turned out to be the least important variable in influencing marketed surplus of rice.

Wheat

Wheat is an important staple crop in India and occupies about 15 per cent of the total cultivated area with a total production of nearly 92 million tonnes (TE 2012–13). Wheat acreage in the country increased from 19.1 million ha in TE 1973–74 to 29.6 million ha in TE 2012–13 and production increased from 24.3 million tonnes to 91.8 million tonnes in TE 2012–13. During the same period, wheat productivity more than doubled from 1,274 kg per ha to 3,094 kg per ha. As a percent of total cropped area, wheat acreage share increased from 11.5 per cent in TE 1973–74 to 15.3 per cent in TE 2012–13. Wheat production increased at an annual compound growth rate of 3.25 per cent during 1971–72 and 2012–13, and this was due to a modest area expansion (0.99 per cent) but a significant yield increase (2.24 per cent). Growth in wheat production was the highest (4.91 per cent) during the seventies which decelerated to 3.39 per cent per year during the 1980s, 3.11 per cent during the 1990s but improved marginally (3.13 per cent) during the last decade. Wheat yield growth rates were particularly rapid during the 1970s and the 1980s. Growth in wheat yield, 2.51 per cent per year in the 1970s and 3.02 per cent per year in the 1980s, slowed down to 1.69 per cent in the 1990s and 1.58 per cent in the first decade of the 2000s. During the last two decades, acreage expansion and yield improvement contributed almost equally to growth in wheat output while yield was the major source of growth in output during the 1980s (Table 34.2).

Government plays an important role in wheat procurement. Wheat procurement which reached a peak of about 21 million tonnes in 2001–02 witnessed a steady decline and touched the lowest level of 9.23 million tonnes in 2006–07. India imported about 5.4 million tonnes of wheat in 2006–07 and about 1.9 million tonnes in 2007–08 which concerned the policymakers, and concerted efforts were made to increase wheat production and procurement. This led to a significant increase in wheat production as well as procurement. Wheat production increased from 75.8 million tonnes to 93.5 million tonnes between 2006–07 and 2012–13, while procurement increased from 9.2 million tonnes to 37.9 million tonnes during the same period. Wheat procurement as a percentage of total production increased from about 12 per cent in 2006–07 to 40.6 per cent in 2012–13 but fell during 2013–14.

In the late 1990s, wheat procurement was mainly concentrated in Punjab and Haryana and share of government procurement as a percentage of production was 59.2 per cent in Punjab and 51.4 per cent in Haryana. The share of Punjab, Haryana, and Uttar Pradesh in total procurement was more than 90 per cent in TE 2003–04, making them almost a monopoly vis-à-vis other states. However, during the last decade, the share of traditional states like Punjab, Haryana, and Uttar Pradesh has declined while the share of Madhya Pradesh and Rajasthan has increased. The share of Madhya Pradesh has increased from less than 2 per cent to over 24 per cent during the last decade. This has happened primarily due to the state policy of additional bonus over the MSP. The procurement trends show that wheat procurement has diversified in terms of coverage of states but at an additional cost. The share of government procurement has been rising over the years in all wheat producing states. Madhya Pradesh has recorded the highest increase of over 30 per cent, from 6 per cent in TE 2001–02 to 37.5 per cent in TE 2011–12. These results indicate that the government has almost a monopsony in wheat procurement and restricted the participation of private sector.

The findings of the study conducted in five major wheat-producing states showed that marketed surplus of wheat was about 81 per cent and ranged from about 61 per cent on marginal farms to 86 per cent on large farms (Table 34.5). The gross marketed surplus was the highest (90.1 per cent) in Punjab, followed by Haryana (82.9 per cent), Madhya Pradesh (82.6 per cent) and the lowest (54.3 per cent) in Rajasthan. The share of various farm size groups in total output, marketed surplus, and area operated as well as farmers' participation in wheat marketing showed that more than two-thirds of the total output of sample households was contributed by medium and large farms while marginal farmers contributed about 5 per cent. A comparison of the shares of respective farm size groups in the total marketed surplus shows that marginal farmers contribute the lowest quantity (4.1 per cent), whereas medium farms offered the highest share (35 per cent) of the total marketed surplus. The share of small and marginal farmers in total output as well as marketed surplus was higher than their share in total area under wheat. More than 96 per cent of the sample households participated in the marketing of wheat, and there was no significant difference among various farm categories.

Table 34.5 Marketable and Marketed Surplus of Wheat on Sample Households

	Marketable Surplus as % of Total Production	Gross Marketed Surplus as % of Total Production	Net Marketed Surplus as % of Total Production
State			
Haryana	82.9	82.9	82.9
Punjab	90.1	90.1	90.1
Uttar Pradesh	68.6	65.1	65.1
Madhya Pradesh	75.2	82.6	80.4
Rajasthan	61.6	54.3	48.4
Total	83.0	80.7	80.1
Farm Size			
Marginal	64.8	61.2	56.3
Small	72.2	69.4	66.9
Semi-medium	79.9	77.7	77.1
Medium	84.7	82.6	82.5
Large	88.1	86.0	86.0
All farms	83.0	80.7	80.1

Source: Field Survey, 2011–12.

The average farm retention (self-consumption, seed, and other purposes) was 15.3 per cent of the total production about 60 per cent of the total retention was for self-consumption, followed by for seed (21.4 per cent), and feed purpose (12.9 per cent).

Farm size, wheat price, awareness of MSP, and access to regulated market have positive influence on marketed surplus while family size and distance to markets have negative impact and most variables are statistically significant, indicating that they significantly influence marketed surplus (Table 34.4). The relationship between farm size and marketed surplus is positive and statistically significant, indicating that with an increase in farm size, marketed surplus ratio also increases. The relative importance of factors influencing marketed surplus indicated that the price received by farmers was the most important factor, followed by an awareness of MSP, farm size, and access to regulated markets. Distance to market was the least important variable in influencing marketed surplus of wheat.

Maize

Maize is the third important cereal in the country with 22.8 million tonnes production, contributing about 9.5 per cent to the country's total cereals production. The area under maize has increased from 5.8 million hectares in TE 1973–74 to about 8.7 million hectares in TE 2012–13, while the production increased by more than 280 per cent from 5.7 million tonnes to about 21.9 million tonnes during the same period, primarily due to a significant increase in yield. The average yield of maize increased from about 990 kg/ha in the early 1970s to 2528 kg/ha during TE 2012–13 but is still much lower compared to the world average and major producers like the United States and China (4.93 t/ha).

Maize production in the country has increased at an annual growth rate of 3.28 per cent during 1971–2012 while area and yield increased at 0.96 per cent and 2.34 per cent, respectively during the same period (Table 34.2). During the 1990s, production of almost all cereals, including rice and wheat witnessed deceleration in growth rates but maize production exhibited an impressive positive and accelerated growth rate (3.74 per cent). During the last two decades, in new non-traditional maize-growing areas, more acreage has been brought under maize cultivation, and the contribution of the area was very close to the contribution of yield in increased production. Maize has experienced a marked regional shift in the production as well as acreage. Traditionally, maize was grown in Uttar Pradesh, Bihar, Madhya Pradesh, and Rajasthan with nearly two-thirds of the total area and over half of the total production in the early 1980s. However, in the recent period, peninsular India has emerged as a dominant maize-growing region and accounts for more than 40 per cent of the total production. Three states, namely Andhra Pradesh, Karnataka, and Tamil Nadu, increased their share in total acreage from less than 10 per cent in TE 1983–84 to 27.4 per cent in TE 2011–12, while production share increased from 15.4 per cent to 42.8 per cent during the same period. Traditional maize-growing states have lost their share in total acreage as well as production during the last three decades.

Table 34.6 Marketable and Marketed Surplus of Maize on Sample Households

	Marketable Surplus as % of Total Production	Gross Marketed Surplus as % of Total Production	Net Marketed Surplus as % of Total Production
State			
Karnataka	90.5	86.5	86.5
Maharashtra	98.3	98.1	98.1
Rajasthan	81.0	86.4	84.4
Total	90.5	88.3	88.1
Farm Size			
Marginal	86.8	79.9	79.9
Small	88.7	83.7	83.4
Semi-medium	91.6	91.1	90.9
Medium	93.4	93.3	93.0
Large	81.1	88.2	88.2
All farms	90.5	88.3	88.1

Source: Field Survey, 2011–12.

The average gross marketed surplus accounted for 88.3 per cent of total maize production in the study area (Table 34.6). In the case of different farm sizes, marketed surplus was the highest (93.35 per cent) on medium households and the lowest on marginal farms (79.9 per cent). Among states, Maharashtra had the highest (98.1 per cent) marketed surplus. The marketed surplus was lower than the marketable surplus in case of small and marginal farmers, thereby indicating distress sale.

The size of farm and maize price had a positive and statistically significant impact on marketed surplus, indicating that with an increase in farm size and higher prices, marketed surplus increases. Family size and number of livestock had a negative impact on marketed surplus, which shows that larger the household family size and livestock herd size, the lower is the marketed surplus of maize (Table 34.4).

Bajra

India is one of the world's leading producers of bajra, both in terms of area (8.54 million ha) and production (9.8 million tonnes) with average productivity of about 1,152 kg per ha during TE 2012–13. The area under bajra declined by about 26 per cent between the early 1980s and TE 2012–13, but production increased by nearly 60 per cent, mainly due to a significant increase in productivity. Bajra, which was the second largest millet in the country after sorghum in terms of area and production till the early 2000s, has thereafter surpassed sorghum and occupied the first position. The share of bajra in total cereals acreage, as well as production, has declined during the last four decades from 12.2 per cent and 6.7 per cent during the TE 1971–72 to 8.6 per cent and 4.2 per cent in the TE 2012–13, respectively. However, the performance of bajra has slightly improved during the last decade mainly due to improvement in yield, from 736 kg per ha to 1149 kg per ha.

Bajra recorded a negative (–0.08 per cent) growth in production during the 1970s before increasing to 1.35 per cent in the 1980s and reaching a level of 2.16 per cent during the last decade. The productivity witnessed an accelerated growth rate during the last four decades. The production (2.16 per cent) and productivity (3.22 per cent) recorded the highest growth rate during the last decade while the growth rates were the lowest during the 1970s (Table 34.2). The variability in both production and productivity has remained fairly high due to extremely low coverage of irrigation facilities.

The marketed surplus of bajra was estimated at 67.7 per cent on all farms and varied from about 60 per cent on small farms to 74 per cent on large farms (Table 34.7). In Haryana, the marketed surplus was higher (83.5 per cent) compared to Uttar Pradesh (61.4 per cent) and Rajasthan (63.4 per cent). The marketed surplus was lower than marketable surplus on all farm categories as well as in Rajasthan and Uttar Pradesh. About 15.6 per cent of the total marketed surplus was procured by government agencies while about 85 per cent was sold to private traders and other buyers. However, large farmers had better access to government agencies than small and marginal farmers. The large farmers received a higher price than small and marginal farmers under both market channels, showing their better bargaining skills.

Table 34.7 Marketable and Marketed Surplus of Bajra on Sample Households

	Marketable Surplus as % of Total Production	Gross Marketed Surplus as % of Total Production	Net Marketed Surplus as % of Total Production
State			
Haryana	83.5	83.5	83.5
Rajasthan	69.5	63.4	60.6
Uttar Pradesh	72.9	61.7	61.7
Total	73.5	67.7	66.1
Farm Size			
Marginal	67.5	60.2	57.8
Small	71.8	66.2	64.4
Semi-medium	74.7	68.7	66.7
Medium	72.8	67.9	66.9
Large	80.4	74.1	73.2
All farms	73.5	67.7	66.1

Source: Field Survey, 2011–12.

Family size and age of head of household had adverse impact on marketed surplus while impact of farm size on marketed surplus was positive and statistically significant indicating that with an increase in farm size, marketed surplus of bajra also increases. Other important factors, which influenced marketed surplus positively, include farmers' awareness of MSP and access to regulated markets (Table 34.4).

Gram

Gram is the most important pulses crop in India and accounts for approximately 35 per cent of total pulses acreage and about 46 per cent of the total production in the country. The area under gram witnessed a declining trend during the post-reforms period as crop acreage declined from 7.5 million hectares in TE 1981–82 to 6.1 million hectares in TE 1993–94 and reached 5.9 million hectares in TE 2001–02 but the trend reversed during the last decade. Although the area under gram cultivation declined during the nineties, production increased from 4.5 million tonnes to 4.8 million tonnes. Gram production reached a record level of 9.53 million tonnes in 2013–14. The gram yield increased from 607 kg/ha in 1971–73 to 913 kg/ha in 2010–12. Gram production increased at an annual compound growth rate of about 1.41 per cent during 1981–2011, while crop acreage and yield recorded 0.34 per cent and 1.06 per cent growth rates, respectively. In the long term, of the 1.41 per cent annual growth in gram production, an increase in yield accounted for about three-fourths of the growth in production while remaining one-fourth came from area expansion. The production recorded a negative growth rate during the 1970s and the 1980s (during green revolution period) but started showing some improvement during the last two decades. The annual growth rate of production became positive (1.19 per cent) during the 1990s and reached 5.51 per cent during the last decade. Gram production (5.51 per cent) and yield (2.10 per cent) recorded the highest growth rate during the last decade (Table 34.1).

The estimates of marketed surplus showed that about 87 per cent of total gram production was sold in the market (Table 34.8). The marketed surplus was highest (88.7 per cent) on the large farms, followed by semi-medium (87.1 per cent) and the lowest on marginal farms

Table 34.8 Marketable and Marketed Surplus of Gram on Sample Households

	Marketable Surplus as % of Total Production	Gross Marketed Surplus as % of Total Production	Net Marketed Surplus as % of Total Production
State			
Maharashtra	90.2	85.2	84.4
Madhya Pradesh	80.7	88.4	88.4
Karnataka	89.6	88.3	88.3
Rajasthan	91.2	84.4	84.3
Total	87.6	87.2	87.0
Farm Size			
Marginal	86.5	85.0	84.7
Small	87.1	86.7	86.5
Semi-medium	87.8	87.1	86.9
Medium	88.4	85.9	85.7
Large	87.2	88.7	88.7
All farms	87.6	87.2	87.0

Source: Field Survey, 2011–12.

Table 34.9 Marketable and Marketed Surplus of Tur on Sample Households

	Marketable Surplus as % of Total Production	Gross Marketed Surplus as % of Total Production	Net Marketed Surplus as % of Total Production
State			
Karnataka	92.7	90.8	90.8
Madhya Pradesh	88.1	96.8	96.8
Maharashtra	83.3	85.7	82.9
Uttar Pradesh	76.0	85.8	85.8
All	90.5	92.5	92.5
Farm Size			
Marginal	84.0	89.5	89.5
Small	85.0	90.5	90.5
Semi-medium	85.9	86.5	86.5
Medium	90.7	99.5	99.5
Large	94.9	85.0	85.0
All farms	90.5	92.5	92.5

Source: Field Survey, 2011–12.

(85 per cent). The average farm retention for self-consumption, seed, and other purposes was 12.2 per cent and varied from 11.7 per cent on medium farms to 13.8 per cent on marginal farms.

Tur

India is the world's largest producer of tur and accounts for about 63 per cent of the global production. Tur is cultivated on about 4.1 million hectares, grown mainly under rainfed conditions (4 per cent area under irrigation), with a production of about 2.8 million tonnes in the country. More than three-fourths of the total cropped area is concentrated in the semi-arid tropics and Maharashtra has the largest acreage (about 31 per cent) under the crop, followed by Karnataka (19.2 per cent), Andhra Pradesh (about 12 per cent), Madhya Pradesh (13.2 per cent), and Uttar Pradesh (about 8 per cent). The compound annual growth rates of area, production, and yield of tur as given in Table 34.2 revealed that growth performance of the crop was not impressive during the last four decades as the crop production registered an annual growth rate of less than 1 per cent (0.97 per cent) while area grew by nearly one per cent, and growth rate in yield was negative. However, performance was relatively better during the last decade when production recorded the highest growth rate (2.22 per cent). Tur yield witnessed a negative growth rate during the 1970s and 1980s before improving marginally during the last two decades but with a less than one per cent growth.

The average marketed surplus in case of tur was quite high (92.5 per cent) and was the highest (99.5 per cent) on medium households and the lowest on semi-medium farms (Table 34.9). The marketed surplus was the highest (96.8 per cent) in Madhya Pradesh and the lowest (85.7 per cent) in Maharashtra.

CONCLUSIONS AND POLICY IMPLICATIONS

Development of efficient and competitive agricultural marketing system is essential for accelerating the growth of agricultural production and marketed surplus and also has the potential to benefit poor consumers. However, marketing structure and organization for agricultural commodities in India varies across different states and commodities and consists of both public and private sectors. For few commodities like rice and wheat government has a direct intervention, while, in most other crops, marketing is dominated by the private sector. The organized marketing of agricultural commodities promoted through a network of regulated markets has helped in ameliorating the market constraints of producers at the wholesale assembling level and protect them from the exploitation of market intermediaries and traders as well as ensured better prices and timely payment for the produce. However, these markets have become restrictive and monopolistic and restricted private investment in the sector, which has led to poor market infrastructure due to lack of investment. To improve the investment in market infrastructure, however, requires undertaking significant investments in technology, institutions, infrastructure and management. Understanding marketed surplus and marketing behaviour can help in designing appropriate policies, technology choices and institutions to facilitate the development of agriculture. Some important policy implications for improving the marketed surplus and infrastructure include strengthening of physical infrastructure, competitive market structure by liberalizing agricultural markets, better reach and quality of market information and extension services, easy access to institutional credit and proper storage at farm household level and improved regulation of markets to avoid exploitation of farmers by market intermediaries particularly in those crops where public procurement is not very effective.

REFERENCES

Bhargava, P.N., and V.S. Rustogi. 1972. 'Study of Marketable Surplus of Paddy in Burdwan District', *Indian Journal of Agricultural Economics*, 27(3) (July–September): 63–8.

Central Statistics Office (CSO). 2014. 'National Accounts Statistics 2014', May. New Delhi: Ministry of Statistics and Programme Implementation, Government of India.

Farruk, M.O. 1970. 'The Structure and Performance of the Rice Marketing System in East Pakistan', Occasional Paper No. 31, Department of Agricultural Economics, Cornell University, Ithaca, New York.

Government of India (GoI). 2015. 'Third Advance Estimates of Food Grains, Oilseeds & Other Commercial Crops for 2014–15', Directorate of Economics & Statistics, Department of Agriculture and Cooperation, Ministry of Agriculture, Government of India, New Delhi, May.

———. 2015a. *Agricultural Statistics at a Glance 2014*. New Delhi: Oxford University Press.

Harris, B. 1982. 'The Marketed Surplus of Paddy in North Arcot District, Tamil Nadu: A Micro-level Causal Model', *Indian Journal of Agricultural Economics*, 37(2): 145–58.

Hussein, M.A., and H.B. Rajbanshi. 1985. 'An Economic Analysis of Marketed Surplus of Paddy in Kathmandu District, Nepal', *Malayan Journal of Agricultural Economics*, 2(1): 54–62.

Krishna, Raj. 1965. 'The Marketable Surplus Function for a Subsistence Crop', *Economic Weekly*, Annual No. 17 (February), pp. 309–20.

Krishnan, T.N. 1965. 'The Marketed Surplus of Foodgrains: Is it Inversely Related to Price', *Economic Weekly* Annual No. 17 (February): 309–20.

Nadkarni, M.V. 1980. 'Marketable Surplus and Market Dependence: A Millet Region of Maharashtra', *Economic and Political Weekly*, 15(13) March 1980, A13–A24.

Narain, Dharm. 1961. 'Distribution of the Marketed Surplus of Agricultural Produce by Size-Level of Holding in India 1950–51'. Institute of Economic Growth Occasional Paper no 2, Asia Publishing House, Bombay, India.

Rahman, A. 1980. 'Marketed Surplus and Product Market in Bangladesh Agriculture', Bangladesh Institute of Development Studies, Dhaka, (mimeo).

Raquibuzzaman, M. 1966. 'Marketed Surplus Function of Major Agricultural Commodities in Pakistan', *The Pakistan Development Review*, 6(3): 376–92.

Sharma, K.L., and M.P. Gupta. 1970. 'Study of Farm Factors Determining Marketed Surplus of Bajra in Jaipur District', *Indian Journal of Agricultural Economics*, 25: 64–68.

Sharma, Vijay Paul. 2011. 'India's Agricultural Development under the New Economic Regime: Policy Perspective and Strategy for the 12th Five Year Plan', *Indian Journal of Agricultural Economics*, 67(1): 46–78.

Sharma, Vijay Paul, and Harsh Wardhan. 2014. 'Assessment of Marketed and Marketable Surplus of Major Foodgrains in India', Report Submitted to the Ministry of Agriculture, Govt. of India, Centre for Management in Agriculture, Indian Institute of Management, Ahmedabad, June 2014.

ANNEXURE 34A

Table 34A The Details of Agro-Economic Research Centres and the Lead Person Involved in the State Report

State	AER Centre/Unit	Authors
Gujarat	AERC Vallabh Vidyanagar	V. D. Shah and Manish Makwana
Haryana	AERC Delhi	Usha Tuteja
Karnataka	ISEC, Bangalore	Parmod Kumar, Elumalai Kannan, Rohit Chaudhary and Kedar Vishnu
Madhya Pradesh	AERC Jabalpur	Hari Om Sharma, Deepak Rathi
Maharashtra	GIPE, Pune	Sangeeta Shroff and Jayanti Kajale
Uttar Pradesh	AERC Allahabad	Ramendu Roy
West Bengal	AERC Shantiniketan	D. Roy, A. Sinha
Punjab	AERC Ludhiana	D.K. Grover, Jasdev Singh and Satwinder Singh

35

Assessment of Pre- and Post-Harvest Losses of Important Crops in India

Elumalai Kannan

India's agricultural production pattern has undergone significant changes overtime. Technology has played an important role in bringing the transformation in agriculture. But, technological developments have altered a multi-commodity production system to a specialized system in different parts of the country. In the process, many traditionally cultivated crops either have lost their area or gone out of cultivation. Further, these developments have increased the incidences of pests and diseases. The increased use of pesticides to maintain the productivity of crops has resulted in developing resistance to insects and pathogens, which unfortunately proved counterproductive by reducing crop yield. The indiscriminate and excessive use of pesticides combined with chemical fertilizers is partly responsible for environmental degradation. Further, this has led to destruction of habitat of beneficial insects and also increased the cost of cultivation of crops.

Estimation of crop losses due to pests and diseases involves complex procedures. It is in fact, difficult to assess the loss caused by the individual pest as a particular crop may be infested by the pest complex in the farmers' field conditions. Further, the extent of crop loss either physical or financial depends upon the type of variety, stage of crop growth, pest population, and weather conditions. Even with these difficulties, crop loss estimates have been made and updated regularly at global as well as national levels by independent researchers.

Generally, crop loss is estimated as the difference between potential (attainable) yield and the actual yield. The potential yield is the yield that would have been obtained in the absence of pests under consideration. By multiplying the area with the estimated yield loss, total loss is obtained. To estimate the crop loss, most of the existing studies have adopted an experimental treatment approach (with or without pest attack through artificial infestation) or fields with natural infestation wherein half of the field is protected against the pest while the other half is not. However, the results obtained from artificial infestation or natural infestation in the selected plots/fields will not be appropriate for extrapolation over a geographical area (Groote 2002). Estimated crop losses under these conditions may not represent the actual field conditions of farmers. Alternatively, crop loss estimates collected directly from the farmers through a sample survey may be more reliable and could be used for extrapolation in similar geographical settings.

Agricultural production is seasonal and exposed to natural environment, but post-production operations play an important role in providing stability in the food supply chain. Losses in food crops occur during harvesting, threshing, drying, storage, transportation, processing, and marketing. In the field and during storage, the products are damaged by insects, rodents, birds, and other pests. Food grain stocks suffer qualitative and quantitative losses while in storage. The quantitative losses are generally caused by factors such as incidences of insect infestation, rodents, birds, and also due to physical changes in temperature, moisture content, etc. The qualitative loss is caused by reduction in nutritive value due to factors such as attack of insect pest, physical changes in the grain, and chemical changes in the fats, carbohydrates, protein, and also by contamination of myco toxins.

As per the available estimates, crop loss caused by pests and diseases are very high (Dhaliwal, Jindal, and Dhawan 2010; Oerke 2006). However, the knowledge on the subject of crop loss at the farm level is very much limited. In addition to losses that occur during the growth period of the crop, there is a huge quantity of grains lost during the process of harvesting, threshing, transportation, and storage. Therefore, the present study makes a comprehensive attempt to estimate the dimension of losses occurring during the pre- and post-harvest stages of the selected crops.

For the pre-harvest losses, generally animal pests (insects, mites, rodents, snails, and birds), plant pathogens (bacteria, fungi, viruses, and nematodes), and weeds are collectively called as pests, which may cause significant economic damage to crops. This broader definition of pests and diseases is followed in the present study. For estimating post-harvest losses, there is a need to establish the extent of losses during storage under different agro-climatic conditions. Causes of storage losses include sprouting, transpiration, respiration, rot due to mould and bacteria, and attack by insects. Sprouting, transpiration, and respiration are physiological activities that depend on the storage environment (mainly temperature and relative humidity). These physiological changes affect the internal composition of the grains and result in destruction of edible material and changes in nutritional quality. However, it would be difficult to measure the loss due to physiological changes at the farm level. Nevertheless, an attempt was made to estimate such losses based on the farmers' estimates.

Keeping in view of importance of the subject, the present study focuses on the following objectives:

1. to estimate the physical losses caused by pests and diseases in rice, wheat, tur, and soybean at the farm level;
2. to examine measures of pest and disease management to reduce crop loss due to pests and diseases at the farm level;
3. to arrive at post-harvest losses in rice, wheat, tur, and soybean under different agro-climatic conditions; and
4. to identify factors responsible for such losses and suggest ways and means to reduce the extent of losses in different operations in order to increase national productivity.

DATA AND METHODOLOGY

The present study was based on the farm-level data collected from 10 major states growing four reference crops, namely rice, wheat, tur, and soybean, during 2011–12. Name of states for these respective crops covered are: rice (Assam, Karnataka, Punjab, Tamil Nadu, Uttar Pradesh, and West Bengal), wheat (Assam, Madhya Pradesh, Punjab, Uttar Pradesh, and West Bengal), tur (Gujarat, Karnataka, and Maharashtra) and soybean (Madhya Pradesh, Maharashtra, and Rajasthan). The crop production constraints particularly infestation by pests and diseases, and losses caused by them were worked out based on the estimates provided by the sample farmers. Pests and diseases, when their population reach beyond a threshold level, cause physical damage to crops, which ultimately result in financial loss to farmers. Besides pests and diseases, other bio-economic factors such as soil fertility, water scarcity, poor seed quality, high input costs, and low output prices lead to considerable financial loss to farmers. Thus, data on these bio-economic variables were also collected from the farmers. The quantification of yield loss was estimated by asking the farmers to identify the pests and diseases by name, frequency of attack, and crop loss by individual pests. Farmers were also asked to mention the actual production with attack of all pests and normal production in the absence of pests.

The post-harvest losses encountered during the process of harvesting, threshing, transportation, and storage were quantified based on the estimates provided by the sample farmers. The study also attempted to identify the storage structure at the farmers' level and enumerate the losses occurring during storage for the reference crops. The control measures adopted by the farmers to minimize the post-harvest losses were also captured through a field survey.

RESULTS AND DISCUSSION

Pre-harvest Losses

In Assam, 92.5 per cent of the sample households reported that low output price was the most important constraint in the cultivation of paddy. In case of pests and diseases, 23.3 per cent of the sample farmers opined it as the most important constraint. Crop losses in high yielding variety (HYV) paddy was estimated to be high as compared to local paddy indicating a higher level of resistance of local paddy against infestation. The loss over the actual production of local paddy ranged between 5.4 per cent and 8.3 per cent while in HYV paddy, it stood between 6.2 per cent and 9.5 per cent across farm size groups. The loss over normal production of local paddy ranged between 5.2 per cent and 7.6 per cent while in HYV paddy, it ranged between 5.8 per cent and 8.7 per cent across farm size groups. Overall, crop loss over normal production was estimated at 6.7 per cent for local and 7.8 per cent for HYV in Assam (Table 35.1).

For paddy cultivation in Karnataka, incidence of pests and diseases emerged as a serious problem with a reporting of 95.6 per cent of the total paddy growing farmers followed by high cost of inputs (90.0 per cent). Paddy yield loss due to all pests ranged from 13.8 per cent among medium farmers to 20.0 per cent among marginal farmers.

Table 35.1 Overall Crop Loss due to Pests, Diseases and Weeds for Sample Crops and States

Crop	Loss over Normal Production (%)	
	Local Variety	HYV
Paddy		
Assam	6.66	7.79
Karnataka	–	16.2
Punjab	–	8.00
Tamil Nadu	–	9.07
Uttar Pradesh	–	2.88
West Bengal	–	15.05
Wheat		
Assam	–	13.76
Punjab	–	7.35
Madhya Pradesh	–	8.89
Uttar Pradesh	–	6.09
West Bengal	–	15.29
Tur		
Karnataka	44.7	43.9
Gujarat	17.11	14.18
Maharashtra	8.12	11.8
Soybean		
Madhya Pradesh	–	11.37
Maharashtra	–	10.48
Rajasthan	23.66	20.74

Note: '–' refers to Not Applicable.
Source: Field survey.

In Punjab, high cost of inputs was reported as the most important constraint by 73 per cent of the households followed by 23 per cent revealing low output price, 14 per cent water deficiency, and 7 per cent pest and disease problem. The loss of paddy output varied from 1.6 to 2.4 quintals per acre, being lowest on small farms and highest on large farm categories due to better management of farms by small farmers as compared to large farmers. The loss over actual production was 7.9 per cent on marginal, 6.1 per cent on small, 8.5 per cent on medium, and 8.9 per cent on large farms categories. In total, magnitude of crop loss due to pests, diseases, and weed infestation in paddy was 8.7 per cent over actual production and about 8.0 per cent over normal production.

Among constraints faced by farmers in the cultivation of paddy in Tamil Nadu, low output price was reported as the most important constraint faced by 57.5 per cent of the sample farmer in Villupuram and 40 per cent in Tiruvarur. About 35 per cent of the sample households informed that pests and diseases problem was an important problem faced by them. Crop loss expressed in terms of actual production varied at 14 per cent for marginal, 12 per cent for small, 14 per cent for medium, and 16 per cent for large farmers in Tiruvarur. The overall loss of paddy was estimated higher among sample farmers of Villupuram.

In Uttar Pradesh, low output price of paddy was reported as the most important constraint by 53 per cent of the sample households. With respect to pest and disease problems, 17 per cent of the sample farmers had reported it as a constraint. The percentage loss over actual production was estimated at 2.96 per cent against 2.88 per cent over normal production. The percentage loss over actual production was estimated the highest at 4.3 per cent for marginal farmers and lowest at 1.5 per cent for large farmers. Similarly, the percentage loss over normal production was estimated at 2.9 per cent for the overall sample. The analysis of data revealed that percentage losses over actual production as well as percentage loss over normal production were highest on marginal farms and lowest on large farms.

Among various constraints, high cost of inputs, low output price, and pest and disease problems ranked most important in the cultivation of paddy in West Bengal. The overall yield loss with attack was estimated at 3.5 quintals per acre. The overall normal production without attack was 23.5 quintals per acre. However, the percentage loss over normal production was less than that of percentage loss over actual production.

In the case of wheat, low output price was the most important constraint for 65.0 per cent of the sample households, and pests and diseases for 90.8 per cent of the households. The yield loss over actual production varied between 16.1 per cent and 21.9 per cent across farm size groups. The loss over the normal production varied from 13.9 per cent to 18.0 per cent. The sample farmers opined that the ruling seed variety and climatic conditions are susceptible to pest and disease attacks. In Punjab, the high cost of inputs was reported as the most important constraint by 76 per cent of the households while 21 per cent informed low output price as the most important constraint. Only 3 per cent of the sample households reported pest and disease problems as the most important constraint. The per cent loss over actual production increased with increase in farm size. Overall, the magnitude of crop loss due to pests, diseases, and weed infestation was 7.9 per cent over actual and 7.35 per cent over normal production.

The incidence of pest and diseases, poor quality of seed, and low price of output were the least important constraints mentioned by the sample farmers in Madhya Pradesh. Crop loss over normal production was 13.9 quintals per acre and over actual production it was 12.6 quintals. There is discernible variation in the loss across farm size groups. In terms of percentages, the loss over normal production was 8.9 and for actual production it was 9.8. The per cent loss over actual and normal production increased with increase in farm size.

In Uttar Pradesh, high cost of inputs and low price of output were considered to be the most important constraints. Low output price was considered most important constraint by over a quarter of the sample households. Interestingly, none of the respondents reported pest and disease problems as a constraint in the cultivation of wheat. The actual yield of wheat with attack varied from 15.2 quintals to 16.7 quintals across farm size groups and for the overall sample it was 16.0 quintals per acre. Normal production without attack varied between 16.3 quintals and 17.7 quintals per acre.

Among various constraints in the cultivation of wheat, high cost of inputs and low output price were considered to be the most important constraints by a significant proportion of the sample households in West Bengal. About half of the sample farmers mentioned that pests and diseases problems, and poor seed quality as the most important problems. The level of crop loss over normal production was lower (15.3 per cent) than that of percentage loss over actual production (18.1 per cent). Among farm size groups, crop loss over actual production was 17.8 per cent for marginal farmers, 18.1 per cent for small farmers and 18.4 per cent for medium farmers. The corresponding figures for crop loss over normal production were 15.1 per cent, 15.3 per cent, and 15.6 per cent.

A high proportion of sample farmers (89.4 per cent) have reported pest and diseases problems as a major constraint affecting the production of tur in Karnataka. The per cent production loss was higher for local varieties than for HYV. In fact, yield loss as percentage of normal production was 44.7 for local varieties and 43.9 for HYV. Similarly, yield loss over actual production was 80.8 per cent for local varieties and 78.3 per cent for HYV.

In Gujarat, pest and disease problems were reported as the most important constraint in cultivation of tur by 80.8 per cent, while 63.3 per cent and 61.7 per cent of sample households mentioned water deficiency and high cost of inputs as the most important constraint, respectively. The per cent loss over actual and normal production was comparatively low in case of HYV in relation to local variety among various categories of farmers. In case of local variety, it varied between 20.0 per cent and 22.8 per cent over actual production and between 16.6 per cent and 18.6 per cent over normal production across farm size groups. However, for HYV it ranged between 16.2 per cent and 17.7 per cent over actual production, and 14.0 per cent and 15.0 per cent over normal production.

Over 20 per cent of sample farmers reported water deficiency, pest and disease problems, high cost of inputs, and low output price as the most important constraints faced in the cultivation of tur in Maharashtra. The proportion of tur production loss in relation to normal production was 8.1 per cent for the overall sample in case of local variety and 11.8 per cent for HYV variety. The per cent loss, both in terms of actual production and normal production, was relatively low for marginal farmers and high for large farmers.

High cost of inputs has been reported as the most important constraint by 60 per cent of the soybean sample households in Madhya Pradesh. Over one-third of the sample farmers mentioned the problem of pests and diseases attack as an important constraint in the cultivation of soybean. The yield loss over actual output increased with increase in farm size with 8.7 per cent for marginal farmers, 10.7 per cent for small farmers, 15.8 per cent for medium farmers, and 16.5 per cent for large farmers. The yield loss over normal production was 14.2 per cent for large farmers, 13.6 per cent for medium farmers, 9.6 per cent for small farmers, and 8.0 per cent for marginal farmers.

In Maharashtra, about 45 per cent of the sample farmers considered pests and disease problem as important as well as most important constraint. Low output price was treated as the most important in the cultivation of soybean by 29.1 per cent of the sample farmers. The proportion of crop loss in relation to actual production was worked out at 10.3 per cent for marginal category,

10.9 per cent for small, 11.7 per cent for medium, and 13.2 per cent for large farmers. The proportion of soybean crop production loss with respect to normal production translated into 9.4 per cent for marginal category, 9.8 per cent for small, 10.4 per cent for medium, and 11.7 per cent for large category.

Pest and disease problem was reported as the most important constraint by 69.2 per cent of the sample households in Rajasthan. The overall loss of output was 31 per cent over actual production and 23.7 per cent over normal production for local variety. Similarly, for HYV, overall loss of output was 26.2 per cent over actual production and 20.7 per cent over normal production.

Post-harvest Losses

Post-harvest losses were captured in the form of quantity of grains lost during different post-harvest operations such as harvesting, threshing, winnowing, transport, handling, and storage undertaken by the sample farmers. The total post-harvest loss varied by crops and states (Table 35.2).

Table 35.2 Overall Post-harvest Loss of the Sample Crops in Select States

Crop	Post-harvest Loss (%)
Paddy	
Assam	7.33
Karnataka	6.87
Punjab	4.43
Tamil Nadu	6.88
Uttar Pradesh	5.57
West Bengal	3.51
Wheat	
Assam	11.71
Madhya Pradesh	8.61
Punjab	1.84
Uttar Pradesh	2.74
West Bengal	7.22
Tur	
Karnataka	11.15
Gujarat	3.05
Maharashtra	6.00
Soybean	
Madhya Pradesh	12.56
Maharashtra	3.66
Rajasthan	3.41

Source: Field survey.

However, there seems to be, by and large, an inverse relationship between post-harvest loss and farm size groups indicating that marginal and small farmers encounter considerable quantity of post-harvest loss due to lack of access to suitable machinery and financial capital.

Among different types of post-harvest losses, quantity of grains lost during harvesting and storage was estimated higher for the reference crops in the select states. In Assam, the overall loss of paddy during storage for the entire sample farmers was estimated at 2.1 kg per quintal, which varied between 1.3 kg for marginal farmers and 2.8 kg for large farmers. The overall loss of paddy in Karnataka during harvesting was 1.9 kg, threshing and winnowing 0.28 kg, transport 0.6 kg, and handling 0.3 kg. However, in Punjab, harvesting loss was estimated at 1.52 kg, transportation loss at 0.06 kg, and storage loss was 2.5 kg per quintal. A similar pattern can be observed in Tamil Nadu, Uttar Pradesh, and West Bengal.

In the case of wheat, total post-harvest loss was computed at 8.5 kg for marginal, 10.7 kg for small, 13.5 kg for medium, and 14.2 kg for large farmers in Assam. In Madhya Pradesh, quantity of wheat grains lost was observed higher during harvesting with 2.9 kg per quintal. The average storage loss for overall sample farmers was 4.8 kg per quintal with the highest amount of loss being estimated for large farmers followed by medium, small, and marginal farmers. In Uttar Pradesh, total post-harvest loss of wheat at the aggregate level was 2.7 kg per quintal, which varied between 3.0 kg among marginal farms and 2.4 kg on large farms. It indicates that total post harvest loss per quintal decreased with increase in size of farms. However, in West Bengal, total post-harvest loss was 7.2 kg with highest being recorded for medium farmers followed by small and marginal farmers.

For tur, among different type of losses, harvesting loss and storage loss was very high at 3.7 kg and 11.2 kg, respectively for the entire farmer sample in Karnataka. In Maharashtra, per quintal average loss of tur was estimated at 6.0 kg, which comprised harvesting loss of 1.6 kg, threshing loss of 1.2 kg, winnowing loss of 0.6 kg, transportation and handling loss of 1.3 kg, and storage loss of 1.3 kg. However, in Gujarat, per quintal loss of tur in various stages post harvest was the highest for marginal farmers and the lowest for large farmers.

In case of soybean, the quantity of soybean lost during harvesting was the highest at 7.1 kg followed by storage (2.6 kg) and threshing and winnowing (2.4 kg) in Madhya Pradesh. In Maharashtra, the total post-harvest loss per quintal was estimated at 3.7 kg, which comprised 1.1 kg of harvesting loss, 0.5 kg of threshing loss, 0.4 kg of winnowing loss, 0.5 kg of transportation loss, 0.6 kg of handling loss, and 0.6 kg of storage loss. Similarly, in Rajasthan it could be observed that quantity of soybean lost was the highest in the harvesting stage followed by threshing and storage.

Total post-harvest loss of paddy was the highest in Assam followed by Tamil Nadu and Karnataka. The post-harvest loss in Uttar Pradesh was 5.6 per cent and in West Bengal it was 3.5 per cent. In case of wheat, the post-harvest loss was relatively high in Assam with 11.7 per cent followed by Madhya Pradesh (8.6 per cent) and West Bengal (7.2 per cent). For tur, the post-harvest loss was relatively high in Karnataka. In case of soybean, the post-harvest loss was the highest in Madhya Pradesh. Maharashtra and Madhya Pradesh registered a post-harvest loss of 3.7 per cent and 3.4 per cent, respectively. It is clear from the analysis that the largest producing states have, by and large, recorded a higher level of post-harvest loss.

CONCLUSIONS AND POLICY SUGGESTIONS

Evidence gained from this all-India study suggests that infusion of new technologies, better practices, coordination, and investment in rural infrastructure are critical for reducing losses. Advantages of information and communication technology (ICT) should be tapped to provide farmers with practical advice for control of insect pests, diseases and weeds. Amount of pre- and post-harvest losses caused by biotic and abiotic factors is found to be substantial.

In order to reduce these losses, scientific knowledge on cultivation practices and post-harvest operations need to be imparted to the farmers. Pests and diseases occur in a complex ways affecting crop yield performances. Evidence shows that adoption of integrated pest and disease management practices is promising for control of pests. Therefore, an integrated approach needs to be promoted for effective control of pests. Concerted efforts should be made to supply agricultural equipment including harvesters and threshing machines on a custom-hiring basis so that resource poor farmers can avail these services at the village level. Local bodies should be facilitated to own and hire out the machinery to the farmers.

Rural infrastructure will play an important role in reducing avoidable post-harvest losses. There is a need to step up not only the amount of public and private investment in building rural agricultural infrastructure, but also quality of such investments. Investments for research on crops such as tur has so far focused on the identification and development of resistant cultivars and on chemical means to control pests and diseases. Biotechnological tools offer greater scope for development of geographically suitable varieties. Farmers will have to be encouraged to follow sensible agronomic practices such as wet and dry system of irrigation and profitable crop rotation. This will reduce the build-up of pests, diseases, and weeds. There is a lack of adequate scientific storage facilities at the village level. Construction of common godowns should be encouraged among local farmers with active support from various agencies. The non-governmental organizations (NGOs) and gram panchayats should play an important role in this regard.

There is a need for rejuvenation of the government extension agencies for approaching the farming community and making themselves indispensable to curtail the dependence of farmers on private input dealers for taking advice regarding farm-related problems. Imparting new training programmes to farmers for timely and cheaper control of insect-pest and disease attack will minimize the production losses due to these constraints.

Finally, a reliable database on crop-loss estimates will help to make proper planning for monitoring and controlling of pests in different crops. Therefore, it is necessary that all the available published estimates are compiled and published regularly for use by different stakeholders.

REFERENCES

Dhaliwal, G.S., Vikas Jindal, and A.K. Dhawan. 2010. 'Insect Pest Problems and Crop Losses: Changing Trends', *Indian Journal of Ecology*, 37(1): 1–7.

Groote, Hugo De. 2002. 'Maize Yield Losses from Stemborers in Kenya', *Insect Science and its Application*, 22(2): 89–96.

Oerke, E.C. 2007. 'Crop Losses to Animal Pests, Plant Pathogens, and Weeds', in David Pimental (ed.), *Encyclopaedia of Pest Management*, pp. 116–20. Florida: CRC Press.

ANNEXURE 35A

Table 35A The Details of Agro-Economic Research Centres and the Lead Person Involved in the State Report

States	AERC Centre/Unit	Authors
Assam	AERC Jorhat	Jatin Bordoloi
Punjab	AERC Ludhiana	D.K. Grover
		J.M. Singh
		Parminder Singh
Madhya Pradesh	AERC Jabalpur	N.K. Raghuwanshi
Maharashtra	AERC Pune	Deepak Shah
West Bengal	AERC Shantiniketan	Debashis Sarkar
		Vivekananda Datta
		Kali Sankar Chattopadhyay
Karnataka	ADRTC Bangalore	Elumalai Kannan and Parmod Kumar
Tamil Nadu	AERC Chennai	K. Jothi Sivagnanam
Uttar Pradesh	AERC Allahabad	Ramendu Roy
Gujarat	AERC VV Nagar	Rajeshree A. Dutta
		Manish Makwana
		Himanshu Parmar
Rajasthan	AERC VV Nagar	Rajeshree A. Dutta
		Manish Makwana
		Himanshu Parmar

36

Fresh Food Supermarket Retailing and Supply Chains in India

Impact on Small Primary Vegetable Producers and Traditional Fruit and Vegetable Retailers

Sukhpal Singh and Naresh Singla

Retailing presently contributes about 10 per cent of India's gross domestic product (GDP) and 6–7 per cent of employment. However, only 4 per cent of retail outlets are bigger than 500 sq. ft and almost all are family owned. Food retail outlets account for one-third of all retail outlets and 63 per cent of total retail sales. Most retail outlets are still small, family-owned operations. During the last decade a number of domestic corporate players have entered the organized food retail sector, including the large food retailers like Spencer's, Reliance Fresh (RF), Aditya Birla Retail Limited (ABRL)'s More, Namdhari Seeds Pvt. Limited (NSPL)'s Namdhari Fresh (NF), and Nilgiris, and others that deal in food and fibre products. Since 1997, 100 per cent foreign direct investment (FDI) in retailing is allowed in single-brand chains, and from 2012, 51 per cent FDI in multi-brand retail trade (MBRT) has been permitted though no foreign player has yet made an entry in the Indian market as the FDI in MBRT remains uncertain with different states and political parties having different perspectives on it and the present Union government opposed to it. Therefore, most of the growth has been driven by the domestic players unlike many other developing countries of Asia and Latin America.

Three major issues of the impact of food supermarkets on local economies include: market concentration and, therefore, producer and consumer interest; downward pressure on producer prices with higher costs and responsibilities; exclusion of small producers and impact on small local retailers. The procurement practices of supermarkets and large processors have a huge impact on farmers and present them with an important challenge. Through their coordinating institutions and mechanisms such as contracts, private standards, sourcing networks, and distribution centres, they are reformulating the rules of the game for farmers and first-stage processors. There is also supplier farmer rationalization due to the larger supplier preference of big retailers. Though supermarkets initially offered higher prices to producers than those offered by traditional channels, farmers incurred extra costs like processing and packaging, marketing, transport, and other transaction costs unlike their counterparts in traditional channels (Boselie, Henson, and Weatherspoon 2003;

Cadlihon et al. 2006; Codron et al. 2004; Ruben, Boselie, and Lu 2007).

An important issue in supply chains is whether small producers can participate and benefit from these chains and markets—which is crucial for their survival as traditional marketing channels weaken or disappear (Harper 2009; Stichele, van de Wal, and Oldenziel 2006). Small farmers have advantages for integrating with the supply chains, as they can supply better quality with intensive management attention to each output unit. However, they lack the size to benefit from economies of scale. The net effect of integrated markets on small farmers depends on the nature of the commodity and its market, as well as the ability of small farmers to coordinate marketing activities. On the other hand, there are also issues of loss of employment and livelihoods due to supermarket expansion. In supermarket stores, low wages, job cuts, long and irregular working hours, and non-contract workers are the abuses reported which are resorted to as strategies to cut labour costs. There can be loss of employment due to supermarket retail chains or higher efficiency of workers. Thus, whereas supermarket chains can lead to new and better employment generation, improvement in food quality, and lower consumer prices and provide new avenues for agricultural development, the negative impacts include exclusion and squeezing out of small producers out of these chains due to high cost and risky investments needed, and the decline of the traditional wholesale markets which may be important for small producers.

Fresh fruit and vegetable (FFV) produce in India is marketed mostly either through the regulated Agricultural Produce Marketing Committee (APMC) markets or totally unregulated local fruit and vegetable markets. Marketing through these traditional channels is characterized by very little attention to grading, sorting and storage, weak regulation, poor handling during loading, unloading, and transport resulting in loss of 30–40 per cent of production. Supply chains for FFVs tend to be multilayered which has implications for the farmers' share in the final consumer price; the quality of produce due to multiple handling; and for the marketing cost as the various agents add their costs. In contrast to fragmented supply chains in traditional markets, supply chains developed by organized retail chains are supposed to be well coordinated.

In this context of global experiences of food retail chain practice and impacts, in India, there have been only a few studies on the impact of food supermarkets on primary producers (Pritchard, Gracy, and Godwin 2010; Sulaiman, Kalaivani, and Handoo 2010) and on traditional retailers (Kalhan, 2007; Kumar, Patwari, and Ayush 2008; Joseph and Soundrarajan 2009).

OBJECTIVES

The present study:

1. explores the procurement channels and practices of major fresh fruit and vegetable retail chains in India and their impact on the primary producers at the procurement end;
2. assesses the likely impact of these chains on traditional small vegetable and fruit retailers at the sales end in India; and
3. examines the possible policy and regulatory provisions to protect and promote livelihoods in the fruit and vegetable sector in the presence of supermarkets in India.

The following sections detail the methodology followed for the study, major findings, and major policy relevant conclusions and ways forward.

METHODOLOGY AND DATA

The study was initiated with a review of relevant literature on the subject and secondary data analysis. The study examines the issues of procurement and sales by retail supermarket chains by examining a few of them across north, west, and south India where all the major players exist. The locations for primary study include Ahmedabad (western region), Chandigarh (northern region), and Bengaluru and Belgaum (southern region). These cities gave access to all the major players like Reliance, ITC, Spencer's, Food World, NF, and More (of the Birlas).

The chains interviewed included Birla's More in Kolar and Belgaum in Karnataka, and Sabarkantha in Gujarat; RF in Sabarkantha in Gujarat; NF in rural Bengaluru in Karnataka; ITC Choupal Fresh in Mohali in Punjab; and Ambala in Haryana. The quantitative and qualitative information about chain operations was obtained from interaction with supply chain managers and front-end mangers of all the retail chains. A farmer's survey was conducted in Malur in Kolar district in Karnataka in the case of Birla's More; and Prantiz in Sabarkantha district of Gujarat for Reliance Fresh. Namdhari Fresh in rural Bengaluru in Karnataka, Lalru in Mohali district of Punjab, and Panjokhra Sahib in Ambala district of Haryana in the case of ITC Choupal Fresh were the other locations. The impact issues were studied through farmer and traditional retailer surveys in the respective locations around the collection centres and in the cities of retail chain outlets respectively. The Collection Centres (CCs) and Distribution Centres (DCs) of all the chains were visited, operations were observed, and the process of collection, processing, and dispatch of vegetables was observed in each case. Supplying

Table 36.1 Retail Chain-wise Number of Farmer Interviewed

State>	Gujarat		Karnataka			Punjab/Haryana
Retail chain>	RF	ABRL	ABRL, Malur	ABRL, Belgaum	Namdhari Fresh	ITC
No. of farmers interviewed	28	22	25	19	33	22

Source: Field survey.

farmers were interviewed at the CCs and/or on their farms/houses in villages. Table 36.1 gives details of chain-wise farmer interviews undertaken in case of each chain.

For traditional retailer interviews, the cities of Ahmedabad, Bengaluru, and Chandigarh each were divided into four zones: north, east, west, and south. The four zones, by and large, represented different segments of the market and coverage of all four types of traditional fruit and vegetable (F&V) retailers—shops, roadside fixed, roadside-cum-home delivery, and only home delivery—ensured that all modes of selling to different segments of buyers are covered. About 15 FFV shops, fixed vendors, mobile vendors, and home delivery vendors each in each zone (a total of 60) in each city were interviewed (Table 36.2). These were in the vicinity of the supermarket outlets to capture the exact impact on them. Not only were vendor's profiles explored, but their perceptions of the impact of modern retail outlets were also investigated and verified with quantitative assessment of the decline in footfalls and sales.

Table 36.2 City-wise Traditional Fruit and Vegetable Retailers Surveyed

Type of Retailer	Ahmedabad	Bengaluru	Chandigarh
Fixed shop owner	21 (33.3)	17 (28.3)	21 (34.4)
Roadside fixed	19 (30.2)	22 (36.7)	16 (26.2)
Home delivery	10 (15.9)	7 (11.7)	12 (19.7)
Roadside-cum-home delivery	13 (20.6)	14 (23.3)	12 (19.7)
All	63 (100)	60 (100)	61 (100)

Note: Figures in brackets are percentage share in total.
Source: Field survey.

RESULTS AND DISCUSSION

Supply Chains of Supermarkets

Aditya Birla Retail Limited and RF in Gujarat, and ABRL in Karnataka procured vegetables from the contact farmers who delivered at the CCs established near the farmer's field on their own. The produce was then sorted, graded, and sent to the DC. After the final sorting and grading, the produce was delivered to the retail stores. ITC's Choupal Fresh in Punjab and Haryana, and ABRL in Belgaum in Karnataka procured vegetables through a consolidator on a commission basis. The consolidator in Belgaum had both contact and contract farmers who delivered the produce at his CC-cum-DC. However, NF in Karnataka had an informal, oral, and non-registered contract with farmers and produce was picked at the farm gate (Figure 36.1). The retail chains in Karnataka were more inclusive (more so in case of NF) than that in Gujarat, Punjab, and Haryana. In Karnataka, the retail chains procured the entire produce of the contract

Gujarat						
Farmer (Contact)	→	Collection Centre (CC)	→	City Processing Centre/District Centre (DC)	→	Store (RF/ABRL)
Karnataka						
Farmer (contact and contract)	→	CC-cum-DC (Consolidator)	→	ABRL Store		
Farmer (Contact)	→	CC	→	DC	→	ABRL store
Farmer	→	DC	→	Namdhari Fresh stores	→	
Punjab and Haryana						
Farmer (Lessee migrant/ Local land holder	→	Consolidator	→	ITC Choupal Fresh		

Figure 36.1 Different Supply Chain Arrangements of Supermarkets in India

farmers, except the rejected produce. However, these chains procured 15 per cent to 59 per cent of the total produce from 'contact' farmers across different locations; based on the day-to-day indent. The rest of the produce was left for the farmers to sell elsewhere. The major problems faced in linking with retail chains were low volume procured due to lower indent and purchase of only A-grade produce by most of these chains.

Retail Chains and Primary Producers

Gujarat

About 75–82 per cent of the farmers were associated with RF and ABRL for only less than a year. ABRL did not have any marginal or small farmers while RF had only 18 per cent small farmers as compared to 27 per cent marginal and 28 per cent small farmers in Gujarat. Further, the average operated landholding size, similar across both chains (15.9 and 14.7 acres respectively) (Tables 36.3 and 36.4), was much higher than the average operated landholding size in Gujarat (6.4 acres) (Table 36.4). Thus, both the chains primarily dealt with larger landholders. The percentage of leased-in land in operated area was 13 per cent in the case of RF and 4 per cent in the case of ABRL farmers. Leasing-out farmers were altogether absent among ABRL farmers while about 14 per cent of RF farmers had leased out land. Farmers across both chains had tube wells and percentage of irrigated area in operated land varied from 83 to 88 per cent. The drip irrigated area and drip irrigated farmers across both categories increased with increase in size of landholding. However, RF farmers had a higher percentage of drip irrigated area to total irrigated area (33 per cent) and a higher percentage of drip irrigating farmers (55 per cent) compared with that in the case of ABRL farmers (9 per cent and 14 per cent respectively). Both chain farmers were

Table 36.3 Category, Retail Chain and Location-wise Distribution of Farmers

State/Retail chain	Average Operated Land	Leased -in Land as % Age of Operated Area*	Net Cultivated Area**	Average Operated Landholding
		Gujarat		
RF	15.90	12.9 (7.2)	12.9 (81.4)	6.45
ABRL	14.74	4 (-)	12.41 (84.2)	
		Karnataka		
ABRL, Malur	7.46	2.9 (-)	5 (67)	4
ABRL, Belgaum	10.76	2 (4.8)	9.39 (87.3)	
NF	4.56	18.9 (-)	4.26 (93.4)	
		Punjab/Haryana		
ITC	9.91	42.2 (17.1)	9.64 (97.3)	9.36(Punjab and 5.26 (Haryana)

Note: * Figures in brackets are for leased out land. **Figures in brackets indicate percentage of net cultivated area in operated area.
Source: Field survey.

Table 36.4 Landholding Profile of Retail Chain Farmer v/s the State Average

(Acres)

Farmer Category	Gujarat		Karnataka			Punjab/Haryana
	RF	ABRL	ABRL, Malur	ABRL, Belgaum	NF	ITC
Marginal	–	–	1 (4)	2 (10.5)	5 (15.2)	–
Small	5 (17.9)	–	14 (56)	13 (68.4)	20 (60.6)	8 (36.4)
Semi-Medium	9 (32.1)	11 (50)	6 (24)	1 (5.3)	8 (24.2)	7 (31.8)
Medium	8 (28.6)	9 (40.9)	4 (16)	–	–	5 (22.7)
Large	6 (21.4)	2 (9.1)	–	3 (15.8)	–	2 (9.1)
All	28 (100)	22 (100)	25 (100)	19 (100)	33 (100)	22 (100)

Note: Figures in parentheses are per cent share of each category in total.
Source: Field survey.

rich in ownership of farm machinery. Farmers across both chains were relatively rich in household asset ownership.

Reliance Fresh procured 41 per cent of the total produce of an average supplying farmer in cauliflower and cabbage each. However, ABRL procured 35 per cent of cauliflower and 39 per cent of tomatoes. Thus, on average, farmers across both chains had to sell 59–65 per cent of their produce in the Jamalpur mandi. The average rejection rate at the CC was 1.7 per cent in the case of RF and 2.5 per cent in the case of ABRL. Farmers realized higher prices in both the retail channels (Rs 7/kg in cauliflower and Rs 4.6/kg in cabbage in case of RF farmers and Rs 3.6/kg in cauliflower and Rs 4.4/kg in tomatoes in case of ABRL farmers) as compared to that in the mandi (Rs 6.4/kg in cauliflower and Rs 4.4/kg in cabbage in the case of RF farmers and Rs 3.5/kg in cauliflower and Rs 3.8/kg in tomatoes in the case of ABRL farmers). The cost of production of cauliflower and cabbage in RF was Rs 2.32/kg each, while that of cauliflower and tomatoes in ABRL was Rs 2.21/kg and Rs 1.99/kg respectively.

Marketing costs were significantly higher in the mandi channel (Re 0.7/kg each in cauliflower and cabbage in the case of RF farmers and Re 0.78/kg in cauliflower and Rs 1.15/kg in tomatoes in the case of ABRL farmers) compared to that in retail channels (Re 0.15/kg cauliflower and Re 0.20/kg in cabbage in the case of RF and Re 0.28/kg in cauliflower and Re 0.41/kg in tomatoes in the case of ABRL). The resulting net income was higher in retail channels (Rs 4.5/kg in cauliflower and Rs 2/kg in cabbage in the case of RF and Rs 1.11/kg in cauliflower and Rs 2/kg in tomatoes in the case of ABRL) as compared to that in the mandi (Rs 3.4/kg in cauliflower and Rs 1.4/kg in cabbage in the case of RF farmers and Re 0.51/kg in cauliflower and Re 0.74/kg in tomatoes in the case of ABRL farmers).

The chains offered market-price-based procurement prices and procured only a limited proportion of the grower's crop without any firm commitment and, more, on a day-to-day basis. They made no provision for any input and did not have any formal contract arrangement. The rejected produce was left for the farmer to dispose of elsewhere as the chains procured only 'A'/'RR'/ABRL grade produce. Lower indent and purchase of only A/RR/ABRL grade were major problems across both the chains. The chains have brought quality consciousness, introduced exotic vegetables and a package of practices for certain vegetables like cucumber and long melon. Farmers also found the chains better on transaction cost as their CCs were located near the farmer's fields which saved farmer's time and cost on selling their produce. The chains, especially RF, also offered somewhat higher prices than market prices in most of the vegetables procured and the coefficient of variation across days and months in the case of RF prices was lower as against those in mandi prices.

Karnataka

Aditya Birla Retail Limited worked with all categories of farmers, except large farmers in Malur. However, NF worked with marginal, small, or semi-medium farmers only (Table 36.3). Small farmers constituted about 56–68 per cent of total farmers across both the retail chains which was higher than that in Karnataka state (26.6 per cent). However, average size of operated landholdings was higher in case of ABRL farmers (10.76 acres in Belgaum and 7.46 acres in Malur) compared to that in the case of NF farmers (4.6 acres), and much higher than average size of operational holdings in Karnataka (4 acres) (Table 36.4).

Only 2–3 per cent of operated land of ABRL farmers across both locations was leased-in as against 19 per cent in the case of NF farmers (Table 36.4). ABRL farmers in Belgaum leased out 5 per cent of the owned land while leasing-out practice was altogether absent among ABRL farmers in Malur and NF farmers. In Malur, ABRL farmers had 70 per cent of operated area as tubewell irrigated compared to 73 per cent and 88 per cent among contact and contract farmers respectively in Belgaum. Similarly, 75 per cent of operated area of NF farmers was tubewell/canal irrigated. No ABRL farmer in Belgaum had any area under drip compared to 48 per cent farmers in Malur who had about 38 per cent area under drip. A total of 15 per cent of NF farmers also had about 17 per cent area under drip. The percentage of farmers with milch animals was higher in the case of NF farmers (76 per cent) compared to 47 per cent in the case of ABRL farmers in Belgaum and 32 per cent in Malur. NF farmers also had a higher number of milch animals/acre of land (0.46) compared to the ABRL farmers (0.33 in Belgaum and 0.12 in Malur). ABRL and NF farmers had similar percentage of gross cropped area (GCA) under contact/contract crops (73–77 per cent).

ABRL in Malur CC procured about 60 per cent and 42 per cent of the total cauliflower and tomatoes respectively as compared to 25 per cent of cauliflower and tomatoes each in the case of contact farmers and 90 per cent of cauliflower and 87.5 per cent of tomatoes in the case of contract farmers in Belgaum. On the other hand, NF procured all the produce of the contract farmers. The rejection rate at Malur CC was only 5 per cent in cauliflower and 6 per cent in tomatoes compared to higher rejection rate of 15 per cent in the case of contact and 10 per cent in the case of contract farmers in the case of cauliflower and 18 per cent in the case of contact and 12.5 per cent in the case of contract farmers in the case of tomatoes at the CC-cum-DC of the consolidator in Belgaum. However, rejection rates in NF at the farm level was only 1–2 per cent.

In ABRL, average price realization was lower in retail channels in the case of contact farmers across both locations (Rs 5.20/unit in cauliflower and Rs 3.1/kg in tomatoes in Malur and Rs 3.8 unit and Rs 3.2/kg in Belgaum) compared with that in non-retail channel (Rs 5.6/unit in cauliflower and Rs 3.55/kg in tomatoes in Malur and Rs 3.9/flower and Rs 3.5/kg in Belgaum) and other retail chains channels (Table 36.5). However, contract farmers in Belgaum realized higher prices in the retail channel (Rs 4.8/unit in cauliflower and Rs 3.75/kg in tomatoes) than that in non-retail channels (Rs 4.4/unit in cauliflower and Rs 3.55/kg in tomatoes). Although NF provided grade-wise prices for bhindi and baby corn, the calculated average price for all grades (Rs 9.69/kg for bhindi and Rs 6.5/kg for baby corn) was lower in NF than that in the mandi (Rs 10/kg for bhindi and Rs 7.8/kg for baby corn).

In ABRL, cost of production was relatively higher among contract farmers in Belgaum (Rs 2.8/flower for cauliflower and Rs 2.3/kg for tomatoes). However, marketing costs were the highest in case of Belgaum contact farmers (Re 0.88/flower in ABRL and Rs 1.52/flower in the mandi for cauliflower; and Re 0.4/kg ABRL and Rs 1.1/kg in the mandi for tomatoes). This was followed by that of the Belgaum contract farmers (Re 0.73/unit in ABRL and Rs 1.41/unit in the mandi for cauliflower; and Re 0.24/kg in ABRL and Re 0.9/kg in the mandi for tomatoes) and Malur contact farmers (Re 0.26/unit in ABRL and Rs 1.33/unit in the mandi for cauliflower; and Re 0.23/kg in ABRL and Re 0.9/kg in the mandi for tomatoes). Thus, ABRL farmers across both locations had lower average cost of marketing in the retail channel than that the in non-retail channel. Hence, all ABRL farmers across both locations had a higher net income in the retail channel compared to the non-retail (mandi) channel. Although, contact farmers had significantly higher yields than that of the contract farmers in Belgaum, lower price realization for contact farmers in both the mandi and retail channels, resulted in the lower net income among contact farmers than that among non-contract farmers. Farmers chose to sell to the retail channel due to cost savings like less time in selling, lower transportation cost, no loading or unloading charge, no sales commission, lower spoilage, and fair and quick weighment and payment, as compared with when they sold the same produce in the mandi. The cost of production was higher among NF contract farmers (Rs 6.67/kg in bhindi and Rs 3.8/kg in baby corn) than that among non-NF farmers (Rs 5.74/kg in bhindi and Rs 3.66/kg in baby corn). NF farmers did not incur marketing costs since the produce was picked from the farm itself while non-NF farmers had to incur a marketing cost of Rs 2/kg in the case of bhindi and Rs 1.78/kg in the case of baby corn. Thus, average cost of production and marketing were higher among non-NF farmers (Rs 7.74/kg in bhindi and Rs 5.4/kg in baby corn) than that among NF farmers (Rs 6.67/kg in bhindi and Rs 3.8/kg in baby corn). The resulting net income was also higher in the case of NF farmers (Rs 3/kg in bhindi and Rs 2.73/kg in baby corn) than that in case of non-NF farmers (Rs 2.26/kg in bhindi and Rs 2.4/kg in baby corn). The major benefits of selling to NF were lower transaction costs, timely supply of good quality inputs at lower than market price, and family labour saving.

However, sometimes, ABRL farmers defaulted due to lower price offered by ABRL compared to mandi prices and higher production due to which farmers preferred to sell the entire produce to the mandi to avoid marketing costs in two different channels. Lower quality produce and lower indent of the company were major problems in supplying to ABRL. In case of NF, 62 per cent farmers reported lower price and non-revision of the price when price in open market increased as their major problem. The other major problems in retail chain linkage were lack of timely supply of agri-inputs and their poor quality, inadequate insurance cover if the crop failed, and delay in procuring the produce.

Table 36.5 Channel-wise Percentage of Cauliflower Sold, Average Price Realized and Rejection Rate in Malur

Channel>	ABRL	Reliance Fresh	HOPCOMS	Heritage@ Fresh	Mandi*
%age of cauliflower sold	46.7 (29.0)#	5.7 (17.0)	7.3 (22.0)	3.0 (9.0)	37.3 (23.0)
Average price (Rs/flower)	5.03	5.10	5.60	5.60	5.60
Rejection rate (%)	5.0	4.5	8	6	2

Note: * three different markets (Kolar, rural Karnataka, and Chennai) were used to sell cauliflower.
Figures in brackets are for farmers who sold through more than one retail channel.
HOPCOMS refers to the Horticultural Producers' Co-operative Marketing and Processing Society.
Source: Field survey.

Punjab and Haryana

Of the total farmers interviewed, 54.5 per cent were local and the rest lessee migrant farmers. The average operated area of retail chain farmers (9.91 acres altogether) and of the lessee and the local farmers separately (8 and 11.5 acres respectively) was higher than the average size of the operational holding at the state level—Punjab 9.36 acres and Haryana 5.26 acres (Table 36.4). Small farmers accounted for only 36 per cent of the total growers of the retail chain (Table 36.3). The proportion of small operators was higher among the lessee category (50 per cent) and only 25 per cent among local landholders compared with the proportion of small and marginal holders in the two states—Punjab 35.4 per cent and Haryana 66.7 per cent. However, lessee migrant farmers, on an average, put about 60 per cent GCA under vegetables compared to only about 23 per cent in the case of local farmers although cropping intensity across both categories was similar (211 and 221).

ITC procured about 23 per cent of cauliflower and bottle gourd each from lessee migrant farmers compared to only 15.5 per cent of cauliflower and bottle gourd each in the case of local farmers. The rejection rate was only 2 per cent. However, the rejection rate of lessee migrant farmers' produce was lower (1.7 per cent) as compared to that of local farmers (2.25 per cent). The average yield was higher in case of lessee migrant farmers (85 quintals in cauliflower and 104 quintals in bottle gourd) than in the case of local farmers (81.11 quintals in cauliflower and 97.8 quintals in bottle gourd). The farmers realized somewhat higher prices in the ITC channel (about Rs 5.5/kg in cauliflower and Rs 4.2/kg in bottle gourd) compared to that in the mandi channel (Rs 5.1–5.4/kg in cauliflower and Rs 3.9–4/kg in bottle gourd). Lessee migrant farmers realized lower prices in the mandi as compared to that realized by local farmers. The average cost of production was higher among local farmers (Rs 3.89/kg) compared to that in the case of lessee migrant farmers (Rs 3.35/kg). The ITC farmers did not incur any marketing cost (except the packing cost in polythene for bottle gourd) since the produce was picked from farm itself. The net income for each crop in each channel was higher for lessee migrant farmers than that for local farmers.

Major reasons for selling to ITC were no transportation costs, time saving, higher price, and supply of crates, free of cost, to pack vegetables. However, about 77 per cent farmers were not satisfied in linking with ITC. The major problems faced were low volumes procured and low price overtime. The lessee farmers being professional vegetable growers had better yields as well as better quality produce. The chain was not able to make an impact on the growers as it was procuring too little because it was not able to sell the procured produce in the market where it faced competition from other retail chains and local vendors and farmer's markets. More recently, the chain has wound up its retailing and procurement operations in the region.

Food Supermarkets and Traditional Retailers

More of traditional sector retailers sold vegetables in Bengaluru (60 per cent) compared with that in Ahmedabad and Chandigarh (46–47 per cent) whereas the proportion of fruit sellers was higher (41 per cent) in Ahmedabad and that of both F&V sellers higher in Chandigarh (36 per cent). The average distance of the retailers from organized retailers was higher in the case of Chandigarh (1.1 km) than that in Bengaluru (0.7 km) and Ahmedabad (0.5 km) which perhaps points to the lower density of modern retail outlets in Chandigarh. The average number of years of presence of the organized retail outlets was two in Bengaluru and only 1.6 years each in Ahmedabad and Chandigarh. Across both Ahmedabad and Bengaluru, RF and More were the nearest organized retail outlets to the local retailers as reported by 48 per cent and 35 per cent retailers respectively in Ahmedabad and 45 per cent and 18 per cent retailers in Bengaluru. In Chandigarh, More, Big Bazaar, RF, and Spencer's were the nearest organized sector outlets in that order.

In Ahmedabad and Bengaluru, about 45–48 per cent each of the retailers sold F&Vs as street/roadside hawkers as compared to only 16 per cent in Chandigarh where 41 per cent retailers sold F&Vs in local neighbourhood or colony markets as compared to only about 21–23 per cent each in Ahmedabad and Bengaluru. About 16–17 per cent retailers each in Ahmedabad and Bengaluru also operated as standalone shops and near small malls respectively. The number of retailers selling in the market popular for special products was higher in Chandigarh (21 per cent) followed by Ahmedabad (11 per cent) and Bengaluru (3 per cent).

The traditional retailers in Bengaluru had the highest number of footfalls both during week days (138) and weekends (155) followed by Ahmedabad (113 during week days and 103 during weekends) and Chandigarh (94 during weekdays and 101 during weekends) before the entry of organized retail chain outlets. However, with the emergence of these new players, the number of footfalls declined across all locations. The percentage decline in footfalls was the highest in Bengaluru (35.5 per cent during week days and 27 per cent during weekends) followed by Ahmedabad (32 per cent during week days and 26.6 per cent during weekends) and Chandigarh (17 per cent during weekdays and 14.9 per cent during weekends).

Further, the number of regular customers visiting the outlets came down everywhere after the entry of modern retail chains; more so in Ahmedabad and Bengaluru and the decline was as much as 20 per cent. In Ahmedabad and Bengaluru, 60 per cent and 45 per cent reported a decline in sales compared with only 33 per cent in Chandigarh. Bengaluru traditional retail sellers reported the largest decline in their turnover (22.5 per cent) and income (31 per cent) followed by Ahmedabad (12.3 per cent and 27.8 per cent respectively) and Chandigarh (9.7 per cent and 19.6 per cent respectively) (Figure 36.2).

There was a marginal increase in employment in case of fixed shop owners and roadside fixed hawkers only after the emergence of organized retail outlets in the neighbourhood. However, the percentage decline in the footfalls was the highest in the case of roadside-cum-home delivery hawkers (49 per cent during week days and 45.5 per cent during weekends) followed by roadside fixed hawkers (47 per cent during week days and 34.3 per cent during weekends), fixed shop owners (18.7 per cent during week days and 19.7 per cent during weekends) and the least in the case of home delivery hawkers (7 per cent each during week days and weekends).

All roadside-cum-home delivery hawkers, 79 per cent of the roadside fixed hawkers, 38 per cent of the fixed shop owners and 20 per cent of home delivery hawkers reported a decline in sales due to the organized retail outlets in their vicinity. The average for all being 60 per cent.The decline in average turnover due to the emergence of organized retail outlets was the highest in case of roadside fixed and roadside-cum-home delivery hawkers (21–23 per cent) followed by fixed shop owners (7 per cent) and the least in case of home delivery hawkers (3.9 per cent); the average for all traditional retailers being 12 per cent. The decline in net income also witnessed a similar trend. The other reasons reported for decline in sales were seasonality, reduced household incomes, high prices, and recession. Therefore,

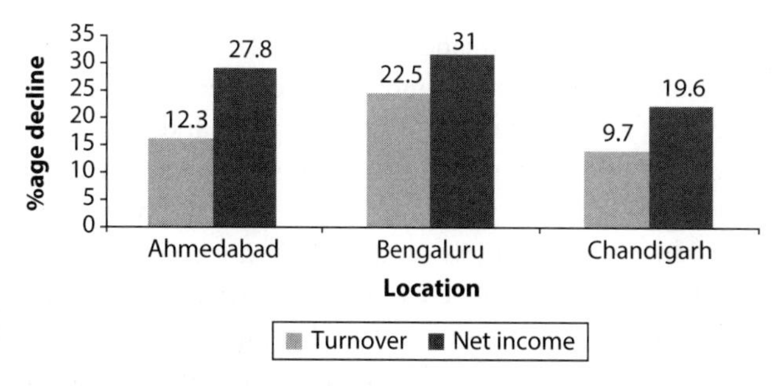

Figure 36.2 City-wise Per Cent Decline in Traditional Outlet Turnover and Net Income after the Entry of Modern Retail Chains

it is important not to attribute the entire sales decline to the entry of modern retail chains.

POLICY SUGGESTIONS

Producer Interface and Policy

Contact farming, as practiced by F&V retail chains in India, is an informal arrangement which does not bind the buyer and the seller. This is not in the interest of the growers, as their market risk is not shared by the agencies. The system of no written contracts and consignments places the financial risks solely with the producers/suppliers and retail chains do not run any financial risk. The retail chains can eliminate all financial risk from their end of the chain due to this direct procurement from growers, as they do not need to maintain stocks, do not bear price risk, and have no commitment to buy.

The functioning of Agricultural Produce Market Committee (APMC) markets needs to be improved to enhance their cost efficiency so that producers could realize better prices. The amended APMC Act allows for the setting up of private markets. It is also necessary to have an open auction system, improve buyer competition in markets, provide better facilities such as cold storage, and improve farmers' access to market information. These markets are important to small farmers and even a significant proportion of medium and large farmers, who still depend on them; they also serve as the main competitors to contract and 'contact' farming (practiced by retail chains) and can improve the terms offered by retail chains to growers.

There are a large number of institutional arrangements to coordinate the small producers which should be assessed for their relevance and effectiveness in a given context. Though, a priori, it seems the cooperative and other similar forms of farmer organization are more relevant and sustainable, especially the new generation cooperatives (NGCs) which are voluntary, more market oriented, member responsive, self-governed, and avoid free riding and horizon problems as they have contractual equity based transaction with grower members.

The government, both the union and the state, should recognize the new institution of producer companies as producer cooperatives and extend all the support as extended to traditional cooperatives. The state governments should instruct their cooperative banks to extend investment and working capital credit to these producer companies. They should also instruct their Agricultural Marketing Boards to extend license to these companies for trading and processing without any conditions. The state governments should ask their Agro Industries Corporations and

Departments of Industries to extend investment capital and working capital grant to these companies for processing and marketing infrastructure creation. The producer companies practicing organic farming can be designated as certifying agencies for third parties and individual growers by the union government agencies like the APEDA. The promotional and NGO bodies promoting and supporting these companies should be given project-based grants by the state/union government.

Role of State and Development Agencies

Governments, NGOs, and donors can facilitate small farmers' access to three key elements in order to have the capacity to supply the supermarket channel: (i) market information identifying the buyer and its requirements, and establishing a market relationship such as having an implicit or explicit contract from the supermarket or the specialized wholesaler, that is, being on the list; (ii) a viable organization/association to reduce coordination costs and enforce delivery from members; and (iii) the requisite physical investments (say in equipment) and managerial improvements to meet the specific product and transaction standards required by the supermarket chain. Moreover, these elements can be mutually reinforcing, for example where having a contract (being on the preferred suppliers' list) acts as a substitute for collateral, inducing a bank to make a loan to a small farmers' group for the purchase of equipment.

Second, governments, NGOs, and donors can facilitate small farmers participating in the supermarket channel by helping the existing wholesale sector adjust to the needs of supermarkets. The European Bank for Reconstruction and Development (EBRD) has a programme in Hungary where it has helped wholesale markets to upgrade their facilities to meet the needs of the supermarket chains. This has the added benefit of keeping alive alternatives for small farmers by increasing the competitiveness of these channels.

Third, governments, NGOs, and donors can facilitate tri- or quadri-partite relationships that facilitate smaller farmer participation. An example can be found in Indonesia, where a combination of a small farmer organization (Makar Buah), a supermarket chain (Carrefour), a seed/chemical company (Syngenta), a government extension programme, and a specialized/dedicated wholesaler (Bimandiri), have formed a fruitful combination to market melons. Carrefour supplies the guaranteed market, Syngenta the financing, and the wholesaler the intermediation and coordination. Therefore, it is crucial that government and donor agencies help small farmers and entrepreneurs to make the investments in equipment, management, technology, commercial practice, and the development of strong and efficient organizations to meet those requirements. There have been such attempts in Brazil and Guatemala as well.

Traditional Retailer Interest Protection and Involvement

As far as the impact on traditional retailers (neighbourhood stores) is concerned, there is a need for a zoning regulation on the pattern of Indonesia. Our findings also show that traditional retailers have suffered a 20–30 per cent decline in sales due after the entry of retail chains in the study areas. Though it is not entirely due to the chain impact as the year 2008–09 also saw a recession but, definitely, there is an impact of the chains on traditional retailers as they do attract their buyers due to the ambience of stores and sometimes cheaper vegetables and fruits as these chains buy in bulk or buy directly from growers and avoid many market charges which small traditional retailers have to pay. Therefore, at least the residential localities of cities could be kept free of supermarket chain outlets.

There is also a need to treat traditional F&V retailers and street vendors in general as part of the city life and protect their interest with adequate policy and legislation. Though national policy on street vendors was framed in 2004, there is not much implementation of the same at the state and local level despite courts' ruling on creating space and facilities for such sellers. Urban planning needs to integrate the interests and concerns of such retailers into city plans. There is also need to organize such retailers on the lines of the Self-Employed Women's Association (SEWA) in Ahmedabad to give them a voice and protect their livelihoods.

Finally, there is a need to combine value chain promotion with a livelihood perspective to enable the resource-poor to enter into and stay into value chains—domestic or global. Innovations in smallholder market linkage are needed in terms of partnerships, use of ICTs, leveraging networks, value chain financing, smallholder policy, and even in contracts which can promote both efficiency and inclusiveness of the linkage. Choosing the right market and market development strategy is a must to scale up and avoid a 'race to the bottom' which can come only by innovation of products and business models. It is not market access or participation but effective market participation which is at the heart of success of any market linkage for primary producers. There is a need to establish mutli-stakeholder initiatives in the chains to protect small producers and traditional retail interest. Support by state/development agencies for small producers and small vendors to enable them to compete with quality and cost efficiency could be quite helpful.

REFERENCES

Boselie, D., S. Henson, and D. Weatherspoon. 2003. 'Supermarket Procurement Practices in Developing Countries: Redefining the Roles of the Public and Private Sectors', *American Journal of Agricultural Economics*, 85(5): 1155–61.

Cadilhon, J-J., P. Moustier, N.D. Poole, P.T.G. Tam, and A.P. Fearne. 2006. 'Traditional vs. Modern Food Systems? Insights from Vegetable Supply Chains to Ho Chi Minh City (Vietnam)'. *Development Policy Review* 24(1): 31–49.

Codron, J.M., Z. Bouhsina, F. Fort, E. Coudel, and A. Puech. 2004. 'Supermarkets in Low-income Mediterranean Countries: Impacts on Horticulture Systems', *Development Policy Review*, 22(5): 587–602.

Harper, M. 2009. 'Development, Value Chains and Exclusion', in M. Harper (ed.), *Inclusive Value Chains in India—Linking the Smallest Producers to Modern Markets*, pp. 1–10. Singapore: World Scientific.

Joseph, M., and N. Soundrarajan. 2009. *Retail in India—A Critical Assessment*. New Delhi: Academic Foundation.

Kalhan, A. 2007. 'Impact of Malls on Small Shops and Hawkers', *Economic and Political Weekly*, 42(22): 2063–6.

Kumar, V., Y. Patwari, and H.N. Ayush. 2008. 'Organized Food Retailing: A Blessing or a Curse?', *Economic and Political Weekly*, 43(20): 67–75.

Pritchard, B., C.P. Gracy, and M. Godwin. 2010. 'The Impacts of Supermarket Procurement on Farming Communities in India: Evidence from Rural Karnataka', *Development Policy Review*, 28(4): 435–56.

Ruben, R., D. Boselie, and H. Lu. 2007. 'Vegetables Procurement by Asian Supermarkets: A Transaction Cost Approach', *Supply Chain Management: An International Journal*, 12(1): 60–8.

Stichele, M.V., S. van der Wal, and J. Oldenziel. 2006. *Who Reaps the Fruit? Critical Issues in the Fresh Fruit and Vegetable Chain*. Amsterdam: Centre for Research on Multinational Corporations (SOMO).

Sulaiman, V.R., N.J. Kalaivani, and J. Handoo. 2010. 'Organised Retailing of Fresh Fruits and Vegetables: Is It Really Helping Producers?', Working Paper: 2010-01. Hyderabad: Centre for Research on Innovation and Science Policy (CRISP).

37

Emergence of New Channels for India's Agricultural Marketing

Nilabja Ghosh and Ananda Vadivelu

Market reforms in agriculture have become an imperative for letting in resources, modern technology, advanced managerial practices, and motivated players for the overall development of the market and the sector in which small farmers are in majority (Ahmed 1996; Sengupta 2013). The new and emerging supply chains are expected to deliver higher prices and returns to the producers even while offering cheaper commodities to consumers and other users. Yet much of these reforms raise apprehensions for the political economy of the intensity that few other policy changes do (Bardhan, Mookherjee, and Tsumagari 2013; Ghosh 2012). In economics, marketing raises highly contentious issues among efficiency, equity and sustainability.

OBJECTIVE

This chapter makes a contemporary assessment of the status and implications of marketing reforms in India when reforms are only at a nascent stage. In particular, the study, based on realities at the state level, inquires if agricultural markets are undergoing reforms and whether the reforms are paving the way to alternate marketing channels and in what ways these new channels differ from the traditional channels they replace in character and performance. The broad method followed is the integration and meta-analysis of evidence brought forward from field surveys conducted by 10 Agro-economic Research Centres (AERCs). Marketing efficiency is assessed both quantitatively by using a modified Shepherd method (Acharya and Agarwal, 2004) and qualitatively, based on perceptions.

MARKETS AND MARKETING CHANNELS: A BACKGROUND

Markets, celebrated in folklore, ballads and history of societies (Yang 1944) are an integral institution of human culture also manifesting specialization and division of labour required for efficiency. The gradual emergence of specialized trading classes evolving from peripatetic to settled modes and from primitive simplistic to capital intensive modern forms replaced the old barter system and diminished the physical distances between the producer and the consumer. Experiences have varied across countries but, by and large, the channels tended to lengthen with the entry of more and more intermediaries. The intermediaries then begin to differ in their skills and investment capacity leading again to a reversion in the length as multiple individual intermediaries finally tend to give way to organized marketing firms endowed with scale, skill, technology, and resources resulting in contests and reduced transaction cost

(Baumol 1982; Jaffe and Ling 2007; Key, Sadoulet, and de Janvry 2000; Young and Hobbs 2002).

Under the Indian Constitution, the state governments have the final say on how marketing of agro-products would operate while the central government can only suggest and advise the states. Existing laws guided by the state's Agricultural Production Marketing Regulation Acts (through APMCs) provide for regulation of agricultural markets where transactions ideally can take place in a fair and transparent manner. The markets also create self-employment opportunities for a fleet of trades in market chains. Open auctions, supervised by democratically elected market bodies would ensure fair practices and competitive prices under idealistic conditions.

The reality is far more complex. Traders who operate in the APMC markets are severely screened and limited through licensing. With hardly any scope for new actors entering the chains, producers enjoy little option in choosing their buyers thus defeating the purpose of the regulation. Moreover, owing to unresolved conflicts in most cases supervision remained poor, bureaucratic, and even corrupt rather than representative. Prices are depressed against the interests of producers owing to the superior power of the traders. Association among traders and alleged collusion between the trader and the state officials make matters worse. Paucity of public funds leave little room for modernizing market facilities. However, the states varied widely in the regulated marketing system and the above description may not apply to all regulated markets.

Globalization in the wake of India's formally joining the World Trade Organization (WTO) made the existence of a vibrant and dynamic marketing system compelling. All state governments were asked to amend the state APMC Acts in tune with a model APMC Act finalized and circulated by the central government in 2003. The Model Act was meant to reform the market by allowing more competition and encouraging innovative new marketing methods to evolve. Yet, marketing of agricultural product in India, labouring from memories of a painful past full of exploitation, rural interlocked markets, and food shortages has become a politically sensitive issue. The proposed new Act meant to override the pre-existing legislation and set the reforms in motion has not been accepted by many state governments. Contract farming, in particular, has been viewed with suspicion drawing from past memories and retail chains calling for external funding and managerial advances have become serious political issues in India.

States vary widely in their nature and pace of reforms in agricultural marketing. In fact, even states that enacted the reforms did not necessarily go by the spirit of the proposal and reforms were at best partial. Contract farming is particularly perceived as 'anti-poor and anti-farmer' in West Bengal but is aggressively promoted in states Punjab, Haryana, and Andhra Pradesh. Private retail chains are becoming popular with customers and farmers in many states but resistance has been stiff in Jharkhand and Uttar Pradesh. Allowing foreign direct investment (FDI) in multi-brand retail is a critical contention in the political economy of the nation. While West Bengal so far only has shown some interest in amending the Act, Bihar repealed the Act in 2006 but failed to enact a new law till date. Only partial amendment of the act is attributed to the two most agriculturally progressive states Punjab and Haryana as well as Madhya Pradesh. Uttar Pradesh amended the act but, facing serious pressure, withdrew the amendment soon after, leaving the old act still in place.

A refusal to amend the act should not be equated with lack of reforms on the ground. In the field studies reported in this report we will find that actual change has been remarkable even in a state that has not legislated reforms while new channels are difficult to come by in the deemed progressive states. Certain states are steadfast in averting changes that are in principle administratively possible even under the existing laws. In fact, the older acts were flexible in principle enabling the states to allow changes through suitable interpretations and 'notifications'. Political will to reform is most important. Regardless of the amendment of the APMC Act early signs of reforms are already apparent in most the states. Development of modern channels such as contact marketing is observable even in states politically averse to these ideas. Entry of new players with modern methods could not be prevented in Uttar Pradesh which did not amend the act. The central government moved towards a national market using persuasion as a means. Fruits and vegetables followed by cereals, pulses, and oilseeds could be gradually dropped from the APMC schedule of regulated commodities but with all the hesitancy shown by the states, Union budget 2015 considered a Constitutional amendment or use of special provisions in the Constitution for setting up a National common market for specified agricultural commodities (Economic Survey 2014–15).

During this period the regular channels, far from being eliminated or phased out were also undergoing transitions in tune with the pressures of competition. Demonstration effects from other states, encouragement from the Centre and overtures of the private companies under the existing regulation are driving the changes on the ground but developments in information and technology were possibly the strongest influence. A flexible Market Intervention Scheme (MIS), E-trading (so far used in trading financial instruments) in the spot and futures market, establishment of derivative exchanges and the ITC e-Choupal deserve mention. Since The Ministry of Agriculture formulated a central sector scheme—Agricultural Research and Marketing Information Network (AGMARKNET) for electronically linking regulated markets spread all over the country. The infrastructure of the APMC markets began to be upgraded.

Integrating the producer with the consumer closely, a modern supply chain can deliver several advantages. By reducing the trader's margin, the farmer's market opportunity and profit can be improved. Besides, managerial efficiency, information dissemination, technologically enabled logistic organization and the market infrastructure at the producer and retail ends can together help in reducing wastage, saving fuel, curtailing greenhouse gas emission, eliminating unfair practices, maintaining quality standards for consumer safety, improve productivity through private extension, and facilitating convenient marketing and shopping.

Yet the reduction of channel length can intensely affect the livelihood of traders whose services remain an unresolved issue. Farmers scattered across underdeveloped rural India are weakly connected with urban and modern society physically. In transacting with poorly informed farmers, traders enjoy the advantage of asymmetric information. Highly maligned for their exploitative practices, the difficult and risky environment in which they operate is given as a rational justification for the low prices they pay producers (Mulky 2008). With lack of competition from the organized sector, their methods remained out of tune with the progress of marketing in the outer world. Trading is perceived to be an easy option in employment.

PRIMARY DATA AND FIELD INFORMATION: METHOD

The 10 Agro-economic Research Centres (AERC) conducted sample survey of participants in desirably two emerging channels in the 11 regions assigned to them. Defining an emerging channel was not easy, but our specification implied a channel that differed from the common traditional channels familiarly seen in the region. Typically, the emerging channel was also shorter than the corresponding traditional channel.

Although the emerging channels studied did not always involve commercial and organized companies in the chain and in some cases even coalesced with the traditional channel at a critical point, in all cases they involved a shorter channel and bypassed the first link which is usually the commission agent or the pre-harvest contractor. For sampling, the participants in the chain included farm households as well as intermediaries at different links in the chain up to the final user in the selected representative emerging channel. For comparison a control sample relating to a most common traditional channel operating in the same area and for the same crop as observed on field exploration is also collected.

This report presents the cases of the following emerging channels (i) Direct marketing (DM) in Andhra Pradesh, Punjab, and Assam; (ii) Contracts (CONTR) in Punjab, Assam, and Uttar Pradesh; (iii) Corporate market intermediation (CMI) in Maharashtra, Himachal Pradesh, and Madhya Pradesh; (iv) Organized retail (RTL) in Jharkhand, Haryana, and Himachal Pradesh; and (v) Trader (TRADER) based channels in slow-moving states West Bengal and Bihar. Whereas in Andhra Pradesh (brinjal, banana) farmers sell to the final consumer at state created venues called the *Rythu Bazaar*, in Assam (orange) the transaction is done between processors and an organized group of farmers that acts as the intermediary association for the constituent farmers. In contrast, in Punjab (kinnow) the farmer sells directly to the next link in the chain who is a wholesale trader in the Farmer Evening Market rather than to the final user as in the other two states and bypasses the traditional pre-harvest contractor (see Table 37A.1).

Contract farming is studied in states Uttar Pradesh (potato, amla), Assam (potato), and Punjab (potato). In two of the cases, the buyer is a multinational processing company and in the other two, a local processor is the buyer. Potato is found to be the studied product in three of the cases. A non-government agency intermediates to protect farmer interest in another case studied in Assam.

The channels are not always distinct. In the emerging channels studied in Maharashtra (pomegranate, onion), Himachal Pradesh (apple), and Madhya Pradesh (soybean), a marketing company that intermediates as a direct buyer from the producer in turn sells to relatively more resourceful trading agents such as malls, processors, and exporters rather than in a typical urban market interface to the final consumer. Thus, the traditional channel is not avoided. In Madhya Pradesh, farmers sell through a much-famed computerized electronic portal hosted by a private company. In the two states that have till now not legislated reforms, West Bengal and Bihar, locating a modern channel was next to impossible but nevertheless models perceptibly variant to traditional ones were found and studied as emerging channels. These channels involved no organized private companies. Rather, local traders who were deemed more reliable by the producers than the market committees stepped into these channels.

Sales by farmers to organized retail chains is also studied in the cases of Haryana, Himachal Pradesh, and Jharkhand but while in Himachal Pradesh a non-profit state-promoted organization intermediates between the producer and the consumer as the retailer, a large commercial retail company is the subject of study in the other two states.

Thus, a profit-oriented company does not participate in a chain in all cases, and in some of these cases the traditional traders are not necessarily excluded. Even where the producer directly sells in the market, the interface may be with the final user (Andhra Pradesh, Assam) or a trader further down in the chain (Punjab). In the direct marketing cases, the farmers cannot avoid marketing costs and in fact except in Punjab the entire marketing cost weighs on them. In emerging channels with a private company participating, the burden of marketing is generally taken over by the company and the producer shares no part of the burden in many cases (Himachal Pradesh [apple], Maharashtra) though in others they are not entirely relieved (Madhya Pradesh, Haryana).

EMPIRICAL FINDINGS

Among the factors influencing the choice of a new channel and making a major shift in traditional practice, higher prices were found to attract many farmers. Security of assured sales also draws participation but the same attraction is reported to retain participation in the traditional channels, so the primacy in this appeal is not established. In contract farming, low marketing cost and superior services are powerful forces that draw farmers while social influence along with habitual acceptance hardly played a role in its rejection. Shorter distances to the disposal site and input support from buyers were mentioned as attractions while hidden costs like demands for bribes and long waiting periods weakened farmers' attachment to the traditional channel.

Our data indicates that efficiency is undoubtedly gained by shifting to the emerging channels. Price magnification from producer to user is higher in the traditional channel and especially large gains are made in Direct Marketing in the Rythu Bazaar of Andhra Pradesh where there are no intermediary margins (Table 37A.2). In the only two aberrant states, West Bengal and Bihar, where the emerging channels studied are also trader operated, the traditional traders remain more efficient in the marketing function. Savings on gross marketing costs for every rupee fetched by the producer are highest in DM followed by RTL. Farmer's price in the emerging channels in most cases is higher. Even after deducting marketing costs and accounting for wastage and rejection, the net price is relatively higher in the emerging channel though with a few exceptions. Profit and the returns from land are also higher in comparison to the corresponding traditional channel functioning in the area and so is productivity in agriculture. The terminal (consumer) price is not necessarily lower in the emerging channel but the quality of the product purchased and the ambience of sale also influence consumer's subjective valuation.

On the qualitative side of efficiency gains, DM denies farmers the gains from specialization. Producers undertake marketing, taking time out of productive activities so that productivity can suffer (seen in the case of Brinjal in Andhra Pradesh). Farmers evince satisfaction with services obtained from new channels in most cases although a hint of suspicion of powerful payers is evident. There is an open interest that the state should support marketing and that building cooperatives could be a better alternative. In all cases of retail marketing, farmers expressed satisfaction at the reduced burden of marketing. In Himachal Pradesh, even the producers excluded from the chain exerted pressure to extricate themselves from their ties with traders to join RTL, thus revealing their preference for retail. Similar appreciation for RTL is observed among consumers who find the electronic displays, air-conditioning of shops and scientific weighing facilities attractive. In the case of DM, the consumers report enjoying farm fresh vegetables in the farmers' markets but alongside there are complaints of 'rude behaviour' of sellers who are untrained in marketing functions.

Relief is obvious among farmers in the corporate intermediated and contract based channels as some of the buyers offer pick-up facilities at farm-gates, some provide storage materials and in most cases the collection centre is within a short distance. In Madhya Pradesh private company ITC provides an electronic platform connecting the two parties. This was a path breaking innovation in bringing modern information technology to farmers. Contracts in CONTR provide a clear protection from price uncertainty. Although contracts with individual traders or pre-harvest contractors made before harvest has already been a common traditional method of marketing of fruits and vegetables, these traditional contracts generated dissatisfaction among farmers over the trader's bargaining strength over producers. The supply of high quality inputs and know-how are specially appreciated in modern contracts but in many cases the unorganized traders too supply input advances and credit for inter-cultural activities necessary for farming.

In the two slower states, while the producers are distressed by their encounters with the more powerful licensed traders, a pool of unemployed youth force potentially step in to replace the vested trading power. Though farmers seem to prefer this alternative, financial poverty and lack of experience of the new agencies relative to the traditional traders is a constraint that depresses efficiency.

In the matter of inclusiveness, the result is mixed with channels showing no bias in any direction if the size of the holding is considered as an indicator of disadvantage. Direct marketing and trader-based channels, in which no organized marketing bodies participate, are significantly more inclusive of small farmers relative to the traditional channels in the regions. Contract farming especially in Uttar Pradesh where a large multinational is the contractor is also more inclusive of small farmers, perhaps as a policy decision, than the corresponding traditional channel in the area. Corporate intermediation has a lower share of small farmers. Participation is less inclusive if social affiliation to majority religion and its forwards castes, ownership of mobile phone, motorcycle, and pump set are considered indicators of privilege.

Discrimination based on quality standards was reported. Preference for higher altitude orchards for apple in Himachal Pradesh is an instance where discrimination is related to geography rather than economic aspects. On-farm storage facilities for onion in Maharashtra is marked as an advantage for inclusion as buyers want to procure at their own convenient time. Ownership of a mobile phone appears to be vitally important for inclusion in the modern chain. A participant in the emerging channel is more often moderately more educated measured by the level of schooling of the heads and the proportion of higher educated members in the family.

On the whole, the corporate intermediation model in Maharashtra and all the contract farming cases involve relatively larger farms. The average farm size seen in Himachal Pradesh is relatively small but there are reports of popular pressure in the state reflected in the higher level of inclusion. At the country level, average farm sizes of participants tend to be higher when profit-oriented private sector companies are the buyers than otherwise.

No significant difference is noted with regard to farm practices between the channels. Use of chemical fertilizer is comparable in all the channels but use of organic manure is considerably less in DM-, CONTR-, and TRADER-based channels relative to the traditional channels. As yet, there is no perceptible shift towards water-saving methods of using sprinkler and drip irrigation which is more of a regional aspect of farming. No farm in either channel was found to be certified as organic. Family labour is used more intensely in most emerging channels.

Producers are mostly found to depend on a single channel although there are instances of channel diversification. The marketing scales in terms of value marketed per farm are higher in the emerging channels. Paradoxically, even in the traditional channels the farmers do not always go through the regulated market. In fact, in Assam, where regulated marketing is yet to take off owing to distances and topography, the APMC market is used more actively by the participants of the emerging channels while the traditional sellers dispose goods in the local and roadside markets. In Bihar too, the practice is similar but larger farmers often carry products to urban regulated markets that are more developed. Producers do not generally avail of other facilities like inputs and credit from buyers in emerging channels although borrowing from traders is reported albeit in rare cases in the traditional channel. In CONTR, farmers however do get inputs, input advances, or technical advice and specifications as well as extension and have expressed high levels of satisfaction with this service.

Direct selling draws small and poorer farmers who get relief from domination from commission agents and other traders but large farmers are rather disinterested because of time constraint. Farmers' market, to which government provides the space and infrastructure, provides an additional avenue of disposing unsold products. The Farmer's Evening Market in Punjab allows a DM platform but without bypassing all traders, though the first link, namely, the Commission agent is avoided. Collective sales by farmer group in Assam overcome the limitations of remoteness, small-sized lots and weak bargaining strength. Contracts intermediated by an NGO, another emerging channel in Assam, also overcomes scale disadvantage while facilitating fair negotiations.

Competition from new marketing agents has helped to improve regulated market functioning in Maharashtra and Madhya Pradesh where there is enough space for traditional and emerging markets to function. The traditional traders are resourceful and operate in larger markets. In other cases, such as Assam, Himachal Pradesh, and Jharkhand there is scope of improving the traditional markets and strengthening the supervision. Punjab too has well-developed and large traditional markets but the system is geared far more for grains than for horticultural products.

Price determination by the forces of demand and supply in an objective manner lies at the heart of the market mechanism, underlined in the theories drawn from Smith and Ricardo. Pricing is seen to be undergoing a change (Young and Hobbs 2002). While auctions, though not often conducted fairly, are still the way of price discovery in traditional marketing, the emerging channels embody a departure and a grey area. Only in DM there are direct simultaneous negotiations leading to the transactions. Sales by farmers' groups in Assam based on information sharing may be another though related example. In all other cases, the negotiation is largely between the transacting parties in isolation especially in CONTR where price is decided even before the transaction. In RTL the price evidently decided unilaterally by the buyer with difficulty. In most cases, mandi prices are the reference prices.

The traditional trader continues to be a prime source of market information to sellers in both channels. The only exceptions are the sellers in RTL who rely only on their buyers' best judgements. Public market intelligence AGMARKET in this study is not found to be effective in enriching the producer directly except in a few sample cases in Haryana and Punjab. The parallel presence of the auction driven traditional marketing channel remains valuable institution for pricing but the futures market can be extremely potent as an objective, and transparent indicator of market price.

Product losses are seen to occur mostly at the stage of harvest and in transit though wastage at retail level is also not insignificant. In nearly all cases, producers have attributed the losses merely to the perishable nature of the crop without considering the possibility of reducing the incidence through technology. Long distances are reported to be an important cause of product wastage associated with the poor condition of roads but waiting for higher prices is also reported to be one of the leading reasons for losses. Although product wastage is reduced in the emerging channels due to the creation of clearing Centres with refrigerated storage facilities, marketing reforms may not be a complete solution as the problem often lies in poor infrastructure that affects both channels. Private companies do not always help as storage function is usually left to farmer's resources. Improved farm practices can reduce product damage but private extension can contribute to their improvement.

POLICY DIRECTIONS

The analysis shows that state governments' political will for reforms is more important than legislation and so the Union government can act only as an advisor and facilitator leaving the actual actions of reforms to the best judgments of the states. In reality, geographic and socio-economic conditions vary widely across states creating differences in political compulsions. Reforms can take different directions in different states for the welfare of the people. However, evidences collected in this study shows that transitions in agro-marketing are in process in almost all states regardless of formal action. The urge to cut down marketing costs and have more options in marketing is manifested in the emergence of a large variety of marketing channels in different parts of the country but an organized corporate entity is not always a player in these channels and sometimes the distinction from traditional channel is not sharp. Producers may or may not be relieved of the entire burden of marketing.

The gain in efficiency from switching to an emerging channel is hard to deny. The new channel is also associated with increases in productivity, profit and returns from farming but distributional implications require attention. When operated by organized companies it tends to draw participation of the larger farmers and may enhance regional disparity. The benefit of specialization is lost in direct marketing. Pricing mechanism is undergoing a change but regulated APMC market prices still set the benchmark and there is a serious need for considering the future of prices. Traditional markets remain large in some regions. APMC markets are improving under competitive pressure but in some states, they function poorly or are even non-starters. While the emerging channels appear to deliver greater efficiency, at present the simultaneous existence of both channels will help to provide options, reduce wastage by diverting rejected products, and cater completely to large markets. The facilities of the APMC market are still useful to players in all channels.

Acknowledgement: The author thanks Agro-economic Research Center's (AERC), Shangeeta Shrof, S. S. Kalamkar, and Jayanti Kajale of AERC, Gokhale Institute of Politics and Economics, Pune, Maharashtra; Gautam Kakaty and Debajit Borah of AERC, Assam Agricultural University, Jorhat Assam; G. Gangadhara Rao and G.M. Jeelani of AERC, Visakhaptnam, Andhra University, Andhra Pradesh; Ranjan Kumar Sinha of AERC for Bihar and Jharkhand (2 states), T. M. Bhagalpur University, Bhagalpur, Bihar; Ranveer Singh, C.S. Vaidya, Meenaakshi, and Pratap Singh of AERC, Himachal Pradesh University, Shimla, Himachal Pradesh; Ramendu Ray, D.K. Singh, and Hasib Ahmad of AERC, University of Allahabad, Allahabad, Uttar Pradesh; Hari Om Sharma and N.K. Raghuwanshi of AERC, Jawaharlal Nehru Krishi Vishwa Vidyalaya, Jabalpur, Madhya Pradesh; Usha Tuteja of Agro-economic Research Centre, Delhi; D.K. Grover, J.M. Singh, Jasdev Singh, and Sanjay Kumar of AERC, Department of Economic and Sociology, Punjab Agricultual University, Ludhiana, Punjab, and Debashis Sarkar and Ramesh Chandra Mondal of AERC, Visva-Bharati, Santiniketan, West Bengal for collection of information. Mr M. Rajeshwor is also thanked for the able research assistance.

REFERENCES

Acharya, S.S., and N.L. Agarwal. 2004. *Agricultural Marketing in India*, New Delhi: Oxford.

Ahmed, Raisuddin. 1996. 'Agricultural Market i\Reforms in South Asia', *American Journal of Agricultural Economics*, 78(3): 815–9.

Bardhan Pranab, Dilip Mookherjee, and Masatoshi Tsumagari. 2013. 'Middlemen Margins and Globalization', *American Economic Journal: Microeconomics*, 5(4): 81–119.

Baumol, W.J. 1982. 'Contestable Markets: An Uprising in the Theory of Industry Structure', *American Economic Review*, 72(1): 1–15.

Ghosh, Jayati. 2012. 'India's Supermarket Moves Show its Tired Government has Run Out of Ideas', *The Guardian*, 20 September.

Government of India. 2015. *Economic Survey 2014–15*, New Delhi: Department of Economic Affairs, Ministry of Finance, Government of India.

Jaffe, Eugene Donald, and Ling Yi. 2007. 'What are the Drivers of Channel Length? Distribution Reform in the People's Republic of China', *International Business Review*, 16(4): 474–93.

Key, Nigel, Elisabeth Sadoulet, and Alain de Janvry. 2000. 'Transactions Costs and Agricultural Household Supply Response', *American Journal of Agricultural Economics*, 82 (May): 245–59.

Mulky, Avinash. 2008. 'Enhancing Marketing Performance: Academic Perspective', *IIMB Management Review*, 20(4): 423–35.

Sengupta, Arjun. 2013. 'Mainstreaming Small Farms: A New Approach Needed', in Nilabja Ghosh and C.S.C. Shekhar (eds), *Future of Indian Agriculture*, New Delhi: Academic Foundation, 33–40.

Shepherd, G.S. 1965. *Marketing Farm Products—Economic Analysis*, Iowa: Iowa State University.

Yang, C.K. 1944. 'A North China Local Market Economy: A Summary of a Periodic Markets in Chowping, Hsien, Shantung', Mimeo, p. 2. New York: Institute of Pacific Relations.

Young, L.M., and J.E. Hobbs. 2002. 'Vertical Linkages in Agri-Food Supply Chains: Chaining Roles for Producers, Commodity Groups, and Government Policy', *Review of Agricultural Economics*, 24(2): 428–41.

ANNEXURE 37A

Table 37A.1 Intermediation in Emerging Marketing Channels in Sample

State	Crop	Channel	Intermediary	Nature	Involvement
Andhra Pradesh	Banana	DM	NONE	Rythu Bazaar, direct to consumer	No private intermediary
Andhra Pradesh	Brinjal	DM	NONE	Rythu Bazaar, direct to consumer	No private intermediary
Assam	Orange	DM	None	FGROUP, Non-profit, sales to processor	No private intermediary, but collective sales
Punjab	Kinnow	DM	Traders	Farmer Evening market, sales to private traders	Private traders only
Himachal Pradesh	Tomato	RTL	Mother dairy	Non-profit, no private intermediary	Non-profit organized
Jharkhand	Cauliflower	RTL	Reliance	Single organized intermediary	Large corporate
Haryana	Muskmelon	RTL	Reliance	Single organized intermediary	Large corporate
Haryana	Tomato	RTL	Reliance	Single organized intermediary	Large corporate
Uttar Pradesh	Potato	CONTR	PepsiCo	Single organized intermediary	Large corporate
Uttar Pradesh	Amla	CONTR	Satkar Foods	Single organized local intermediary	Local corporate
Assam	Potato	CONTR	Kishalaya Food	Single organized local intermediary but NGO intermediated	Local corporate
Punjab	Potato	CONTR	PepsiCo	Single organized local intermediary but public intermediation	Large corporate
Himachal Pradesh	Apple	CMI	Adani	Sales too private traders via single organized corporate intermediary	Large corporate with private traders
Madhya Pradesh	Soybean	CMI	ITC	Sales to traders via e-portal of organized corporate intermediary	Large corporate with private traders
Maharashtra	Onion	CMI	DFPCL	Sales to traders via organized intermediary	Large corporate
Maharashtra	Pomegranate	CMI	DFPCL	Sales to traders via organized intermediary	Large corporate
Bihar	Mango	TRADER	Local	Sales to traders via local trader group	Private traders only
West Bengal	Arum	TRADER	Local	Sales to traders via local trader group	Private traders only

Source: Field survey.

Table 37A.2 Gross Marketing Cost Reduction in the Emerging Channels per Farmer's Rupee

Channel	Quantum (Rs)	Relative (%)
Direct marketing (DM)	0.83	69.25
Corporate marketing intermediation (CMI)	0.24	28.79
Marketing to processors on contract (CONTR)	0.27	63.90
Marketing to organized retailer (RTL)	0.37	23.22
Marketing by local traders* (TRADER)	−0.33	−44.59

Source: Computed from survey data.
Note: Marketing cost includes trader's margins and is expressed in value as well as value relative (%) to that in traditional channel. The figures are averages of channels.
* includes West Bengal where the traditional channel and emerging channel are not strictly comparable as two different crops are reported.

Table 37A.3 The Details of Agro-Economic Research Centres and the Lead Person Involved in the State Report

State	AERC/Unit	Authors
Maharashtra	AERC, Gokhale Institute of Politics and Economics, Pune	Shangeeta Shrof, S.S. Kalamkar and Jayanti Kajale
Assam	AERC, Assam Agricultural University, Jorhat	Gautam Kakaty and Debajit Borah
Andhra Pradesh	AERC, Visakhapatnam, Andhra University	G. Gangadhara Rao and G. M. Jeelani
Bihar and Jharkhand	AERC for Bihar and Jharkhand, T. M. Bhagalpur University, Bhagalpur, Bihar	Ranjan Kumar Sinha
Himachal Pradesh	AERC, Himachal Pradesh University, Shimla	Ranveer Singh, C.S. Vaidya, Meenakshi and Pratap Singh
Uttar Pradesh	AERC, University of Allahabad, Allahabad	Ramendu Ray, D.K. Singh and Hasib Ahmad
Madhya Pradesh	AERC, Jawaharlal Nehru Krishi Vishwa Vidyalaya, Jabalpur	Hari Om Sharma and N.K. Raghuwanshi
Delhi	Agro-Economic Research Centre, Delhi University	Usha Tuteja
Punjab	AERC, Department of Economic and Sociology, Punjab Agricultural University, Ludhiana	D.K. Grover, J.M. Singh, Jasdev Singh and Sanjay Kumar
West Bengal	AERC, Visva-Bharati, Santiniketan	Debashis Sarkar and Ramesh Chandra Mondal

38

Integration of Major Agriculture Markets in Karnataka

An Empirical Study of Selected Commodities

M.J. Bhende

Many developing country policymakers are currently considering more ambitious regional integration initiatives that include reform and integration of food and agriculture markets to enhance domestic production and to tap export potential. Under the transitional process of globalization characterized by liberalization, privatization, free market economy, aggressive consumerism, and entry of private players, both national and multinational in the trade of agricultural commodities, it is of paramount importance to put in place an efficient market system. An efficient agricultural marketing system is seen as an important means for raising the income level of farmers and for promoting the economic development of a country. The farmers allocate their resources according to their comparative advantage and invest in modern farm inputs to obtain enhanced productivity and production. The overall market performance may be indicated by spatial price behaviour in regional markets. Spatial market performance may be evaluated in terms of its price relationships.

A key premise of several arguments in economics is that markets allow for price signals to be transmitted both spatially and vertically. An obvious example is the assessment of the relative merits of alternative trade and/or policy environments: potential losses for a country or a group of economic agents and benefits crucially depend, inter alia, upon markets receiving price signals, which, in turn, depends upon a number of market features, including their very existence. The extent to which a price shock at one point affects a price at another point can broadly indicate whether efficient arbitrage exists in the space that includes the two points. At two extremes, one may assume that a full transmission of price shocks can indicate the presence of a frictionless and well-functioning market, while at the other extreme a total absence of transmission may make the very existence of a market questionable. Therefore, the degree of price transmission can provide at least a broad assessment of the extent to which markets are functioning in a predictable way and price signals are passing-through consistently between different markets.

A wide economic literature has studied the relationship between prices, either spatial or vertical. Concerning the former, a recent wide critical review is in Fackler and Goodwin (2002). The premises of full price transmission and market integration correspond to those of the standard competition model: in a frictionless undistorted world, the Law of One Price (LOP) is supposed to regulate spatial price relations, while pricing along production chains will depend exclusively on production costs. The absence of market integration or of complete pass-through of price changes from one market to another has important implications for economic welfare. Incomplete price transmission arising either due to trade and other policies, or due to transaction costs such as poor transport and communication infrastructure, results in a reduction in the price information available to economic agents and consequently may lead to decisions that contribute to inefficient outcomes.

In integrated markets, an arbitrage process operates that limits price differences in time, form, and space, to the marketing costs. Markets that are not integrated may convey inaccurate price information, distorting the marketing decisions of the producers and contributing to inefficient product movements (Tomek and Robinson 1990). Sexton, Kling, and Carman (1991) identified three reasons for a lack of market integration: prohibitive transaction costs, different kinds of trading barriers, and imperfect competition. This research aims at a better understanding of the different aspects of regional integration of markets for agriculture commodities. Specifically the study is envisaged with the following objectives:

1. to study the structure and behaviour of prices in selected agricultural commodities in Karnataka;
2. to analyse the spatial pricing efficiency of commodities selected markets of Karnataka; and
3. to assess the prices risk for different commodities in the selected markets of Karnataka.

METHODOLOGY AND DATA

Time series data on the daily arrivals and prices of concerned agricultural commodities, that is, rice, maize, cotton, groundnut, chick pea, and pigeon pea were downloaded from the Karnataka State Agricultural Marketing Board (KSAMB), Bangalore website (www.krishimaratavani.com). KSAMB receives daily arrivals and prices data from the respective Agricultural Produce Market Committees (APMCs) of various markets throughout the state. The data on arrivals refer to the total arrivals during the day in quintals in a market place. The data on prices used refer to the modal price for the day's transaction. Modal price is considered to be superior to the daily average price as it represents the major proportion of the commodity marketed on the particular day in a particular market. The data on daily arrivals and modal prices were

Table 38.1 Five Markets Selected by Commodities for the Study

S. No.	Crop	Markets Selected
1	Rice	Bangalore, Bangarpet, Mangalore, Mysore, Gulbarga
2	Maize	Bellary, Davangere, Hassan, Honnali, and Koppal
3	Chick pea	Bidar, Bangalore, Dharwad, Gadag, Raichur
4	Pigeon pea	Bidar, Gulbarga, Sedam, Laxmeshwara, Basavakalyan
5	Cotton	Bijapur, Chitradurga, Haveri, Raichur, Ranebennur
6	Groundnut	Bagalkot, Chalakere, Gadag, Hubli, Raichur

Source: Karnataka State Agriculture Marketing Board (http://krishimaratavahini.kar.nic.in/)

collected for the period of January 2004 to November 2010 (seven years). The necessary data were collected from five major markets for each of the six commodities (Table 38.1) based on the maximum annual arrivals during 2010.

Several authors have studied price transmission within the context of the LOP (Ardeni 1989; Baffes 1991) or within the context of market integration (Blauch 1997; Gardner and Brooks 1994; Palaskas and Harriss 1993; Ravallion 1986; Sexton, Kling, and Carman 1991; Zanias 1991, 1999). The concept and the analytical techniques have also been used to evaluate policy reform, such as ex post assessment of market integration in the context of the implementation of the structural adjustment programmes (Alexander and Wyeth 1994; Dercon 1995; Goletti and Babu 1994). Another vein of research focuses on vertical price transmission along the supply chain from the consumer to the producer level (see for example Brorsen et al. 1985; Kinnucan and Forker 1987; Kumar and Sharma 2003; Prakash 1998; Schroeter and Azzam, 1991; von Cramon-Taubadel, 1999).

The large body of research on market integration and price transmission, both spatially and vertically, has applied different quantitative techniques and has highlighted several factors that impede the pass-through of price signals. In theory, spatial price determination models suggest that, if two markets are linked by trade in a free market regime, excess demand, or supply shocks in one market will have an equal impact on price in both markets.

This research focuses on spatial price differences in the major agricultural markets in Karnataka for important food and cash crops. Taking into account the characteristics of the markets under study, an empirical test is developed to verify whether price patterns in different locations cohere. In an efficient market system, prices move together:

'trade takes place if price in the importing region equals price in the exporting region plus the unit transport cost incurred by moving between the two' (Ravallion 1986). In such cases, trade literature postulates the existence of the representative price, that is, a price which prevails at all markets, which is known as LOP.

ANALYSIS OF TIME SERIES

Database for time series analysis was the data on price and arrival in chronological order. Such a database is usually conceived of as being generated by certain regulated base causative sources such as long-time forces, cyclical forces usually with a periodicity of longer duration than a year, annual fluctuation associated with season of a year and many other sources of fluctuations producing random effects. These are described below.

The graph of such a time series data gives a rough idea about the nature of fluctuations in the value of the variable with time. The fluctuations have the combined effect of various causes such as the ones mentioned above which sometimes induce a sharp rise and fall. Segregating these various types of fluctuations in the time series is known as analysis of time series. The important basic components of time series are (i) secular trend; (ii) seasonal variation/periodic movements; (iii) cyclical movements; and (iv) irregular variations.

Estimation of Seasonal Indices of Monthly Data

As a first step, to estimate the seasonal index a 12-month centred moving averages as calculated removes a large part of fluctuation due to seasonal effects so that what remains is mainly attributable to other sources, namely, long-term effects (*Tt*) and cyclical effect (*Ct*); the irregular variation (*It*) due to random causes is also minimized as process of smoothing out effect. Thus, this affords a means of not only estimating trend cycle effect but also estimating seasonal components. These indices are used using appropriate techniques.

RELATIONSHIP BETWEEN TWO MARKETS

It is of interest to know whether the variations in prices in one market are transmitted to other markets or variation in prices do not impinge upon prices in other markets. In other words whether the markets are integrated or disintegrated. The most convenient and simple method for measuring the relationship between two markets is conventionally through computation of correlation coefficient between unadjusted series of two markets. However, while correlating price series of two markets, of a given period, it is necessary to adjust for trend, otherwise there will be a biased measure of market integration, because price contains trend effects also. Therefore, price series adjusted for trend is recommended as a better measure. A significant correlation implies that the markets are well integrated, that is, markets are mutually dependent on price.

Cointegration analysis will allow us to verify the degree of association between the markets. George Blyn (1973) also discusses Price series correlation as a measure of market integration. This research proposes to adopt cointegraion method. A substantially improved procedure is now available for conducting Box-Jenkins ARIMA analysis which relieves the requirement for a seasoned perspective in evaluating the sometimes ambiguous autocorrelation and partial autocorrelation residual patterns to determine an appropriate Box-Jenkins model for use in developing a forecast model. We have used Auto Regressive Integrated Moving Average (ARIMA) Model to study the spatial price efficiency of major commodity markets in Karnataka.

For co-integration analysis, at first the order of integration of the price series is to be determined. The number of times a series has to be differenced before it comes stationary is itself the order of integration. A series, which has to differenced once to become stationary, has an order of integration one, denoted as I(1). A series, which has been differed twice to become stationary has an order of integration two, denoted as I(2). In determining the order of integration under study, the Dickey-Fuller test was employed. The Dickey-Fuller test is based on the regression.

$$Pt = bPt\text{-}1 + e$$

Here, 'b' will be zero if 'P' follows random walk. It will be negative and significantly different from zero if Pt is stationary.

Without going into too much detail, there is a "duality" between a given time series and the autoregressive model (ARM) representing it; that is, the equivalent time series can be generated by the model. The AR models are always invertible. However, analogous to the stationarity condition described above, there are certain conditions for the Box-Jenkins MA parameters to be invertible.

Trend Measures the overall movement of a series. It usually estimated by fitting a linear trend equation to the data of the form Y = a + bt + e where Y is the variable under study and T is the time variable. 'a' and 'b' are the constants of the equation where b is the estimated trend. Seasonal Index is estimated using the ratio to moving average method. Finally, the residual of the fitted ARIMA models are cross-correlated to identify possible lead or lag effects between the series.

The data collected on arrivals and prices of six crops for five markets each from the Karnataka State Agricultural Marketing Board (KSAMB) are subjected to statistical analysis using techniques outlined above.

MARKET RISK

Market risk is a dominant source of income fluctuations in agriculture sector all over the world. A concept discussed in this context is Value-at-Risk (VaR). VaR has been established as a standard tool among financial institutions to depict the downside risk of a market portfolio. It measures the maximum loss of the portfolio value that will occur over some period at some specific confidence level due to risky market factors. Though VaR has been primarily designed for addressing the needs of financial institutions, it also has the potential for applications in agribusiness. Value at Risk (VaR) measures the maximum expected loss over a given time period at a given confidence level that may arise from uncertain market factors.

The general techniques commonly used for VaR estimates include:

1. **Parametric**
   - Delta-Normal method (Variance-Covariance Method)
   - Monte Carlo simulation
2. **Non-parametric**
   - Historical Simulation

RESULTS AND DISCUSSION

Spatial market integration refers to a situation in which prices of a commodity in separated markets move together and price signals and information are transmitted smoothly across the markets, hence, spatial market performance may be evaluated in terms of the relationship between the prices of spatially separated markets and spatial behaviour in the markets may be used as a measure of overall market performance. The present study empirically evaluates spatial integration of selected agriculture commodities among major markets. The market integration was assessed by employing co-integration technique. In the present context, co-integration analysis is employed to examine whether the one market is integrated with the other market. This is studied by testing whether the Law of one Price (LOP) holds in these markets.

Seasonality

Rice

The arrivals into Bangalore, Mangalore and Mysore had a fairly uniform flow of arrivals of rice. Gulbaraga and Mysore had highly seasonal arrivals. The prices of rice showed seasonal variation only in Mysore and Gulbaraga and in the other markets notably Bangalore and Mangalore and Bangarapet the prices were stable. Arrivals in all the markets have witnessed an over 15 per cent increase with the exception of Bangalore and Mangalore where there has been no quantitative increase in the quantity of arrivals. Prices of rice in all the markets have increased at rates varying from 6 per cent to 15 per cent per annum and the increase has been significant.

Maize

The behaviour of prices across markets for maize was not the same. In Davangere, Belgaum, and Bangalore the prices were largely stable, whereas in Honali and Kopal it peaked in July to October. On the other hand the arrivals showed marked seasonality which varied from market to market. Though the arrivals to all the markets have been increasing modestly, a huge increase in arrivals was observed in Koppal market of around 18 per cent per annum. The prices of maize on an average have been increasing at around 10 per cent per annum in all the markets.

Chick Pea

The results reveal that in the case of chickpea Bangalore. Gadag and Raichur markets showed an inverse relationship between arrivals and prices. In the case of Dharwad and Bidar no clear relationship was observed. In general price was higher during August to November as compared to the earlier period. Gadag and Raichur witnessed the highest rate of growth in arrivals of over 15 per cent per annum. Arrivals to Bidara and Dharwad were not significant.

Pigeon Pea

Pigeon pea prices were seasonally stable both in Gulbarga and Bidar which is not true of the arrivals which is principally peaks during the first part of the year. Arrivals of pigeon pea to Bidar, Sedam, and Basavakalyan markets have all decreased significantly and showed a marked increase in Gulbarga and Laxmeshwara where the arrivals has risen each year by 16 and 12 per cent respectively. Prices of pigeon pea have increased sharply by over 25 per cent across markets.

Cotton

The seasonal behaviour of prices was not consistent across markets. In some markets prices are higher during the second half of the year in some markets like Chitradurga, Haveri, and Ranebennur whereas, in Bijapur the first half of the year had higher prices. The explanation could not be found in arrivals entirely. Arrivals of cotton to most markets were significant of over 15 per cent in all the markets studied except Raichur where it is under 5 annum.

Groundnut

Most of the markets for onion with the sole exception of Gadag and Honali were seasonal markets where arrivals take place only during the production season in the market hinterland. The price of groundnut did not witness display any marked seasonality and was fairly uniform across markets. The arrivals to Bagalkot and Raichur showed a significant decline of about 15 per cent per annum. In Challakere and Gadag, the increase in arrivals has been significant at around 15 per cent per annum.

Market Price Integration

Market price integration is characterized by three situations: (a) Inter-dependence, (b) Independence, and (c) the leadership. Interdependence is a sign of market integration whereas independence and price leadership do not augur well for market integration

Rice

Five important markets of rice, namely Bangarapet, Bangalore, Gulbarga, Mangalore, and Mysore, were analysed for spatial price efficiency. The results indicate that Bangarpet market leads Bangalore by 5–6 days, whereas Bangalore is leading Bangarpet market by 2 and 3 days. Mysore market leads Bangarpet market by 6 and 7 days whereas it leads Gulbarga market by 6 days. The results indicated that Bangarpet and Gulbarga; Bangarpet and Mangalore; Bangalore and Gulbarga; Bangalore and Mysore; Bangalore and Mangalore; Gulbarga and Mangalore as well as Mangalore and Mysore markets were independent of each other.

Maize

Five markets viz., Davangere, Honnalli, Hassan, Bellary, and Koppal were selected to study the spatial efficiency. The analysis of prices at Davangere and Honnali; Davangere and Koppal; Davangere and Hassan markets indicated that these markets are independent of each other i.e., the markets were not related to each other. The results indicate that Davangere leads Bellary market by 4 days. Honnali and Koppal markets are independent of each other. The results indicate that prices of maize in Bellary market lead Honnali market prices by 4 days. Similarly, Hassan market leads Honnali prices by 5 days whereas later has a tendency to lead by 6 days. The results indicate that Bellary market leads Koppal market prices by 5 and 6 days, whereas Koppal market has a tendency to lead by 1, 2, 3, 4, and 6 days. Hassan market is leading Koppal market by 5 and 6 days, whereas Koppal is found to lead Hassan market by 3 days. The analysis of relationship between maize prices at Bellary and Hassan prices indicated that Hassan is leading by a day and Bellary leads by 1 and 3 days.

Chick Pea

The spatial integration between the markets for chickpea prices was studied for Bangalore, Bidar, Dharwar, Gadag, and Raichur markets. The analysis of chickpea prices at Bangalore and Bidar market indicated that Bidar had a tendency to lead Bangalore market by 1, 2, and 6 days. The interrelationship between Bangalore market price and Dharwar market price revealed that Bangalore market led Dharwad by 4 and 5 days. This is a case of unidirectional causality. With regard to Bangalore and Gadag markets the results indicated that Gadag is leading by 7 days. The analysis of relationship between Bangalore and Raichur; Bidar and Dharwar; Bidar and Gadag; Bidar and Raichur; Dharwar and Gadag; as well as between Dharwar and Raichur indicated that these markets operate independently from each other and no price transmission takes place between the pair of markets listed above. The relationship between prices at Gadag and Raichur markets indicated that Raichur market leads Gadag prices by a day.

Pigeon Pea

The markets studied for Pigeon pea were Bidar, Gulbarga, Sedam, Basavakalyana, and Laxmeshwara. Despite the proximity, Gulbaraga and Bidar markets were not integrated. Similarly Bidar and Laxmeshwara markets were found independent of each other. The relationship between the prices of Bidar and Sedam were not strong, however, Bidar market leads by 6 days. Analysis of prices at Bidar and Laxmeshwara shows that Bidar is leads Basvakalyan prices by 1, 2, and 3 days. Similarly, Sedam was found to be leading the Gulbarga market by 4 and 5 days. Gulbarga and Laxmeshwara markets were independent of each other. The same was true of Gulbarga and Basavakalyan,

Sedam and Laxmeshwara markets. The Sedam market has a tendency to lead the Basavakalyan market by a day, whereas Sedam market led Basvakalyan market by 5 and 6 days. Basavakalyan market has a tendency to lead Laxmeshwara market by 1, 2, and 3 days.

Cotton

The study of integration of cotton prices between Bijapur and Chitradurga market revealed that Bijapur is leading by 2 and 7 days, which is a sign of market leadership. Similarly, Bijapur market is leading Haveri prices by 2 and 3 days. The interrelationship between Bijapur market and Raichur market indicated that both the markets are independent of each other. Analysis of prices from Bijapur and Ranibennur markets indicated that Bijapur market leads Ranebennur by 7 days. Between Chitradurga market price and Haveri market, Haveri led by 4 and 5 days whereas Chitradurga led by 7 days. The results indicate that Chitradurga has a tendency to lead Raichur market by a day whereas Raichur has a tendency to lead by 2, 3, 4, and 5 days. The interrelationship between cotton prices at Chitradurga and Ranibennur; Haveri and Raichur; Haveri and Ranebennur as well as between Raichur and Ranibennur market revealed that the above listed pairs of markets are independent of each.

Groundnut

The analysis of interrelationship between groundnut prices to study spatial integration was carried out for five major markets in Karnataka. The five markets selected for the study are Bagalkot, Challakere, Gadag, Hubali, and Raichur. The results about interrelationship between groundnut prices at Bagalkot and Challakere market indicate that Challakere market is leading Bagalkot market by 1 and 2 days. Bagalkot market was found to be leading Gadag by 6 days, whereas Gadag is leading Bagalkot market by 7 days. Bagalkot and Hubli markets are interdependent as indicated by the fact that Bagalkot leads Hubli market by 7 days. The study of Bagalkot market and Raichur market shows that Bagalkot leads by 1 and 2 days. Between Challakere and Gadag market, Challakere leads by 2 and 3 days. The study of Challakere and Raichur market indicated that Challakere market is leading by 2, 6 and 7 days, similarly, Challakere market leads Hubli by 2 and 7 days. The results about interdependence between Gadag and Hubli as well as Gadag and Raichur market indicates that Gadag is leading Hubli market by a day and Raichur market by 4 days. The results indicate that Hubli and Raichur markets have the tendency to lead each other by 6 and 7 days.

From the foregoing it is clear while at the macro level markets for commodities are integrated, some of the regional markets are not so well integrated and some even behave independent of each other due to the poor communication facilities between markets. This gives opportunities for making abnormal arbitrage profits which does not augur well for market efficiency. Efforts should be made to make markets for each commodity to behave like one large market where price signals from one market is quickly transmitted to the other.

Market Risk Analysis

The price risk of the commodities has been studied in the Value at Risk (VaR) framework. The price risk for rice at Bangarapet, Gulbarga, and Mysore markets was high as compared to Mangalore and Bangalore where the price risk in rice was moderate. The risk for maize was by and large moderate in most of the markets studied. Price risk for chickpea was found low in Bangalore, Bidar, Dharwar and Gadag markets and moderate in Raichur market. Similarly, Bidar and Sedam markets indicated low price risk for pigeon pea whereas, Gulbarga, Laxmeshwara and Basavakalyan have moderate risk. All the major markets selected for groundnut indicated moderate price risk. In case of cotton, Bijapur market experienced low price risk whereas, Raichur and Ranebennur markets markets faced high price risk. Chitradurga and Haveri reported moderate price risk in the marketing of cotton.

CONCLUSIONS AND POLICY IMPLICATIONS

The results of the study reveal that there is a degree of pricing inefficiency with regard to the smaller markets of each of the commodities. While the major markets are by and large price efficient as indicated by the degree of spatial price integration, some market prices tend to move independently of these markets. The markets for major cash crops, namely cotton and groundnut were found better integrated whereas, most of the markets for cereals and pulses that is, for rice, maize, chickpea and pigeon pea were functioning independent of each other. The distance between two markets was not an important criterion for integration of two markets. However, there is a need to improve both the transportation and other infrastructure facilities in these markets and perhaps improve the competition so that price integration is facilitated.

Volatility in these markets is high and in some commodities the volatility is so high that it could lead to undermining the production of these commodities due to the uncertainty that these markets prices induce. There is a need to introduce stability in prices after understanding the reasons for the instability and controlling these factors. The operation of the price stabilisation fund could be streamlined to ensure that volatility is checked.

REFERENCES

Alexander, Carlo, and John Wyeth. 1994. 'Cointegration and Market Integration: The Indonesian Rice Markets', *The Journal of Development Studies*, 30(2): 303–28.

Ardeni, P.G. 1989. 'Does the Law of One Price Really Hold for Commodity Prices?', *American Journal of Agricultural Economics*, 71(3): 661–9.

Baffes, J. 1991. 'Some Further Evidence on the Law of One Price', *American Journal of Agricultural Economics*, 73(4): 21–7.

Baffes J. and M. Ajwad. 2001. 'Identifying Price Linkages: A Review of the Literature and an Application to the World Market of Cotton', *Applied Economics*, 33(15), 1927–41.

Blauch, B. 1997. 'Testing for Food Market Integration Revisited', *The Journal of Development Studies*, 33(4): 512–34.

Blyn, George. 1973. 'Price Series Correlations as a Measure of Market Integration', *Indian Journal of Agricultural Economics*, 18(2): 56–9.

Brorsen, B.W., J.P. Chavas, W.R. Grant, and L.D. Schnake. 1985. 'Marketing Margins and Price Uncertainty: The Case of US Wheat Market', *American Journal of Agricultural Economics*, 67(3): 521–8.

Dercon, S. 1995. 'On Market Integration and Liberalization: Method and Application to Ethiopia', *The Journal of Development Studies*, 32(1): 112–43.

Fackler, P.L., and B.K. Goodwin. 2002. 'Spatial Price Analysis', in B.L. Gardner, and G.C. Rausser (eds), *Handbook of Agricultural Economics*. Vol. 1, Part 2. Amsterdam: Elsevier Science, 971–1024.

Gardner, B.L., and K.M. Brooks. 1994. 'Food Prices and Market Integration in Russia: 1992–93', *American Journal of Agricultural Economics*, 76(3): 641–6.

Goletti, Francsco, and Suresh Babu. 1994. 'Market Liberalization and Integration of Maize Markets in Malawi', *Agricultural Economics* 11(2–3): 311–24.

Kinnucan, H.W., and O.D. Forker. 1987. 'Asymmetry in Farm Retail Price Transmission for Major Dairy Products', *American Journal of Agricultural Economics*, 69(2): 285–92.

Kumar, Parmod, and R.K. Sharma. 2003. 'Spatial Price Integration and Pricing Efficiency at the Farm Level: A Study of Paddy in Haryana', *Indian Journal of Agricultural Economics* 58(2): 201–17.

Palaskas, T., and B. Harriss. 1993. 'Testing Market Integration: New Approaches with Case Material from the West Bengal Food Economy'. *Journal of Development Studies* 30(1): 1–57.

Prakash, A. 1998. 'The Transmission of Signals in a Decentralised Commodity Marketing System: The Case of UK Pork Market', Unpublished PhD thesis, University of London.

Ravallion, M. 1986. 'Testing Market Integration', *American Journal of Agricultural Economics*, 68(2): 292–307.

Schroeter, J.R., and A. Azzam. 1991. 'Marketing Margins, Market Power, and Price Uncertainty', *American Journal of Agricultural Economics*, 73(4): 990–9.

Sexton, R.J., C.L. Kling, and H.F. Carman. 1991. 'Market Integration, Efficiency of Arbitrage, and Imperfect Competition: Methodology and Application to U.S. Celery', *American Journal of Agricultural Economics*, 73(3), 568–80.

Tomek, W., and K. Robinson. 1990. *Agricultural Product Prices*. Ithaca, NY: Cornell University Press.

Von Cramon-Taubadel, S. 1999. 'Estimating Asymmetric Price Response with the Error Correction Representation: An Application to German Pork Market', *European Review of Agricultural Economics* 25(1): 1–18.

Zanias, G. 1991. 'Testing for Integration in the EC Agricultural Product Markets'. *Journal of Agricultural Economics* 44(3): 418–27.

———. 1999. 'Seasonality and Spatial Integration in Agricultural (Product) Markets'. *Agricultural Economics* 20(3): 253–62.

39

Agricultural Market Integration and Trade Competitiveness of Indian Agriculture

D. Kumara Charyulu and M. Prahadeeswaran

Indian Agriculture is characterized by small holdings, limited off-farm employment opportunities, and inadequate institutional and infrastructure support. Therefore, farmers are unable to withstand major structural shifts if required. Any adverse impact on their income and/or increase in risk will have a significant impact on their livelihood. Farmers do respond to market prices and adjust to market requirements by changing cropping patterns and input use. However, this is all right as long as these crops are remunerative in each region. If all crops suitable in an area are not competitive in the international market, then the very livelihood of the farmers will be adversely affected. Even if there is a remunerative crop, farmers may not be able to take advantage of the crop if they are exposed completely to the volatility of the international market, as their risk-bearing ability is poor. In the absence of efficient futures and options market which are prevalent in developed countries and some newly emerging economies, our farmers/agribusiness are at a disadvantage in managing volatility in world market prices. Therefore it is essential that the Agreement on Agriculture (AoA) implications in terms of both profitability and volatility are not too drastic for farmers to cope up with.

In the early 1990s, India had emphasized the need for exports. Till then there was no consistent policy towards exports. The country's export basket in the past was heavily dependent on items such as tea, mates and coffee, which only recently has widened to include other commodities like rice, marine products, fruits and vegetables, etc. However, dedicated systems to cater to the export market are yet to develop. Exporters have to take special care in maintaining the quality at all stages of exports. Still, our current information flow along the value chain is very weak. These often lead to additional costs. Therefore, there are asymmetric trade opportunities for Indian exporters which constraining our exports. In addition, the importers in other countries are also putting additional restrictions or raise quality issue to prevent imports from certain countries. There are likely to affect Indian exports to a large extent.

Although India enjoys advantage in exporting some commodities, in the period post the General Agreement on Tariffs and Trade (GATT) international trade has become highly competitive and the competitive advantages of some of these commodities would be lost due to the infrastructural advantages prevalent in the competing countries. Therefore, infrastructure development for efficient movement, handling, grading, packing, processing, trade network, and information dissemination systems is needs to be developed on a priority basis. So, building of these backward/forward linkages is also important to improve production efficiency.

However, past studies conducted by some researchers on the impact of globalization and liberalization on Indian agriculture concluded that agriculture sector witnessed sharp improvement in terms of trade (ToT) during initial years of reforms. In the post-World Trade Organization (WTO) period though ToT remained favourable compared to the period before reforms but there is decline in them. Growth rate in GDP of the agriculture sector showed almost no change during the pre-reform decade and post-reform period. But, the advantage India has in production of labour-intensive crops such as fruits and vegetables, and other crops such as basmati rice may not be adequate to compensate for the likely imports of other commodities and larger fluctuations in prices.

Nevertheless, the demands and trends in world markets will increasingly influence both the patterns of production and price expectations. On the other hand, the price could be the best incentive to give a strong boost to investment in agriculture as well as adoption of modern technologies and thereby to the raising of agricultural production and productivity. Similarly, the rise in domestic prices could put pressure on the public distribution system and accentuate the problem of food subsidy. The nature and character of state intervention and state support will have to undergo qualitative changes in order to not only realize the opportunities for exports, but also to cope with the implications of our agriculture coming into increasing alignment with the international marketplace.

With this background, the chapter attempts to highlight the following researchable issues:

1. How to widen India's trade base in international markets? (India has remained a marginal player in world agricultural trade despite being a big country and a major producer of several agricultural commodities.)
2. Is India able to influence the world market in terms of price, quantity, and quality?
3. Whether Indian farmers and exports benefit out of any rise in international prices or able to cope up the price shocks/volatility?
4. How to improve the farmers' knowledge and organizational capabilities to benefit from trade opportunities?

To address the above issues in the context of growing global competition, it is crucial to assess the trade competitiveness of Indian agricultural commodities in terms of price, quantity, and quality in comparison with major global players; and understanding the market integration and price transmission between markets so as to build the capacity of Indian farmers who lack knowledge and technology, financial and organization capacity in order to gain from trade opportunities.

OBJECTIVES

Broadly, the present study has been planned to cover four major objectives:

1. Understanding the structure of India's major agricultural exports and imports compared with major global players
2. Assessing trade competitiveness of selected agricultural commodities in terms of price, quantity, and quality
3. Examining the price realization of farmers and assessing the influence of price changes on domestic and international markets
4. Understanding the existing institutional support and suggestions for building the capacity of farmers for better price realization and improving competitiveness

DATA AND METHODOLOGY

The issue of price linkages in product markets both at local and international levels has been studied in the literature extensively either under the notion of the law of one price (Ardeni 1989; Protopapadakis and Stoll 1983, 1986) or under the notion of market integration (Baulch 1997; Gardner and Brooks 1994; Ravallion 1986). Integration between markets, either national or international, is one of the important phenomena which needs to be checked through co-integration tests. Price transmission is expected to take place between integrated markets (domestic–domestic or international–domestic) over short- and long-term basis. Transmission of changes in world prices into various domestic prices of agricultural commodities can be estimated through a Vector Error Correction Model (VECM) framework and estimation of long-run as well as short-run impacts on prices are also possible. To accomplish this, various prices, namely, international prices (reference market prices), country's export and import prices, wholesale prices, and producer prices were collected from major regulated markets in the country and various other sources.

Vector Error Correction Model

On the basis of the properties, the test for unit roots was also applied to the residuals of the static regression between each pair of prices, in order to test for co-integration following the Engle and Granger (1987) procedure. Where co-integration arose, a set of Auto Regressive Distributed Lag (ARDL) models were specified and estimated as follows:

$$pd_t = a + \tau T + \sum_{j=1}^{J} \beta_j pd_{t-j} + \sum_{k=0}^{K} \gamma_k pw_{t-k} + e_t \quad \text{Equation 1}$$

where, *pd* are the countries' (logarithm of the) import unit values in time *t*, *pw* is the (log) world reference price, *a* is an intercept, *T* is a time trend, *e* is the error term, and *t* is the period index. Where the null of absence of co-integration is rejected in the Engle and Granger (1987) procedure, the adjustment taking place around the long-run equilibrium can be modelled through an Error Correction (ECM) specification, such as:

$$\Delta pd_t = a + \delta T + \rho[pd_{t-1} - \lambda_1 pw_{t-1}] + \sum_{j=1}^{J} \beta_j^* \Delta pd_{t-j} + \sum_{k=0}^{K} \gamma_j^* \Delta pw_{t-k} + h_t \quad \text{Equation 2}$$

in which the coefficient = (1 − Σ) $j\ \rho\ \beta$ usually named 'ECM coefficient', indicates the short-run adjustment of prices toward the long-run equilibrium, and λ_1 is the same as the one calculated from the ARDL model in Equation 1.

Results reported here include for each commodity the parameters and the *t* statistics for the long-run equilibrium, together with the results of the estimation of the corresponding ECM specifications. In order to test for Granger non-causality between the pairs of prices, Equation 1 and its reverse form have been estimated by dropping the contemporaneous coefficients, according to:

$$pd_t = a + \tau T + \sum_{j=1}^{J} \beta_j pd_{t-j} + \sum_{k=1}^{K} \gamma_k pw_{t-k} + e_t$$

$$pw_t = a + \tau T + \sum_{j=1}^{J} \beta_j' pd_{t-j} + \sum_{k=1}^{K} \gamma_k' pw_{t-k} + z_t \quad \text{Equation 3}$$

Both equations were tested for $\gamma_k \beta_j\ \gamma_k'\ \beta_j'$ significantly different from zero for any *j*, *k*. Acceptance of the null implies that past values of the series on the right hand side are not adding information on the actual values of the series on the left hand side, on top of what is provided by its own past values. If this happens in both equations, then neither of the two series is Granger-causing the other, while if the null can be rejected in one of them, the price appearing on the left hand side will be Granger-causing the other. Given that a co-integrating relation must exist between the two series involved if Granger non-causality is rejected in at least one of the two equations, this test has been used here first, as a confirmation of the test for the long-run equilibrium; second, to understand which of the two prices acts as a source of information for the other; and third, to gain qualitative elements to understand the results, in terms of the causality direction. Rejection of the null in both the equations is to be considered as indicating a model misspecification or incompleteness, as it implies that both series are being Granger-caused by some third unknown variable. This test was performed, on monthly data, for those pairs of prices showing the presence of long-run equilibrium.

Assessment of the change in trade pattern was analysed using the time series data on quantity and value of exports and imports of various principal agricultural commodities. Secondary data was obtained from *Agricultural Statistics at a Glance*, web data sources like Food and Agriculture Organization of the United Nations (FAO), Centre for Monitoring Indian Economy (CMIE), and *IndiaTrade* for the periods between 1985–86 and 2008–09. Secondary data was also collected to examine the trend/pattern of trade during the above mentioned years and calculated the annual growth rates. Commodity-wise comparison between India and the major global players was done using the strengths, weaknesses, opportunities, and threats (SWOT) framework.

Overall, the study broadly covered eight commodities under six categorical groups. They are paddy (foodgrains), groundnut and castor (oilseeds), sugarcane (sugar crops), cumin (spices), cashew nut and tea (plantation crops), and mango and its pulp (fruits). For the selected agricultural commodities, data on cost of production and output prices was collected for the selected region/state from different sources which could supplement primary data collected from farmers and traders in the respective domestic markets. Value of tradable and non-tradable inputs and outputs was used to assess the trade competitiveness of selected commodities using the Domestic Resource Cost Ratio (DRCR) methodology. Further, the Extended DRCR framework was also adopted to address environmental issues. Secondary data on quantity and quality were collected for the selected commodities to assess the robustness of trade and quality competitiveness.

RESULTS AND DISCUSSION

The broad commodity-wise results of the study have been summarized in Table 39.1.

Despite the degree of competitiveness, export and import of rice is largely influenced by policy (export ban on non-basmati rice) and participation of large export firms. Fluctuations in world prices, especially export price is found to influence the producer price and vice-versa. Both producer and export prices were influencing each other; the

Table 39.1 Trade Competitiveness of Different Agricultural Commodities (Under Exportable Hypothesis)

Commodity	NPC	EPC	ESC	DRCR
Basmati rice	0.91	0.89	0.90	0.32
Groundnut	1.00	1.01	1.07	0.03
Castor	1.11	1.19	1.21	0.09
Sugar*	1.28	–	–	0.807
Cumin	0.80	0.70	0.71	−0.08
Cashewnut	1.21	1.35	1.39	−0.13
Tea	N.A	N.A	N.A	N.A
Mango	N.A	N.A	N.A	N.A

* 1995 estimates.
Note: NPC = nominal protection coefficient; EPC = effective protection coefficient; and ESC = effective subsidy coefficient.
Source: Authors' computation.

significance of influence of producer price on export price was found high (see Table 39.2). Even though the measures of nominal protection coefficient (NPC), effective protection coefficient (EPC), and effective subsidy coefficient (ESC) are close to one, the estimated DRCR is as low as 0.32; which indicates that export of Indian basmati rice is much more competitive. Export firms aim to bring quality produce (organic or untraceable level of chemicals) and they promote such activities by providing technical and input support to farmers. Linking rice farmers with these firms would benefit both the parties. Basmati rice is a special product and its export is not affected by price advantage or quality.

India is the second largest producer groundnut in the world, which is the most produced oilseed in India. The trade competitiveness of groundnut has been declining over the past six-year period. But, it is still competitive in the international market. The export prices are influencing the producer prices and the opposite was not significant (see Table 39.3). International prices are slightly influenced by our major wholesale markets in the country. All

Table 39.2 Granger-Causality Wald Test for Rice Long-term Price Integration

Dependent	Independent	Chi-Square	Pr > ChiSq
Producer Price	Export Price	6.65	0.0360
Export Price	Producer Price	26.03	<0.0001
Producer Price	Import Price	5.73	0.0570
Import Price	Producer Price	2.96	0.2281
Export Price	Import Price	3.96	0.1380
Import Price	Export Price	0.92	0.6301

Source: Authors' computation.

Table 39.3 Granger-Causality Wald Test for Cumin Long-term Price Integration

Dependant	Independent	Chi-Square	Pr > ChiSq
International price	Junagadh price	16.07	0.0003
Junagadh price	International price	4.35	0.1135
International price	Gondel price	28.76	<0.0001
Gondel price	International price	1.41	0.4951
International price	Rajkot price	19.93	<0.0001
Rajkot price	International price	0.36	0.8365
Junagadh price	Gondel price	24.06	<0.0001
Gondel price	Junagadh price	0.75	0.6884
Junagadh price	Rajkot price	30.27	<0.0001
Rajkot price	Junagadh price	21.59	<0.0001
Gondel price	Rajkot price	17.80	0.0001
Rajkot price	Gondel price	49.36	<0.0001

Source: Authors' computation.

the three major domestic markets (Junagadh, Rajkot, and Gondel) are well integrated.

India is the largest producer of castor in the world. India faces stiff competition from exporting countries, primarily from Brazil and China. At the moment the NPC value of castor was more than one. We are less competitive in the international market under exportable hypothesis. But, still India has a strong comparative advantage in the production of castor beans. Just like groundnut, export prices are influencing the producer prices and the opposite was not significant. Our domestic whole markets like Mumbai, Rajkot, and Unjha have a little influence on the international market prices. All major markets in Gujarat (Junagadh, Rajkot, Gondel, and Unjha) showed significant integration in price transmission between them.

Sugar is highly a controlled commodity and prices showed total disintegration among them. Policy support towards increasing the efficiency of the sugar mills is the immediate need and strategies of successful sugar mills (such as Warnanagar sugar mill) should be replicated. It is necessity for a dramatic change in sugar policy and efficiency promotion is the need of this hour.

India is the largest producer of cumin in the world and accounts for 70 per cent of world's production followed by Syria, Iran, and Turkey. India also consumes 66 per cent of the total world production. Production of cumin is highly competitive in the international market. The calculated NPC value was less than one under exportable hypothesis. Our major wholesale markets (Unjha, Rajkot, and Gondel) were able to slightly influence the international prices. But, the opposite was not significant (see Table 39.4). The price

Table 39.4 Granger-Causality Wald Test for Cumin Long-term Price Integration

Dependant	Independent	Chi-Square	Pr > ChiSq
International price	Mumbai price	7.83	0.0200
Mumbai price	International price	3.44	0.1794
International price	Unjha price	6.90	0.0317
Unjha price	International price	0.82	0.6622
International price	Rajkot price	9.77	0.0076
Rajkot price	International price	5.59	0.0611
International price	Gondel price	9.90	0.0071
Gondel price	International price	4.93	0.0849
International price	Junagadh price	5.71	0.0576
Junagadh price	International price	1.48	0.4771
Mumbai price	Unjha price	0.78	0.6781
Unjha price	Mumbai price	15.43	0.0004
Mumbai price	Rajkot price	2.27	0.3218
Rajkot price	Mumbai price	0.26	0.8789
Mumbai price	Gondel price	6.71	0.0349
Gondel price	Mumbai price	0.84	0.6574
Mumbai price	Junagadh price	0.56	0.7542
Junagadh price	Mumbai price	7.30	0.0260
Unjha price	Rajkot price	16.34	0.0003
Rajkot price	Unjha price	1.58	0.4546
Unjha price	Gondel price	26.48	<.0001
Gondel price	Unjha price	0.15	0.9300
Unjha price	Junagadh price	21.24	<.0001
Junagadh price	Unjha price	1.24	0.5392
Rajkot price	Gondel price	6.57	0.0375
Gondel price	Rajkot price	3.38	0.1849
Rajkot price	Junagadh price	3.03	0.2198
Junagadh price	Rajkot price	10.68	0.0048
Gondel price	Junagadh price	0.54	0.7627
Junagadh price	Gondel price	6.48	0.0392

Source: Authors' computation.

transmission mechanism is functioning well among the all domestic markets along with the Mumbai market.

Cashewnut is the important plantation crop in India. India is the second largest producer of cashewnuts in the world after Vietnam. Brazil, Nigeria, Tanzania, Indonesia, and Mozambique are other major producers in the world. The degree of competitiveness in cashewnut was low under importable hypothesis for raw cashewnuts. India has strong competitive and comparative advantage in case of cashewnut kernel exports in the international market. Our producer prices are independent with import prices. The domestic wholesale market prices were not integrated with international prices. However, significant integration was observed among domestic markets (Kerala, Andhra Pradesh, Goa, and Karnataka).

Tea producers in the southern part of the country face problem of less price realization and a little share in the consumer rupee. Farmers' price realization is only 10–12 per cent of the market price. Various blends of tea may be introduced to win the consumers choice. Export of tea from India is declining over the years due to competitors like Sri Lanka. However, introduction of new blends, tea products (ice tea, lemon tea, etc.) and promotional measures are needed to increase/keep India's market share in the world market. Geographical indication like 'Darjeeling Tea' would help to boost the performance of the Indian tea sector. Encouraging domestic consumption of tea through tie-ups with Amul/other dairy outlets may also be attempted.

Supply of mango is largely influenced by the bearing habit of the (variety) trees and climatic fluctuations. Existing processing units provide a strong forward linkage to the mango growers and price fluctuations are also brought under control. The Gandevi Model may be replicated for establishing strong forward linkage to the mango producers in the state/country. The Kesar variety is highly preferable for export both as fruit and pulp, not only because of the price competitiveness but also due to its taste and consumers' preference. The Hazard Analysis Critical Control Point (HACCP) system may be advocated to the processing units in order to get premium prices and to have a significant share in European markets.

CONCLUSIONS AND POLICY IMPLICATIONS

Overall, the study brought out the following conclusions:

1. Growing internal demand due to increase in population and income. India will have to strive hard to meet its domestic consumption. Among the different commodities analysed, India has trade competitiveness only in the case of cumin followed by groundnut. All the eight commodities covered in the study exhibited huge domestic demands. Therefore, the Government of India (GoI) has to develop a comprehensive plan and strategy while dealing with the export of these commodities to international markets.
2. The price transmission analysis concluded that major wholesale markets influence international prices to a small extent. However, integration between the domestic markets was observed only in a few commodities. Modernization of the agricultural marketing

system and introduction of information and communications technology (ICT) will fill up these gaps and enhance the price transmission process.

3. The real impact of future markets/future trading needs to be analysed critically. Many traders, processors, and farmers opined high negative perceptions about that system.
4. Huge fluctuations in currency exchange between the rupee and dollar create a lot of impact on the export business. There should be some short-term stability mechanism for a certain period of transactions (at least a week) for smoothening of the agricultural trade.
5. Concerted efforts would be needed to increase the production through productivity enhancement technologies as well as post-harvest handling of commodities. All the commodities covered in the study showed that the productivity levels in India are far below when compared with key global players.
6. Protection or insulation in many developed markets remains high and allowable export subsidies and domestic support still threaten the stability of agricultural markets in the developing countries like India. This issue needs to be addressed well in forthcoming deliberations of the WTO.
7. High tariffs and other non-tariff barriers did not allow us to enter developed country markets. It has clear impacts on populations of developing countries whose dependency on agriculture is very high. These issues should be resolved in future negotiations.
8. Commodity-wise improved and safe package of practices, networking of processing facilities, creation of supply and value chains, and quality testing labs should be developed to boost the exports further in that specific region/state.
9. Capacity building of farmers in production and processing aspects, improved extension communication system, and well-integrated marketing system will play a key role in the promotion of agricultural exports.
10. Lack of proper transportation is an important market constraint that effects market integration. The existing infrastructure is highly inadequate, outdated, and inefficient. Therefore, there is a need to introduce an integrated system of bulk handling and transportation of agricultural commodities.
11. An important bottleneck is the lack sufficient storage and warehousing facilities, especially cold storage for perishable commodities. Reforms are needed in the functioning of Agricultural Produce and Marketing Committee (APMC) markets to hasten up the process of marketing and trading.
12. The flow of real time information across various markets that are interlinked throughout the country will create tremendous demand and efficiency in the system.

REFERENCES

Ardeni, P.G. 1989. 'Does the Law of One Price Really Hold for Commodity Prices?', *American Journal of Agricultural Economics*, 71(3): 661–9.

Baulch, B. 1997. 'Transfer Costs, Spatial Arbitrage, and Testing for Food Market Integration', *American Journal of Agricultural Economics*, 79(2): 477–87.

Charyulu, D. Kumara, and M. Prahadeeswaran. 2013. *Agricultural Market Integration and Trade Competitiveness of Indian Agriculture*. CMA Publication no. 243. New Delhi: Allied Publishers.

Engle, R.F., and C.W.J. Granger. 1987. 'Cointegration and Error Correction: Representation, Estimation and Testing', *Econometrica*, 55(2): 251–76.

Gardner, B., and K.M. Brooks. 1994. 'Food Prices and Market Integration in Russia: 1992–1993', *American Journal of Agricultural Economics*, 76(3): 641–6.

Protopapadakis, A.A., and H.R. Stoll. 1983. 'Spot and Futures Prices and the Law of One Price'. *Journal of Finance* 38(5): 1431–55.

———. 1986. 'Some Empirical Evidence on Commodity Arbitrage and the Law of One Price', *Journal of International Money and Finance*, 8(2): 341–51.

Ravallion, M. 1986. 'Testing Market Integration', *American Journal of Agricultural Economics*, 68(1): 102–9.

40

Market Access and Constraints in Marketing of Goats and Their Products

Ramendu Roy and D.K. Singh

Agriculture alone cannot provide sufficient income and employment to the ever-growing population of the country. At present, a majority of the farmers of the country are not willing to remain in agriculture because it is not a profitable business as compared to other occupations. The suicide reports among the farmers have become a cause of alarm. In order to improve the income of farmers, there is a need for diversification in agriculture. The allied activities such as livestock, poultry, fishing, bee keeping, etc., have sufficient potential to generate adequate income and employment for target groups among the weaker sections of the society.

Animal husbandry supports the farmers at time of failure of crops. Agriculture and animal husbandry are complementary and supplementary to each other. Livestock not only provide milk, manure, meat, skin, etc., but it is also source of energy. Since the availability of cultivated land has been decreasing rapidly, therefore, keeping livestock has become necessary to compensate for the decreasing income from crops. In order to get regular income, the farmers belonging to economically weaker section of the society, keep the different species of livestock. The medium and large farmers prefer to keep the cows and buffaloes while the marginal and small farmers and landless always prefer to keep the goats, sheep, poultry, and others. Among livestock, goats require very low investment and their maintenance costs are also very low. Goats are usually associated with the poorest of the poor. Goats mostly depend on shrubs and trees, reared in different agro-climatic conditions and are also used in ceremonial feasting. On account of these, goats are among the most popular domesticated animals. Goat meat is widely consumed. The skin of a goat is a basic raw material for tanneries. Since the goat has a short gestation period, it generates adequate returns in a short period. They also require less stall feeding than other livestock.

In spite of these benefits, poor access to veterinary support, inadequate availability of feed and fodder, less availability of space, etc., are basic reasons for higher mortalities in the case of goats. Goats are still reared in the traditional way along with other occupations. Indigenous breeds of goats are kept in most part of the country. On account of these, the productivity of goat products is low in comparison to China. It does not have a commercialized status in most parts of the country. The rearing of goats, in fact, provides livelihood and subsistence to millions of small, marginal farmers and landless across the country. The goats are kept for meat, kid's fibre, and milk production (Kumar 2003; CLE 2008). Goats require less investment and less maintenance cost. Goats are recognized as a liquid asset because they generate reasonable income in a short period of time. Now-a-days, the demand for goat products is significantly increasing in domestic markets and side by side the export demand is also gaining momentum. Most of the livestock markets have neither any basic amenities nor infrastructural facilities. The weight of goats is measured by physical assessment. In most of the cases, the price of a goat is fixed whatever the traders want. There is much harassment and exploitation of goat sellers during the marketing of goats in the unorganized markets. The un-remunerative price of goats followed by delayed payment are also important constraints in marketing of goats. The marketing channels are also very wide and complicated. The structure of markets is mostly unsystematic and unscientific (Thomsen, 1951, Dixit, 2006). The goats are mostly sold in livestock markets (Mondal and Sinha, 2009, Sharma, 2009, Thaware, 2009). A limited number of exclusive goat markets are available across the country. On account of these, the sale of goats take place at unfavourable places, unfavourable times, and unfavourable terms. The entire process of the goat market is biased in favour of traders/butchers. This leads to goat keepers to sell their goats at disadvantageous price. In order to learn the ways and means to improve market access of live goats and their products in Uttar Pradesh (UP), Madhya Pradesh (MP), West Bengal (WB), and Maharashtra, the present study was undertaken. The following objectives were framed for this study:

1. to study the goat marketing system and marketing access to goat farmers;
2. to ascertain the constraints in marketing of goats;
3. to identify leverage points for developing a system ensuring fair marketing access and prices of goats;
4. to study the marketing system of goat products; and
5. to suggest suitable development strategies for the efficient marketing of goats and their products.

METHODOLOGY AND DATA

Keeping in view the objectives of the study, a multistage sampling technique was adopted for the selection of agro-climatic regions, districts, markets, villages, and goat-keepers by the four participating centres from their respective states. First, each participating centre identified two agro-climatic regions on the basis of maximum concentration of population of goats from the state. From each region, one district having the maximum number of goats was selected by each centre. From each district, two important goats/livestock markets were selected purposively. Two villages, one in the vicinity of the market and other far from the markets were selected on the basis of availability of sufficient number of goat-keepers in the villages. A list of goat-keepers was prepared and classified into four groups, namely 1–5 goats, 6–15 goats, 16–30 goats, and above 30 goats. From this list, 50 goat-keepers were selected from four villages of two markets of a district on the basis of probability proportion to number in the groups by Allahabad, Jabalpur, and Santiniketan centres while the Pune centre selected 52 sample goat-keepers instead of 50 sample goat-keepers by adopting the same selection procedures, thus, a total number of 402 goat keepers from eight districts of four states comprised the sample size (Table 40.1).

In order to know the marketing system, 20 sellers and 20 buyers were also selected from four markets of two districts. Thus, in all 80 sellers and 80 buyers from 16 markets of four states constituted the sample for the study. Apart from these, the information regarding goat products (meat and skin) were also required. For this purpose, each participating centre had also selected 20 butchers, 20 petty skin traders, and eight wholesale skin merchants from four selected markets of their respective state. Thus, 80 butchers, 80 petty skin traders, and 32 wholesale skin traders formed the sample for the study. The reference year for the study was from July 2007 to June 2008.

RESULTS AND DISCUSSION

General Findings Based on Secondary Information

The population of goats has been continuously increasing from 1950 to 2005 in four selected states. The growth rate in goat population was also positive in the selected states during the study period. Goat products, such as milk, dung, meat, skin, etc., were also showed an increasing trend across the selected states. The percentage share of goat population to total livestock was highest being 34.03 per cent in WB, followed by 28.73 per cent, 22.56 per cent, and 22.11 per cent in Maharashtra, MP, and UP respectively in 2003. The contribution of goat milk in the total production of milk was the highest, being 40.13 per cent, in MP while it was only 6.28 per cent and 4.71 per cent in UP and Maharashtra respectively during 2005–06. The price of goat milk was around Rs 10 per litre in the selected states in 2007–08.

The production of goat meat was more in MP and WB than that of other livestock during 2005–06, while the production of buffalo meat was highest in UP in comparison to goat meat during the corresponding year. The growth rate in production of goat meat has witnessed a positive trend in the selected states. The skin markets were totally governed by the tanneries in UP. The width and length

Table 40.1 Details of Selected Goat-keepers from Villages of UP, MP, WB, and Mahrashtra

Name of the States	Name of the Selected Regions	Name of Districts	Name of Markets		Nos. of Selected Villages		Selected Samples (Nos.)		Total Sample
			I	II	Market (I)	Market (II)	Near the Vicinity the Market	Off the Market	
UP	Western	Etah	Padawa	Dholna	2	2	25	25	50
	Eastern	Bahraich	Bahraich Sadar	Nonpara	2	2	25	25	50
Sub Total	**2**	**2**	**2**	**2**	**4**	**4**	**50**	**50**	**100**
MP	Malwa	Dhar	Kukshi	Dahi	2	2	25	25	50
	Kymore	Sidhi	Waidhan	Deosar	2	2	25	25	50
Sub Total	**2**	**2**	**2**	**2**	**4**	**4**	**50**	**50**	**100**
WB	Rarh	Burdwan	Natunhat	Gueskora	2	2	25	25	50
	Gangatic Plain	Murshidabad	Dakbangla	Baldanga	2	2	25	25	50
Sub Total	**2**	**2**	**2**	**2**	**4**	**4**	**50**	**50**	**100**
Maharashtra	Vidarbh	Yavatmal	Darvha	Pusad	2	2	26	26	52
	Western Maharashtra	Ahemad Nagar	Supa	Kashti	2	2	26	26	52
Sub Total	**2**	**2**	**2**	**2**	**4**	**4**	**52**	**52**	**104**
All	**8**	**8**	**8**	**8**	**16**	**16**	**202**	**202**	**404**
			16		**32**				

Source: Field survey.

of skins were the main criteria in the fixation of its price. The infrastructure and other facilities of slaughterhouses were extremely poor in the selected markets of the study areas. The goats were slaughtered in open places, residential premises, and shops. Most of slaughterhouses of UP were ill maintained and also found in very deplorable conditions. The number of slaughterhouses were inadequate in the four selected states. The livestock markets of UP and WB were unorganized while these were mostly regulated in Maharashtra.

It was also governed under Agricultural Produce Market Committee Regulation Act (APMC Act). The basic amenities were also absent in goat markets of UP, while these were fully available in the markets of Maharashtra.

Findings Based on Primary Information

Of the total 404 selected goat-keepers, the marginal farmers and landless were main goat-keepers of the sampled farms across the four selected states (Table 40.2).

The other backward classes (OBC) and scheduled caste/ scheduled tribe (SC/ST) castes were the main sampled goat-keepers in the study areas. The sample goat-keepers of Maharashtra and MP had kept more goats on their farms than that of UP and WB. The sample goat-keepers of UP and WB had kept a limited number of goats on their farms ranging between one to five goats. Agriculture, dairy, goat rearing, and labour were main occupations of sample goat-keepers of the study areas. There was no exclusive occupation of goat rearing of the majority of sampled goat-keepers of four selected states. The number of goats on the sample farms of four states was higher than other livestock. The population of female goats and kids was higher than

Table 40.2 Distribution of Sampled Households by Ownership of Land (No.)

Size class of Operated Land (hect.)	States				
	UP	MP	WB	Maharashtra	Total
Landless	18 (18.00)	7 (7.00)	60 (60.00)	41 (39.42)	126 (31.19)
Up to 100	59 (59.00)	37 (37.00)	33 (33.00)	30 (28.85)	159 (39.36)
1–2	15 (15.00)	24 (24.00)	7 (7.00)	15 (14.42)	61 (15.10)
2–4	5 (5.00)	23 (23.00)	–	15 (14.42)	43 (10.64)
Above 4	3 (3.00)	9 (9.00)	–	3 (2.89)	15 (3.71)
Total	100 (100.00)	100 (100.00)	100 (100.00)	104 (100.00)	404 (100.00)

Note: Figures in brackets are percentage of total.
Source: Field survey.

the population of male goats on the sample farms. Male goats were generally kept only for two years by sample goat-keepers. The goat-keeping had generated employment for 134 days per annum to the sample goat-keepers. The females and children were more involved in rearing of goats than the male members. Agriculture followed by labour and goat-rearing were main sources of income of sampled goat-keepers. The per household income of sampled goat-keepers of Maharashtra was highest being Rs 43,096 per annum followed by Rs 36,537, Rs 36,098 and Rs 24,265 on the sampled farms of UP, WB, and MP respectively. The sampled goat-keepers of Maharashtra were much economically well off than their counterparts of the selected states. The sampled goat-keepers of UP were more conscious about keeping different types of breeds of goats on their farms than the sampled goat-keepers of other selected states. The traditional way of goat-rearing practice was more prevalent on the sampled farms across the study areas. The 'Breed Improved Programme' was not actively launched in the study areas. An extensive production system of rearing of goats was prevalent across sampled farms.

Marketing of Live Goats

Of the total number of 5,493 goats, only 43.49 per cent goats were sold by sampled goat-keepers of the study areas during the reference year. Of the total 2,389 goats sold off by 404 sample goat-keepers, 57.35 per cent goats were sold at doorsteps against 42.65 per cent in the markets. The sample goat-keepers of Maharashtra and UP sold the maximum number of goats at the door than in the markets which was found to be the reverse in MP and WB (Table 40.3). The landless and marginal sampled goat-keepers had sold the maximum number of goats at the door rather than in the market. Large sampled goat-keepers had better market access as compared to other categories of farmers. The distance of market from village played a significant role in the choice of marketing place of goats. The size of farms did not play significant role in the choice of marketing place. The sale of goats at the door or in the markets did not have a positive relation with size of flocks. It differs from state to state which depends on the convenience of goat-keepers at particular period and time.

Table 40.3 Marketing of Goats at the Door and in the Market (Nos.)

State	Population of Goats	Sold		
		At Door	In the Market	Total Sold
UP	756	185 (51.97)	171 (48.03)	356 (100.00)
MP	1872	226 (36.87)	387 (63.13)	613 (100.00)
WB	1083	73 (32.88)	149 (67.12)	222 (100.00)
Maharashtra	1782	886 (73.96)	312 (26.04)	1198 (100.00)
All	5493	1370 (57.35)	1019 (42.65)	2389 (100.00)

Note: Figures in brackets are percentages to total.
Source: Field survey.

Out of total goats sold, 57.35 per cent goats were sold at the door through the following channels:

1. Goat-keepers → Goat-rearers
   Goat-keepers → Local Consumers
2. Goat-keepers → Butchers
3. Goat-keepers → Traders → Wholesalers → Butchers/Consumers
4. Goat-keepers → Butchers → Traders → Retailers → Butchers/Consumers

Among the buyers, the butchers and professional traders were much involved in purchasing of goats at the door as well as in the markets of UP, MP, and WB, while in Maharashtra, butchers and local consumers purchased the maximum number of goats at the door (see Figure 40.1). The type of buyers of goats differs from state to state. The demand, supply, and price of goats on a particular market day also influenced the choice of buyers with regard to sex-wise purchase of goats. The average value of a goat sold at the door was Rs 2,718, Rs 1,794, and Rs 1,326 of a male, female, and kid respectively on the selected farms of Maharashtra followed by Rs 2,697, Rs 1,759, and Rs 870 for a male, female, and kid respectively on the sample farms of UP The average price of a male, female, and kid goat was Rs 907, Rs 592, and Rs 389, respectively on the selected farms of WB, which was very low in comparison to UP and Maharashtra states. This was probably due to low weight of maximum number of goats sold. The price of a goat sold in the market was Rs 2,614, Rs 2,237, and Rs 712 for a male, female, and kid, respectively in the selected markets of UP. The main reasons for sale of the live goats at the door were dominance of traders in the markets, less bargaining capacity of the seller of goats and long distance from home to market. Net gain was more for villages near the market than off the market. There were no separate goat markets in the selected districts of WB and Maharashtra. The goats were sold and purchased in livestock/general markets along with other livestock. The variation in price of different sex of goats in the selected

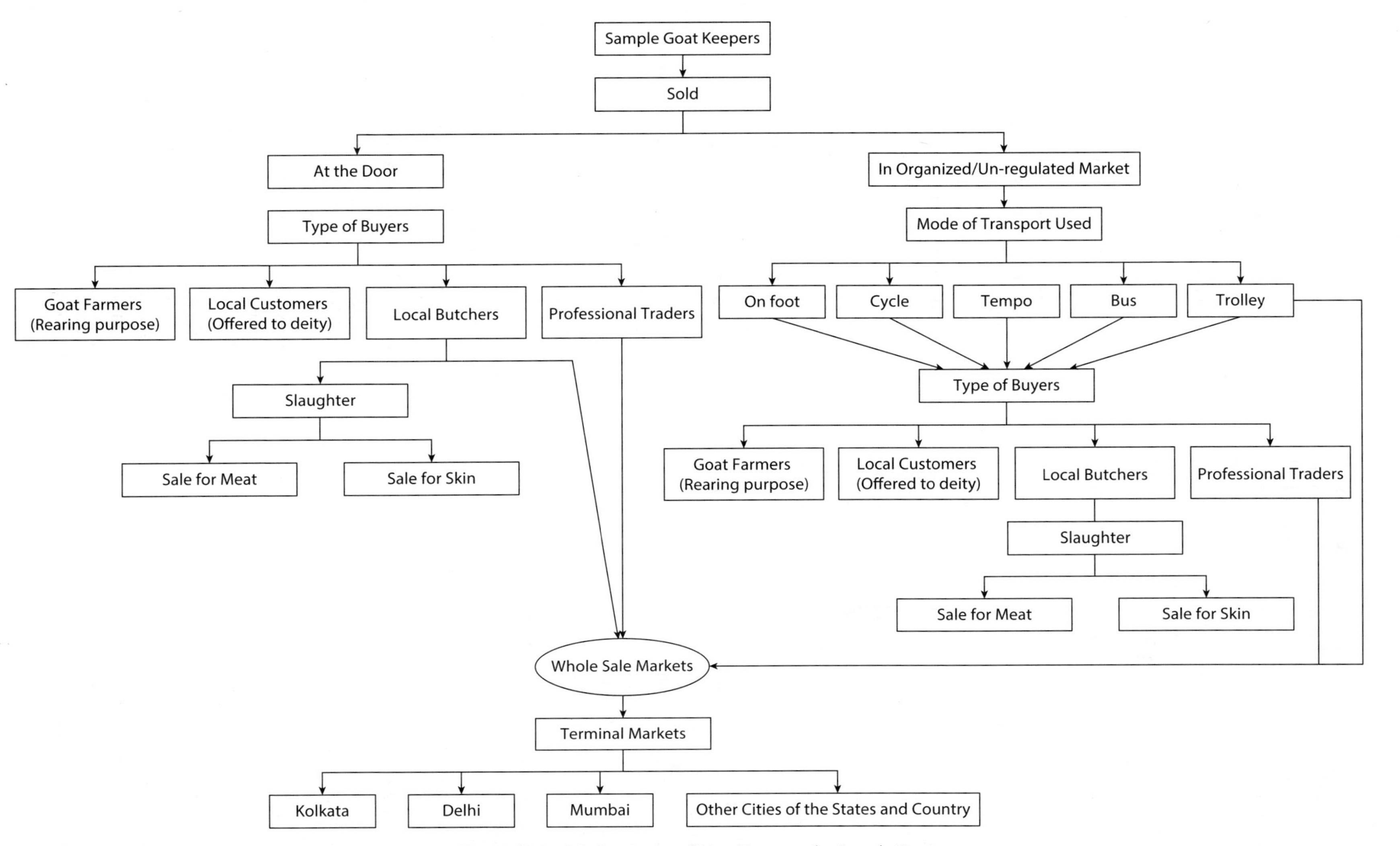

Figure 40.1 Market Access of Live Goats on the Sample Farms

Source: Generated by Author.

markets of UP, MP, and Maharashtra was not so wide. It was above Rs 2,000 per male goat across the selected markets of UP, MP, and Maharashtra. The only exception was in WB which was below Rs 1,000. The average price of all types of goats was higher in the markets of UP, MP, and WB than the price at the door while it was the reverse in Maharashtra. The goats sold at the door do not incur any type of marketing costs and no exploitation occurs. The goat-keepers belonging to villages near the markets had better market access than those far from the markets. The maximum sale of goats at the door was noticed in villages that were situated far from the markets.

Marketing of Goat Products

Among the goat products, milk and dung were marketed by goat-keepers themselves while meat and skins were sold by butchers. The average production of goat milk per annum was worked out to be 95.96 kg, 50.59 kg, and 31.61 kg for MP, Maharashtra, and UP, respectively. Per litre price of goat milk was estimated at Rs 10 across the selected areas of the study. Out of total production of goat milk, 62 per cent was consumed by sampled goat-keepers themselves and the remaining 38 per cent was sold. The sample goat-keepers of MP sold about 50 per cent of total production of goat milk. The butchers are one of the most important functionaries in the channel of the goat marketing system. The 75 selected butchers from 16 markets of four states had purchased 29914 goats in the reference year of which 97.06 per cent were slaughtered during the same year. The average production of meat of slaughtered goat was estimated at 14.06 kg, 9.82 kg, 9.82 kg, and 6.7 kg for MP, UP, Maharashtra, and WB, respectively. The value of goat meat per kg was worked out to Rs 181 in West Bengal followed by Rs 163 per kg in Maharashtra, while it was only Rs 139 and 133 per kg in UP and MP, respectively. The meat was sold to local consumers and hotels. Out of total production of goat meat of 362,184 kg, the maximum share being 84.68 per cent was purchased by local consumers and remaining was supplied to local hotels. Two marketing channels were involved in marketing of meat in the study areas, namely:

1. Butchers → Consumers
2. Butchers → Hotels

The butchers also earned handsome income by selling the skin of slaughtered goats. The highest price of a skin being Rs 64 in Maharashtra followed by Rs 61 in MP. The price of skin depends on its size, colour, softness, etc. The skin markets in UP were totally managed by private contractors. There was a total monopoly of the processors in the skin markets of UP.

Constraints in Marketing of Live Goats and their Products

Constraints Expressed by Goat-keepers

The market infrastructural facilities were not up to the mark in the selected markets of UP, WB, and MP while it was much better in the selected markets of Maharashtra. Apart from this, basic amenities such as water, electricity, rest rooms, boundary walls, platforms, etc., were also absent in selected markets of UP and WB, while these were available in the selected markets of Maharashtra. The un-remunerative price of goats and delay in payment were common phenomena across the selected markets of selected states. The exploitation and foul practices were much prevalent in the unorganized livestock markets of UP and WB. Market intelligence services in the livestock markets of UP, MP, and WB were not in operation. The price of goats was fixed with mutual negotiation. A scientific weighing process was not adopted in the markets. The visual assessment was only recognized as criterion in the buying and selling of goats. This type of tendency was prevailing more or less across the selected markets of the four states. The price chart was also not displayed in the selected markets of UP and WB. The goat-keepers do not have direct access to the consumers. A number of hands are involved before it reaches the ultimate consumers. The marketing system of goats was totally dominated by butchers and professional traders across the selected markets of four states. The attitude of owners of livestock markets of UP was much favourable towards the buyers of goats than actual seller of goats. Since the majority of sample goat-keepers were poor, illiterate, and innocent, therefore, they had poor access to banks to get adequate loans to carry on goat farming in a better way. The majority of the sampled goat-keepers of UP, WB, and MP were not satisfied with the availability of loans from financial institutions. The informal financial institutions have still a strong hold at the grass-root level. The majority of the sampled goat-keepers were not satisfied with prevailing rate of interest of loans. The goat-keepers emphasized that it should not be more than 4 per cent per annum.

The professional buyers of goats faced a lot of problems during transit of goats from primary/secondary markets to terminal markets. The buyers had to make unauthorized payments at a number of places during the export of goats from primary markets to terminal markets. Apart from this, loaded trucks were unnecessarily held up on the route. On account of this, a number of loaded goats died in the trucks. This leads to a decrease in the margin of profit of buyers of UP, while the buyers of Maharashtra had no problem in marketing goats. Most of the selected buyers of Maharashtra were satisfied with fair dealing in purchasing and selling of goats in different markets of the state and out of the state.

Constraints Expressed by Butchers and Skin Traders

Almost all the selected butchers of the selected markets of four states had expressed their views that they had no proper slaughterhouses to slaughter the goats. The goats were slaughtered either in meat shops or in their homes. There were no modern abattoirs in the selected districts of UP to produce hygienic meat. Most of the selected butchers of UP were under clutch of moneylenders. They had poor access to banks. The support of veterinary doctors was not so helpful for butchers. Most of slaughterhouses of UP were ill maintained and found in deplorable conditions. Most of the butchers of selected areas were not equipped with better equipment to slaughter goats in a scientific manner. On account of this, the goat meat was not produced at global standards, particularly in UP.

The grade, size, colour, weight, etc., are considered in fixing the price of goat skin. Since the majority of skin traders were illiterate, therefore, they were not well acquainted with the quality of skin. They were exploited by wholesalers/commission agents of tanneries. The value of skin was also not properly estimated. The small traders had no adequate facility to retain skins up to a month which forced them to sell the goat skin at lower prices. Most of the sampled skin traders of UP were economically poor, hence they had taken loans from moneylenders/big skin traders at high rates of interest to meet their financial obligations. Hence, they were bound to sell the skin to moneylenders on their terms and conditions. Goat milk was not popularized among consumers. Most of production of goat milk was consumed by goat-keepers themselves. It was not commercialized in UP. Low demand of goat milk in the markets and cities was the main constraint in marketing of goat milk.

POLICY SUGGESTIONS

Small and marginal farmers and landless are very much involved in goat farming. Among the livestock, cows and buffaloes have been getting due attention while small ruminants such as goats, sheep, pigs, etc., have been mostly neglected. The rearing and marketing of live goats are still managed by individual farmers. Goat farming has not been taken as commercialized farming as poultry farming, bee-keeping, and fish farming. Therefore, there is a need to make an effort for sustain development of goats through better marketing with due returns. If this enterprise is managed properly, it could be a driving force in the overall development of goat-keepers. The goat-keepers only require a reasonable price for their goats. Goat-keeping is economically effective to those who rear goats in the rural areas of the country.

The goat-keepers face a number of difficulties from rearing to marketing of live goats. The marketing of goat products namely milk, dung, meat, skin, etc., are also not properly channelized. There are many hurdles in the way of its proper marketing. The question arises how we would overcome the difficulties and constraints to leverage the marketing efficiency in favour of goat-keepers and other market functionaries. The following suggestions are made on the basis of observations and information of primary and secondary sources of four participating centres. Since, the livestock markets of Maharashtra are governed by the APMC, Act 1963, therefore, the sampled goat-keepers of this state had not faced much difficulties and problems in marketing of goats in the markets. Most of the livestock markets of Maharashtra are also organized and well equipped with modern amenities; hence, they faced less exploitation and cheating in marketing of live goats as compared to goat sellers of UP, West Bengal, and MP states.

Suggestions for Betterment of Marketing of Goats

The state governments should give high priority for efficient road and communication networks across the length and breadth of the states. The elimination of multiple layers of intermediaries between goat-keepers and final consumers is the most important steps to realize better prices for goats. There is also a need to eliminate malpractices from livestock markets through the introduction of Emerging Marketing Channels. This would be helpful in improving the bargaining power of goat-keepers. The goat-keepers will definitely bring their goats to markets without any fear to get higher prices for goats rather than selling at the door. All the livestock/goat markets of UP and West Bengal are still unorganized/unregulated. Therefore, the state governments should take step to regulate the livestock markets to safeguard the interests of goat-keepers. These markets should be covered under APMC Act like it is in Maharashtra state. The goat-keepers should be motivated to form goat-keepers' societies/self-help groups in villages to reduce the length of the marketing channel. These societies will safeguard the interest of goat-keepers and also prevent the exploitation from different type of agents. This would be an efficient supply chain enabling them to significantly reduce the cost of delivery and get better returns on their goats. The infrastructural facilities, basic amenities, communication networks, etc., were very poor in unregulated as well as in regulated markets of livestock. These need to be modernized in the interest of sellers and buyers of goats. The goats/livestock markets are very limited in number. There was also a very limited exclusive goat markets in UP. Therefore, block-level markets should be established in potential areas to provide better market access to goat-keepers. The National Bank for Agriculture and Rural Development (NABARD) and Regional Rural Banks (RRBs) should devise policies to provide loans to goat-keepers as and when financial crises arise to avoid distress sale of goats. The present credit and delivery system is unsuitable for goat-keepers. It requires rejuvenation for poor goat-keepers. Weighing machines should be available in livestock markets to estimate the correct weight of a goat. The traditional practice of weight estimation by hand lifting should not be allowed in goat markets. Outsider buyers of goats should get all types of assistance and help in the goat markets and outside the markets. This would be helpful in the promotion of export of goats from primary markets to terminal markets.

Suggestions for Improvement of Better Production and Marketing of Goat Products

In order to get global standards of production of goat meat, modern slaughterhouses will have to be constructed in potential pockets of the country. The cleaning, washing, packing, refrigerating, transporting, etc., will have to be undertaken at par with international standards. These facilities will also facilitate the hygienic production of goat meat. The private investors should also be mobilized to establish modern slaughterhouses in potential pockets of the country. The state as well as central governments should provide adequate amounts to the district administration for the construction of modern slaughterhouses. The remodelling of old meat shops is also needed to produce hygienic goat meat. Nationalized banks should provide loans to butchers to construct modern meat shops, only then can mini cold storage facilities be developed to preserve the unsold meat. The state governments should not allow the butchers to slaughter goats at unauthorized and unhygienic places. The butchers should get all types of modern infrastructural facilities to produce goat meat at global standards. This would facilitate exports of huge quantities of goat meat from potential areas of the country.

Suggestions for the Betterment of the Skin Trade

The goatskin is the basic raw material for tanneries. The country gets a considerable amount of foreign exchange every year through the export of raw skin and its processed goods. Butchers are main suppliers of goatskin to tanneries but they do not get reasonable value of goatskin from processors/wholesalers. Therefore, it is suggested that the credit delivery should be prompt and adequate for petty skin traders to avoid distress sale of skins. The skin markets should also be regularized to protect the exploitation of skin traders from big skin merchants. The directorate of the industries of the different states should give license and registration to private sectors for establishment of small skin processing units by providing them technical assistance, skill, and credit to stimulate better market access of goat products. It would be helpful in reducing the monopoly of big tanneries. There should be a one-window facility for finance, technical expertize, market intelligence, etc., to make goat rearing a really profitable enterprise for the target groups of the society.

REFERENCES

Council for Leather Export. 2008. *Export of Leather and Leather Products: Facts and Figures 2007–2008*. Chennai, India: Council for Leather Export.

Dixit, R.S. 2006. *Agricultural Marketing in India*. Gurgaon: Shubhi Publication.

Kumar, Shalander, and P.K. Pant. 2003. 'Development Perspective of Goat Rearing in India: Status, Issues and Strategies'. *Indian Journal of Agricultural Economics* 58 (4): 752–67.

Mondal, R. C and A. Sinha. 2009. 'Market Access and Constraints in Marketing of Goats and their Products in West Bengal'. AER Centre, Visva-Bharati, Santiniketan, Study No. 158, West Bengal.

Sharma, Hari Om. 2009. 'Market Access and Constraints in Marketing of Goats and their Products in M.P'. AER Centre, JNKVV, Jabalpur, Study No. 98.

Thaware, Kailash. 2009. 'Market Access and Constraints in Goat Marketing and their Products in Maharashtra'. AER Centre, Gokhale Institute of Politics and Economics, Pune.

Thomsen, Fredrick Lundy. 1951. *Agricultural Marketing*. New York: Mcgraw Hill-Hill Book Company.

41

Producer Companies in India

A Study of Organization and Performance

Sukhpal Singh and Tarunvir Singh

Primary producers' organizations or collectivities are being argued to be the only institutions which can protect small farmers from the ill-effects of globalization or make them participate successfully in modern competitive markets (Trebbin and Hassler 2012). Producers' organizations not only help farmers buy or sell better due to scale benefits but also provide lower transaction costs for sellers and buyers, besides providing technical help in production and creating social capital. It is also argued that cooperatives or such collectivities are needed for small farmers as they help realize better output prices and credit terms and, thus, can help eliminate interlocking of factor and product markets into which small farmers are generally trapped (Patibandla and Sastry 2004).

In India, there are many legal forms of organizations into which a primary producer can organize themselves but cooperatives have been the most prevalent. The cooperative form of organization has been perceived and seen as a means to achieve reduction in poverty and an increase in wellbeing of local people in the presence of other structural constraints like small holdings, lack of bargaining power of small sellers of produce or services, and competition from other forms of organizations. However, cooperatives across the developing world have been more of a failure than success and are alleged to have led to exclusion of really poor, elite capture of such bodies, and promoting differentiation instead of equity in rural communities like in case of sugar cooperatives in Gujarat. In India, the only exceptions to the failure have been sugar and milk cooperatives in some states especially in Maharashtra and Gujarat. However, even in Gujarat, there are as many cases of failure as there are of success of cooperation which include chicory, tobacco, cotton, vegetables, grains, and canal irrigation cooperatives.

In this context, there has been a constant search for alternative forms of collectivization or cooperation to achieve the objectives of development of poor people. Though some researchers also differentiate between collectivization and cooperation in the sense that whereas former refers to organizing to avoid exploitation in markets and the latter as organizing in situations of missing markets.

In India, the alternative ways of registering farmer producer organizations (FPOs) include societies and trusts, cooperatives, Mutually Aided Cooperative Societies (MACS). Self-reliant Cooperatives, private limited companies, public limited companies, and Producer Companies (PCs). There could also be Mutual Benefit Trusts (MBT) under the Trust Act. Farmer organizations registered under the Societies Act cannot legally involve in any trading of inputs or outputs as they are not trading entities. However, until recently, in India and many other developing countries, traditional cooperatives were mostly organized under the cooperative structure, like the State Cooperative Societies Acts in India. However, due to political interference, corruption, elite capture, and similar issues, the cooperatives soon lost their vibrancy and became known for their poor efficiency and loss-making ways. The major problems of traditional cooperatives have been capital constraint due to the withdrawal of financial support by the government, high competition from other players in the market, and access to credit (capital) and technology, besides free riding by members. The government support to these cooperatives has declined, though gradually and selectively. At the same time, they face higher competition due to privatization and liberalization policies. The new environment, however, provides new opportunities for cooperatives due to state withdrawal and deregulation. In addition, there is an increased need and relevance of cooperatives due to the structural adjustment programme, and globalization policies, which are marginalizing the resource-poor producers. The new and potential role of cooperatives in the new economic regime includes provision of inputs, economies of scale, fine-tuning of produce to the market, facilitating more competition in primary markets, and capturing surplus in adjoining stages of the value chain.

A PC is a relatively new legal entity of the producers of any kind, namely, agricultural produce, forest produce, artisanal products, or any other local produce, where the members are primary producers. As a legal entity, PC was enacted in 2003 as per section IXA of the Indian Companies Act, 1956. Since the above enactment, the PC has been hailed as the organizational form that will empower and improve the bargaining power, net incomes, and quality of life of small and marginal farmers/producers in India. Table 41.1 compares PCs with the traditional cooperative structure in India.

The PCs in India come close to the new generation cooperatives (NGCs) and cooperative companies (CCs) in various developed and developing countries as they come close to the PC structure in terms of organization and policy. An NGC is one, which has restricted or limited membership, links product delivery rights to producer member equity, raises capital through tradable equity shares among membership, enforces contractual delivery of produce by members, distributes returns based on patronage, goes for value addition through processing or marketing, and makes use of information efficiently throughout the vertical system. However, it retains a one-member, one-vote principle for major policy decisions. The advantages of delivery rights shares for members are: assured procurement prices and market share of profits due to value

Table 41.1 Differences between a Cooperative and a PC in India

Feature	Cooperative	PC
Registration under	Coop societies Act	Companies Act
Membership	Open to any individual or cooperative	Only to producer members and their agencies
Professionals on Board	Not provided	Can be co-opted
Area of operation	Restricted	Throughout India
Relation with other entities	Only transaction based	Can form joint ventures and alliances
Shares	Not tradable	Tradable within membership only
Member stakes	No linkage with number of shares held	Articles of association can provide for linking shares and delivery rights
Voting rights	One person one vote but RoC and government have veto power	Only one member one vote and non-producer cannot vote
Reserves	Can be created if profit made	Mandatory to create reserves
Profit sharing	Limited dividend on capital	Based on patronage but reserves must and limit on dividend
Role of government	Significant	Minimal
Disclosure and audit requirements	Annual report to regulator	Very strict as per the Companies Act
Administrative control	Excessive	None
External equity	No provision	No provision
Borrowing power	Restricted	Many options
Dispute settlement	Through co-op system	Through arbitration

Source: Kumar et al. 2007; Mondal 2010; and NABCONS 2011.

addition (residue claims), and appreciation of share prices due to better performance of the cooperative. This kind of restructuring, especially equity linked delivery shares and contractual delivery of produce, helps cooperatives to tackle problems of free riding by the membership horizon, which is at the root of financial constraint; and opportunism, both of members as well as of the cooperative. This arrangement by cooperatives has helped them become economically efficient, financially viable, and obtain member loyalty wherever it has been tried. The problems with the NGC concept and its practice as pointed out by various critics are: (i) preferred shares provision compromises the principle of user ownership, though it protects the user control principle; (ii) in practice, member control may operate by the control of delivery rights rather than by the one-member–one-vote principle; (iii) it is more suited for large growers who can afford large upfront investment in processing/marketing; (iv) they are more like closely held companies; and (v) have the potential danger to turn into investor oriented companies (IOC) instead of a user-oriented cooperatives (UOC). In practice, though the NGCs have been able to raise 30–50 per cent of their total capital through delivery rights issues, the problems include: (i) off-market purchases to meet contract terms by the growers; (ii) leasing of delivery rights by members; and (iii) dependence on non-producer member equity and business (Harris, Stefanson, and Fulton 1996; Nilsson 1997).

In the Asian region, there are not many documented cases of CCs or NGCs. However, more recently, there has been a spurt in their numbers. A PC (NorminCrop) of the northern Mindanao vegetable producers' association (NorminVeggies) in the Philippines has been successful in interfacing with large buyers for its small farmers of vegetables by working on cooperative lines. It plans production at the cluster level with the help of cluster leaders and provides all the marketing facilitation for a fee. The farmer and the buyer are responsible for quality and delivery, and purchase conditions agreed respectively (Vorley, Lundy, and MacGregor 2009). However, the unique experience of organizing small farmers under a more business-like entity called farmer companies in Sri Lanka since the 1990s is quite relevant for India.

In Sri Lanka, farmer companies are investor-owned companies established under the Companies Act as people's companies registered with the Registrar of Companies and follow rules and regulations of a private company. They were established to accelerate commercialization in non-plantation agriculture as recommended by the National Development Council (NDC) of Sri Lanka in 1995. They are registered with a minimum of 50 members to safeguard against possible private ownership by imposing restrictions on membership and share trading. Only farmers and other stakeholders involved in agriculture living within a particular geographical region can become shareholders and shares cannot be traded except among farmers eligible for membership. In addition, the maximum number of shares a farmer can own is limited to 10 per cent of shares issued at a given time according to the relevant provision of the Act. These companies were organized by different government agencies and membership ranged from 200–2,200 each and they were involved in different stages of the agricultural value chains like input supply, procurement, selling, packaging, credit supply, crop/produce purchase, contract farming, and manufacturing of tea. Most of them suffered from poor capital base, lack of farmer participation, restriction on shareholding, and poor perception of these entities by farmers as service providers.

In Sri Lanka, the PCs failed to achieve expected objectives, due to various reasons like: (i) politicization of farmer companies; (ii) lack of managerial and entrepreneurial skills due to poor recruitment of management staff; (iii) lack of sound plans and poor management by incompetent boards of directors without professional advice; (iv) lack of proper mechanisms to monitor and evaluate; (v) mistrust between farmer company management and farmers; (vi) farmer perception of the farmer company as a service provider; (vii) awareness gap between the shareholders and the farmer company; and (viii) restriction on share capital ownership (Esham and Usami 2007).

In India too, like in Sri Lanka, the first set of PCs were promoted and supported by a state government (Madhya Pradesh) under a World Bank (WB) poverty reduction project since 2005. In the case of PCs in MP, the state government which was also the promoting body provided a one-time grant of Rs 25 lakh to each PC as fixed-deposit revolving fund to obtain a bank loan against it, and also another annual grant of maximum Rs 7 lakh per year for five years for administrative and other expenses in the manner of 100 per cent in first year, 85 per cent in second year (Rs 590,000), 70 per cent in third year (Rs 490,000), 55 per cent in fourth year (Rs 385,000) and 40 per cent in fifth year (Rs 280,000). Further, interest subsidy up to a limit of Rs 2 lakh was provided on any term loan taken by the PC and a grant of up to 75 per cent of the cost up to a maximum of Rs 2 lakh was given for any certification expenses like Food Products Order (FPO), Global Good Agricultural Practices (Globalgap), etc.

The major research questions regarding the role of PCs include: how far are PCs an improvement over the existing cooperative or other models of producer organization? How relevant and appropriate are the PCs in the context of globalized markets? What is the competitive edge of PCs over other modes of farmer or primary producer organization? What kind of policy treatment do the PCs need to grow as vibrant producer entities and to make an impact on the livelihoods of small producers?

While there are some unresolved questions in the current design and context, the PC as an enterprise of small and marginal farmers/producers nevertheless appears to be a powerful vehicle to empower small farmers/producers and improve their quality of life leading to better rural development in India. It is an appropriate time to assess the functioning of the PCs and their impact on the small and marginal farmers/producers in India as they have been in existence for almost a decade now.

Within the above context, the study examines:

1. The current status of the PCs in India in terms of the ownership and the management structure.
2. Business performance of the existing PCs on various parameters.
3. Differences, if any, among PCs organized/facilitated by different external stakeholders like the private sector, NGOs, government, and reasons thereof.
4. Problems faced by these PCs and mechanisms to address such constraints.

The study attempts to understand the current mode of operation and effectiveness of the PCs with reference to the small and marginal farmers/producers in India. This, in turn, helps review the design of the PC and the amendments and policy mechanisms that may be necessary to make the PC an effective institutional arrangement of the small and marginal farmers/producers leading to development of small producers in rural India.

METHODOLOGY AND DATA

As of mid-2011, there were over 156 PCs in India. Of these registered PCs, the PCs of District Poverty Initiative Project-Madhya Pradesh (MPDPIP) are the most cited. The above PCs sell their produce to any large national and international buyers/processors or to their promoters. In its attempt to aggregate the produce from the marginal producers, the above PC model focuses on the common interest groups (CIGs) or self-help groups (SHGs) as the basic units for aggregation with no limit on the size of membership and size of cluster/operational area.

As of mid-2011, there were only 25 PCs in India which were registered before March 2008 (NABCONS 2011). Further, 34 per cent of all PCs were in western India, 24 per cent each in south and east India, and 17 per cent in north India. More of the PCs during 2008–11 were registered in western, southern, and eastern India. About 74 per cent of all the PCs (139) were in agriculture or allied sectors which included animal husbandry and fisheries and 64 per cent in agriculture alone which included crop agriculture, horticulture, organic, herbal, and medicinal plants. Another 10 per cent were into non-farm business, 4 per cent in power, and 12 per cent in various other businesses. Thus, we have effectively 133 PCs including 17 from MPDPIP (not part of the NABARD-supplied list) from which we ended up choosing our sample of PCs and that too, given the existence of older and functional PCs, from the central and western Indian states of MP, Gujarat, Maharashtra, and Rajasthan. Incidentally, the states of MP, Maharashtra, Rajasthan, and Gujarat each have more than two-thirds of their operated holdings as marginal or small, the highest being in Gujarat (75 per cent) and around 67 per cent each in the other three states.

The study includes both the survey and the case study methodologies. The data on ownership and management structure of the PCs was collected through a sample study of PCs across major locations, commodities, and types of promoters of such entities like NGOs, government, private-sector entities, and farmers and their organizations themselves. Multiple case studies of PCs (3–4 in each category and in major states) have been undertaken to understand the operational modalities and the challenges in the functioning of the PCs. For both survey and case study, sampling of PCs was based on geographic area, length of operation, type of promoter, number of members, type of ownership, etc., but all of them belong to farm produce as the dynamics of other sectors are quite different in terms of markets and production systems and risks involved. Therefore, we excluded non-farm PCs from this study.

This study covers only farm business-related and allied PCs to maintain the uniformity in comparisons and understand the dynamics of such entities. Further, we have covered only those PCs which were completely new as PCs and were not converted into PCs from other entities like cooperatives and societies (for example Vasundhara Agri-horti PC (VAPCOLS) in Maharashtra and Umang Mahila PC in Uttarakhand, respectively) as it would be difficult to assess the performance in such cases due to the earlier business being under another entity and present performance also being result of earlier set-up to a large extent. The study covered 24 PCs across four states and four types of promoters, that is, government, NGOs, farmers' organizations, and private corporate entities (Table 41.2).

In addition, the study attempted to cover mostly those PCs which were at least three year old so as to get a good data base to assess the performance, but many of those contacted were not that old or had been only registered

Table 41.2 Distribution of PCs by Type of Promoter

State	NGO-promoted	State-promoted	Farmer Group-promoted	Corporate-promoted
Punjab	–	–	–	6 (all defunct)
Rajasthan	4	–	–	–
MP	5	6	–	–
Gujarat	2	–	2	–
Maharashtra	2	–	1	2 (spurious)
All	13	6	3	8

Source: Primary Survey.

but not operationalized. Thus, we have 48 per cent that are at least three years old, another 48 per cent that are more than one or two years old and one (4 per cent) which is only one year old in its legal existence.

RESULTS AND DISCUSSION

The membership/shareholding of PCs in India ranges from individual producers to informal SHGs and individual producers, registered SHGs and individual members, and only institutional members. The number of members ranged from 11–220 in Maharashtra, 30–6,000 in Gujarat, 344–1,200 in Rajasthan, and 10–6,500 in MP. Though authorized capital ranged from Rs 3–25 lakh across PCs, the paid-up capital remained within Rs 1–5 lakh with only one touching Rs 10 lakh (Table 41.3). However, most of the DPIP PCs in MP had given out shares to a large number of farmers ranging from 1,200–6,500. The number of users was even higher ranging from 2,460–8,000. Except one PC, most of the PCs mainly did business with membership and non-member business was only around 20 per cent (Table 41.4). Most of them represented really small farmers with land holdings averaging around one hectare. Most of them had professional managers, though in many cases the turnover of professionals was high. A significant support to start the PCs had come from promoting agencies or projects, especially in MP. Most of the MPDPIP PCs were into seed production and farm input supply as their main business. The financial performance of most of the PCs was weak with most making losses and other very low profits (Table 41.5).

The NGO–ASA promoted PCs had an even lower capital base (20–40 per cent of authorized capital) with a very small membership and user base, and low professional support. They too worked with small holders and largely member farmers for business transactions as they were into better cotton programme including input supply. They were of relatively recent origin and one of them did make profits within two years despite the fact that they did not get any state support unlike MPDPIP and other PCs in the state. The other NGO (Srijan and PRADAN) promoted PCs had much larger capital base (Rs 3–51 lakh) and all of them had received support from MPDPIP as by the DPIP PCs. Most of them had only institutional members and some professional support/staff. However, their user base was much larger (2,200–4,000) as two of them were into the dairy and poultry business which required scale. For this, they also (two of them) depended on non-member business up to 40 per cent of their turnover. However, only the poultry PC (of PRADAN) was consistently into profits.

In Gujarat, NGO-promoted PCs were not able to raise authorized capital to a large extent which was anyway small (Rs 1–5 lakh) and shareholding was restricted to a few groups and farmers. They did not have any professional managers. The user member base was small, farm holdings larger than in MP, and sold mostly inputs and facilitated producer selling. However, both were not active now and were being restructured and re-energized. On the other hand, the farmer organization (Bhartiya Kissan Sangh [BKS] and Onion Growers Co-operative Federation [OGCF]) promoted PCs were doing better in terms of business volumes as well as profits. The most striking case was that of BKS leader promoted PC which had touched a farmer member base of 6,000 across six districts and achieved turnover of the order of Rs 25 crore within two years and was making good profits without any external support. However, its farmer base was large and medium farmers and it had high value produce for exports and domestic markets. The other farmers organization-based PC was doing good sales and business management but seemed more of an outfit of a few people with large non-member sales.

The PCs in Rajasthan were relatively very new and had modest farmer base (300–1,200) with mostly individual shareholders (100–500) but had large number of farmer groups associated with them. They had fairly good professional support from the promoter NGO (Access Development Services [ADS]). In some of them, non-member dependence was high (20–60 per cent) though the farmer base was really made up of marginal and small farmers, that too, in tribal areas. Though most of them were also into input supply, two of them also ventured into facilitation of seed contract farming and ginger production and marketing. Their annual turnover was in the range of Rs 10–30 lakh and all of them were into profits, though modest.

Table 41.3 State-wise Basic Profile of PCs in India

State>MP											
Promoter> Producer Co.> Parameters	MPDPIP (all from CIGs)						ASA (from SHGs)		SRIJAN (from CIG)	PRADAN (First from CIG and Second from Co-ops)	
	Khajurao (Chattarpur)	Nowgaon Agri (Tikamgarh)	Sagar Samridhi Crop (Sagar)	Khujner Agi (Rajgarh)	Samrath Kissan (Shajjapur)	Ramraja Crop (Tikamgarh)	Nimad farmers (Ojhar, Barwani)	Khargaon (Khargaon)	Sagar Shri Mahila Dugh Utpadak (Sagar)	Sironj Crop (Sironj, Vidisha)	MP Women's Poultry (Bhopal)
Authorised capital (Rs Lakh)	5	3	5	5	15	25	5	5	25	5	25
Share capital (Rs Lakh)	4.6	1	1	1.8	9.5	1.67	2.14	1	5	3.42	25
Shareholding Pattern Types of holder											
Initial	12	10	200	450	3000	650	367	5	10	10	5 (Coop.)
–Individual	–	–	–	–	–	–					
–Group											
Now	4625	1000	1203	1860	6500	1647	514	100	10	1910	6 (Coop.)
–Individual	140 (23% of all)								142		
–Groups											
Shares per member	10	10	10	10	12–22	100	20	50	10	10–100	1–10
FBG/SHG/Coop. Associated	140	No	No	200	No	No	No	1000	142	No	6
No. of directors	12	5	11	5	14	7	15	5	8	10	6
No. of Prof. Managers (who pays them)	3 (DPIP and PC)	1(DPIP)	1 (DPIP and PC)	3 (PC and DPIP)	2(DPIP and PC)	2 (DPIP and PC)	1 (ASA)	1(ASA)	3 (DPIP and SRIJAN)	1 (DPIP)	1 (PC)
Total No. of employees	10	3	4	9	5	4	4	2	11	5	NA

Main business	Seed contract farming and responsible soya production facilitation and input supply including soil testing	Seed contract farming facilitation and input supply	Seed contract farming facilitation and input supply besides seed processing	Seed contract farming and responsible soya production facilitation with input supply and collective sale, and sale of daily necessities	Seed contract farming and better soya production facilitation and input supply	Seed contract farming and responsible soya production facilitation and input supply	(Better) Cotton production facilitation and input supply	(Better) Cotton production facilitation and input supply	Milk procurement and sale and cattle feed manufacture and sale	Seed contract farming facilitation and input supply	Poultry production facilitation and sale of poultry inputs and birds (meat)
State >		Maharashtra				Rajasthan				Gujarat	
Producer Co. > Parameters	Baliraja krishak (Sangamner)	Waghad Agri (Dindori, Nashik)	Devnadi Valley Agri (Sinnar)	Udaipur Agro (Jhadol)	Jhambukh and Kissan Agro (Banswara)	Vijwa Agro (Dungarpur)	Dungaria (Dungarpur)	Dhari Krishik Vikas PC (Dhari)	Mahagujrat Agricotton PC (Amreli)	Farmer Crop Care PC (Budhel, Bhavnagar)	North Gujarat agro PC (Palanpur)
Established from	Krishak panchayat Sangathan	WUA	WUAs	4 kissan sanghs	Kissan Sangh	Kissan Sangh	Kissan Sangh	WUAs	–	–	–
Authorised capital (Rs lakh)	5	5	5	10	10	1	1	1	100	11	5
Share capital (Rs lakh)	2.25	1.7	2.3 lakh	300000	110000	58300	30000	1	88	10.42	3.205
Main promoter/ facilitator*	NGO	WUA Federation	Yuva Mtra (NGO)	ADS (NGO)	ADS (NGO)	ADS)NGO)	ADS (NGO)	DSC (NGO)	BKS (FO)	BOGCF/ SFWA (Co-ops)	Farmers Union (FO) and 3 NGOs
Shareholding Pattern											
Types of holder											
Initial	100	24	103	1200	600	588	344		800	11	200
–Individual				99				10			
–Group											
Now	220	107	173	1200	800	588	344		6000	30	641
–Individual				99				10			
–Groups											

(*Cont'd*)

Table 41.3 (*Cont'd*)

	State > MP										
Promoter> Producer Co.> Parameters	MPDPIP (all from CIGs)						ASA (from SHGs)		SRIJAN (from CIG)	PRADAN (first from CIG and Second from Co-ops)	
	Khajurao (Chattarpur)	Nowgaon Agri (Tikamgarh)	Sagar Samridhi Crop (Sagar)	Khujner Agi (Rajgarh)	Samrath Kissan (Shajjapur)	Ramraja Crop (Tikamgarh)	Nimad Farmers (Ojhar, Barwani)	Khargaon (Khargaon)	Sagar Shri Mahila Dugh Utpadak (Sagar)	Sironj Crop (Sironj, Vidisha)	MP Women's Poultry (Bhopal)
Share Range	50–1000	1000	Not fixed	1–5	1–200	10–100	10–100	1000	1–25	500–5500	5
FBG/SHG/Coop. Associated	1	1	1	99	62	38	26	10 (1)	13 (1)	6	13
No. of directors	11	7	11	8 (6)	9 (2)	7 (3)	10	1 (DSC)	6 (PC)	No	No
No. of Prof. Managers (who pays them)	2 (Lok panchayat)	0	NGO staff	3 (ADS)	2 (ADS)	2 (ADS) & 1 (KVK)	1 (ADS)	1 (DSC)	6 (PC)		–
No. of employees	2	1 WUA Staff	NGO staff	2	2	1	1	1	10	3	–
Main business	Bio-input supply and marketing of organic produce	Input supply and grape sales	Vegetable selling	Input supply and purchase of produce	Input supply and collective marketing of produce	Input and grain supply	Cotton seed contract farming facilitation and input supply	Input supply	Cotton and mango selling and input supply thru own retail shops at local level	Input selling	Input supply and fruit selling

Note: *ALL PCs in MP were promoted by DPIP. Sahyadri Farmers (Nashik) and Sahyadri Agri (Pune) PCs from Maharashtra are removed from this table as they are fake PCs.
Source: Field survey.

Table 41.4 Membership Profile of PCs in India

	State > MP										
Producer Co.> Parameters	Khajurao (Chattarpur)	Nowgaon Agri (Tikamgarh)	Sagar Samridhi Crop (Reori, Sagar)	Khujner Agiculture (Rajgarh)	Samrath Kissan (Shajjapur)	Ramraja Crop (Tikamgarh)	Nimad farmers (Ojhar, Barwani)	Khargaon (Khargaon)	Sagar Shri Mahila Dugh Utpadak (Jaishee nagar, Sagar)	Sironj Crop (Sironj, Vidisha)	MP Women's Poultry PC (Bhopal)
Total users	6600	5000	3000	2460	8000	3000	767	148	2200	3000	4000
– Member	4600	1000	1200	1860	6500	1647	517	73	1500 (SHG)	1000	4000
– Non-member	2000	4000	1800(inputs only)	600	1500	1353	250	75	700	2000	0
% of total business from non-members	15	30	20	20	20	50	5	50	40	40	0
Avg. Size of holding of member (Range) in Ha	1 (0.2–3.4)	0.8 (1–10)	0.8 (0.4–4)	1 (0.25–10)	1.5 (0.3–7)	1 (0.4–1.6)	1 (0.25–2.5)	2 (0.5–5)	2 (1–4) Buffaloes	1.5 (0.5–3)	400 (300–600) (birds)
State>		Maharashtra			Rajasthan				Gujarat		
PC> Parameters	Baliraja krishak PC (Sangamner)	Waghad Agri PC (Dindori, Nashik)	Devnadi Valley Agri PC (Sinnar)	Udaipur Agro PC (Jhadol)	Jhambukhand Kissan Agro PC (Banswara)	Vijwa Agro PC (Dungarpur)	Dungaria PC (Dungarpur)	Dhari Krishik Vikas PC (Dhari)	Mahagujrat Agricotton PC (Amreli)	Farmer Crop Care PC (Budhel, Bhavnagar)	North Gujarat agro PC (Palanpur)
Total users	750	107	173	3000	1500	788	382	1500	6000	Everyone but prefer member first	700
– Member	220	107	173	1200	800	588	344	1150	6000		641
– Non-member	530			1800	700	200	38	350			60
% of total business from non-members	70	0	0	60	20–30	10	10	25	0	90	10
Size of holding of member (Range) in Ha	1 (0.4–6)	2.4 (2–4)	1.5 (0.1–4.1)	0.8 (0.2–4)	0.8 (0.16–1.6)	0.72 (0.16–2.4)	0.48 (0.08–1.12)	2.74 (0.45–4.57)	10 (0.4–16)	4 (0.8–6)	18.88 (2.2–34.28)

Note: Sahyadri Farmers (Nashik) and Sahyadri Agri (Pune) PCs from Maharashtra are removed from this table as they are fake PCs.
Source: Field survey.

Table 41.5 Business Performance and Profile of PCs in India

	MP										
Producer Co.> Parameters	Khajurao PC (Chattarpur)	Nowgaon Agri PC (Tikamgarh)	Sagar Samridhi Crop PC (Reori, Sagar)	Khujner Agiculture PC (Rajgarh)	Samrath Kissan PC (Shajjapur)	Ramraja Crop PC (Tikamgarh)	Nimad Farmers (Ojhar, Barwani)	Khargaon (Khargaon)	Sagar Shri Mahila Dugh (Sagar)	Sironj Crop (Sironj, Vidisha)	MP Women's PPC (Bhopal)
Year (2006–07)										(2005–06)	(2007–08)
Turnover (Rs Lakh)	3.63	2	44.60	4.68	4	18.42	–	–	4.46	91	389.65
Profit (loss) in Rs	30362	25	–	(99000)	(4000)	(240000)			(4000)	27000	5.02 Lakh
Year (2009–10)											
Turnover (Rs Lakh)	29.95	30	28.10	40.22	158	15.38	4.19	66000	45.29	150	958.25
Profit (loss) in Rs	(34198)	(1.74 lakh)	(10000)	59000	(5.2 lakh)	(4.5 lakh)	9068	(7000)	(1300000)	80000	8.48 Lakh
Year (2010–11)											
Turnover (Rs Lakh)	22.57	54.70	62	100.58	189	15.47	5.27	No business in 2010–11	49.57	127	518.47
Profit (loss) in Rs	(14600)	15000	(2.3 lakh)	26536	50000	(150000)	62000		(458000)	(12,00,000)	1613000
State>	Maharashtra			Rajasthan				Gujarat			
Producer Co.> Parameters	Baliraja krishak (Sangamner)	Waghad Agri (Dindori, Nashik)	Devnadi Valley Agri (Sinnar)	Udaipur Agro (Jhadol)	Jhambukhand Kissan (Banswara)	Vijwa Agro PC (Dungarpur)	Dungaria PC (Dungarpur)	Dhari Krishik Vikas PC (Dhari)	Mahagujrat Agricotton (Amreli)	Farmer Crop Care Bhavnagar)	North Gujarat agro (Palanpur)
Year (2009–10)			–	–	–	–	–				
Turnover (Rs Lakh)	No turnover	39.25	–					15.59	1000	230.61	–
Profit (loss) in Rs	(10840)	(691000)	28					29533	–	61320	–
Year (2010–11)			–								
Turnover (Rs Lakh)	5.59	108					9.506	10.04	2500	289.34	21
Profit (loss) in Rs	(15530)	(400000)		13.854006	15.5913847	28.9114458	60000	(206000)	26,98000	278485	200000

Note: Sahyadri Farmers (Nashik) and Sahyadri Agri (Pune) PCs from Maharashtra are removed from this table as they are fake PCs.
Source: Field survey.

The PCs in Maharashtra presented a mixed bag with some being extremely genuine and others completely fake as there were also non-NGO (corporate) promoters in some cases. Of the two NGO-promoted PCs, the capital base was small (Rs 2–3 lakh), the number of shareholders was small (200), and professional help was missing. Similar was the case of one farmer group promoted PC which had similar profile. Though in all three cases, the farmer base comprised of small farmers, in one case, non-member dependence was very high (70 per cent of business). All of them made losses and suffered from capital shortage. On the other hand, both the exporter and the business-person-promoted PCs were more on paper, did not have any farmer base, and shareholding was restricted to small numbers (11 and 70). The PCs were managed by staff of the other businesses of the promoters.

In terms of profitability, the majority of DPIP PCs in MP were into losses, the majority of non-DPIP were into profits, all except one in Gujarat were into profits as was the case of all four PCs in Rajasthan. All three in Maharashtra were into losses. However, most of the PCs did benefit the members in some way or the other like employment generation, increase in income, higher market price, dividends and fair and prompt payments and self-respect and identity for small producers.

So far as the viability of the PCs is concerned, there were different factors in different contexts. In the case of Maha Gujarat Agricotton, it was more of scale, nature of farmers, and crops handled. On the other hand, in the case of poultry cooperatives, it was the scale and the professional management of the value chain besides women-member involvement which made the difference. In case of Nimad and Khargone, it was again the high value crop—cotton, that too, better cotton market and the ASA support which made them sustain as was the case with most of the ADS promoted PCs in Rajasthan. Another explanation for most PCs being in loss could be that as PC income is taxable, the PCs tend to pass on the surplus generated to members as price benefit to avoid taxation.

The PCs have been founded on the strength of pre-existing organizations like water users' associations and SHGs or other such entities in most cases though there are also completely fresh origins of PCs in India like the one in north Gujarat and a few in Maharashtra and MP.

Across states, problems have been reported and seen in the functioning of PCs. Major policy issues for PCs in MP were: lack of working capital support other than the DPIP initially, poor professional management due to inability to afford professionals as well as high turnover of professionals, besides the difficulty of managing the accounts and paper work of the PCs. In Gujarat, the Development Support Centre (DSC) reported registration process- and compliance-related problems. Further, since cooperatives have been in existence in Gujarat for a long time, the PC is seen as a competitive entity rather than a complementing one by the bureaucracy and other stakeholders do not have orientation on PCs. The PC also had difficulty in obtaining various licenses.

In Rajasthan, the major issues in PC management include working capital shortage and lack of access to loans as banks ask for 3-year balance sheet, as the PCs are not yet recognised by the union or state government for any incentive or support. Second, banks refuse to lend to the PCs due to the lack of state or government guarantees. Besides this, professionals are costly to hire and sustain. Some of the PCs managers are also being lured by the private sector interested in promoting collectivities of producers. There has also been some resistance from local traders.

In Maharashtra, most of the studied PCs are very commodity specific in their business like grapes or organic produce or vegetables, which places additional pressure to perform and be viable as individual crop or produce markets can be very volatile. The PCs in India, in general, appear to be product focused rather than producer/farmer focused.

The major hurdles for a PC, in general, are: getting registration and digital signatures of board of directors who are small farmers and illiterate villagers who do not have any identity proofs, accessing capital from outside; and not being able to access grants as they are commercial entities. The PCs also suffer from tax on income (30.2 per cent) unlike cooperatives which can show income under tax-free heads. There have also been cases of hijacking of PCs. In MP, the government appointed its own Chief Executive Officers (CEOs) as it had given grants to PCs. In Gujarat, a PC was hijacked by the promoters. However, the PC Act provides for handling such malpractices.

A comparison of cooperatives and PCs in policy treatment in India shows that income tax exemption, non-taxable welfare income exemption, land lease at nominal rates or free, fertilizer allocation to Primary Agricultural Co-operative Societies (PACS), foundation seed supply and marketing support to seed cooperatives, state agency grants to cooperatives, export incentives and provision of distribution outlets for selling products which is available to cooperatives is not available to PCs.

CONCLUSIONS AND POLICY IMPLICATIONS

The PCs suffer from lack of finance in the formative years, as the kind of support provided by the MP government has not come from any other state government or the union government until recently. The union government has already made provision for PCs in the 2013–14 budget for matching grants up to Rs 10 lakh per FPO/PC with a provision of Rs 50 crore and for a credit guarantee fund for FPOs through the Small Farmer Agribusiness Consortium (SFAC) with allocation of Rs 100 crore. The Reserve Bank of India has put PCs under priority sector lending up to Rs 5 crore per PC. The National Bank for Agriculture and Rural Development (NABARD) has a fund for promotion of producer organizations, which provides for business plan based loans to PCs as well as capacity building grants to promoting agencies. Recently, the Ministry of Agriculture and Farmers Welfare has advised all the state governments to treat PCs on par with cooperatives for policy incentives.

A working group on PCs, convened by the National Resource Centre for Rural Livelihoods of PRADAN, an NGO promoting such enterprises, and chaired by Nitin Desai, in its report in 2009 suggested a few mechanisms to deal with some of the problems: The first was to provide a mechanism to assess the value of shares held by the farmers and declare a fair value periodically. This is expected to incentivize members to acquire more shares. The second was to enable transfer of shares within membership at freshly assessed value rather than at par value as well as issue new shares at new fair value of a share. It also recommended provision for issuance of preference shares, bonds and debentures to raise external risk capital from non-producer members. However, if dividend is not paid to such members for three years continuously, then they acquire voting rights automatically as per existing law. It also recommended buy back of shares by the PC, if needed. Its other recommendations included lowering the start up authorized capital from Rs 1 lakh to Rs 10000, tax exemption to PCs for a few years, and restriction on membership of the PC for new members by the existing members for reasons of business (NRCRL 2009).

Governmental support in the form of grants during the early stages of the PC should be made available. Exemption from corporate tax at least for initial few years and the inclusion of financing agency on the board of PC can also help. In India, banks give collateral free loans to small and medium enterprises (SMEs) which can also cover PCs. Similarly, a PC can be treated as a non-banking financial company (NBFC) to provide loans to farmer members. It is also possible to mobilize more equity from within the membership. For example, some PCs have attempted variation in shareholding related patronage to mobilize capital. One had voting rights linked to the patronage and another linked patronage to shareholding. Others had minimum patronage in terms of sale or purchase transactions annually with the PC to remain members.

The PCs also need to choose their activity portfolio carefully keeping in mind the member centrality. It is possible to identify new activities in local areas which are valuable for small farmers, for example, custom hiring of farm machinery and equipment which they cannot afford to buy but can rent in. This is being done in some parts of India viably by private entities and PACS.

It was found in the case of NGCs in the USA that planning and development and financing and costs were the two major factors in success of the NGCs across various agro processing sectors (Carlberg, Ward, and Holocomb 2006). In fact, it is also argued that effective farmer producer organizations need to have clarity of mission, sound governance, strong responsive and accountable leadership, social inclusion, be demand and service delivery driven, have high technical and managerial capacity, and effective engagement with external actors like government, donors, or the private sector. However, even this needs to be supplemented by supportive and enabling a legal, regulatory, and policy environment that guarantees autonomy and level playing field (Thompson et al. 2009).

There is also need for a central agency to promote PCs with grants and disseminate awareness about the concept and practices of PCs among farmer producers and other stakeholders. Both the state and the union governments in India should recognize PCs as producer cooperatives and extend all support as extended to traditional cooperatives in terms of credit, licenses for inputs and output sale and purchase. They should be considered eligible for investment and working capital grants for processing and marketing infrastructure creation. The PCs practicing organic farming can be designated as certifying agencies for third parties and individual growers by the union government agencies like the Agricultural and Processed Foods Products Export Development Authority (APEDA) and Food Safety Standards Authority of India (FSSAI). The promotional and non-governmental organisations supporting these PCs should be given project-based grants by the state/union government.

REFERENCES

Carlberg, J.G., C.E. Ward, and R.B. Holocomb. 2006. 'Success Factors for New Generation Co-operatives', *International Food and Agribusiness Management Review*, 9(1): 33–52.

Esham Mand K Usmi. 2007. 'Evaluating the Performance of Farmer Companies in Sri Lanka: A case study of Ridi Bendi Ela Farmer Company'. *The Journal of Agricultural Sciences*, 3(2): 86–100.

Harris, Andrea, B. Stefanson, and M. Fulton. 1996. 'New Generation Cooperatives and Cooperative Theory', *Journal of Cooperatives*, 11(2): 15–28.

Kumar, A, H K Deka, P Das, and P Ojha. 2007. Livelihood opportunities in broiler farming livelihood resource book. PRADAN, New Delhi.

Mondal, A. 2010. 'Farmers' producer company (FPC): concept, practice and learning—a case from action for social advancement'. *Financing Agriculture*, 42(7): 29–33.

NABARD Consultancy Services (NABCONS). 2011. *Integration of Small Producers into Producer Companies: Status and Scope*. Hyderabad: NABARD Consultancy Services.

National Resource Centre for Rural Livelihoods (NRCRL) 2009. *Report of the Working Group on Producer Companies*. New Delhi: NRCRL, PRADAN, January.

Nilsson, J. 1997. 'New Generation Farmer Co-ops', *Review of International Cooperation*, 90(1): 32–8.

Patibandla, M., and T. Sastry. 2004. 'Capitalism and Co-operation: Co-operative Institutions in a Developing Economy', *Economic and Political Weekly*, 39(27), July: 2997–3004.

Thompson, J., A. Teshome, T.A. Hughes, E. Chirwa, and J. Omiti. 2009. 'The Seven Habits of Highly Effective Farmers' Organisations', Future Agricultures Policy Brief 032, Futures of Agriculture Consortium, Brighton, Sussex, June.

Trebbin, A., and M. Hassler. 2012. 'Farmers' Producer Companies in India: A New Concept for Collective Action?', *Environment and Planning*, 44(2): 411–27.

Vorley, B., M. Lundy, and J. MacGregor. 2009 'Business Models that are Inclusive of Small Farmers', in C.A. da Silva, D. Baker, A.W. Shepherd, L. Jenane, and S. Miranda-da-Cruiz (eds), *Agro-industries for Development*, pp. 186–222. Oxfordshire: FAO, UNIDO, and CABI.

42

Evaluation of Price Support and Market Intervention Schemes

D.S. Bhupal

In the neo-liberal framework of economic development, agricultural input costs have been rising. There has been a demand and supply mismatch of agricultural commodities mainly due to half-hearted reforms in agriculture which did not bring in much needed private investment and the public sector investment in irrigation, fertilizers, pesticides, seeds, research, and extension services could not maintain its earlier tempo. The edible oilseeds sector is a classic example of imbalance in supply and demand. In 1993–94 due to impact of Technology Mission on Oilseeds and Pulses (TMOP), the country became almost self-sufficient in edible oils. However, once the sector was put on Open General License, imports have gone beyond 67 per cent of total consumption. In the current year out of a total consumption of 18 million tonnes, about 12 million tonnes were imported.

Environmental degradation has also been a concern and is manifested in the form of diminishing soil fertility and water table and water logging. The need for change in cropping patterns from low-value crops to high-value crops has been in focus to increase rural income.

The market plays an important role in the determination of farmers' income, change in land use, overall production, change in cropping patterns, etc. Approximately one-third of all persons gainfully employed in the country are engaged in the field of marketing and about one-forth of the national income is earned by the marketing profession (MOSPI 2010: 8). However, in an economy like India where about two-thirds of the people are dependent upon agriculture and where a huge disparity in income and wealth exists, government intervention becomes indispensable. Moreover as farmers have the least control over input and output markets, their emphasis has been on increasing production by using more inputs, particularly fertilizers, pesticides, and irrigation, resulting many times in a crash in output prices, thus necessitating government intervention through the Price Support Scheme (PSS). Other important intervention takes place through the Market Intervention Scheme (MIS).

In the PSS, the government besides announcing the minimum support price (MSP) for 24 major agricultural commodities defends the price by procurement. Whereas in the case of MIS, particularly apple 'C' grade, no MSP is announced. The state government in consultation with central government announces the procurement price for its agency/s to buy at that price. The MIS is applicable in two situations: one, when production is more than 10 per cent of the preceding year; and two, when the price of a commodity falls below 10 per cent of the preceding year. The MIS, unlike the PSS is an ad hoc arrangement for a selected period of the year. This in fact is one of the attempts to affect market price (Centre for Business Planning). This chapter examines the PSS with regard to sunflower in Haryana and MIS with regard to apple 'C' grade in Uttarakhand.

AREA BACKGROUND

Haryana has seen tremendous growth in agricultural production, intensive land use, farm machinery utilization, surface and ground water irrigation, and excessive use of fertilizers and pesticides. It has 106 main market yards, 178 sub-market yards, and a huge number of Village Purchase Centres. Thus, each market yard covers about 152 square kilometres and serves 64 villages. Still it faces a heavy rush of peak season arrivals, thereby reluctance on the part of private buyers due to obvious reasons. The role of public sector procurement agencies under PSS therefore becomes important. In addition to wheat, paddy, millet, etc., sunflower has been covered under PSS in the state.

In Uttarakhand, due to limited local demand marketing of farm produce is difficult due to lack of infrastructure like collection centres, storage, proper transport, roads, etc. There are 36 wholesale markets, 30 rural and primary markets, and 58 regulated markets. The average area served is about 962 square kilometres, which is about 7 times more than in Haryana. Though the population density in Uttarakhand is much less, per market population in Uttarakhand is 146,368, almost double in comparison to about 74,453 in Haryana. Thus, government intervention in the agricultural marketing becomes sine qua non (Bhupal 2009). The central government shares the burden of intervention to the extent of 15 per cent of MSP under PSS and 50 per cent under MIS.

This exercise evaluates PSS in the marketing of sunflower in Haryana and MIS with regard to apple 'C' grade in Uttarakhand.

OBJECTIVES

The specific objectives of the study were:

1. to analyse the extent of coverage of MIS for apple 'C' grade and PSS for sunflower with respect to farmers.
2. to ascertain the socio-economic factors that influence coverage of villages and farmers with regard to MIS and PSS.
3. to understand problems of stakeholders in the operation of MIS and PSS.
4. to study the effect of MIS and PSS on the market price and suggest policy measures.

DATA AND METHODOLOGY

As the area covered under sunflower in Haryana is too little to find place in any published document of the state government, data with regard to area, production, marketing, etc., were collected from the offices of the Directorate of Economics and Statistics, Directorate of Agriculture, and the Haryana State Co-operative Supply and Marketing Federation Limited (HAFED). Sunflower is grown and mostly marketed only in two districts, Ambala and Kurukshetra. Most of the oil extraction mills are also located in Shahabad (Kurukshetra) and Ambala; therefore the two districts were selected and from the two districts four blocks: Shahabad and Thanesar from Kurukshetra and Barara and Saha from Ambala were selected. At the next stage, four villages from each district, namely, Padlu, Damli, Bir Mathana, and Kaulapur from Kurukshetra and villages Barara, Jamalmajra, Nahoni, and Ugala from Ambala were selected.

Household questionnaires were canvassed for the collection of information from farmers of all categories, village schedules were completed with the help of village heads or patwaris, and district schedule by the department of agriculture. Data for market arrivals and prices were collected from the Agricultural Produce Market Committees (APMCs). In the case of apple 'C' grade, in Uttarakhand, little information of one figure of 1.86 lakh with regard to coverage under MIS was noticed in the available literature. Hence, from the offices of the Directorate of Economics and Statistics and Directorate of Horticulture details of area, production, and marketing of apple were obtained. It was noticed that only in district Uttarkashi MIS for apple 'C' grade was operationalized, though apple was grown in other districts. In fact, the MIS was operational in one block, Mori, so block Mori was a natural choice. Therefore, district Uttarkashi alone was selected at the first stage, Block Mori at the second stage and from there eight villages and 'Toks' (small hamlets) namely Thunara, Kiranu, Arakot, Bhutanu, Gokool, Jhatodee, Kaleech, and Makuri were selected. In the entire state, a large part of MIS for apple 'C' grade was implemented only in these villages/hamlets. However, among these villages/toks, gram sabhas were formed only in Bhutanu, Arakot, and Gokool. We have to opt for more villages/toks because the number of households without MIS was not enough to select an adequate sample.

Nature also worked against us during the process of data collection. In June a shocking message of at least three of the respondents being eliminated by floods and sludge was received. The Uttarakhand government's webpage also does not provide much information, thus we were handicapped

Table 42.1 Sample Size

Item	Haryana	Total	Uttarakhand	Total
Selected Distt.	Ambala, Kurukshetra	2	Uttarkashi	1
Tehsil/block	Barara, Saha; Shahabad, Thanesar	4	Mori	1
Crop	Sunflower		Apple 'C' grade	
Beneficiary farmers @			30 (8)	30 (8)
Non-beneficiaries @	96 (8)	96 (8)	39 (11)	39 (11)
District schedules	2	2	1	1
Village schedules	8	8	8	8

Notes:@ None can be described as beneficiary or non-beneficiary Villages/toks in parentheses.
Source: DES, Haryana, HAFED Ltd, DES, Uttarakhand, Directorate of Horticulture, Uttarakhand.

in getting the required information. The details of sample are given in Table 42.1.

Categorizing the farmers into beneficiary and non-beneficiary in Haryana is confusing, because there were no such farmers who directly sold to HAFED (the only agency to implement PSS). Second, no marginal farmer grows sunflower.

RESULTS AND DISCUSSION

Case of Sunflower

Area and Production

Sunflower has high adaptability to diverse agro-climatic conditions and requires less irrigation, hence is most suitable in water shortage conditions. In Haryana it is grown in summer when other crops do not compete with it and can give a good combination with potato and groundnut. The state needs a change in cropping patterns.

In India, it gains importance due to acute shortage of edible oils. The government took various measures to implement the recommendations of TMOP. Consequently, by 1993–94 the area under sunflower increased to 2.5 million hectares and production to about 2 million metric tonnes. However, the impact of liberal imports became visible sooner. The average area under sunflower in the country has come down to 1.5 million hectares and production to less than 1 million metric tonnes. The average yield is still under 600 kg/hectare. Close examination of data shows that during the last decade (2000–01 to 2010–11) the area and production of sunflower in the country have been going down at compound annual rates of 0.045 per cent and 0.036 per cent respectively with a variation of 40 per cent and there is a negligible increase in yield rate. Some serious efforts will be needed to stabilize area and production and to bring yield to the level in Haryana if not more or to the world standards. The yield in Haryana is more than two times of national average.

Haryana plays a negligible role in the contribution of sunflower. Haryana's contribution to average production of sunflower in India is around 3.5 per cent, while it covers only 1.2 per cent of the area.

As there is no surety of market price and, like other edible oils, sunflower is also susceptible to weather and pests, it has not become the preferable crop of the farmers. Therefore, marketing and price also need detailed examination.

Price and Marketing of Sunflower

Though agriculture is a state subject, the fact remains that the centre, through a number of interventions, procurement, storage (CWC), export/import policy, inputs and infrastructure, direct monetary benefits in the form of subsidies, physical restrictions, release of grants with conditions (NHM grants to states with condition of change in APMC Act for example). etc., affects the level and standard of marketing of agricultural produce in the country. The banning of cotton export, removal of Guar seed from futures, banning of onion exports are few examples. The states' role through physical facilities and marketing infrastructure as well as monetary incentives/disincentives in the form of bonus on MSP, concessions on power, fuel, and waiver of interest on loans also affects in many ways.

Prices of competing crops, instability in returns of potato—a most suitable compensatory crop in North India—and liberal imports at lower tariffs, etc., of other oils are important factors causing a decline. Pricing of the commodity in fact is a tactical tool to effect huge changes (Li, Sexton, and Xia 2006).

With regard to MSP and procurement two–three issues emerge. MSP of edible oils should be based along with cost of cultivation on other factors like soil health and need of change in cropping pattern, water availability and requirement, foreign exchange outgo on imports vis-à-vis returns to the farmers and most importantly weather and climatic risks. A statutory provision should be made so that policy ad-hocism does not become a rule. Third, it is observed that MSP works as pivot for the market price. The market price during the peak season revolves around MSP. Hence, incentives to farmers to promote crop rotation should be kept under consideration. Lastly, determination of cost of cultivation (starting from primary data collection to calculations) needs a thorough review (MOSPI 2010) as there are instances when MSP was routinely increased by Rs 5–10 for years or Rs 50 later or there was no change at all though cost of cultivation increased. On the contrary, when the general price level increased, cost of cultivation of sunflower for the years 1996 through 1998 shows decline of around Rs 2000/ha.

Between 2001 and 2011, MSP of sunflower increased at a negligible rate of 0.08 per cent, which is almost not compatible with prices of inputs determining cost of cultivation. Like determination of MSP, procurement has never been consistent and compatible with production or market arrivals. Many times, it was not introduced even when market prices were lower than the MSP. Procurement of sunflower was 7 per cent, the highest ever of production during 2000–01. After that, it has never reached even 1 per cent of production. Trend of procurement vis-à-vis production is negative (*Agricultural Statistics at a Glance*).

SUNFLOWER MARKETING IN HARYANA

Haryana is well placed as far as provision of marketing and infrastructure facilities are concerned. Most of the sale takes place in local purchase centres or market yards. There is not a single centre exclusively established for the purpose of PSS. The nodal agency has to make purchases from the regulated market system under the supervision of the concerned APMC. Sunflower is sold like any other commodity in the regulated markets.

Farmers and traders from Punjab also sell sunflower in the Ambala and Shahabad regulated markets due to large-scale processing of sunflower in Shahabad and Ambala.

Like many other districts in Haryana, both Ambala and Kurukshetra are well developed in infrastructure and well connected with the catchment villages through roads. Details of market-wise arrivals show that overall market arrivals in Thanesar have increased by 0.05 per cent compounded annually during the last five years but with a huge variation in arrivals of individual commodities. Many have gone down substantially while others have increased. For example, arrivals of oilseeds have increased by about 0.5 per cent annually whereas there is a significant decline, about 0.4 per cent, in arrivals of sunflower (Table 42.2).

The arrivals in Shahabad another important market in district Kurukshetra have similarly increased overall by 0.02 per cent annually but with a huge variation commodity wise. Arrivals of oilseeds have gone up by about 0.15 per cent, but there is decline in arrivals of sunflower by about 0.1 per cent annually. The decline in arrivals of sunflower in the district is associated with the declining pattern of production. However, Shahabad is known for receiving sunflower even from the Punjab, decrease in arrivals in this market might be indicating decreasing production of sunflower in the Punjab. In the Punjab, the area under sunflower has come down from 70,000 hectares to 15,000 hectares. In district Ambala, another important sunflower producing district, the department of agriculture due to miniscule arrivals has stopped even to enumerate the crop under a separate head.

Procurement under the PSS by the HAFED/National Agricultural Cooperative Marketing Federation of India (NAFED) has never been substantial, neither in number of years it was under taken and nor as a proportion of production or arrivals. In Haryana, total procurement of sunflower on behalf of NAFED was undertaken by HAFED for two years, 1648 quintals in 2009–10 valued Rs 36.5 lakh and 811 quintals in 2010–11 worth Rs 19.06 lakh. Thus these negligible purchases made under PSS cannot make any dent on area, price, farmers' income, etc.

In fact, the purchases made under the PSS were not only insufficient, but the process adopted was also not as per the objectives of the scheme. Even during 2009–10, the price paid was below the MSP. The purpose of PSS was to buy directly from the farmers so that distress sales could be avoided or the farmers should be saved from depressed market prices. In the case of sunflower in

Table 42.2 Sunflower Arrivals (Qutls) in All the Markets of Kurukshetra

	2009	2010	2011	2012	2013	CGR
Thanesar	9,506	9,169	4,815	2,938	1,869	–0.35539
Ladwa	39,066	30,114	7,878	5,920	5,764	-0.42038
Shahabad	114,749	118811	95,974	82,815	92,030	-0.07709
Ismailabad	17,008	21,746	14,666	8,557	13,060	-0.13592
Pipli	14,980	12,853	7,289	6,336	6,498	-0.21162
Babain	21,608	13,588	8,193	8,463	10,378	-0.17636
Distt.	216,917	206,281	138,815	115,029	129,599	-0.14906

Source: DES, Haryana, HAFED Ltd.

Haryana, the government issues a notification and NAFED requests HAFED to buy on its behalf. The representatives of the corporation go to the market and buy the commodity without bothering whether it was from the farmers or not.

The produce is then stored in the godowns by HAFED on behalf of NAFED and it was up to NAFED when to dispose the produce off. Generally, it has to wait for the instructions from the government for disposal. The loss due to difference in prices paid and received is made good by the government up to 15 per cent of the MSP. However, payment creates problems as the corporation has to make payment immediately but the money from the government comes after months, sometime after years. As far as HAFED's cost benefit is concerned, it takes a fixed commission from NAFED, which is included in latter's total expenditure.

The process and operation, thus, does not appear to meet the objectives. The whole process starting from the issuance of notification, procurement, quantity to be purchased, storage, disposal, and release of payment, etc., therefore, need a thorough review and modification to achieve the targets of self-sufficiency in edible oils.

SURVEY RESULTS

Data from respondents show that leasing in and out of land takes place in each size group. However, none of the size groups is without any sort of debt. In all about 40 per cent households took loans, 40 per cent of small-size and 50 per cent of large-size households were in debt. The loan was taken mainly for production. The loan was mostly from commercial banks. Loans taken from commission agents or from moneylenders, relatives, etc., was not revealed. Lastly, the average loan ranges between Rs 2 and 4 lakhs.

Data for two years 2010–11 and 2011–12 show that small-size households have devoted the highest share 16 per cent and 11 per cent of gross cropped area (GCA) to sunflower during the two years. The share of sunflower in medium- and large-size farms remains 8–9 per cent for both the years. No marginal farmer was sowing sunflower. A substantial part of paid-out costs goes in the production process whereas marketing costs are almost negligible. The price received by the households in 2009–10 was less than the MSP despite the procurement by HAFED under PSS and a little over the MSP in the year 2011–12, though there was no PSS operation in that year. As far as marketing channels and market charges are concerned, for the producers there is only one channel in the market: seller–commission agent–buyer.

CONCLUSION AND POLICY OPTIONS

It is thus obvious that PSS operations in the case of sunflower in Haryana are very limited. However, as the PSS directly affects the farmers' income, to some extent quality of their living, and cropping pattern, they support it. Considering the miniscule level of PSS, district agricultural officials, APMC officials, and nodal agencies did not have much to say. Based upon the information available about the level of PSS, the relevance of the scheme becomes questionable. In fact, by looking at the coverage of area under sunflower, yield, and production of sunflower in the state one cannot find any impact of the two years' intervention. However, looking at the need of the edible oilseeds and needed change in cropping pattern, measures to improve yield, area, and production will be essential, and for that price factors in the form of MSP and PSS will play a major role. Hence, the above discussion will lead to suggest following policy options.

Suggested Action

1. Limited current level of PSS in sunflower maximum could do was to motivate the other buyers to offer higher prices. Had there been no intervention and consequent uplift in the market mood, the farmers' returns could have fallen further.
2. Considering the importance and need of edible oils in the country, it is necessary that area under oilseeds in general and under sunflower in Haryana in particular must increase, because yield of sunflower in Haryana is about twice of the all-India average. For that, price factors are important; therefore MSP of sunflower needs to cover the risk factors related with production and market instability and also offer attractive returns, keeping in mind all costs of imports. To maintain that higher level of MSP, PSS operations need to be made almost regular and more quantity needs to be purchased.
3. In the case of sunflower, MSP should be made a tool to promote the crop.
4. HAFED can introduce its own processing of sunflower either by taking over the existing mill/s or establishing new ones. That will help in regular procurement and will also assure the farmers of stable returns from sunflower. Moreover, new mill/s with upgraded technology will have a higher milling efficiency.
5. As sunflower has remained concentrated in a few blocks of district Kurukshetra and Ambala, emphatic extension services will help promote the crop in other areas.

6. Sunflower is also susceptible to pests, therefore, attractive crop insurance can be helpful in its promotion.
7. Adequate credit supply needs to be maintained at reasonable rates of interest.
8. As far as infrastructure is concerned, marketing, transport, roads, etc., are well established in the state. Only crushing of oil seeds needs modern mills and equipment.
9. Promotion of sunflower is also important from the point of view of spoiled soil health in the state due to wheat paddy rotation, specifically in Kurukshetra and Ambala, the districts where sunflower is struggling for survival.
10. Potato and groundnut, other complementary crops, are equally meaningful and suitable. Therefore, marketing of these crops, with assured returns through price factors need encouragement, which ultimately will help in promoting sunflower.

Case of Apple 'C' Grade

In India apples are categorized into three grades 'A', 'B', and 'C'. Along with other specifications such as colour, maturity, freshness, unpunctured skin, brands, and varieties, diameter is an important criterion of categorization. 'A' grade apples have more than 80 mm diameter, 'B' grade apples have between 65 and 80 mm, and all those less than 65 mm are graded as 'C'.

In Himachal Pradesh, the Himachal Pradesh Horticultural Produce Marketing and Processing Corporation (HPMC) buys at the stipulated price, processes, and sells. However, in Uttarakhand, the horticulture department is entrusted by the government to buy 'C' grade apples at the stipulated price. MIS in Uttarakhand is not a regular feature. Kumaun Mandal Vikas Nigam and Garhwal Mandal Vikas Nigam are not at all involved in processing or procurement.

Coverage under MIS

Uttarakhand plays a minor role in area and production of apple in the country. During three years (2008–09 to 2010–11), the share of Uttarakhand in area under apples has been between 5 and 6 per cent while in production only 2–3 per cent. Naturally, in yield it is behind other states.

However, in the state, the share of district Uttarkashi has been about 23 per cent in area under apples and about 32 per cent in production during the three years. Further, in block Mori, about 83 per cent of area under fruits is used for apple. The share of apple in the production of fruits is about 87 per cent. Moreover, in the district, Mori block covers about 42 per cent area under apple and about 44 per cent of production. Thus, a little less than half of apple in district Uttarkashi is produced in block Mori and overall about 14 per cent of apple produced in the state comes from this block.

Uttarakhand in general and being the top north district Uttarkashi in particular, suffers from the lack of good marketing infrastructure. There are 66 wholesale markets in the state. However, number of regulated markets is 58 only, with 25 principal regulated market yards and 33 sub-market yards. Out of 25 markets, 20 are functional. In district Uttarkashi, the sole regulated market is non-functional. Therefore, most of the fruits and vegetables (including from Mori) are sold in Dehradun and Kanpur, mostly through contractors.

Marketing Practices

After the intervention of Mother Dairy and other private players like Reliance, Birla, Chirag, Shree Jagdamba Samiti (SJS), etc., horticultural produce which was marketed and consumed locally earlier has access to distant markets like Kanpur, Dehradun, and Delhi (Bhupal 2009).

As far as 'C' grade apple is concerned, if not bought by private processing units it is generally sold in the nearby market at throw-away prices. Many times farmers, not sure of recovery of even transport costs, do not bring produce to the market. In absence of a local market, MIS becomes important. The scheme takes shape after state government's orders for procurement, which is supposed to take place at the MIS purchase centres, established for the purpose. In district Uttarkashi, five such centres have been established. For Mori block, apple 'C' grade is generally purchased at the Arakot Centre. In Uttarakhand only for 3–4 years, apple 'C' grade was purchased under MIS and that too not in substantial quantity. In other words, MIS has not played any effective role so far in coverage of apple 'C' grade.

A detailed discussion with farmers suggests that about 18 to 20 per cent of fruit turns into 'C' grade, depending upon the snowfall, rainfall, setting of the fruit, pollination, etc. If we roughly take 15 per cent average, we can say that during the years 2008–09 and 2010–11 when only 86.46 and 33.25 metric tonnes of apple 'C' grade was procured out of a production of 4,000 to 5,000 metric tonnes and 6,000 to 8,000 metric tonnes of apple 'C' grade during these years respectively, which works out 2.1 per cent to 1.7 per cent in 2008–09 and between 0.55 per cent to 0.41 per cent in the year 2010–11. Therefore, much more

Table 42.3 Apple Marketing under Market Intervention Scheme

Year	Rate, Rs/kg	Quantity (MT)	Amt. Lakh Rs	Agency*
2007–08	4.5	114.95	5.17	HMT/ KGMVN
2008–09	4.5	86.46 (4-5k)@	3.89	HMT
2009–10	0	0	0	Na
2010–11	6	33.25 (6-8k)@	1.99	HMT
2011–12	0	0	0	Na

Notes: *HTM: Horticulture Mobile Team; KGMVN: Kumaun Garhwal Mandal Vikas Nigam; NK: not known; NA: not applicable @ rough estimates of apple 'C' grade as 15–18% of total apple production during the years.
Source: DES, Uttarakhand, Directorate of Horticulture, Uttarakhand.

needs to be done. Table 42.3 gives details of procurement under MIS during the years.

Disposal by Respondents

About the pattern of disposal under the MIS, the following points are made. Only one-year data could be used because there was no MIS for apple 'C' grade during the year 2009–10. Second, during the year 2010–11 when the apple 'C' grade was purchased under MIS, a total of 33.25 MT of apple 'C' grade was purchased. In addition, out of that 152 quintals or about 46 per cent were sold by our respondents. Therefore, the sample covers about 50 per cent of the targeted crop. Hence, the results, namely, opinions, difficulties, perceptions, etc., can be treated with confidence. The largest share of produce sold under MIS comes from marginal farmers, followed by small farmers. There is six times difference in price received through MIS and that received for other category of the produce that is for 'A' and 'B' category apples.

Another important issue is that though respondents sold about 50 per cent of the total procurement made under the MIS, but that covers only 5.4 per cent of their total production. If 15–18 per cent of the produce turns out to be 'C' grade then even the respondents were left with two-thirds of the produce still to be marketed (Table 42.4). One can consider the position of other farmers of the Mori block who were not our respondents, other blocks in the district, and other districts in the state. In other words, the fate of their 'C' grade produce. Thus nutritious food goes to the waste basket (Srinivas 2013; Yasmeen 2013). Overall, thus at current levels of procurement, MIS covers almost nothing and fails to make any effect. Therefore, the argument that in the absence of MIS even these returns will not be possible, may hold.

However, for MIS to make a significant effect it should be regular and reasonable quantity needs to be procured. The procurement price should be considered in terms of utility and value of processed 'C' grade apples. For example, for *murabba* (preserve), only small size apples are most suited/used, and that too of any quality. The murabba is sold at Rs 140–150 per kg, then paying Rs 6 or 7 per kg of apple 'C' grade seems unjustified. It can be easily understood if Patanjali Yogpeeth buys 'C' grade apples at a higher price than MIS. It underlines the need of processing of 'C' grade apples into jams, jellies, squashes, juices, murabbas, etc.

Looking at huge margins, an NGO like SJS can earn profits even keeping the margin very low and with all liberal expenses on its staff and payment to farmers. For example, in Delhi one kg. of apple Murabba is being sold for Rs 150 per kg, which contains hardly 300 grams of apple, the rest is sugar and water. Thus with 1 kg of 'C' grade apple, 2.5 to 3 kg murrabba worth Rs 400–450 is prepared. In other words, there is lot of scope to increase the procurement price of apples.

Limited availability of credit with the households is another issue. A loan of Rs 26 lakh was available to total 69 sample households (beneficiary and non-beneficiary) and with that they were running their economies, agriculture, horticulture, animal husbandry—all put together an annual economy of lakhs of rupees. With this small loan,

Table 42.4 Apple Produced by Farmers and Its Disposal Pattern Uttarkashi, Beneficiary

Crops	Production (qtls)	Kept for Home Consumption (qtls)		Marketed (qts) under			Price (Rs/kg) through	
	2010–11	2010–11	% of Prod.	Other	MIS	% Sold under MIS	MIS	Other
Marginal	1,127	28	2.48	1,027	73	7.11	6.5	36.4
Small	1,067	8	0.75	999	61	6.11	6.5	37.5
Medium	816	7	0.86	790	19	2.41	6.5	36.3
Large	0	0	0	0	0	0	0	0
All	3,010	43	1.43	2,816	152	5.40	6.5	36.75

Source: DES, Uttarakhand, Directorate of Horticulture, Uttarakhand.

they were providing food, shelter, health, education, etc., to 1,115 persons. After deducting the number of children and senior citizens, employment to about 800 persons was provided. The amount of loan per household in both types of households works out to less than Rs 33 thousand in the case of non-beneficiary households and about Rs 44 thousand in the case of beneficiary households. Hence, time, quantity, and cost of availability of loan are other issues. Moreover, this loan was for production and not for marketing. Clearly, efforts are needed for arranging loans from public sector institutions.

CONCLUSIONS AND POLICY OPTIONS

In the light of the above following points will be helpful in improving the production and marketing of apple in the region.

1. Production of apple, particularly through yield enhancement needs to be improved. For that, agronomical efforts (quality plants, proper care, and nursing, etc.) along with provision of easy and adequate credit need to be made.
2. High quality seeds and extension services for the proper care of the plants need to be emphasized, so that ratio of 'C' grade apple to that of 'A' and 'B' is reduced.
3. The state lacks in marketing infrastructure, particularly in number of required regulated markets, which need consideration.
4. Along with markets, proper storage, transportation, and packing, etc., need to be improved.
5. As production, per se cannot improve the income and living standard of the producers unless it is efficiently marketed, processing in the area needs to be taken up speedily.
6. If 'C' grade apple is not bought by private processing units and under the MIS, it is either sold in the market at throw-away prices or it turns into waste. Therefore, processing facilities under private public partnership along with one like HPMC needs to be considered (Reddy 2001). There is no dearth of demand of processed apple with handsome margins; hence, it would be beneficial to the economy of the state as well.
7. As far as MIS is concerned, with this negligible intervention in the market, the role of MIS in influencing cropping pattern, farmers' income, market price, etc., cannot be significant. However, that cannot be construed that it might not have affected the farmers' returns (*Report of Task Force*). Hence, the concept of MIS needs to be emphasized keeping in mind the total production of 'C' grade apples and its purchase by private agencies. It would be worthwhile that after the purchases made by the private agencies, the entire left over produce should be procured at a reasonable rate under MIS and processed by the government/ cooperative body.
8. It would be worthwhile if the minimum price of apple 'C' grade is determined by keeping in mind the market value of its processed products along with the cost of cultivation.

REFERENCES

Agricultural Statistics at a Glance (various issues). Ministry of Agriculture, GoI.

Bhupal, D.S. 2009. *Impact Assessment Study of Agricultural Market Reforms.* Delhi: AERC.

Government of India (GoI). 2002. 'Report of Task Force on Agricultural Marketing Reforms'. Ministry of Agriculture Department of Agriculture & Cooperation Krishi Bhawan, New Delhi.

Li, Lan, Richard J. Sexton, and Tian Xia. 2006. 'Food Retailers' Pricing and Marketing Strategies, with Implications for Producers'. California: Agricultural Issues Center, University of California.

Ministry of Statistics and Programme Implementation (MOSPI). 2010. *Manual on Agricultural Prices and Marketing.* New Delhi: Government of India, Central Statistics Office Sansad Marg.

Reddy, Y.V. 2001. *Advantages of Processing.* Address at Conference of Indian Society of Agriculture Marketing at Vizag on 3 February.

Srinivas, Tulasi. 2013. 'As Mothers Made it…', in *Food and Culture: A Reading* (3rd Edition), edited by Carole Courihan and Panny Van Esteriks, 355–75. UK: Routledge.

Yasmeen, Gisele. 2013. 'Not from Scratch: Thai Food', in Carole Courihan and Panny van Esteriks (eds), *Food and Culture: A Reading* (3rd Edition), pp. 320–9. UK: Routledge.

43

Impact Assessment of Agricultural Market Reforms

Evidence from Uttarakhand and Haryana

D.S. Bhupal

To enhance private investment in agriculture, some important policy measures, such as a Model Act to change the Agricultural Produce Marketing Committee Regulation Acts (APMC Acts), contract farming, liberalization of agricultural exports/imports, and futures trade in agricultural commodities were brought about. These were necessitated after signing of international trade agreements like WTO and Free Trade Area Pacts. It was considered necessary to allow the corporate sector to have captive production, marketing arrangements, storage, processing, transportation facilities, and export and import of agricultural commodities. It was envisaged that with free trade a lot of opportunities would emerge for Indian agriculture in the international markets due to removal of trade barriers, quantitative restrictions, and reduction in tariffs by developed nations, where the demand for competitive commodities from the developing economies was much higher. It was emphasized that once the agricultural subsidies were withdrawn, agricultural trade would expand, leaving export markets open for countries like India (Gulati and Kelly 1994), where labour costs were less and agricultural production would be cheaper and competitive. The farmers would be able to share benefits if they were linked with international agriculture through trade and industry. If commercial crops replace staple crops threatening food security, it was argued that why produce everything locally and store for fear of shortages when it could be imported at much lower costs, even lower than carry over costs (Parikh). This was to counter arguments by Nayyar and Sen (1994) who emphatically pointed out that many times it might not be possible even to meet the fraction of requirement of India. Whenever she, a huge market in herself, entered the international market as a buyer or seller, the international prices significantly shot up or crashed respectively, consequently threatening the food security of the nation and livelihood of the farmers.

With the entry of the corporate sector after implementation of these policies, it was expected that high-value crops like fruits, vegetables, and plantations would get preference in place of low-margin crops like cereals (Acharya 2001). Therefore, there will be enhanced employment opportunities because of higher production of high-value crops which require more labour, and due to more investment in transport, processing, and subsidiary activities like road-side small marketing, transport repair and accessory shops, and due to overall development of infrastructure like roads, warehouses, processing units, banks, etc. Enhanced employment and more production of high-value crops will push up income which in the hands of population will ensure food security by easy access to food and nutrition and ultimately be helpful in removal of poverty and hunger. The increased income would mean more demand for industrial goods. Thus, reforms in agricultural marketing would result in overall agricultural growth which ultimately will take the growth of the economy to a higher trajectory (Reddy 2001).

With this vision the central government conditioned certain grants, like benefits under National Horticulture Mission with amendments of APMC acts, for the motivation of states. The Model Act paved the way for direct purchase by the corporate sector from the farmers. Contract farming was introduced to help the sector build up their captive supply. The other benefits of contract farming were envisaged in the form of enhanced income to small and marginal farmers due to assured improved seeds and other inputs and buy-back arrangements of produce by the corporate sector.

To promote the regular flow of raw materials and to meet trade commitments of processed products uninterruptedly, the corporate sector was allowed strategically located huge areas in the form of special economic zones (SEZs); additionally, to cover risks and discover prices, future trading was permitted. For that commodity exchanges were established. One of the strongest arguments in favour of future trading was that farmers would benefit from the price discovery as well as by minimizing price risks.

Finally, to increase the share in international markets trade restrictions were removed and about 1,400 product lines were put on the open general licence (OGL) in the very first year and substantial reduction in import duties was effected, which was followed by lowering the quantity of buffer food-grain stocks.

All these developments brought changes in India's pattern of exports and imports. They might have helped achieve a higher export-led growth trajectory and higher income through competitive pricing (Bawcutt 1996), but agricultural growth did not see a spectacular upward shift. This may be due to the fact that the crucial investment in irrigation, infrastructure, marketing, transportation, and storage was neither made nor facilitated to the expectations of the corporate sector. In addition, other administrative barriers like unrestricted movement of agricultural produce throughout the country, choice to sell sugarcane free of area restrictions, etc., were not removed. If higher growth in agriculture could not be achieved, how far were cropping pattern, rural income, employment, etc., affected by these changes was not seriously assessed. This study was planned to find out the impact of changes in agricultural marketing policies at the grass root level, particularly, regarding change in cropping pattern, farmers' income, and rural employment pattern.

OBJECTIVES

The specific objectives of the study were:

1. to find out changes in the cropping pattern such as replacement of lower value crops by higher value commercial crops;
2. to examine changes in agricultural production and farmers' income;
3. to analyse changes in the form and nature of employment; and
4. to suggest modifications in policy.

DATA AND METHODOLOGY

To pursue the above objectives primary data from two important states Haryana and Uttarakhand, totally divergent in topography, agro-ecology, and production practices, were collected. With the objective of commonalty of crops, vegetables from Haryana and fruits and vegetables from Uttarakhand were chosen. With the help of agricultural departments of the state governments, tentative lists of private parties buying fruits and vegetables from both the states were obtained. Reliance, Birla Group, and Mother Dairy were important players in both states. Chirag was active in Uttarakhand. However, as the Mother Dairy, a semi-cooperative body, was the key and oldest player and data with it were well documented, we chose only that area where Mother Dairy and other players were active. Our focus was on the farmers selling to them. We also selected other farmers, the control group of sample, from the same area who were selling their produce in the market, in the case of Haryana in Sonepat or Azadpur market, Delhi, and in the case of Uttarakhand in the nearby markets such as Nainital, Bhimtal, Bhowali, and Haldwani.

The selection of respondents was based upon a stratified four-stage random selection. In the first stage after consultation with Mother Dairy and the state agricultural departments, Sonepat district from Haryana and Nainital district from Uttarakhand were selected. Sonepat and Nainital tehsils were selected at the second stage. At the third stage villages (both types—selling to Mother Dairy and other private buyers) from each tehsil were selected. In Sonepat four villages where farmers were selling either to Mother

Dairy or to other groups in the villages and three villages were the control group were selected. In Uttarakhand four villages were selected—two with farmers selling to Mother Dairy and the other two with farmers selling in the market in one case and in the other selling to both, Mother Dairy, other players, and in the market as well.

With the help of Mother Dairy, a list of local farmers selling to Mother Dairy was prepared and with information gathered from these farmers and local heads, lists of other farmers were prepared and then the selection of the requisite number of respondents from both the sets of farmers was made for collection of data on the pre-tested questionnaires. As the agricultural practices differed vastly in the two states, there was a little difference in the two sets of questionnaires canvassed for data collection in the two states. The data were collected from 90 sample households in Haryana and 50 in Uttarakhand. The list of villages and respondents is given in Tables 43.1 and 43.2.

LIMITATIONS

As the data have been purposively collected from areas growing horticultural crops and also wherein the private marketing agencies were involved in both the states, it imposes a big limitation that the results cannot be interpreted to represent the overall scenario of each state. Second, due to resource constraints, the aggregate situation of food availability or supply could not be taken into account. For example, we know that there is ample scope to increase food supply from other than traditional states. Hence, reduction in area and production of wheat in the sample area needs not to be interpreted as an alarming signal about food security. The results, therefore, are to be interpreted with these limitations.

Table 43.1 List of Selected Villages

State	District	Tehsil	Villages	Control Villages
Uttarakhand	Nainital	Nainital/ Dhari	1 Budiwana	1 Malla Ramgarh
			2 Sunkia	2 Simyal
			3 Simyal	
Haryana	Sonepat	Sonepat	1 Jhundpur	1 Khewda
			2 Jakhauli	2 Garh Mirakpur
			3 Jajel	3 Palda
			4 Manoli	

Source: Field Survey.

BACKGROUND OF THE AREA AND SOCIO-ECONOMIC PROFILE OF THE RESPONDENTS

Haryana and Uttarakhand are vastly different in its four aspects of agriculture, namely, natural endowments, human resources to depend upon, infrastructure, and market access.

In Haryana, sample district Sonepat is adjoining the national capital Delhi, which has a huge demand potential for high-value crops like flowers, fruits, and vegetables and dairy and poultry products. Better road and rail links provide a quick delivery mechanism of the produce with relatively little loss/wastages and lower costs, whereas district Nainital does not have that location advantage, road and rail connectivity, for quick disposal of perishable commodities. On the other hand, climate, weather, and moisture content in the atmosphere in the hill regions are more advantageous for these crops, which Haryana lacks. In fact, to take advantage of demand potential and market accessibility, production of high-value crops seems to be demand-oriented in Haryana, which in Uttarakhand should be supply-pushed.

LAND DETAILS

The difference in land ownership and cultivated land emerges clearly between the two states. For example, in Uttarakhand, because of obvious reasons we did not observe any leasing in or leasing out of land whereas in Haryana, in both categories of farmers we find cultivated area more than owned area due to leasing in by the households growing vegetables. The reasons were obvious: growing vegetables was more profitable than cereal crops and most of the cultivation of vegetables is done by small and marginal farmers who, not finding other work in the village, generally go for vegetable cultivation and for that have to lease in land. This practice was noted in and around Delhi and UP as well.

Table 43.2 Number of Respondents

State ⟶	Uttarakhand				Haryana						
Villages ⟶	Sunkia	Budiwana	Simyal	Malla Ramgarh	Jhundpur	Jakhauli	Jajel	Manoli	Khewda	G. Mirkpur	Palda
Control	–	–	6	10	5	6	–	10	9	8	12
Others	14	12	8	–	8	12	14	6	–	–	–
TOTAL	14	12	14	10	13	18	14	16	9	8	12

Source: Field Survey.

In Uttarakhand, modern technology, namely, use of tractors, irrigation, chemical fertilizers, and pesticides can hardly be applied. Most of the operations are carried out manually just like in kitchen gardens. Therefore, the question of leasing in and out of land does not arise. Second, as the male members from the area come down to the plains for want of work, most of the agricultural operations are carried out by female members who can do that to a limited scale or on the owned land. Lastly, almost everyone in the area, due to lack of other economic activities, does farming. Therefore, leasing in and out is almost ruled out.

Most of the area under both types of sample households is rain-fed. Irrigation takes place only with the natural rainwater collected in small ponds by a few households. Whereas in Haryana, almost the entire area is irrigated, and for that every irrigation source—canals, tubewells/bore wells, and pump sets, is used. Even the leased in area which has increased during the three-year period of study by about 6 per cent is fully irrigated.

LIVELIHOOD AND SOURCE OF INCOME OF THE RESPONDENTS

In both the states, the source of livelihood of all the respondents was agriculture. However, in a few cases male members have gone for other income sources, for example, service and their own subsidiary business of transportation—tractor/trolley, in the case of Haryana and commercial passenger vehicles in Uttarakhand. However, such households are not more than 2 per cent. Female members in both the states were found working on fields, more aggressively in Uttarakhand (98 per cent) compared to 68 per cent in Haryana. Second, in Haryana, most of the harshest operations like ploughing, levelling, harvesting, threshing, etc., were performed mechanically as compared to manually and that too mostly by women in Uttarakhand. In Uttarakhand, in a few cases, operations of fruit collection, bringing it to the collection centre, sorting out, etc., were contracted out to persons from border areas/Nepal.

In both the states, the entire family income was under female custody, but spending was not their domain. All major spending decisions were taken by the male members except in rare cases in consultation with dominant female members.

RESULTS AND DISCUSSION

Cropping Pattern

With regard to cropping pattern, some interesting results emerge. First, there is difference in cropping pattern between the plains and hill areas. In the hills, fruit trees are grown on almost every part, and vegetables and other crops are intermixed. Therefore, change in area under fruit trees may be only due to replacement of trees or addition of new area, which however was not the case during the period of the study.

In Haryana a few observations are: first, in the sample region most of the area is devoted to vegetables and other crops. Unlike Uttarakhand, there is no area under fruits.

During the three years, 2005–06 to 2007–08, for which data were collected, it is observed that in the control group of households in Uttarakhand, the area under fruits as percentage of GCA went up from 57 per cent to 58 per cent. The area under vegetables was down from 41 per cent to 39 per cent, under maize it was only 1.3 per cent to 1.4 per cent of GCA. Finally it is wheat under which the area was less than 1 per cent, coming down from 0.81 per cent to 0.62 per cent during the period.

Two conclusions can be drawn from this trend: one, that as the farmers from this group of households sell generally in the market on their own, sometimes in the nearby markets like Bhimtal, Nainital, Bhowali, and mostly in Haldwani, which is about 75–80 km away from the villages, carrying fresh and perishable produce like vegetables daily to such a distance may not be economical in comparison to fruits, which can be considered as semi-perishable, particularly in comparison with vegetables. That is why the area under fruits is on the rise and that under vegetables is declining. Second, area under cereals like wheat is also on the descending path. This may be due to the fact that as fruits and vegetables from the area are now moving down to the plains in larger quantities due to the entry of some NGOs like Chirag, Birla, and the Reliance groups and Mother Dairy, returns from their cultivation may be becoming remunerative in comparison to crops like wheat, oilseeds, pulses, etc.

In comparison, in the non-control group, area under fruits is about 37–38 per cent and under vegetables about 55–56 per cent. In addition, in this group we find the trends of area just reversing, in the case of fruits on the decline, maybe marginally, from 37.44 per cent to 36.84 per cent, and under vegetables increasing from 55.1 per cent to 56 per cent. Possibly, because vegetables are taken away by the agencies to their retail outlets, thus creating enough market space for these perishable commodities. As far as area under wheat is concerned, there is not much difference in both types of farmers. It is on the decline. However, under maize mostly being cultivated for the purpose of sweet corn and baby corn, is on the increase. Thus area under maize has increased by about 12.5 per cent in the control group households and by about 7 per cent in non-control group sample households (Table 43.3). However, the most important point is about the decline in area under cereal crops, particularly wheat, which declined

Table 43.3a Area (Nalies) Under Fruit Crops (Control Group, Uttarakhand)

Crops	2005–06	2006–07	2007–08	2006–07/2005–06	2007–08/2006–07	2007–08/2005–06
All Fruits	351	377	377	7.4	0	7.4
All Vegetables	253	259	253	2.37	−2.31	0
Wheat	5	4	4	−20	0	−20
Maize	8	8	9	0	12.5	12.5
Other Crops	2	1	2	−50	100	0

Note: One Nali is equal to a 20th part of an acre.
Source: Field Survey.

Table 43.3b Area (Nalies) Under Fruit Crops (Non-Control Group, Uttarakhand)

Crops	2005–06	2006–07	2007–08	2006–07/2005–06	2007–08/2006–07	2007–08/2005–06
All Fruits	708	693	703	−2.12	1.44	−0.7
All Vegetables	1042	1060	1067	1.73	0.66	2.4
Wheat	39	35	29	−10.26	−17.14	−25.64
Maize	102	103	109	0.98	5.83	6.86
Other Crops	32	25	21	−21.88	−16.00	−34.38

Source: Field Survey.

by about 20 per cent in the control group households and by about 26 per cent in non-control households.

It has serious implications on food security of the area. It was noted that about a decade back, when there were no private agencies to buy horticultural crops from the area people were growing wheat, pulses, oilseeds, etc. Now they have to purchase wheat and even edible oils to meet their household consumption needs. Almost their total foodgrain demand could be met locally then, and now about 25 per cent of wheat is purchased from markets like Haldwani.

The increase in area under maize could have neutralized the shortfall in area under wheat and thereby the availability of cereals, had the area under maize been used to produce corn. However, it is being used to produce sweet corn and baby corn for supply to Delhi and other vegetable markets, which will enhance farmers' income but would not help meet food needs.

In Haryana about 19 per cent area of the control group households has increased under vegetables during the last three years, and by 7 per cent for the non-control group whereas area under wheat of the control group households decreased by about 2 per cent and that for the non-control group by about 11 per cent, which is a huge area. If the pattern continues, it may be a serious cause for food security, particularly foodgrains (Table 43.4).

Table 43.4a Area (Acres) Under Crops and % Change, Haryana (Control)

Area	2005–06	2006–07	2007–08	2006–07/2005–06	2007–08/2006–07	2007–08/2005–06
All Vegetables	106.8	117	127.3	9.55	8.80	19.19
Wheat	186.2	176	182.7	−5.48	3.81	−1.88
Mustard	2	2	2	0.00	0.00	0.00
Total	188.2	178	184.7	−5.42	3.76	−1.86

Source: Field Survey.

Table 43.4b Area (Acres) Under Crops and % Change, Haryana (Non-Control)

Area	2005–06	2006–07	2007–08	2006–07/2005–06	2007–08/2006–07	2007–08/2005–06
All Vegetables	228.5 (40)	247 (40)	245 (40)	8.10	−0.81	7.22
Wheat	164.5	145.5	146.5	−11.55	0.69	−10.94
Mustard	12	9	10	−25.00	11.11	−16.67
Total	176.5	154.5	156.5	−12.46	1.29	−11.33

Source: Field Survey.

We have seen area shifting towards horticultural crops in Uttarakhand, and there are reports about area under foodgrains, particularly under coarse cereals in Rajasthan also shifting towards commercial crops like Jathropa. As far as area under wheat in Haryana is concerned, that has gone down by about 13 per cent in both groups. There was no area under maize.

Production

The production of fruits in the control group households increased by about 9 per cent during the period, whereas that in the non-control group households marginally by less than 2 per cent. However, production of vegetables in both the sample groups increased by about 11.5 per cent in the control group and by about 8 per cent in the non-control group, thus leading to overall increase in production by about 20 per cent. However, production of food grains, area under which has been declining, particularly of wheat which fell by about 18 per cent in both sample groups. As far as production of maize in the control group households is concerned, it increased by about 23 per cent but fell by about 21 per cent in non-control group households. Thus, overall there may not be a major change in the production of maize. However, certainly there is shortfall in wheat production.

As far as production in Haryana is concerned, area and production of vegetables in both sample groups have increased, and production increased significantly by more than 17 per cent in the control group and by more than 33 per cent in the non-control group. Thus aggregate production increased by about 51 per cent in the three-year period. This not only led to an increase in farmers' income and employment, but also provided nutritious food to city consumers where the produce was sold. However, there was a shortfall in the production of wheat by about 2 per cent in the control group and by about 9 per cent in the non-control group. That needs to be addressed.

In sum, the impact on the cropping pattern and production is visible, there is an increase in area and production of fruits and vegetables, the increase in production is more than the increase in area. This shows an increase in productivity as well, which should be beneficial to the farmers and consumers. On the field, employment in the case of horticultural crops, particularly vegetables, has also increased. This extra requirement of labour is met by women mostly in the case of Uttarakhand, who are already overburdened. Growing of horticultural crops is beneficial in comparison to growing of cereals or pulses or edible oil crops. That is why there is a shift in area. The marketed surplus of fruit and vegetables is going up due to the intervention by Mother Dairy and other players in the sample area. Not only the marketed surplus of sellers to these players has gone up, but also that is true in the case of the other group of sample households, who are left with more market space in the absence of sellers to Mother Dairy and other players whose main markets are far off.

Income

The very fact, that the area under cereal crops is being shifted towards horticultural crops which are more prone to weather, price fluctuations, and moreover are not as much crucial as food grains, is not for nothing. There cannot be any other reason for that except a major difference in returns both due to difference in production (quantity) and margins (difference in costs and prices) in both the states. The data in terms of per unit area, except potato, show a positive variation in production of vegetables: cauliflower, cabbage, peas, and other vegetables. In other words, along with area expansion, there is an increase in yield of these vegetables. The generation of extra income, in comparison to traditional crops is supported by the benefit cost ratio also. The benefit–cost ratio of some selected vegetables was found to the tune of 1.87 in the case of tomato followed by potato (1.79), and ginger (1.68) under field conditions. Nonetheless, highest benefit–cost ratio was recorded in the case of capsicum (2.20) under protected cultivation (Dixit nd Pandey 2009). Earlier, (Bhupal, 1989, 1999, 2001, 2003, 2004, 2006, 2009) we recorded positive benefit cost ratios in the case of vegetable production and marketing in Delhi.

Employment

The production process, starting from bed preparation to harvesting of fruit and vegetables, sorting, marketing, etc., is labour intensive and requires more labour power per unit of area and per unit of output in comparison to other crops. It requires the whole family to work in fields to complete the operations in time. From both the states all the respondents were of the opinion that more labour days were needed in the cultivation of fruits and vegetables. In comparison to that labour requirement for the production of wheat, maize, and a few pulses was much less. However, in Uttarakhand about 98 per cent respondents were of the opinion that most of the extra labour was provided by the household females. The remaining 2 per cent hired labour from the North, even people from Nepal. Employment opportunities thus significantly increase with change in cropping pattern towards horticultural crops.

In states like Haryana, where cropping pattern is shifting in favour of cash crops, specifically in the sample area, extra labour absorption is useful to only those families who have family labour but not enough land. However, most of the agricultural labour in Haryana is from outside. The labour cost in Haryana is about the highest in the country. It helps the immigrant labour to earn some extra income. Overall, in vegetable cultivation extra employment is generated and mostly female members benefit.

In sum, with the intervention of Mother Dairy and others, few points from the data emerge, first, there is increase in area under horticultural crops, particularly under vegetables, in both the states and production of vegetables has increased. The increase in production has taken place both due to area expansion as well as due to increase in yield. Second, the marketed surplus has increased in the case of both types of producers/sellers, mainly because Mother Dairy and other players have increased the demand from their suppliers leaving the local market space to be filled by others. Third, there is an increase in on-field employment for the producers of vegetables and thereby in income because horticultural crops fetch better returns. Fourth, there is decrease in area under cereals in both type of respondents in both the states and production has gone down. Moreover, the production has gone down despite the fact that the production per unit of area has gone up, meaning the decline in area under cereals is nullifying the increase in yield too. The demand for cereals in the case of Uttarakhand respondents is being met by buying from the market. Therefore, the issue of food security of these areas will have to be addressed if the production of horticulture crops is to be stepped up. In Uttarakhand we have seen the yield rate of wheat is equal to almost pre-green revolution yield rate, which needs to be addressed by technological improvements.

Thus, the entry of Mother Dairy in the marketing of fruits and vegetables has had a positive effect on the farmers' income and employment through the change in the cropping pattern in both the states Haryana and Uttarakhand. In Uttarakhand, the contribution of the direct purchase from the farmers seems more as they do not have any other alternative. Therefore, the role of Mother Dairy in improving farmers' conditions in remote areas of Uttarakhand through enhancement of horticulture production needs to be strengthened further. It would be more so in the case of entry of other private players. Even some of the conscious farmers were aware of the gap in lower prices being offered by Mother Dairy in comparison to the prevailing market prices at their sale points, Delhi in this case. However, they were also aware of the situation that would emerge in the absence of Mother Dairy. We are also aware of the vast difference in prices the consumers pay in Delhi and farmers in Uttarakhand receive. For example, we bought apples at Rs 15 per kg, plums Rs 5 per kg, pears Rs 2 per kg, the highest price we paid for each item in the sample area, whereas the same quality of fruits were selling in the Delhi retail market, Dari retail fruit and vegetable market, at Rs 60, Rs 40, and Rs 12 per kg respectively at that time. Whatever be the transport costs, overhead charges, wastages, and other marketing charges and margins, the four, eight, and six times difference of price cannot be justified. This does not show that the entire price difference is pocketed by Mother Dairy. One should compare the prices with wholesale prices prevailing in the market. In that case, the difference is reduced by almost 50 per cent. In fact, we have seen during our earlier studies that it is the retailer in Delhi who pockets between 34 per cent to 54 per cent consumers' price depending upon the locality, time, and type of fruit and vegetables (Bhupal 2003).

BROAD CONCLUSIONS AND SUGGESTED ACTION

In light of the above, it can be concluded that with the intervention of Mother Dairy and other players, cropping patterns and production are changing in favour of horticultural crops. This change should be helpful to enhance farmers' income, generate employment opportunities, and increase availability of nutritious food to the consumers. The promotion of horticultural crops at least in states like Uttarakhand should find some extra support as it most suits the conditions. Therefore, marketing facilities and infrastructure such as roads, transportation, storage, and on-site processing need to be upgraded and supported. A network of Small and Medium Processing Units spread over vast areas will be most advisable. In addition, efforts to increase yield in hill areas by upgrading technology need to be made. In fact, the overall yield in the sample households of 6–7 quintals per acre in the case of maize and about 8–9 quintals per acre in the case of wheat is equal to almost pre-green revolution yield of these crops in plains. This needs serious exercise on the part of agronomists to enhance yield rate. To meet these challenges, more research units with the public sector and private capital need to be established and encouraged through financial incentives and policy intervention.

Increased yield rate will also address the problem of food security as area under and production of main cereal crops, wheat and maize, are showing a declining trend in the sample households. In fact, the technological intervention in the case of horticultural crops is also needed to spare land for other uses, both in hill areas as well as in plains.

Most importantly in Haryana, where a sizeable portion of land adjoining the National Highway No. 1 has already been devoted for the construction of malls, shopping and housing complexes, and in future land for such activities will be needed more aggressively, efforts to increase productivity become essential. One can recall that increased expenditure on agricultural research, extension services, and marketing measures helped the country to achieve self-sufficiency in food production in the green revolution period. A big boost to support agriculture through the Private Public Partnership (PPP) model with regard to enforcing changes in cropping pattern, land-use changes, production technology, and distribution, instead of patchwork through government missions, is the need of the hour to take agricultural growth to a higher trajectory, enhance farmers' income, and thereby enhance overall demand for industrial goods and other services which will be helpful to increase the overall growth of the economy.

REFERENCES

Acharya, S.S. 2001. 'Domestic Agricultural Marketing: Policies, Incentives and Integration', in S.S. Acharya, and D.P. Chaudhri (eds), *Indian Agricultural Policy at the Crossroads*. Jaipur: Rawat Publications, 129–212.

Bawcutt, D.E. 1996. 'Agricultural Marketing in a Highly Competitive and Customer Responsive Food Chain-experience in United Kingdom', *Indian Journal of Agricultural Marketing*, 10(3): 1–10.

Bhupal, D.S. 1989. *Price Spread in the Marketing of Vegetables in Delhi*. Delhi: AERC.

———. 1999. *Origin of Vegetables in Delhi*. Delhi: AERC.

———. 2001. *Changing Pattern of Vegetable Marketing in Delhi*. IJAM.

———. 2003. *Price Spread in the Marketing of Cauliflower, Okra and Palak in Delhi*. Sao Paulo: PENSA–V.

———. 2004. *Fruit and Vegetable Markets in and around Delhi*. Delhi: AERC.

———. 2006. *Production and Marketing of Vegetables in Delhi*. University of Sussex: STEPS Centre.

———. 2009. *Price Spread in the Marketing of Vegetables in and around Delhi*. ICH.

Dixit, A.K., and B.M. Pandey. 2009. 'Horticulture Technology Mission (MM-I) Project "Status of Horticulture and Market Opportunities in the State of Uttarakhand"', Paper presented in the Annual conference of ISAM at Ludhiana, Feb. 2009

Gulati, A., and Tim Kelly. 1994. *Trade Liberalization and Indian Agriculture*. Oxford University Press.

Nayyar, D., and A. Sen. 1994. 'Agriculture under Trade Policy Regime', *Economic and Political Weekly*, 29(20): 1187–203.

Parikh, K. *Writings and Lectures about Food Security and Storage Costs*.

Reddy, Y.V. 2001. 'Advantages of Processing', Address at Conference of Indian Society of Agriculture Marketing, Vizag, 3 February.

44

Understanding the Growth and Prospects of Agro-processing Industries in West Bengal, Bihar, and Maharashtra

Jiban Kumar Ghosh, Fazlul Haque Khan, and Vivekananda Datta

Dependence on agricultural sector, particularly on crop cultivation has resulted in widespread unemployment and underemployment in the country. The agricultural sector is characterized by ever-declining land–man ratio, predominance of small and fragmented land holdings and increasing application of labour-saving production technologies. It is thus being increasingly realized now that the very capacity of the agricultural sector is not enough to absorb the growing labour force. On the other, the organized industry sector, due to its capital-intensive nature cannot offer much scope for absorption of additional labour force. The environment of liberalization, privatization, and globalization has also thrown up newer challenges for employment. Obviously, all these have aggravated the unemployment and underemployment situation in India which underscores the need for alternative avenues for employment generation. This brings the development of agro-processing industries into sharp focus.

Agro-processing industry in India is largely a house of small-scale enterprises. They are highly heterogeneous in terms of capital investment, technology in use, scale of operation, quality and quantum of output, and composition and level of employment. More importantly, levels of productivity among tiny and small enterprises are low. There must be a lot of constraining factors covering institutional, technological, and marketing that are holding up productivity of the agro-processing units to low levels. There is, therefore, a need to address these constraints so that productivity of the agro-processing sector may be improved. As a whole, the strength of agro-based industry is comparatively less than those of non-agro-based industries. Moreover, the growth profile of the number of agro-based enterprises is uneven across the regions of India. It is this trend in the growth of agro-based manufacturing enterprises calls for undertaking this study. The main objective of the study is to study the problems and prospects of agro-processing industries.

METHODOLOGY AND DATA

The study was undertaken by three Agro-economic Research Centres (AERCs) in three states, namely West Bengal, Bihar, and Maharashtra. The present study attempts to consolidate the findings obtained from individual centre studies. The study makes use of both secondary and primary data. For secondary data, the study draws upon sources like the quinquennial National Sample Survey data for unorganized manufacturing and Annual Survey of Industries data for organized manufacturing. Secondary data relates to the select years, namely, 1994–95 and 2000–01, the latest available year being 2000–01. In India, the bulk of the units in the agro-processing sector are small and unregistered. Considering this, primary level data from the selected processing units are collected in order to capture the problems at the grass root level so that policy recommendation could be made for the promotion of agro-based industries.

Primary data was collected from the selected agro-processing units in each selected state. As the products of agro-industries are both edible and non-edible, the agro-based industries are classified into agro-food industries (or food processing industries) and agro-non-food industries. Thus, primary data are thus collected from the selected processing units chosen from both agro-food industries and agro-non-food industries. Altogether, 30 sample processing units are studied in each state. In Bihar, however, the study is based on 27 sample processing units. The sample units are selected at random proportionately spread over food and non-food processing segments of agro-based enterprises.

In selecting processing units, the food processing activities are broadly divided into three categories, namely primary food processing units, mainly grain processing units; spice and horticultural products and livestock-based processing units including fish processing. Similarly, non-food processing units are broadly divided into four categories namely, textile products, wood and its products, paper and its products, and leather and its products. For each category of enterprise, the dominant processing activity was selected consulting available secondary data. Sample districts are identified on the basis of the concentration of units of activities. In the case of food processing enterprises, for each selected processing enterprise, six units of different sizes, namely, OAMEs, NDMEs, and DMEs[1] with their distribution as 3:2:1, are covered. Within the non-food processing segment of the agro-based industry, for each selected processing unit, three units of different sizes, namely OAMEs, NDMEs, and DMEs in the ratio of 1:1:1, are selected. The units in the suggested proportion could not be selected in Maharashtra due to non-availability of entrepreneurs of a particular category at the time of survey. In West Bengal, the selected processing units are paddy processing, fruit (mango processing), fish processing, leather and its products, paper and its products, textile products (jute), and wood and its products. In Bihar, the selected units are paddy based processing activity, horticultural products (litchi), livestock-based processing units, textile products, wood and its products, and leather and its products. In Maharashtra, the selected processing units are cashew, fish, rice mill, leather, paper, textile, and wood. Primary data from the selected processing units are collected through canvassing structured schedule and questionnaire prepared for the purpose of the study. Data are analysed through simple tabular analysis.

[1] Own Account Manufacturing Enterprises (OAMEs) are those units which are run without the help of any hired worker. An establishment that employs less than six workers is known as a Non-Directory Manufacturing Establishment (NDME) while the one employing a total of six or more workers is categorized as a Directory Manufacturing Establishment (DME).

RESULTS AND DISCUSSION

Status of Agro-based Industry in the Selected States: Analysis Based on Secondary Data

West Bengal

As evidenced by Annual Survey of Industries data, the strength of agro-based industry is comparatively less than those of non-agro-based industries in the organized sector of manufacturing enterprises of the state. Evidently, however, in the concerned period between 1994–95 and 2000–01, the organized segment has tended to concentrate more and more on agro-based industrial enterprises. In the unorganized segment of manufacturing enterprises, the dominance of agro-based industry is clearly noticed. The unorganized segment of agro-industrial sector had as many as 86.30 per cent of total manufacturing enterprises, 81.54 per cent of employment of workers, and 69.09 per cent of gross value added. During the reference period, agro-based enterprises (both food and non-food) witnessed increase in the number of units leading to an increase in their share in units from 80.51 per cent in 1994–95 to 86.30 per cent in 2000–01. Importantly, agro-based industry is largely a house of household-based tiny and small enterprises, the proportion of OAMEs in the unorganized segment of manufacturing enterprises being 89.59 per cent.

Bihar

In Bihar, the unorganized manufacturing sector is characterized by the dominance of agro-based industries (including agro-food and agro-non-food) sharing 53.00 per cent in the number of total working units in 1994–95. Among the agro-based industries, the share of agro-food processing industries was estimated to be higher (28.45 per cent) than agro-non-food processing industries (24.55 per cent). Data for the year 2000–01 shows a significant decline in the number of working units under the groups of 'agro-food', 'agro-non-food', and 'non-agro-based industries' as compared to that of 1994–95.

Maharashtra

In Maharashtra, the unorganized sector clearly dominates the organized sector as far as the number of the units is concerned in both the years—1994–95 and 2000–01. In the organized sector, non-agro-based industries dominate with their share being around 70 per cent. However, in the unorganized sector, the agro-based industries are seen to be dominating the non-agro-based industries and their number has greatly increased (92.87 per cent) over the concerned period whereas that of non-agro-based industries has fallen (the percentage change being –19.98 over the period). Further, in the organized sector, the share of food processing industries in total agro-based enterprises has increased in the reference period while in the unorganized sector their share has declined.

Status of the Sample Units: Analysis Based on Primary Data

West Bengal

Status of the units were ascertained in terms of years of existence, average age of the units, and registration status. In West Bengal, all the sample processing units were existing ones, the average age of the unit being varied from 10 to 20 years in case of food processing units and from 3 to 22 years in the case of non-food processing units. It is observed, that investors are not keen on registering their units. On the aggregate, in about 50 per cent of the cases, entrepreneurs of the processing units are found to have registered their units. Notably, OAME units in all the category of enterprises are entirely unregistered. Average area of working place occupied by the DME units of manufacturing activity is seen to be more than the other category of manufacturing enterprises.

Bihar

In Bihar, most of the units are existing ones. Further, most of the surveyed processing units have been working in the unorganized sector are tiny, small, and artisan-based enterprises and so they are mostly unregistered. The average age of the sample processing units ranged between 8 to 35 years. DMEs under cereal-based processing activity occupied the largest area followed by DMEs of horticultural-product-based activity, wood-based processing activity, and livestock-based processing activity.

Maharashtra

In Maharashtra, the majority of the units are existing ones. It is the cashew processing units and the rice mills which are seen to be the new units. In the state, most of the units are registered. Four fish processing units and one OAME each from leather, textile, and wood category are the unregistered units. The fish units carry out their activity outside the house near the beach. The area covered by the units using machinery—cashew units, rice mills, and binding units are seen to be more than the other business units.

Production and Operation Cycle of the Activities

West Bengal

In West Bengal, the level of working of the units varied from activity to activity depending on the availability of working capital and seasonality of the activity in terms of input availability and demand for output. For all the activities, it is seen that monthly working days ranged between 26 to 30 days. The difference is noted in the case of per year working days. The number of working days per year for food processing units is observed to be relatively less than those of non-food processing units. Depending on the time taken for processing of the unit, the number of production cycles each unit completes is seen to be different being varied according to the type and size of the activity. Notably, within the category of food processing enterprises, the number cycles completed in a year increased with the size of the unit which is not observable uniformly across the category of enterprises in the non-food processing segment.

Bihar

In Bihar, number of working days per month as well as working hours per day were seen uniform in most of the cases, except in horticultural-crop-based (litchi), dairy-products-based and textile-products-based processing activities. As litchi-based processing activity is run hardly for 22 days to one month, so, in case of DME of this, double-shift work is undertaken. In regard to textile processing activity also, two production activities in two shifts, or more than eight hours are undertaken. Therefore, in these cases, working hours per day is longer. Livestock based processing activity is everyday business without fail on priority basis; however, its working hours is shorter (five hours). The number of production cycles, which the unit completes in a year, also differs with the type and size of the processing unit. It was quite higher in cases of livestock (300) and leather-based processing activity (ranging from 312 to 355). In all other activities, the number of production cycles were quite lower depending upon the availability of raw materials, time taken for processing the same and scale of operation.

Maharashtra

In Maharashtra, the number of working days per month as well as working hours per day are seen to be uniform for all the units. The difference is noted as far as working days per year are concerned. As the food processing units are located in the costal district of Ratnagiri, all the activities come to a halt because of heavy rains during June–September. Therefore, working days per year for these units are less than the non-food processing units in Pune and Mumbai. Depending upon the nature of activity, number of days required for other components of the operation cycle (stocking period, marketing, and credit realization period) are seen to be different for different activities. The number of production cycles which a unit completes in a year also differs with the type and size of activity. Normally within a category, the number of cycles completed increases with the size of the unit. Depending upon the time taken for processing of the unit, the number of production cycles each unit completes is seen to be different.

Net Income from Investments in Processing Units

West Bengal

In West Bengal, all the activities gave a positive net income, being varied among the activities depending upon the size of the investment. This is uniformly observable in the case of food processing units. Within the group of food processing units, paddy processing activity gave the maximum net income at Rs 185,718 per year followed by fish processing activity at Rs 161,583 and fruit processing activity at Rs 145,666. Small investments in units, such as fruit processing, yielded net income of smaller amounts in comparison with other units in the food processing category. For the group of non-food processing units, this particular pattern is not uniformly observed, although, paper-based processing units with maximum investment among non-food processing units accrued maximum net income of Rs 115,333 followed by wood-based processing units at Rs 89,583, leather-based processing units at Rs 74,133, and jute-based textile units at Rs 68,800. For all the processing activities (food and non-food), net income increased with the size of the unit.

Bihar

In Bihar, all the activities and units yielded positive net returns. The data revealed that except DME category of livestock-based processing activity, in all other cases under agro-food processing activities net returns increased with the size of the unit. In the non-food processing segment, similar pattern is observed except in case of net income earned by DME of textile-based processing activity (Rs 46,600), which is a bit lower than its NDME (Rs 51,850). It thus simply indicates the efficiency of the investments in bigger units.

Maharashtra

In Maharashtra, all the activities and units show a positive net return. For the food processing activities, the net return increases with the size of the unit. Among these activities, the highest net income is earned by the cashew processing unit (DME) followed by fish processing (DME). Among non-food processing units, this particular pattern, that is, increasing income with increasing size is not observed. This might be because of the heterogeneous nature of these non-food processing activities. Even within the categories, in some cases, the products of these units are differing slightly from the other units.

Employment Generation

West Bengal

Employment generation by the processing units covered in West Bengal showed wide variation. In the food processing category of enterprises, maximum employment generation from the investment was observed in the case of fish processing units with 7,662 man-days per unit per year followed by fruit processing (4,195 man-days) and paddy processing (1,550 man-days). Among the non-food processing units, maximum employment generation by the activity was observed in the case of wood-based product manufacturing units (2,150 man-days) followed by paper-based units (2,100 man-days), leather-based units (1,760 man-days), and jute-based textile product units (1,730 man-days). As expected, labour employment in the units increased with the increase in the size of the unit. With regard to employment across sexes, fruit processing units in the food processing sector and jute-based textile units in the non-food sector are seen to be female-dominated ones.

Bihar

With regard to employment opportunities created by the sample processing units in Bihar, it is observed that the number of total labour days engaged in the units increased with the size. The highest number of total man-days employed was seen in case of DME of horticultural-product-based activity figured at 24,200. It was followed by DMEs of cereal-based, wood-based, textile-based, leather-based, and livestock-based processing activities at 7,796, 4,050, 3,000, 2,700, and 2,000 respectively. It could also be observed that only OAMEs of cereal-based, horticulture, and textile-based processing activities engaged female family labourers. In regard to employment of hired female workers, it is found that only the DME and NDME categories of two agro-food processing activities, namely, cereal- and horticultural-product-based activities employed female workers on a hiring basis.

Maharashtra

In Maharashtra, the number of total labour employment in the units is increasing with the size as is expected. The highest number of workers is found in cashew processing DME units. It is also observable that all the categories in the food processing sector except one have engaged female family labourers. Thus, food processing (which can be carried out along with the domestic chores) is seen to be a female-dominated activity. As against this, in all, only three categories in the non-food sector (wherein work is carried out within the household) have engaged female family labourers. A similar pattern is found as far as hired female labourers are concerned. Leather as well as wood processing units are seen to be basically male-dominated units.

Problems Faced by Manufacturing Enterprises

West Bengal

Reportedly, in West Bengal, the problem of non-availability of raw materials throughout the year, variability of prices of raw materials and absence of information network to keep track of raw materials prices and availability came to be featured prominently in the array of problems faced by the entrepreneurs of sample processing units in West Bengal. For food processing units, the major problem in procuring raw materials reported to be variability of prices of raw materials (cent per cent) followed by absence of information network (72.22 per cent) and non-availability of raw materials (66.67 per cent) throughout the year. As far as the non-food processing units are concerned, the specific problem faced by the enterprises in procuring raw materials was reported to be variability of prices of raw materials (cent per cent). The next important problem faced by the non-food units reported to be absence of information networks (50 per cent) to keep track of raw material prices and availability. Thus, both food and non-food units faced the variability of prices of raw materials and in procuring the same face difficulties in fixing prices of products, having a bearing on the marketability of their products.

In the field of marketing of processed products, reportedly for food processing units, the main problem was lack of a proper domestic market of processed products (72.22 per cent), followed by the absence of a good network purveying market information (66.67 per cent), and dependence on middlemen for marketing the processed products (66.67 per cent). Notably, all the OAME units in the food processing segment reported these three problems uniformly across the category of enterprises. For non-food processing units, the major problems were reported to be the absence of a strong network for obtaining market information (58.33 per cent), followed by lack of a proper market for processed products (50 per cent) in the domestic market, and dependence on middlemen for marketing the processed products (41.67 per cent). Here again, OAME units in all categories of enterprises reported the above three problems in the sphere of marketing of their products.

Bihar

In Bihar, problems of non-availability of adequate raw materials due to lack of capital, supporting machines/equipments, and absence of required infrastructural facilities were reported by the majority of the food processing units. Fluctuations in prices of raw materials, absence of information networks and circumstantial purchase of raw materials from middlemen at higher rates were also prominently reported by the sample food processing units. Non-availability of skilled labourers, availability of raw materials (litchi) for a very short period and difficulty in determining prices of value added products were specifically felt by DMEs of agro-food processing activities. As far as agro non-food processing activities are concerned, lack of capital, poor quality of raw materials, and no easy availability of bank credit were reported to have been faced by OAMEs. Like agro-food processing activities, NDMEs of agro-non-food processing activities did come across the problems of poor electricity supply position, variability of prices of raw materials, and purchasing of raw materials from distant markets (Kolkata in the case of leather). Procurement of raw materials sometimes from informal trade channels, non-availability of strong supporting infrastructure, and raw materials (in the case of leather), were the main constraints faced by DMEs of agro-non-food processing activities.

With regard to the marketing of processed products in the domestic market, long market channels causing lower net income, non-existing support by NGOs/cooperative marketing societies and selling the value added products to local middlemen at lower prices were found to have been faced by entrepreneurs of small units in case of agro food processing activities. Seasonality of demand for the products, lack of mutual understanding among enterprises for preparing common marketing strategy, and quality consciousness of consumers compelling the entrepreneurs to sell their products in distant markets for higher prices were experienced by all NDMEs. Existence of tough competition, absence of widespread networks for marketing the products in the state (particularly litchi juice, syrup, and pulp), and transport-related problems in taking the value added products to terminal markets were the constraints faced mainly by DMEs of agro-food processing activities. Regarding constraints faced by sample entrepreneurs of agro-non-food processing activities, it is revealing that determination of the price of the products by middlemen or big traders, no option to choose profitable markets but to market the product mainly in the local market preferred generally by low-income groups of people yielding lower returns were reported to be the main problems faced by OAMEs. Marketing through middlemen resulting in lower net profit, and demand for quality product based on design-oriented preferences by the consumers were felt by NDMEs. Lower returns as a result of scattered markets, existence of tough competition, and uncertainty of ready demand were reported to be the main problems faced by DMEs of agro non-food processing category of enterprises.

Maharashtra

In Maharashtra, within the food processing segment, the majority of the cashew and fish units have reported non-availability of raw materials throughout the year. As far as the cashew units are concerned, non-availability of good quality cashews is mainly due to inability of the small units to find agents or sellers supplying good quality raw material. In the absence of information/resources to find the same, these units are often at a disadvantage if the cashews supplied are not of good quality. The units have also reported non-availability of labourers during the peak season and variability of prices. The fish units also face this problem, as during the months of monsoon, fishing does not take place. It was reported by these units that over a period of time, the supply of good quality fish has been reducing due to various reasons such as entry of large firms. Another major problem faced by the fish units is the absence of any government schemes/promotional agencies (unlike in the case of cashew units covered under the District Industries Centre [DIC]/Khadi and Village Industries Commission [KVIC]) which would provide support to these units. In case of the rice mills, the main problem reported was irregularity in the electricity supply. They have also reported that there is a tough competition from other rice mills as due to the liberal policy of the government regarding licensing, many new rice mills are being established. As far as the non-food processing units are concerned, majority of the units reported non-availability of cheap labour as one of the important problems. Similarly, the absence of any promotional agencies is another problem reported by the units.

Prospects of the Units

West Bengal

As revealed by primary data, within the group of food processing industries, paddy processing activity gave maximum net return in West Bengal. The state of West Bengal being blessed with the largest production of paddy has the potential for investing in the paddy processing industry. The industrial units in the future can take advantage of the growing demand for the value-added processed product in India as well as abroad. However, as observed in the study, this would be possible if the units have access to information networks to keep track of raw material prices and availability.

Within the group of non-food processing industries, textile and leather units yielded lower net income, although, they have shown relatively better performance in terms of growth in number of units. The common problem faced by the entrepreneurs of these units was reported to be the absence of network for the marketing of their products. Obviously, these units could enhance their earning capacity if they are provided with better infrastructure purveying market information for their processed products. Paper-based manufacturing units gave the highest net return amongst the non-food processing units and thus offer scope for investing in units manufacturing paper-based products.

Bihar

The development of agro-based industry in the state of Bihar is largely dependent on the importance attached to fruits and vegetables vis-à-vis other crops. In Bihar, significantly large areas are under different top qualities of fruits, namely mango, banana, litchi, guava, lemon, and pineapple. The quantum of production of these fruits is quite large. However, in the absence of required storage, preservation, and proper marketing facilities within and outside the state, a good quantum of these fruits are wasted and sometimes sold at unremunerative prices. Hence, there is great potential for installation of agro-processing industries based on these fruits in areas/regions with their production in abundance. Among cereal-based processing activities, apart from paddy and wheat, there is a high prospect for agro-processing industries based on maize in Bihar. With regard to livestock-based processing activity, the dairy industry in the cooperative sector under the brand name Sudha has achieved marked success in Bihar. In the unorganized sector, also there is great potential and bright prospect for processing of milk into khowa, ghee, butter, cream, paneer, lassi, etc.

As far as non-food processing units are concerned, it is revealing that leather and leather products in Bihar have a small share (0.56 per cent) in total production of agro-based industries. However, considering the magnitude and quality of livestock wealth and traditional expertise of leather men in Bihar, there appears to be a good potential for industries relating to leather and leather products in the state. Similarly, if the traditional expertise of weavers are utilized properly by providing them necessary inputs and infrastructure then the prospects of textile-based processing industries in Bihar is undoubtedly bright. Wood-based processing activities have also great potential in Bihar as the demand for value added products based on wood is on the increase with the growth of urbanization in the state.

Maharashtra

The analysis of the data collected from the sample processing units in Maharashtra shows that cashew units (DME) have earned highest net income followed by fish units (DME). The cashew units are newly established units under DIC/KVIC schemes. Due to the increasing demand for the cashew nuts in the domestic as well as international markets and due to the existence of huge untapped potential for processing of the fruit, the units cantake advantage of the expanding markets in the future as well. Fish processing units are existing units working since 10 to 20 years. Most of these have their business on the beaches under unhygienic conditions. The major problem faced by these units is the irregular supply of good quality fish. Similarly, these units are not covered under any scheme as they do not require heavy capital investments. Assistance to these units in terms of market information and value addition techniques would increase their earning capacity. The major problem for the rice mills is the contraction of the business due to opening up of many rice mills in the vicinity under the regime of liberal licensing policy of the government. The mills thus need to be more competitive and modernized.

With regard to the non-food sector, overall, there has been a contraction in the number of non-food units. As far as the sample units are concerned, these are located in the developed districts of Pune and Mumbai. The units do not exhibit uniformly increasing net income with size within a category. Thus, the DME units are not necessarily the units with highest net income. This may be indicative of efficiency of small units which depend mainly on family labour. The units mainly have reported non-availability of labour, absence of governmental support, and existence of rivalry as the main problems. The units would expand their earning capacity and be more competitive if they are provided information regarding market conditions and various existing schemes and extension services.

CONCLUSIONS AND POLICY IMPLICATIONS

On the basis of findings the following are the major policy recommendations that emerged from the study.

West Bengal

Adequate infrastructure like marketing infrastructure, storehouse, and cold storage facilities assume great significance in the context of the growth of agro-based enterprises. This is particularly evidenced by primary-level data analysis of sample food processing units. Thus, public investment in developing the required infrastructure needs to be stepped up for the growth of agro-based enterprises.

Pricing of products is an important element of marketing of agro-based products. In the present study, sample processing units experienced one major problem of variable prices of raw materials varying over the seasons. In the face of variable prices of raw materials, the processing units find difficulty in fixing prices of their products in advance. This has deterred these units from entering into forward contracts with the customers who can purchase their products at reasonable prices, thus ensuring the marketability of the products. Moreover, for want of information network infrastructure, the processing units are unable to assess the supply demand conditions of raw materials and thus prices of raw materials. They are also unable to forecast market demand for the product. This calls for creating infrastructure in the form of developing network linkages.

The constraint/problem common to the OAME and NDME categories of fish processing enterprises is the absence of information network both in the sphere of availing raw materials and marketing of the product. Therefore, assistance to these units in terms creating access to information networks would enhance the efficiency of these units.

As for the non-food processing units, the textile units have faced the basic problem of low market demand for the products. Similar to textile units, leather-based activities also face the problem of marketing of their products. However, the common problem faced by the entrepreneurs of leather and textile units reported to be the absence of network for the marketing of their products. Obviously, these units could enhance their numerical strength if they are backed by better infrastructural support providing market information for their processed product.

Bihar

Information Centres should be established. These can give information relating not only to market prices, availability of raw materials, technical know- how in connection with concerned activities, but also about various government schemes meant for promoting agro-processing activities.

There exists large scope for expanding livestock-based processing activity (milk processing). However, it will require proper input supply, a free marketing mechanism, scientific preservation facilities, milk chilling plants at different places in the private sector, infrastructural facilities, marketing intelligence and information systems, packaging facilities at producers' level, and skill development training programmes for the entrepreneurs and workers of such processing units.

For the development of textile-based processing enterprises, handloom parks should be established in and around potential districts. In view of larger concentration of *tasar* and silk units in and around Bhagalpur, expansionary measures by the DIC should be undertaken.

Maharashtra

It is important that the information is accessible and reaches various regions. For easy accessibility of this information, information centres should be established. These can give information relating not only about various government schemes but also markets, prices, and the technical know-how relating to the concerned activities. This should guide the entrepreneurs in adding value to their products and in reducing costs.

Establishment of cooperative marketing for cashews and fish units can be promoted. This is important as the units mainly rely on agents for marketing of the produce as well as for procuring raw material.

The sample non-food processing units reported absence of any promotional agency or government help in running the activity. Hence, it is felt that efforts should be made to encourage establishment of self-help groups or cooperative production/marketing units which would also act as information centres for the units.

The potential of small-scale agro-based industries to expand can be improved if they are provided with good quality infrastructure, information about the market and prices, and technical knowhow. Establishment of institutions for procuring raw material and marketing of produce will help them in taking advantage of the scale economies and getting directly in touch with the terminal market and getting a better price for their products.

To conclude, tasks are many to perform for reducing uneven growth of agro-processing industries across the regions of India. Apart from easing of infrastructural bottlenecks in the form of developing market infrastructure, roads communication, storehouse and cold storage facilities what is important is that the information is accessible and reaches to the entrepreneurs of processing units. This calls for creating infrastructure in the form of developing network linkages. However, performing of tasks enumerated above would require coordinated efforts among different departments of the government as well as amongst government and non-government agencies.

45

Agro-processing Industries in Bihar

Rajiv Kumar Sinha and Ranjan Kumar Sinha

Corresponding with the objectives of making food more digestible, nutritious, and extending the shelf life, food processing offers an opportunity for the creation of sustainable livelihoods and economic development for rural communities. Food processing includes; value-added forms of almost all agro-food-based and agro-non-food-based commodities. It covers all the processes that food items go through from the farm to the time it arrives on the consumer's plate. Food processing has come a long way in the last few decades. Dynamics of the industry have been altered by the ever-changing lifestyles, food habits, and tastes of the consumers globally. The food processing sector (FPS) forms an important segment of the Indian economy in terms of its contribution to GDP. As far the status related to contribution and growth of food processing industries (FPIs) in India are concerned, the GDP of FPI had increased by 1.21 times during the period from 2006–07 to 2010–11. It increased from Rs 52,164 crore to Rs 62,933 crore during the period. In percentage terms, the GDP of FPI increased from 9.40 per cent in 2006–07 to 14.90 per cent in the year 2010–11. The sector contributes as much as 9.00 to 10.00 per cent of GDP in agriculture and manufacturing sectors. Encouraging increases in the number, of registered food processing units (RFPUs) belonging to the sectors, namely: (i) meat, fish, fruits, vegetables, and oils (15.77 per cent); (ii) dairy products (102.58 per cent); (iii) grain mill products (52.49 per cent); (iv) other food products (59.64 per cent); (v) beverages (76.38 per cent); and (vi) total (50.24 per cent) were observed during the period of 1998–99 to 2010–11. An increase of 1.13 times is also visible, in number of persons employed under registered RFPUs in India during the period 2006–07 to 2010–11. The number increased from 14.76 lakh persons in 2006–07 to 16.75 lakh in the year 2010–11. Though the growth percentage declined by 1.79 during the period, that is, it came down from 6.09 to 4.30. Agro processing is defined as set of techno-economic activities, applied to all the produces, originating from agricultural farm, livestock, and aquaculture sources and forests for their conservation, handling, and value addition to make them usable as food, feed, fibre, fuel, or industrial raw materials. The agro-processing sector has experienced expansion during the last five decades, starting with a handful of facilities which were mainly operating at the domestic/cottage level. Appreciably, of late, agro-processing has been recognized as the sunrise sector of the Indian economy in view of its large potential for growth and likely socio-economic impact, specifically on employment and income generation. Some estimates suggest that in developed countries, up to 14.00 per cent of the total work force is engaged in agro-processing sector directly or indirectly. However, in India, only about 3.00 per cent of the work force finds employment in this sector revealing its under-developed state and vast untapped potential for employment. Properly developed, the agro-processing sector can make India a major player at the global level for marketing and supply of processed food, feed, and a wide range of other plant and animal products.

Agro-processing Industries (APIs), based on both food-products and non-food products, are faced with various constraints/problems. However, it has been more important in the wake of increasing emphasis on nutritional food security to ward off the 'silent-hunger.' It was also suggested that the quality consciousness and preference for health food by the high-income domestic consumers should also be capitalized up through development of agro-based high-value processed and branded products. In India, the processing units based on grains, horticultural products, livestock products, and fish have ample opportunities. India with 2.5 per cent of global area, supporting 16.7 per cent of the world population produces 22.4 per cent of world paddy production, 11.7 per cent of wheat, 15.00 per cent of rapeseed, 7.70 per cent of potato, 24.60 per cent of sugarcane, 43.30 per cent of jute, and 27.28 per cent of pulses produced globally. In the current scenario, India contributes 10.30 per cent and 9.20 per cent to the global production of fruits and vegetables respectively. In spite of strong base in horticultural products, a very negligible percentage of fruits and 0.5 per cent of vegetables are processed as against 70.00 per cent in Brazil. Moreover, India loses over 30.00 per cent of its produce of fruits and vegetables annually in the absence of proper infrastructural facilities. Agro-processing industries have, thus, vast potential in India.

In view of the above, this chapter is based on the study entitled 'Understanding the Growth and Prospects of Agro-Processing Industries in Bihar' with following objectives: (i) to present a profile of the agro-processing industries and recent trends; (ii) to examine the existing location pattern of selected agro industries; (iii) to study the impact of the agro-processing industry on agriculture; (iv) to study the economics of agro-processing units; (v) to analyse the marketing behaviour of agro-processed products; (vi) to study the employment potential from agro-processing industries; (vii) to analyse the constraints on acceleration of production; and (viii) to review the export performance of various agro-based commodities and constraints faced in accelerating the growth of export from the sector.

METHODOLOGY AND DATA

Secondary data, such as the quinquennial National Sample Survey data (NSS data) on unorganized manufacturing and Annual Survey of Industries (ASI) data for the organized segment have been used to gain a comprehensive view of the agro-processing sector. In view of tiny and small-scale agro-based industrial enterprises, being highly heterogeneous, it was worthwhile to look into each of the three layers namely: Own Account Manufacturing Enterprises (OAMEs), Non-Directory Manufacturing Establishments (NDMEs), and Directory Manufacturing Establishments (DMEs) and thus, covered in the sample. Primary data was collected from different units through a duly structured schedule.

The primary data have been collected from the selected processing units chosen from both agro-food industries and agro-non-food industries. As per the suggested design, all together 27 sample processing units were selected at random proportionately spread over food and non-food processing segments of agro-based enterprises. Of these, 18 processing units have been selected within the group of food processing and 9 from the non-food processing segment of agro-based enterprises. In the non-food processing group, paper and its products enterprises could not be taken up for a detailed study due to its non-existence or very poor concentration. The food processing activities were broadly divided in three categories, namely (i) primary food processing units, mainly grain processing units; (ii) spice and horticultural products; and (iii) livestock-based processing units including fish processing. Non-food processing units were broadly divided into three categories, namely (i) textile products; (ii) wood and its products; and (iii) leather and its products. Within the group of food processing and non-food processing agro-based activities, for each category of enterprise, the dominant processing activity was selected considering the concentration of units in the state. In the case of the food processing component of agro-based enterprises, for each selected processing enterprise, six units of different sizes: OAMEs run without the help of any hired worker, NDMEs employing less than six workers, and DMEs employing a total of six or more workers, but less than 10 workers (along with power supply) with their distribution as 3:2:1 were covered.

Within the non-food processing segment of agro-based industry, for each selected processing unit, three units of different sizes, namely: OAMEs, NDMEs, and DMEs in the ratio of 1:1:1 were selected. Having chosen processing activities, the districts were selected according to the concentration of selected agro-based enterprises. As such, the selected districts became more than one (five districts) depending on the location of the specific agro-processing activity chosen for the study. These have been listed in Table 45.1.

Table 45.1 Sample Processing Units and Selected Districts for the State of Bihar

S. No.	Processing Activity	Selected District	Number of Sample Units
Food Processing			
a.	Paddy Processing	Rohtas	06
b.	Fruit (Litchi) Processing	Muzaffarpur	06
c.	Milk Processing	Khagaria	06
	Sub-total		**18**
Non-Food Processing			
a.	Textile Products	Bhagalpur	03
b.	Wood and its Products	Patna	03
c.	Leather and its Products	Patna	03
	Sub-total		**09**
	Total		27

RESULTS AND DISCUSSIONS

Status of Agro-based Industries in the State

In regard to working units, non-agro-based industries dominated (75.08 per cent) over total agro-based industries (24.92 per cent) as per the data of 1994–95. Besetting fact in this regard is that the share of food processing industries and 'agro-non-food processing industries' in Bihar in the referred year were quite lower at 12.03 per cent and 12.90 per cent respectively. As regards investment, a very low share of total agro-based industries in the state (3.61 per cent) was found in comparison to total non-agro-based industries. However, within the category of agro-based industries, agro-food processing industries were cornered to only 3.05 per cent of total investment in that sector. Agro-non-food processing industries shared very negligible proportion (0.56 per cent) in Bihar. As per 1994–95 data, major investment could be seen in favour of non-agro-based industries (96.40 per cent). Non-agro-based industries provided employment to 222,172 workers, with gross output of Rs 1,777,327 lakh, and Rs 408,507 lakh as net value added. In comparison to these, the same figures for total agro-based industries in Bihar were 41,779, Rs 139,601 lakh, and Rs 33,444 lakh, respectively. The share and contribution of agro-food processing industries on the parameters of employment (number of workers), 'gross output' and 'net value added' in the state were 1.85, 4.30, and 3.61 times more than those of agro-non-food processing industries. As far the latest scenario of Bihar in regard to food processing industries is concerned, up to December 2013, a total of 191 projects were sanctioned with a total project cost of Rs 2,606 crore, and a grant amounting to Rs 202 crore was released. The employment generation was 15,181. Out of the sanctioned projects, only 111 (58.11 per cent) had gone into commercial production. By September 2014, the total number of sanctioned projects increased to 328 with a total cost of Rs 3,871 crore. Of these, 180 (54.88 per cent) units started the commercial production. The grant released amounted to Rs 294 crore and the estimated employment generation also increased to 21,240. It is also observed that the three principal types of food processing industries are rice milling, wheat milling, and maize milling. Between December 2013 and September 2014 a period of nine months, as many as 30 rice milling, 6 wheat milling, and 11 maize milling units had started in Bihar. This was indeed a substantial addition to the state's industrial scenario.

Economic Structure and Status of Unorganized Manufacturing Sector

Data-based picture of the status (share) and changes in agro-food, agro-non-food based processing industries and non-agro-based industries in Bihar have been discussed here. The change in the number of working units for different categories of industries has been measured by taking into account data available for the years 1994–95 and 2000–01 meant for unorganized manufacturing sector and provided by the National Sample Survey Organization. It is clear that in the year 1994–95; agro-based industries (including agro–food- and agro-non-food-based processing industries) dominated sharing 53.00 per cent (711279) number of the total working units. Non-agro-based industries shared a little less 47.00 per cent than total agro-based industries. Among the agro-based industries, the number of units related to agro-food-based processing activities was higher estimated at 381,810 (28.45 per cent) than agro-non-food processing industries 329,469 (24.55 per cent).

Sample Districts and Selected Processing Activities

In the case of the food processing component of agro-based enterprises for (i) primary food processing units, mainly grain processing units (rice mills in Rohtas district); (ii) horticultural product-based enterprises (that is, litchi-based processing activities in Muzaffarpur district; and (iii) livestock-based processing units (operational in Khagaria district under unorganized sector) were selected. Within the non-food processing segment of agro-based industries: (i) Bhagalpur district for textile product-based processing activities; (ii) Patna district for wood and its products processing enterprises; and (iii) again Patna district for leather and its products processing activities were selected for the study. As paper and its products processing industries were not functional in Bihar, in the organized and unorganized sector as well, no paper processing unit was taken into consideration. Here as such five districts were selected for the study depending on the location of the specific agro-processing activity. Data based structure of socio-economic profile of sample entrepreneurs infers that processing activity IV (textile-based processing activity) is run by members of the Momin community (100 per cent). Processing activity–V (that is, wood-based processing activity) has been undertaken by entrepreneurs of carpenters belonging to the other backward classes (OBC) group. Processing activity VI (that is, leather-based processing activity) was run by entrepreneurs of the Scheduled Caste class only. Processing activity Nos I, II, and III (that is, cereal-based processing enterprises, horticultural crop-based processing activity and livestock-based processing activities) were run by the entrepreneurs of OBC and others (50.00 per cent, 33.34 per cent, and 66.66 per cent for activity II and 16.66 per cent and 83.34 per cent for activity III), respectively.

Regarding the age of sample entrepreneurs, data embraces only 16.66 and 33.33 per cent of them in the senior citizen's age group of above 60 years to be engaged in processing activities I, II, and IV, respectively. Most of the sample entrepreneurs belonged to the age group of 45–60 years. Almost equal percentages of 66.67 of sample entrepreneurs in this age group were found running processing activities IV, V, and VI.

As far educational status scenario of the sample entrepreneurs is concerned, only 16.66 per cent of them were found to be illiterate undertaking processing activity III (that is, dairy-based activity). About 16.66 per cent and 33.34 per cent of the sample entrepreneurs engaged in processing activities I and II respectively were technically qualified. The majority of the sample entrepreneurs undertaking different processing activities were educated up to 10th standard. With regard to economic status in the form of landholding, it is revealed that 66.67, 100, and 33.33 per cent of the sample entrepreneurs owning processing activities IV, V, and VI respectively owned <1 ha and 1–2 ha of land. Only 16.66 per cent and 50.00 per cent of them running processing activities I and II (that is, cereal-based and horticultural crop-based processing activities), were found owning more than 10 ha of land area. The remaining sample entrepreneurs running agro-food-based processing activities I, II, and III were in the landholding groups of 2–4 ha and 4–10 ha. For previous experience in selected activities, the table inscribes all of the sample entrepreneurs to have more than five years' experience. Only 16.67 per cent and 33.34 per cent of the entrepreneurs running activities I and IV (that is, cereal-based processing activity and textile-based processing activity) possessed more than 30 years' experience.

Cost of Investment and Its Financing

It is revealed that most of the units are existing ones. It can also be seen that most of the surveyed processing units have been working in unorganized sector tiny, small, and artisan-based enterprises and so they are mostly unregistered. Paddy based processing industry, that is, rice mill in Rohtas district under the category of DME, litchi-based processing industry at Muzaffarpur 'Litchika International,' are registered under agro-food processing activities, DME, in the fields of leather and 'wood-based' processing enterprises both in Patna district are registered. Except for one OAME under food processing activity and one NDME under horticultural products-based processing activity, all the sample units were existing ones. The average age of the sample processing units ranged between 8 to 35 years.

DMEs under cereal-based processing activity in Rohtas district occupied the largest area (35,865 sq. ft). It was followed by DMEs of horticultural product-based activity (8,000 sq. ft), wood-based processing activity (1,400 sq. ft) and livestock-based processing activity also spread in a significantly larger area of 1500 sq. ft. It is revealed that generally within a particular group of processing activity, investment increased with the size of the unit. The size of the total investments went on increasing with the size of the enterprises. OAMEs showed lower total investments in comparison to that of NDMEs and DMEs. The percentages of working capital were found lower in case of OAMEs than NDMEs and DMEs except in the case of livestock-based processing activity and textile product enterprises. Data also suggests that size of investments were higher in case of DMEs meant for primary food-based processing units, that is, rice mills (Rs 7,796,000), litchi-based processing activity (Rs 15,960,000), livestock-based processing activity (Rs 1,050,000), wood-based and leather-based DMEs (Rs 1,600,000) and (Rs 900,000), respectively. Percentages of block capital have remained much higher than respective working capitals in all the three groups (OAMEs, NDMEs, and DMEs) in case of both agro-food processing activities and agro-non-food processing enterprises except NDME and DME of wood-based processing activities. The possible reason for this may be that the machinery, tools, and equipment do not require heavy expenditure for this enterprise. Besides, comparatively higher expenses are incurred in procuring raw materials (wood from Assam, Bettiah, and distant remote areas of Jharkhand and West Bengal). The share of block capital is seen to be very high in all cases. It varied from 86.96 per cent to 57.15 per cent with regard to agro-food processing activities and from 72.23 per cent to 39.66 per cent in case of agro-non-food processing activities.

Economics of Investment in Agro-processing Units

The number of working days per month as well as working hours per day were seen as uniform in most of the cases, except in horticultural crop-based (litchi), dairy products-based and textile products-based processing activities. As litchi-based processing activity is run hardly for 22 days to one month, therefore, in the case of DME of this, double shift work is undertaken. With regard to textile processing activity also, production activities in two shifts, or more than 8 hours are undertaken. Therefore, in these cases, working hours/day is longer. Livestock based processing activity is an everyday business without fail on a priority basis; however, its work hour is shorter (5 hours). Depending upon the nature of activities, the number of days required for other components of the whole operation cycle (namely, input stock, production process, output stocking, marketing, and credit realization) was seen to be different for different processing activities. The number of production cycles, which the unit completes in a year, also differs with the type and size of the processing unit. Except for cereal-based processing activity and litchi-based processing activity, the numbers of working days/year were quite higher (ranging from 300 to 355) for all other activities. The reason being the fact that litchi is a very short duration crop (available for processing from 22 days to a maximum of 30–35 days). Paddy is also not available for continuous processing in abundant quantum for more than three to four months. The number of days taken for credit realization was lower in case of OAMEs and NDMEs of most of the processing activities as they generally took loans from non-formal agencies. With regard to production cycle/year, the data in table discloses that it was quite higher only in cases of livestock (300) and leather and its products based processing activity (ranging from 312 to 355). In all other activities, numbers of production cycle were quite low depending upon the availabilities of raw materials, time taken in bringing them to consumable form, and scale of operation.

Employment Generation

It can be seen that the number of total labour in the units increased with the size. The highest number of total labourers' (who got employment in a year at 8 hours/day), was seen in the case of DME of horticultural product-based activity figured at 24,200. It was followed by DMEs of cereal-based, wood-based, textile-based, leather-based and livestock-based processing activities at 7,796, 4,050, 3,000, 2,700, and 2,000 respectively. It could also be observed that only OAMEs of cereal-based, horticulture-, and textile-based processing activities engaged female family labourers. With regard to hired female workers, their involvement could be seen only in the case of DMEs and NDME of two agro-food processing activities, namely, cereal- and horticultural product-based activities. It reveals that most of the processing activities (under both agro-food and agro-non-food categories) did not prefer to employ female workers.

Conclusively, it may be inferred that though processing activities of larger size provide greater employment opportunities, the contribution of NDMEs and OAMEs cannot be underestimated.

Problems and Prospects of Agro-processing Industries

Non-availability of adequate raw materials due to lack of capital, supporting machines/equipment, and absence of required infrastructural facilities were reported by the majority of the agro-food processing units. Fluctuations in prices of raw materials, absence of information networks and circumstantial need of purchasing raw materials from middlemen at higher rates were also prominently reported by the sample agro-food processing units. Non-availability of skilled labour, availability of raw materials (litchi) for a very short period, and difficulty in determining prices of value added products were felt by DMEs of agro-food processing activities. As far as agro-non-food processing activities are concerned, lack of capital, poor quality of raw materials, and no easy availability of bank credit were the main problems faced by OAMEs. Like agro-food processing activities, NDMEs of agro-non-food processing activities did come across the problems of poor electricity supply, variability of prices of raw materials, and purchasing of raw materials from distant markets (Kolkata in the case of leather). Procurement of raw materials, sometimes from informal trade channels, non-availability of strong supporting infrastructure, and raw materials not adequately available in the state (in the case of leather), were the main constraints faced by DMEs of agro-non-food processing activities.

In case of agro-food processing activities, long market channels causing lower net income and non-existent support by NGOs/cooperative marketing societies are limiting factors causing processors to sell value added products to local middlemen at lower prices. Seasonality of demand for the products, lack of mutual understanding among enterprises for preparing a common marketing strategy, and quality consciousness of consumers compelling the entrepreneurs to sell their products in distant markets for higher prices were experienced by all NDMEs. Existence of cut-throat competition, absence of a widespread network for marketing the products in Bihar (particularly litchi juice, syrup, and pulp), and transport-related problems in taking the value added products to terminal markets were the constraints faced mainly by DMEs of agro-food processing activities.

Regarding constraints faced by sample entrepreneurs of agro-non-food processing activities, determination of the price of the products by middlemen or big traders, no option for other profitable markets, and demand for product mainly in the local market preferred generally by low-income group of people yielding lower returns were observed as the main problems faced by OAMEs. Marketing through middlemen resulting in lower net profit, demand influenced by design-oriented preferences and more quality, design and brand consciousness of consumers were felt by NDMEs. Lower returns as a result of scattered markets, existence of tough competition due to presence of a good number of entrepreneurs and uncertainty of ready demand, and availability of value added products is generally ensured on order basis, could be found as main problems faced by DMEs of agro-non-food processing activities category.

With regard to prospects of agro-food-based processing industries, it is to be mentioned here that growing incomes, changing food habits, and lack of employment opportunities during post-harvest and after-sowing periods have encouraged the growth of the agro-processing sector. Based on the industrial units covered by ASI, the agro-based industries in Bihar accounted for nearly half of the gross value added. If the remaining smaller units are also taken into account the share of agro-based industries (ABIs) will be still higher. However, the potential of agro-based industries is not fully utilized. The development of ABIs is largely dependent on the importance assigned to fruits and vegetables vis-à-vis other crops. It is to be noted here that significantly large areas are under different top qualities of fruits in Bihar. Mango, banana, litchi, guava, lemon, pineapple, and others' occupy 140,786 ha, 29,013 ha, 28,758 ha, 27,994 ha, 17,122 ha, 4,454 ha, and 31,284 ha respectively. The quantum of production of these fruits is quite high. However, in the absence of required storage, preservation, and proper marketing facilities within and outside the state, a good quantum of these fruits are wasted and sometimes sold at un-remunerative prices. Hence, there is great potential for installation of agro-processing industries based on these fruits and vegetables too in areas/regions with their production in abundance. The processing of mango, litchi, banana, etc., will also take care of seasonal gluts, storage, and retention of their nutritive values, apart from providing income and employment. Among cereal-based processing activities, apart from paddy and wheat, there is high prospect for APIs based on maize in Bihar having a total area under autumn and rabi maize estimated at 472.90 thousand ha, and total production is estimated at 1076.30 thousand tonnes. The bulk of maize is produced mainly in the north-eastern districts of Bihar. As per a rough estimate, nearly 25.00 per cent of maize produced is used for human consumption locally. About 20.00 to 25.00 per cent is used in feeding milch animals and other domestic animals. As much as 50.00 to 55.00 per cent of the total quantum of maize produced is sent to Andhra Pradesh and other states from Bihar, which is processed there as value added products for human consumption, poultry feed, fish feed, animal feed, starch-making, etc. If processing industries based on maize are installed at different points in the districts of its surplus production, it will not only make proper and optimum use of this cereal crop but also be instrumental in a big way in creation of additional employment opportunities in rural and urban areas both and help in enhancing the income of the farmers and the people in general. It is, thus, important to record here that maize-based APIs can be effectively established for producing/manufacturing various value added products for human being, as animal feed and bio-diesel, etc. As regards livestock-based processing activity, the dairy industry in the cooperative sector under the brand name Sudha has achieved marked success in Bihar. In the unorganized sector also, there are great potential and bright prospects for processing of milk into khowa, ghee, butter, cream, paneer, lassi, etc. It will, however, require proper input supply, an exploitation-free marketing mechanism, scientific preservation facilities, milk chilling plants at different places in the private sector, infrastructural facilities, marketing intelligence and information systems, packaging facilities at producers' level, and skill development training programmes for the entrepreneurs and workers of such processing units. Of course, there are some problems, weaknesses, and lacunae in procurement of raw materials, operational aspects, and marketing of value added products of agro-food-based processing industries. However, if these constraints could be removed strategically with vision and determination, the prospect of APIs in Bihar is undoubtedly bright.

With regard to agro-non-food processing units, as a result of an increase in income, urbanization, and the demonstration effect causing a change in preferences, there has been a rise in the demand for 'value added products' based on agro-non-food processing, such as textiles, wood, and leather. According to the ASI data of 2004–05, leather and leather products in Bihar has a small share (0.56 per cent) in its total production of Rs 1922 thousand crore from the agro-based industries. However, considering the magnitude and quality of livestock wealth and traditional expertise of leather men in Bihar there appears to be a good potential for industries relating to leather and leather products in the state. Similarly, if the traditional expertise of weavers and their presence in significantly good numbers in some districts of Bihar (particularly, Bhagalpur, Gaya, Aurangabad, Patna, Banka, Madhubani, Siwan, and Nawada districts) are utilized properly by providing them necessary inputs, technical supervision, training on latest machines, remunerative marketing facilities with reduced number of middlemen, and better power supply, then the prospects of textile based processing industries in Bihar is undoubtedly bright. As regards wood-based processing activities, it also has great potential in Bihar. Urbanization has been promoting the use of varieties of 'value added products of wood'. As far as the sample units of agro-non-food category are concerned, these are located in comparatively developed districts of Patna and Bhagalpur. If the problems/constraints faced by sample processing units at different stages of production process are suitably addressed, and the factors making agro-non-food processing activities flabby are removed with vision, the prospects of agro-non-food- and agro-food-based processing industries in Bihar are sure to be very bright. What is needed is to use the inherent potential and available resources in different areas/fields.

CONCLUSIONS AND POLICY IMPLICATIONS

Keeping in view the vast potential for expanding agro-processing activities in the state, prevailing problems and existing potentials (prospects) of sample agro-processing units, the following action points could be suggested:

1. Arrangements should be made for making capital available to the potential entrepreneurs engaged in agro-processing activities.
2. Information centres should be established. These can give information relating to not only market prices, availability of raw materials, technical knowhow in connection with concerned activities, but also about various government schemes meant for promoting agro-processing activities.
3. Deficiency of supporting infrastructure should be removed by ensuring quality all-weather roads to rural and urban areas, regular power supply, means of communication, and strengthening formal credit institutions.
4. With a view to ensuring the supply of raw materials at reasonable prices and in time, and marketing of the produce, cooperatives be made instrumental and strengthened.
5. Locally available raw materials' based processing units related to fruits and vegetables have to be promoted.
6. Emphasis should be given on formation and strengthening of Self-Help Groups (SHGs).
7. There is a need to revive the handloom and power loom industry of the state, particularly in potential districts. This can be effectively done by providing credit and capital (machinery) on easy terms without many procedural complexities. Skill Development Training Programmes for weavers can also help achieve the goal.
8. For the development of textile-based processing enterprises, handloom parks should be established in and around potential districts. In view of larger concentration of tasar and silk units in and around Bhagalpur, expansion measures by DIC should be taken.
9. Technological and infrastructural backup be provided for preservation of raw materials and the agro-food-based processed products under PPP endeavours.
10. With a view to ensure quality standardization of the produce, particularly of agro-food products, Certification Centres/Laboratories should be opened, preferably at KVC, at PUSA and BAC, Sabour under RAU.
11. Exhibition of agro-food- and agro-non-food-based processed products should be arranged in all the government sponsored *melas*/fairs with a view to ensure wide publicity.

REFERENCES

Central Statistics Office (CSO). 2005. Annual Survey of Industries, 2004–05. MOSPI, Government of India.

Chadha, G. K., and Sahu. 2003. 'Small Scale Agro Industry in India: Low Productivity in its Achilles' Heel'. *Indian Journal of Agricultural Economics* 58(3): 518–43.

Chengappa, P. G. 2004. 'Emerging Trends in Agro Processing in India'. *Indian Journal of Agricultural Economics* 59(1): 55–74.

DAC, Ministry of Agriculture. 'State of Indian Agriculture (2012–13)'. New Delhi: Govt. of India.

Department of Industry, Govt. of Bihar.

Directorate of Statistics and Evaluation, Bihar. 2003. *Bihar through Figures*. Patna.

Finance Department, Government of Bihar. 2008. *Economic Survey (2007–08)*, pp 85–96.

———. 2009. *Economic Survey (2008–09)*, pp. 64–5, 91–8.

———. *Economic Survey (2014–15)*.

Govt. of Bihar. *Economic Survey (2008–09)*, Finance Department. p. 66.

Index of Industrial Production. *State of Indian Agriculture (2012–13)*.

Jha, U.M. 2008. 'Growth, Status and Prospects of Agro Processing Industries: An Overview: Workshop Papers', Vol on Understanding the Growth and Prospects of Agro Processing Industries in Bihar, 15–16 April 2008, pp 17–18.

Ministry of Agriculture and Farmers Welfare (MoA&FW). 'State of Indian Agriculture 2012–13'. Department of Agriculture and Cooperation Directorate of Economics and Statistics New Delhi.

———, MOSPI. *State of Indian Agriculture (2012–13)*.

National Accounts Statistics, MoSPI. *State of Indian Agriculture (2012–13)*.

National Sample Survey Organization. 1995. *Report No. 433, Unorganized Manufacturing Sector in India, Its Size, Employment and Some Key Estimates*. NSS 51st Round, July 1994-June, 1995.

———. 1995. *Report No. 434, Unorganized Manufacturing Enterprises in India: Salient Features*. NSS 51st Round.

———. 2002. *Report No. 477, Unorganized Manufacturing Sector in India (2000–01), Key Results*. NSS 56th Round.

———. 2002. *Report No. 478, Unorganized Manufacturing Sector in India (2000–01), Characteristics of Enterprises*. NSS 56th Round.

———. 2002. *Report No. 480. Unorganized Manufacture Sector in India (2000–01) Input, Output and Value Added NSS 56*th *Round (July, 2000-June, 2001)*. New Delhi: Ministry of Statistics and Programme Implementation, Govt. of India.

Rahim, Kazi M.B., and Jiban Kumar Ghosh. 'Status of Agro based Industries in Bihar and West Bengal: A Comparative Study', Vol. on Understanding the Growth and Prospects of Agro Processing Industries in Bihar, 15–16 April 2008, pp 1–2.

46

Agro-processing Industries in Maharashtra

Status and Prospects

Jayanti Kajale

The agricultural sector of India has moved away from scarcity to surplus food production. In response to the changing pattern of demand in the domestic as well as international markets and the need for moving away from traditional crop production, the agricultural sector has been diversifying into production of non-foodgrain crops and processing of the agro products. With this background, the agro-processing sector has come to occupy an important position as the activities involved are being looked at as growth engines generating income and employment for the agricultural sector as a whole. It is widely recognized that due to the diverse agro-climatic conditions, Indian soil can grow a wide variety of crops and a wide range of allied activities can also be pursued. On the other hand, with growing population and increasing incomes, the demand for agro-based food as well as non-food products are expected to rise. The liberalization of the Indian economy and world trade as well as rising consumer prosperity has thrown up new opportunities for diversification in the agro-based food as well as non-food processing sector and opened up new avenues for growth.

Recognizing the importance of food processing for the country as a whole, the government set up the Ministry of Food Processing Industries in 1988 and the food processing industry was identified as a thrust area for development. The Indian food processing industry has become an attractive destination for investors all over the world. It needs to be noted however, that the food processing sector of India is still at a nascent stage. Processing of agricultural commodities constitutes a small proportion of the raw material available. In 2004–05, food processing sector contributed about 14 per cent of manufacturing GDP. Of this, the unorganized sector accounted for more than 70 per cent of production in terms of volume and 50 per cent in terms of value. On the export front, India contributes 1.5 per cent (2003–04) to the global agricultural exports despite its leadership in agricultural production (D&B Information Services India Private Limited 2007). The share of this sector in GDP has almost remained the same for the last 10 years (Ganesh Kumar, Panda, and Burfisher 2006). Processing of fruits and vegetables is around 2 per cent, around 35 per cent in milk, 21 per cent in meat, and 6 per cent in poultry products. By international comparison, these levels are significantly low. Processing of agriculture produce is around 40 per cent in China, 30 per cent in Thailand, 70 per cent in Brazil, 78 per cent in the Philippines, and 80 per cent in Malaysia. According to one estimate, due to inadequate processing facilities, fruits and vegetables worth Rs 40000 crore were being wasted annually. Thus, food processing is also necessary for reducing wastage which normally take place in the post-harvest period or during period of a bumper crop.

The agro-based non-food processing sector can be sub divided into segments like textile-, wood-, paper-, and leather-based products. With increasing urbanization, increasing incomes, and spread of education, demand for these products has been increasing. As compared to the pre-reform period, the export of agro-based processed products from these segments especially textile and leather have been increasing in the post-reform period. The Economic Survey of India, 2007–08, reports impressive growth of leather, wood, and textile sectors and a fluctuating growth for the paper industry during April–November 2007 (http://indiabudget.nic.in).

In view of the current status and prospects of the agro-processing sector, it is essential to understand the nature of growth and constraints faced by the agro-processing sector. At the disaggregated level, states have exhibited varying performance of this sector. Maharashtra is one of the states with high potential for growth of food as well as non-food processing sector. It is a leading industrial state and is known for the advantage it has in the horticultural sector. As a result, it is one of the important states as far as development of agro-based industries is concerned. The data shows that Maharashtra had the highest number (856 or 19 per cent of the total in India) of units manufacturing fruit and vegetable products with 10 per cent of the total installed capacity in 2007 (www.indiastat.com). It was also the state with highest value of gross output generated by the registered small-scale industries (www.indiastat.com). Maharashtra was among the front-runners to receive the highest share of foreign direct investment (FDI) in food processing during the last five years. The dairy and consumer industries received FDI worth Rs 2.7 billion each. Within the non-food industries, Maharashtra is one of the important states as far as textile and textile based products and the leather-based products are concerned. These products have high export potential also. As the data for Maharashtra shows, these sectors have recorded impressive growth. Similarly, it is equally important to note contraction in the paper-based and wood-based sectors.

With this background, the chapter analyses working of the agro-processing industries in Maharashtra. Based on the secondary data, it examines profile of the agro-processing industries/units in the organized as well as unorganized sector of Maharashtra during 1994–95 and 2000–01. Further, this chapter analyses primary data collected from the field and discusses category-wise working of various agro-processing units. It also studies problems and prospects of agro-processing units and makes policy suggestions.

This section introduction is followed by a discussion on the methodology and sampling design of the study. This is followed by major findings of the study. The last section discusses conclusions and policy implications emerging from the study.

METHODOLOGY AND DATA

The study is based on both secondary and primary data. As far as secondary data is concerned, the study has used the quinquennial National Sample Survey Orgnization data relating to the unorganized manufacturing sector and the Annual Survey of Industries data for the organized sector. With the help of this data, the profile of agro-based industries in the state in terms of various characteristics such as employment, output, investment, etc., is studied for the years 1994–95 and 2000–01.

The primary data was collected from various units through canvassing a structured schedule. The information so collected from selected units was used to analyse the economics behind the working of these units, various problems faced by them, and their prospects.

Agro-based industries are classified as food- and non-food-based/processing units. Therefore, the sample consists of these two types of units. Altogether, 30 sample processing units were selected at random proportionately spread over the food and non-food processing segment of agro-based enterprises. As the share of food processing units is almost 60 per cent of the total agro-based units at all-India level (as is shown by Annual Survey of Industries data), out of the total of 30 units, 18 food processing units were selected and the remaining 12 were from the non-food processing segment of agro-based enterprises.

The food-processing activities are usually broadly divided into three categories, namely, primary food processing units mainly grain processing units; spice and horticultural products and livestock-based processing units including fish processing. The non-food processing units are broadly divided into four categories namely, textile products, wood and its products, paper and its products, and leather and its products. Within the group of food processing and non-food processing agro-based activity, for each category of enterprise, the dominant processing activity was selected considering available data relating to state area under the concerned crop and also the number of registered units engaged in the concerned activity in the state. Based on the concentration of selected agro-based enterprises, sample districts were chosen. In the case of the food-processing component of agro-based enterprises, for each selected processing enterprise, six units of different sizes namely Own Account Enterprises (OAMEs),

Non-Directory Manufacturing Enterprises (NDMEs) and Directory Manufacturing Enterprises (DMEs) with their distribution as 3:2:1 (subject to availability) were to be selected.[1] The units that could not be selected in the requisite proportion were from sub-categories-cashew units (2:3:1) and textile units (0:2:1).

Within non-food processing segment of agro-based industry, for each selected processing activity, three units of different sizes namely OAMEs, NDMEs, and DMEs in the ratio 1:1:1 were selected. The units selected were those operating on a small scale, that is, family enterprises and also those running with the help of hired workers. A field survey was conducted during the months of May–July 2008. The data relating to income relates to financial year 2007–08.

RESULTS AND DISCUSSION

Status of the Agro-based Industries in Maharashtra

It is observed from the secondary data that the unorganized sector constitutes a major segment of the agro-processing sector in terms of number of units. This is indicative of expansion of unregulated activity and easy entry or access of small units into the market. The data shows that that the food processing industries are growing at a faster pace over the concerned time period considering both the organized as well as unorganized sectors. Among the non-food agro-based industries, wood and paper units have been adversely affected in terms of numbers in both the sectors. However, the leather units have registered a positive increase in both the sectors.

In the organized sector, the share of agro-processing industries in total investment, employment, output and also the net value has increased over the period as compared to the non agro based industries. The increased share of agro-processing units in total employment has ensured increased share of total agro-based industries (38 per cent in 2000–01 as against 31 per cent in 1994–95). The data also indicates changing composition of the agro-based industries—expansion of the food processing sector and contraction of the non-food agro-based sector over the concerned period.

In the unorganized sector, the share of OAMEs has increased and that of other classes, that is, NDME and DME has fallen. Thus, more and more households engaged themselves in food-processing activities. This may be indicative of higher efficiency of small-scale household units which rely mainly on family/female labour for conducting the activity. It was observed that the share of food processing industries in total industries, their share in the total workers, and in the gross value added (GVA) almost stayed constant over the years though there was a rise in value added per worker and per enterprise. In the case of non-food agro-based industries however, the number of units and workers showed a sharp rise during the concerned period. This sharp rise led to an increase in the share of total agro-based industries in 2000–01 in terms of number of units (75 per cent) and number of workers (64 per cent). The main driving force had been the textile products sector. Though wood and its products was the dominating sector in the year 1994–95, the textile expanded at a faster pace over the concerned period and it accounted for almost one-third of all industries and one-fifth of the total workers engaged in the unorganized sector in the latter period. The data also shows that the net value added per worker as well as per enterprise registered a decline. Thus, the productivity of the units declined over the years. A disturbing trend observed is the fall in the employment per enterprise for almost all agro-based enterprises. This was in compliance with the economy wide phenomenon of falling elasticity of employment and increasing intensity of unemployment.

Profile of the Sample Districts and Selected Entrepreneurs/Processing Activities: *Food Processing Units*

Primary Milling of Grains—Rice Milling

In Maharashtra, among the foodgrains that undergo primary processing, paddy occupies largest area contributing around 12 per cent to the area and therefore it was selected as the dominant activity in this category. Ratnagiri is an important coastal district in paddy production in the state which currently contributes around 5 per cent and 9 per cent to state paddy area and production, respectively and is known for the prevalence of rice mills. Hence, Ratnagiri district which lies in the Konkan division was selected as a sample district.

Horticultural Products: Cashew Nuts

Among various horticultural crops, cashew occupies around 11 per cent of the total state horticultural area. The area under this crop was 19,913 ha in 1994–95, which increased to 90,382 ha in 2004–05 registering an increase of 78 per cent (CMIE 2007). Maharashtra is the major cashew producing state of India. The yield of cashew nuts in Maharashtra is very high and the state contributes around 32 per cent to

[1] OAME = ENTP employing only the owner-worker or his or her family workers; NDME = ENTP with at least one and utmost five workers of whom one is hired worker; DME = ENTP with at least six workers including at least one who is hired.

the state cashew production. Cashew processing units were therefore chosen for the field work. Among the cashew nut growing coastal districts, Ratnagiri contributes around 42 per cent to the total state area under cashew nuts and has a number of cashew nut processing units. Therefore, district Ratnagiri was selected as the sample district.

Fish-based Processing

As primary processing of fish is carried out at the household level by a large number of families in coastal areas of Maharashtra, this activity was selected as the sample activity. Ratnagiri, located on the west coast of Maharashtra was selected as the sample district for the survey of fish processing units. The district has a coastline of 167 km and a number of families are dependent on fishing and fish processing activities.

Profile of the Sample Districts and Selected Entrepreneurs/Processing Activities: *Non–food Processing Units*

As far as non food processing sector is concerned, within each of the categories, that is, textile, wood, leather, and paper, a number of activities are carried out. However, for selecting the sample, only those activities were considered which are carried out on a smaller scale. Therefore, tailoring units, notebook-making units, furniture-making units, and units making leather goods were selected. The available secondary data clearly shows that these non-food agro-industries are working on a large scale in developed, highly urbanized districts. Therefore, the two most developed districts, namely Mumbai and Pune, were selected for the non-food-based agro-units. Mumbai city district was selected for selecting sample units for leather processing activity. Units in other categories, that is, paper, textile, and wood products were selected from Pune district.

Profile of the Sample Entrepreneurs of Agro-processing Activities

The socio-economic profile of the sample entrepreneurs shows that most of the entrepreneurs belonged to the category 'others' which also included people from other religions (for example, Muslims who are involved in fish processing). Around 67 per cent of the leather entrepreneurs belonged to the SC/ST category as tanning of the animal skin has been the traditional business of this community. The average size of the family varied between 3–7 for food and 4–9 for non-food agro units. It was observed that for the majority of the categories, women were also the earning members along with the male members. The average dependency ratio was higher for non-food units (49.91 per cent) than the food units (32.50 per cent). This possibly indicated higher participation of household members in the processing activity in case of the food processing units.

As far as education is concerned, it was observed that majority of the entrepreneurs were educated up to the 10th standard. However, the entrepreneurs possessing cashew units, rice mills, and paper-based (binding) units were educated above 10th standard. Thus, the entrepreneurs engaged in activities which need technical knowhow and relatively heavy investments in terms of machinery, were better educated. As far as landholding size is concerned, it was noted that urban-based households engaged in non-food processing agro-based activities did not possess any land. The households however, in rural areas of Konkan engaged in cashew processing and rice milling owned land. These households were dependent on agriculture and agro-based activities for their survival. Families engaged in fish processing also did not possess any land.

It was also observed that majority of the units were existing units and had experience of more than 5 to10 years. This is specifically true in the case of fish and leather units as the business is carried on traditionally and hence the household members have learnt the business traditionally. It is noted, that the cashew units are the newly established units and all the entrepreneurs have been trained, as running the business needs technical training and knowledge about the machinery.

It can also be observed that most of the units were registered except four fish processing units and one unit each from leather, textile, and wood categories which were unregistered units. The fish units had not availed of any loan facilities and were not covered by any scheme. They carried out their activity outside the house near the beach. The area covered by the units using machinery—cashew units, rice mills, and binding units was more than the other business units.

Economics of the Agro-processing Units

Cost of Investment and Its Financing

It was observed that generally, within a category, investment increased with the size of the unit. This means that size of the investment by OAMEs was lower than that of NDMEs which in turn had lower investment than DMEs. The size of the investment was higher in case of rice mills and paper-based activity of binding which depended upon costly machinery for processing. The quantum of investment was lower for fish processing units. For other units, the share of block capital was very high; it varied from 64 per cent to 99 per cent for food units. For non-food units, it ranged from as low as 27 per cent to a high of 97 per cent. The size of the working capital was lower for the OAME units as these units did not have to incur expenditure on wages/salaries.

Majority of the units had financed the activity using their own funds. Units engaged in cashew processing and rice milling and one (DME) in book binding and leather took loans to finance their own investments. The share of loan ranged from 37 per cent to 80 per cent. The former two received subsidies as well under the District Industries Centre (DIC) or Khadi and Village Boards (KVIC) schemes which helped them to finance the investment needed thereby reducing their reliance on other sources like loan/own fund. It can be seen that units engaged in fish processing and majority units in non-food processing activities utilized their own funds to finance investment. Only two units out of the sample of 30 units have taken non-institutional loans.

Production and Operation Cycles

The data relating to production and operation cycles shows that the units had production cycles of different durations. The majority of the units did not face any problems as far as marketing of the produce and credit realization was concerned. As far as procuring the raw material and marketing of the produce is concerned, it was observed that as the units were small in size often working only with family labour, the capacity of the units to reach out to various markets was limited and they worked through agents. Hence, the unit owners directly did not come in contact with the terminal consumers/markets. Units like rice mills, leather units, textile mills, and furniture units, which processed only the raw material provided to them by the customers at their doorstep, did not have strong linkages with either input or output markets. All the units except the cashew units have reported that they have only one source (market) for procuring raw material as well as selling their product.

Investment and the Net Income from Investments

Table 46.1 shows the gross value of the output (GVO), investment, and net income earned by the sample units.

Table 46.1 Per Unit Net Income from the Investment of the Sample Units (Rs)

Processing Activity	Type of Enterprise	GVO of the Processed Product	Expenditure			Net Income
			Fixed	Variable	Total	
A) Food						
1. Cashew	OAME	265000.00	45690.00	162520.00	208210.00	56790.00
	NDME	491666.67	47317.41	351446.67	398764.08	92902.59
	DME	1120000.00	142048.33	304100.00	446148.33	673851.67
2. Fish	OAME	616960.00	3028.26	369353.00	372381.26	244578.74
	NDME	1152000.00	30222.50	751370.00	781592.50	370407.50
	DME	5000000.00	12116.70	4121250.00	4133366.70	866633.30
3. Rice Mill	OAME	83520.00	37025.79	6150.00	43175.79	40344.21
	NDME	198560.00	88329.19	40640.00	128969.19	69590.81
	DME	520000.00	174058.30	213740.00	387798.30	132201.70
B) Non-Food						
1. Leather	OAME	475200.00	62066.66	22350.00	84416.66	390783.34
	NDME	270000.00	74500.00	92400.00	166900.00	103100.00
	DME	1650000.00	502942.58	926400.00	1429342.58	220657.42
2. Paper	OAME	238200.00	21233.34	30000.00	51233.34	186966.66
	NDME	4400550.00	72905.70	3880300.00	3953205.70	447344.30
	DME	988080.00	291016.67	570400.00	861416.67	126663.33
3. Textile	OAME	40200.00	2866.67	11300.00	14166.67	26033.33
	NDME	748800.00	167125.00	216000.00	383125.00	365675.00
	DME	0.00	0.00	0.00	0.00	0.00
4. Wood	OAME	984720.00	36683.33	658600.00	695283.33	289436.67
	NDME	1495690.00	125766.67	1026400.00	1152166.67	343523.33
	DME	1045235.00	282382.50	679900.00	962282.50	82952.50

Source: Field Survey.

First, a comparison of variable and fixed cost shows that the share of variable costs was higher than that of fixed costs for most of the units. On an average, 71 per cent of the investment was variable investment. In case of fish processing households, the share of fixed costs was marginal. This was because the activity was labour intensive and essentially formed a part of the unorganized sector of the economy with no expenditure on salaries, tax, or insurance premium.

Size of the total investment was seen to be increasing with size of the units specifically the food processing units. Similarly, in absolute terms, the paper and leather processing units were seen to have had heavy investments as compared to other units.

For all the units, the total investment was compared with the GVO to find the net return. All the activities and units showed a positive net return. It can be seen that for the food processing activities, the net return increased with the size of the unit. Among these activities, the highest net income was earned by the cashew processing unit (DME) followed by fish processing (DME). Among non-food processing units, this particular pattern, that is, increasing income with increasing size was not observed. This might be because of the heterogeneous nature of these non-food processing activities. Even within the categories, in some cases, variations in products produced/sold were observed.

Table 46.2 shows values not only of net income and expenditure, but also of net income per labourer and the ratio of net income to expenditure. To get an idea about the productivity of the unit, net income per unit of labour (hired plus family) was calculated. It was observed that this indicator followed a pattern similar to that of net income. For cashew and fish processing units, productivity was seen to be increasing with the size of the unit within the category. These are female-labour-intensive units.

Table 46.2 Income and Expenditure of the Sample Units (Rs)

Processing Activity	Type of Enterprise	Total Expenditure	Net Income	Net Income/Total Expenditure (%)	Net Income Per Labourer
Food					
1. Cashew	OAME	208210	56790	27.28	14198
	NDME	398764	92902.6	23.3	18581
	DME	446148	673852	151.04	74872
2. Fish	OAME	372381	244579	65.68	48916
	NDME	781593	370408	47.39	52915
	DME	4133367	866633	20.97	123805
3. Rice Mill	OAME	43175.8	40344.2	93.44	40344
	NDME	128969	69590.8	53.96	13918
	DME	387798	132202	34.09	22034
Non Food					
1. Leather	OAME	84416.66	390783	462.92	130261.11
	NDME	166900	103100	61.77	25775
	DME	1429342.6	220657	15.44	36776.24
2. Binding	OAME	51233.34	186967	364.93	93483.33
	NDME	3953205.7	447344	11.32	89468.86
	DME	861416.67	126663	14.7	21110.56
3. Textile	OAME	14166.67	26033.3	183.76	26033.33
	NDME	383125	365675	95.45	52239.29
	DME	–	–	–	–
4. Wood	OAME	695283.33	289437	41.63	144718.34
	NDME	1152166.7	343523	29.82	114507.78
	DME	962282.5	82952.5	8.62	13825.42

Source: Field Survey.

However, this was not necessarily true of the non-food units. For leather-, paper-, and wood-based units, productivity of the OAME units in each category was very high. These were the units working only with family labour. Higher labour productivity of these units possibly indicated efficiency of the family labour and also pointed out that indicators like technology used, skill of the labourers, and type of labour used also could have influenced output per labourer.

Again for standardizing and comparing the net income across the classes, the ratio of net income to total expenditure was found out. It was generally observed that the ratio was declining within a particular category with increase in the size of the unit. This shows that as the size increases, the expenditure rises more than proportionately. This points out at the efficiency of the investment of the smaller units.

Employment Generation

Table 46.3 presents the details of the employment generation in the processing units. It can be seen that the number of total labour in the units was increasing with the size as was expected. The highest number of workers, that is, nine was found in cashew processing DME unit.

It can also be seen that all the categories in the food processing sector except one had engaged female family labourers. Thus, food processing (which can be carried out along with the domestic chores) was seen to be a female-dominated activity. As against this, in all, only three categories in the non-food sector (wherein work was carried out within the household) had engaged female family labourers. A similar pattern was found as far as hired female labourers were concerned. Leather as well as wood processing units were basically male-dominated units.

Table 46.3 Employment Generation under in the Sample Units (Standard Man Days of 8 hrs Per Day)

Processing Activity	Type of Enterprise	Family Labour		Hired Labour		Average Total Labour
		Male	Female	Male	Female	
A) Food						
1. Cashew	OAME	2	2	0	0	4
	NDME	1	2	0	2	5
	DME	0	1	0	8	9
2. Fish	OAME	2	3	0	0	5
	NDME	2	3	1	1	7
	DME	2	3	2	0	7
3. Rice Mill	OAME	1	0	0	0	1
	NDME	2	1	2	0	5
	DME	2	2	2	0	6
B) Non-Food						
1. Leather	OAME	3	0	0	0	3
	NDME	2	1	1	0	4
	DME	0	0	6	0	6
2. Binding	OAME	2	0	0	0	2
	NDME	2	2	1	0	5
	DME	1	1	4	0	6
3. Textile	OAME	1	0	0	0	1
	NDME	2	2	1	2	7
	DME	0	0	0	0	0
4. Wood	OAME	2	0	0	0	2
	NDME	2	0	1	0	3
	DME	1	0	5	0	6

Source: Field Survey.

Problems and Prospects of Agro-Processing Industries

The majority of the cashew and fish units reported non-availability of raw material throughout the year. For the cashew units, non-availability of good quality cashews was mainly due to inability of the small units to find agents or sellers supplying good quality raw material. The fish units also faced this problem. It was reported by these units that over a period of time, the supply of good quality fish had been reducing due to various reasons such reason being the entry of large firms. Another major problem faced by the fish units was the absence of any government schemes/promotional agencies (unlike in case of cashew units covered under DIC/KVIC) which would provide support to these units. In case of the rice mills, the main problem reported was irregularity in the electricity supply. These units also reported that there was tough competition from other rice mills as due to the liberal policy of the government regarding licensing, many new rice mills were being established.

The non-food units mainly reported non-availability of labour, absence of government support, and existence of rivalry as the main problems.

With increasing urbanization and income, changing food habits and consumption pattern, the demand for agro-based food as well as non-food products is increasing. The units would be able to compete and increase their earning capacity if they have sound knowledge of market conditions and various government schemes.

CONCLUSIONS AND POLICY SUGGESTIONS

In view of the potential for expanding the agro-processing activities in the state and the existing problems and prospects of the sample agro-processing units, the following policy suggestions emerge from the study.

The agricultural department, through the extension services disseminates information regarding various government schemes. It is important that the information is accessible and reaches various regions. For easy accessibility of this information, information centres should be established. These can give information relating not only about various government schemes but also markets, prices, and the technical knowhow relating to the concerned activities. This should guide the entrepreneurs in adding value to their products and in reducing costs.

Ensuring good quality infrastructure including roads and supply of electricity is very important. This is specifically important for the hilly parts of the Konkan area where some of the sample units are located. Good quality infrastructure would help in reducing the costs of the units. Uninterrupted supply of electricity would help the rice mills in increasing their net return.

Establishment of cooperative marketing for cashews and fish units can be promoted. This is important as the units mainly rely solely on agents for marketing of the produce as well as for procuring raw material. In case of lack of information regarding the market functionaries, the units may have to compromise as far as the quality of the raw material and price of the produce is concerned.

The administration of the schemes (for example, those by KVIC/DIC) should be made simple and transparent so that the borrowers or the units availing of the scheme do not face administrative hurdles.

Online trading for cashews can be encouraged.

In the wake of non-availability of fish throughout the year and reduction in the supply over the years, aquaculture should be promoted in the state.

Diversification of the agricultural sector through the agro-processing sector and creation of non-farm employment attains utmost importance in the wake of the agrarian crisis experienced by the state economy in the post-2000 period. Therefore, promotion of agro-based non-food processing units also assumes great significance. The following policy suggestions are made as far as non-food processing units are concerned.

The sample units reported absence of any promotional agency or government help in running the activity. Hence, it is felt that efforts should be made to encourage establishment of self-help groups or cooperative production/marketing units which would also act as information centres for the units.

As in case of food processing units, information regarding various government schemes (giving loans/subsidies) should be easily accessible to the public so that the entrepreneurs take advantage of it. Small-scale agro-based industries have been playing an important role in serving the market. It can be said that the potential of these industries to expand can be improved if they are provided with good quality infrastructure, information about the market and prices, and the technical knowhow. This would also help them in taking advantage of the wider markets. Establishment of institutions for procuring raw material and marketing of produce will help them in taking advantage of the scale economies and getting directly in touch with the terminal market and getting a better price for their products.

REFERENCES

Centre for Monitoring Indian Economy. 2007. *Agriculture Database*. Mumbai: Centre for Monitoring Indian Economy.

Chadha, G.K., and P.P. Sahu. 2003. 'Small Scale Agro-Industry in India: Low Productivity is its Achilles' Heel', *Indian Journal of Agricultural Economics*, 58(3): 518–43.

Dun & Bradsheet Information Services India Private Limited. 2007. *Emerging Food Processing SMEs of India*. Mumbai. http://test.dnb.co.in/Food%20Processing/default.asp.

Ganesh Kumar, A., Manoj Panda, and Mary E. Burfisher. 2006. *Reforms in Indian Agro-processing and Agriculture Sectors in the context of Unilateral and Multilateral Trade Agreements*. Report submitted to the Economic Research Service/ United States Department of Agriculture, Washington, DC WP-2006–011, Indira Gandhi Institute of Development Research, Mumbai, http://www.igidr.ac.in/~agk/AGK_Publications.htm.

Government of Maharashtra (GoM). 2005–06. *District Socio Economic Abstract: Pune*. Mumbai: Directorate of Economics and Statistics, Mumbai.

———. 2006–07. *District Socio Economic Abstract: Ratnagiri*. Mumbai: Directorate of Economics and Statistics, Mumbai.

———. (various years). *Economic Survey of Maharashtra* Mumbai: Directorate of Economics and Statistics, Planning Department.

Government of India (GoI). 2005–06. *Annual Report*. New Delhi: Ministry of Textiles.

———. various years. *Annual Survey of Industries*, 1994–95 and 2000–01. Kolkata: Ministry of Statistics and Programme Implementation, CSO.

———. various years. *Economic Survey of India*. New Delhi: Ministry of Finance and Company Affairs, Economic Division.

———. various years. *General Economic Tables, Maharashtra*. Census of India, Series 14.

———. 2002. *Unorganised Manufacturing Sector in India*. 56th round, July 2000–June 2001, NSSO, Ministry of Statistics and Programme Implementation.

Kachru, R.P. 2006. 'Agro-Processing Industries in India: Growth, Status and Prospects'. agricoop.nic.in.

Patil, Vikas. 2006. *Cashew Guidance Booklet*. Agricultural Technology Management Administration, Department of Agriculture, GoM.

47

Economics of Paddy Processing Sector in India

An Assessment

Komol Singha and Rohi Choudhary

The present chapter is an extract from a consolidated research report, conducted on hulling and milling ratios in major paddy growing states in India. Paddy or rice is the staple food for more than half of humanity in the world (Razavi and Farahmandfar 2008; Singha 2012b), around two-thirds of the World's population (Roy et al. 2011). Rice is the seed or kernel of paddy, covered by two different layers, namely, bran (inner layer) and husk (outer layer). Hereafter, rice means 'husked paddy', ready for cooking; while paddy means 'un-husked grain'. Literally, paddy becomes rice only when the husk and bran are removed properly through different milling processes and milling performance is largely measured by the head rice yield or quality of kernel (Dauda et al. 2012; Razavi and Farahmandfar 2008). Its milling operation is nothing but a process of separation of the husk (de-husking) and bran (polishing), to produce the edible portion (endosperm) for human consumption (Trade Development Authority of Pakistan [TDAP] 2010). The former process is known as 'hulling' and the latter one is known as 'milling' of paddy. However, often, rice kernels are susceptible to breakage and its yield reduces due to inefficient milling processes. In addition, the basic objective of modern rice milling is to remove the hull, bran and germ with minimum breakage of endosperms. Therefore, an efficient milling process is highly recommended for a better recovery ratio and quality rice kernel production (Dauda et al. 2012). Therefore, an efficient milling system is also identified as one of the most important remedies to prevent post-harvest paddy losses, and its need is widely realized in the world now (Appiah, Guisse, and Dartey 2011; Joshi 2004; Lele 1970; TDAP 2010).

The Asian continent is identified as the highest agricultural density population bloc, supported by rice cultivation in the world (Boyce 1988; Joshi 2004). When we traced back historically, grain or paddy has shaped the culture, diet, and economy of the people of this continent (Gomez 2001). Compared to other cereals, rice is more capable of sustaining land productivity without manure or fertilizers, or limited application of fertilizer (Boyce 1988). Since the 1980s, some studies identified rice as the staple food for more than 50 per cent of India's population (Joshi 2004; Saunders et al. 1980). Besides, the country has got the largest share of its land area under rice cultivation (Thiyagarajan and Gujja 2010), and became the second largest producer of rice, approximately 21 per cent of global rice production, next to China (Joshi 2004; Nayak 1996). Since the hulling and milling process is the oldest agro-processing industry in the country, it provides ample opportunities for the development of rice-based value-added products. Apart from rice milling, processing of rice bran for oil extraction, energy generation from husk, etc., are important by-products of the industry, and it generates income and employment to a large extent. As such, the hulling and milling of paddy is the oldest and the largest agro-processing industry in India. Almost the entire produce (90 per cent) of paddy is converted into rice every year by paddy processing units of varying sizes and capacities spread across the country, especially by traditional and inefficient processing mills. The remaining 10 per cent of it is stored as seed for next season's crop (Singha 2012a). In India, the major problem faced by the rice processing industry is that only about a half of the entire paddy production is processed by modern mills and the rest are being processed by inefficient traditional hullers, leading to a considerable post-harvest loss or processing loss. Consequently, there is reasonable scope for augmenting processing industry and reducing post-harvest loss with interventions and modernization.

METHODOLOGY AND DATA

Having understood the research problem discussed above, the present study tries to assess the economics of paddy processing units in India. Why has the larger share of the paddy in the country still been processed by traditional and inefficient units? It further explores a relative share of benefit generated by the different milling techniques (modern and traditional). For the purpose, both primary and secondary data were used. The primary data were collected directly from the respondents (millers) of three major paddy producing states of India—West Bengal, Punjab, and Karnataka, one each from the eastern, northern, and southern zones respectively. The secondary data were gathered from the government's published works by: the Ministry of Agriculture and Farmers Welfare (MoAFW), Food Processing Industries, Directorate of Economics and Statistics of the respective states, Agricultural Marketing and Food and Civil Supplies, etc. The year 2009–10 is taken as the reference year, to evaluate the hulling and milling ratios of paddy. However, the data pertaining to 2007–08 and 2008–09 were also collected to validate yearly growth trends and fluctuations of paddy conversion ratios. As of the sample size, altogether 282 mills from the three states were interviewed (Karnataka 92 sample mills, Punjab 90 sample mills and West Bengal 100 sample mills). Of the 282 mills, 157 were modern mills (Karnataka 67 mills, Punjab 40 mills, and West Bengal 50 mills) and 125 were traditional and huller mills (these two traditional and huller mills may be used interchangeably, Karnataka 25, Punjab 50, and West Bengal 50 huller mills). The sample districts were selected based on the concentration of rice mills and area under paddy cultivation in the respective states. They are: (i) Karnataka—Mandya, Davanagere, and Tumkur; (ii) West Bengal—Burdwan and Birbhum; and (iii) Punjab—Sangrur and Patiala. The data gathered through schedules were analysed by using descriptive statistics, especially the ratios and averages as per requirements of the study.

RESULTS AND DISCUSSION

Area, Production, and Yield of Paddy in India

As rice constitutes major food supply in the country, the crop's growth and performance of area, production, and yield have become major issues of attention. As per the secondary data, the overall growth rate of paddy production in the country in the last two decades (1991–92 to 2009–10) has been recorded at 0.93 per cent. However, the growth rate of area has turned out to be negative at the tune of -0.09 per cent, following urbanization and development of secondary and tertiary sectors. In the case of yield level, due to the growth of technology, it has increased at the tune of 1.02 per cent during the same period (Annexure 47A, Table 47A.2).

As for the state-wise contribution, the overall growth of the paddy sector in Punjab exhibited quite an impressive trend and recorded above the national level. In particular, the total contributions of area, production, and yield in the last two decades in Punjab were found to be at 1.6 per cent, 2.7 per cent, and 1.1 per cent respectively. For West Bengal, the growth rates of area, production, and yield were -0.08 per cent, 0.96 per cent, and 1.05 per cent respectively. In the case of Karnataka, growth rates of area, production, and yield were 0.84 per cent, 1.42 per cent, and 0.57 per cent respectively during the same period. The overall growth of paddy area and production in the last two decades in the three states has been above the national level (refer to Annexure 47A, Table 47A.2). However, yield being the indicator of performance, its growth rate in the two states—Karnataka and West Bengal—was not found to be statistically significant (Figure 47.1).

The growth performance of the districts within the selected states was also found to be impressive. While comparing among the selected states, Punjab performed better than the other two states in terms of area, production, and yield of the crop. Understandably, the bigger states

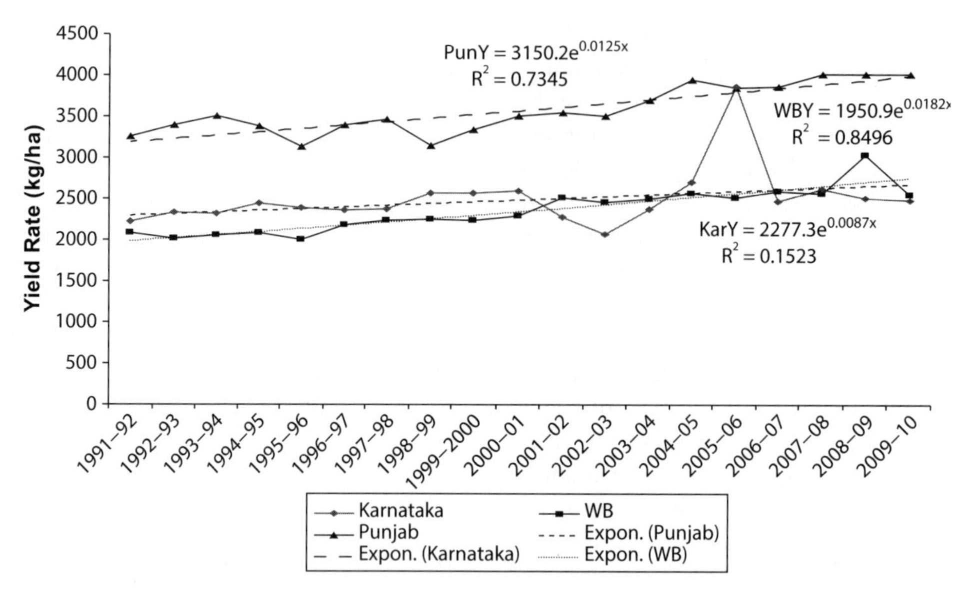

Figure 47.1 Trend in the Yield of Rice

Source: Ministry of Agriculture (MoA) (2012).

with larger area under paddy cultivation produced larger quantity of paddy. In terms of yield, the district of Sangrur in Punjab with merely 267 thousand hectares of area for paddy cultivation, showed a yield level of 4,617 kg/ha, the highest rate when compared to any other districts taken for study across the states in 2008–09. This literally showed that the state of Punjab has adopted better technology in the field of agriculture compared to that of the other two states (MoA, 2012; Planning Commission 2012; Reserve Bank of India (RBI) 2011).

Modernization of Paddy Mills in India

With the emergence of modernization and development of technology, the traditional hand pounding or foot pounding (*Dhenki*) rice mechanism became obsolete. The rice hullers, shellers, and modern rice mills became popular (Singha 2012a). Recently, with the coming out of efficient modern mills, the traditional hullers and shellers have become redundant. According to Joshi (2004), the hullers seldom give about 65 per cent of yield with 20–30 per cent broken rice. Besides, it does not give completely cleaned rice. However, the modern rice mills (single pass) have the capacity to process 2–4 tons per hour with yield recovery of 70 per cent and grain breakage of 10 per cent only. According to Lele (1970), modern rice mills have out-turned advantage of 1.6 per cent of parboiled rice more over the traditional hullers. However, in case of non-parboiled rice, the out-turn ratio of modern mills was 2.5 per cent higher than that of the shellers and 6.6 per cent higher than the hullers (Lele 1970; Saunders et al. 1980). Besides, the modern rice mill has many advantages in terms of mechanical adjustments (for example rubber roll clearance, separator bed inclination, feed rates) that are automated for maximum efficiency and ease of operation. The whitener-polishers are provided with gauges that sense the current load on the motor drives, which indicates the pressure on the grain. The new India Licensing Rules has made modernization of rice mills compulsory. Figure 47.2 depicts the processes and operational mechanism of modern rice mill.

The enforcement of 'Rice Milling Industry (Regulation and Licensing) Act' provided the following measures (Nayak 1996), to develop rice mills in India.

1. The new rice mills will undertake de-husking of paddy separately by rubber roll sheller or centrifugal de-husker and shall have paddy separators and cleaners in addition to polishers.
2. All the existing mills except single hullers shall be modernized.
3. Promotional efforts in the form of technical assistance, concessional finance, subsidy for modernization, extension programmes, training, research, development, etc.

Status of Rice Milling in the Selected States

Karnataka has got quite a large concentration of rice mills—about 1,755 modern rice mills as on 2008–09

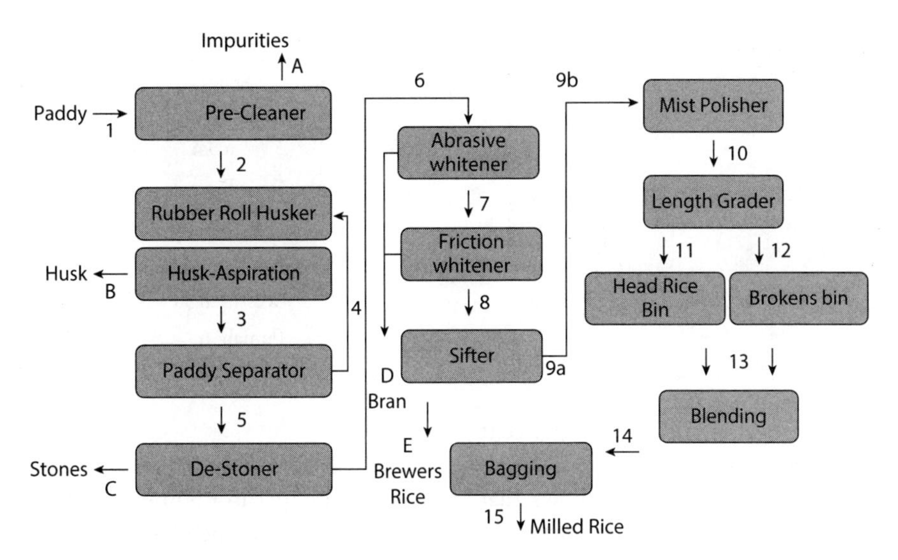

Figure 47.2 Flow Diagram of Modern Rice Milling Process

Source: Singha (2012a).

In Punjab, over the years, there has been a steady growth of improved/modern rice mills from 1,965 in the year 1993 to 3,163 numbers in 2009–10, further to 3,778 units in the year 2011–12. Most of these have got milling capacity ranging from 0.5 tonnes per hour (TPH) to 10 TPH. Sangrur district of the state has got the highest concentration of paddy mills in the state with around 16 per cent of the total capacity in the state in year 2009–10. Ludhiana, Patiala, and Moga are the other important districts that have got a large share of modern paddy mills in the state with about 14 per cent, 11 per cent, and 8 per cent respectively.

West Bengal also has got quite a large concentration of rice mills with more than 1,100 modern rice mills at present, but these mills are usually of low capacities where the capacity in terms of paddy throughout varies between 0.25 TPH and 16.0 TPH. The average paddy processing capacity of these modern rice mills in the state stands at about 2.5 TPH based on operation in two shifts (16 hour per day), thereby making an aggregated annual paddy processing capacity of around 15 million tonnes. Over the years, the rice milling industry in India has undergone different phases of technological transformations related to winning, parboiling, and drying systems; although it lags far behind the countries like USA, UK, Germany, Japan, Taiwan, etc. There are only a few fully automatic plants in India, which have installed colour cortex machines imported from Japan, USA, UK, etc. In West Bengal, the Burdwan and Bolpur (Birbhum) clusters have installed colour cortex machines and silky polishing machines, and the parboiled rice produced by these units may compete in national and international markets.

Presently, the major players in the industry are the age-old huller units as against the ever-improving modern rice mills in the country. On the one hand, the huller mills have the advantage of being cheap and simple to operate, but are very inefficient in converting paddy to rice. On the other, the modern mills give the highest yield of rice with the least breakage and better quality of by-product like bran. Normally, the huller mills yield bran having the lowest oil content as it contains appreciable amount of husk and broken rice. However, the oil content in bran from modern mills is far better than that of the traditional mills in this respect.

The traditional mills being small and running primarily on 'custom hiring' basis, requires a fewer number of labourers. Machines are operated when customers come for processing/hulling paddy and pay custom charges. It thus turned out that the modern rice mills create better employment opportunities in the agro-processing industry than the traditional mills. It is intuitive to note that the share of by-product decreases as technology of mills upgrade through the different phases. Conversely, the out-turn ratio of rice increases as the technology improves. Unlike traditional mills, apart from the main product—fine rice or polished rice, the by-products like, broken rice, bran, and husk are being produced separately by the modern mills. Figure 47.3 presents the much-needed attention towards modernization of the mills in respect to achieving high yield levels and reducing brokerage to the maximum extent. In terms of out-turn ratio (paddy to rice), the state of Punjab registered the highest with an average of 68.7 per cent in the last three-year study period. It was followed by Karnataka with an average of 58.7 per cent and, at the

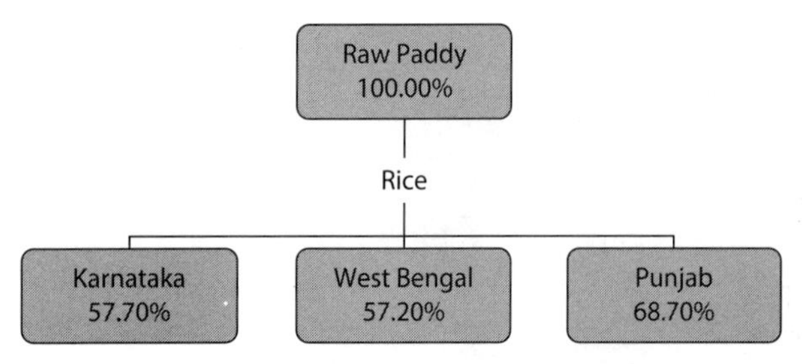

Figure 47.3 Paddy Processed Ratio by Traditional Mill

Source: Primary Survey.

bottom, the state of West Bengal registered 57.2 per cent of conversion ratio during the same period.

In the case of modern mills, Figure 47.4 shows that the average out-turn ratio of Karnataka in three years was found to be 63 per cent and was almost same for the West Bengal as well during the same period. In Punjab, it was relatively higher than that of the former two states, and stood at 69.5 per cent, but was not much different from the traditional mills of the state.

The share of the main product and the by-products can be seen from the Table 47.1. As of the modern mills, the conversion ratio of paddy to fine rice (the the milling ratio) was 62.73 per cent for Karnataka, 63.16 per cent for West Bengal, and 69.45 for Punjab in the three-year study period. In the case of ratio for broken rice, it stood at 2 per cent each for Karnataka and West Bengal, while it turned out to be 3.4 per cent for the state of Punjab. The proportion of husk per quintal of paddy turned out to be about 20.23 per cent for Karnataka, and 20 per cent each for West Bengal and Punjab. As for the bran, on an average, the out-turn ratio of Karnataka was turned out to be 4.4 per cent, 4.68 per cent for the state of West Bengal, and 3.1 per cent for Punjab.

As of the traditional mills, the ratio of recovery of fine rice (out-turn ratio) turned out to be 58.70 per cent in Karnataka, 57.22 per cent in West Bengal and as for Punjab, it was 68.70 per cent per quintals of paddy processed—which is about 10 to 11 per cent higher than the other two states. In addition, it is almost 10 per cent less

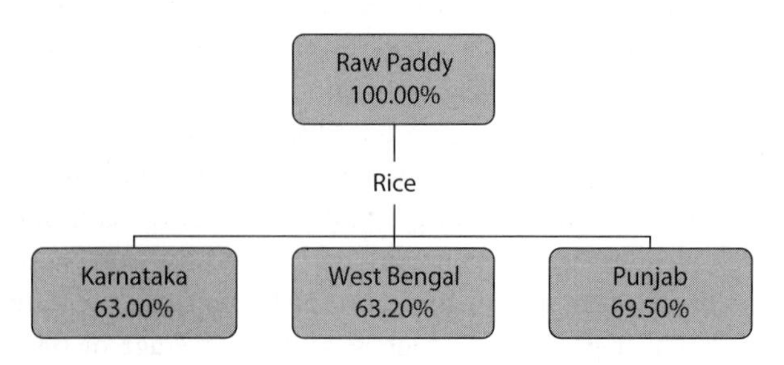

Figure 47.4 Paddy Processed Ratio by Modern Mills

Source: Primary Survey.

Table 47.1 Average Out-turn Ratio of Rice and By-Products

Types of Unit	States	Fine Rice	Broken Rice	Husk	Bran
Modern Rice Mills	Karnataka	62.73	2.00	20.23	4.40
	West Bengal	63.16	2.02	20.02	4.68
	Punjab	69.45	3.40	19.90	3.10
Traditional Rice Mills	Karnataka	58.70	10.20	26.60	
	West Bengal	57.22	4.87	36.63	
	Punjab	68.70	3.40	23.00	

Source: Primary Survey.

than that of modern rice mills in all states. It should also be noted here that in traditional huller units, the recovery of broken-rice was much higher than the modern rice mills.

Processing Cost of Modern and Traditional Rice Mills

We broadly subdivide the processing costs into two major components—Variable Costs and Fixed Cost. On the one hand, the fixed costs include costs like insurance of the processing unit, depreciation of capital goods and machinery, administrative expenses, etc., that include fixed rent or lease rent amount, rent of hired machinery, etc. On the other hand, the variable costs for processing activities include costs like labour costs, energy charges (namely, electricity, fuel, etc.), storage, and maintenance charges, etc., including the other variable costs like packing, transportation, handling, etc. As has been mentioned earlier, there exist significant differences within the modern rice mills depending upon the advancement or the milling technology adopted. For instance, depreciation charges may vary considerably according to the age of machine used, even the two mills using same technology. In addition, this component is more significant for the modern mills than the traditional ones because the modern mills use advanced technology involving huge investment (Lele 1970). As different milling techniques demand different combinations of inputs, there are fair amounts of variations in the input application of modern rice mills, which understandably influence different costs in the processing activity. In the findings of Afzalinia, Shaker, and Zare (2002), there has been wide difference in the milling costs ranging from Rs 75.17 per quintal (Three Abrasive Whitener with Friction Whitener as a Polisher Mill) to 574.40 per quintal (Three Abrasive Whitener with Rubber Roll Polisher) of paddy processed among the different milling techniques in Iran.[1]

[1] Value is converted into Rs per quintal of paddy processed from the CAN$ per kg of paddy (exchange rate 1 CAN$ = Rs 53.6903).

Table 47.2 Average Paddy Processing Cost by Modern Mills (Owner-cum-Trader)

States	Costs (Rs/qtl.)	2007–08	2008–09	2009–10	Average
Karnataka	Variable	42.37	44.53	48.37	45.09
	Fixed	22.69	23.13	24.39	23.40
West Bengal	Variable	47.29	53.48	58.61	53.13
	Fixed	11.69	13.22	14.07	12.99
Punjab	Variable	43.70	47.00	51.80	47.50
	Fixed	20.00	21.40	22.70	21.37
Karnataka	Fixed + Variable	65.06	67.66	72.76	68.49
West Bengal	Fixed + Variable	58.98	66.70	72.68	66.12
Punjab	Fixed + Variable	63.70	68.40	74.50	68.87
Three States	Fixed + Variable	62.58	67.59	73.31	67.83

Source: Primary Survey.

From Table 47.2 we can summarize that the average paddy processing cost (three states) of modern mills in three states turns out to be Rs 67.83 per quintal of paddy processed in the three-year study period. It is slightly higher than Bangladesh as given by Zaman et al. (2001) in 2000 at the tune of Rs 48 per quintal of paddy processed, but lower than the figures given by Shwetha, Mahajanashetti, and Kerur (2011) in Karnataka (India) of Rs 196.40 per quintal of paddy processed for modern units in the five years from 2005–06 to 2009–10.

When we analyse the difference between the types of cost—variable and fixed, the former is an all-time high. In addition, the gap between the two was quite high in the case of West Bengal compared to the other two states. However, one thing is to be kept in mind is that the processing cost analysed in our present study excludes some major components such as staff salary, administrative cost, and working capital interest, as Shwetha, Mahajanashetti, and Kerur (2011) included in their study. Among the states, the paddy processing cost of Punjab was at the highest at an average of Rs 68.87 per quintal in the three years. It was followed by Karnataka with Rs 68.49 per quintal and the state of West Bengal falls at the bottom with Rs 66.12 per quintal of paddy processed during the same period.

To maintain uniformity with the modern mills in terms components included in processing cost, Table 47.3 depicts paddy processing cost incurred by the traditional mills in three states with limited cost components under the two broader cost components—variable and fixed cost. As mentioned in the methodology section, the present study of traditional mills concentrates on the non-parboiled rice for the two states—Karnataka and Punjab and custom hiring basis of ownership. However, the state of West Bengal covers parboiled rice but like other two states, custom hiring basis of ownership of mills is studied.

From Table 47.3, we can summarize that on an average, the paddy processing costs by the traditional mills are found to be Rs 15.12 per quintal, Rs 12.18 per quintal, and Rs 17.83 per quintal of paddy processed in Karnataka, West Bengal, and Punjab respectively in the three years of the study period from 2007–08 to 2009–10. Of the states, as has been observed, Punjab appears on the top in terms of cost incurred and West Bengal turns out to be at the bottom. Like modern mills, the variable cost is higher than the fixed cost in all the states. It is also found that the fixed cost incurred by West Bengal is extremely low compared to other two states.

In contrast to the modern rice mills, the paddy processed by the traditional mills (hullers in this case) is

Table 47.3 Average Paddy Processing Cost by Traditional Mills (Custom Hiring) in Rs/qtl

Costs	States	2007–08	2008–09	2009–10	Average
Variable Cost	Karnataka	10.36	12.67	14.88	12.64
	West Bengal	9.27	11.42	14.77	11.82
	Punjab	13.65	15.15	21.10	15.60
Fixed Cost	Karnataka	2.31	2.51	2.73	2.58
	West Bengal	0.30	0.36	0.43	0.36
	Punjab	2.12	2.19	2.40	2.23
Total (Fixed and Variable Cost)	Karnataka	12.66	15.08	17.61	15.12
	West Bengal	9.57	11.78	15.19	12.18
	Punjab	15.77	17.34	23.50	17.83
Average	All Three States	12.67	14.73	18.77	15.04

Source: Primary Survey.

exclusively under the custom hiring basis. The custom hiring category of mills per se incurs relatively lower cost than the modern mills. The cost of machines is also much lower compared to the modern mills. As a result, the depreciation charges and insurance cost of the traditional mills are also relatively much lower compared to modern mills. However, for comparison, it is necessary to consider the processing costs borne by the traditional rice mills. According to Shwetha, Mahajanashetti, and Kerur (2011), the overall paddy processing cost by the traditional mills was Rs 127 per quintal of paddy processed in 2009–10 in Karnataka. In the finding of Singha (2012b), it was around Rs 15 per quintal in the same state in the same period. However, the difference between the two sources might be the components included in the cost estimation. The latter did not include some of the major cost heads like staff salary, working capital expenditure head, and administrative cost, etc. Similarly, the study finding of paddy processing cost by Afzalinia, Shaker, and Zare (2002) in Iran was estimated at around Rs 75 per quintal of paddy processed (0.014 CAN$/kg = 0.75 INR/kg × 100 kg = Rs 75/qtl) for the low technology mills (may not be equivalent specification of the *Huller mills* of our present study). Based on the total quantity of rice produced, an attempt has been made here to explore the type of mill that is dominating the rice processing industry in the study areas in three states. Table 47.4 portrays the relative shares of paddy processed by different mills belonging to traditional and modern mills.

From Table 47.4 we can see that on an average, a total of 91.22 per cent of paddy of the sample collected for the present study in West Bengal was processed by modern mills in three years. Hardly 9 per cent of the total paddy was left to traditional mills for processing in the state. In the case of Karnataka, altogether 81 per cent of the paddy was processed by the modern mills and 19 per cent was processed by the traditional mills. Though the major share of Punjab's rice production is exported, 86 per cent of the paddy was processed by the modern mills and the remaining 14 per cent of the total paddy was processed by the traditional mills in the state.

Table 47.4 Relative Shares of different Milling Techniques

States	Units	Paddy Processed Per Mill	% of Paddy Processed
Karnataka	Modern	47.55	80.88
	Traditional	11.24	19.12
	Total	58.79	100.00
West Bengal	Modern	51.55	91.22
	Traditional	4.96	8.78
	Total	56.52	100.00
Punjab	Modern	3.65	86.23
	Traditional	0.58	13.77
	Total	4.23	100.00

Note: Quantity in '000 qtl. per mill.
Source: Primary Survey.

In absolute terms, the state of Karnataka could process 47.55 thousand quintals of paddy per mill under the modern mills category and 11.24 thousand quintals of paddy under traditional mills in the three years from 2007–08 to 2009–10. Both mills put together, the state could process almost 59 thousand quintals of paddy per mill on an average in the three-year period. As for West Bengal, a total of 51.55 thousand quintals of paddy processed per mill under the modern mills category and 4.96 thousand quintals of paddy under traditional mills in the three years from 2007–08 to 2009–10. Both mills put together, the state could process almost 57 thousand quintals of paddy per mill on an average in the three-year period. In the case of Punjab, a total of 3.65 thousand quintals of paddy was processed per mill under the modern mills category and 0.58 thousand quintals of paddy under traditional mills in the three years from 2007–08 to 2009–10. Both mills put together, the state could process 4.23 thousand quintals of paddy per mill on an average in the three-year period.

Hurdles in the Process of Paddy Processing

This section puts forward the problems and prospects faced by the paddy processing industry in the country. Most of the milling units in the country faced certain difficulties. With the help of primary data, Tables 47.5a and 47.5b specifically reflect the reasons for underutilization of the installed capacity both in modern and traditional mills in the three states.

Reasons for Underutilization of Installed Capacity

According to Lele (1970), the decline in the raw paddy production and government levies are the major factors for underutilization of the paddy processing mills. Singha (2012b) revealed that labour shortage was identified as one of the most important factors for underutilization of rice mills in Karnataka (India).

It can be clearly observed from the Table 47.5a, in the case of modern mills, labour problem was the main obstacle faced by the millers of Karnataka, whereas supply

Table 47.5a Reasons for Underutilisation of Modern Mills

Reasons	No. of Respondent (Mill)	% of Total Sample*
Karnataka		
Labour Problem	28	42
Power-interruption	11	16
Raw Material	13	19
Technical Problem	8	12
Market fluctuation & others	7	10
West Bengal		
Shortage/Inconsistent Supply of Raw Material	41	82
Break-down of Machinery	18	36
Unavailability of Skilled Labourers	15	30
Power-interruption	14	28
Punjab		
Increase in Number of Rice Mills (competition)	24	60
Technology Limitation	19	47
Delay Process of Official Work	12	28
Storage Space (infrastructure)	11	27

Note: % is estimated out of the total sample 67, (No. of Mills/Sample*100).
Source: Primary Survey.

Table 47.5b Reasons for Underutilisation of Traditional Mills

Reasons	No. of Respondent (Mill)	% of Total Sample*
Karnataka		
Labour Problem	10	40
Technical Problem	8	32
Power-interruption	7	28
West Bengal		
Local Paddy being Channelized to Modern Mills	31	62
Shortage of Supply from Farmer Households	28	56
Unavailability of Labourers to Run Hullers	24	48
High Concentration of Hullers in the Same Locality	13	26
Punjab		
Local Paddy being Channelized to Modern Mills	29	58
Shortage of Supply from Farmer Households	34	68
Unavailability of Labourers to Run Hullers	42	84
High Concentration of Hullers in the Same Locality	23	46

Note: % is estimated out of the total sample 67, (No. of Mill/Sample*100).
Source: Primary Survey.

of raw materials was the main problem faced by West Bengal. Since, Punjab is considered to be a better performer in terms of the adoption of technology, competition of the mills was identified as the major factor for underutilization of the mills. In the case of traditional mills, the labour problem was again the major problem faced by Karnataka. While in West Bengal and Punjab, the local paddy being transferred to modern mills was identified as the major threat to traditional mills (Table 47.5b).

Brief Analysis

Compared to the modern rice mills, the processing cost per quintal by the traditional rice mills (hullers), run on custom hiring basis, have been found to be very low. Traditional mills are being used for small-scale production, especially in the rural areas. For the large-scale operations, modern rice mills enjoy economies of scale. This is what evident from Punjab, the state exports a major share of rice produced. The modern rice mills function as individual commercial business units that purchase paddy from the farmers, traders, local stockists, etc., and convert it into fine rice. The fine rice is then sold to the wholesalers, retailers, and even to the government as levy. In the process, profit is accrued by the millers for value addition for processing from raw paddy to fine rice. The modern rice mills are expected to give higher yields of rice, a small share of by-products like, broken-rice, bran, husk, etc., and process the by-products separately. The economics of modernization is essentially based on better ratio of recovery, not solely on the quantum of production. However, in the large-scale business, little low out-turn ratio does not make much difference. From the primary survey, it is evident that the net return per quintals of paddy processed by the modern mills turned out be, on an average, Rs 17.01 for Karnataka, Rs 16.23 for West Bengal, and Rs 22.87 for Punjab. Though the state of Karnataka and West Bengal accrued lower profits compared to Punjab, they still benefit from the economies of scale. It was

Rs 25.00, Rs 21.18, and Rs 34.36 for Karnataka, West Bengal, and Punjab respectively.

The traditional huller units do not produce fine rice, broken rice, and other by-products separately, a mixture of broken-bran-husk is produced. Therefore, a perfect comparison between modern and traditional mills cannot be made. Unlike the modern mills, the output of by-products cannot be owned by the miller because the mills are run under the custom hiring basis. Therefore, the efficiency and quantum of paddy processed or capacity of paddy processed by the traditional mills cannot be compared with the advanced modern mills. However, the huller units remain very important for the poor farmers in the rural areas. As they do not buy rice from the market, they need to get their paddy processed in their locality. Thus, it appears that the huller units running on a custom hiring basis perform similar tasks as the modern rice mills at much greater convenience in rural areas. In addition, the issues relating to the marketing of rice did not arise at all for the custom hiring mills or traditional type. In terms of net revenue, paddy to rice out-turn ratio and cost of production, the traditional mills seem to be better than the modern mills, but both have certain advantages over the other. Modern mills have got advantages in large-scale and commercial production, while the traditional mills cannot be easily ignored in the rural areas of the country. Nevertheless, the need for modernization of the paddy processing mills is at the crossroads in the country now.

CONCLUSION AND POLICY SUGGESTIONS

The study thus highlighted the main issue of the improvement of the mills in order to yield good quality rice. Upgradation of the existing hullers or traditional rice mills with advanced techniques into modern rice mills can improve the milling ratio across the country. Rice being a staple food for all of us, its production, processing, and consumption are also closely connected to the issue of poverty alleviation in the country. When focused on the economics of paddy processing, better out-turn ratio and sophisticated paddy mills contribute a significant role in the development of the country. The rice-milling industry not only processes rice for consumption but also ensures development of entrepreneurship and generates employment at the grass root level. For further development, some of the key policy implications are suggested:

1. As shortage of paddy is one of the major problems in the country, a strong extension service needs to be developed to increase the area under paddy cultivation.
2. Ensuring appropriate technology transfer to produce parboiled rice is highly recommended and financial institutions should extend financial support to the needy millers to produce parboiled rice.
3. Training programmes to be conducted on public-private partnership model to tackle the problem of inefficient and unskilled labour force in this industry especially among rural youths, aiming the enhancement of technical skills.

REFERENCES

Afzalinia, S., M. Shaker, and E. Zare. 2002. *Comparison of Different Rice Milling Methods*. Presented on 27–28 September 2002 at '2002 ASAE/CSAE North-Central Intersectional Meeting', Saskatoon, Saskatchewan (Canada).

Appiah, F., R. Guisse, and P. Dartey. 2011. 'Post-Harvest Losses of Rice from Harvesting to Milling in Ghana', *Journal of Stored Products and Post-harvest Research*, 2(4): 64–71.

Boyce, James K. 1988. 'Technological and Institutional Alternatives in Asian Rice Irrigation', *Economic and Political Weekly*, 23(13): A6–A22.

Dauda, Solomon Musa, Peter Aderemi Adeoye, Kehinde Bello, and Abdulfatai Adesina Agboola. 2012. 'Performance Evaluation of a Locally Developed Rice De-hulling Machine', *International Journal of Agronomy and Agricultural Research*, 2(1): 15–21.

Gomez, Kwanchai A. 2001. *Rice, the Grain of Culture*. Bangkok: Thai Rice Foundation.

Joshi, Bhavesh Kumar. 2004. *Post-Harvest Profile of Paddy/Rice*. New Delhi: Directorate of Marketing and Inspection, Dept. of Agriculture and Cooperation, Ministry of Agriculture, Government of India.

Lele, Uma J. 1970. 'Modernisation of Rice Milling Industry: Lesson from Past Experience'. *Economic and Political Weekly*, 5(28): 1081–90.

Ministry of Agriculture (MoA). 2012. *Agricultural Statistics at a Glance 2012*. New Delhi: Directorate of Economics and Statistics (Department of Agriculture and Cooperation), Government of India.

Nayak, Purusottam. 1996. 'Problems and Prospects of Rice Mill Modernization: A Case Study'. *Journal of Assam University* 1(1): 22–8.

Planning Commission. 2012. *Economic Survey 2011–12*. New Delhi: Government of India.

Razavi, S.M.A., and R. Farahmandfar. 2008. 'Effect of Hulling and Milling on the Physical Properties of Rice Grains'. *International Agro-physics* 22: 353–9.

Reserve Bank of India (RBI). 2011. *Handbook of Statistics on the Indian Economy for the year 2010–11*. Mumbai: Govt. of India.

Roy, Poritosh, Takahiro Orikasa, Hiroshi Okadome, Nobutaka Nakamura, and Takeo Shiina. 2011. 'Processing Conditions, Rice Properties, Health and Environment'. *International Journal of Environmental Research and Public Health* 8 (June 3): 1957–76.

Saunders, R.M., A.P. Mossman, T. Wasserman, and B.C. Beagle. 1980. *Rice Post-harvest Losses in Developing Countries*. US Department of Agriculture Science and Education Administration. Agricultural Reviews and Manuals 9 ARM-W-12 April.

Singha, Komol. 2012a. *Structure and Performance of Paddy Processing Industry in India: A Case of Karnataka*. CA, USA: Scientific & Academic Publishing.

———. 2012b. 'Economics of Paddy Processing Industry in India: A Case of Karnataka', *Scientific Journal of Agriculture*, 1(4): 80–91.

Shwetha, M.K., S.B. Mahajanashetti, and N.M. Kerur. 2011. 'Economics of Paddy Processing: A Comparative Analysis of Conventional and Modern Rice Mills', *Karnataka Journal of Agricultural Science*, 24(3): 331–5.

Thiyagarajan, T.M., and Biksham Gujja. 2010. SRI: Produce More Rice with Higher Resource Use Efficiency and Minimal Environmental Degradation. National Workshop on SRI in India, Stock, Taking and Future Directions in the context of Food Security and Climate Change, on 21–22 December at ICRISAT, Patancheru, Andhra Pradesh (India).

Trade Development Authority of Pakistan (TDAP). 2010. *Post-Harvest Losses of Rice*. Pakistan: Agro Food Division, Trade Development Authority of Pakistan, Government of Pakistan.

Zaman, Z.U., T. Mishima, S. Hisano, and M. Gergely. 2001. 'The Role of Rice Processing Industries in Bangladesh: A Case Study of the Sherpur District', *The Review of Agricultural Economics*, 57 (March): 121–33.

ANNEXURE 47A

Table 47A.1 Growth of Area, Production and Yield of Paddy in India

Year	Karnataka			West Bengal			Punjab			India		
	A	P	Y	A	P	Y	A	P	Y	A	P	Y
1991–92	1269	2826	2227	5719	11954	2090	2074	6755	3257	42649	74678	1751
1992–93	1317	3069	2331	5695	11445	2010	2065	7002	3391	41775	72868	1744
1993–94	1374	3183	2317	5876	12111	2061	2179	7642	3507	42541	80298	1888
1994–95	1296	3168	2445	5890	12236	2078	2277	7703	3383	42814	81814	1911
1995–96	1265	3024	2390	5953	11887	1997	2161	6768	3132	42837	76975	1797
1996–97	1359	3212	2364	5801	12637	2179	2159	7334	3397	43433	81617	1879
1997–98	1353	3213	2374	5900	13237	2243	2281	7904	3465	43446	82422	1897
1998–99	1427	3657	2563	5904	13317	2255	2519	7940	3152	44802	85973	1919
1999–2000	1450	3717	2564	6150	13760	2237	2604	8716	3347	45162	89683	1986
2000–01	1483	3847	2593	5435	12428	2287	2611	9154	3506	44712	84977	1901
2001–02	1418	3234	2281	6069	15257	2514	2487	8816	3545	44904	93340	2079
2002–03	1155	2390	2070	5842	14389	2463	2530	8880	3510	41176	71820	1744
2003–04	1074	2550	2375	5857	14662	2504	2614	9656	3694	42593	88526	2078
2004–05	1308	3547	2712	5784	14885	2574	2647	10437	3943	41907	83132	1984
2005–06	1485	5744	3868	5783	14511	2509	2642	10193	3858	43660	91793	2102
2006–07	1395	3446	2470	5687	14746	2593	2621	10138	3868	43814	93355	2131
2007–08	1416	3717	2625	5720	14720	2574	2610	10489	4019	43914	96693	2202
2008–09	1511	3802	2516	5936	18037	3039	2735	11000	4022	45537	99182	2178
2009–10	1487	3691	2482	5630	14341	2547	2802	11236	4010	41918	89093	2125
CAGR	0.84	1.42	0.57	-0.08	0.96	1.05	1.60	2.71	1.10	-0.09	0.93	1.02
2010–11*	1420	3057	2153	4760	8883	1866	2820	10837	3843	42214	80651	1911

* Advance Estimate; A, P and Y imply area in '000 ha, production in '000 ton and yield in kg/ha

Source: MoA (2012); Economic Survey (2012); RBI (2011)

Table 47A.2 The Details of Agro-Economic Research Centres and the Lead Person Involved in the State Report

S. No.	States	AERC Who Carried the Study in the State	The Lead Person Who Carried Out the Study
1.	Punjab	AERC Ludhiana	D.K. Grover
2.	Karnataka	ADRTC Bangalore	Komol Singha
3.	West Bengal	AERC Shantiniketan	Kali Sankar Chattopadhayay Debajit Roy

48

An Economic Analysis of Chickpea and Its Value Added Products in Agri-Export Zones for Pulses in Madhya Pradesh

Hari Om Sharma and Deepak Rathi

On the agricultural front, India has moved from chronic food security to food surplus in the last four decades. This became possible due to cutting edge science coupled with the fast adoption of improved production technologies by the farmers and government policies according high priority to agricultural production by making large planned investment in infrastructure, irrigation, power, credit, research, and extension. As a result of technological breakthroughs and their effective dissemination, foodgrain production increased from 50.82 (1950–51) to 217.3 million tonnes (2006–07), oilseeds production increased from 5.16 (1950–51) to 24.29 million tonnes (2006–07) and pulses production also increased from 8.41 (1950–51) to 14.02 million tonnes (2006–07). Now foodgrain, oilseed, and pulse production has reached a plateau and stagnates to only 208.80, 18.05, and 13.57 million tonnes (triennium ending [TE] 2007) respectively.

At this juncture, not only productivity of crops is needed to be increased but at the same time the quantum upgrade of value addition activities of international quality standards are also required to provide sustainable income and employment to farmers throughout the year. The farmers of the country are not getting remunerative prices for the products they produce and as a result, they fight with poverty. Therefore, at this stage it becomes necessary for the farmers to know the value addition technology available for the products and by-products they produce. Value added activities also enhance income and employment of market intermediaries involved in the marketing chains.

Value addition technology includes all post-harvest practices and marketing technology and creates the time, place, form, and possession utilities in a particular product. These include, winnowing, threshing, grading, standardization, quality control, storage, processing, packaging, transportation, brand name, publicity, selling of produce through suitable market channels, etc.

This value addition for sustainability also gave answers to all these questions: (i) what will be the best alternative channel of product marketing?; (ii) what will be the best alternative technology for grading/standardization/storage/processing/packaging/transportation/publicity of their products?; (iii) how can farmers get sustainable income from marketing of a particular product and become entrepreneurs?; and (iv) how can the traders/processors become entrepreneurs? Chickpea has all prominent properties required for value addition. People have known this quality since 10,000 BC. Because they could not harvest the value addition properties properly, they were unable to get remunerative price for their products.

The world's total production of chickpeas is around 8.5 million tonnes annually and is grown approximately over 10 million hectares of land. The desi type (Traditional) chickpea contributes around 80 per cent and the Kabuli type around 20 per cent of the total production. India is the largest producer of chickpea, contributing around 70 per cent (around 6 million tonnes) of the world's total production. Apart from India, Turkey (7 per cent), Pakistan (5 per cent), Iran (3 per cent), Mexico (3 per cent), Australia (2 per cent) Canada (2 per cent), and Ethiopia (2 per cent) are the other major chickpea producing countries of the world.

In India, chickpea is grown in the rain-fed areas as they are best suited for its production. Chickpea-producing states in India are Madhya Pradesh (MP), Uttar Pradesh, Rajasthan, Maharashtra, and Andhra Pradesh. Madhya Pradesh produces the major share of 42 per cent of Indian's production of around 6 million tonnes. Andhra Pradesh, Uttar Pradesh, Maharashtra, and Rajasthan follow Madhya Pradesh and contribute around 11, 12, 13, and 9 per cent of total production, respectively.

Madhya Pradesh contributed 2474.6 thousand tonnes chickpea cultivated in an area of 2692.6 thousand hectare. However, an average farmer harvested a yield of only up to 926 kg/ha (2006). The districts Vidisha, Raisen, Hoshangabad, Rajgarh, Chhindwara, Narsighpur, Shivpuri, and Guna were identified for Agri-Export Zones (AEZs) for pulses in the state. These districts contributed about 33.94 per cent of area for chickpea in the state. The area and production of this particular crop in the state showed an increasing trend, but the yield remained constant mostly for the past 10 to 15 years. The productivity levels of these districts (1082 kg/ha) are found to be more than the state average (927 kg/ha) and less than the potential yield (3000 kg/ha) of the chickpea harvested in front-line demonstrations in the state. This might be due to the poor adoption of the recommended crop production and marketing technologies and the various constraints associated with these functions. Aimed at analysing the cost and return structure of chickpea and its value added products and marketing aspects, this study was framed to find out the exact solution for increasing the production as well as income and employment of cultivators and market functionaries in AEZs for pulses in MP. Theobjectives of this study are:

1. to determine the growth of area, production, and productivity of chickpea in the last 15 years (1992–2006) in AEZs for pulses in MP;
2. to analyse the yield and expenditure gap of chickpea at different levels of adoption;
3. to analyse the structure of cost and return and resource use efficiency in chickpea production at different levels of adoption;
4. to examine the pattern of marketing of chickpea;
5. to examine the nature and extent of value addition (primary processing) and its profitability over chickpea and assess the cost of processing of chickpea at the miller level; and
6. to assess marketing patterns and trade-related issues in chickpea products under AEZ for pulses.

METHODOLOGY AND DATA

Both primary and secondary data were collected for the study. The district-wise time series data on area, production, and productivity of chickpea were collected for the period 1992 to 2006. The average of triennium ending (TE) 1995 and triennium ending 2006 were treated as the base and current year respectively for the study.

The primary data were collected from the chickpea growers, village merchants, wholesalers, and processors. The study is confined to two districts, that is, Vidisha (7.96 per cent) and Narsinghpur (5.75 per cent) of MP for collection of primary data. These districts were selected purposively for the study because districts were identified by the Government as AEZs for pulses and have the highest area under chickpea in MP. A list of all the blocks in each selected district was prepared in descending order of chickpea production and a block having the highest production of the crop was selected for the investigation. Similarly, a list was prepared of the all villages in the each selected block with their production of chickpea listed in descending order. The villages were categorized into three categories (low, medium, and high chickpea producing villages) by using the cumulative frequency technique. A village under each of these three categories was randomly selected for the investigation. Further, a complete list of farmers of all the selected villages and area under chickpea was prepared and 40 farmers from each selected village were randomly selected by using a Random Number Table. Thus, 240 households from six villages were considered for the investigation (Table 48.1). The selected respondents were further classified into three categories (low, moderate, and high) by using Mean +/−1 standard deviation (SD) of yield of chickpea, as yield of crop is directly related to adoption of recommended technologies (Sharma, Nahatkar, and Patel 2000).

A pre-tested interview schedule was used for collection of required data from the respondents. The interview

Table 48.1 Number of Selected Chickpea Growers in Different Levels of Adoption of Recommended Technologies

Level of Adoption of Recommended Technologies	Yield Levels (qtl/ha)	No. of Chickpea Growers
1. Low (Mean – SD)	Below 9.37	27
2. Moderate	9.37 to 15.35	182
3. High (Mean + SD)	Above 15.35	31
Mean	12.36	
Standard Deviation (SD)	2.99	

Source: Field survey.

schedule having all the information about the sample farmer, namely, land utilization pattern, cropping pattern, farm assets and household assets, and expenses on input used, high-yielding variety (HYV) seeds, seed treatment fungicides, fertilizer and manures, micronutrients, bio-fertilizer, insect pest control, disease control, labour used (human labour, bullock labour, machine labour), and output (yield of main product and by-product) expected constraints related to crop production, processing, and marketing, etc. The primary data were collected from the individual sample respondents through the survey method by personal contact. The Statistical Package for Social Science (SPSS) was used for classification, analysis, and tabulation of collected data. The collected data were analysed with mean, percentage, coefficient of variance, correlation, regression analysis, etc. The yield gap analysis was performed with the help of following formulae:

Yield gap I = Maximum Farm Yield – Potential Yield*

Yield gap II = Average Farm Yield – Maximum Farm Yield

Yield gap III = Average Farm Yield – Potential Yield

* Potential Yield considered as Maximum Farm Yield obtained by a farmer in adoptive trails conducted by All India Coordinated Research Project on Chickpea, Sehore (MP) in the area under the study.

A multiple regression model including following variables was used for analyzing resource use efficiency of chickpea production.

$$Y = a + b_1 X_1 + b_2 X_2 + b_3 X_3 + b_4 X_4 + b_5 X_5 + b_6 X_6 + b_7 X_7 + b_8 X_8 + b_9 X_{10}$$

where: Y = Yield of Chickpea (q/ha), a = intercept value, X_1 = Expenses incurred in field preparation, X_2 = Expenses incurred in High Yield Varieties Seed, X_3 = Expenses incurred in sowing, X_4 = Expenses incurred in fertilizers, X_5 = Expenses incurred in bio-fertilizer, X_6 = Expenses incurred in seed treatment, X_7 = Expenses incurred in weed management, X_8 = Expenses incurred in insect pest management, X_9 = Expenses incurred in harvesting and b_1 to b_9 = Regression Coefficient.

The primary data were also collected from sufficient numbers of village merchants, wholesalers, and processors/millers for analysis of cost of marketing and processing.

RESULTS AND DISCUSSION

Trend and Growth of Area Production and Productivity Based on Time Series Data

Chickpea was found to be grown by the cultivators almost in all the districts of the state, although its intensity was found different in different districts. In the state there were 28 districts found in the low-intensity area (<50 thousand hectares) covering about 20 per cent of area and production of the state, while 12 and 8 districts were found to have moderate (50 to 100 thousand hectares), and high-intensity areas (>100 thousand hectares) respectively which covers about 80 per cent (32 per cent in moderate and 48 per cent in high) of total production. It is surprising to note that Chhindwara, Rajgarh, Shivpuri, Guna, and Hoshangabad districts come under low and moderate intensity area and were found to be on the list of the AEZs for pulses, while the districts, namely, Rewa, Datia, Satna, Ujjain, Shajapur, Jabalpur, Panna, and Sehore (moderate intensity areas), and Dewas, Chattarpur, Damoh, Sagar (high intensity area) were not found in the list of the AEZ districts for pulses declared by the government although the production of chickpea, infrastructural and other facilities were found to be similar in all these districts of the state.

Amongst the different AEZ districts, the Raisen (41.91 per cent), Vidisha(31.59 per cent), Rajgarh (16.51 per cent), Narsinghpur, (16.51 per cent), Shivpuri (15.33 per cent) and Chhindwara (9.59 per cent) showed positive relative change, while Hoshangabad (–17.94 per cent), and Guna (–26.93 per cent) showed negative relative change during the period under study. The area of chickpea was found to be fluctuate more in Rajgarh district (28.79 per cent) followed by Guna (27.56 per cent), Shivpuri (21.21 per cent), Hoshangabad (17.36 per cent), Raisen (13.92 per cent), Vidisha (13.72 per cent), Narsinghpur (9.25 per cent), and Chhindwara (4.58 per cent) districts of AEZ for pulses in MP during the period under study. As regards to growth of area of chickpea in MP, it has increased with the linear and compound growth of 1.10 per cent and 2.30 per cent per annum. The growth of area of chickpea was also found more in other districts (1.29 per cent and 1.25 per cent per annum) as compared to AEZ districts (0.82 per cent and 0.83 per cent per annum). The highest positive linear and compound growth in area was observed in Raisen district (2.78 per cent and 2.76 per cent per annum) followed by Vidisha (2.50 per cent 2.53 per cent per annum), Shivpuri (1.98 per cent and 2.02 per cent per annum), Chhindwara (1.60 per cent and 1.59 per cent per annum), Narsinghpur (1.04 per cent and 0.98 per cent per annum), and Rajgarh (0.65 per cent and 0.06 per cent per annum), while negative and linear and compound growth was observed in Hoshangabad (−2.36 per cent and 2.34 per cent per annum) and Guna (−1.14 per cent and −2.47 per cent per annum) districts in AEZ districts of MP.

In MP, the production of chickpea increased by 505.97 thousand tonnes (30.13 per cent) from 1,679.53 thousand tonnes (base year) to 2,185.50 thousand tonnes (current year) with the fluctuation of 16.83 per cent (348.30 thousand tonnes) during the period under study. The increase in production was found to be more in others districts (34.57 per cent) as compared to AEZ districts (22.84 per cent). Amongst the different AEZ districts, the Raisen (76.85 per cent) showed highest positive relative change followed by Vidisha (60.36 per cent), Rajgarh (46.68 per cent), Shivpuri (30.95 per cent), Chhindwara (17.79 per cent), and Narsinghpur (9.50 per cent), while Hoshangabad (−2.52 per cent) and Guna (−37.02 per cent) showed negative relative change during the period under study.

The productivity of chickpea fluctuated between 13.50 per cent (Raisen) to 29.82 per cent (Guna) in different districts AEZ for pulses in MP during the period under study. As regards to growth in productivity of chickpea is concerned for MP, it was increased with the linear and compound growth of 0.84 and 0.81 per cent per annum respectively. The growth in area of chickpea was also found more in AEZ districts (1.21 and 1.19 per cent per annum) as compared to MP and other districts (0.59 and 0.55 per cent per annum). The highest positive linear and compound growth in productivity was observed in Vidisha district (1.94 and 1.98 per cent per annum) followed by Raisen (1.91 and 2.02 per cent per annum), Hoshangabad (1.49 and 1.50 per cent per annum), Rajgarh (1.16 and 1.08 per cent per annum), Guna (0.68 and 0.47 per cent per annum), Shivpuri (0.62 and 0.53 per cent per annum), Narsinghpur (0.55 and 0.48 per cent per annum), and Chhindwara (0.51 and 0.38 per cent per annum) districts. The growth rate of productivity of chickpea in all the AEZ districts was found positive.

Yield and Expenditure Gap

Analysis of primary data collected from the chickpea growers indicates that there was a considerable yield gap (Yield Gap III) of 17.64 qtl/ha (142.72 per cent) between the potential yield (30.00 qtl/ha) and average farm yield (12.36 qtl/ha) on an average chickpea grower's farm. Out of this total gap (Yield Gap III), a gap of 13.46 qtl/ha (Yield Gap I) and 4.18 qtl/ha (Yield Gap II) was found between the potential yield and average farm yield, and maximum farm yield (16.54 qtl/ha) and average farm yield respectively. The Yield Gap I denoted that the Recommended Package of Practices (RPP) of chickpea are not transferred fully to chickpea growers from lab to land, and there is a difference in soil and climatic conditions of the experimental and farmer's field, while the Yield Gap II was found due to the socio-economic constraints present in study area. The Yield Gap I was found more than the Yield Gap II revealed that lacuna in transfer of technology is more than the socio economic constraints. The farmers are not able to adopt the RPP due to lack of knowledge rather than the socio economic constraints present in the study area. It is also observed from the data that as the level of adoption increases from low to high the yield gap decreases from 282.65 to 15.92 per cent (Yield Gap I), 37.18 to 35.46 per cent (Yield Gap II), and 424.93 to 57.03 per cent (Yield Gap III) (see Table 48.2).

The total cost of cultivation of chickpea at RPP was estimated to Rs 32880.12/ha, in which total variable cost (50.30 per cent) was found to be highest followed by total fixed cost (49.30 per cent). In the total variable cost the total input cost (Rs 6952.57/ha) was found to be lesser than the total labour cost (Rs 8889.74/ha) at RPP, while in average farmer's field the total labour cost (31.51 per cent) was found to be highest than the total input cost (24.60 per cent). An average farmer was found to invest Rs 18716.14/ha to cultivate chickpea in his field, in which he invested Rs 4605.03/ha as the input cost and Rs 5897.19/ha as total labour cost. The gap of 75.68 per cent was found in average farmer's field than the RPP. It is also observed that the total input cost, total cost of cultivation, total fixed cost, and

Table 48.2 Yield Gap at Different Level of Adoption (q/hectare)

Particulars	Level of Adoption			
	Low	Moderate	High	Average
1. Average Farm Yield	5.72	12.26	19.11	12.36
2. Maximum Farm Yield	7.84	15.91	25.88	16.54
3. Potential Yield	30.00	30.00	30.00	30.00
Yield Gap I (2–3)	22.16 **(282.65)**	14.09 **(88.56)**	4.12 **(15.92)**	13.46 **(81.34)**
Yield Gap II (1–2)	2.13 **(37.18)**	3.65 **(29.77)**	6.78 **(35.46)**	4.18 **(33.85)**
Yield Gap III (1–3)	24.29 **(424.93)**	17.74 **(144.70)**	10.90 **(57.03)**	17.64 **(142.72)**

Note: Figures in the parenthesis show percentage yield gap
Source: Primary data.

total indirect variable cost were found to be increased with levels of adoption, while the total labour decreased with levels of adoption.

An average chickpea grower of the study area found to receive a gross income of Rs 27,452.87/ha. He got a net income of Rs 16,480.74/ha and Rs 8,736.73/ha at total variable and total cost of cultivation respectively. On investment of Rs 1.00 an average chickpea grower got a return of Rs 2.50 and Rs 1.47 at total variable cost and total cost of cultivation respectively which revealed that the chickpea production was found to be a profitable enterprise in the study area. However, it may be increased manifolds by adoption of RPP and by removal of constraints that came across in the adoption of RPP by the chickpea growers. The data also revealed that the cost of production shows increasing trend with the level of adoption. An average farmer was found to invest Rs 882.61 and Rs 641.75 as total cost and total variable cost respectively to produce a quintal of Chickpea.

Resource Use Efficiency

Regression analysis found that expenses on field preparation, high yielding variety seed, fertilizer, bio-fertilizer, seed treatment, weed management, insect pest management, and harvesting showed a positive effect on yield, expenses on sowing method being an exception. The yield of chickpea showed positive response to variable expenses on field preparation (0.323), weed management (0.213), harvesting (0.34) while expenses on fertilizer and high yielding variety seeds also showed positive and significant response over gross income. The negative response of gross income to sowing methods reveals that excessive use of labour reduces gross income of cultivators.

Pattern of Marketing

The majority of chickpea growers of the study area disposed of their produce through three marketing channels, namely, Channel I: Producer—Village Merchant—Wholesaler at Regulated Market–Processor/Miller, Channel II: Producer—Wholesaler at Regulated Market–Processor/Miller, and Channel III: Producer—Local Traders—Processor/Miller. Amongst all these channels, Channel II was found be more popular amongst chickpea growers. The maximum (88.25 per cent) of the total produce was found to be disposed off from this particular channel followed by Channel III (10.69 per cent) and Channel I (6.88 per cent). The average marketing cost incurred in different marketing channels was found to be Rs 61.67/qtl comprises cost of bags (35.64 per cent); loading/unloading (11.66 per cent); storage (25.15 per cent); transportation (12.55 per cent); market fee (7.67 per cent); establishment charges (1.25 per cent); weighing, filling, and stacking (3.18 per cent); commission (1.59 per cent); and spoilage (2.58 per cent). Amongst the different marketing channels the highest cost was found to be incurred in Channel II (Rs 86.09/qtl) followed by Channel I (Rs 52.69 qtl/ha), and Channel III (Rs 46.24/qtl). The average price spread was found to be Rs 209.33/qtl (9.71 per cent) in the marketing of chickpea and producer got 91.15 per cent share in the processor's rupee. The maximum price spread was found to be Channel III (18.26 per cent) followed by Channel I (13.23 per cent), and Channel II (7.44 per cent), while the highest producer share in processor's rupee was found to be in Channel II (93.08 per cent) followed by Channel I (88.32 per cent), and Channel III (84.56 per cent). Channel II (14.45 per cent) was found to be more efficient followed by Channel I (8.56 per cent), and Channel III (6.48 per cent) in the study area.

Constraints in Production and Marketing

The non-availability of good quality input specially HYVs seeds, fertilizers (specially muriate of potash and zinc sulphate), insecticides, fungicides (92.92 per cent), lack of demonstrations/field trails in farmers' fields (90.00 per cent), lack of knowledge about the composition of different nutrients in chemicals and the preparation of required concentration of the chemical through the branded chemicals and fertilizers (87.92 per cent), high cost of inputs including labours (82.50 per cent), high wages of labour at peak operation period, namely, sowing, harvesting, threshing, etc. (82.92 per cent), lack of knowledge, soil testing facilities (77.50 per cent), non-availability of input in time (75.00 per cent), irregular power supply/power cut at peak operation period (74.58 per cent), lack of Irrigation facilities/low water table (51.70 per cent), lack of knowledge about RPP viz. soil treatment, seed treatment, rhizobium and PSB culture treatment, Integrated Plant Nutrients Management, Integrated Pest Management technologies, application of fertilizers and micronutrients (Zn and S), plant protection chemicals, etc. (55.00 per cent) and Inadequate scale of crop loan (34.17 per cent) were found be major constraints in production of chickpea in the area under study.

The lack of market intelligence services (90.00 per cent), lack of knowledge about warehousing facilities present in the regulated markets (87.92 per cent), lack of knowledge about value addition technologies of chickpea (82.92 per cent), lack of market news at village level (82.50 per cent), lack of market credit facilities (92.92 per cent), low price of grain (77.50 per cent), lack of storage facilities (75.00 per cent), lack of knowledge about proper grading technology

(74.58 per cent), and lack of all weather roads (55.00 per cent), were identified as major constraint in efficient marketing of soybean.

Value Addition

As regards value addition technologies preformed by the chickpea growers in the area under study, the majority of chickpea growers preformed various value added activities (Table 48.3) such as picking of green leaves (96.67 per cent) and green pods for sale in the local market, preparation of *dal* (35.83 per cent), roasted grains (17.92 per cent), chickpea flour (97.08 per cent), namkeen (97.92 per cent), and sweet (97.50 per cent). Amongst all these activities, the preparation of dal was found to be more economical viable as an average chickpea grower got an additional return of Rs 9.08 on investment of Re 1.00, followed by preparation of *namkeen* (Rs 7.39), sweet (Rs 7.20), roasted grain (Rs 7.60), roasted dal (Rs 7.20), selling of green pods (Rs 2.30), and green leaves (Rs 2.24). The value addition activities, that is, cleaning/grading of grains and packing of grains in small packets were not preformed in the area under study.

Processing Cost

The total processing cost incurred to process 1.00 quintal of grain to dal was found to be Rs 63.20 per quintal, in which the share of fixed and variable cost was respectively of 6.96 and 93.04 per cent. The cost of bags was found to be the main item of total variable cost (42.20 per cent) followed by filling of bags, that is, (*palledari*) (12.66 per cent), expenditure on labour (9.22 per cent), commission (7.91 per cent), interest or working capital (9.59 per cent), electricity charges (5.81 per cent), operating of machine (3.69 per cent). The main item of the total fixed cost (Rs 4.40/quintal) was found to be rent on buildings (Rs 3.67/quintal) followed by depreciation on machine (Rs 0.33/quintal) and interest on fixed capital (Rs 0.40/quintal). It is interesting to note that the by-product of dal, that is, *chuni*, which is the main ingredient of milch cattle was not only found to recovered the total cost of processing of dal, but the miller also comes in profit by selling this to cattle owners (Rs 140/quintal). An average processor got a net profit of Rs 88.32/q from the processing of grains.

CONCLUSIONS AND POLICY IMPLICATIONS

The following conclusion and policy implications are made from the above findings:

1. It was found during the course of investigation that after the declaration of the districts under AEZ for pulses in MP, the farmers, traders, processors, and the general people were of the opinion that

Table 48.3 Processing of Chickpea at Farmers Level (Rs/qtl)

Value Added Products	Number of Respondents	Additional Cost	Additional Income Over Grains	Additional Net Return Over Grains	Additional Cost Additional Benefit Ratio
Green Leaves	232 (96.67)	356.45	800	443.55	2.24
Green Pods	87 (36.25)	241.67	556	315.22	2.3
Cleaning/Grading of grains	0	60	156	96	2.6
Packaging of grains in 1 Kg pack	0	130	244	114	1.88
Split of grains (dal)	86 (35.83)	115	1,044	929	9.08
Roasted Grains	43 (17.92)	367	2,644	2,277	7.2
Roasted dal	26 (10.83)	367	2,789	2,422	7.6
Chickpea Flour	233 (97.08)	400	802.11	402.11	2.01
Namkeens	235 (97.92)	487	3,600	3,113	7.39
Sweets	234 (97.5)	543	3,844	3,301	708

Note: Figures in parentheses show percentage to total respondents.
Source: Field survey.

the developmental activities, employment opportunities, etc., will be increased manifold in these districts. But, it is surprising to note that no change was observed in this regard, even the traders of the districts were not aware that their district fall under the AEZ for pulses by the government.

2. As yield instability was found to be major source of instability in chickpea production in the districts under AEZ for pulses and the spread of improved technology was found to be associated with decline variability in production. Hence, there is need to pay extra special attention to production and distribution of improved package recommended package of practices to break the yield barriers and bring sustainability in production. Expansion of area under irrigation, development of water sheds, development of varieties resistance to insect pest and climatic stress were found to be other major factors for reducing the variability in area production and yield of chickpea in the area under study.
3. An effective channel for transfer of production and marketing technology in the farmers' field is needed as wide yield and expenditure gap was observed with recommended package of practices of chickpea, namely, improved varieties, balance use of fertilizer, bio-fertilizer, micro-nutrients, weedicides, pesticides, etc., in the area. The transfer of technology may be effectively done by the processors' by providing extra incentives and motivation to them. As, it is clear that the ultimate profit marker in the production and marketing of chickpea was found to be processor. Whatever the cultivator harvested in their field the major portion of this was found to be in the hands of the processors. Hence, it is the duty of the processors to provide the full package of the quality input to farmers at subsidized rates. If the government provides special motivation and facilities to the processors they will come forward for doing so. This will become an effective measure for the removal of production as well as marketing constraints prevailing in the area under study.
4. The resources were not been found to be fully utilized by the chickpea growers and processors. Hence, there is tremendous scope for increasing the income and employment of chickpea growers and traders/processors by adoption of modern technology and by removing the constraints present in the production and marketing of chickpea in the area under study. It is also observed during the course of investigation that chickpea growers use chemicals specially hormones, which were not recommended by the Agricultural Universities or Department of Agriculture. Hence, efforts should be made to stop these activities as early as possible. There is also an urgent need to enact a law through which the business of spurious quality input being used in agriculture sector can be checked effectively.
5. The involvement of cooperative sector in the production, processing, and marketing of chickpea in the area under study was not found. However, in near future such possibilities cannot be ignored as the market of value added products are found to be increased in near future. Hence, in order to accelerate the pace of chickpea marketing the state government should come forward to facilitate the processors, traders, regulated markets, etc., to meet the global challenges of forward marketing.
6. As chickpea is a miracle crop, a number of value added products were found prepared by the farm women for their home consumption and saved lot of money. Hence, there is a possibility to convert into a cottage industry though creation of Self-help Groups performing economic activities. The Self-help Group can easily be formed by providing training and motivation to women on farms.
7. The majority of processors were found to be processing chickpea with outdated and old machines. This should be replaced with modern ones. The traders and processors will also be required to know the world market, trade centres of chickpea, and other AEZs of the country. Hence, orientation programmes for the development of processors need to be launched by the state government for the development of the processing sector in the state.

REFERENCES

Sharma, H.O., S.B. Nahatkar, and M.M. Patel. 2000. 'Economics of Soybean Production Technology at Different Levels of Technological Adoption in Sehore Development Block of M.P.', *Indian Journal of Agricultural Economics*, 55(3): 534–5.

ANNEXURE 48A

Table 48A The Details of AERCs and the Lead Person Involved in the State Report

S. No.	States	AERC Who Carried the Study in the State	The Lead Person Who Carried Out the Study
1.	Andhra Pradesh	AERC Waltare	M. Nageswara Rao
2.	Bihar	AERC Bhagalpur	Ranjan Kumar Sinha
3.	Haryana	AERC Delhi	Usha Tuteja
4.	Rajasthan	AERC Vidyanagar	Mrutyunjay Swain Ramesh H. Patel Manish Kant Ojha
5.	West Bengal	AERC Shantiniketan	Kali Sankar Chattopadhayay Debajit Roy
6.	Karnataka	ADRTC Bangalore	Parmod Kumar
7.	Tamil Nadu	AERC Chennai	K. Jothi Sivagnanam A. Abdul Salam S.R. Muthusamy

49

Economics of Commercial Silk Weavers of Assam

Bharati Gogoi

Silk fibre is made of the protein secreted from the silk glands of silkworms. It is a high-valued low-volume commodity. India ranks as the second largest producer accounting for nearly 16 per cent of the world's raw silk production. There are four major types of silkworms—Mulberry, Tasar, Muga, and Eri. The Indian silk industry has a unique distinction of producing all four major types of silk. Mulberry silk is produced throughout the country from Kashmir to Kerala and Gujarat to Assam, while Orissa, Bihar, Madhya Pradesh, and Uttar Pradesh produce Tasar silk on a large scale. Muga and Eri silks are produced only in the North-Eastern states.

Silk culture in Assam probably originated in the Vedic age, as there are references of rearing of silkworms by the Assamese people for production of various silk cloths even in the age of 'Ramayana'. Such references are also found in Kautilya's *Arthasastra*.

Silk weaving was once a household affair in Assam. Nowadays it is a sustainable farm-based economic enterprise positively favouring the rural poor in the unorganized sector. Rearing of mulberry, muga, and eri silkworms has been playing an important role in the economic development of a large section of rural population of the state. Oak tasar silk production in Assam was introduced in 1972.

The average production of handloom in the state is very poor in comparison to the rest of the country. Most of the weavers are still using traditional looms including loin and throw shuttle looms. In spite of the existence of a large number looms, the actual production forms only a small share of the total handloom production in the country. The average productivity per day per loom in Assam is only 0.63 metres against the all-India average of 1.29 metres. Standard of looms, level of utilization, and operating conditions of the looms are considered as some of the critical factors for low productivity.

In Assam, at present about 2.8 lakh looms run commercially. About 5.70 lakh looms run on a semi-commercial basis to earn a subsidiary income, the remaining being domestic looms, run at leisure hours to meet the family requirement of a few items of fabrics. The handloom industry of Assam is basically silk oriented.

Although the state is enriched with all four varieties of silk products, the present study is confined to only mulberry and muga silk weavers of Sualkuchi which is the centre of the silk weaving of the state. Sualkuchi is better known as the silk village of Assam. History has references of Sualkuchi as a production centre of silk even in the reign of King Dharam Pal in the eleventh century. With this backdrop, the study is designed with the following objectives:

1. to study the social and economic status of commercial weaver families in the society;
2. to study the sources of collection of Muga and Mulberry silk yarn/cocoon;
3. to study the cost and return of per unit of cloth produced by weaver families;
4. to study the marketing system, marketing channels, and marketing costs of the finished products; and
5. to identify the problems faced by the weavers families.

METHODOLOGY AND DATA

The study is confined to Pat (Mulberry) and Muga commercial weavers. The silk weavers-dominated area of Sualkuchi was purposively selected for the study. A list of weavers having Muga and Pat looms with their numbers was collected from the concerned area. The list was further stratified on the basis of number of looms possessed by each family as: up to five looms, five to ten looms, ten to twenty looms, twenty to thirty looms, thirty to forty looms, and forty and above looms. One hundred (100) weaver households were drawn randomly as samples of the study with the help of the ratio proportionate technique from each stratum. In order to study the marketing channels of finished products of Muga and Pat silk, three wholesalers were purposively selected from the Sualkuchi market on the criteria of having retailers under them in three different city/towns, namely, Guwahati city, Pathsala town, and Jorhat town. The study relates to the year 2006–07. Accordingly, the relevant data were collected from the selected respondents.

RESULTS AND DISCUSSION

The total number of family members of the sample households was 531 comprising 55.93 per cent (297) males and 44.07 per cent (234) females; the sex ratio being 788 females per 1000 males. There were 29.57 per cent below 15 years of age and 3.77 per cent in the 65 years and above age group. The remaining 66.66 per cent constituted the main work force of the sample households of which the highest number was found in the age group of 25 years to 35 years (21.28 per cent).

The literacy rate of the family members of the sample households was found at 87.76 per cent, which was much higher than that of the state's literacy rate of 62.25 per cent as per the 2001 Census. Among the sample households, only 12.24 per cent were illiterate; of which 8.66 per cent were in the age group of 0–15 years which means that a major portion of the illiterate persons were either infants or small kids. There were 13.18 per cent who fell in the category of just literate and 18.83 per cent were educated up to the lower primary level. It was found that though the literacy rate of sample households was as high as 87.76 per cent, only 0.76 per cent family members of the sample households were postgraduate degree holders and 1.88 per cent had graduate degrees. The highest numbers of literate persons were found in the literacy level of up to high school leaving certificate (HSLC) at 39.36 per cent followed by 10.55 per cent HSLC passed and 3.20 per cent higher secondary school leaving certificate (HSSLC) passed. It was observed that as weaving was the family enterprise of most of the sample households since generations, the inclination towards family business starts in the minds of the young population quite early. As a result, generally they prefer to be actively involved in the weaving business rather than pursuing higher education, considering the fact that weaving was an economically gainful enterprise. Therefore, they usually opted for the family business after obtaining HSLC-level education.

In the sample, the highest number was found as workers comprising 37.85 per cent of the total population. There were 121 male workers (22.78 per cent of total population) and 60 female workers (15.07 per cent of total population) in sample families. In the category of helpers, there were 68 male helpers (12.81 per cent of total population) and 78 female helpers (14.69 per cent of total population), that is, 27.50 per cent of total population in the sample households. The figure in the non-worker category was 34.65 per cent of the total population. The higher percentage of non-workers can be attributed to inclusion of all infants, school-/college-going students and old persons of the sample population.

One of the important features of the weaving industry of Sualkuchi was the use of hired workers on a contractual basis. The common and prominent practice among the weaver households of the area was that the young girls and boys in batches, mostly from the Bodo community, were employed as weavers on annual/piece work on wage basis.

The landholding pattern of the sample households depicts a somewhat contrasting picture with rest of the rural Assam. Assam being an agricultural state, agriculture is the source of livelihood for more than 70.00 per cent of the state's total population. However, Sualkuchi was found to be an industrial village and agriculture did not play any role in the village economy. All 100 sample households do not cultivate any crop. They were totally dependent on others for their basic needs of food items.

It was found that the total net income from weaving of 100 sample households was Rs 37,844,207.91 during the year under reference (2006–07). Apart from weaving,

12 sample households earned Rs 252,000.00 from business, trade, commerce, and transport. Most of these activities were related directly or indirectly to silk production. However, it was also reported that some family members of the sample households had government or private jobs as their source of income. The annual earnings of those service holders stood at Rs 668,404.00 in the reference year. The livestock (poultry birds and ducks) farms earned the sample households an annual income of Rs 21,600.00 from that source. Total income from subsidiary sources of the sample households was Rs 941,404.00 during the year 2006–07.

In the sample area, various types of handlooms were used. The handloom was a manually operated simple machine which was locally called *Tat Shal*. Among the sample weaver households, there were a total of 1,725 numbers of looms comprising of 63.48 per cent Pat looms (1,095 numbers) and 36.52 per cent Muga looms (630 numbers). Pat looms were exclusively used for production of Mulberry silk and Muga looms were used for production of Muga silk. Predominance of Dobby looms was clearly visible in the sample households. Of the total 1,725 numbers of looms, 92.58 per cent were Fly shuttle looms fitted with a Dobby machine followed by 5.68 per cent Fly shuttle looms with a Drawbuoy. The remaining 1.74 per cent looms were other varieties of Fly shuttle looms.

Handloom weaving is a labour-intensive enterprise. Per loom number of man days required in the year for production of 51,519 number of silk items of the sample households was 212.99 man-days, the shares of family labour and hired labour being 42.60 man days (20.00 per cent of total man days) and 170.39 man-days (80.00 per cent of total man days) respectively. The proportion of male and female family labour was 40:60 where as it was 20:80 in the case of hired labour.

The estimated wages to the family labour was Rs 4,259.79 per loom of which Rs 1,703.91 (40.00 per cent) was for male and Rs 2,555.88 (60.00 per cent) for female family labour. The wages paid to the hired labour in the year stood at Rs 17,039.18 per loom of which Rs 3,407.84 per loom (20.00 per cent) was for male weavers and Rs 13,631.34 (80.00 per cent) for female weavers. The cost, combining all workers (both family labour and hired labour) was Rs 21,298.97 per loom; amounting 24.00 per cent (Rs 5,111.75 per loom) was for males and 76.00 per cent (Rs 16,187.22 per loom) for female weavers. The average costs of production of silk of sample households were Rs 54,450.79. Out of the total cost of production, 39.12 per cent was spent on labour wages, 38.18 per cent was incurred on purchasing raw materials, 20.90 per cent was on accessories and provisions and the remaining 1.80 per cent was attributed to managerial costs (Table 49.1).

Muga silk is costlier than Pat silk. As such, the demand for Pat silk was higher than Muga silk for economic reasons. That is why the silk weaver households give more emphasis on the production of the Pat rather than Muga silk. The total number of Pat and Muga products inclusive of all varieties of items produced by the sample households was 51,519 during the reference year.

Overall, the average return from silk produced by the sample households was Rs 76,389.46 per loom in 2006–07 of which 56.00 per cent came from sale proceeds of Mulberry silk and the remaining 44.00 per cent was contributed by Muga silk. The annual average income per loom was calculated by deducting the cost of production from the value of all silk woven by the sample households and was Rs 21,938.67, consisting of 66.02 per cent (Rs 22,818.56) from Mulberry silk and 33.98 per cent (Rs 20,409.33) from Muga silk (Table 49.2).

It was found that silk weaving of the sample households was an economically viable enterprise as the Benefit–Cost Ratio (BCR) was found at 1.40:1. Category wise BCR against Pat silk production of the sample households was 1.51:1, whereas the BCR of Muga silk production was recorded at 1.29:1.

Table 49.1 Annual Cost of Production of Silk Clothes of Sample Households (Per Loom)

Items	Muga Weavers	Pat Weavers	Average Cost	P.C. to Total Cost
Numbers of Loom	630	1,095	1,725	
Equipments, Provisions Cost	11,406.85	11,361.84	11,378.28	20.90
Labour Costs:				
(i) Family Labour	4,315.58	4,227.70	4,259.79	7.82
(ii) Hired Labour	17,831.50	16,583.32	17,039.18	31.29
(iii) Total Lab. Cost	22,147.08	20,811.02	21,298.97	39.12
Raw Material Cost	36,681.11	11,649.56	20,791.52	38.18
Misc. Expenditure@ 2%	1,316.93	789.34	982.02	1.80
Total Cost	**71,523.40**	**44,628.19**	**54,450.79**	**100.00**

Table 49.2 Return to Sample Households from Silk Weaving (Per Loom)

Variety of Silk	Numbers of Loom	Average Return	Average Cost of Production	Average Net Return	BCR
Pat	1,095	67,446.75	44,628.19	22,818.56	1.51:1
Muga	630	91,932.72	71,523.40	20,409.33	1.29:1
Total	1,725	76,389.46	54,450.79	21,938.67	1.40:1

Note: BCR: Benefit–cost ratio.

An analysis of marketing channels indicated that there were six marketing channels operating in silk trading in Sualkuchi, namely, (i) Producer–Customer; (ii) Producer–Retailer–Customer; (iii) Producer–Wholesaler–Retailer–Customer; (iv) Producer–Commission Agent–Retailer–Customer; (v) Producer–Cooperative Society–State Govt.–Customer; and (vi) Producer–Cooperative Society–Wholesaler–Retailer–Customer.

The market efficiency of six different marketing channels in four sample markets, that is, Sualkuchi, Pathsala, Guwahati, and Jorhat on Pat and Muga silk was examined against the sample households. The sample silk producers reported that 49.50 per cent of their produce was sold to cooperative societies, of which 46.00 per cent was purchased by the state government and the remaining 3.50 per cent was purchased by wholesalers. The sample households directly sold 28.00 per cent of their produce to wholesalers and 15.00 per cent to retailers. The sample households also sold 7.00 per cent of their produce to commission agents, who were local traders. It was noticed that only 0.50 per cent of the produce of the sample weaver households was directly sold to the customers who visited Sualkuchi market to purchase some Pat Muga silk either for personal use or for some ceremonies, especially for marriage ceremonies.

PROBLEMS OF SILK CLOTH WEAVING

The economic activities of the sample households of Sualkuchi were centred on silk production. In the course of silk production, numerous difficulties were faced by the sample weaver households. Some of these problems were:

1. Declining trend in the number of local weavers since the last decade. As a result, most of the sample households had to depend upon migrant weavers. They have to pay higher wages to these hired weavers. Bodo young girls from neighbouring Goreswar, Tamulpur, and Tangla who are experts in silk cloth weaving demand higher wages for their work. So, availability of skilled weavers and high financial involvement associated with it were identified to be the major problems faced by the sample households.
2. The price of Muga yarn has escalated so much that pure fabric has become unaffordable to the common man/woman. The Tasar yarn that is dyed in Muga colours has very similar outward physical properties; in reality Tasar yarn is inferior to Muga yarn so far as the inner qualities are concerned. It is very difficult even for an expert to find out the differences. Being cheaper in price, a trend of gradual replacement of Muga yarn by Tasar yarn is slowly taking place in fabric production. The sample weaver families opined that the day is not very far when entire Muga culture will be jeopardized, if a time-bound work plan is not prepared to increase its production substantially.
3. Most of the raw materials, namely, mulberry silk yarn, golden and silver *guna* (*Zari*) and cotton art threads, etc., are imported from outside the State. Mulberry yarn mainly comes from Bangalore and golden/silver guna and cotton art threads used in making motifs and designs in silk clothes are brought from Surat, while the Muga cocoons are brought from Upper Assam and the Garo Hills areas. Sporadic price due to artificial short supply during the peak season makes the cost of production of Muga much higher. It has very often been observed that the non-availability of good yarn and relatively higher price of the same quality yarn because of unfair profit motives of the traders, middlemen, etc., create hurdles for the weavers to run the industry smoothly and to have desired economic returns.
4. Notwithstanding some degree of changes toward modernization that have taken place in the silk industry of Sualkuchi, there is still immense scope of improvement. No sustainable attempt from any quarter for upgrading the technologies with sophistication in reeling, spinning, winding, and weaving as well as in diversifying the products to match the today's global market has been made for the sample households.
5. It is observed that though the weavers of Sualkuchi have all the skills and urges of producing high-quality world-class silk clothes with eye-catching designs, they still are lagging far behind in entering the international market. Lack of proper guidance and marketing channels to outside the country are the hurdles yet to be overcome by the sample weavers of the area in particular.

6. Another irritating problem faced by the sample households was the erratic power supply in the area, which interrupts the working hours of the weavers. Though the handloom industries of Sualkuchi do not use electrical equipments, the weavers work on their looms up to 10 pm to 11 pm at night. Frequent power cuts by the Assam State Electricity Board has compelled many of the sample households to look for alternative power supplies like purchasing generator sets or hiring power from generator sets of neighbouring co-weavers.

PROSPECTS OF SILK WEAVING

The silk industry of Sualkuchi has undergone many changes. The same is true for the age-old silk culture of Assam as well. To cope with the challenges of time, the silk industry of Assam must try to explore advanced ways to upgrade the traditional silk culture as a whole. There are a number of issues to be looked into for uplifting the economic condition of the sample weaver households in particular and for further progress of the silk industry of the state in general.

1. Introduction of Dobby machines has brought remarkable changes in traditional designs of Assamese silk textiles produced in Sualkuchi. However, to boost the volume of production, there is ample scope of introducing simple power looms besides semi-automatic handlooms, sizing plants, etc. A few years back, the Central Silk Board of India demonstrated some sophisticated reeling and spinning machines for Muga, Mulberry and Eri yarns in order to impart training to local artisans. Introduction of semi-automatic handlooms and simple power looms together with in-plant training can bring about marked changes in the silk industry of Sualkuchi. There is immense scope of such a technological revolution. Concerted efforts in terms of effective extension services and monitoring programmes can change the attitude of the people for adoption of available technologies, including indoor rearing of muga and may help the silk industry of Sualkuchi in particular and the state as a whole.
2. A revolutionary change in the history of silk culture of Sualkuchi textile took place during the third decade of the twentieth century. Weaving of cloth on Fly Shuttle was started with the help of Jacquard machines. Designing floral decoration was made easy. It was found that although some of the weaver families of Sualkuchi had adopted Jacquard machine technology in their looms, the sample households were yet to go for Jacquard machines. Use of Jacquard machines can bring in much improvement in the designing of silk cloth of the sample households.
3. The initiative taken by the Central Silk Board in popularizing relatively more efficient machines in place of traditional equipment 'Bhir' for reeling Muga yarn should be escalated and steps may be taken to expand this culture among the Muga weavers of Sualkuchi. The sample households, in spite of their awareness about the efficiency of the new machine, had expressed their financial inability to go for such an expensive machine. The cost of such machine is reported to be between Rs 5000 to Rs 10000 whereas the cost of 'Bhir' is less than Rs 500. Such a machine, if provided at subsidized prices, will have many of the weaver households willing to go for it. They also opined that even if a subsidized rate is not feasible, provision of finance will be of great help for the weavers of the area in adopting more efficient machines. Necessary assistance in procuring the costly Muga reeling machine will be a great impetus to the Muga cloth weavers of the study area.
4. There is ample scope for a harmonious combination of traditional designs with new ideas by taking the help of computer technologies to work out brilliant and more appealing modern patterns in the silk cloth woven by the expert weavers of Sualkuchi. This will help in impressing more customers both in local and outside markets. The urgent need for initiating some coordinated research and training in this direction will help the silk cloth weaving industry of Sualkuchi to a great extent.
5. The handloom and textile industry of Sualkuchi is one of the most valuable assets of Assamese culture and tradition. The increased need of modernization in terms of infrastructural facilities demands the introduction of more improved varieties of looms that are at present in operation in advanced textile towns like Salen (Tamil Nadu) and Panipat (Haryana).
6. Establishment of a textile museum at Sualkuchi by Government patronage will be an appropriate step to keep alive the unique tradition and culture of silk industry of Assam which in turn may attract tourists into this area.
7. Some of the accessories made of bamboo and wood used in handloom for weaving various types

of cloth at Sualkuchi are now replaced mostly by similar accessories made out of materials like plastic and iron. All these changes have contributed positively towards the growth and development of the industry. This welcoming trend of transformation may be encouraged by ensuring ample supply of such items.

8. Cooperation is one of the important priorities for bringing about changes in the economy in a country like ours. The present well-organized and efficiently functioning Co-operative Societies of Sualkuchi have a role to accept the challenges of rural development by reaching the weaker section of the society. The grassroots level Weaving Co-operative Societies as an umbrella organization should function as a prime mover for the purpose of rural development as well. The Co-operative Societies have to see that the hired weavers get the actual wages for their labour. On the other hand, remunerative returns to the poor weavers of the area for selling their finished silk cloth items through the Co-operative Societies must be ensured.
9. Availability of Bank finance to the willing weaver families will open up the path of speedy adoption of modern technologies. Lack of finance which is considered to be a major impediment for the poor weaver families can be mitigated by providing institutional finance to the willing weaver families and that in turn will improve both quality and quantity of the silk clothes produced in the study area.

CONCLUSION AND POLICY IMPLICATIONS

Keeping in view the emerging process of transformation in the age-old silk industry of Sualkuchi, following suggestions are offered:

1. Introduction of semi-automatic handlooms and simple power looms can bring about remarkable changes in the silk industry of Sualkuchi as well as in the state.
2. Use of Jacquard machines will bring much improvement in the design-making of the silk cloth produced by the sample households.
3. Relatively more efficient machines in place of traditional equipment 'Bhir' for reeling Muga yarn should be escalated and steps may be taken to expand this culture among the Muga weavers of Sualkuchi.
4. A harmonious combination of traditional designs with new ideas by taking the help of computer technologies should be worked out to evolve brilliant and more appealing patterns in silk cloth. Establishment of a research cum training centre in the line can meet a long-pending necessity of the area.
5. Adoption of modern technologies, especially use of sophisticated tools and equipments requires sufficient amount of money. Lack of finance in case of poor weaver families should be mitigated by providing institutional finance to the willing weaver families of the area.
6. Appropriate steps should be taken to safeguard the golden yarn, Muga. Unless Muga yarn production is increased substantially, the price of already much costlier Muga yarn will increase further and there is every possibility that entire Muga culture may be jeopardized through its replacement by low-cost Tasar yarn.
7. Supply of electricity to the silk villages should be improved as it will help the weavers of Sualkuchi to get uninterrupted working hours.
8. The Co-operative Societies should ensure that the hired weavers get proper wages for their labour. At the same time, remunerative returns must be ensured to the poor weavers of the area for selling their finished silk clothes items through the Co-operative Societies.
9. Establishment of a Yarn Bank for Pat and Muga yarns with reasonable prices at Sualkuchi will fulfil a much-needed requirement of the silk cloth weavers of the area. It will rescue the silk weavers from the clutches of the traders who very often exploit them by raising the rate of yarns.
10. To keep alive the unique tradition and culture of silk industry of Assam and to attract tourists into the area, a textile museum should be established at Sualkuchi.

Everything made of hand is precious and has excellent market value. The age-old traditional fabrics have the highest status in handloom textiles. Silk cloth produced in Assam are milestones in this regard. It would be very encouraging and reassuring if efforts are made to preserve different textile items traditionally produced in Sualkuchi with the unique colourful patterns. However, in spite of vast potentials in terms of skills, artistic designs, beautiful crafts, and colour combinations, the silk industry in Assam is lagging behind mainly because of half-hearted efforts by the development agencies, financial shortage, lack of suitable markets for finished products, and shortage of proper training facilities. There is no doubt that if both state and central governments join hands with various handloom development organizations and undertake the responsibilities of providing necessary support for marketing outlets, production activities, and other infrastructure for the weavers, the silk products of Assam can make a mark in the national as well as international handloom scenario.

REFERENCES

Baishya, P. 1989. *Small and Cottage Industries—A Study in Assam*. Assam: Manas Publication.

———. 1969. *A Cultural History of Assam*. Guwahati: Lawyer's Book Stall.

Baruah, K.L. 1933. *Early History of Kamrup*. Guwahati: Lawyer's Book Stall.

Borgohain, J.N. 2006. 'Silk Industry in Assam', *Assam Tribune*, 14 May.

Chetia, Sanjita. 2006. 'The Assamese Handloom Textiles Tradition'. Parbotia, Tinsukia: The Assam Computers.

Chowdhury, N.S. 1982. *Eri Silk Industry*. Guwahati: Directorate of Information and Public Relations.

Chowdhury, P.C. 1959. *The History of Civilization to the 12th Century AD*. Guwahati: Dept. of Historical and Antiquarian Studies in Assam.

Goswami, A.D. 1980. *Sahitya Samanya*. Kokrajhar: Sahitya Sabha Bhawan.

Goswami, P.C. 1988. *The Economic Development of Assam*. New Delhi: Kalyani Publishers.

Konwar, K.R. 1981. *Practical Hand Book of Weaving*. Assam: Directorate of Sericulture.

Medhi, D.N. 1956. *Handloom Weaving and Designing*. Guwahati: Lawyer's Book Stall.

Nair, G., and K.R. Babu. 1993. *Demand and Supply Prospects for High Quality Raw Silk*. New Delhi: Oxford & IBH Publishing.

Sarof, D.N. 1982. *Indian Craft Development and Potential*. New Delhi: Vikas Publishing House.

Semon, H.F. 1897. *Monograph of Cotton Fabrics of Assam*. Calcutta.

50

Economics of Production, Processing, and Marketing of Fodder Crops in Selected States of India

D.K. Grover and Sanjay Kumar

The livestock sector in India contributes in the range of 30 to 35 per cent of the total agricultural output. The desired annual growth of the agriculture sector can be accomplished only through enhancing overall productivity of the livestock sector. This would require a steady and adequate supply of quality fodder for supporting the livestock population. Having only 4 to 5 per cent of total cropping area under fodder cultivation and low productivity of fodder crops has resulted in a severe deficit of green fodder, dry fodder, and concentrates (Dikshit and Birthal 2010). For development of the livestock sector, the need of the hour is, therefore, to meet this shortfall of fodder (which is over 55 per cent) by adopting suitable measures for increasing the production of crop residues, green fodder, and agricultural by-products (Government of India 2007). Fodder deficit can mainly be attributed to our limitations in increasing the area under fodder crops, limited availability of good high yielding fodder varieties, lack of quality seeds of improved hybrids/varieties, poor quality of dry fodder like paddy/wheat straw, changing crop pattern in favour of cash crops, etc. Besides, low priority accorded to investment in fodder production, lack of post-harvest management for surplus fodder, poor management of grazing/pasture lands and inadequate research, extension, and manpower support have also aggravated the shortfall situation of fodder. The importance of feeds and fodders in dairy farming needs no emphasis. With increase in the pressure on land due to urbanization and industrialization and decrease in the area under fodder and food crops coupled with increasing demand for milk and milk products, the dependency of livestock/dairy farmers on external or purchased inputs has also increased and it is putting pressure especially on the resource poor dairy farmers. Efforts are being made and are underway to reduce the gap between the requirement and availability of feeds and fodders through technological interventions to increase the yields, bringing more area under fodder crops, conservation of feeds and fodders, improving the nutritive value of the poor quality roughage, formulation of balanced rations, feeding of unconventional feeds, etc. However, 'fodder scarcity' continues and it has become a challenging issue in most of the developing countries including India, where dairying is largely the avocation of the poor, especially women. The study was carried out to accomplish the following objectives:

1. to study the status of fodder crops cultivated in selected states;

2. to estimate the costs of production and returns associated with the cultivation of important fodder crops;
3. to identify the processing and marketing system and to estimate the costs and returns at each link for these fodder crops; and
4. to study the problems faced by the producers in production, marketing, and processing of these fodder crops.

METHODOLOGY AND DATA

The study was conducted in the Gujarat, Madhya Pradesh, Karnataka, and Punjab states of India. Important fodder crops in the India include berseem, sorghum, guar, maize, cowpea, oats, chari, bajra, moth, lucerne, jowar, etc. In the present study, one most important fodder crop each in the kharif, rabi, and summer seasons of the selected states were selected for the in-depth analysis. Amongst different districts of each state, three districts with the highest area of fodder in the state were selected purposively. Amongst the selected districts, two blocks from each district, one block near and one distant to the periphery of district headquarter were selected randomly to realize the effect of distance factor in the findings. From each block, a cluster of three to five villages were randomly chosen. Finally, a sample of 25 farmers was selected randomly from each selected cluster, making a total sample of 150 households. The primary data pertaining to the year 2008–09 were collected by the personal interview method. However, fodders marketing and processing practices were not commonly found in India, yet the hay/silage method of fodder processing was used by a few farmers. A sample of marketing intermediaries involved in the marketing and processors associated with the processing of fodder were randomly chosen from the selected blocks to know the different stages of the fodder marketing and processing and to assess the costs involved at each stage.

RESULTS AND DISCUSSION

Status of Livestock Population

The size of livestock herd in Gujarat increased from 196.7 lakh in 1992 to 237.9 lakh in 2007 indicating a spectacular average annual growth rate (AAGR) of 1.28 per cent during the period 1992–2007. Similarly, the livestock population in Madhya Pradesh showed an increasing trend over the years and the total livestock population was found to have increased with the annual growth of 1.90 per cent in the year 2007 as compared to the year 1992 (0.32 lakh). Likewise, the total livestock population in the state of Karnataka has increased from 295.7 lakh in 1992 to 30.86 million in 2007 with compound annual growth rate (CAGR) of 0.29 per cent. On the contrary, the livestock population in Punjab has been decreasing continuously since 1990, and showed a tremendous decline from about 97 lakh in 1990 to only 71 lakh during 2007, at the rate of 1.5 per cent per annum.

Status of Fodder Crops Cultivation

Fodder cultivation is still found to be in a nascent stage in Madhya Pradesh. Out of the total fodder area (0.74 lakh ha), the cultivators of Madhya Pradesh devoted their maximum area under the cultivation of bajra (20 per cent) followed by Jowar (4 per cent), berseem (2 per cent), and maize (1 per cent). The area of fodder was found to have declined over the years from 0.97 lakh ha (1990–94) to 0.75 lakh ha (2006–09) in Madhya Pradesh during the last 20 years. In Punjab, on an average, about 5.83 lakh ha were under fodder crops during the period 2005–09, which comes out to be about 7 per cent of gross cropped area of the state. The area under fodder crops was found to decrease continuously from the average area of 7.8 lakh ha during the period 1990–94. The fodder crops occupied about 2.64 lakh ha area in the kharif season and about 2.97 lakh ha during the rabi season. Maize fodder was also cultivated during the summer season covering about 21 thousand hectare area during the season. Sorghum, bajra, and guara were important kharif fodders covering about 24, 14, and 3 per cent of the total area under fodder cultivation in the state during the period 2005–09. Berseem and oats were the important rabi fodders covering about 34 and 12 per cent of the total area under total fodder cultivation in the state. Maize fodder covered about 4 per cent of the total area under fodder cultivation in the state during the period 2005–09.

Agro-socio-economic Characteristics of Fodder Growers

The majority of households in each selected state had between four to eight family members. Most of the sample households had a young head with age above 30 years except in Karnataka where about 46 per cent of the sample households had a head of age up to 30 years. Heads of about 82 per cent sample households were literate in Gujarat and Karnataka. Illiteracy was found to be significantly higher among the Madhya Pradesh farmers, that is, 52 per cent. The majority of farmers of the selected states had a net annual income below Rs 1 lakh, except in Punjab, where most of the sample households (about 55 per cent) had an annual income of more than Rs 5 lakh. The average landholding was the highest for Madhya Pradesh farmers (6.19 ha) and the least for Karnataka farmers (3.14 ha). In Gujarat, 82.23 per cent of operational land was irrigated. In Gujarat, nearly 99 per cent of the total livestock were bovines and about 86 per cent of the total adult cows in milk were crossbred. The value of total livestock per sample household was found to be Rs 1.73 lakh. In Madhya Pradesh, the present value of indigenous cows, crossbred cows, and buffalo were found to be Rs 0.10 lakh, Rs 0.20 lakh, and Rs 0.23 lakh respectively. In Karnataka, among indigenous cattle, for overall sample farmers the average price per female dry was the highest with Rs 18,857, followed by female in milk (Rs 16,576), and female not calved (Rs 16,500). Similarly for crossbred cattle the average price for female in milk was the highest (Rs 21,835) followed by female dry (Rs 13,750), and male (Rs 12,857). In Punjab, the average sample household was found to rear about six buffaloes and about to cattle on the farm.

Economics of Production of Fodder Crops

In Gujarat, the average cost of cultivation per hectare for kharif maize (cereal) comes to Rs 15,107. Human labour (33 per cent), machine labour (20 per cent), farmyard manure (FYM) (15 per cent), and chemical fertilizer (15 per cent) were the major contributors in the total cost of cultivation. In Madhya Pradesh, the total cost of cultivation of maize fodder was Rs 9,264.64/ha in the cultivation of maize. The FYM (37 per cent), machine labour (16 per cent), seed (13 per cent), chemical fertilizer (12 per cent), hired human labour (10 per cent) and family labour (7 per cent) were found to be major components of cost of cultivation of maize in the area under study. Jowar is an important food/fodder crop in Karnataka. The overall estimated variable cost was Rs 223/ha. Family labour accounted for the highest proportion of total cost. In Punjab, the total variable cost on per hectare basis for the most important fodder crop during kharif season (sorghum) was found to be Rs 11,946. Amongst variable cost components, the share of human labour was more than 71 per cent. The total cost of cultivation per hectare for lucerne, being the most important forage crop of Gujarat during rabi season, was Rs 31,372. The item-wise examination of cost shows that in the total cost of cultivation, the share of seed cost was highest at about 34 per cent. In Madhya Pradesh, berseem is found to be a major fodder crop cultivated by the majority of fodder growers in the winter season and an average fodder grower invested Rs 13,835.66/ha in the cultivation of berseem. The FYM (33 per cent), seed (26 per cent), machine labour (11 per cent), irrigation (9 per cent), chemical fertilizer (8 per cent), hired human labour (7 per cent), and family labour (4 per cent) were found to be the main components of cost of cultivation of berseem in the area under study. In Punjab, the total variable cost on a per hectare basis for the most important fodder crop during rabi season (berseem) was found to be Rs 18,231. Human labour was found to take a larger proportion of the cost as its share was about 66 per cent. In Gujarat, the total cost of cultivation per hectare for lucerne (summer) was Rs 25,075. The item-wise examination of cost data shows that in total cost of cultivation, share of seed cost was the highest at about 35 per cent. In Madhya Pradesh, jowar was found to be a major fodder crop cultivated by the majority of fodder growers in the summer season and an average fodder grower invested Rs 9,264.64/ha in the cultivation of jowar. The FYM (32 per cent), machine labour (16 per cent), seed (11 per cent), hired human labour (11 per cent), chemical fertilizer (10 per cent), irrigation (9 per cent), and family labour (8 per cent) were found to be the main components of cost of cultivation of maize the area under study. In Punjab, the total variable cost on per hectare basis for most important fodder crop during summer season (maize fodder) was found to be Rs 8,948. About 60 per cent of the operational cost was incurred on human labour, most of which is required during the harvesting of the crop.

Table 50.1 shows that in Gujarat, during the kharif season, net returns per hectare for maize cereal crop comes to Rs 32,775 which was higher by Rs 10,821 compared to a net return of Rs 21,954 for maize grown as pure green fodder. Paddy is a crop competing with maize. Overall, gross value of production (MP + BP) and total variable cost of paddy were Rs 34,375 and Rs 16,444 respectively. Overall, net return per hectare for paddy was Rs 18,291. In the rabi season, the net return per hectare was Rs 13,828 for lucerne whereas it was Rs 33,922 for the competing crop—wheat. In the summer, the net return for study crop lucerne was only Rs 6,569 whereas it was Rs 16,246 for competing crop jowar/sorghum grown as green fodder crop (Table 50.1). In Madhya Pradesh, there was no competition of fodder crops with other crops in the area under study. The comparative picture of fodder crops showed that the cultivation of beseem was more profitable in the area under study in which an average fodder grower invested Rs 13,835/ha and received Rs 52,521/ha revealed that on an investment of Rupee 1, he got Rs 3.80 as benefit over the variable cost, while he received only Rs 1.80 and 1.69 on investment of Rupee 1 respectively on the cultivation of maize and jowar. He also got maximum net returns from the cultivation of berseem (Rs 52,521.47/ha) as compared to the cultivation of maize (Rs 16,664.92/ha) and jowar (Rs 16,092/ha). In Karnataka, the returns over variable cost

Table 50.1 Economics of Fodder Crop vis-à-vis Competing Crop, Sample Households, Selected States, India, 2008–09

(Rs/ha)

Particulars	Yield (Qtls/ha)	Price (Rs/qtl)	Gross Returns	Total Variable Costs	Return Over Variable Costs
			Kharif season		
			Gujarat		
Maize (MP+BP)	NR	NR	48,905	16,130	32,775
Paddy (MP+BP)	NR	NR	34,735	16,444	18,291
			Madhya Pradesh		
Maize chari	269.37	96.26	25,929	9,264	16,665
Competing crop	NR	NR	NR	NR	NR
			Karnataka		
Jowar	3.9	200	775.1	222.5	552.6
Paddy	15.8	890.0	15,774	5,474	10,300
			Punjab		
Sorghum	448	56	25,082	11,946	13,136
Paddy	59	775	45,725	15,635	30,090
			Rabi season		
			Gujarat		
Lucerne	397.82	126	50,221	36,393	13,828
Wheat	NR	NR	50,079	16,158	33,922
			Madhya Pradesh		
Berseem	649.73	102.13	66,357.13	13,835.66	52,521.47
Competing crop	NR	NR	NR	NR	NR
Karnataka	NR	NR	NR	NR	NR
			Punjab		
Berseem	855	49	41,895	18,231	23,664
Wheat	47	1,080	56,635	17,129	39,506
			Summer Season		
			Gujarat		
Lucerne	260.24	125	32,418	25,850	6,569
Bajra	190.03	146	27,731	18,646	9,085
			Madhya Pradesh		
Jowar chari	253.37	101.03	25,597.97	9,505.85	16,092.12
Competing crop	NR	NR	NR	NR	NR
Karnataka	NR	NR	NR	NR	NR
			Punjab		
Maize fodder	361	56	20,220	8,948	11,272
Maize grain	37	725	26,825	11,285	15,540

Note: NR is not reported. MP- Main product; BP- By-product.
Source: Field survey.

fetched from paddy on a per hectare basis were Rs 10,300 as compared to Rs 552 for the jowar fodder. Farmers in Karnataka did not allocate area under fodder crops during rabi and summer seasons due to low profitability in relation to their competing crops. In Punjab, the returns over variable cost fetched from paddy on a per hectare basis were more than double as compared to sorghum. Berseem was found to be more remunerative as compared to sorghum but still the returns over variable cost were only 65 per cent as compared to the most important competing crop during the rabi season (wheat). Likewise, during the summer season, maize fodder was found to be less remunerative as compared to most important competing crop during the season, that is, maize grain. The returns over variable cost for maize fodder were only 70 per cent as compared to maize fodder during the season.

Processing and Marketing System for Fodder Crops

Amongst the selected states of India, only Gujarat and Punjab commercially cultivated fodder crops. In Gujarat, fodder is generally sold by producers through one marketing channel, namely Producer–Local Trader–Consumer (Table 50.2). In this channel, the local trader incurred marketing expenses mainly on transportation and loading/unloading of fodder and the marketing cost remained around Rs 23/qtl in all the three seasons. The consumer's price was Rs 300/qtl in kharif and it touched to Rs 400/qtl in the summer. The net profit margin of the local trader on consumer's price was the highest at Rs 52 (9.17 per cent) in rabi season and the lowest at Rs 27 (8.9 per cent) in the kharif season. In Punjab, in Channel I (Producer–Forwarding agent/Commission agent–Dairy owner/Consumer), the produce was directly taken by the producer to the forwarding/commission agent, who were forwarding the produce to the big dairy owners keeping in view the fodder demanded, through the chaff cutters. In Channel II (Producer–Forwarding agent/Commission agent–Chaff cutter–Consumer), the chaff cutter purchases the produce from forwarding/commission agent, who charges their commission from the producer as well as buyer. In Channel III, the produce is directly disposed of to the consumers in the village itself. In Channel-I for the sale of sorghum, the producer's share in the consumer's rupee was found to vary from 74 to 77 per cent for the different fodder crops. In Channel-II, the producer's share in consumer's rupee was about 65 to 70 per cent for different crops.

Amongst the selected states of India, the processing of fodder was done only in Gujarat (through haymaking) and Punjab states (through silage). In Gujarat, the cost of harvesting, packing, loading, unloading, etc., are operational costs for haymaking (Table 50.3). Overall, post-harvesting operational costs of processing one quintal fodder was found to be the highest for the summer season. It was Rs 27 for lucerne and Rs 27 for bajra fodder. On the other hand, processing cost was the lowest at Rs 21 for wheat in the rabi season. It was Rs 24 and Rs 25 for kharif maize fodder and kharif bajra fodder respectively. Among various operational costs, the share of harvesting in total cost was more than 50 per cent. In Punjab, silage or ensilage is a method of preservation of green fodder through fermentation to retard spoiling and this method of processing is more popular as compared to haymaking. This is practiced during the kharif season when sorghum, bajra, and chari are mixed, chaffed, and put in an underground pit. The average storage capacity of the pit was found to be 1,935 quintals. The storage period was up to one year from the time of storage (July to August). Less than 1 per cent of the produce was found to be spoiled as the rainwater enters from the corners through the sheets used. Regarding the post-harvest operational cost involved in silage making, it was about Rs 11/qtl. About 74 per cent of the operational cost has to be incurred during chaffing followed by transportation (18 per cent) and pit-making (about 6 per cent).

Problems Faced by Fodder Growers

In Gujarat, inferior quality of seeds of fodder crops, non-availability of adequate quantity of required brand high-yield variety (HYV) seeds, problems related to insects/pests and plant diseases, and the lack of technical knowledge were the major problems in production of fodder crops. In Madhya Pradesh, lack of technical knowhow was found to be the biggest problem observed during the course of investigation and reported by the maximum numbers of respondents in the area under study. The inferior quality of seed, faulty input delivery system, high expenditure in production due power cuts, non-availability of skilled labour in time and high cost of labour, faulty government policy as distribution of mini kits of fodder seeds from veterinary department instead of agriculture department, were the other major problems found in the study area reported by the majority of the respondents in production of fodder crops. In Karnataka, the highest percentage of farmers reported problems with respect to access to credit, labour availability, high expenditure on production, seed quality, and access to technical knowledge. In Punjab, supply of poor quality and un-recommended varieties of seed, shortage of labour especially during harvesting of the crop, lack the technical knowledge, and acquisition of credit were the major problems faced by the fodder growers during production of these crops in the study area.

In Gujarat, fodder growers reported problems in respect of non-availability of market information in time and transport facilities at reasonable rates. In Punjab, low price in the market, lack of market information, and delayed payment for the produce by the commission agents in the market were reported as the major marketing problems confronted by fodder growers of the study area.

CONCLUSIONS AND POLICY IMPLICATIONS

In Gujarat, fodder markets being unorganized and unregulated, fodder production become low priority enterprise in

Table 50.2 Marketing Costs, Margins and Price Spread Analysis of Various Fodder Crops During Peak Seasons in Different Channels, Sample Households, Selected States, India, 2008–09

(Rs/qtl)

Particulars/Channels	Punjab									Gujarat		
	Sorghum			Berseem			Maize			Kharif	Rabi	Summer
	Ch 1	Ch 2	Ch 3	Ch 1	Ch 2	Ch 3	Ch 1	Ch 2	Ch 3			
1. Net price received by the producer	47.4 (73.8)	47.4 (70.5)	44.0 (100.0)	41.9 (73.5)	41.9 (65.0)	40.5 (100.0)	50.9 (76.7)	50.9 (70.7)	48.5 (100.0)	250 (83.3)	275 (78.6)	340 (85.0)
2. Marketing costs of producer/local trader												
(i) Weighing charges	0.4 (0.6)	0.4 (0.6)	–	0.5 (0.9)	0.5 (0.8)	–	0.3 (0.5)	0.3 (0.4)	–	–	–	–
(ii) Loading/ unloading	2.3 (3.6)	2.3 (3.4)	–	2.0 (3.5)	2.0 (3.1)	–	2.1 (3.2)	2.1 (2.9)	–	15.00 (5.0)	15.00 (4.3)	15.00 (3.8)
(iii) Transportation	4.8 (7.5)	4.8 (7.1)	–	3.6 (6.3)	3.6 (5.6)	–	3.5 (5.3)	3.5 (4.9)	–	8.33 (2.8)	7.69 (2.2)	8.33 (2.1)
(iv) Commission charges	1.1 (1.7)	1.1 (1.6)	–	1.0 (1.8)	1.0 (1.6)	–	1.2 (1.8)	1.2 (1.7)	–	–	–	–
Sub-total	8.6 (13.4)	8.6 (12.8)	–	7.1 (12.5)	7.1 (11.00	–	7.1 (10.7)	7.1 (10.0)	–	23.33 (7.8)	22.69 (6.5)	23.33 (5.8)
3. Selling price of Producer	56.0 (87.2)	56.0 (83.3)	–	49.0 (86.0)	49.0 (76.0)	–	58.0 (87.3)	58.0 (80.6)	–	–	–	–
Net margin of local trader	–	–	–	–	–	–	–	–	–	26.67 (8.9)	52.31 (14.9)	36.67 (9.2)
4. Purchase price of chaff cutter	–	56.0 (83.3)	–	–	49.0 (76.0)	–	–	56.0 (83.3)	–	–	–	–
5. Costs incurred by Chaff cutter												
(i) Chaffing, Weighing etc.	–	4.0 (5.9)	–	–	4.0 (6.2)	–	–	4.7 (6.5)	–	–	–	–
Sub-total	–	4.0 (5.9)	–	–	4.0 (6.2)	–	–	4.7 (6.5)	–	–	–	–
Net margins of chaff cutter	–	7.2 (10.7)	–	–	11.5 (17.8)	–	–	10.1 (14.0)	–	–	–	–
6. Costs incurred by Dairy owner (Consumer)												
(i) Chaff cutter charges*	6.0 (9.3)	–	–	6.0 (10.5)	–	–	6.0 (9.0)	–	–	–	–	–
(ii) Commission charges	2.2 (3.4)	–	–	2.0 (3.5)	–	–	2.4 (3.6)	–	–	–	–	–
Sub-total	8.2 (12.8)	–	–	8.0 (14.0)	–	–	8.4 (12.7)	–	–	–	–	–
7. Consumer's price	64.2 (100.0)	67.2 (100.0)	44.0 (100.0)	57.0 (100.0)	64.5 (100.0)	40.5 (100.0)	66.4 (100.0)	72.0 (100.0)	48.5 (100.0)	300 (100.0)	350 (100.0)	300 (100.0)

*Includes Chaffing, Weighing, Packing, Loading/unloading, Transportation etc charges

1. Figures in parentheses show percentage of consumers' price
2. In Punjab, the prevalent marketing channels are:
   Channel-1: Producer–Forwarding/commission agent–Dairy Owner (consumer)
   Channel-2: Producer–Forwarding/commission agent–Chaff cutter–Consumer
   Channel-3: Producer–Consumer
3. In Gujarat, the prevalent marketing channel is Producer–local trader–consumer

Source: Field survey.

Table 50.3 Details Regarding Processing of Fodder Crops, Sample Households, Selected States, India, 2008–09

Particulars	Gujarat			Punjab
	Kharif Fodder	Rabi Fodder	Summer Fodder	Kharif Fodder
1. Processing method adopted (% Households)				
Haymaking	43.33	5.33	23.33	–
Silage making	–	–	–	3.0
2. Average Storage capacity (qtl)	NR	NR	NR	2100
3. Average quantity of produce stored (qtl)	57.91	100.63	28.50	1935
4. Percent capacity utilized	NR	NR	NR	92.0
5. Material used for storage (%)				
Sheet	100	100	54.29	3.0
Chemical	24.62	37.50	20.00	1.0
6. Produce lost during storage (%)	14.14	14.18	NR	0.7
7. Post-Harvest operational costs (Rs/qtl)				
Harvesting	9.28 (36.4)	5.65 (24.1)	6.25 (22.9)	–
Packing	3.21 (12.6)	2.61 (11.1)	4.69 (17.1)	–
Loading/Unloading	2.79 (11.0)	4.78 (20.4)	5.47 (20.0)	–
Transportation	3.98 (15.6)	5.22 (22.2)	4.38 (16.0)	2.0 (17.7)
Chaffing	1.18 (4.6)	–	–	8.3 (73.5)
Pit making	–	–	–	0.7 (6.2)
Storage	2.65 (10.4)	2.61 (11.1)	3.13 (11.4)	–
Chemical Used	0.74 (2.9)	0.87 (3.7)	1.56 (5.7)	0.1 (0.9)
Sheet Used	–	–	–	0.2 (1.8)
Any Other	1.67 (6.5)	1.74 (7.4)	1.88 (6.9)	–
Total	25.50 (100.00)	23.48 (100.00)	27.34 (100.00)	11.3 (100.0)

Note: Figures in parentheses show percentage to total cost, NR is Not reported.
Source: Field survey.

potential fodder production areas. In addition, dry fodder being mainly by-products from cereal crops, their economics linked with demand and price realization of main products. In a normal year, there were surplus productions of fodder/grass. Hence, organizing fodder banks in these areas is suggested. Fodder/grass from surplus production areas may be stored in these fodder banks in normal years. It is suggested that government must evolve an arrangement to produce HYV seeds for fodder crops in adequate quantity and these should be made available at reasonable rate in adequate quantity to the farmers. There is a need to adopt a price mechanism which ensures higher net returns from cereal crops and prevents a shift in crop patterns from cereal crops to cash crops. Create an organized marketing structure in surplus fodder/grass production areas. In addition, arrange to provide uninterrupted market information to farmers. The production of grass/fodder can be increase by regeneration of wastelands through controlled exploitation and growing grass in a systematic manner. The problematic lands may be treated to make them fit for growing grass. Cultivation of fodder trees on marginal land and degraded forest areas will be helpful in increasing forage production. In addition, encourage silvi-pasture in waste lands. Government may provide organizational and financial support to individuals for making investment in such treatments. Large producers of fodder/grass should be encouraged to create godowns by providing institutional credit at reasonable rates. They should also be provided bank credit for growing fodder. A separate feed and fodder development authority should be established within the Directorate of Animal Husbandry with necessary technical manpower to undertake inter-agency co-ordination in fodder production, fodder seed production, conservation, and transportation of fodder. Forest grass should be harvested during the monsoon season and converted into hay and packed, compressed, and transported to other destinations. This would be helpful in reducing the fodder deficit. The state should develop and maintain pasture and fodder patches along water reservoir, canals and rivers. Gram panchayat should be encouraged for development of pasture lands.

In Madhya Pradesh, fodder cultivation has not shown too much progress in the state since 1990. The cultivators are still growing fodder in the line of crop cultivation and the majority of them do not know the recommended package of practices of fodder cultivation. The fodder growers were also not performing any fodder preservation techniques, namely, hay and silage making for the lean period. They were not cultivating fodder on commercial lines as none of them were involved in marketing fodder in the state. Hence, it is right time that the state government re-intensifies their efforts in progress of fodder in the state because without introducing a dairy based faming system approach on the farmers' farms, their income will not double, which is the ultimate target of the state government. It is the only activity that farmers have done for a long time. It not only generates income but also enhances employment at own farms. The mini kit of fodder crops were found to be distributed by the animal husbandry department and they were not taking interest in the extension activities regarding fodder, due to lack of training in it, it lacks the aura of being a doctor, and fodder is more inclined towards agriculture. The animal husbandry department in the state is only concerned with the treatment aspect and improvement of breeds because the money lies here. Investing interest in the fodder sector will benefit the livestock owners, but who cares? Hence, there is an urgent need to create a separate department for fodder development separate from the animal husbandry department, or to merge the fodder development sector in the agriculture department for better extension activities and distribution of fodder mini kits with technical knowhow because the cultivation of fodder is more or less similar to the cultivation of crops.

In Karnataka, concerted efforts should be made to encourage the farmers to cultivate green fodder crops to enhance the quality of livestock rearing across districts in the state. This may be attempted initially by providing subsidized seed materials and fertilizers to a group of potential farmers at the hobli level. This can be replicated to others through these successful farmer groups. It is thus, necessary to conduct farmers' training periodically by the officials of the Department of Agriculture to impart skill and technical knowledge to the farmers. In this regard, coordination between the Department of Agriculture and Department of Animal Husbandry and Veterinary Services is necessary for better sharing of technical knowledge including on feeding practices with the farmers. There is huge scope for increasing the yield of napier and jowar through the adoption of better technology and field management. For this, good quality seed material and other inputs should be made available. Local institutions should be encouraged to play an active role in protecting the common property resources, which not only will help in the development of livestock enterprises but also in the maintenance of ecological balance. Efforts should be made to popularize improved breeds of different livestock that are adaptable to different agro-climatic conditions. Karnataka has relatively a large area under dry land. The livestock species suitable to dry land areas should be promoted so that they perform better in those areas. Efforts should also be made to promote rearing of high quality buffaloes for improving the dairy development. This assumes importance in the context of decline or stagnant cattle population in the state. Availability of reliable data on fodder cultivation will be useful for better planning of livestock development in the state. Concerted efforts should be made by the government departments to systematically collect and publish data on fodder cultivation.

In Punjab, due to heavy pressure of growing wheat and paddy, the area under fodder has been decreasing, and so as the composition of the livestock population. As a viable means of diversification, cultivation of fodder should be increased along with increase in livestock population, in order to make it more productive. Farmers suggested improving the quality of seedlings and frequent checks by the department officials can help in this direction. More emphasis is needed to evolve the high yielding varieties for various fodder crops as presently these are regarded as lesser important crops. The centre government grant of Rs 6 crores to the state government during 2009–10 and 2010–11 for providing subsidies to purchase quality berseem seed to cattle farmers, need to be increased keeping in view the serious problem of non-availability of quality seed for various fodder crops in the state. Further, the state government needs to use such subsidies more effectively for right cause and concern. The primary agricultural credit cooperative societies and other funding agencies should be persuaded to provide adequate short-term credit facilities to cover the operational cost. There is need to make more efforts for effective extension for these hitherto neglected crops so that the farmers may be able to know the latest know how regarding these crops. On the marketing front, most of the fodder growers were in favour of establishment of regulated markets in the region. To stabilize the prices, the farmers were in favour of establishment of better market infrastructure by the government so that the prices may not go down by the certain minimum level and they may come out of the clutches of the commission agents. The state has abundant roughage (wheat and rice crop), which can be used in making silage through processes developed and recommended by Punjab Agricultural University, Ludhiana. The centre provides a subsidy of 80 per cent for making silo pits with automatic loader. To promote the processing of fodder, these facilities are needed to be spread to more number of farmers.

REFERENCES

Acharya, S.S., and N.L. Agarwal. 2005. *Agricultural Marketing in India*. New Delhi: Oxford & IBH Publishing.

Bhende, M.J., R.S. Deshpande, and P. Thippaiah. 2004. *Evaluation of Feed and Fodder Development under the Centrally Sponsored Schemes in Karnataka*. Bangalore: Agricultural Development and Rural Transformation Centre.

Birthal, P.S., and P.P. Rao. 2004. 'Intensification of Livestock Production in India: Patterns, Trends and Determinants'. *Indian Journal of Agricultural Economics* 59(3): 555–65.

Dikshit, A.K., and P.S. Birthal. 2010. 'India's Livestock Feed Demand: Estimates and Projections'. *Agricultural Economics Research Review*, 23: 15–28.

Erenstein, O., W. Thorpe, J. Singh, and A. Verma. 2007. *Crop–Livestock Interactions and Livelihoods in the Indo-Gangetic Plains, India: A Regional Synthesis*. Mexico: International Maize and Wheat Improvement Centre (CIMMYT).

Government of India. 2002. *Report of the Working Group on Animal Husbandry and Dairying for the 10th Five Year Plan (2002–2007)*. New Delhi: Planning Commission.

———. 2007. *Report of the Working Group on Animal Husbandry and Dairying for the 11th Five Year Plan (2007–2012)*. New Delhi: Planning Commission.

Indian Council of Agricultural Research. 2011. *Handbook of Agriculture*. New Delhi: Indian Council of Agricultural Research (ICAR), New Delhi.

Kumar, S., and S. A. Faruqui. 2009. 'Production Potential and Economic Viability of Food Forage Based Cropping System under Irrigated conditions'. *Indian Journal of Agronomy*, 54(1): 46–51.

Mishra, A.K., D.B.V. Raman, M.S. Prasad, and Y.S. Ramakrishna. *Strategies for Forage Production and Utilisation*. Hyderabad: KVK Central Research Institute for Dryland Agriculture.

Sharma, R.P., K.R. Raman, and A.K. Singh. 2009. 'Fodder Productivity and Economics of Pearlmillet with Legumes Intercropping under Various Row Proportions', *Indian Journal of Agronomy*, 54(3): 301–05.

Wylie, P. 2007. 'Economics of Pastures versus Grain or Forage Crops', *Tropical Grasslands*, 41: 229–33.

ANNEXURE 50A

TABLE 50A The Details of Agro-Economic Research Centres and the Lead Person Involved in the State Report

S No	State	Authors
1	Gujarat	V.D. Shah
		Manish Makwana
		Shreekant Sharma
2	Karnataka	Elumalai Kannan
3	M.P.	Hari Om Sharma
		N.K. Raghuwanshi

Section IV
Incentives, Agricultural Prices, and Food Security

51

Food Policy Transition in India

Nilabja Ghosh

An overwhelming presence of the State in the food market has been a hallmark of India's food policy. The hysteric fear of food insecurity was a natural reminiscence of the famines and privation that the people had gone through in the past. After a successful transition through the green revolution (GR) in the 1960s and the 1970s and further through two decades of economic liberalization starting in the 1990s the food market finally shows signs of an imminent move towards a market-based regime too. Such a transition would undoubtedly present one of the toughest challenges to the political economy of the country and a test on all the optimisms and misgivings attached on the market mechanism by various groups.

OBJECTIVES

This chapter studies the character of India's journey towards an idealistically free food market by hypothesizing the counterfactual of the exit of the state in the recent decades. Food market liberalization is alternatively visualized as greater integration with global market measured by the place of international trade in the disposal of the production and as a shrinking of the place occupied by public operations in the larger food market comprised of the free market and the administered market. The government has been intervening in the larger market since independence through its own operations. Using data collected from secondary sources and treating Rice and Wheat as the two subject food grains of concern, this chapter inquires whether over the period of economic reforms public operations in procurement, distribution, and stocking have grown and whether the food market has become more open to global demand and supply by offering an increasing share of domestic production for international trading. The role of minimum support price (MSP), the main tool for administering food price, in shaping market prices and in deciding the volumes traded in the free market vis-á-vis the administered market is also examined using econometric regressions.

BACKGROUND

The Bengal famine of 1770s[1] that influenced economic theory and policy over two centuries, the report of the Royal Commission on Agriculture in 1928, the colonial strategies taken before and during the World War II and the experience of the 1943 Bengal famine arguably described as a classic case of market failure as against a wartime

[1] Popularly known as the 'monnontor' of 1976 this Bengal Famine was consequent of a catastrophe of two successive years of poor rainfall supported by wrong and high-handed policies of the East India Company. About 10 million people of Bengal, Bihar, and Odisha are estimated to have died from starvation (Wikipedia, website).

compulsion (Anirban 2010; Ó Gráda 2015; Sen 1981) all reflected grounds for the poor confidence placed on the market. The sceptical outlook on a free food market continued in the decades after independence, reinforced by the impact of partition and the dire condition of national food shortage. The Food Grain Procurement Committee 1950, the Food Grain Enquiry in 1957, the Agricultural Prices Commission (APC) of 1965–66, and even the Food Policy Committee 1983 found enough justification for control on the food market. The lack of confidence on the rule of price to take care of the food problem and clear the demand–supply disequilibrium in the market was not surprising when agriculture was in a state of stagnation and complete exhaustion (Krishna 1963) even while the population was growing at over 2 per cent per annum. Both physical availability and affordability being requirements to gain food security, public policy addressed production as well as distribution.

The structural adjustment program (SAP) 1991 established the intention of orienting the overall economic policy towards the market mechanism and the signing of the World Trade Organization (WTO) treaty in 1995 manifested a further commitment to remove trade barriers. Yet opening up the food market in particular still remains a delayed option fraught with trepidation and controversies. In fact, food had come in the way of the General Agreement on Tariffs and Trade (GATT) from the beginning and even the WTO gave relaxation to developing countries as negotiations continued over latitudes given to states on procurement at MSP, stocking and distribution at administered issue prices, all of which are suspected to 'distort' prices and have repercussions on the larger world food market. Free market, alternatively laissez faire market economy is based on the rule of price drawing from notions of a competitive market derived from the works of Adam Smith and Ricardo but the role of the state as a regulator, though perhaps not an inhibitor of the market,[2] is largely accepted. The continual vigilance of the state is even more vital in food which as an essential item cannot be substituted by other goods without causing malnutrition, starvation, and unrest. Food supply responses to price changes are also restrained by limitations of area, soils, and weather specificities, technological bounds of the time, and a host of non-price factors such as the lack of resources, risk, poor transport, and lack of information.

Food policy is the sum total of all proactive interventions, regulations, and state-promoted institutions that directly or indirectly help to shape prices. Marketing policy is that part of the food policy that aims to facilitate markets to operate in a fair manner to determine competitive prices rather than to directly change prices in any particular way. On the contrary, price policy is a set of strategies that seeks to guide the price in any pre-selected direction (Kahlon and Tyagi 1983). The Food Corporation of India (FCI) was established in 1964 to execute government's food policy but it began to deal with rice and wheat, the production of which was technologically promoted by the green revolution (GR). The Agricultural Prices Commission (APC) was set up in 1965 to advise government on food price, setting in motion a strong price policy as the distinction between procurement price and MSP blurred from the 1970s. The APC renamed Commission for Agricultural Cost and Price (CACP) in 1985 took cost of production into consideration to ensure remunerative price to farmers. To ensure food security, the government continued the colonial practice of statutory food rationing which went through an evolutionary process as it was replaced by Fair Price Shops (FPS) in 1953 and the announcement of Central Issue Price (CIP) to which state governments added bonuses. The resultant nation-wide Public Distribution System (PDS) was later modified to a targeted PDS (TPDS) to correct for bias, reduce leakages, and to target the poor and the momentous recent enactment of the legally enforceable universal Food Security Act (FSA)[3] in 2013 to make food security a right. The gap between public demand for cheap grain and sale of grains by farmers determined by an open ended and voluntary procurement system is bridged by public stocking, international import and export trade, and budgetary allocations for funding the operation (Ghosh 2004). Meanwhile a National Food Security Mission (NFSM), was launched in October 2007 to escalate food production to meet food security needs.

India's food policy confronted a new paradigm since the 1990s dotted with mistrust of the market and a clamour that food should be treated as an entitlement (Drèze and Sen 1989) but the road was uneven and direction indeterminate. The Agricultural Produce Market Committee (APMC) Act was amended in 2003 to allow and strengthen private channels. There were proposals of dismantling the FCI or converting it into a competitive player (Gulati, Sharma, and Kahhkonen 1996; World Bank 1999), the

[2] In 1776, Adam Smith, wrote on the Bengal famine of 1976 in *The Wealth of Nations* (vol 2, p. 110), 'The drought in Bengal, a few years ago, might probably have occasioned a very great dearth. Some improper regulations, some injudicious restraints imposed by the servants of the East India Company upon the rice trade, contributed, perhaps, to turn that dearth into a famine.'

[3] National Food Security Act, 2013 (also Right to Food Act) is an act of Parliament of India.

PDS was revamped a number of times (Ghosh and Guha-Khasnobis 2008), the universal FSA was legislated under controversy and futures trading in grain was allowed and disallowed successively (Ghosh and Chakravaty 2010). Recently a report published on food policy reforms emphasizing the reorientation of the FCI (FCI 2015) was a further step in food market liberalization. Negotiations on a Trade Facilitation Agreement and a possible Permanent Solution (Ghosh 2014; Narayan 2012) created fresh challenges and complexities for the nation surrounding procurement and stocking of grains requiring politically tough stands. The post 2000 years witnessed many developments in the path to reforming India's food market.

DATA AND METHODS

To revisit and re-examine the path taken by the Indian food policy in liberalization this paper traces the course of India's journey towards the ideals of free market by measuring the space yielded by state to the private traders and to international traders in particular. However, to the extent that the state agencies continue to operate merely as competitive forces in the market without inhibiting other traders overtly or indirectly through instruments such as administration of prices, the presence of the state may not imply the inhibition of market regime. The purpose of the transition would be to give greater role to the signal of market price that is determined purely by forces of demand and supply. The chapter seeks to confirm the diminishing role of the administered prices in determining prices in the market. The analysis made in the chapter relates to the two main foodgrains—rice and wheat—and using price, production, and public operation data obtained from official central government sources (MoA various; MoA, website) measures the share of the state and the autarchic economy and estimates the effect of price administration on not only market prices but also on the purchases of the state which hypothetically behaves as another market actor.

Annual data from 1980–81 to 2012–13 are used. Production, procurement, distribution through the PDS, import and export data are for financial years (FY) and public stock relates to the last month March of the FY. Monthly wholesale price (WSP) reported as indices at the all India level are used for creating the variables on market prices. Harvest time WSP are only considered to denote marketing season prices that farmers receive. The administered prices considered are the current MSPs generally announced prior to sowing. All market variables are taken for the relevant season specified keeping in view the specific crop calendars of the two crops. Rice is a kharif crop sown in June–July and its harvest starts in October so that the harvest time is taken to carry on from October to December. Wheat is a rabi crop harvested from April. Therefore, the harvesting period, April to May overlaps with the succeeding financial year. The analysis constructs and portrays relevant measures of government's place in market through graphs, tables, and econometric regression equations in log-linear specifications.

THE RETREAT OF THE STATE: ANALYSIS

The shrinking of the state in the economy is an essential element of liberalization. In an idealistic case, the government is visualized to be relieved of the task of procuring and carrying over grains in anticipation of emergencies since imports can be the answer in such a context. The argument treats food as any other commodity that is freely traded. With reforms running their course for over two decades it is useful to assess the revealed response of the government in being able to step away gradually from the market and to allow space to agents of the free market. The basic equations underlying public operations in grains are given by the identities.

$$\text{Market Availability} = \text{Production} + \text{Imports} - \text{Exports} - \text{Public Stock Addition} \quad (1a)$$

$$\text{Public Stock Addition} = \text{Closing Stocks}(-1) + \text{Procurement} - \text{Distribution} - \text{Opening Stock} \quad (1b)$$

The assessment is made by using two indicator variables relating to both grains together namely PUBOP or public operation and OPEN or openness measuring government's retraction from the domestic market as a buyer and a seller at administered prices and by allowing an exporter or an importer to step in. The variables are given as follows:

$$\text{PUBOP} = (\text{Procurement} + \text{Distribution from PDS})/\text{Domestic Production} \quad (2a)$$

$$\text{OPEN} = (\text{Export} + \text{Import})/\text{Domestic production} \quad (2b)$$

In Equation 2a PUBOP relates to that part of domestic production of rice and wheat that passes through the public channel expressed as an index with base 1980–81. The place of trade in the foodgrain economy is measured in Equation 2b by the variable OPEN indexed with the same base as PUBOP. A rising value of OPEN and falling PUBOP indicate withdrawal of the state from the market. The plot of PUBOP and OPEN in Figure 51.1 shows the vacillations in the policy. After an initial sign of depression after 1990, PUBOP remained stagnant but only until 1998–99 after which it rose to a new high but again slumped in 2006 probably a reaction to successive good monsoon years but then onwards the state's share in the food economy continued growing. Openness of the food

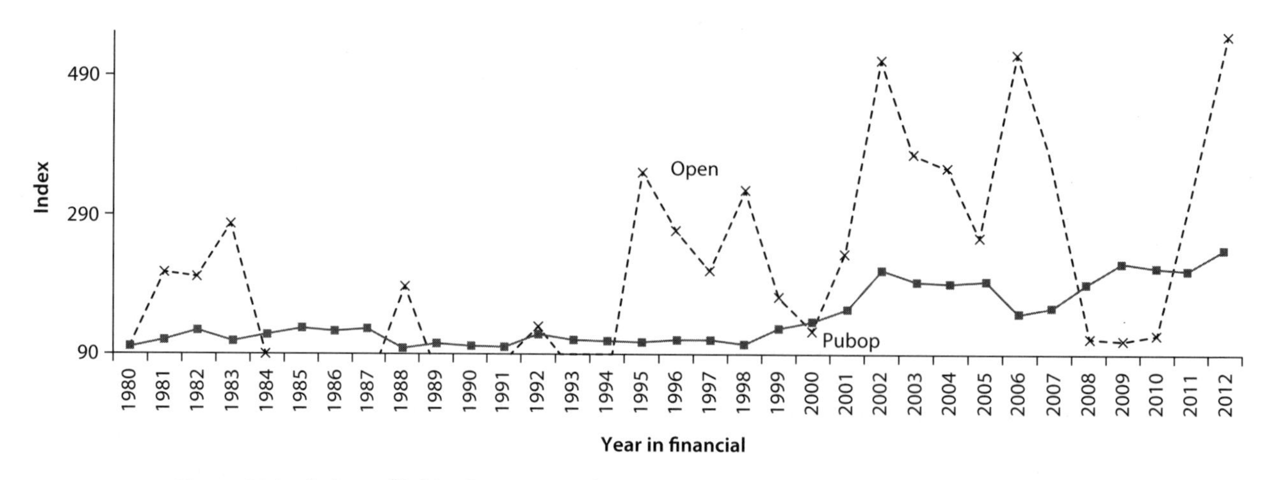

Figure 51.1 Indices of Public Operations and Openness to Trade of Rice and Wheat (Base 1980–81)

market measured by OPEN has been far more volatile than PUBOP in the post-liberalization period especially from the mid-1990s. Trading picked up in the wake of the WTO treaty but witnessed wide swings, the most notable being the slump in the period 2008–10 though improvement has again followed.

Table 51.1 shows that in comparison to a 32 per cent share of the public channels in production in triennium ending (TE) 1983–84, public operations in grains comprised 62.5 per cent of production in 2012 of which 33 per cent was procurement. The larger share of procurement compared to distribution is a sign of accretion to public stocks. The gap widened in TE 2000 to be reversed in TE 2005 but remained wide after TE 2010. The rise of exports of rice and wheat is the principal factor behind the movements in openness, as imports diminished. The WTO regime appears to have improved the openness of the food market though the degree of openness at 4 per cent remains small as the domestic capability overwhelmingly retains its importance. Table 51.1 also reveals that till TE 1990 India's food imports exceeded exports but the position reversed thereafter when India became a net exporter. Openness moved down from TE 1983 but expanded four times in TE 2005–06 and again in TE 2012.

Table 51.2 shows that procurements grew by 6.2 per cent per annum, distribution by 12.3 per cent and stock by 8.2 per cent in the period 2008–12. As can be expected movement of procurement and distribution have not been in the same direction but stock grew in six out of the eight periods presented. Public stocking is an issue of much contention being felt as alternatively a necessity for the national food security, a wastage of public resources, and an opportunity for farmers to sell. It is also seen as a disincentive creating an element of uncertainty for farmers in other countries as stock will either prevent imports or trigger sudden bulk exports (WTO website). In other words, a relation between public stock and international trade is expected though the direction is indeterminate.

Table 51.1 Public Intervention and Free Trade in Food Grain Market: Rice and Wheat

T.E.	Procurement	Distribution	Public Operations	Exports	Imports	Openness
1983	16.5	15.5	32.0	0.6	2.3	2.80
1990	15.8	13.0	28.8	0.4	0.8	1.20
1995	16.1	14.7	30.8	1.8	0.1	1.90
2000	21.3	13.2	34.5	2.0	0.6	2.60
2005	26.1	28.2	54.3	4.1	0.0	4.10
2010	32.1	26.5	58.7	1.3	0.1	1.40
2011	32.2	28.6	60.7	2.2	0.1	2.26
2012	33.0	29.5	62.5	4.6	0.0	4.60

Note: TE is triennium ending. All figures are as percentage of production.
Source: Computed.

Table 51.2 Annual Growth Rates (%) Per Annum of Rice and Wheat in Public Operations

	Production	Procurement	Distribution	Stock
1980–84	3.3	14.0	−0.1	21.2
1984–88	5.0	−4.7	7.4	−26.4
1988–92	1.1	11.3	0.3	19.7
1992–96	3.8	−4.6	9.3	6.7
1996–2000	0.6	17.8	−8.2	28.5
2000–04	0.0	0.5	22.7	−21.0
2004–08	3.8	8.1	−1.1	5.6
2008–12	2.5	6.2	12.3	8.2

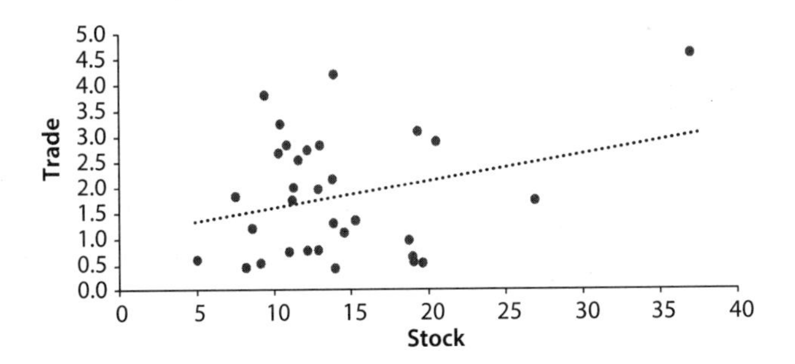

Figure 51.2 Trade and Public Stocks in Rice and Wheat

Note: Trade = Export + Import, Stock is opening stock. Both Trade and Stock are deflated by production.

This is affirmed in Figure 51.2 which, however, hints at a positive relation. Since imports have reduced in significance the relation may be a sign that exports are responsive to stock as apprehended. In practice, the food stock becomes extremely useful in occasional years of poor monsoon.

It may be pertinent to keep in view that the population of India has been rising all along. Public operation being meant for feeding the people, the government's share in the market would be expected to keep pace with the population growth. Figure 51.3 plots production and procurements both as totals and at per capita levels. The per capita procurement showed better growth performance than per capita production probably reflecting the growing commitment of the government to the food security through various distribution and food based welfare schemes which, in turn, urged to the government to seek supplies in the market in competition to other buyers. A break in the upward trajectory of procurement from 2008 may be indicative of the impact of the National Food Security Mission (NFSM).

Intercrop Dimensions

Notable changes are becoming apparent between the two staples as rice seems to have gained prominence in both operations and trade. In the three periods 1980–92, 1992–2000, and 2000–12 the public operations grew by 138 per cent, 58 per cent, and 69 per cent and distribution grew by 68 per cent, 5 per cent, and 213 per cent in rice and operation by 94 per cent, 61 per cent, and 85 per cent, and distribution by 7 per cent, 3 per cent, and 287 per cent in the case of wheat. Similar improvement is marked in respect of openness. There has been a tendency of catching up first by rice and then by wheat which was by far the less important crop in public demand and international market. Figures 51.4 and 51.5 are suggestive of this convergence reflected by the intersection of the graphs for public operation and openness.

The Role of Minimum Support Price

The word support in MSP suggests the minimum safety net function of the instrument for farmers against prices falling

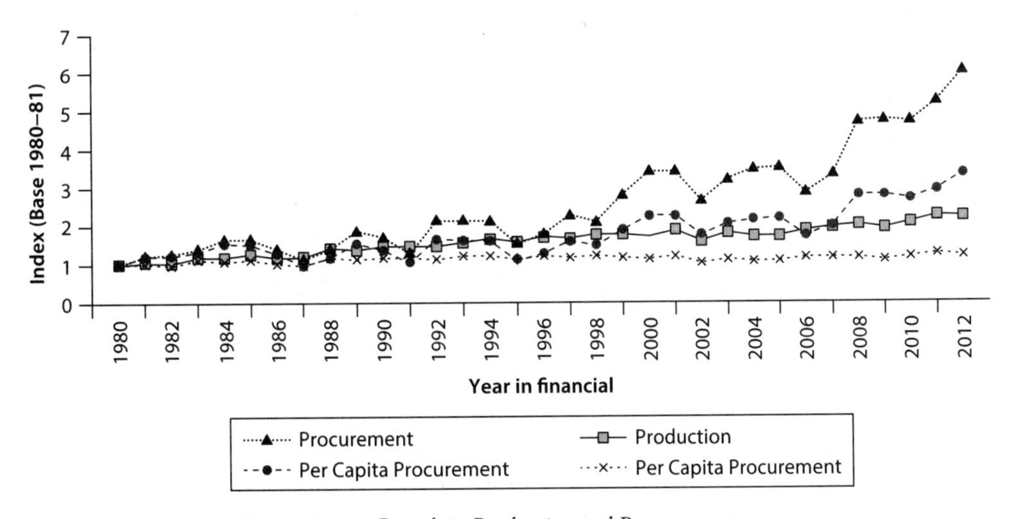

Figure 51.3 Growth in Production and Procurements

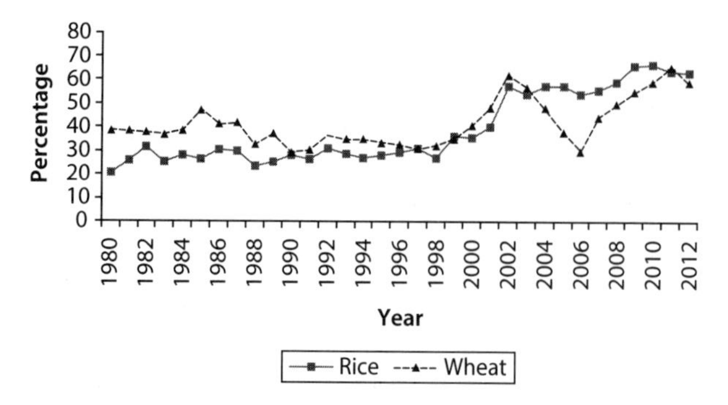

Figure 51.4 Public Operations in Rice and Wheat

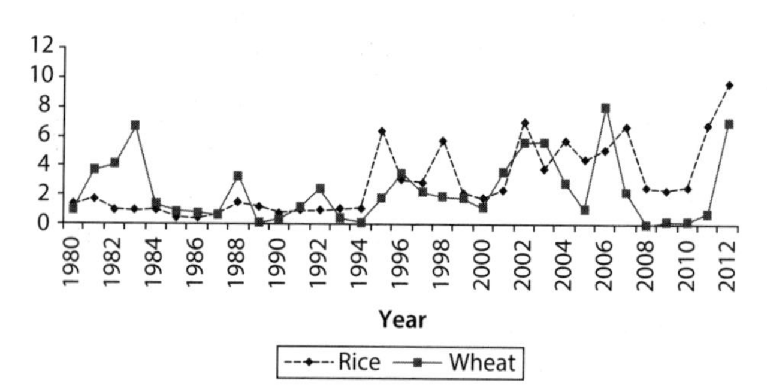

Figure 51.5 Openness of Rice and Wheat

below certain distress levels. In reality, it has served as a key instrument for influencing farmers to redirect production efforts towards socially desirable crops. With market forces becoming more active, MSP is expected to become less useful and even redundant. Figure 51.6 plots the ratio of the harvest time wholesale price relative to MSP expressed as index which was higher for rice till the year 2000 while wheat price was closer to the MSP. The ratio then declined for both rice and wheat and converged as rice too gained significance in public operations. The MSP and the market price are gradually coming closer.

Do MSPs influence market prices and do they direct grains away from the free market toward public channels? To answer this in Equations 3a and 3b, based on the sample 1980–81 to 2012–13 with dependent variables RPRICE and WPRICE, expressed as indices with common base (1980–81), being respectively the harvest period WSP of kharif rice and wheat are regressed on exogenous variables.

The market prices are regressed on their one period lagged values and the MSPs represented by RMSP and WMSP respectively for rice and wheat announced for the current season prior to sowing time and predating the harvest price realization. In addition, it is presumed that the total stock available with the government of both foodgrains rice and wheat (FGST) determine with intensity with which the government conducts its procurement operations. Larger stock would mitigate the effort and urgency of procurement, and perhaps impose budgetary and managerial incompatibility even though the procurement system in India in practice happens to be open ended which means that all voluntary offers at declared MSP would be accepted.

In both cases of rice and wheat the short-run dynamics played a dominant role evident from the significant parameters of the lagged prices. However, prices are also driven by the MSP though the response appears to be less intense for rice. Given the past price behaviour a 1 per cent rise in MSP is seen to increase the harvest time price by point 0.27 per cent in the case of rice and 0.54 per cent in the case of wheat. Public stock of the two crops has a weak negative influence on wheat price with insignificant coefficient but in the case of rice the t-statistics of the coefficient being less than 1, the stock variable FGST is not considered in the equation presented.

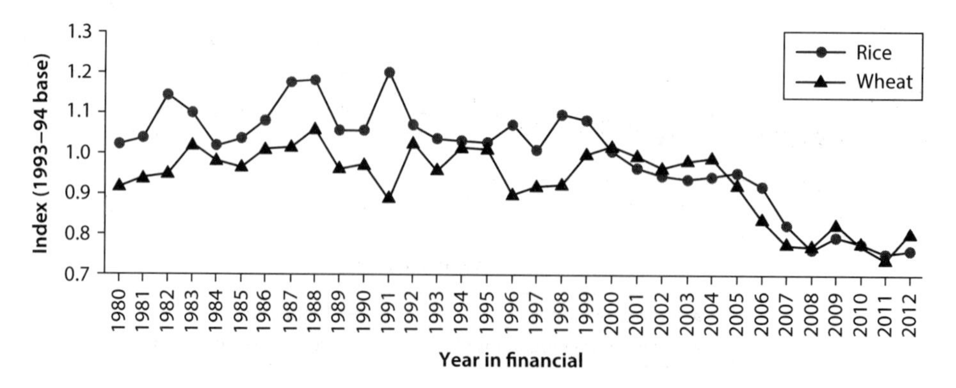

Figure 51.6 Index for Harvest Time Prices of Rice and Wheat Deflated by MSP

Table 51.3 Effect of Minimum Support Price (MSP) on Public Procurement

Variable	Procurement Rice	Procurement Rice	Procurement Wheat	Procurement Wheat
Equation	4	5	6	7
Constant	−3.39	−2.87	−1.43	−1.58
	(−0.71)	(−0.61)	(−0.97)	(−1.10)
MSP	1.99		1.24	–
	(4.26***)		(1.56)	
MSP/Market Price	–	2.06	–	1.15
		(4.54***)		(1.51)
Market Price	−2.20	–	−0.68	–
	(−4.35***)		(−0.63)	
Production	0.35	0.31	0.88	0.99
	(1.32)	(1.22)	(2.12**)	(2.86***)
R–Square	0.936	0.937	0.639	0.648
D.W.	2.2	2.2	1.7	1.7

Note: All variables are in logarithms, Sample period: 1980–81 to 2012–13, all price indices are deflated by wholesale price index of all commodities. ***, ** and * denotes level of significance at 1%, 5% and 10% respectively.

$$\text{RPRICE} = 0.29 + 0.67 * \text{RPRICE}(-1) + 0.27 * \text{RMSP} \quad (3a)$$

(3.44***) (6.38***) (2.99***)

$$\text{WPRICE} = 0.40 + 0.40 * \text{WPRICE}(-1) + 0.54 * \text{WMSP} - 0.04 * \text{FGST} \quad (3b)$$

(1.96*) (2.48**) (3.94***) (−1.40)

How far the MSP is effective in diverting the food supply in the free market is examined by looking for the response of the alternative public channel to changes in prices. In Table 51.3 regression Equations 4 and 5 explain procurement of rice using two different specifications. Similarly, Equations 6 and 7 explain wheat procurement. Besides prices, the current production is also treated as a variable influencing the procurement. In each case, the first specification considers the MSP and the market price both deflated by the wholesale price of all commodities as separate variables and the second equation takes both prices as a comprehensive variable measuring the ratio between them. Based on the DW-statistics all the equations are corrected for the autocorrelation.

All the price variables appear with the expected signs showing a positive response to MSP, the MSP relative to market price and production, and a negative response to the market price indicating the competition between the public and private channels. In Equations 4 and 6 the MSP declared earlier has a positive effect significant at the 1 per cent case of rice but insignificant in the case of wheat; however, the market price has a significant negative effect in case of rice but its coefficient is insignificant in wheat. Production of wheat has a greater effect on its procurement than the price variables, appearing with highly significant coefficients while the rice procurement does not respond to production significantly. Equations 5 and 7 with differing specifications perform marginally better going by the adjusted R^2. The relative price measured by the ratio of MSP to market price is a significant positive influence in rice procurement. The response of wheat procurement relative to MSP is positive but weaker with t-statistics not significant at 10 per cent. As in Equations 4 and 6, wheat procurement is influenced strongly by its production and weakly by the price competition among public and private agents whereas the case is reverse for rice.

DISCUSSION

Public operations in food market remain unabated, the market has become more open and market prices are coming close to MSPs. The price relative to MSP was higher for rice than wheat but, over time, they converged and while both rice and wheat competed for a place in public channels and trade, the market prices and the administered price are currently moving together. Prices of both rice and wheat are strongly influenced by their MSPs. In the case of rice, public procurement is strongly determined by MSP and by a public–private competition but production rather than MSP is the driver in the case of wheat. The open-ended procurement policy makes wheat procurement, made mostly in surplus northern states, more responsive to the volume of the production whereas the administered price is a useful tool for directing rice between the two channels.

CONCLUSION

Over time, the share of the public sector in the food market has not shown any sign of contracting but the market became more open in the international market. Rice and wheat, the two staples of the Indian population have gained relative significance in public operations and trade in succession. While the Indian food market is now more open in the international market internally, the traditional food security concerns and misgivings about markets still weigh on the government, sustaining the policy reliant on price administration and public procurement created decades ago. In the current paradigm India needs to do more to work out a mechanism that is responsive to supply and demand forces in the market with a medium term purview for safeguards against short term production fluctuations. A policy for ensuring food security in the current paradigm needs to be less reliant on administration by the state on a regular basis.

REFERENCES

Anirban. 2010. 'A Gruesome Firsthand Account of the Bengal Famine of 1770–1772. It is a Miracle'. December 10. http://milkmiracle.net/2010/12/20/bengal-famine.

Ó Gráda, Cormac. 2015. *Eating People Is Wrong, and Other Essays on Famine, Its Past, and Its Future.* Princeton University Press.

Drèze, Jean, and Amartya Sen. 1989. *Hunger and Public Action.* United Kingdom: Clarendon Press.

Food Corporation of India (FCI). 2015. *Report of the High Level Committee on Reorienting the Role and Restructuring of Food Corporation of India.* January.

Ghosh, Nilabja. 2004. 'Impact of Trade Liberalization on Returns from Land: A Regional Study of Indian Agriculture', in *The WTO Developing Countries and the Doha Development Agenda,* edited by Basudev Guha-Khasnobis. New York: Palgrave Macmillan.

———. 2014. 'Why India Should or Should not Ratify the TFA?', *Yojana,* September.

Ghosh, Nilabja, and B. Guha-Khasnobis. 2008. 'Measuring the Efficiency of Targeted Schemes: Public Works Programmes in India', in *Food Security: Indicators, Measurement and the Impact of Trade Openness,* edited by Basudeb Guha-Khasnobis, Shabd S. Acharya, and Benjamis Davis. Oxford.

Ghosh, Nilabja, and Sangeeta Chakravarty. 2010. 'Agricultural Prices and Futures Trading: Interactions and the Transfer of News', in *Commodity Derivative Markets: Opportunities and Challenges,* edited by Madhoo Pavaskar and Nilanjan Ghosh. TAER.

Gulati, Ashok, Pradeep Sharma, and Satu Kahhkonen. 1996. 'The Food Corporation of India: Successes and Failures in Indian Food Grain Marketing', IRIS-India Working Paper No. 18.

Kahlon, A.S., and D.S. Tyagi. 1983. *Agricultural Price Policy in India.* New Delhi.

Krishna, R. 1963. 'Farm Supply Response in India–Pakistan: A Case Study of the Punjab Region', *The Economic Journal,* 73(291): 477–87.

Ministry of Agriculture (MoA). various. *Agriculture Statistics at a Glance.* Directorate of Economics and Statistics, Department of Agriculture and Cooperation, Ministry of Agriculture, Government of India.

———. website. http://eands.dacnet.nic.in/. Directorate of Economics and Statistics, Department of Agriculture and Cooperation, Directorate of Economics and Statistics, Department of Agriculture and Cooperation, Government of India.

Narayan, Sudha. 2012. 'The National Food Security Act vis-à-vis the WTO Agreement on Agriculture', *Economic & Political Weekly,* 49(1): 40–6.

Sen, Amartya. 1981. *Poverty and Famines: An Essay on Entitlement and Deprivation.* Oxford: Oxford University Press.

World Trade Organization (WTO). website. http://www.wto.org/englishlthewtoe/ministe/mc9 e/briefe.htm.

Wikipedia. website. Bengal Famine of 1770. https://en.wikipedia.org/wiki/Bengal_famine_of_1770.

52

Food Insecurity and Vulnerability
A Study of Agricultural Sector in Himachal Pradesh

M.L. Sharma, C.S. Vaidya, and N.K. Sharma

Food security has been a consistent theme raised in specific contexts in a number of world conferences convened by the United Nations in the 1990s. The World Conference on Human Rights (Vienna 1993) emphasized the need to ensure that everyone enjoyed a right to food. The concept of food security as understood now has been considered at a number of levels: global, regional, national, state, household, and individual. While the ultimate concern may be at the household and individual levels, it is important to realize that food security at the levels outside the household has a strong bearing on the performance at the household level.

The Food and Agriculture Organization of the United Nations (FAO) (1983) had formulated that the basic concept of food security implied that 'all people at all times have both physical and economic access to the basic food they need'. The World Bank (1986) has modified this formulation to indicate that food security is 'access by all people at all times to enough food for an active, healthy life'. Now that we have grown so much food, is every Indian getting more to eat? Not according to the Economic Survey. Only 390.6 grams of wheat and rice were available for every Indian per day in 2000–01, down from 426 grams per day in 1999. On the production front, during the 1990s the growth of agricultural productivity and production decelerated as compared to the 1980s. The per capita food production failed to keep pace with the population growth. During 1990s, while population increased at the rate of 1.84 per cent per annum, food grain production per year increased by only 0.90 per cent. Per capita availability of food grains declined from 177 kgs per year in 1991–92 to 163.2 kgs during 2000–01. It was estimated that decline of cereal intake varied across regions by as little as 1 per cent in Himachal Pradesh to 11 per cent in Rajasthan.

Himachal Pradesh has attained the distinction of being regarded as model of hill development. There are still doubts about the equitable distribution of gains of development and less privileged sections of farming community may not have gained significantly and their status may have remained more or less stagnant. Himachal Pradesh has a large predominance of marginal holdings (more than 80 per cent), where agriculture is practiced on subsistence level. But the declining size of, particularly, marginal holdings has further aggravated the problem of food, feed, and nutritional insecurity in the state.

The increasing number of holdings and ever increasing trend towards marginal holdings combined with poor farm yields is leading towards unsustainability in terms of providing means of sustenance. The apprehension is that the present trend may be leading to a situation insufficiency of farm outputs in meeting out the food requirements and may be leading to hunger gap. Even if some groups are not food insecure, they are vulnerable to food insecurity with a slight shock which can be in the form of bad weather, adverse input and output prices, labour shortages, or changes in policy regime. One of the most welcome changes in mitigating food insecurity in Himachal Pradesh is the diversification of cropping pattern towards high pay of commercial crops, the returns from such crops are higher than the traditional subsistence crops, but the miniscule farm size offsets the benefits of such cropping pattern and the marginal commercial farms may also be suffering from food insecurity. This fact is also substantiated by the fact that the marginal commercial farmers migrate during harvesting season to work on large orchards.

The above facts indicate that there may be prevalence of food insecurity in, both, subsistence as well as in commercial farming scenarios. It becomes, therefore, important to identify the farming groups which are food insecure or may be vulnerable to food insecurity. In the light of foregoing discussion the study is based on following specific objectives:

1. to identify the groups vulnerable to food insecurity and to quantify the extent of food insecurity in terms of hunger gap;
2. to highlight the socio-economic conditions of food insecure groups in the state;
3. to highlight coping mechanism adopted by food insecure and vulnerable groups to mitigate the food insecurity;
4. to analyse the public/private and financial capital of food insecure groups and their dependence on such capital for avoiding the food insecurity; and
5. to examine the role of government intervention through policy and programmes for providing food security net to vulnerable groups.

METHODOLOGY AND DATA

The issue of food security is most relevant for the poorer section of the society and hence, population below poverty line has been taken as a basic sampling parameter. Consequently, two districts with highest concentration of below poverty line families in the state (Shimla and Chamba) were selected for the study and to find out the availability of different food items, home produced as well as supplied through public distribution system (PDS) to categories of population belonging to Below Poverty Line (BPL), Antodaya Anna Yojna (AAY), Above Poverty Line (APL), Additional PDS families, and Annapurna Scheme. Among selected districts, two blocks in each district, comprising highest poverty within the blocks, were chosen. The selected blocks were Tissa and Salooni in district Chamba and Chirgaon and Nerwa in district Shimla. Among selected blocks thirty farm families from each district comprising different categories were selected for the detail study through multi-stage random sampling technique. The study is based on 120 sampled households, located in four blocks of two districts.

RESULTS AND DISCUSSION

The study has come out with following findings.

Identification and Assistance to Vulnerable Groups

One of the main constituents of the government strategy for poverty alleviation has been the Targeted Public Distribution system which ensures the availability of essential commodities to families belonging to APL, BPL, Antyodaya (poorest) and Annapurna (Indigent) through a net work of 4,335 fair price shops. During the year 2009 a total of 9,33,438 families were covered as APL and were getting 15 kg wheat flour and 20 kg rice per family under specially subsidized rates of Rs 8 and Rs 9 per kg respectively on the landed cost policy prescribed by Government of India (GoI) with effect from 28 July 2007. A total of 5,14,000 families were targeted for placement in BPL Scheme out of which 1,97,100, the poorest of the poor families, were further transferred to AAY Scheme. The GoI was allocating 3,965 Mettric Tonnes wheat and 7130 MT rice per month for these families. Under AAY, 1,97,100 families out of the BPL list were identified and were allocated 2,955 MT wheat and 3940 MT rice per month. The scale of issue per family was as per GoI norms, 15 kg of wheat and 20 kg of rice which has been enhanced by the Himachal Pradesh Govt. with effect from 28 July 2007. The Annapurna Scheme was introduced in the State with effect from 1st April 2000. The numerical ceiling of the beneficiaries was fixed by the GoI at 6,373 persons for this State for providing food security to the old destitute/indigent citizens of 65 years of age or above. Out of the fixed target, 4,997 persons were indentified in the Pradesh till February 2008 under this scheme. Rice is issued to beneficiaries free of cost at the scale of 10 kg per person per month.

Socio-economic Profile of Sample Households

The average family size of sample was about five persons with women lagging behind in educational levels. Government of Himachal Pradesh has been providing assistance in the shape of mid day meal and provision of free books to primary class children. This system helped children in both ways, that is, encouragement in getting education as well as financial assistance. Agriculture is the main occupation of about 60 and 70 per cent households in Chamba and Shimla districts, respectively. Female participation in agriculture was more than 90 per cent and it is highest in AAY category. Major share of the area under agriculture and horticulture was with APL category of households. The area under horticulture was less than 0.03 hectare per farm with BPL and AAY categories in Chamba but it was slightly better in district Shimla. Apple was contributing cash income of about Rs 3,500 per farm in Chamba but in Shimla this was Rs 8,950 per farm in Chirgaon and whopping Rs 1,33,333 in Nerwa block. Out of total value of farm produce about half was from apple production providing great boost to the household economies. Low value of foodgrains indicated vulnerability to food insecurity among different categories.

Income from Different Sources of Sample Households

Horticulture, agriculture, animal husbandry, labour, government services, and business were different sources from where the rural household were drawing income for their livelihood. In agriculture maize and wheat and in horticulture apple were the main crops grown by the farmers in study districts. The total income from these sources was Rs 54,047 and Rs 91,173 per farm per year in Tissa and Salooni blocks of Chamba district, respectively. These figures were Rs 1,49,949 and Rs 1,87,417 for Chirgaon and Nerwa Blocks for district Shimla (Table 52.1). The pattern of income from different sources indicates that at farm level the households were drawing highest income of Rs 17,816 and Rs 46,440 from government services in Tissa and Salooni blocks of Chamba district. Whereas, in Shimla district the main source of income was horticulture generating a total income of Rs 89,500 and Rs 46,440 in Chirgaon and Nerwa blocks, respectively.

These levels of income have been used in identifying the food secure and insecure groups. The criterion adopted is that the category generating low income from average income of all the categories has been identified as insecure category. On this basis the BPL households have been found to be food insecure in all the four study blocks. Same situation prevailed among the households belonging to AAY category with the exception of Chirgaon block in Shimla. The APL households were generally food secure whereas a no definite trend was present in case of APDS group households.

ANNUAL CONSUMPTION AND NET AVAILABILITY OF CEREALS

Consumption behaviour of sampled households is the main indicator of food security and this is true other way round also. It is envisaged here to analyse as to what proportion of the total consumption is met out from own farm production and how the gap between the two is bridged. In district Chamba, on an average, total annual consumption requirement of cereal was 234 qtls in Tissa block and it was 198 qtls per annum in Salooni block (Table 52.2). Though rice was the major food in their consumption pattern, it was cultivated only in Shimla district and the supply of rice from PDS was the only source of its availability. The net availability of cereals from home produce was 53 and 63 per cent in Tissa and Salooni blocks, respectively indicating a deficit of about 47 and 37 per cent in these blocks.

In Shimla district, average annual consumption of cereals was about 231 and 241 qtls in Chirgaon and Nerwa blocks, respectively. On an average, the net availability of home produce was 38 and 61 per cent in these blocks, indicating a deficit of 62 and 39 per cent, respectively. This indicates higher availability of home

Table 52.1 Total Income of Sampled Household from All Sources (Rs/farm/year)

Chamba											
Tissa						Salooni					
BPL	AAY	APL	APDS	Annapurna	Total	BPL	AAY	APL	APDS	Annapurna	Total
22658	36134	107886	54047	0	56755	73425	55516	134342	103720	0	91173
Shimla											
Chirgaon						Nerwa					
114964	74971	247613	210196	0	149949	127033	181944	112360	133500	0	187417

Source: Field survey.

Table: 52.2 Annual Cereal Consumption and Net Availability (Qtls)

Blocks	Annual Cereal Consumption	Home Produce				Surplus/ Deficit	Deficit Percentage
		Maize	Wheat		Total		
Chamba							
(a) Tissa							
BPL	43.54	16.63	2.63		19.26	–24.28	56.22
AAY	77.38	23.62	9.19		32.18	–44.57	57.60
APL	88.30	39.37	14.00		53.37	–34.93	39.56
APDS	24.35	10.50	8.75		19.25	–5.10	20.94
Annapurna	–	–	–		–	–	–
Total	233.57	90.12	34.57		124.69	–108.88	46.61
(b) Salooni							
BPL	13.26	4.37	7.44		11.81	–1.45	10.93
AAY	83.31	25.82	15.75		41.57	–41.74	50.10
APL	51.16	11.37	23.62		34.99	–16.17	31.60
APDS	50.68	16.19	21.00		37.19	–13.49	26.42
Annapurna	–	–	–		–	–	–
Total	198.41	57.75	67.81		125.56	–72.85	36.72
Shimla							
(a) Chirgaon							
BPL	89.96	12.68	16.19	4.77	37.64	–52.32	58.15
AAY	65.38	9.62	8.75	6.12	24.49	–40.89	62.54
APL	41.62	6.13	6.56	4.37	17.06	–24.56	59.01
APDS	32.54	1.75	6.12	5.25	13.12	–19.42	59.68
Annapurna	1.35	–	–	–	–	–1.35	100.00
Total	230.85	30.18	37.62	20.11	87.91	–142.94	61.92
(b) Nerwa							
BPL	22.72	5.25	6.12	4.37	15.74	–6.98	30.72
AAY	179.02	42.35	28.00	26.25	96.60	–82.42	46.03
APL	34.14	7.00	7.88	9.63	24.51	–9.63	28.20
APDS	5.32	2.63	2.63	5.25	10.51	+5.19	+97.55
Annapurna	–	–	–	–	–	–	–
Total	241.20	57.23	44.63	45.50	147.46	–93.74	38.86

Source: Field survey.

produce in Nerwa block. Thus, Tissa block in Chamba and Nerwa in Shimla district were comparatively poor in respect of net availability of cereals.

This indicates a need of government assistance for poverty alleviation as well as ensuring food security, especially for the vulnerable groups like BPL and AAY. It was found that the provisions made for the supply of cereals on subsidized rates to different categories of households under PDS of state Government has proved to be very efficient. It is heartening to note that PDS has been able to fulfil the cereal requirements as indicated by deficit and none of the sampled farmer had to purchase cereals from open market.

STATUS OF FOOD SECURITY

The cereals are the staple diet of sampled households and shortfall in its on-farm production is an indicator of food insecure nature of these households. However, they have enough resources for purchasing the requisite quantities of cereals and as such, by definition, may not be food insecure. Presently, an attempt has been made to make a reflection on status of food security based on average production of cereals on farm. The categories of farms/farmers groups having production less than the average farm production of different categories have been termed as food insecure groups. It was assumed that the gap between the average

Table 52.3 Food Secure and Unsecured Categories of Cereals in Study Districts

Districts/ Blocks	Net Availabilities of Cereals			Total Consumption of Cereals
	Maize	Wheat	Rice	
Chamba				
(a) Tissa				
BPL	Secure	Insecure	–	Secure
AAY	Secure	Insecure	–	Secure
APL	Secure	Insecure	–	Insecure
APDS	Secure	Secure	–	Insecure
(b) Salooni				
BPL	Secure	Secure	–	Insecure
AAY	Secure	Insecure	–	Secure
APL	Insecure	Secure	–	Insecure
APDS	Secure	Secure	–	Insecure
Shimla				
(a) Chirgaon				
BPL	Secure	Insecure	Insecure	Insecure
AAY	Secure	Insecure	Secure	Secure
APL	Secure	Insecure	Secure	Insecure
APDS	Insecure	Secure	Secure	Insecure
Annapurna	Insecure	Insecure	Insecure	Insecure
(b) Nerwa				
BPL	Secure	Insecure	Insecure	Insecure
AAY	Secure	Insecure	Insecure	Secure
APL	Insecure	Insecure	Secure	Secure
APDS	Insecure	Secure	Secure	Insecure

Source: Field survey.

and actual production is the assessment of food secure and insecure categories of households. The results of analysis have been presented in Table 52.3. It was found that BPL and AAY categories in Tissa block of Chamba district were food secure as far as cereal production and consumption are concerned whereas only AAY households were food secure in Salooni block of this district. But in district Shimla, only AAY households were secure in Chirgaon block and AAY and APL were secure in Nerwa block.

CONSUMPTION AND AVAILABILITY OF PULSES

Pulses are the essential items in consumption pattern of the people but its production has been decreasing since last two decades. In such a scenario, the highly subsidized pulses provided through PDS are proving very helpful in improving food security status of poor households. It was found that except Tissa, pulses were grown in all other blocks of both the districts. In pulses growing blocks, the percentage of deficit over the net availability from farm produce was 84, 72, and 60 per cent in Salooni, Chirgaon, and Nerwa blocks, respectively. In case of annual purchase of pulses from PDS and open market, only APL category of Tissa and Salooni and BPL category of Nerwa has purchased pulses from open market at the rate of 0.43, 0.76, and 0.07 qtls per annum, respectively.

PER CAPITA CALORIE INTAKE AND NUTRITIONAL SECURITY

On an average, per capita per day calorie intake in Chamba district was 2,141 and 2,502 in Tissa and Salooni blocks, respectively (Table 52.4). In both the blocks, highest calorie intake was in case of APL category and was 2,201 and 2,602 calories per day per capita. In case of Shimla district, on an average per capita calorie intake was 2,482 and 2,157 in Chirgaon and Nerwa blocks, respectively with APL category persons topping with 2,601 and 2,378 calories per day per capita. On the basis of this analysis, the food secure and insecure groups have been identified and results presented in Table 52.5. In this table the categories consuming less than the total average calorie intake of the block have been marked as insecure and those consuming equal or more than the average were categorized as secure categories. The results indicate that BPL category was nutritionally secure except in Chirgaon block of Shimla. The persons belonging to AAY category were nutritionally secure only in Nerwa block of Shimla. APL category was invariably nutritionally secure. APDS category was secure only in Chirgaon block of Shimla.

Table 52.4 Average Calories Intake by of Family Members of Different Group (Calorie/intake/per day)

Districts	Calorie Intake											
	BPL	AAY	APL	APDS	ANNAPURNA	TOTAL	BPL	AAY	APL	APDS	ANNAPURNA	TOTAL
Chamba	Tissa						Salooni					
	2161	2090	2201	2123	–	2141	2545	2439	2602	2483	–	2502
Shimla	Chirgaon						Nerwa					
	2445	2429	2601	2531	2249	2482	2235	2269	2378	2135	–	2157

Source: Field survey.

Table 52.5 Nutritional Status of Different Groups

District	Nutritional Status									
	BPL	AAY	APL	APDS	AN	BPL	AAY	APL	APDS	AN
Chamba	Tissa					Salooni				
	Secure	Insecure	Secure	Insecure	–	Secure	Insecure	Secure	Insecure	–
Shimla	Chirgaon					Nerwa				
	Insecure	Insecure	Secure	Secure	Insecure	Secure	Secure	Secure	Insecure	–

Source: Field survey.

CONCLUSIONS AND POLICY IMPLICATIONS

There is a large scope for cultivation of fruit and vegetables, including off season vegetables, in all the blocks but due lack of irrigation facilities, especially in Tissa and Salooni blocks, the farm families are unable to diversify cropping pattern at their desired extent.

Therefore, for alleviating hunger and poverty, it is important to support more public investment in rainfed and backward hilly areas. In hilly topography, there exists a scope for augmenting irrigation facilities through lift and flow irrigation schemes. Presently, cultivation of herbs and medicinal plants is emerging in a very big way. Therefore, availability of irrigation facilities will not only become helpful for cultivation of fruit and vegetable but also generate higher returns from diversification towards herb and medicinal plants cultivation due to suitable agro-climatic conditions of hilly topography in Himachal Pradesh.

In order to be effective the food security policy must evolve as a basic element of a social security policy with proper coordination among the various government departments, private sector, and non-government organizations. The direct food and nutrition support for the poor through a minimum safety net should be properly balanced with improvements in the quality of life of local people through investments in education, drinking water, sanitation, and healthcare.

Further, future food security programmes should have a broad objective of increased agricultural production and enhanced access to food through a participatory approach of local people with emphasis on resources efficiency, social equity, and preservation of the environment. Centralized anti-poverty programmes should give way to local initiative and local participation based on the principles of efficiency, equity, and environmental conservation.

It is very interesting to note that in some categories, quantity of cereals supplied through PDS is higher as compared to the quantity produced at home. This indicates that reliance on PDS is increasing steadily leading to neglect of farm production. This is eroding the very production base and pushing local agriculture towards unsustainability. This is an alarming situation and the farm neglect is further aggravated by the menace of stray and wild animals.

It has been observed that poor quality seeds lead to variation and low level in productivity. There are many sources from which farmers obtain seeds—retained crop produce, other farmers, traders, cooperatives, government departments, agricultural universities, public and private seed companies and since certification is not compulsion, it is difficult to monitor the quality of seeds from informal sources. Therefore, it is suggested that extension services should educate farmers in the importance of using quality seed.

Considering the overwhelming importance of the rural sector, additional emphasis has to be placed on rural development and non-farm activities to increase income and employment as demand for agricultural goods slows down. This addresses the domestic policy orientation aspect. This is especially valid in the case of study districts, where both the income and employment generating capacity of agriculture and the growth of the non-farm sector have slowed down considerably.

This study makes a strong case for a drastic change in the food management policy and, therefore, in the overall agricultural strategy. Apart from reforming procurement and price support policies, the government should seek to increase production of non-cereal foods like fruits, vegetables and animal products including milk, their storage facilities and the processing of agricultural produce. There is good of scope to increase production of these products in study districts as well as in hill agriculture of Himachal Pradesh.

The impact of changes in climatic factors on crop productivity raises several researchable issues. The need for short duration wheat variety is felt necessary to address the problem of shorter winter as well as rising temperatures, which affect the productivity adversely. Since the food security is the primary concern of the Government, the declining production and productivity has to be effectively taken care of. This is a challenge to agricultural research system, which also necessitate revamping of the crop breeding programme suitably to fit the regionally differentiated strategy. Technology policy needs to stress more on development of short duration variety to escape climatic aberrations.

REFERENCES

Adil, Mohommad. 2001. 'Decline in Cereal Consumption Varies across States', *Economic Times*, 19 December 2001.

Barah, B.C. 2007. 'Criticality of Rice and Wheat System in Sustainable Food Security in India: An Analysis', Agricultural Situation in India.

Chand, Ramesh. 2000. 'Trends in Food Consumption and Nutrition—Food Security Concern (Reporter's Report)', *Indian Journal of Agriculture Economics*, 61 (3): 571–81.

Datta, K.K., S. Mandal, A.K. Tripathi, S.B. Singh, M.R. Verma, and S. Mohanty. 2000. 'Retrospect and Prospect of Food Security in North Eastern Hilly Region of India', *Agricultural Situation in India*. August 2000.

George, P.S. 1996. 'Public Distribution System, Food Subsidy and Production Incentives', *Economic and Political Weekly*, 31(39), 28 September: A140–A144.

———.1999. Some Reflections on Food Security, Presidential Address 59th Annual Conference of Indian Society of Agricultural Economics, Jawahar Lal Nenru Krishi Vishwa Vidyalaya Jabalpur, 1–3 December.

Minhas, B.S. 1976. 'Towards National Food Security, Presidential Address', *Indian Journal of Agricultural Economics*, 31(4): 8–19.

Panagariya, Arvind. 2002. 'Stamping in Nutrition', *Economic Times*, 24 February 2002, 6.

Ram, G.S. 1996. 'Food Security System, Poverty Issue and Rural Development, Presidential Address', *Agricultural Economics Research Review*, 9(2): 121–7.

53

Public Policies and Sustainable Agricultural Development

A Case Study of Commercialized Agriculture

Brajesh Jha

Though India by virtue of its diverse natural resource endowment and surplus labour has potential to produce large number of agricultural commodities; there is dearth of adequate exportable surpluses on sustained basis. The self-sufficiency in foodgrain has also been dented in the recent period since country had to import considerable amount of agricultural commodities at a relatively higher prices. Though stock of food grains in the recent years is comfortable, prices of fruits and vegetables increased suggesting burgeoning gap between production and consumption of horticulture commodities. Perverse aggregate supply response of agriculture and inadequate post-harvest infrastructure are some of the important reasons for the above. The breaking of this perverseness requires investment, infrastructure, technology, institutions, and similar other measures; the same can be achieved through reforms in agriculture.

A significant proportion of public expenditure in agriculture is consumed in subsidizing agricultural inputs. Though subsidized farm inputs have been instrumental in increasing agricultural production in the 1970s, continuance of same has discouraged efficient use of many of these resources. This has also caused degradation and depletion of natural resources in certain regions of the country (Jha 2000a). Rationalization of price is widely recognized as an important element of reforms in agriculture. In an opening economy rationalization requires that prices of tradable input and outputs are aligned to the long-term trend in international prices, whereas prices of non-tradable inputs would reflect scarcity of resources for the society.

Such changes in agricultural prices depending on the technological options and resource endowment of the region would influence farmers' resource utilization pattern; this will subsequently affect the long-term growth of agriculture in a region. The present study attempted to study the above linkage in a farm economy with following specific objectives:

1. to assess changes in public policies towards agriculture in the recent period and its likely effect on the prices of agricultural commodities;
2. to study the extent of sustainability of agricultural growth in the Northwest India and factors responsible for same; and

3. to evaluate responses of alternate policies: price, technology, and institution on various elements of sustainability in a farming system framework.

METHODOLOGY AND DATA

Since factors that constrain long-term growth of agriculture vary across regions, the present study has chosen Northwest India for the analysis as this region emerged as the grain basket of India and quite in news because of disconcerting trends in agricultural development of the country. Besides, the region is in the forefront of adoption of commercial agricultural practices, optimization would present more realistic picture for such region. The changes in public policies besides discussing price, technology, and institution related policies also measure extent of protection to domestic agricultural commodities. Protection is measured with Nominal Protection Coefficient (NPC); the NPC is the ratio of domestic to world prices at common reference point. The NPCs measure divergence between domestic and international prices; the NPC greater than one indicates that country is protecting the commodity while NPC lesser than one suggests otherwise (Jha 2000b). Behaviour of farm outputs prices was discussed with trend growth and stability in these prices. The extent of integration of agricultural output prices was studied with the correlation of prices across states.

The sustainability of agricultural system in the Northwest India was evaluated with the help of Conway's indicators (Conway 1985) that considers productivity, stability, low-vulnerability, and equity as yardsticks for measuring sustainability. The importance of these indicators, however, varies across regions; farmers in a resource rich region like Haryana as compared to a resource-starved region are less vulnerable to natural calamities. The analysis assumed that risk especially downside risk in return would reflect vulnerability of farmers in the northwest India. The study considers growth as an essential element of sustainable agricultural development of a region especially when population is growing. The present study, therefore, evaluated sustainability on the basis of following indicators, that is, growth, stability, productivity, and vulnerability.

The growth and stability in time series data were worked out with the help of linear and exponential trend equations (see Jha 2008, 2011). The coefficient of variation around trend in productivity and return of important crops was computed to assess stability of agricultural system in the country. Vulnerability was assessed with the down side risk; the frequency and magnitude of down side risk was measured with the probability of failure and expected negative deviation from trend return and yield of important crops. The growth in productivity has been discussed with total factor productivity growth of important crops like paddy and wheat. Though there are several ways of measuring factor productivity growth, the present study worked out trans-log index, developed by Tornquist (Jha 2008). Most of the above indices suggest that agricultural system in the Northwest India was constrained to the extent that the present rate of growth in agriculture is also suffering. The study also delineated factors responsible for the above trend.

The present study further evaluated the effect of alternate policies on factors that constrain sustainable growth in the Northwest India; the same has been undertaken on an average farm of Kurukshetra district of Haryana using linear programming framework (Hazell 1971, Jha 1995). The effect of policies has been discussed with a mix of economic, ecological, and social indicators. Some of the ecological parameters relevant to field condition and also incorporated in the present evaluation were: soil degradation, ground water depletion, deterioration in quantity and quality of biomass further causing stress on common property resources (for constraints and activity-related details of methodology see Jha 2008).

RESULTS AND DISCUSSION

A temporal trend in NPC suggested that with trade liberalization the NPCs for most of the agricultural commodities approached towards one (Table 53.1). A decrease in protection clearly suggested towards integration of domestic and international market. The inter–commodity differences in protection were also reducing.

There have been attempts to reorient domestic agricultural policies as well. Restrictions on the movement of many agricultural commodities were abolished. The special provision of the Essential Commodities Act, which has come in force in 1981, was repealed towards the end of

Table 53.1 Decreasing Protection in Indian Agriculture

Commodities	Nominal Protection Coefficients					
	1992	1996	2000	2001	2002	2003
Paddy	0.90	0.82	1.10	1.14	1.28	1.20
Wheat	1.02	0.94	1.14	1.04	1.10	1.12
Maize	1.14	0.86	1.18	1.3	1.24	1.05
Sorghum	1.10	0.86	1.10	1.16	1.25	1.22
Cotton	0.81	0.83	1.06	1.15	1.12	0.91

Note: These estimates are for leading producing state of the commodity; for example, Punjab for paddy and wheat; Uttar Pradesh, Rajasthan, Gujarat, West Bengal, and Andhra Pradesh for maize, sorghum, and cotton, respectively.
Source: Author's computation.

the 1990s, primarily to encourage private trading activities. Future market in a large number of commodities was initiated. Agriculture Produce Market Regulation (APMR) Act has been redesigned in the Model APMR Act to address most of the concerns of marketing agricultural commodities. The state governments are being encouraged to adopt the model APMR Act. An integrated Food Legislation is being enforced to address multiplicity of legislation for processing sector. Further, to encourage large-scale investments in processing sector, many agro-processing industries have been de-reserved, the scale of investment is also increased for some commodities. The intensity of compulsory procurement levies for paddy and sugarcane has also been reduced gradually. All these attempt to improve efficiency in the post-harvest operations.

Although minimum support price (MSP) has been instrumental in agricultural development of the country; also created certain kind of problems in an opening economy (Jha and Mohapatra 2004). A relatively higher MSP and its downward rigidity has resulted in situation as that of high food stocks with hardly any taker of same in the domestic and international market. The opening up of trade has started showing its impact on market prices of many farm outputs. The MSP was largely influencing Farm Harvest Prices (FHP) in the 1980s; divergence in these prices has started only after mid-1990s. The world prices have started influencing domestic prices of many agricultural outputs. Interestingly, the FHP has declined for many commodities during the period (2000–03).

With the removal of restrictions from domestic markets, it is hypothesized that market across space would integrate; whereas, price integration, as measured with the correlation coefficient between FHP in states, has decreased in the recent decade. The spatial integration as an inverse of coefficient of variation in FHPs across states has weakened consistently over the years (2002–05). Removal of certain restrictions from domestic market does not appear to have affected private trade significantly; informal restriction on domestic market is reported to have persisted. Uncertainty associated with the fumbling of some of the domestic market restrictions has also discouraged private trading activities. The cost of domestic transport remains high; in such situation external trade, generally in small amount, affects market price of commodity only in regions adjacent to a port. This possibly widens disparity in market prices across states.

With the liberalization of trade of farm inputs administered prices of some of the farm inputs such as pesticides, fertilizers (especially phosphatic and potassic) have started reflecting world prices. The above changes in relative prices will affect resource utilization pattern of farmers. The farmers who actually implement an agricultural policy attempt to maximize their farm return; the externalities associated with the stated objectives are often ignored. At times one finds that particular farm policy in spite of certain positive outcomes has been detrimental to the long-term growth of agriculture in a region. In this context the much-acclaimed biochemical technology and policy regime of the late 1960s is an example. The negative externalities associated with the adoption of biochemical technology were ignored in light of the broader objective of achieving self-sufficiency. The farm policies are, however, politically so much sensitive that frequent alteration to correct imbalances becomes difficult. Prioritization of the long-term concerns over the short run objectives becomes even more difficult. The long-run effects of policy changes, therefore, need to be evaluated a-priori. This evaluation has to be region-specific since the externalities, which constrain long-term growth of agriculture in a region vary across regions depending on the status of natural resources in the region. The northwest India was selected purposively for the present analysis.

MEASURING SUSTAINABILITY OF AGRICULTURAL SYSTEM IN NORTHWEST INDIA

The sustainability of agricultural system in the northwest India is evaluated on the basis of growth, stability, and productivity of agriculture at regional level and also vulnerability of farmers of the region. In the northwest India, paddy and wheat occupy more than 80 per cent of cropped area; this is the main source of agricultural growth of the region. The growth in productivity of these crops has, however, tapered off in the recent decade. A decreasing trend in factor productivity of paddy is also because of data related anomaly, productivity of paddy as available from 'Statistical Abstract of Haryana', pools basmati and non-basmati paddy data together; productivity of these crops are significantly different, proportion of these varieties in the aggregate (state/district level) production of paddy varies over the years, therefore, some of fluctuation in productivity of paddy as obtained from pooled data for paddy is also on the above account.

In wheat, total factor productivity is decreasing for the state as a whole. Further, to observe regional pattern in agricultural productivity, total factor productivity of wheat was calculated for selected districts of Haryana: Kurukshetra and Mahindergarh. Though downward trend in productivity was established for both the districts; the decreasing trend in factor productivity in a progressive district like Kurukshetra has started earlier than the late-adopting district like Mahindergarh. The above findings, to some extent, substantiate the theory of technological maturity in agriculture as propounded by several researchers.

Further enquiry into decreasing trend in total factor productivity of the above crops suggests that the increase in the cost of production and stagnation in yield of these crops are some of the important reasons for the above trend in factor productivity. A comparative account of cost structure suggests that share of labour has increased significantly during the reference period (1988–90, 1997–99, and 2003–05). The hike in administered wages for agricultural labour, which cannot be justified on the basis of decreasing total factor productivity of important crop of the region has accounted for a higher share of labour in total cost. Interestingly, real wages for agriculture workers has also decreased in the subsequent years: 2000–03 (Jha 2006). A higher share of chemicals in total cost of paddy and wheat suggest higher application of pesticides in Kurukshetra; the district has been an early adopter of biochemical technology. Again, decline in micronutrient status of soil has led to increase in cost of production; since farmers now has to apply higher doses of micronutrients containing fertilizers to receive same levels of production.

The ramifications of degradation of some of these resources go beyond the agricultural system. A higher dose of nitrogenous fertilizers on account of degradation of soil has reduced potable quality of ground water. It may be noted that depletion of ground water resources and introduction of submersible technology has not only threatened the ground water aquifer but have also introduced a new source of inequity in the rural area of the semi-arid region of the country (Jha 2000a). The disturbing trend in agricultural productivity owes it to various biotic and abiotic stresses; these are often reported to have been aggravated by policy variables. Present study assesses several technological and policy options for reversing the unsustainable trend in the northwest India. Technological options can be segregated into land and non-land based options; the non-land-based options are primarily resource-saving and resource-intensive farm practices, at times adopted by farmers for crops already in cultivation in the region.

Some of the negative externalities of the existing paddy-wheat crop enterprises could also be minimized by encouraging utilization of land for crops other than paddy and wheat. Field visit suggests that on large farms, there is limited possibility of diversification towards fruits and agro-forestry whereas on small farms, these are vegetables. The cultivation of many of these crops is at the suboptimal level; though reasons for suboptimal levels could be many, inadequate infrastructures for storage and processing have been the most important. Dearth of suitable institution has also constrained area under certain crops; for example, acreage under sunflower is affected because of non-availability of quality seeds. The problem of spurious seed is more conspicuous in the hybrid variety of seeds of different crops. In this regard an institution that effectively regulates quality of seeds will be important. Field visits to area also indicate towards technological constraints; farmers report that the short durational pulse (summer season) varieties, which can suitably be adjusted between wheat and paddy crops during the month of April–May–June, is highly sensitive to the occasional high temperature during the above period. The present study suggests that incorporation of certain attributes into the existing variety of summer pulse will stabilize yield of the above crop. The institutional and technological constraints mentioned above hold good for several other crops of the region.

The non-land-based technological options in the present analysis are the farm practices used by the selected farmers of the region that have potential to reduce the negative externalities of intensive agriculture in the region. Certain agricultural practices based on the wisdom of conventional agricultural practices (like organic farming) are often suggested to improve the existing state of resource degradation. In many of such practices crop yield in the short run is often less than the yield as obtained from crops using biochemical technology; though crop yield in the long run is always reported to have been higher in the earlier case. Unfortunately, farmers, most of the time prioritize their short-term interest over the long-term goal.

Some resource conserving agricultural practices as that of Systems of Rice Intensification (SRI) is also suggested, its adoption in the northwest India is, however, restricted. The Integrated Pest and Nutrient Management (IPNM): a compromise between intensive use of chemicals and traditional agricultural practices is being practiced by the selected few farmers. Though benefits and costs of IPNM vary across studies, the present study assumes that IPNM will increase agricultural production on an average/synthetic farm by 15 per cent and decrease cost of cultivation by around 5 per cent. The existing regime of subsidization of farm input prices is widely believed as important reasons for limited adoption of many of the above practices.

In the era of liberalization price rationalization is often argued as panacea for many ills that Indian agriculture is plagued with. Rationalization of price in the present study requires that the existing market price of farm input and output should be aligned with its shadow prices. For tradable farm outputs and inputs shadow price is close to world prices. The above assumption holds true for most of the tradable farm inputs with the exception of urea. In urea, domestic market is still insulated from the world price. In non-tradable farm inputs, rationalization of price should reflect cost of production of these inputs. The power tariff for agriculture in most of the states is not related to actual cost of producing electricity. Though a consensus amongst the state chief ministers that power tariff in no case should be less than 50 per cent of the cost of power produced in the state was arrived in the year 1996. About canal irrigation pricing, Vaidyanathan Committee suggests that price of water from canal should vary from a minimum of 5 per cent of gross output for cereals to a maximum of 12 per cent for commercial crops. Price rationalization in the present analysis is based on above recommendation.

The impact of alternate price policies and technological options on sustainable agricultural development of the region is evaluated in a farming system framework with different indicators that explain economic, ecological, and social environment around a farm. Some of these indicators are farm return, variability in return, variable cost, amount of irrigation water, synthetic farm inputs (chemical fertilizers, plant protection chemicals), crop and livestock residues for fuel and fodder, aggregate employment and distribution of employment. Analysis of these indicators suggests that some crops, like, basmati paddy, even though highly profitable, are not cultivated by farmers to its potential as returns are highly unstable. The unstable return was on account of various institutional and technological factors. Similarly, crops like sugarcane, though profitable, are highly water intensive and could only be recommended hesitatingly in the semi-arid region of the country. Information on the above indicators and questions related to resource allocation pertains to the survey year (2000–01).

Most of the farms in the study area were mixed farms wherein dairy plays an important role. The existing crop enterprise-mix on an average farm was dominated by paddy and wheat crops; there was no significant difference across farms in the above context. The next most important crop rotation in the study area was fodder consisting of jowar (chari) and berseem crops. The share of fodder in total cropped area decreased with the increase of farm size; this suggests higher intensity of livestock on small farm. Similar trend across farm size was evident in vegetables; vegetable on commercial scale was cultivated on some small farms only; whereas, area under agroforestry, at considerable level, was cultivated on some of the large farms only. These crops unlike paddy, wheat, and fodders were cultivated on the selected small and large farms, respectively; therefore, share of these crops in synthetic farm was extremely low.

The farm-level impacts in a linear programming framework has been studied by ascertaining changes in the levels of various activities/farm outputs following changes in the price of farm input, outputs (alternate price policy), and input–output coefficients (alternate farm practices). The impact was evaluated by assessing changes in the above indicators in alternate situations. Alternate situations were; Economically-Ecologically-Efficient (E-E-Efficient) plans that present crop combination when farmers are trying to use their resources optimally on the basis of two economic criteria: maximization of gross margin (return) and parameterization of negative deviation from mean (risk). Though availability of resources and its utilization varies across different sizes of farm; the E-E-Efficient plans in the present study were based on average synthetic farm of 2.7 hectare; the evaluation results have been derived for three situations, (E-E-Efficient plan I, II, and III).

In E-E-Efficient plan I, resources were used optimally with the existing enterprise-mixes, input and output prices were the average price which prevailed during the last three years of survey. Field visit suggested possibility of various land-based technological options; suitability of these options often varies across farm; fodder and vegetable-based crop rotation was more suitable on small farm; while fruits (kinnow, a citrus fruit) and agroforestry (eucalyptus, papular, subabul/acacia) were more suitable on large farm. Farmers across farm-sizes were of the opinion that summer pulse can be incorporated between wheat and paddy crop rotation, provided it is less sensitive to the occasional high temperature during the month of May.

In E-E-Efficient plan II, resources were used optimally with the existing enterprise-mixes, while input prices were rationalized (Jha 2009). The present study presumed that prices of farm outputs were almost rationalized during the study period (2000–02). World price would reflect shadow prices of tradable farm inputs while prices of non-tradable inputs should be reflected in the scarce price of the said resource. Several experts have opined about pricing of non-tradable farm inputs like canal irrigation water, tariff of power for its use in agriculture. In E-E-Efficient plan III, resources were used optimally with the changed farm practices whereas enterprise-mixes were as it existed on an average farm. The above evaluation used rationalized prices of farm inputs as per the previous plan (E-E-Efficient plan II). The existing non-land based resource utilization pattern reflected the situation as it would be with the adoption of IPNM practices.

The linear programming based optimization has led to specialization in favour of crops like sugarcane, wheat-paddy, and potato-based crop rotation (maize-potato-sunflower/rapeseed). Vegetable and fodder-based crop rotation had been more profitable on small farms due to regular supply of labour. The optimal levels of these crops were, however, restricted on account of market-related constraints in the study area. Some of such crop specific constraints were having insufficient storage facility in potato, limited processing facility in sugarcane, inadequate chilling centres, and air-conditioned vans for fruits and vegetables. Besides infrastructure, the resource-based constraints also restricted acreage under certain crops. The present study presumed that any optimal plan must not consume water significantly higher than the existing plan. Sugarcane, like paddy and wheat, is also a water intensive crop and acreage under such crops has, therefore, been restricted in the present evaluation. On one of the above accounts, the maximum constraint was imposed for crops in the present evaluation.

As was apparent from the E-E-Efficient plan I, there was limited scope of improvement in the utilization of existing resource (land) with the maximum constraints as it was in the above crops. The difference between highly specialized enterprise mix and the one obtained after imposing the maximum constraints suggests losses on an average farm due to institutional and infrastructure related constraints in the region. On an average farm of 2.7 hectare, there has been considerable amount of return foregone on account of infrastructure, institution, and resource constraints.

Table 53.2 Evaluation of Alternate Situations on the Basis of Different Economic, Ecological, and Social Parameters

Parameters	Existing	E-E-Effi-I	E-E-Effi-II	E-E-Effi-III
Economic parameters				
Gross Margin ('00Rs)	567.87	577.74	438.76	616.86
Negative Deviation (in Rs)	139.16	110.67	211.42	182.16
Ecological parameters				
Irrigation-monsoon (July–Oct)	239.4	244.2	107.8	184.1
Irrigation-otherseas. (Nov–June)	90.2	86.4	82.3	80.8
Nutrient mining (kg)				
Nitrogen	514.2	536.7	421.5	523.1
Phosphate	106.1	110.4	106.9	131.0
Potash	596.5	629.9	467.6	613.3
Fuel derived ('000 kcals)	6138.8	6198.8	3596.6	4454.5
Fodder obtained				
(a) Dry crude protein	143.5	153.5	176.1	146.6
Total digestible nutrients	2582.4	2839.4	2544.1	2493.0
Social parameters				
Employment-total (man days)	432.7	439.2	390.5	449.6
Employment-Sowing and harvesting	291.5	298.9	242.3	284.5
Employment-other operations	141.2	140.3	148.2	165.1

Source: Author's computation.

The rationalization of farm input and output prices have resulted in decrease of area under paddy and wheat crops and increase in area under maize- and potato-based crop rotation. The above reallocation has resulted in loss of farm return (more than 10 per cent). The loss in the above plan was caused because of increase in the cost of agriculture; though there was significant respite on account of use of natural resource, especially water. Since rationalization has squeezed farmers' return; there were less chances of it getting adopted by farmers in the existing price regime. The findings also highlighted that too much of emphasis on price or market-based instruments may not encourage sustainable agricultural development; this has to be supplemented with the technological innovations. In the study area, IPNM appeared to be an important option to counter the existing stress on natural resources and simultaneously increase farm income. In the E-E-Efficient plan III, paddy and wheat regained its status, maize-potato-mustard and cotton-mustard emerged as another important crop rotations. As a result of adoption of improved farm practices, farm return in E-E-Efficient plan III has increased considerably over the E-E-Efficient plan II.

POLICY SUGGESTIONS

In the era of globalization, it is trendy to highlight the role of price-corrections in increasing growth; the present analysis, however, suggests that merely rationalization of price would not lead to a sustained growth in agriculture. The sustained growth in fact requires suitable technology, adequate infrastructure facilities, and a set of institutions. The desired institution is not only in the sense of aggregator or facilitator to stem out problems associated with the small and scattered production of agriculture but also effective-enforcer of State laws as in the case of sale of spurious seeds. Institution is also desired in the form of interventionist policies like limiting the use of groundwater aquifer. Since factors that influence future growth of agriculture vary across regions the present study may promote similar studies in other regions of the country. Such evaluation study would generate evidences on the role of alternate policy instruments: price, technology, and institution for sustainable agricultural development of the region.

REFERENCES

Christeensen, L.R. 1975. 'Concept and Measurement of Agricultural Productivity', *American Journal of Agricultural Economics*, 57(5): 910–15.

Conway, G.R. 1985. 'Agro-ecosystems Analysis', *Agricultural Administration* 20: 31–55.

Hazell, P.B.R. 1971. 'A Linear Alternative to Quadratic and Semi-variance Programming for Farm Planning under Uncertainty', *American Journal of Agricultural Economics* 53: 53–62.

Jha, Brajesh. 1995. 'The Trade-off between Return and Risk in Farm Planning, MOTAD and Target MOTAD Approach', *Indian Journal of Agricultural Economics* 50(2): 193–99.

———. 1996. 'Market Induced Risk in the Greenbelt Farms of India', *Journal of Rural Development* 14(3): 409–15.

———. 2000a. 'Implications of Intensive Agriculture on Soil and Groundwater Resources', *Indian Jl. of Agricultural Economics*, 55(2): 182–93.

———. 2000b. *Indian Agriculture and the Multilateral Trading System*. New Delhi: Bookwell.

———. 2008. *Public Policies and Sustainable Agricultural Development: A Case Study of the Commercialized Agriculture*, a report submitted to Ministry of Agriculture Government of India, Delhi.

———. 2009. *Evaluating Agricultural Policy in a Farming System Framework: A Case Study from North West India*. Institute of Economic Growth (IEG) Working Paper Series No. E/299/2009.

Jha, Brajesh and B.B. Mohapatra. 2003. 'Liberalisation and Agricultural Prices: Some Disconcerting Trends.' *Indian Journal of Agricultural Economics*, 58(3): 375–86.

———. 2004. 'Liberalisation and Agricultural Prices: Some Disconcerting Trends.' *Indian Journal of Agricultural Economics*, 59(3): 375–86.

54

Impact of Diesel/Power Subsidy Withdrawal on Production Cost of Important Crops in Punjab

D.K. Grover, J.M. Singh, Jasdev Singh, and Sanjay Kumar

The diesel retail prices continue to be regulated by the government of India since early 1970s and hence, contributed in a major way towards the buildup of fuel subsidies over the years. Government of India has taken a number of measures to reform its fuel subsidy system. In June 2010, petrol pricing was liberalized and the intention to liberalize diesel prices announced. In its 2012–13 budget speech, the government stated its intention to limit all central subsidies (including those on fuels) to less than 2 per cent of gross domestic product (GDP) in 2012–13, and reducing them to less than 1.75 per cent of GDP over three years. In January 2013, the government announced that Oil Marketing Companies (OMCs) would have greater flexibility in setting diesel prices and that bulk users of diesel would pay unsubsidized prices. (International Monetary Fund 2013). Subsidizing consumers of petroleum products including diesel has been a common phenomenon in many developing and emerging economies. Fuel subsidies generally arise out of desire to shield consumers, especially poor households, from high and often volatile fuel costs for lighting, cooking, and transportation. However, fuel subsidies are both inefficient and inequitable (Anand et al. 2013). Such subsidies encourage overconsumption of fuel, delay the adoption of energy-efficient technologies, and crowd out high-priority public spending on physical infrastructure, education, health, and social protection. Most of the benefits of fuel subsidies also go to higher income groups who tend to consume more fuel (Arze del Granado, Coady, and Gillingham 2012). Recognition of these shortcomings has led to an active debate in India as to the merits of replacing these subsidies with better targeted safety net measures. Fuel subsidy reforms have been on Indian government's policy reform agenda over the last decade.

Oil Marketing Companies increased per litre diesel prices by Rs 0.5 per month from January 2013 onwards with some exceptions. Since then, the diesel prices had risen by a cumulative Rs 10.12 per litre in sixteen installments and the diesel subsidy is likely to be completely removed with automatic deregulation of fuel in next few months if the rupee continue to strengthen and monthly price hikes continue (Daily Post 2014). The subsidy on diesel is financed by the government of India, while electric power subsidy to agriculture is financed by the state governments. During the recent decade impact of electricity subsidy to agriculture on financial health of state economy has been the most debated issue in Punjab. Keeping this in view, the present study has been undertaken to estimate the use of diesel in crop production and analyse the likely impact of diesel/energy price policy on the cost-profitability relationship of major crops in Punjab. The specific objectives undertaken under the study were as follows:

1. to study the status of electricity and diesel use for various crop-production activities in the state;
2. to examine the impact of squeezing diesel subsidy/ enhancing diesel price on the cost of agricultural production and profitability; and
3. to estimate the likely impact of power subsidy withdrawal (Hypothetical) on the cost of cultivation/ production and profitability of major crops in the state.

METHODOLOGY AND DATA

To meet the specific objectives of the study, at first stage of sampling six districts namely Hoshiarpur, Amritsar, Jalandhar, Ludhiana, Bathinda, and Fazilika, representing all the major agro-climatic regions of the state, were selected purposively. From each of the selected district, two blocks were selected randomly. Thus, overall twelve blocks from the sample districts were selected. From each selected block a cluster of villages was selected randomly for the farm household survey. Finally, from each of the selected village cluster, twenty-five cultivators with five cultivators representing each farm size category, as per standard national-level definition of operational holdings, were selected randomly. Thus, overall from the state, total sample of 300 farmer households, comprising sixty farmers each, from marginal, small, semi-medium, medium, and large categories forms the basis for the present enquiry. To work out the components of returns over variable cost of selected crops, data on variable costs and output obtained were collected on an especially prepared schedule for important crops. Simple tabular analysis was conducted to analyse the results. Simulations by changing diesel prices were undertaken to work out the cost of production/ cultivation of crops by keeping prices of all other inputs at constant level as at the time of data collection and just varying diesel prices only in order to see the impact of diesel subsidy withdrawal on cost of production of crops and also worked out 'diesel price hike coefficient'. The changing diesel prices and its impact on cost of production of major crops was seen at I (base) level (1 February 2013), secondly at II (current) level (1 June 2014), and thirdly at III (proposed) level with zero diesel subsidy. Impact of de-subsidized electricity supply to agriculture sector on cost of production of important crops in Punjab was also worked out. For this purpose per hectare electricity consumption for irrigation of a crop on sample farms was worked out using following formula:

Electricity use in irrigation (Kwh/ha) = Use of electric pump (electric motor or submersible) for irrigation of crop (hours/ha) × HP of electric pump × 0.746 kwh + 20% inefficiency

Cost of electricity consumption for each crop was estimated through multiplying the per hectare electricity consumption with the subsidy amount of Rs 4.18/kwh, which was approved by the Punjab State Electricity Regulatory Commission for electricity supplied to the farm sector in Punjab during year 2012–13 (PSERC-Tariff Order for FY 2013–14).

RESULTS AND DISCUSSION

The results have been discussed under following sub-heads:

1. Agro-socio-economic characters of the respondents
2. Cost-return structure of major crops
3. Diesel price hike coefficient and its impact on cost of production of major crops
4. Diesel/power subsidy withdrawal impact on cost of production of crops

Agro-socio-economic Characters of the Respondents

The agro-socio-economic characters of the sample respondents revealed that there were about 40 per cent adult males in the families of respondent farmers followed by nearly 35 per cent adult females and 25 per cent minors. The farm category-wise analysis revealed that relative proportion of adult males as compared to adult females was more among marginal, small, and semi-medium categories as compared other farm categories. About 38 per cent of the family heads among sample households were more than 50 years old while nearly 35 per cent were aged between 36 and 50 years and remaining 27 per cent were quite young and aged up to 35 years. As far as education level of the respondents was concerned, nearly 50 per cent of the respondents were educated up to matric level while remaining were educated up to secondary, primary, graduation, and post-graduation level. The operational holding size per farm was 5.62 hectares with land owned being 3.61 hectares; land leased-in at 2.13 hectares and land leased-out at 0.12 hectares. The entire area under cultivation was irrigated with average rental value of land leased-in being Rs 68,180 and that of land leased-out Rs 71,430 per hectare. Electric motor, submersible pump, diesel engine, generator, and canal were the various sources of irrigation on the sample farms. The number of submersible pumps per farm in an overall scenario were highest (0.76) followed by generator (0.27), electric motor (0.23), and diesel engine (0.17). The total income from farming, dairy farming, service, business, and other sources was Rs 8,01,364 per farm. The relative share of farming in the total income was 99.20 per cent followed by minor share from dairy farming, service sector, business, and other sources. The category-wise analysis revealed that the relative share of farming in total income was lowest on marginal farms and highest on the large farms.

The average number of tractors per farm was 0.79 with 43.94 H.P. and present value being Rs 1,94,106. The category-wise analysis revealed that the numbers of tractors per farm were more on large farms as compared to other farm categories and all the respondents on medium and large farms owned at least one tractor The average number of tractor drawn implements was 5.70 on large, 4.20 on medium, 3.70 on semi-medium, 1.32 on small, 0.18 on marginal, and 2.89 in an overall scenario. Therefore, the present value of tractor drawn implements was more on large farms due to their higher number as compared to other farm categories. Cropping pattern and cropping intensity on the sample farms revealed that during kharif season in overall scenario, paddy was the major crop occupying 2.32 hectares (18.83 per cent of gross cropped area) of the operational holding followed by Bt cotton, basmati, sugarcane, maize, guara, fodder, and vegetables. During rabi season wheat was the major crop sown on 4.58 hectares (37.20 per cent of the gross cropped area) followed by fodder, potato, and other minor crops. The crops sown during zaid season were; potato, sunflower, spring maize, vegetables, and mentha. The percentage share of zaid season crops was 8.51 per cent of the gross cropped area on the sample farms. In an overall scenario, the cropping intensity on the sample farms was 219.24 per cent with highest on large farms (222.81 per cent) and lowest on marginal (206.84 per cent) farms. The source of diesel purchase was asked from the respondent farmers and results of the study showed that about 95 per cent of the respondent farmers purchased diesel from the private petrol pumps and about 5 per cent from cooperative societies. The distance of the petrol pump from the farmer's house was within 5 km as reported by 95 per cent of the respondents. Mostly farmers purchased diesel seasonally (61.13 per cent) and one time purchase of diesel was about 50 litres as reported by about 50 per cent of the respondents.

Cost-return Structure of Major Crops

Paddy: The cost-return structure of paddy cultivation revealed that in an overall scenario, human labour use per hectare of paddy cultivation worked out to be 313.85 hours while category-wise analysis revealed that highest number of human labour hours were spent on medium farms followed by marginal farms which were lowest. In case of machine labour (tractor), overall 13.14 hours were utilized in various field operations while combine harvesting hours for paddy crop were estimated as 1.77 hours per hectare. The estimated irrigation hours using electric motor/submersible pump were 255.97 hours per hectare. Diesel engine use for irrigating paddy crop was more on marginal farms while generator use was more on large farms. Similarly, diesel consumption per hectare on owned and hired machinery taken together was 155.50 litres in paddy cultivation. In an overall scenario, the proportionate share of diesel consumption in total variable cost was 23.49 per cent while according to farm categories its share was lowest on small farms and highest on large farms. The returns over variable cost (ROVC) in paddy cultivation were about Rs 49,627 per hectare in an overall scenario while it was highest (Rs 50,702) on semi-medium farms and lowest (Rs 50,034) on medium farms.

Basmati: In case of basmati crop, 462.47 human labour hours were spent on various farm related operations per hectare while these were highest on small farms and lowest on medium farm category. The tractor use on sample farms was estimated at 16.22 hours per hectare while combine harvester use was 0.97 hours. The total irrigation hours using electric motor/submersible pump in basmati cultivation were 230.83 hours per hectare. Diesel engine and generator use for irrigating basmati crop were estimated to be 34.37 hours and 0.84 hours, respectively. In overall scenario, diesel consumption per hectare, including owned and hired machinery, was 123.41 litres. The category-wise analysis revealed that diesel consumption per hectare was highest on marginal (131.59 litres) farms and lowest on medium (111.34 litres) farms. The relative share of diesel consumption in total variable cost was worked out to be 17.26 per cent. The ROVC in basmati cultivation were about Rs 56,898 per hectare in an overall situation while it was highest (Rs 64,061) on small farms and lowest (Rs 55,725) on semi-medium farms.

Cotton: In cotton crop, 547.95 human labour hours were utilized for raising this crop on sample farms while according to farm category, highest number of labour hours were spent on large farms and lowest on the marginal farms. The tractor use was 25.15 hours per hectare for various farm operations in cotton cultivation. Total irrigation hours estimated for the use of electric motor/submersible pump were 27.03 while estimated diesel consumption on the sample farms in overall scenario was 118.64 litres. The relative share of diesel use was 11.70 per cent of total variable cost constituting 8.99 per cent from owned sources and 2.71 per cent from hired ones. The ROVC from cotton cultivation were Rs 24,883 per hectare with highest (Rs 30,048) on marginal farms and lowest (Rs 21,809) on large farms.

Sugarcane: In case of sugarcane crop, 1122.92 human labour hours per hectare were spent on undertaking various cultivation related farm operations in an overall situation. The tractor hours utilized for various farm operations were 22.23 per hectare. For irrigating sugarcane crop, 138.06 hours were spent while using electric motor/submersible pump. Also, 192 litres of diesel was consumed per hectare in sugarcane cultivation with more on medium farms and less on small farms. The relative share of diesel use in total variable cost was found to be 8.60 per cent with 7.91 per cent from owned sources and 0.69 per cent from hired machinery. The ROVC were estimated to be Rs 1,10,258 per hectare in sugarcane cultivation while according to farm category; it was Rs 1,18,775 on small farms, which were highest among farm categories and Rs 1,12,783 on marginal farms, which were lowest.

Maize: In maize crop, 311.84 human labour hours were utilized while raising the crop on the sample farms. The tractor use for various farm related operations was 11.19 hours while combine harvester use for harvesting maize crop was 1.29 hours. The irrigation hours estimated for the use of electric motor/submersible pump were 33.88 hours. The total diesel consumption in maize cultivation was 125.83 litres constituting 94.41 litres from owned sources and 31.42 litres from hired machinery. In overall scenario, the relative share of diesel use was 19.39 per cent of total variable cost. The ROVC were Rs 21,019 per hectare for maize cultivation while farm category-wise analysis revealed that returns were highest (Rs 23,869) on semi-medium farms and lowest (Rs 16,372) on medium farms.

Wheat: In case of wheat crop, 124.53 human labour hours per hectare were utilized to raise this crop on the sample farms. The tractor use was 20.13 hours for undertaking various field operations while for harvesting, using combine harvester, 2.33 hours per hectare were estimated to have been utilized. In an overall scenario, the irrigation hours worked out on the basis of electric motor/submersible pump use were 54.14 hours. The total diesel consumption in wheat crop on the sample farms was 144.13 litres. The relative share of diesel use in total variable cost was 26.28 per cent with 13.70 per cent share from owned machinery and 12.58 per cent from hired machine use. In overall scenario, the ROVC in wheat cultivation were estimated to be Rs 39,806 per hectare and according to farm size category; it was about Rs 40,330 on large farms, which was highest among farm categories, and Rs 35,422 on marginal farms, which was lowest.

Sunflower: In overall scenario, 218.48 human labour hours per hectare were estimated to have been used in sunflower cultivation on the sample farms. The tractor use in various farm operations worked out to be 18.95 hours. Also, combine harvester was used to harvest sunflower crop and 2.5 hours per hectare were spent on this particular operation. For irrigating sunflower crop also, electric motor and submersible pumps were used and in overall situation, 82.36 hours were spent while irrigating sunflower crop. The total consumption of diesel for various farm operations in sunflower crop was estimated to be 175.73 litres. The share of diesel use in total variable cost in sunflower cultivation was 29.32 per cent. The ROVC in sunflower cultivation worked out to be Rs 16,322 per hectare while on medium farms the returns were Rs 15,597 and Rs 17,868 per hectare on large farms.

Diesel Price Hike Coefficient and Its Impact on Cost of Production of Major Crops

The changing diesel prices and its impact on cost of production of paddy, basmati, cotton, sugarcane, maize, wheat, and sunflower was seen at I (base) level (1 February 2013), secondly at II (current) level (1 June 2014), and thirdly at III (proposed) level with zero diesel subsidy. The cost of production of paddy increased from Rs 494 per quintal at I level to Rs 517 and Rs 528 per quintal at II and III level, respectively. The diesel price hike coefficient was worked out to be 2.53 showing that with one rupee increase in diesel price, the cost of production of paddy increased by 2.53 rupees. In basmati crop, cost of production increased from Rs 890 per quintal at I level to Rs 921 at II and further Rs 936 at III level and the diesel price hike coefficient was worked out at 3.36. In cotton crop, the cost of production was estimated as Rs 2,859 per quintal at I level which increased to Rs 2,967 at II level and Rs 2,959 at III level if there was complete withdrawal of diesel subsidy. The diesel price hike coefficient was worked out at 7.31 in case of cotton crop. The cost of production of sugarcane worked out to be Rs 125 per quintal at I level which increased to Rs 127 at II level and further Rs 128 per quintal at III level. The diesel price hike coefficient was estimated to be 0.24 showing that with one rupee increase in diesel price, the cost of production of sugarcane increased by 0.24 rupees.

Similarly, the cost of production of maize was Rs 714 at I level which increased to Rs 742 and Rs 756 at II and III level, respectively with the total abolition of diesel subsidy and also, diesel price hike coefficient was worked out to be 3.03 for maize crop. In case of wheat crop, the cost of production increased from Rs 572 per quintal at I level to Rs 603 at II level and further Rs 618 at III level. The diesel price hike coefficient for wheat crop was worked out at 3.29 showing that with one rupee increase in diesel price the cost of production of wheat increased by 3.29 rupees. Lastly, the cost of production of sunflower worked out to be Rs 1,631 per quintal at I level which increased to Rs 1,729 and Rs 1,775 at II and III level, respectively while the diesel price hike coefficient was worked out to be 10.45.

Diesel/Power Subsidy Withdrawal Impact on Cost of Production of Crops

Diesel: It was found out that due to withdrawal of diesel subsidy (Table 54.1), the cost of production of sunflower increased by 8.81 per cent followed by 7.90 per cent in case of wheat, 7.06 per cent in paddy, 5.83 per cent in maize, 5.18 per cent in basmati, 3.52 per cent in cotton, and 2.63 per cent in sugarcane. However, the increase in cost of production of different crops according to various farm size categories due to withdrawal of diesel subsidy did not show any specific trend of increase or decline.

Power: The major impact of power subsidy withdrawal (Table 54.2) was seen on increase in cost of production of paddy (25.30 per cent) due to more number of irrigations applied to this crop followed by basmati (21.24 per cent), sunflower (9.07 per cent), wheat (6.64 per cent), maize (3.50 per cent), sugarcane (2.63 per cent), and cotton (1.75 per cent). The impact of power subsidy withdrawal was more on semi-medium, medium, and large farm categories as compared to marginal and small farms.

Aggregate impact of diesel and power subsidy: Further, the impact of both diesel and power subsidy withdrawal (Table 54.3) on cost of production of major crops in Punjab was dwelled upon and it was found that the cost of production of paddy increased by 32.35 per cent due to withdrawal of both diesel and power subsidy. The increase in cost of production of basmati was by 26.42 per cent followed by sunflower (17.88 per cent), wheat (14.55 per cent), maize (9.33 per cent), sugarcane (5.97 per cent), and cotton (5.26 per cent). The farm category-wise analysis revealed that increase in cost of production of major crops was more pronounced on semi-medium, medium, and large categories as compared to marginal and small farms. Thus, the impact of power and diesel subsidy withdrawal was estimated to have been more on large and medium farmers as compared to marginal and small farmers.

CONCLUSIONS AND POLICY IMPLICATIONS

The above discussion brings us to the conclusion regarding change in cost of cultivation/production of important crops in Punjab which was worked out after taking into account the impact of withdrawal of diesel and power subsidies with emphasis on bringing out specific policy implications

Table 54.1 Impact of Diesel Subsidy Withdrawal on Cost of Production, Important Crops, Punjab, 2012–13 (Rs/qtl)

Crops	Marginal	Small	Semi Medium	Medium	Large	Overall	Marginal	Small	Semi Medium	Medium	Large	Overall
	With Subsidy						Without Subsidy					
Paddy	505	490	490	505	489	494	537 (6.22)	530 (8.18)	526 (7.41)	539 (6.76)	529 (8.37)	528 (7.06)*
Basmati	895	853	871	802	882	890	944 (5.46)	897 (5.16)	916 (5.22)	839 (4.61)	923 (4.57)	936 (5.18)
Cotton	2728	2756	2740	2921	3056	2859	2826 (3.58)	2846 (3.24)	2836 (3.50)	3030 (3.75)	3168 (3.65)	2959 (3.52)
Sugarcane	120	118	119	123	121	125	124 (2.79)	121 (2.73)	122 (2.78)	126 (2.78)	124 (2.65)	128 (2.63)
Maize	725	702	658	799	697	714	744 (2.64)	749 (6.68)	694 (5.55)	847 (5.96)	739 (6.04)	756 (5.83)
Wheat	623	568	551	561	538	572	665 (6.73)	610 (7.43)	596 (8.23)	605 (7.83)	583 (8.27)	618 (7.90)
Sunflower	–	–	–	1664	1560	1631	–	–	–	1810 (8.74)	1703 (9.21)	1775 (8.81)

Note: *Figures in parentheses indicate per cent increase in cost of production due to withdrawal of diesel subsidy.
Source: Field survey.

Table 54.2: Impact of Power Subsidy Withdrawal on Cost of Production, Important Crops, Punjab, 2012–13 (Rs/qtl)

Crops	Marginal	Small	Semi Medium	Medium	Large	Overall	Marginal	Small	Semi Medium	Medium	Large	Overall
	With Subsidy						Without Subsidy					
Paddy	505	490	490	505	489	494	606 (20.00)	606 (23.67)	618 (26.12)	630 (24.75)	618 (26.38)	619* (25.30)
Basmati	895	853	871	802	882	890	1032 (15.31)	1034 (21.22)	1074 (23.31)	1001 (24.81)	1097 (24.38)	1079 (21.24)
Cotton	2728	2756	2740	2921	3056	2859	2765 (1.36)	2804 (1.74)	2790 (1.82)	2973 (1.78)	3115 (1.93)	2909 (1.75)
Sugarcane	120	118	119	123	121	125	124 (3.33)	122 (3.39)	124 (4.20)	128 (4.07)	126 (4.13)	126 (3.28)
Maize	725	702	658	799	697	714	744 (2.62)	724 (3.13)	684 (3.95)	824 (3.13)	728 (4.45)	739 (3.50)
Wheat	623	568	551	561	538	572	655 (5.14)	604 (6.34)	588 (6.72)	601 (7.13)	577 (7.25)	610 (6.64)
Sunflower	–	–	–	1664	1560	1631	–	–	–	1818 (9.25)	1708 (9.49)	1779 (9.07)

Note: *Figures in parentheses indicate per cent increase in cost of production due to withdrawal of power subsidy.
Source: Field survey.

Table 54.3 Impact of Diesel and Power Subsidy Withdrawal on Cost of Production, Important Crops, Punjab, 2012–13 (Rs/qtl)

Crops	Marginal	Small	Semi Medium	Medium	Large	Overall	Marginal	Small	Semi Medium	Medium	Large	Overall
	With Diesel and Power Subsidy						Without Diesel and Power Subsidy					
Paddy	505	490	490	505	489	494	637 (26.23)	646 (31.85)	654 (33.53)	664 (31.51)	659 (34.75)	654 (32.35)*
Basmati	895	853	871	802	882	890	1081 (20.77)	1078 (26.38)	1119 (28.53)	1038 (29.42)	1137 (28.95)	1125 (26.42)
Cotton	2728	2756	2740	2921	3056	2859	2863 (4.94)	2893 (4.98)	2886 (5.33)	3083 (5.53)	3226 (5.58)	3009 (5.26)
Sugarcane	120	118	119	123	121	125	127 (6.13)	125 (6.11)	127 (6.98)	131 (6.84)	129 (6.78)	129 (5.97)
Maize	725	702	658	799	697	714	763 (5.26)	771 (9.82)	721 (9.50)	872 (9.09)	770 (10.49)	781 (9.33)
Wheat	623	568	551	561	538	572	697 (11.87)	646 (13.77)	633 (14.94)	645 (14.95)	621 (15.52)	655 (14.55)
Sunflower	–	–	–	1664	1560	1631	–	–	–	1964 (18.00)	1852 (18.70)	1923 (17.88)

Note: *Figures in parentheses indicate per cent increase in cost of production due to withdrawal of diesel and power subsidy.
Source: Field survey.

affecting the state agriculture due to withdrawal of diesel and power subsidy. It was found, after undertaking various simulations regarding change in diesel prices keeping prices of all other inputs at constant level (cetris-paribus), that the cost of production of paddy increased by 7.06 per cent with the withdrawal of diesel subsidy. Similarly, the increase in cost of production of basmati was by 5.18 per cent; while in other crops it was 3.52 per cent in cotton, 2.63 per cent in sugarcane, 5.83 per cent in maize, 7.90 per cent in wheat, and 8.81 per cent in sunflower. The diesel price hike coefficient showed that with one rupee increase in diesel price, the resultant cost of production of paddy increased by Rs 2.53; while in other crops such as basmati, increase in cost of production was by Rs 3.36, in cotton by Rs 7.31, in sugarcane by Rs 0.24, in maize by Rs 3.03, in wheat by Rs 3.29, and in case of sunflower by Rs 10.45.

The increase in cost of production of different crops under various farm categories due to withdrawal of diesel subsidy did not show any specific trend of increase or decline according to size of the farm category. The major impact of power subsidy withdrawal was seen on increase in cost of production of paddy (25.30 per cent) crop due to more number of irrigations applied to this crop followed by basmati (21.24 per cent), sunflower (9.07 per cent), wheat (6.64 per cent), maize (3.50 per cent), sugarcane (3.28 per cent), and cotton (1.75 per cent). The impact of power subsidy withdrawal was more on semi-medium, medium, and large farm categories as compared to marginal and small farms. In aggregate, the cost of production of paddy increased by 32.35 per cent due to withdrawal of both diesel and power subsidy. Similarly, the increase in cost of production of basmati was by 25.42 per cent followed by sunflower (17.88 per cent), wheat (14.55 per cent), maize (9.33 per cent), sugarcane (5.97 per cent), and cotton (5.26 per cent). The farm category-wise analysis showed that the impact of power and diesel subsidy withdrawal would be more on large and medium farmers as compared to their smaller counterparts.

The major policy issues drawn from the discussion was that Punjab government should emphasize the union government to increase the minimum support price (MSP) of paddy and wheat, which are the crops for which state farmers get assured price, in commensurate with the diesel price hike coefficient. For other crops also, MSP should be enhanced in proportionate to the diesel price hike coefficient, for which MSP is announced but is not actually implemented. In case, power subsidy is withdrawn by the state government, farmers, especially marginal and small one's, should be compensated according to the electricity usage bill generated for irrigating various crops on their farms. Thus, for keeping marginal and small farmers in farming business, subsidies, especially power subsidy, should not be withdrawn, however, their form can be changed for the benefit of these farmers in general and farming community in particular.

REFERENCES

Anand, Rahul, David Coady, Adil Mohommad, Vimal Thakoor, and James P. Walsh. 2013. 'The Fiscal and Welfare Impacts of Reforming Fuel Subsidies in India'. IMF Working Paper, Asia and Pacific Department, WP 13/128, International Monetary Fund, Washington, DC.

Anonymous. 2012. Agriculture at a glance, Information Serivce, Department of Agriculture, Punjab, Chandigarh.

———. 2013 PSERC-Tariff Order PSPCL, Punjab State Electricity Regulatory Commission, Chandigarh.

———. 2014. "Losses on Sale of Diesel Dips to New Low," *Daily Post*, New Delhi, 2 June 2014.

Arze del Granado, J., D. Coady and R. Gillingham. 2012. "The Unequal Benefits of Fuel Subsidies: A Review of Evidence for Developing Countries." *World Development* 40(11): 2234–48.

International Monetary Fund. 2013. *Energy Subsidy Reform: Lessons and Implications*. Washington DC.

55

Impact of MGNREGA on Wage Rate, Food Security, and Rural Urban Migration

Parmod Kumar

National Rural Employment Guarantee Act, now the Mahatma Gandhi National Rural Employment Guarantee Act (MGNREGA) from 2 October 2009, was passed in the year 2005. The basic objective of the Act is to ensure livelihood and food security by providing unskilled work to people through creation of sustainable assets. Under the provisions of the Act, the state has to ensure enhancement of livelihood security to the households in rural areas by providing at least one hundred days of guaranteed wage employment to every household whose adult members volunteer to do unskilled work. In-built with various transparency and accountability measures and provisions for social audits, this Act, for the first time, brings the role of the state as provider of livelihood. The MGNREGA Scheme has high potential in terms of employment generation, alleviation of poverty, food security, halting migration, and overall rural development. Based on this background this study is conceptualized with the following objectives:

1. to measure the extent of manpower employment generated under MGNREGA, their various socio-economic characteristics, and gender variability in implementing MGNREGA since its inception in the selected states;
2. to compare wage differentials between MGNREGA activities and other wage employment activities;
3. to understand the effect of MGNREGA on the pattern of migration from rural to urban areas;
4. to find out the nature of assets created under MGNREGA and their durability;
5. to identify factors determining the participation of people in MGNREGA scheme and whether MGNREGA has been successful in ensuring better food security to the beneficiaries; and
6. to assess the implementation of MGNREGA, its functioning, and to suggest suitable policy measures to further strengthen the programme.

METHODOLOGY AND DATA

The study is based on both primary and secondary data. Primary data was collected from the selected villages and households in 16 states. From each selected state, five districts were selected, one each from the north, south, east, west, and central locations of the state. From each districts, two villages were selected, keeping into account their distance from the location of the district or the main city/town. From each selected village, primary survey was carried out on twenty participants in MGNREGA and five nonparticipants working as wage employed. In this fashion, from each state, ten villages were selected and a total number of 250 households were surveyed in detail with the help of structured household questionnaire. In this way around 200 participants and 50 non-participants were selected from each state and data was collected in 16 states. The total sample consists of 3,166 participants and 839 non-participants. The selected states were: Karnataka, Andhra Pradesh, and Kerala in the south; Himachal Pradesh, Uttar Pradesh, Haryana, and Punjab in the north; Madhya Pradesh and Chhattisgarh in the central; Maharashtra, Gujarat, and Rajasthan in the west; Bihar and West Bengal in the east; and Sikkim and Assam in the northeast. The data was collected through structured questionnaires. The data pertain to the reference period of January to December 2009.

In addition to household questionnaire, a village schedule was also designed to capture the general changes that have taken place in the village during the last one decade and to take note of increase in labour charges for agricultural operations after the implementation of MGNREGA. The village schedule also has qualitative questions related to change in lifestyle of the villagers taking place during the last one decade. One village schedule in each village was filled up with the help of a 'Group Discussion' with the Pachayat Members, Officials, educated and other well informed people available in the village being surveyed.

RESULTS AND DISCUSSION

Total Employment Generated and Its Socio-Economic Characteristics

In the three phases of MGNREGA implementation in India from 2006–07 to 2013–14 (up to October), 81 crore households were issued job cards at the country as a whole, out of which around 34 crore households were provided employment averaging around 4.5 crore households working in MGNREGA per annum, that constitutes roughly around 30 per cent of the rural households in the country as a whole. Andhra Pradesh, Uttar Pradesh, and Rajasthan each employed more than 3 crore households during this period. A total number of 1.5 thousand crore man days of employment was generated by MGNREGA during the above mentioned time period. The share of Scheduled Castes (SCs) and Scheduled Tribes (STs) in the total person days generated was 26.9 and 22.0 per cent, respectively while share of women in the total employment was 48.0 per cent.

At the aggregate, a total number of 45 person days of employment was provided by MGNREGA whereas the target set under the programme is 100 days of employment per household. Highest number of 54 days of employment, that is slightly above 50 per cent of the target, was achieved only in the year 2009–10. Among the states, highest numbers of days of employment (60 to 70 days) was provided by the northeastern states of Mizoram, Nagaland, Tripura, Sikkim, and Manipur. Rajasthan, Madhya Pradesh, and Andhra Pradesh provided between 50 to 60 days of employment. The other states like Chhattisgarh, Himachal Pradesh, Tamil Nadu, Karnataka, Maharashtra, Uttar Pradesh, Jharkhand, and Odisha provided 40 to 50 days of employment while Haryana, Jammu and Kashmir, Uttarakhand, Gujarat, Kerala, and Assam provided 30 to 40 days of employment. The states that lied at the bottom included Bihar (31 days), Arunachal Pradesh, West Bengal, and Punjab (28 days each), and Goa, only 25 days of employment.

Out of the total 34 crore households working in MGNREGA during its full tenure, only 2.9 crore households completed 100 days of employment. Around 25 per cent households working in MGNREGA completed 100 days in Mizoram, 20 per cent in Tripura, 18 per cent in Sikkim and Nagaland each, 16 percent in Rajasthan, and 14 per cent in Manipur. Tamil Nadu and Andhra Pradesh were the other states where around 10 to 13 per cent households completed 100 days of employment. Goa, Punjab, and West Bengal were at the bottom where only less than 2 per cent households completed 100 days of employment. At the all India aggregate, only 8.4 per cent households completed 100 days of employment during the entire period of MGNREGA in operation up till October 2013.

Number of Projects Completed and Total Amount Spent

Water conservation was the leading activity which occupied around 24 per cent projects under MGNREGA, followed by rural connectivity projects: 17 per cent, provision of irrigation: 14 per cent, drought proofing: 13 per cent, land development: 10 per cent each, renovation of traditional water bodies, and micro irrigation: 6 per cent, and flood control: 3 per cent. During the entire period of MGNREGA, a total number of 1 crore projects were completed and around 2.9 crore were ongoing. Thus, out of total 4 crore projects taken up under MGNREGA, around 30 per cent were completed and rest of 70 per cent were in progress. A total amount of Rs 2,35,084 crore was spent on the MGNREGA with an average of slightly less than Rs 30 thousand crore every year. Working out the total expenditure incurred per project, it turns out around Rs 59 thousand per project for all MGNREGA works undertaken so far at the aggregate.

During the whole period of implementation of MGNREGA a total amount of Rs 75 thousand crore was spent on rural connectivity, Rs 45 thousand crore on water conservation, Rs 27 and Rs 25 thousand crore on renovation of traditional water bodies and drought proofing, respectively, Rs 17 thousand crore on provision of irrigation, Rs 16 thousand crore on land development, Rs 12 thousand crore on micro irrigation, Rs 11 thousand crore on flood control, and around Rs 6 thousand crore on other activities. At the aggregate, the highest amount per project was spent on renovation of traditional water bodies, Rs 121 thousand per project that was closely followed by Rs 112 thousand per project on rural connectivity. Expenditure on flood control lied on the third place with an expenditure of Rs 79 thousand per project. Micro irrigation had a spending of Rs 53 thousand per project, followed by drought proofing, Rs 49 thousand per project, water conservation, Rs 47 thousand per project, land development, Rs 40 thousand per project, and provision of irrigation, Rs 29 thousand per project. Thus, even though water conservation topped in the total numbers of projects undertaken, but spending on per project was much less on water conservation compared to rural connectivity that topped among all projects, not only in the total amount spent but also amount spent per project. State-wise highest amount per project was spent in Manipur, Rs 297 thousand, followed by Nagaland (Rs 245 thousand), Mizoram (Rs 269 thousand), Tamil Nadu (Rs 255 thousand), Assam (Rs 191 thousand), and Maharashtra (Rs 160 thousand). The states that lied at the bottom in spending per project were Andhra Pradesh (Rs 18 thousand), Gujarat (Rs 41 thousand), Karnataka and Goa (Rs 48 thousand), Kerala (Rs 49 thousand), and Uttar Pradesh (Rs 54 thousand).

Qualitative Indicators of MGNREGA Performance

During 2008–09 to 2013–14 (up to October), a total number of 10.52 crore muster rolls were opened in the country, out of which around 85 per cent were verified by the authorities who carried out the auditing work. Social auditing of MGNREGA work of the Gram Panchayats (GP) was held in around 87 per cent of the GPs during the above mentioned period. The social audit was held in above 90 per cent GPs in Tamil Nadu, Madhya Pradesh, Kerala, and Nagaland, whereas it was held in less than 60 per cent GPs in Arunachal Pradesh, around 60 to 65 per cent GPs in Jammu and Kashmir and Karnataka. The percentage of works inspected at the district level was very low, only 12 per cent, whereas the works inspected at the block level was as high as 81 per cent. Almost half of the works were inspected at the district level in Arunachal Pradesh while proportion of inspected works was half to 1/3rd in Assam, Sikkim, Nagaland, and Kerala. In rest of the states, less than 1/3rd works were being inspected at the district level. Complaint redressal system was adopted under MGNREGA and a total number of 2,15,542 complaints were registered in all the states, out of which around 84 per cent were redressed. Complaint redressal was 100 per cent in Goa, Arunachal Pradesh, and Mizoram. It was less than 80 per cent in Madhya Pradesh, Maharashtra, Odisha, West Bengal, and Gujarat, while in rest of the states, above 80 per cent complaints were redressed during the above mentioned period. According to the legislation on MGNREGA, if a member of a household has not been provided employment after issuing him/her a job card after a lapse of fifteen days, the GPs are supposed to provide unemployment allowance and such amount would be borne by the concerned state government. During the period 2007–08 to 2013–14 (up to October), unemployment allowance was due for 4.83 crore person days for which employment was not provided to the job card holders but only 2,478 days of allowance was paid, that makes only 0.01 per cent days of unemployment allowance paid and it was not more than 0.04 per cent in any state.

MAIN FINDINGS—PRIMARY SURVEY

Household Characteristics, their Income and Consumption Pattern

The average household size was 4.75 with participants having average family size of 4.7 and non-participants, 4.9. The average numbers of earners in the family were 2.2 members among participating families and 2.6 members among the non-participating families. Similarly, the number of members in working age (that is, 16 to 60 years) was 74.4 per cent among participants and 73.7 per cent among non-participants. Looking at the education status among the selected households, the percentage of illiterate was around 1/3rd among the participants and less than 1/3rd among the non-participants. On the overall, non-participants were better educated compared to participant household members. Looking at the caste distribution among the participating households, the percentage of households belonging to SC, ST, and Other Backward Castes (OBC) was 34, 17, and 34 per cent, respectively while General category had only 16 per cent proportion among the selected households.

The trends in occupation depict that among the participating households, the proportion of work provided by MGNREGA was only a small proportion of their aggregate employment. Out of the total man days employed per household including all the working members, the share of MGNREGA varied between 12 to 32 per cent among different states. It was less than 15 per cent in Karnataka, Kerala, Assam, Gujarat, and West Bengal. Its proportion was between 15 to 25 per cent in Uttar Pradesh, Sikkim, Madhya Pradesh, Chhattisgarh, Maharashtra, Himachal Pradesh, Rajasthan, Haryana, and Punjab. The share of MGNREGA in total employment was above 25 per cent only in two states namely Bihar and Andhra Pradesh. At the aggregate, MGNREGA provided 18 per cent share in the total employment among our selected households. Casual labour in agriculture and non-agriculture sector constituted more than 40 per cent share in employment. Self-employment in agriculture and livestock constituted around 20 per cent share and self-employment in business and regular salary had around 5 and 10 per cent share, respectively in the total employment among the selected participants.

A glance on the household income statistics reveals that the estimated per household income of non-participant households was higher compared to participant households. On an average, the selected non-participant households earned Rs 70 thousand per annum compared to Rs 59 thousand earned by the participating households. Comparing the sources of income across different activities, wage income constituted a lion's share in the income of both participating as well as non-participating households. Earnings from agricultural wages contributed around 17 per cent, followed by wage earnings from non-agricultural activities, 22 per cent, while wage earnings in MGNREGA activities contributed only 12 per cent share in the total household income of participants. In addition to wage earnings, income from self-employment in agriculture and livestock constituted around 17 per cent share of their household income while regular salaried job contributed around 14 per cent share in the household income of the participating households. Trends in share of various sources were somewhat similar in the case of non-participating households.

Work Profile under MGNREGA, Wage Structure, and Migration Issues

According to our survey data, on an average, less than two members (1.7) per family were employed under MGNREGA. Among the selected states, the average exceeded two members per family working in MGNREGA in Sikkim, Gujarat, Chhattisgarh, and Andhra Pradesh. It was between 1.5 and two members in Karnataka, Haryana, Madhya Pradesh, Maharashtra, and West Bengal. The states that employed less than 1.5 members per family were Uttar Pradesh, Bihar, Kerala, Assam, Punjab, Himachal Pradesh, and Rajasthan. The highest numbers of members employed under MGNREGA among the selected households was found 2.8 members in Sikkim and lowest, 1.07, in Kerala. Out of 1.68 members employed under MGNREGA at the aggregate, 0.98 members belonged to male gender and 0.70 members belonged to female gender. Only in Gujarat and Rajasthan, the numbers of female member per household working in MGNREGA exceeded that of male and in Sikkim and Maharashtra their percentage was same. Against the average size of 1.68 members per family at the aggregate, the average size was 1.47 for the SCs, 1.67 for STs, and 1.53 for the OBCs. The SC and ST households' average size was highest, 2.63 and 2.53 members in Gujarat and lowest, 0.22 and 0.19 members in Bihar, respectively.

On an average, 68 days per household employment was generated among our selected participants. The states that topped in employment generation among our selected participants included Maharashtra (100 days), Haryana (94 days), Himachal Pradesh (92 days), and Rajasthan, Sikkim, and Gujarat (slightly above 80 days). The states that were slightly above or below the national average were Madhya Pradesh, Karnataka, Kerala, and Uttar Pradesh (between 80 to 60 days). The states that lied at the bottom were Bihar (32 days), Andhra Pradesh (43 days), and Assam (48 days). Looking at the ratio of employment among the male and female workers, numbers of days of employment was shared by male (37 days) and female (30 days) with a per cent share of 56 for male and 44 for female.

Out of 16 states for which analysis is done, only in 10 states information about households completing 100 days of employment was available. Among these ten states, the percentage of households that completed 100 days, their percentage in Himachal Pradesh was exceptionally high (85 per cent). In Haryana and Rajasthan, 48.5 and 44.5 per cent households completed 100 days. In Karnataka and Sikkim around 1/4th of the participant households completed 100 days of employment. In Bihar, Assam, Gujarat, and West Bengal less than 5 per cent households completed 100 days and in Uttar Pradesh, around 10 per cent households completed 100 days. At the aggregate, only 1/4th of the selected participants in these ten states completed 100 days. In other words, MGNREGA was not quite successful in providing social security to the households as households had to depend on other activities for earning their livelihood as MGNREGA provided only 18 per cent share of the total employment to the selected households.

Looking at the wage rate on which employment was provided, average wage rate at the aggregate was recorded at Rs 100 and it was not particularly different among male and female. The highest wage was recorded in Haryana (Rs 150), followed by Kerala (Rs 125), Punjab (Rs 123), and Himachal (Rs 110). Among the selected states lowest wage rate was paid in Rajasthan (Rs 80), Chhattisgarh (Rs 83), West Bengal (Rs 84), and Karnataka (Rs 86). However, in most of the states actual wage rate obtained under MGNREGA was below the stipulated minimum wage rate fixed by the states under the Minimum Wages Act 1948. The difference between the actual payment and minimum stipulated wages was specifically high in Karnataka (Rs 33), Maharashtra (Rs 22), Rajasthan and Assam (Rs 21), Madhya Pradesh (Rs 19), Andhra Pradesh and Punjab (Rs 14), Gujarat and Haryana (Rs 12), and Bihar (Rs 10). Last but not the least, the average distance of workplace form the residence or village of the households was less than 2 kms in all the states with few exceptions.

Among the surveyed households, the highest work under MGNREGA was concentrated on rural connectivity which shared around 40 per cent of the total employment followed by water conservation and water harvesting which shared 17 per cent of employment under MGNREGA. Land development (12 per cent), renovation of traditional water bodies (11 per cent), flood control and protection (8 per cent), and micro irrigation (5 per cent) were the other major activities of employment under MGNREGA. On the question of how was the quality of the assets created through MGNREGA work, a little less than half of the households indicated that the assets created were very good while another half of them indicated that assets created were of the good quality. Only less than 3 per cent households pointed out that the assets created were bad or worst in quality. We enquired the selected households whether after registration, if they did not get employment, did they receive any unemployment allowance; households indicated that they did not receive any such allowance except in Maharashtra and West Bengal where households received only a poultry amount as unemployment allowance.

Our statistics on migration indicates that around 0.20 members per family (with average size of 4.7 members) migrated because of not getting work under MGNREGA. Out of the selected states, the numbers of per family members migrated because of not getting work averaged at 0.54 in Assam, 0.44 in Rajasthan, 0.31 in Madhya Pradesh and Maharashtra each, 0.20 in Andhra Pradesh, Chhattisgarh, and Himachal Pradesh, each, and less than 0.1 members in rest of the selected states. Thus, incidences of villagers' migration in search of work despite having been registered for MGNREGA were still recorded in the surveyed villages. However, there were also incidences whereby around 0.12 members per family among the participant households returned back to the village to work under MGNREGA at the aggregate who hitherto were working elsewhere before the implementation of this programme. The members retuning back to work under MGNREGA was highest in the state of Bihar where around 0.65 members per family returned back to work under MGNREGA after the implementation of the Act. Among other states, the incidence was recorded in Andhra Pradesh, Rajasthan, Madhya Pradesh, and Maharashtra where, on an average, 0.1 to 0.2 members per family returned back to work in MGNREGA after implementation of the Act. Punjab, Haryana, and Assam were the only states where no such reverse migration incidences were recorded. On the overall, it is difficult to say whether the MGNREGA programme has been successful in cutting down the incidences of labour migration from villages in search of job. The majority of the households who returned back to work in MGNREGA pointed out that they were now better off compared to earlier working as a migrant labourer.

DETERMINANTS OF PARTICIPATION IN MGNREGA

The logit function provided us the probabilities of the participation of a household in MGNREGA activities. State-level regression results showed that the households who had alternate employment opportunities and those who had higher income contribution from other activities had less incentive to work in MGNREGA. The coefficient for employment other than MGNREGA was negative and significant in Sikkim, Haryana, Madhya Pradesh, and Chhattisgarh. Coefficient of income other than MGNREGA was significant and negative in Madhya Pradesh, Chhattisgarh, Punjab, Maharashtra, and Himachal Pradesh. The household size had significant and positive sign in Karnataka, Andhra Pradesh, Kerala, Haryana, Madhya Pradesh, Chhattisgarh, Punjab, Maharashtra, and West Bengal indicating with increase in family size there was more probability of household members working in MGNREGA among the selected households. Household size had significant but negative relationship in Uttar Pradesh and Himachal Pradesh, indicating low participation at higher family size in these two states.

The value of assets and land ownership had negative sign in the regression, indicating household members with land ownership or better assets accumulation had less probability of participating in MGNREGA activities. The coefficient was significant with a negative sign in Karnataka, Uttar Pradesh, Sikkim, Madhya Pradesh, Assam, Punjab, and West Bengal. On the opposite, if a household owned an Antodaya Anna Yojna (AAY) or Below Poverty Line (BPL) card or if they belonged to SC or ST community they had higher possibility of entering into MGNREGA work. The coefficient of dummy BPL was found positive and significant in Karnataka, Sikkim, and in Haryana. Similarly, coefficient of social characteristics (household belonging to SC, ST, and OBC) was found significant and positive in Sikkim, Andhra Pradesh, Chhattisgarh, and Maharashtra. From the household OLS regression (ordinary least squares), the most important and significant variable emerged was wage rate in MGNREGA with a positive sign in almost all the states indicating that with higher wage rate households preferred to work in MGNREGA.

Some interesting relations were observed in the member level logit regression. Among the members in a household, those who worked in MGNREGA had a direct and significant relationship with age and negative relationship with education. The implication is that older age and less educated people preferred to work in MGNREGA as the latter is known providing soft wages. Similarly, the dummy on sex indicates that the male members had higher probability of working in MGNREGA compared to female members, although female proportion in total work force constituted around 45 per cent, varying in its degree from state to state. The members with BPL and AAY cards and members belonging to SC and ST community showed higher probability of getting employed in MGNREGA. The above findings were generally true across the states.

THE FUNCTIONING OF MGNREGA—QUALITATIVE ASPECTS—(FIELD SURVEY)

On the qualitative questions, a majority of the households indicated that they did not have to pay any bribe to get a job card issued. Regarding irregularities in the job card, around 15 per cent households at the aggregate indicated that either no entry was made in the job card about the work performed under MGNREGA or entries were missing or fake; entries were overwritten or signature column was blank, while clear cut majority observed no such irregularities. Around 80 per cent of the household were given employment in response to their application for work. All households who did not get work within 15 days indicated that they did not get any unemployment allowances in lieu of not getting work within the period of 15 days after putting up their application for work under MGNREGA.

On the system of payment of wages, almost all participating households agreed that wage rate for male and female was same. The payment system was both daily-wage basis and piece rate/task wage basis. In majority of cases, work was measured on collective or team management basis while in a thin majority, it was measured on individual work basis. A majority of participant households pointed out that wages were paid either fortnightly or monthly basis but around 12 per cent participants pointed out that they had to wait for a longer period or at least more than a month to realize their wages from MGNREGA work. Among the irregularities in wage payments, the participant households indicated that there was delay in wage payments after the work was finished; the wage paid was less than the task performed and the participants faced problem in accessing post office or bank account; and lastly, they were not aware on what basis wages were determined in case of those to whom wages were not paid on daily wage basis. Delay in wage payment was reported by highest numbers of participants in Andhra Pradesh, Chhattisgarh, Madhya Pradesh, Gujarat, and Rajasthan.

Regarding information about the work to be performed and facilities available at the worksite, around 2/3rd majority of participants pointed out that they were given requisite details of the work to be performed. About the facilities available at the worksite, around 3/4th of the participants agreed that drinking water facility was provided at the worksite. About the facilities like shade for period of rest; childcare facilities; first-aid kit; and primary medicines available at the worksite, around 40 to 50 per cent participants replied that these facilities were not available on the worksite. Lack of drinking water, childcare, and medicine facility at the workplace was mostly reported by participants in Karnataka, Haryana, Madhya Pradesh, and Punjab.

On the monitoring of the MGNREGA functioning, more than 80 per cent participants indicated that the work was being monitored through some authority but majority of them did not know whether any auditing of the accounts take place or not. In Haryana around 80 per cent participants indicated that there was no monitoring taking place while 16 per cent expressed their unawareness and only 4 per cent participants indicated that monitoring of MGNREGA work was being held. In all other states more than 60 per cent participants indicated that the work was being monitored. Very few participants lodged any complaint, and even who indicated that they lodged a complaint, only 7 per cent of them said that their complaints were taken care of.

Some incidents of migration out of the village as well as migration back to the village (to work under MGNREGA) were cited, but the extent of the same was only miniscule, not leading to the conclusion that MGNREGA had any conclusive evidence of affecting labour migration into any particular direction. Some household members migrating out for job after implementation of MGNREGA among the selected states was observed comparatively higher in Bihar, Gujarat, Assam, Rajasthan, and Maharashtra. However, in

Bihar and Maharashtra the incidence of family members migrating back to village to work under MGNREGA was also found higher than the other states, indicating the reverse migration occurring along with the incidence of migration among the participant households. Regarding the question of villagers' awareness about 'Mahatma Gandhi National Rural Employment Guarantee Act' under implementation in the village, a clear 2/3rd majority of the respondents pointed out that people in the village were aware about the same. However, households were hardly aware about the provision of unemployment allowance under MGNREGA. Similarly, majority of the respondents were not aware about provision of the worksite facilities, mandatory availability of muster rolls at the worksite, and list of permissible works under the MGNREGA.

To understand how the MGNREGA programme has affected the general life of villagers, we enquired few questions related to participants' day-to-day life. Around 67 per cent participants were of the view that MGNREGA has enhanced food security of the villagers by providing them employment, and thus, purchasing power to have better access to food. Around 60 per cent participants pointed out that MGNREGA has given greater independence to women. Around 65 per cent agreed that MGNREGA provided protection against extreme poverty. On the migration issues, around 49 per cent indicated that MGNREGA has helped to reduce distress migration from the village to cities. Similarly, around 50 to 60 per cent pointed out that MGNREGA has reduced indebtedness by generating purchasing power at the local economy.

We further probed the food security issues among the participants. To our question: did your family get full two square meals throughout the reference year, around 24 per cent households answered in negative. If the households did not have sufficient food how did they cope up with the situation? Around 37 per cent affected households indicated that they borrowed from some sources to cope up with the situation. Around 13 per cent pointed out that they reduced the numbers of meals during the crisis period while others took other measures like catching fishes or rats and so on. The states where maximum number of households indicated not having two square meals among the selected states were the poor states of Assam and Bihar, while in the states of Haryana and Andhra Pradesh, no household reported not having sufficient meal during any month of the reference year.

MGNREGA IMPACT ON VILLAGE ECONOMY

The surveyed villages had mixed picture with some villages having perfect infrastructure like road, post office, bank, self-help group, school, primary health centre, Fair Price Shop (FPS), and so on, while others had to travel some distance to approach the same. During the last ten years there has been a slight change in the occupation structure in the selected villages. The prevailing wage rates in agriculture were fluctuating widely. Prevailing wage rate in non-agricultural sector were much higher compared to the agricultural sector and the level of skilled wages were almost double that of unskilled wages.

Comparing the prevailing wage rate in the selected villages over the last five years, that is, since the time MGNREGA has come into implementation, the wage rate in agriculture sector has increased by slightly less than 50 per cent for male and slightly above 50 per cent for the female. By the same estimates, wage rate for unskilled as well as skilled labour in the non-agricultural sector increased by slightly less amount compared to agriculture labour except the wage rate in mining during the same time period. The wage rate for unskilled labour in non-agriculture and construction work increased slightly less than the wage rate increase in agriculture while wage rate for skilled labour in mining increased slightly more than agriculture. The wage rate for technical work like electrician, plumber, and pump set boring increased by less than that of agriculture (between 35 to 47 per cent). Thus, increase in wage rate in agriculture more than most of the other activities within the village indicate the enhanced demand for wage labourers due to employment works in MGNREGA that goes parallel with the agriculture sector, thereby causing a competition in the labour market for the agriculture sector. Increases in charges for agricultural operations per acre on an average were almost similar to increase in agricultural wages as overall wages observed an increase of around 49 per cent compared to around 46 per cent increase in cost of per acre agricultural operations as per our group discussion data.

A majority of the villages indicated that shortage of agricultural labour has increased after the implementation of MGNREGA. In majority of the villages the shortage of labour was observed during the sowing and harvesting months of kharif and rabi seasons, especially in the months of July, August, and September and March and April. This was more so after the implementation of MGNREGA. A majority of villagers were of the view that after MGNREGA implementation cost of production in agriculture has increased by 10 to 20 per cent because of scarcity of labour.

On the question, whether workers who earlier migrated out of the village to work in city are now coming back to work in MGNREGA, the trend of villagers returning back to the village to work in MGNREGA was found more prevalent in Andhra Pradesh, Himachal Pradesh, West Bengal, Bihar, and Karnataka while reverse was the case in Gujarat and Kerala. But a majority of participants in the discussion indicated that MGNREGA has not made any significant changes in the migration pattern in the village.

Another point of debate was how the MGNREGA has affected living standards of villagers. A clear majority indicated that MGNREGA has not been successful in raising their living standards or their consumption level and the reasons was quoted that the programme has not provided enough numbers of days of work to make a significant dent on the poverty level, although a minority of them were of the view that MGNREGA has been successful in doing so, to some extent. The latter ones indicated that MGNREGA has improved living standards by providing work within the village and by ensuring same wage rate to female as equal to that of male. To another question, whether MGNREGA has changed the trend of attached labour in agriculture, a significant majority said yes as people were getting better payments within the village compared to agricultural work so the trends of attached labour for the agricultural work were declining. However, MGNREGA has certainly increased people awareness towards Government schemes through increase in the showcasing by television, newspaper, Gram Panchayat and Gram Sabhas and by other means. Among the selected states, in Sikkim, Andhra Pradesh, Kerala, Rajasthan, West Bengal, Uttar Pradesh, Maharashtra, and Gujarat, a clear majority of the discussants expressed that the household consumption as well as enrollment of children in the school have increased after implementation of MGNREGA that has provided extra purchasing power in the hands of the villagers. On the question of awareness, almost all states observed increased awareness of the households towards existing government schemes because of their participation in the Gram Sabha and also because of joint working opportunities in MGNREGA.

VILLAGERS' SUGGESTIONS TO RAISE EFFICACY OF MGNREGA

Among the steps needed to ensure better implementation of MGNREGA, the major ones suggested by the discussants included: increasing working days and wage rate; providing food within the programme; allowing private land development through MGNREGA work for longevity of the programme; and by providing proper information on various aspects of the programme; implementation should be carried out though local bodies and job card should be given in the hands of the workers; quick payment after work.

POLICY SUGGESTIONS

In the light of above discussion following policy suggestions can be made to improve the functioning of MGNREGA.

The MGNREGA has not been successful in providing stipulated 100 days employment to all the registered persons. The reasons expressed by the Panchayat and district officials were many, including lack of funds; money not being provided from the Central authorities on time; the gap with which money reaches to the Panchayat officials; and money being provided only for few months and not the whole year. The results of the household survey clearly indicate that unless participants are given work for the stipulated 100 days, MGNREGA shall not be able to make any significant dent on the rural poverty and would fail in its basic objective. Therefore, provision of 100 days employment to all the participants should be made mandatory and strict action should be taken against the Panchayats which fail in fulfilling this target. The issue of timely provision of money to the Panchayats should be looked into so that MGNREGA work does not suffer because of lack of funds with the Panchayats.

Another big anomaly was found in the wage rate paid under MGNREGA. Under the MGNREGA Act, Panchayats are ordained to pay at least equal to the minimum wage determined for the state during a particular period. However, the actual wages paid under MGNREGA were found much lower. Among participants in Karnataka, those who were paid equal to or above the stipulated minimum wage, their percentage was only 1.4. Those who were paid Rs 100 or above constituted only 22 per cent and those who were paid between Rs 80 and Rs 100, their percentage was 63, while the percentage of those paid less than Rs 80 was around 15. Thus, above 40 per cent of the selected participants were paid less wages by 50 per cent or more, compared to the stipulated minimum wage in the state during the reference period. Among the corrections suggested by the households, almost all of them wanted that the minimum stipulated wages should be ensured for all participants irrespective to the nature of work they were involved in.

In the village analysis it was observed that there seems to be a conflicting interest between the MGNREGA and the farming community. Farmers across the board are feeling that they are facing labour shortage for agricultural activities because of the diversion of labour caused by MGNREGA activities. With a meticulous planning, this problem can be solved without affecting anyone adversely. In our secondary analysis, we saw that MGNREGA has provided not more than 45 days of employment per household at the all India level and all states failed in providing stipulated 100 days of employment to all households working in the programme. Even if the stipulated 100 days employment is provided by the MGNREGA, still there is enough scope for the labour force to work in the agricultural sector. There is, however, need to plan the MGNREGA work at the Panchayat level in such a way that it does not clash with the sowing and harvesting season in agriculture when the demand for agriculture labour is highest. The projects taken up under MGNREGA should be planned in such a way that labour is strictly employed for the project after the sowing and harvesting season of main rabi and kharif crops is over. This planning has to be done at the Panchayat/Block and District level, depending upon the cropping pattern of the respective regions. It would not only provide necessary labour force for agricultural operations but also would increase employment and income opportunities for the villagers during the off-season, including that of marginal and small farmers who do not have enough work at the farm in the off-season.

Proper punishment system should be put up in place for the unscrupulous officials who are found guilty of indulging in corruption and other untoward activities. Similarly, those Gram Panchayats that work efficiently in running the MGNREGA system should be rewarded and felicitated appropriately.

The provision of food/grain at the workplace and easy institutional credit can attract more villagers, especially the poor ones, towards working in MGNREGA and also ensures better food security to the participants.

The Unique Identification (UID) should be used for the better functioning of MGNREGA (Anderson et al. 2013). Bank accounts for MGNREGA workers will be linked to the unique biometric ID. As a result, the actual transfer of payments will immediately reach the hands of those it is intended for. This would drastically reduce the alleged inherent corruption in the current system and increase the amounts and reliability of payments to the workers.

REFERENCES

Anderson, Siwan, Ashok Kotwal, Ashwini Kulkarni, and Bharat Ramaswami. 2013. *Measuring the Impacts of Linking NREGA Payments to UID*. Working Paper, International Growth Centre, London: London School of Economics and Political Science.

Basu Arnab K. 2011. *Impact of Rural Employment Guarantee Schemes on Seasonal Labor Markets: Optimum Compensation and Workers' Welfare*. Discussion Paper No. 5701, *ZEF*, University of Bonn Germany *and IZA*, May 2011.

Dey, Subhasish. 2010. *Evaluating India's National Rural Employment Guarantee Scheme: The Case of Birbhum District, West Bengal*, Working Paper No. 490, Netherlands: International Institute of Social Studies.

Erlend Berg, Sambit Bhattacharyya, Rajasekhar Durgam, and Manjula Ramachandra. 2012. *Can Rural Public Works affect Agricultural Wages: Evidence from India*? CSAE Working Paper WPS/2012-05, Centre for the Study of African Economies, Department of Economics, Oxford: University of Oxford.

Kareemulla, K., S. Kumar, K.S. Reddy, C.A. Rama Rao, and B. Venketeswarlu. 2010. "Impact of NREGS on Rural Livelihoods and Agricultural Capital Formation." *Indian Journal of Agricultural Economics* 65(3): 524–39.

Khera, R. and N. Nayak. 2009. "Women Workers and Perceptions of the NREGA." *Economic and Political Weekly*, XLIV (43), 12 February.

NREGA. 2013. "Mahatma Gandhi National Rural Employment Guarantee Scheme (MNREGA)," Ministry of Rural Development (MRD), Government of India, New Delhi, available at http://nrega.nic.in/netnrega/home.aspx

Planning Commission. 2001. *Evaluation Study on Employment Assurance Scheme (EAS)*, Programme Evaluation Organization, Planning Commission, Government of India, New Delhi.

Shankar, S., R. Gaiha, and R. Jha. 2011. "Information, Access and Targeting: The National Rural Employment Guarantee Scheme in India." *Oxford Development Studies* 39(1): 69–95.

ANNEXURE 55A

Table 55A.1 The Details of AERCs and the Lead Person Involved in the State Report

S. No.	States	AERC Who Carried the Study in the State	The Lead Person Who Carried Out the Study
1.	Andhra Pradesh	AERC Waltare	G. Gangadhara Rao and K. Adiseshu
2.	Assam	AERC Guwahati	Jotin Bordoloi
3.	Bihar	AERC Bhagalpur	Rajiv Kumar Sinha and Rosline K. Marandi
4.	Chhattisgarh	AERC Jabalpur	Hari Om Sharma and Deepak Rathi
5.	Gujarat	AERC Vidyanagar	V.D. Shah and Manish Makwana
6.	Haryana	AERC Delhi	D.S. Bhupal

7.	Himachal Pradesh	AERC Shimla	C.S. Vaidya and Ranveer Singh
8.	Kerala	AERC Chennai	R. Arunachalam and A. Abdul Salam
9.	Madhya Pradesh	AERC Jabalpur	Hari Om Sharma and Deepak Rathi
10.	Maharashtra	AERC Pune	Jayanti Kajale and Sangeeta Shroff
11.	Punjab	AERC Ludhiana	Kamal Vatta; D.K. Grover and Tinku Grover
12.	Rajasthan	AERC Vidyanagar	Mrutyunjay Swain and Shreekant Sharma
13.	Sikkim	AERC Shantiniketan	Jiban Kumar Ghosh and Snehasish Karmakar
14.	Uttar Pradesh	AERC Allahabad	Ramendu Roy and Ramji Pandey
15.	West Bengal	AERC Shantiniketan	Jiban Kumar Ghosh
16.	Karnataka	ADRTC Bangalore	Parmod Kumar and I. Maruthi

Table 55A.2 The Details of Agro-Economic Research Centres and the Lead Person Involved in the State Report

States	AERC Centre/Unitthe	Authors
Andhra Pradesh	AERC Waltare	G. Gangadhara Rao and K. Adiseshu
Assam	AERC Jorhat	Jotin Bordoloi
Bihar	AERC Bhagalpur	Rajiv Kumar Sinha and Rosline K. Marandi
Chhattisgarh	AERC Jabalpur	Hari Om Sharma and Deepak Rathi
Gujarat	AERC Vidyanagar	V.D. Shah and Manish Makwana
Haryana	AERC Delhi	D.S. Bhupal
Himachal Pradesh	AERC Shimla	C.S. Vaidya and Ranveer Singh
Kerala	AERC Chennai	R. Arunachalam and A. Abdul Salam
Madhya Pradesh	AERC Jabalpur	Hari Om Sharma and Deepak Rathi
Maharashtra	AERC Pune	Jayanti Kajale and Sangeeta Shroff
Punjab	AERC Ludhiana	Kamal Vatta; D.K. Grover and Tinku Grover
Rajasthan	AERC Vidyanagar	Mrutyunjay Swain and Shreekant Sharma
Sikkim	AERC Shantiniketan	Jiban Kumar Ghosh and Snehasish Karmakar
Uttar Pradesh	AERC Allahabad	Ramendu Roy and Ramji Pandey
West Bengal	AERC Shantiniketan	Jiban Kumar Ghosh
Karnataka	ADRTC Bangalore	Parmod Kumar and I. Maruthi

56

An Evaluation Study of Prime Minister's Rehabilitation Package for Farmers in Suicide-Prone Districts of Andhra Pradesh, Karnataka, Kerala, and Maharashtra

M.J. Bhende and P. Thippaiah

Although Indian agriculture has been facing a serious crisis since the late 1980s, it has assumed an alarming dimension since the middle of the 1990s. The diversification of agriculture from subsistence to commercial farming resulted in increased dependence on purchased (market) inputs. The rise in market dependence had its toll on profitability while the lack of safety nets made things worse for the farming community as a whole. Moreover, uncertain monsoons, presence of spurious inputs like seeds, fertilizers, and plant protection chemicals, fluctuations in the production, and imperfect markets have contributed to the distress and the frustration of farmers. The important manifestation of the crisis in agriculture was the stagnating, if not deteriorating, terms of trade for agriculture (Government of India 2007).

The agrarian distress reached a climax at the end of the twentieth century manifested by a large number of suicides committed by farmers in some parts of India. The Situation Assessment Surveys of the National Sample Survey Organization (NSSO) has reconfirmed the worsening situation of farming households, indicating that 48.6 per cent of the farmer households in India are indebted and about 40 per cent farmer households in the country did not like farming because it is not profitable, is risky, and it lacks social status and felt that, given a choice, they would take up some other career (NSSO 2005).

Farmers' suicides have been receiving a lot of social and public policy attention.[1] Suicides were mainly concentrated in Karnataka, Andhra Pradesh, and Maharashtra. A large number of suicides were reported in Karnataka in the first three years of the decade starting 2000–01, while Andhra Pradesh had the maximum number in 2004–05. In 2006, there was virtually a suicide epidemic in Maharashtra. These incidents raised serious questions about the state

[1] Some of the studies are Deshpande (2002), Mohanty and Shroff (2004), Sarma (2004), Deshpande and Prabhu (2005), Gill and Singh (2006), Mishra (2006, 2006a), Mohankumar and Sharma (2006), Satish (2006), Singh (2006), Sridhar (2006), Mitra and Shroff (2007), Vaidyanathan (2007), Shroff (2008), Padhi (2009), Parthasarathy and Shameem (1998), Mohan Rao (1998), Vasavi (1999), Narasimha Rao and Suri (2006). Srijit Mishra (2006).

of the agrarian economy and the economic hardships faced by farmers.

Concerned with farmers' suicides in some parts of the country, the Hon'ble Prime Minister of India Dr Manmohan Singh, after having visited some parts of the Vidarbha region in Maharashtra, announced a rehabilitation package on 1 July 2006 to mitigate the distress of farmers in the identified districts. On 29 September 2006, the Union Cabinet approved the Rehabilitation Package for 31 identified districts in Andhra Pradesh, Karnataka, Kerala, and Maharashtra (Table 56.1) called the Prime Minister's Rehabilitation Package (PMRP) for Farmers in Suicide-Prone Districts of Andhra Pradesh, Karnataka, Kerala and Maharashtra'. The implementation period of the PMRP is fixed for three years and included both short (immediate) and medium-term measures.

PRIME MINISTER'S REHABILITATION PACKAGE (PMRP)

The PMRP aimed at establishing a sustainable and viable farming and livelihood support system through debt relief to farmers, complete institutional credit coverage, crop-centric approach to support agriculture and other subsidiary activities. In order to alleviate the hardships faced by the debt-stressed families of farmers, ex gratia assistance from the Prime Minister's National Relief Fund (PMNRF) at Rs 50 lakh per district has also been provided. The package covers the following:

1. Complete credit cover through institutional credit sources, including debt relief to farmers by restructuring overdue loans, interest waiver, and provision of fresh loans
2. Provision of assured irrigation facilities
3. Watershed management, which includes development of participatory watershed, construction of check-dams, and rainwater harvesting structures
4. Seed replacement programme
5. Subsidiary income-generating activities like livestock, dairying, fisheries
6. Diversification of activities and value addition through horticulture
7. Extension support services

The PMRP involves a total amount of Rs 16978.69 crore, consisting of Rs 10,579.43 crore (62.31 per cent) as subsidy/grants and Rs 6399.26 crore (37.69 per cent) as loans. The state-wise financial summary of the rehabilitation package is given in Table 56.1.

As stated earlier, the PMRP has several financial components which provide not only short-term benefits but also envisage development of suicide-prone regions by strengthening the infrastructure and institutions. It is evident from Table 56.2 that an amount of Rs 16,978.69 crore was sanctioned under this relief package for the four identified states, while 16.01 per cent of the amount is earmarked for waiving of overdue interest and 57.35 per cent is for assured irrigation under the Accelerated Irrigation Benefit Programme (AIBP). The rest of the amount was allocated for watershed activities (10.96 per cent), seed replacement (4.89 per cent), micro irrigation (4.72 per cent), diversification of agriculture and horticulture development, cattle and fisheries development, and so on (22.08 per cent).

The main objective of this study is to understand the ground reality and check whether the benefits of the package are reaching to the intended beneficiaries or not and to suggest measures to achieve the goals envisaged in the PMRP. The specific objectives of the study are:

1. to explore whether the benefits of the package are reaching the intended beneficiaries;
2. to assess the overall social and economic impact of the rehabilitation package; and
3. to study the constraints involved in the implementation of the package and suggest improvement.

Table 56.1 State-wise Financial Summary of Rehabilitation Package (in Crores)

State	Rehabilitation Package					
	Subsidy/Grant		Loan		Total	
	Amount (Rs)	Per cent in Total	Amount (Rs)	Per cent in Total	Amount (Rs)	Per cent in Total
Andhra Pradesh	5,943.31	56.18	3707.24	57.94	9650.55	56.84
Karnataka	1,568.07	14.82	1121.57	17.53	2689.64	15.84
Kerala	577.21	5.46	188.03	2.93	765.24	4.51
Maharashtra	2,490.84	23.54	1382.42	21.60	3873.26	22.81
Total	10,579.43	100.00	6,399.26	100.00	16,978.69	100.00

Source: P.M. Package, concept note, Government of India 2009.

Table 56.2 Financial Summary of the Prime Minister's Rehabilitation Package (in Crores)

S.No.	Item of Assistance	Total Amount	% in Grand Total	Subsidy/Loan/Grant	Subsidy/Grant	Loan
1.	Ex-gratia assistance from PMNRF	15.50	0.09	Only grant	15.50	–
2.	Restructuring/Rescheduling of loans	9,051.81		Only loan	–	9,051.81
3.	Credit Flow	20,114.05		Only loan	–	20,114.05
4.	Interest Waiver*	2,718.25	16.01	Only subsidy	2,718.25	–
5.	*Assured Irrigation*					
	1. Major & Medium Irrigation	6,530.29	38.46	AIBP norms	4,268.84	2,261.45
	2. Minor Irrigation	3,207.81	18.89	Only loan	–	3,207.81
	Total:	9,738.10	57.35		4,268.84	5,469.26
6.	Seed Replacement	830.10	4.89	Only subsidy	830.10	
7.	*Watershed Development*					
	1. Participatory Watershed	837.00	4.93	Only grant	837.00	–
	2. Check Dams	930.00	5.48	Only loan	–	930.00
	3. Rain Water harvesting	93.00	0.55	50:50	93.00	–
	Total:	1,860.00	10.96	–	930.00	930.00
8.	*Micro Irrigation***	801.53	4.72	Only subsidy	801.53	–
9.	*Horticulture Development (NHM)*	452.78	2.67	Only subsidy	452.78	–
10.	*Extension Services*	15.50	0.09	Only grant	15.50	–
11.	*Subsidiary Income Activities*	546.93	3.22	Only Subsidy	546.93	–
	Grand Total	16,978.69	100.00		10,579.43 (62.31)	6,399.26 (37.69)

Notes: *Burden of interest waiver will be shared equally by the Central and State Government in the ratio 50:50.
** Scheme provides for 50 per cent subsidy and 50 per cent loan. of 50 per cent subsidy, 40 per cent is Central share and 10 per cent is from State Government.
Amount under Restructuring/Rescheduling of loans and credit flow has not been included in the grand total.
Source: Concept note, Government of India

METHODOLOGY AND SELECTION OF SAMPLES

The PMRP for farmers in suicide-prone districts is being implemented in six districts each of Karnataka and Maharashtra, three districts of Kerala, and sixteen districts of Andhra Pradesh. To study the process of implementation of the PM's rehabilitation package and get the feedback from the beneficiaries, three districts are selected from each state on the basis of the highest cumulative suicide cases registered in the district in the past. The districts selected for primary data collection for the present study are:

1. **Andhra Pradesh**: Anantapur, Guntur, and Warangal
2. **Karnataka**: Belgaum, Chitradurga, and Hassan
3. **Kerala:** Kasaragod, Palakkad, and Wayanad
4. **Maharashtra:** Amravati, Buldhana, and Yavatmal

RESULTS AND DISCUSSION

The PMRP includes both credit and non-credit components and there are some schemes directed towards individual beneficiaries like waiving off overdue interest, rescheduling of loans, expansion of micro irrigation, distribution of subsidized seed, and so on, whereas others are community based related to infrastructure like development of irrigation, watershed development, and so on. PMRP components were implemented in the identified districts with varying levels of achievements. As stated earlier, the data are collected from forty households, each from three districts of Andhra Pradesh, Karnataka, Kerala, and Maharashtra. Total sample size thus comprises 120 households from each state. The data are analysed and the findings are summarized below.

EX GRATIA PAYMENT

The Government of India released Rs 50 lakh per district to facilitate payment of ex gratia amount to the families of distressed farmers for meeting contingent expenditure on healthcare and education needs of their children. The amount was disbursed by the district collector on the recommendation of the district committee. There is a variation in disbursement of ex-gratia payment to individual farmers across states as well as within a state.

Andhra Pradesh government disbursed Rs 10000 each as ex-gratia payment to the distressed families to meet contingent expenditure needs for health, education, and others and utilized the full amount of Rs 8 crore received from the Government of India. The Government of Karnataka has disbursed less than 60 per cent of the total allocation of Rs 3 crore. The average amount of ex-gratia payment disbursed ranged from Rs 2,420 in Chitradurga district to Rs 10,000 per beneficiary in Hassan district. In Kerala, Rs 1.445 crore is disbursed to 4,913 farm families as ex-gratia payment. The average amount of ex gratia payment per beneficiary varied from Rs 2,844 in Kasaragod to Rs 3,003 in Wayanad. In Maharashtra, the entire amount of Rs 3 crore provided to six District Collectors was disbursed to 3,974 distressed families by the end of March 2008 itself.

CREDIT COMPONENT

The credit component includes waiving off the entire interest on overdue loans as on 1 July 2006, rescheduling of overdue loans over a period of three to five years with a one year moratorium, and provision of fresh credit by the banking system. The actual loan amount rescheduled, the number of accounts benefited by rescheduling of loans and also the actual interest waived off are higher than the estimated overdue loan amount and the number of farmers both in Karnataka and Maharashtra. On the contrary, the actual amount of overdue loans rescheduled as well as interest actually waived off was less than the estimated overdue loan amount and interest in Andhra Pradesh and Kerala. Surprisingly, none of the farmers interviewed were aware of either the overdue interest waived off or the principal amount due. Most of the farmers were reluctant to borrow after getting their overdue loans rescheduled expecting the government to waive off the entire loan (principal) amount.

SEED REPLACEMENT

In order to provide immediate assistance for the farmers, a massive seed replacement programme was launched with 50 per cent subsidy in all the 31 identified districts. The total allocation for the seed replacement programme was Rs 830.10 crore for a period of three years. The allocation made for distributing subsidized seeds was fully utilized by Andhra Pradesh, Maharashtra, and Kerala. The Government of Karnataka could spend less than 30 per cent of the allocation made for seed distribution till March 2009.

The certified seeds distributed represent major crops grown in the respective district. Subsidized seed is provided to cover a maximum area of 2 hectare only. In Andhra Pradesh, groundnut, chickpea, maize, and soybean together shared more than 95 per cent of the total quantity of seed and 84 per cent of the total subsidy under the PMRP. In Karnataka, paddy, maize, ragi, sunflower, and chickpea dominated the subsidized seed. The subsidy amount per beneficiary farmer ranged from Rs 185 in Hassan to Rs 504 in Belgaum. In Kerala, about 10 per cent of the farmers from Wayanad and an equal number of farmers from Palakkad benefited from the seed replacement scheme. The beneficiary farmers received an average subsidy of Rs 50 per head. In Maharashtra, soybean seed was the most sought after seed by the sample farmers. Farmers reported modest increase in the yield levels due use of certified seed.

MICRO IRRIGATION: SPRINKLER AND DRIP IRRIGATION

The Department of agriculture and horticulture distributed drip and sprinkler irrigation sets to the farmers with varying levels of subsidy. The subsidy provided varies across districts and among the states. It was as low as 30.8 per cent in Chitradurga (Karnataka) to 90 per cent in Andhra Pradesh. Andhra Pradesh, surpassed the target and spent Rs 283.10 crore against the release of Rs 234.00 crore by the Government of India. The Government of Kerala could use a very small amount, that is, 17 per cent of the total Rs 6.36 crore released by the Government of India. The Government of Karnataka spent Rs 74 per cent of the released amount whereas the Government of Maharashtra could utilize 92 per cent of the total amount released till March 2009.

There also appears some bias against the suppliers of micro irrigation systems in Kerala. The farmers were given 37 per cent and 50 per cent subsidy on irrigation set, depending on the supplier or the dealer. Majority (than 93 per cent) of the farmers opined that there is a modest increase in the area covered under irrigation due to introduction of sprinkler or drip irrigation system.

EXTENSION SERVICES

Strengthening of extension services was one of the components of the PMRP and is implemented by Agriculture Technology Management Agencies (ATMA) in all the districts. The activities include training programmes, seminars and workshops, visits to demonstration fields, research stations, and study tours of the farmers for exposing themselves to new farming systems and technologies. The Government of Andhra Pradesh could spend 46 per cent of the total amount of Rs 26.52 crore released by the government of India till 30 June 2008. The department of agriculture, Government of Karnataka, has used about two-third of the total released amount of Rs 2.99 crore. In Maharashtra, Rs 10.44 crore (91 per cent) is spent by ATMA for various activities against the release of Rs 11.50 crore by the State Government. Most of the funds released for extension in Maharashtra were spent on strengthening of self-help groups.

More than 60 per cent of the farmers form Anantapur and Warangal districts have said that they have attended crop demonstration and all of them benefited from the programme. On the contrary, no extension activity is reported in Guntur district. In Karnataka, overall, 30 per cent of the sample households have attended crop demonstrations and only a handful (1 to 2 per cent) of the sample farmers benefited either from exposure trips arranged by the department of agriculture or a daylong training programme cum workshop on improved agricultural practices organized at the KVKs. In Kerala, it was proposed to establish Farmer Counselling Centres at the existing Krishi Bhavans at Wayanad, Palakkad, and Kasaragod districts. Only 12 per cent of the beneficiaries have confirmed that they have attended the counselling programme.

SUBSIDIARY INCOME-GENERATING ACTIVITIES

Animal Husbandry

In order to support the subsidiary income activities of the farmers, the PMRP included components like supply of high yielding animals, calf-rearing, feed and fodder supply, animal healthcare, setting up of bulk milk chilling plants, fodder block-making units, provision of breeding services, and estrus synchronization, and so on. On an average, 50 per cent subsidy is provided for purchasing milk animals. The animal husbandry department of Andhra Pradesh, an implementing agency of the package, could spend 74 per cent of Rs 195 crore released by the Government of Andhra Pradesh. In Karnataka, Livestock Development Agency along with Karnataka Milk Federation (KMF) has been implementing the programme and could spend 36 per cent of the total Rs 25.48 crore released. The implementing agencies in Kerala utilized little more than 78 per cent (Rs 33.39 crore) of the total grants of Rs 42.52 crore released under PMRP. In case of Maharashtra, the package involved a total investment of Rs 98.87 crore in six districts over a period of three years. The department of animal husbandry utilized almost 96 per cent of the amount released (Rs 50.98 crore) by the government from 2006–07 through 2008–09.

The beneficiary household from Anantapur purchased, on an average, one milch animal as against roughly two milch animals by the beneficiaries from Guntur and Warangal districts. The average annual income from dairy activity ranged from Rs 17,400 per beneficiary in Warangal district to Rs 29,000 in Guntur district, with an average income of Rs 24,533 for all the beneficiaries.

In Karnataka, around 70 per cent of the beneficiary households purchased two milch animals and rest one each. Almost all the beneficiaries from Karnataka have reported that they have not received payments for the purchase of feed and concentrate and for rearing calves. The average net income per annum from milch animals ranged from Rs 8,200 per beneficiary household in Hassan district to Rs 13,400 per beneficiary in Belgaum district.

In Kerala, a little more than 15 per cent of the farmers have taken advantage of the animal husbandry and dairying component of the PMRP and they have purchased milch cattle, goats, and sheep. The average income per household from dairy (sale of milk) is Rs 20,572 for the Wayanad farmers, Rs 14,150 for the Palakkad farmers, and Rs 12,900 for the Kasaragod farmers. The farmers interviewed have also hinted at paying some bribe to officials for accessing and obtaining the assistance and relief from the package.

In Maharashtra, only 12.5 per cent of the sample beneficiaries purchased milch cows and bullocks under the subsidy scheme The Government of Maharashtra also provided subsidy of about Rs 10,851 per farmer household for construction of cattle shed/purchase of feed and for calf-rearing under the PMRP. The average net income from the livestock was Rs 3,860 per household. Most of the beneficiary farmers from Maharashtra opined that the high yielding cross-bred cows purchased under this scheme could not be sustained under hot climatic conditions of the regions and it affected the milk productivity.

Fisheries

The Programme envisaged development of fishponds in 100 hectares per district by providing 40 per cent of capital and input costs, the rest planned to be sourced through bank credit. The Government of Andhra Pradesh spent Rs 7 crore for the development of fisheries in the identified sixteen districts during the last three years beginning with 2006–07 and has constructed 699 hectares of fishponds under the PMRP. On an average, income per beneficiary from fish cultivation ranged from Rs 30,000 in Guntur district to Rs 40,000 in Warangal. Overall, the scheme was successful in Andhra Pradesh.

As far as Karnataka is concerned, an amount of Rs 6.24 crore was earmarked for the fisheries developmental activities in the six districts. Fishery department in Karnataka could spend Rs 1.264 crore during 2007–08 and 2008–09 taken together. The average expenditure incurred for development of fishponds ranged from Rs 0.75 lakh in Belgaum district to Rs 2.28 lakh in Hassan district. The average subsidy amount works out to Rs 50,075 per beneficiary or Rs 89,420 per ha. The fisheries activity helped to augment household income ranging from Rs 8,200 in Hassan to Rs 22,222 in Chitradurga district. The supply of good fingerlings (fish seed) is a major constraint affecting the fish yield. The better-off farmers and those who already had small tanks or ponds in and around their fields have availed of the subsidy under the fishery component of the PM's Package. The farmers who constructed new ponds are rare.

The Fish Farmers Development Agency (FFDA) was identified as the nodal agency for implementing the programme in Kerala. FFDA utilized the entire amount (Rs 1.03 crore) released by the government of Kerala during the first year to develop aquaculture on 180.4 ha of area. The Government of Kerala released the remaining Rs 2.30 crore during 2007–08. However, the area covered during the second year is not known as scrutiny of applications as well as survey of the area was in progress. A total of 1,184 households from Kasaragod and 454 from Wayanad and 214 from Palakkad benefited from the fisheries programme during 2006–07. The major activity taken so far is stocking of fingerlings.

In Maharashtra, total allocation for fisheries component was Rs 6.21 crore. However, Rs 23 lakh was released and of this only Rs 18.93 lakh could be spent for the development/repair of fishponds in the identified six districts of Maharashtra. A total of ninety-one fishponds covering 102.34 ha area were constructed and ninety-one farmers benefited by this scheme.

NATIONAL HORTICULTURE MISSION (NHM)

NHM was included as one of the component of PMRP. Andhra Pradesh utilized 84 per cent of the total grants of Rs 484.47 received from the Government of India for NHM activities. On an average, Rs 66,833 was invested by each beneficiary farmer, of which Rs 34,923 was subsidy component provided under the NHM programme. A majority, that is, 73 per cent of beneficiaries were satisfied with the scheme.

The Government of Karnataka could spend only 60 per cent of the total 97.77 crore released by the government of India for NHM for all the six districts in Karnataka. The area covered under NHM was roughly 60 per cent of the targeted area of 1,44,817 ha. Most of the farmers availed the subsidy for development and extension/rejuvenation of mango, banana, pomegranate, and sapota plantations as well as for purchase of micro irrigation system.

In Kerala, during the three-year period (2006–07 to 2009), about 54 per cent of the total earmarked amount of Rs 79.07 crore was released and 92 per cent of the released amount was utilized for implementing the various components under the horticulture mission. Overall, 18 per cent of the farmers from the three identified districts of Kerala participated in NHM programme.

In the case of Maharashtra, Rs 225 crore has been sanctioned by the government towards the NHM programme. Of this, Rs 115.26 crore (51.26 per cent) was received by 31 March 2009. On an average, two-thirds of the sample farmers benefited from the NHM in the study districts of Maharashtra. The average area brought under horticulture varied from 0.94 hectare per beneficiary in Amravati district to 1.24 ha in Buldhana district.

ASSURED IRRIGATION

Assured irrigation component accounts for 57.35 per cent of the total PMRP of Rs 16,978.69 crore earmarked for the 31 districts from four states. The funds were provided to complete ongoing major and medium irrigation projects in the selected districts to augment area under irrigation and also for activities such as construction of new tanks, modernization of tanks, pickups and barrages, flood control works, and lift irrigation schemes were taken up under the minor irrigation component.

The Government of Andhra Pradesh could utilize Rs 6,442.87 crore for irrigation development against the target of Rs 8,984.23 crore. Minor irrigation works are not undertaken in any of the three selected districts of Andhra Pradesh. The Government of India sanctioned Rs 1208.71 crore for major and medium irrigation projects in Karnataka. Most of the funds are used for lining of canals and strengthening of embankments of major and medium irrigation projects. Funds from National Bank for Agriculture and Rural Development (NABARD) are used mostly for rehabilitation and rejuvenation of irrigation tanks. Since most of the works under the minor irrigation are in various stages of completion, the actual benefits from these programmes cannot be ascertained at this point of time.

In Maharashtra, a total of Rs 2,679 crore was released for irrigation projects, of which Rs 2,641 crore is utilized by 31 March 2009. However, the irrigation potential created is only 53 per cent of the targeted area of 1,67,871 ha. The average area irrigated increased from 1.55 ha to 2.73 ha after its implementation. In the case of Kerala, the Government of India provided a total of Rs 105.03 crore over a period of three years for completion of all major, medium, and minor irrigation projects in the three selected districts. A total of

Rs 6.65 crore have been released under Rural Infrastructure Development Fund (RIDF) to undertake 132 minor irrigation projects during the last three years. The projects are incomplete, and hence, its impact cannot be ascertained.

PARTICIPATORY WATERSHED PROGRAMME

Drought-proofing is one of the important components of the PMRP for farmers in the suicide-prone districts and hence, the watershed development programme was included in order to increase the production and productivity of crops. The scheme is subdivided into three sub-schemes, namely, check-dams, participatory watershed development programme, and rainwater harvesting scheme.

PARTICIPATORY WATERSHED DEVELOPMENT PROGRAMME

The Prime Minister's Rehabilitation Package provided Rs 837 crore as grants to the state governments for implementation of participatory watersheds in the 31 identified districts of four states and the rest of the financial requirement was to be met from loans raised under RIDF from NABARD. It was envisaged that about 15,000 ha per district would be treated under the participatory watershed development programme for which a grant support of Rs 60 lakh per watershed of 1,000 hectare each was provided.

In Andhra Pradesh, watershed development programme covered an area of 4.83 lakh hectares with an expenditure of Rs 4,756.26 crore as on 31 January 2010. It was observed that on an average, 2.16 hectares of land per beneficiary household were treated under watershed programme. Farm-ponds were dug to conserve rainwater and bring more area under irrigation by the selected sample households. The average cost for construction of farm pond was Rs 18,055 and farmers contributed on average Rs 8,572. All the beneficiary farmers agreed that the scheme would benefit them in increasing the irrigated area and help them realize higher yields.

In Maharashtra, about 90188 hectare of area is treated under the watershed development programme with an expenditure of Rs 18.65 crore, against the sanctioned amount of Rs 54 crore, by 31 May 2009. On an average Rs 37,073 per farm-pond/check-dam is spent on construction in the selected areas and sample households shared about Rs 11,282 for the structure. Only 40 per cent of the farmers anticipated that the scheme would benefit them. A package of Rs 81 crore was earmarked for watershed development in the three districts of Kerala. However, as per available the data Rs 1.53 crore was spent to treat 15,600 ha of area under the watershed development programme in the selected districts of Kerala.

Implementation of the participatory watershed development component in Karnataka is delayed as the Government of Karnataka had to seek permission from the Government of India to modify the watershed component to follow the SUJALA model in implementation of watershed programme in the selected districts. A sum of Rs 162 crore is provided by NABARD for implementation of participatory watershed development in Karnataka. The Government of Karnataka could spend Rs 7.15 crore till the end of March 2009. As per the information made available, a substantial amount (Rs 40 lakh per district) was disbursed to non-government organizations for community mobilization and training. Land treatment and other field activities were not initiated till March 2009, and hence, no progress is reported.

CONSTRUCTION OF CHECK-DAMS

Check-dams are constructed at the end of the watershed to impound the excess water after allowing water to seep into the aquifer through the entire course of the drainage line of the watershed. As per the guidelines of the PMRP, on an average, 500 check-dams are to be built every year at an average cost of Rs 2 lakh per check-dam in each district over three years.

Under the PM's package the total allocation for Andhra Pradesh is Rs 480 crore. However, no amount has been spent so far during this period. In Karnataka, all previous check-dam projects sanctioned/kept in abeyance were withdrawn and the amount of Rs 180 crore meant for check-dams would be utilized for the development of watershed projects based on SUJALA watershed model wherein a check-dam is one of the components of watershed development. The Government of Kerala has in principle accorded sanction for the construction of ninety check-dams under the PMRP in the identified districts involving an amount of Rs 82 crore. The progress of the construction of check-dams in Kerala is not clear. The Government of Maharashtra received Rs 180 crore towards construction of check-dams and could construct 7,970 check-dams against the target of 9,000 check-dams by 2008–09 by spending Rs 162.53 crore.

RAINWATER HARVESTING STRUCTURES

The rainwater harvesting scheme aiming at accelerated growth of irrigation potential for ensuring agricultural development of Schedule Caste (SC)/Schedule Tribe (ST) beneficiaries including small and marginal farmers was provided with 50 per cent back-ended capital subsidy along with 50 per cent bank loan, covering 1,000 beneficiaries a year in each of the identified district. Rs 93.00 crore was provided as subsidy/grant for constructing rainwater harvesting structures to be shared by the Central and State Governments equally.

Andhra Pradesh received Rs 48 crore allocations for construction of rainwater harvesting structures under the package. However, no amount has been spent so far on the rainwater harvesting scheme in Andhra Pradesh (status as on 28 February 2009). In Karnataka, it was decided that an allocation of Rs 18 crore made for construction of rainwater harvesting structures would be used for the participatory watershed development programme. The Government of Kerala has a target to construct 1,000 rainwater harvesting structures per year per district for SC/ST and small and marginal farmers. Funds are yet to be released by the Government of India. However, the Government of Kerala is implementing another rainwater harvesting scheme called 'Jalanidhi' and requested NABARD for release of Rs 9 crore. The Government of Maharashtra implemented the rainwater harvesting scheme. The beneficiaries were provided 50 per cent back-ended capital subsidy along with 50 per cent bank loan, covering 1,000 beneficiaries a year in each of the identified districts.

CONCLUSIONS AND SUGGESTIONS

State governments are implementing parallel programme along with the PMRP to assist distressed farmers in the suicide-prone districts (with identical components). For example, some state governments provide monetary compensation to the farm families whose members have committed suicide. A few states have also resorted to waiver of loans from cooperative societies, supplying seed with subsidy, promoting micro irrigation and horticulture with subsidy, construction of farm-ponds, rain water harvesting, and so on. The parallel implementation of state schemes along with the PM package creates confusion among the beneficiaries. However, it was observed that most of the farmers in the suicide-prone districts benefited from one or the other scheme. Some farmers got benefit from two schemes and a few others from more than even three schemes also.

The average amount of ex-gratia received by the beneficiaries varied not only across the states but also within the states. Collusion between local leaders and officials is also evident as some well-off farmers got the ex-gratia payments whereas poor households were left out.

Overdue credit was rescheduled and interest was waived off under the credit component of the PMRP, however, it was observed that many beneficiaries were not aware about the quantum of interest waived off or the yearly instalment of principal they have to repay. Very few borrowers opted for fresh loans.

Distribution of certified seeds with 50 per cent subsidy in the identified districts is appreciated by all the farmers. This is one of the important components of the PM's package which helped farmers immensely.

Under the micro irrigation scheme, sprinkler and drip irrigation sets are supplied to the farmers at 35 to 50 per cent subsidy with the exception of Andhra Pradesh where subsidy accounted for 90 per cent of the cost of the equipment. It was observed that the subsidy amount received by the beneficiaries varied across the districts (Karnataka) as well as within the districts among the beneficiaries (Kerala). There is a need to provide clear-cut guidelines on the subsidy component, so that all the beneficiaries would get the same benefit without any room for leakages or corruption. All the beneficiaries reported that the adoption of micro irrigation system has helped in efficient use of irrigation water leading to modest expansion in the irrigated area.

Agriculture Technology Management Agencies is identified as the nodal agency in all the districts to ensure extension support and convergence at district level. A few farmers benefited from the activities identified under extension component in four states. Close monitoring of extension activities is necessary to achieve the desired goals.

There is a provision in the package to provide assistance for feed and fodder for one year for milch animals and also support for rearing a calf. However, none of the respondents has received assistance for feed and fodder (Karnataka). Periodic monitoring and surprise checks by the competent authorities are necessary to control the pilferage of funds. The full potential of milk yield could not be attained in Maharashtra due to heat stress and poor availability of fodder. It was felt that local breeds may be more suitable as they can sustain the heat and require less fodder.

Small and marginal farmers benefited the least from the fisheries programme as the initial investment for development of fishponds is very high and the subsidy component is only 40 per cent. Availability of quality seed and infrastructure (availability of ice, transport, markets) are major constraints faced by the beneficiaries in the identified districts.

Over the years, horticulture has emerged as a sunrise sector having the potential to accelerate the growth of our agrarian economy. National Health Mission (NHM) is being implemented in most of the identified districts. Most of the beneficiaries availed subsidies for purchase of micro irrigation equipments, plantation of fruit crops, and rejuvenation of old orchards.

The benefits from development of minor irrigation are shared by the community as a whole. In Andhra Pradesh, work on minor irrigation projects is not initiated in any of the identified districts. In Karnataka and Kerala, most of the works under the minor irrigation were in various stages of completion, and therefore, the actual benefits (expansion of irrigated area, quantity and quality of irrigation water supply, and so on) from these programmes cannot be ascertained at this point of time. It was observed that the quality of civil work done in the rejuvenation/rehabilitation of tanks under the PM's package in Karnataka was of poor quality and the farmers have complained to the authorities. It is suggested that the Gram Panchayat should have the authority to check and supervise the work related to minor irrigation under their jurisdiction.

Accelerated Irrigation Development Programme, watershed development programme, and micro irrigation are aimed at increasing the area under irrigation. These schemes did increase the area under irrigation and improved the productivity of all major crops cultivated in the districts. Thus, irrigation projects have to be completed and watershed activities have to be further promoted to recharge groundwater. The farmers will be in a position to face the droughts only when protective irrigation is available.

Overall, the study concludes that the farmers benefited from various components of PMRP such as interest waiver, rescheduling of loans and subsidy given under various schemes which enabled them to be eligible for fresh loans and also helped them augment their incomes through subsidiary activities. However, their capacity to cope with drought conditions whenever monsoons failed was still weak and the PM's package through its multiple schemes had a limited impact on this front.

REFERENCES

Deshpande, R.S. 2002. 'Suicide by Farmers in Karnataka: Agrarian Distress and Possible Alleviatory Steps', *Economic and Political Weekly*, 37(26): 2601–10.

Deshpande, R.S. and Nagesh Prabhu. 2005. 'Farmers' Distress: Proof Beyond Question', *Economic and Political Weekly*, 44(45): 4663–5.

Gill, Anitha, and Lakhwinder Singh. 2006. 'Farmers' Suicides and Response of Public Policy.' *Economic and Political Weekly*, 41(26): 2762–8.

Government of India. 2007. *Report of the Expert Group on Agricultural Indebtedness*. New Delhi: Ministry of Finance, Government of India, July.

Mishra, Srijit. 2006. 'Farmers' Suicides in Maharashtra', *Economic and Political Weekly*, 41(16): 1538–45.

———. 2006a. 'Suicides Mortality Rates across States of India', *Economic and Political Weekly*, 41(16): 1566–9.

Mohankumar, S. and R.K. Sharma. 2006. 'Analysis of Farmer Suicides in Kerala', *Economic and Political Weekly*, 41(16): 1553–8.

Mohan Rao, R.M. 1998. 'Study on Causes for Suicides among Cotton Farmers in Andhra Pradesh.' NABARD, Department of Co-operation and Applied Economics, Andhra University, Visakhapatnam.

Mohana Rao, L.K., G. Gangadhara Rao, K. Adiseshu, N. Ram Gopal, M. Nageswara Rao, G.M. Jeelani, K.V. Giribabu, K. Ramesh, and P. Malathi. 2010. Report on 'Prime Minister's Rehabilitation Package for Suicide-Prone Districts of Andhra Pradesh', Report No. 131, Agro-economic Research Centre, Andhra University, Visakhapatnam, July 2010.

Mohanty, B.B. and Sangeeta Shroff. 2004. 'Farmers' Suicides in Maharashtra.' *Economic and Political Weekly* 39 (52), 25 December: 5599–606.

Narasimha Rao and P. Suri, K.C. 2006. 'Dimensions of Agrarian Distress in Andhra Pradesh.' *Economic and Political Weekly*, 22 April: 1546–52.

National Sample Survey Organization (NSSO). 2005. Situation Assessment Survey of Farmers: Indebtedness of Farmer Households, 59th Round, (January–December).

Padhi, Ranjana. 2009. 'On Women Surviving Farmer Suicides in Punjab.' *Economic and Political Weekly*, 44(19): 53–59.

Parthasarathy and Shameem. 1998. 'Suicides of Cotton Farmers in Andhra Pradesh, An Exploratory Study.' *Economic and Political Weekly*, 33(13): 720–26.

Raj Kumar, R., S.R. Muthusamy, and A. Abdul Salam. 2010. Report on 'Prime Minister's Rehabilitation Package for Suicide-Prone Districts of Andhra Pradesh', Report No. 144, Agro-economic Research Centre, University of Madras, Chennai, Tamil Nadu.

Satish, P. 2006. 'Agricultural Institutional Credit, Indebtedness and Suicides in Punjab.' *Economic and Political Weekly*, 41(26): 2754–61.

Shroff, Sangeeta. 2008. 'Agrarian Distress and Farmers' Suicides in India: Focus on Maharashtra.' *The India Economy Review*, 5: 204–9.

Singh, Sukhpal. 2006. 'Credit, Indebtedness and Farmer Suicides in Punjab.' *Economic and Political Weekly*, 41(30): 3330–1.

Sridhar, V. 2006. 'Why Do Farmers Commit Suicide?', *Economic and Political Weekly*, 41(16): 1565–9.

Vaidyanathan, A. 2007. 'Farmers' Suicides and the Agrarian Crisis', *Economic and Political Weekly*, 41(38): 4009–13.

Vasavi, R. 1999. Agrarian Distress in Bidar, 'Market State and Suicides', *Economic and Political Weekly*, 34(22): 2263–8.

ANNEXURE 56A

Table 56A The Details of Agro-Economic Research Centres and the Lead Person Involved in the State Report

States	AERC Centre/Unit	Authors
Andhra Pradesh	AERC Waltare	L.K. Mohana Rao and Gangadhar Rao
Kerala	AERC Chennai	R. Raj Kumar
Maharashtra	AERC Pune	S.S. Kalamkar and Sangeeta Shroff

Section V
Constraints and Emerging Solutions

57

Pro-Poor Policy Contemplation

P.C. Bodh

THE PRO-POOR POLICY

Food and Agricultural Organization of the United Nations (FAO) had long been struggling to mitigate rural poverty with the participation of the poverty-stricken developing nations. It took various steps to gain deeper insight into the development-resistant rural poverty. One such action was initiated in 2007 by undertaking a series of FAO-funded studies in India with the collaboration of Ministry of Agriculture and Farmers Welfare. These were typical perception gathering field research studies executed through the Ministry of Agriculture and Cooperation's field research network of Agro Economic Research Units at the Institute of Economic Growth (IEG) of Delhi; the Institute of Social and Economic Change (ISEC), Bengaluru; and National Centre for Agricultural Economics and Policy Research (NCAP). The novelty of this research initiative was that it was driven by the global development experience that developmental investment in the non-farm activities in the rural areas is most cost-effective and outcome-oriented in eradicating poverty. As the Director incharge of the project in the Department of Agriculture, Cooperation and Farmers Welfare, I was associated with this project till its conclusion at the Regional Round Table at Bangkok after completion of the Indian part of the research projects at IEG, Delhi in July 2010. These research projects and the associated institutes were as follows:

1. Policy Analysis of Rural Non-Farm Employment in India—by IEG, Delhi;
2. Managing Common Pool Resources (CPR) of non-Timber Forests Produces (NTFP) in Tribal Areas of Eastern India with Reference to Non-Timber Forest Produces in Chhattisgarh; and Small Scale Culture Fisheries in Odisha by ISEC—ISEC, Bengaluru;
3. Infrastructure Development for Agricultural Growth and Poverty Alleviation—National Centre for Applied Agricultural Policy, New Delhi.

A brief review of the research project in view of the ever growing concern for eradicating rural poverty and the constraints of farming in putting farmers on higher levels of welfare and prosperity is given here. Provided here is a bird's eye-view of the research approach followed in these studies; their findings and country results, and finally the eight implementable policy actions thrown up by the research projects.

A relook at the outcomes of this important research initiative is all the more relevant today when development resistant socio-economic deprivation and indebtedness is culminating into farmers suicide. Rural poverty keeps the Indian state and central Governments and international developmental agencies engaged in making better socio-economic policy. It was this development concern that a policy analysis of this scale was thought necessary.

This Policy Analysis Study was assigned to a team of academicians of India's three institutes of repute.

RESEARCH APPROACH AND FINDINGS

Each of the three studies had different methodology to ensure their results. First, study on non-farm employment under guidance of Brajesh Jha followed diagnostic approach to the issue of rural employment, its outcome, and policy measures required for non-farm employment. The study used data available from India's National Sample Survey on employment and unorganized sector and Central Statistical Organization's Economic Census data. The study highlights trends in the non-farm sector in rural work force between 1999–2000 to 2004–15. It throws light on such important issues as trends in Non-Farm Employment, pattern of rural diversification, and its impact on non-farm employment and review of Government's policy related to non-farm employment.

The second study, Managing Common Pool Resources (CPR) for poverty eradication in the tribal areas of Eastern India was carried out by Indian Institute for Social and Economic Change under the guidance of R.S. Despande and D.K. Marothia. The investigators reviewed existing literature and looked into the working of existing common pool resource activities in NTFPs and small-scale culture fisheries. The case study of NTFP in Chattisgarh probed into the working of forests based common pool resource management for poverty eradication; and the second study, that is, small scale culture fisheries case of Odisha looked into the working of multiuse common water bodies.

The selection of common pool resource for the study was made on the basis of the extent of population of the poor and their exclusive dependence on these resources, namely non-timber forest products and multi-use water-body-based fisheries. The first CPR led the research team to the case study of Bastar and Sarguja districts of Chhattisgarh. The second resource, that is, multi-use water-body-based small culture fisheries favoured the case study of Jajpur and Cuttak districts of the state Odisha.

In this study, as a basic method, primary and secondary data analysis was combined with the ethnographic details. The ethnographic information was collected by personal field visits. The study focused on bringing out interlinkages among poverty, livelihood system and common property resources, and interface between poverty and common pool resources across different states.

Considering the special needs of this subject, the usual survey method was not followed. Rather, participatory appraisal and focus group discussion techniques were used for gathering information.

The third study was carried out by NCAP under the overall guidance of ISEC and Smita Sirohi was the researcher. This study looks into the impact of infrastructure on agricultural development and poverty alleviation to gauge the overall impact on employment and poverty and looked at the importance of new forms of infrastructure as automatic weather station, information and communication technology, marketing infrastructure, and rural mobile phone, and so on.

The study methodology was guided by its objective of understanding the impact of rural infrastructure developed so far, issues related to it, and effective delivery approaches. For this the researchers found it necessary to study various aspects of infrastructure and its linkages with poverty eradication through its developmental linkage.

The researchers, among other things, examined specifically those infrastructures which have stronger rural poverty eradication characteristics. This includes basic amenities such as roads, power, irrigation, drinking water, sanitation, education, health, and housing.

The study has also analysed policy approach towards infrastructure development in detail by studying latest initiatives made during the period of the study. These included Rural Infrastructure Development Fund, Bharat Nirman Yojana, Accelerated Irrigation Benefit Programme (AIBP), new forms of infrastructures such as information communication technology initiatives of establishing common service centres covering all six lakhs villages in the country, agricultural marketing information system rural bazars; and rural area specific infrastructure such as automatic weather stations, cold chains and retail chains, modern terminal markets, and rural mobile phones.

The study also talks of a new approach towards prioritizing rural infrastructure of roads, electricity, irrigation, agricultural extension, and information communication technology through a novel participatory approach of public private partnership to ensure fund availability, better management, effective utilization of resources available in the public sector and private sector domains.

Through the analysis of data available and experience with various macroeconomic policies dealing with various sectors, institutional incentives for increasing employment, and direct employment generation programmes and the focus on the decentralization of rural development, the study indicates that only a labour intensive primary resource based growth of manufacturing has the potential for rural non-farm employment generation. This has the production linkage of assured supply of agricultural produce for manufacturing and consumption linkage of spending on their goods.

For this to happen a cluster approach of development of large and medium enterprises linking with micro, small, and medium enterprises is suggested in the study. The approach specifically suggests the necessity for encouraging micro, small, and medium enterprises based on organic manufacturing with large and medium enterprises. Suggested to be of key importance was necessity to ensure institutional support for fiscal and monetary incentive to village resource based manufacturing and service activities to offset the disadvantages of poor infrastructure and institutional support in the rural areas. This necessitates infrastructure development measures of developing rural industrial estates Common Facility Centres (CFC) and providing subsidized land and cheap electricity.

To make possible full utilization of labour-intensive primary-resource-based manufacturing potential in the rural areas with full backward and forward linkages of production and consumption, all out effort for development of work skills as well as entrepreneurial capacity in the rural areas were found of key importance. Encouraging the participation of the voluntary sector and the private sector in this endeavour was found to be crucial. Not only this, the content of the training should focus on both conventional and information and communication technology based skills.

FINDINGS OF THE STUDIES

Convergent and Integrated Approach

One of the problems these studies suggested was the multiplicity of schemes related to poverty eradication and employment generation under implementation by various concerned developmental Ministries and Departments on both Central and State Government levels. To ensure that all these schemes move towards delivering their goals for complementing and strengthening each other, what was indicated to be of key importance was the need for ensuring convergence in their concepts, plan formulation, funding, implementation, and monitoring. This, it suggested, would ensure cost effectiveness, timely delivery, and results of an integrated development approach.

Common Pool Resources

Common Pool Resources have been traditionally in community ownership in India with very effective practices to manage them. This has been found true of NTFPs in Chattisgarh as well as small culture fisheries of Odisha. In both cases the resources were found important for the livelihood of the people.

The highlight of the success of the initiative for eradicating poverty and generating employment and income through effective management on all levels of economic activities involved in NTFP was the emergence of a strong chain of both nationalized and other NTFPs primary and cooperative societies, such as Joint Forestry Management Committees (JFMCs), self-help groups, and Marts and Sanjivinis. The problem, however, according to this study, was with the collection of non-nationalized NTFPs which are collected under usufruct rights due to which prices received by the collectors is far lower than the hard work put in for collecting the products. The solution suggested for this problem was extension of the present marketing network designed by Chattisgargh, Forest Produce (Trading and Development) Cooperative Society/Federation (MFPCF) in the remote villages. The success of marketing network was found to have been made possible by JFMC and Forest Development Authority (FDA).

Shared management principle has been followed by the government in implementing various forest based programmes in the tribal regions of Chhattisgarh. Joint forest management is involved in all the programmes initiated. A huge net saving of Rs 90 million have been accumulated by the JFMCs operational in the state. But the problem is the lack of institutional arrangements that can provide it an element of robust institutional management strength.

Institutional shortcomings faced by the common water bodies in Odisha have rendered this common pool resource ineffective in benefitting the local population in the way it should have. The identified shortcomings, the study suggested, were required to be removed to address the issues of sustainable use of water bodies for efficient production with no social and economic inequity. The solutions are to be found in public-private-community partnership governance in fish production with workable institutional arrangements supported by effective authority system.

Assigning the responsibility of the management to the Panchayatiraj Institutions with clear arrangements of harvesting, consumption rights was suggested to be the institutional way out. Similarly, designing a financial plan for utilization of available funds through JFMCs and FDA for value addition activities for NTFPs was suggested to be another way forwarded. Similarly, a proper participatory management of multi-use common water bodies for promotion of pisciculture to ensure the conversion of this resource into common pool resource and evolve responsible fishing conduct was suggested to be the most desired approach for managing fisheries better.

AVAILABILITY OF INFRASTRUCTURE

The availability of core infrastructure such as all weather roads, electricity, irrigation, rural credit infrastructure in various states of India indicate that 13 states of India are in low infrastructure development category. These are: Jharkhand, Assam, Odisha, Meghalaya, Uttaranchal, Rajasthan, Madhya Pradesh, Bihar, Chhattisgarh, Tripura, Uttar Pradesh, Himachal Pradesh and Jammu and Kashmir.

The approach suggested by the authors of the study was to use innovative means of funding, management, and work allocation such as public private partnership model in developing new forms of infrastructure. In addition to it, community participation, agri-business and marketing, Information and Communication Technology (ICT), investment, and improving regulatory structure are other ways to ensure new forms of infrastructure.

The new forms of infrastructure, beyond the conventional infrastructure, that has immense potential in non-farm employment and poverty alleviation were automatic weather stations, rural mobile, modern marketing, storage facilities, retail chains, commodity future and forward markets, and centre for perishable goods.

The study suggested that ensuring modern infrastructure development was key to poverty eradication schemes and development programmes in view of their growing importance in higher agricultural growth and rural poverty eradication and employment generation.

NON-FARM INCOME AND EMPLOYMENT

The NSS data on employment shows a sharp fall in share of agriculture in total workforce. Though manufacturing has traditionally been the most important industry in the non-farm sector, its share has reduced significantly. Nonetheless, productivity in rural non-farm sector is significantly lower than that of productivity in urban sectors. Enterprise survey showed that share of own account enterprises have reduced and that of establishment with hired workers has increased. Analysis shows that both distress and development related factors have led to the growth of rural non-farm employment in the country. The study argues that growth in agriculture, manufacturing, and tourism would trigger growth of productive employment in other sectors of non-farm economy. Though there have been numerous public institutions to encourage rural manufacturing but the gap between the productivity in rural and urban sector continues. This has also failed to check flight of manufacturing enterprises away from rural sector. The study, therefore, argues for sufficient incentives to encourage manufacturing in rural vicinity. Considering the higher proportion of workers in unorganized sector, the government has to come forward for provisioning of social security benefits.

The most important result of the Pro-Poor Policy programme was the churning out of rich research findings consisted in the three studies mentioned in the previous sections. These studies look into Pro-Poor Policy issues in the areas of non-farm employment, managing common pool resources, new forms of rural infrastructure. Findings of these studies are likely to provide a firm base for way forward in eradicating poverty by giving a thrust to non-farm employment generation.

Second important outcome of FAO's Pro-Poor Policy project is the recommendation about mobilization of the stakeholders that has taken place through various international, national, and departmental conferences, workshops, and discussions. The project process was initiated by convening an inception workshop, the same was organized for the Pro-Poor Policy Formulation, Dialogue and Implementation at the Country Level and was held at New Delhi during 14–15 December 2007 by Department of Agriculture and Cooperation. It involved all the important stakeholders ministries and departments, organizations, and representatives of FAO.

Third, the next important event was participation of the country in the Regional Consultation and High Level Round Table held during 4–6 March 2008 at Bangkok, Thailand, to give final shape to the Pro-Poor Policy Programme. This was followed by a country level Pro-Poor Policy Dialogue held at NCAP, Delhi on 22 December 2008. This was aimed at finding out the initial results of the studies being carried out by the identified institutes. This workshop was attended by all stakeholders including representatives of FAO.

Fourth activity organized by Pro-Poor Policy India was an internal meeting of the institutes involved in the three studies and the concerned Divisions of Department of Agriculture and Cooperation and experts on the thematic areas. This internal meeting was followed by a national workshop of the stakeholders held at ISEC, Bangalore on 25 May 2010, to discuss the findings of the studies on Managing Common Pool Resources in the tribal areas of Chattisgarh and Odisha. This workshop was attended by all the stakeholders, including progressive farmers and representatives of non-government organizations. The final national workshop of the stakeholders was held on 14 July 2010 at IEG, Delhi to finalize the studies on Policy Analysis of Rural Non-Farm Employment in India; and Infrastructure Development for Agricultural Growth and Poverty Alleviation. This workshop was also attended by all the stakeholder ministries and departments, representatives from NGOs and Progressive Farmers Organizations.

The officials involved in the Pro-Poor Policy India participated in capacity building trainings and workshops at Bangkok and Siem Reap which prepared them for their effective roles in implementation of the country's non-farm employment generation programmes.

EIGHT PRO-POOR ACTIONABLES

A three-pronged Pro-Poor Policy was the policy suggestion outcome of Pro-Poor FAO Project of Ministry of Agriculture and Farmers Welfare. This considers the commonality that has emerged from the major findings of three studies.

Village-Resource-Based Manufacturing

The policy analysis study very strongly suggests, from Indian economy's experience, that an all out effort for developing village-resource-based manufacturing is the key to pushing non-farm employment and income generation, which necessitates cluster approach of developing small, micro, and medium organic manufacturing enterprises linked to large and medium enterprises and ensuring massive skill and entrepreneurial capabilities, facilitating rural land and electricity availability.

Conventional and New Forms of Infrastructure

Second, the study of infrastructure development has suggested the need for giving a massive push to both conventional and new forms of infrastructure to ensure rapid poverty eradication and non-farm employment generation by employing novel funding and management model of PPP and involvement of NGO and SHG in the fields.

Common Pool Resources Crucial

Third, the case studies of NTFP in Chhattisgarh, and multi-use common water bodies in Odisha indicate that common pool resource play crucial role in poverty eradication, and all that they need is robust institutional and regulatory arrangements.

Considering these converging findings of the three studies, a three-pronged strategy was suggested for implementing all major recommendations. The first plank of this strategy is CPR and skill development to provide:

1. Water and Forest Resource Development Assistance.
2. Modern markets development assistance in rural areas with modern facilities
3. Automatic weather stations development assistance.
4. Developing resource centres with linkages to goods, services, markets, and enterprises.

Rural Employment Skill Development

For CPR development there has to be rural employment skill development relevant to NTFP, multi-use common water bodies, mountain and desert resources of tourism, and so on, information communication technology and entrepreneurial capacity building relevant to village resource based manufacturing.

Awareness of Pro-Poor Policy

The second plank of the three-pronged strategy is to increase awareness of Pro-Poor Policy for non-farm employment and income generation. This is achievable through workshops and discussions in all high priority zones indicated according to conventional infrastructure development status and poverty and common pool resource distribution and incidence of poverty. There are states and regions with low infrastructure development accompanied by low rural employment and development and inefficient use of common pool resources.

High Priority to Low Infrastructure Development States

The priority of three-pronged Pro-Poor Policy action plan discussed in this presentation was considered to be the states with low infrastructure development and states with high and low CPRs but high rural poverty with high rural tribal poverty as mentioned in the slide. With regard to development of cluster approach development of organic manufacturing, skill development, and providing promotional facilities populations close to water bodies, forest, and rural vicinities will be given priority.

Policy Prescriptions

A major way forward is the strategy to imbibe Pro-Poor Policy programme in the country's planning process, plan formulation, implementation, and monitoring with a sharp focus on outcome of poverty eradication programmes.

Since eventually the project's aim was assisting the participating countries in reducing rural poverty and supporting implementation of these policies at country level, arriving at implementable/actionable points of recommendations is the most important. Secondly, in view of FAO's focus on not letting United Nations' Millennium Development Goal's (MDG) of poverty eradication diluted due to persistence of hidden hunger, broad-based growth with an appropriate Pro-Poor Policy was an essential prerequisite. Thirdly, this also means ensuring reasonably high growth rate of 8 per cent supported by 4 per cent annual growth rate in agriculture.

Identifying eight implementable policy recommendations out of the given sets of recommendations from each study was made by first identifying common areas of important implementable recommendations in each sphere. Then implementable recommendations of holistic and mutually supplementary nature were picked up to the reach actionable points.

Specific areas of common policy issues emerged when certain recommendations from each of these studies were selected and the actionable points under each study were arranged into three groups of recommendations. This exercise finally provided complete body of implementable recommendations with clear lines of connectivity, actions, and relevance to the existing government policies and programmes was arrived at. These were compatible with MDGs and Government of India's flagship programmes of inclusive growth.

THREE-PRONGED PRO-POOR POLICY

A three-pronged implementable action plan for CPR Development and Skill Development (Resource & Skill Development Scheme (RSDS)) for Rural Employment; concept promotion and policy imbibing schemes is suggested:

1. Common Pool Resource Development (CPRD) and Skill Development (SD).
2. Awareness Promotion Workshops/Discussions.
3. Pro-Poor Policy Imbibing Strategy.

Each of these is briefly explained below.

Common Pool Resource Development and Skill Development

This was the main plank of implementable actions emerging from a close study of findings, analysis, and action plans of the three studies. It has the following two components:

CPRD and Participatory Maintenance

(i) Water and forest Resource Development Assistance to local institutions.
(ii) Development of modern markets in rural areas with modern facilities.
(iii) Development of Automatic Weather Stations.
(iv) Common(Pool) Resource Villages for developing resource centers with linkages to goods/service, markets, and enterprises.

Rural Employment Skill Development (RESD)

Rural non-farm employment skill development being most important component of poverty eradication in the rural areas needs to be ensured in the following ways:

1. SD for NTFP collection, grading, and primary processing.
2. SD for water bodies related economic activities.
3. SD for mountain/deserts resources, including tourism.
4. SD for ICT.
5. Entrepreneurial resource development linking small rural enterprise to the medium and large enterprises.

In addition to the suggested implementable recommendations in the line of findings of the studies, the Project considered other forms of CPRs and systems for poor development. First, CPRs include pastures, community lands, and institutes and institutions, for which Project needs to devise systems and procedures for ensuring equitable access to these common resources and develop skills to ensure their efficient utilization. Second, skill development for improving methods of gathering minor produces through efficiency enhancing tools and methods of productivity, increasing value-addition to minor forest produces; water harvesting for homes, fields, and training in harnessing of such resources, is important. Thirdly, other common pool resource activities that need project support are gathering information on irrigation, time, and methods; skill development in seed production, compost production, for increasing efficiency; skill development in post-harvesting and marketing activities; and agriculture support services skills development in workmanship of smiths, and so on, in transportation and construction services.

Awareness Promotion Workshops and Discussions

Awareness promotion workshops and discussions forums are required to be organized in the targeted beneficiary states that fall in the priority zone according to the findings of these three studies. The studies on infrastructure, non-farm employment study, as well as the study on CPR, indicate certain priority states and regions with low infrastructure development, low rural employment, and development, efficient use of CPRs. However, in most of the states—awareness promotion is required to be focused on priority.

This may need a programme for covering all identified states or some selected states representing adequately all important zones of India which need special efforts for making the stakeholders aware of the need for well thought out pro-poor rural development approach.

The list of stake holders should include: (i) state governments; (ii) central government; (iii) progressive farmers and local level institutions; (iv) business and industry engaged in targeted rural areas; (v) state agriculture and rural development universities institutes; (vi) policy advocacy groups from NGOs; (vii) Panchayati Raj Institutions.

Policy Imbibing Programme

Pro-poor policy interventions have to be devised in such a way that these are internalized by all the stakeholders. Therefore, the policy suggestions should be imbibed in a methodical way. Along with the effort towards widespreading the pro-poor policy and development concepts among various stakeholders in the entire government, private and NGO sectors—imbibing the concepts of pro-poor policy and development in the plan formulation, implementation and monitoring is most crucial. This is possible by strongly impressing upon the key government organizations responsible for formulating, approving, and funding plans for poverty eradication and employment generation.

These policy interventions have to be elaborated to get them implemented by the Union Planning Bodies and State Planning Bodies. The best way to do this is to organize presentations to the planning bodies. This will expound the concept of pro-poor policy and elaborate on the findings of these three studies along with their final recommendations. The presentation should address to the decision making group in the planning bodies. The bottom line of such an approach should be inclusion of a section on pro-poor policy focus of Asia Pacific Region in the plan documents of the state and central governments.

COVERAGE AND PRIORITIZATION

The development approach should be guided by a clear sense of status of development and abundance of resources.

1. For the new forms of infrastructure, such as AWS, MTM, and ICT, states with low infrastructure development as Jharkhand, Assam, Odisha, Meghalaya, Uttaranchal, Rajasthan, Madhya Pradesh, Bihar, Chhattisgarh, Tripura, Uttar Pradesh, Himachal Pradesh and Jammu and Kashmir need to be covered under capacity building projects.
2. States with both high and low CPRs but with high levels of rural poverty with high rural travel poverty need to be accorded priority in projects and schemes related to CPRs of NTFP and fisheries, and so on. These states include Odisha, Madhya Pradesh, Tamil Nadu, Meghalaya, Mizoram, West Bengal, Bihar, Arunachal, Assam, Manipur, Nagaland, Sikkim, Tripura, and Maharashtra.
3. For non-farm employment, populations close to rural industrial clusters/and rural development projects belonging to water, forests, energy should be focused.

THE RELEVANCE IN THE CURRENT DEVELOPMENT STRATEGY AND THE WAY FORWARD

Though the timeline suggested for implementation of the pro-poor policy recommendations was their implementation plan formulation, implementation and monitoring co-terminus with the formulating its 12 Five-Year Plan, new concepts and ideas of development can be promoted and imbibed continuously. Pro-poor policy findings of India need to be effectively advocated to the union and state government departments and organizations, their planners and policymakers, and all other stakeholders. Other stakeholders who need to be made aware of these schemes are: (i) progressive farmers and local level institutions; (ii) business and industry engaged in agriculture; (iii) state agriculture and rural development universities and institutes; (iv) policy advocacy groups from NGOs; and, (v) Panchayati Raj Institutions.

This needs not only high level workshops and conferences involving Planning Bodies and funding agencies, but also advocating inclusion of the implementable/actionable points of recommendations with assured funding support of FAO and the International Fund for Agricultural Development (IFAD) or the government departments.

The bottom line of these programmes was initiating four CPRs based development schemes and five rural employment skill development schemes in fifteen states of India; completion of concept promotion workshop and conferences in five representative zones of India; and advocating for inclusion of a pro-poor policy section in the plan documents. Though various schemes on these have been formulated and implemented what is needed to be contemplated on is the issue that large number of India states is still facing the problem of chronic poverty and indebtedness-triggered suicides. This hints that the pro-poor policy research project's findings and the policy prescriptions need to be pursued further in terms of both further studies and implementation of the recommendations.

In the context of the present National Skill Development Programme and Doubling of Farmers Income and the focus on higher welfare and prosperity of the farmers, the findings and recommendations and the eight implementable actionable points of the pro-poor policy gain more importance. The importance of this insight lies in the fact that, in addition to the newly realized potential of livestock and allied sector activities in raising farm income and employment, the thematic areas of agricultural development has proven to most investment efficient in eradicating poverty.

58

Climate Change and Indian Agriculture

A Review

Nilabja Ghosh

Human beings are as much a victim of the ominous global threat called climate change (CC) as they are the cause of it. Climate change refers to a change in the statistical distribution of the weather conditions over a reasonably long period, at least a decade. The United Nations Framework Convention on Climate Change (UNFCCC) adds a further dimension by defining CC as a change in climate that is attributable directly or indirectly to human activity that alters the composition of the global atmosphere and which is, in addition to natural climate variability, observed over comparable times. The UNFCCC actually ties the concept of CC with the emission of greenhouse gases (GHG) without confusing it with shorter period weather fluctuations and patterns like the El Nino although they can well be influenced by CC (Nadolnyak, Vedenov, and Novak 2008).

Agriculture, with its total exposure to nature, is at a higher risk to adverse weather events than other economic activities. India's agriculture is especially vulnerable to CC that manifests in weather events and their consequences, not just for her tropical geography, but because a large section of poor people employed in farming are indefensibly exposed to natural calamities. CC is a complex and cross-cutting subject. On the other hand, agriculture is also seen as a major source of emissions that cause CC but interestingly if planned properly, it can also provide a vital solution to the problem. This study focused on the relationships, dialogues, and research directions implied by CC in context of agriculture. The chapter was initiated with the following specific objectives:

1. to provide a comprehensive overview of CC in context of Indian agriculture; and
2. to facilitate understanding of India's policy options based primarily on existing literature reviewed in a multi-disciplinary framework.

METHODOLOGY AND DATA

The chapter has been based on the in-depth review of literature available to comprehend the concept of CC in the context of Indian agriculture. The overview of various studies conducted during the last decades and other available information on the subject was quite helpful to understand different policy options before the country to tackle this alarming issue.

RESULTS AND DISCUSSION

Much of south Asian agriculture remains dependent on uncertain monsoon which makes understanding and predicting monsoon important for food security and livelihood. Indian farmers are a victim of weather events on a regular basis as droughts, floods, waterlogging, cyclones, and thunder storms, all make farming unproductive and risky. They have developed indigenous methods of coping with such adversities but many of these mechanisms are either distressful such as migration, indebtedness, and curtailment of consumption while others are viewed as economically inefficient such as diversification towards less rewarding crops and reduced dependence on market purchased inputs. Farmers in coastal regions are vulnerable to floods and tidal waves and ingress of the sea, while in mountains, droughts are common due to rapid water run-off. Even plain areas lying in downstream locations are at a disadvantage when they are negatively affected by rain at higher altitudes. Some of the districts of Bihar, drained by the Himalayan Rivers, are most vulnerable to floods and waterlogging leading to migration as a way of coping. Climate change can make things even more difficult.

Evidences suggest that climates in this planet have changed repeatedly in geological time scales leading to the changes in life forms, rise of agriculture, and collapse of civilizations. While such devastating adjustments of climate were caused solely by natural factors, in the last two centuries, human beings, by burning fuels, releasing aerosols, manufacturing cement, and altering land use patterns, have added substances collectively called GHG to the atmosphere leading to the menace of CC. Anthropogenic interventions tend to raise global temperature over time disturbing nature's balance. The invention of the Stevenson Screen had enabled regular and scientific measurement and recording of temperatures over a century ago and developments in satellite imagery further facilitated the process. First observed by an amateur climatologist Callendar in 1938, human beings were however slow to recognize the association that fossil fuel burning has with carbon dioxide concentration in the atmosphere and the rise in global temperatures. An acknowledgement of the relation in a report of the giant oil company Exxon Mobil, reports of the unprecedented rise in estimated temperature in the last century by National Research Council in 2006, and the celebrated Keeling curve produced by an observatory in 2009 made the world community rise up to the looming threat of CC. The plot of global land and water temperature from 1880 to 2000 in Figure 58.1 is a clear indication of global warming.

GREENHOUSE GASES AND A SINK

A GHG is a chemical that is naturally present in earth's atmosphere. It maintains world temperatures at levels conducive for life but beyond a level it can warm the world to an intolerable level. Fortunately, chemical reactions limit the residence of the GHG in the atmosphere and transform them to their benign forms called sinks. When human activities generate excessive accumulation of GHG in the atmosphere, reflection of solar radiation to the space is reduced as in a greenhouse for plants. As a result, the

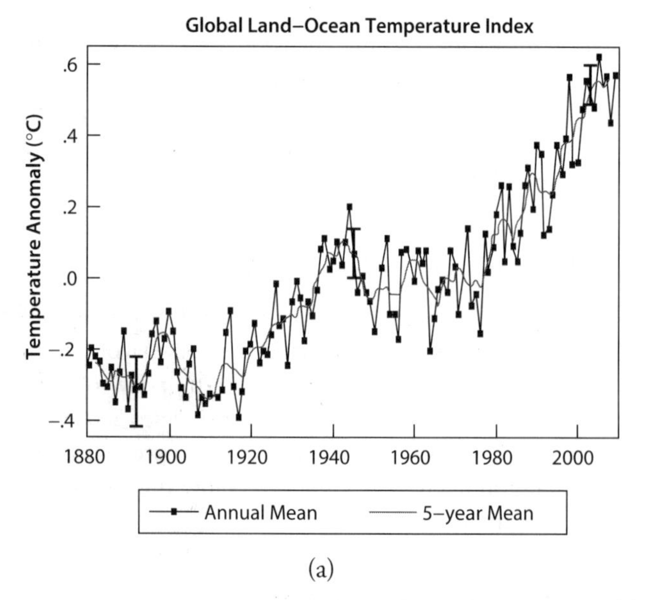

(a)

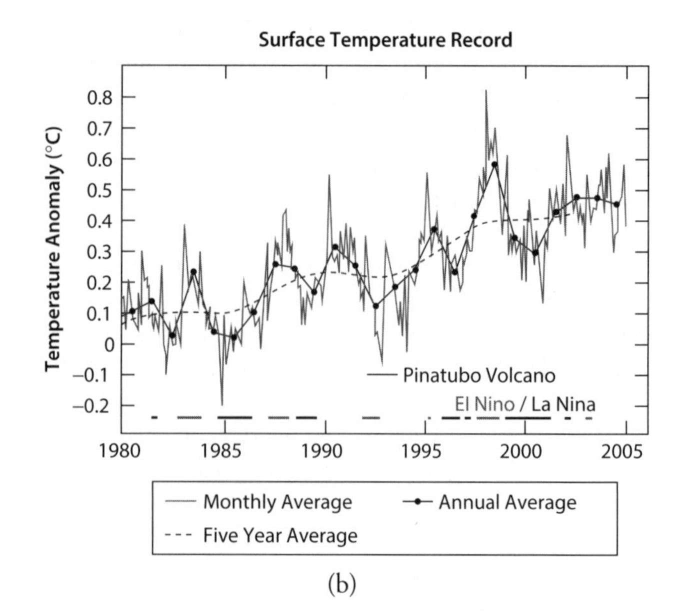

(b)

Figure 58.1 Global Temperature Graph

Source: http://en.wikipedia.org/wiki/Image:Instrumental_Temperature_Record.png
http://en.wikipedia.org/wiki/Image:Short_Instrumental_Temperature_Record.png

heat trapped in earth's atmosphere can transcend tolerable limits. Concentration of carbon dioxide, methane, and nitrous oxide, also historically associated with natural climatic changes and also all related to fossil fuel burning for human use, increased since the industrial revolution. They are the three most important GHG. Besides, other gases resulting from industrial activities, fluorinated gases like CFC, ozone, and water vapour also act as GHG but the emission of some of these are already regulated by the Montreal Protocol known to be a highly successful example of international agreement.

Sinks are reservoirs that trap GHG. They are also bodies that absorb GHG and reduce their concentration in the atmosphere. Historically, carbon losses resulted as native ecosystems were converted to farmlands to secure food by human beings. Trees sequester or capture and store carbon for years which otherwise would reach the environment till the time they are felled and destroyed when the same carbon is released back. Further, roots, litters, and decayed parts of trees generate biomasses that are transferred to the soil as soil organic matter (SOM). Both forests and soils are, therefore, carbon sinks while open soils are prone to carbon loss depending on temperature, moisture, and acidity. Conservation of SOM depends on the type of vegetation and more often its root structure. Similarly, methane and nitrous oxide can under certain situations undergo further chemical reaction to become harmless sink.

PROJECTIONS AND POLICY

Global warming, however, is not just the temperature rise due to GHG emission but also its far reaching effects making up CC. Heating of land and water have complex effects, interactions, and feedback effects on the biosphere. It alters the rainfall patterns in unknown ways. Hydrology of rivers, sequestration of water in ice-caps and glaciers, and water levels in seas and oceans are also affected. Weather forecasting is a traditional practice and custom in most societies but when the climate itself gets altered, forecasting weather assumes a challenging dimension. Understanding and projecting climate, which is a long-term phenomenon, in a dynamic scenario is no less challenging. It involves identifying the drivers, measuring them and their mutual interactions with reasonable insights about future. Emissions too become drivers but predicting them means predicting human behavior. General Circulation Models (GCM) were initial physical models that gave way to more sophisticated climate, economic, and agronomic models with the advent of powerful computers to project global and regional climate under alternative scenarios.

CC raises two types of policy issues—Mitigation and Adaptation. Atmosphere is a global common property without political borders and its interconnectedness suggests that the United Nations is the ideal umbrella body to work out acceptable ways to arrest the menace. The World Meteorological Organization held the first World Climate Conference in 1979, establishing the World Climate Programme and setting in process motivated activities in the 1980s. A number of conventions which were proof of intentions and protocols that became binding when ratified by the members followed to control the emissions. Regardless of where it is made, emissions affect the planet and developed countries that took the lead in industrialization have to take responsibility of past contributions to problems that are borne by poor nations now trying to develop. The UNFCCC explicitly recognized the *common and* differentiated responsibilities of the countries but arriving at consensus is not easy as is exemplified amply by the vacillations surrounding ratification of Kyoto Protocol (KP) and as manifested by other negotiations including the most recent twenty-first yearly session of the Conference of the Parties (COP) of UNFCCC in Paris 2015 where no detailed timetable or country-specific goals for emissions could be produced despite the universal agreement on zero carbon as soon as possible.

Mitigation or setting controls on emissions also can mean slowing down economic growth which can go against the interest of developing countries and the economic conditions of the world's poor. However, greater efficiency of energy use, shifts to cleaner energy, and strengthening of conducive institutions and incentives are ways to address CC without compromising growth. Making use of flexible and market compatible mechanisms provided by the KP such as the carbon market and the clean development mechanism can facilitate the shift and even help develop new economic sectors. On the other hand, adaptation, as a way to tolerate global warming which remains unavoidable even if mitigation is possible, is a far broader subject as it brings under fold all individuals and communities of the world affected by emissions made within certain sectors. Adaptation involves understanding how natural systems respond to actual or expected changes in climate or environment (Mitchel and Taneer 2007) and is a process that moderates the harm or exploits the benefits of CC. Both mitigation and adaptation are costly processes and, with limited resources, conflicts are bound to arise about their primacy. While many advocate treating them as twin processes with enough synergies to be exploited, some disagree drawing attention to their divergences of focus and their temporal and spatial reaches (Klein et al. 2005).

HOW CLIMATE CHANGE AFFECTS INDIAN AGRICULTURE

Indian climate is predominated by monsoon that involves seasonal winds bringing rain due to global circulations caused by temperature gradients. Moreover, northern India also gets rainfall from another global circulation known as the western disturbance bringing rain in end winter. The performance of India's agriculture has always been determined by the vagaries of monsoon though expansion of irrigation has reduced this monsoon dependence over time. Broadly crops in India are grown in two seasons. The kharif season is associated with the main monsoon while rabi crops like wheat and mustard in north India and commercial and food crops in the south depend respectively on western disturbances and the returning monsoon. Models indicate that CC's effect on Indian agriculture comes not just from higher temperatures but also from its effects on rainfall patterns. CC may make monsoon more variable with delayed onsets though there is no conclusive suggestion on the quantum of rainfall. Intensified monsoon circulation may even increase rainfall but possibilities of droughts and excess rainfall are enhanced besides other extreme events. Hot weather may become a cause of human distress, pests, crop damage, and even more emissions.

Crops have specific biological demands at various stages of their growth. Moisture, heat, and nutrients are major needs for plant growth and crop production. CC is likely to affect agriculture primarily by its effect on water availability and temperature. The impact of CC on rainfall is not known with certainty. Frequent droughts, floods, and extreme events and greater uncertainty are expected. Climatic processes being complex and interrelated across wide stretches in the world, linking any climate event specifically to CC is difficult but agriculture can be adversely affected by lack of rainfall and excess or untimely rainfall. The problem can be particularly serious in geographical regions like mountains where rivers have faster hydrology and in coastal and low lying areas more vulnerable to floods, waterlogging, and land degradation (Rasul 2014). Crops have specific biological needs for heat at various stages of grain formation and although higher temperature is not necessarily harmful, hotter weather related to crop demand can lead to greater transpiration, wilting, and untimely grain maturation. Higher temperatures can also undermine human health and thereby labour quality. Apart from higher temperature and variable rainfall, rising sea level, particularly affecting coastal and island agriculture, furious storms created by rising temperature gradients, forest fires caused by hotter and drier weather, and feedback effects with a spiraling effect on CC are other aspects of CC that have implication for Indian agriculture. Besides the chief causal factor which is increased carbon dioxide in air also has a direct effect on crop growth.

It is useful to note that the effect of CC can be both adverse and favourable for agriculture and some countries may even gain from CC (Easterling et al. 2007; Parry, Rosenzweig, and Livermore 2005; Tol 2002). Higher temperature helps in germination and grain maturation but the desired heat at any point of time differs across different crops dictated by biology. In cold and frost affected areas cultivation may in fact be enabled and rendered more productive by CC when the number of snow-free days increases. In general, the nature of effect will depend on the crops in question and the reference region. Agro-ecological models (FAO/IIASA) predict that CC will, by and large, affect agricultural yields in negative ways, though with regional variations, because of increased respiration and shortened vegetative and grain-filling periods and reduced floral reproduction (Satake and Yoshida 1978, Nishiyama and Satake 1981). It may hurt cold weather crops like wheat and mustard in tropical winters but may help in the higher altitudes in India. Rise in night-time temperature and spikelet sterility may also hurt yield of rice, known to be a hot weather crop (Yoshida and Parao 1976, Welch et al. 2010). Besides the quantitative implications, high temperature can also interfere with chemical processes and undermine nutrient quality in a crop even if yield is maintained. Similarly, while water stress is a serious problem for agriculture, intensified monsoon circulations can bring rainfall in dry areas and help agriculture while contrarily, excess rainfall can destroy crops by flooding. Carbon fertilization, which raises photosynthetic activities, is also a possible favourable effect, but the responses of plants vary by types and ecosystems and the net effect of CC is rather unresolved (Ainsworth and Long 2005; Kimball, Kobayashi, Bindi 2002; Long et al. 2006).

The possible effect of CC on crop yields is drawing particular research interest. While effect of rainfall on crop yield has long been recognized, over time special focus on the role of rainfall and the specific optima of rainfall also received only limited attention (Lahiri and Roy 1985). Important food crops are often promoted by public policy. The impact of economic incentives like price support, subsidies, insurance, grants, and investments can be better studied with econometric methods only when the rainfall effects are factored in with care. Incorporation of temperature data can make the models more reliable for inference and forecasting especially in context of CC. Panel data based models have been common for studying effects of rainfall (Kathy, Nicholas, and Gianfranco 2011) by capturing both temporal and spatial differences and recently spatial autocorrelation coefficients and spatial models are gaining popularity even in India (Kumar 2013, Kumari and Ghosh 2014, Mythili 2008). Besides yield, CC also affects cropping practices such as the choice of crops, their rotation, and changes of crop calendar as also the practices of irrigation. Dynamic crop growth models such as INFOCROP and WHTGROWS suggested that while rice and wheat yield in north India can be adversely affected by a rise in mean temperature the optimal sowing periods for rabi crops are likely to be advanced to avoid higher temperature arising later in the season which will in turn disturb the entire cropping pattern (Aggarwal and Sinha 1993; Hundal and Kaur 2007; Kalra et al. 2008). Shifts from and even elimination of certain crops are also not impossible if habitats for some cold weather crops cease to be conducive.

WEATHER EVENTS AND RESPONSES OF INDIAN AGRICULTURE

India has faced several extreme weather events in the last decade though not conclusively associated with GHG emissions. Late monsoon (2012), unseasonal rainfall in October (2010), devastating flash flood in Uttarakhand (2013), unexpected hailstorms (2015), drought like situation in succession (2014, 2015), warm winter (2009, 2015), unprecedented flood in Chennai (2015) are events that socially and economically affected farmers, putting them to distress and even suicides and imposing immense pressure of public finance. Timely relief from calamity, disaster management, and ameliorating farmer distress have, for the ruling government, become regular challenges, both economic and political in essence that only improvements of adaptation measures can help. Adaptation is a method by which farmers and the agricultural system cope with climatic events including disasters and reduce damages. Such mechanisms can be evolved through historical practices and innovations strengthened and supplemented by public initiatives in technology and institutions. The response of the government to the recent climatic experiences and the growing awareness about CC has increased research interest in and the development of drought resistant seeds, weather, and crop forecasting, both for early warning to farmers and for advance alert to the administration for planning policies and for remodeling of the public crop insurance scheme NAIS to MNAIS and further to PMFBY in 2016.

Investments on and modernization of space technology and its application, expansion of weather stations, intense drought management, and accommodation of advanced econometric methods for forecasting are some of the support services taken by government to enable adaptation (Ghosh, Chakravarty, and Bhatt 2008). Timely dissemination of weather alerts and exchanges of information are facilitated by information technology including the use of smart phones and the farmer portals in electronic media on real time basis. Interactive and holistic public extension is geared to encourage farmers to adopt technology consistent with adaptation and to exchange mutual experiences and ideas.

Crop insurance is a way to compensate farmers' losses due to weather events while conserving their incentive for production (Ghosh, Bhatt, and Yadav 2016). The government has been trying out various models of crop insurance with the support of banks, keeping in view the poverty and distress of the large farm community though success till date is limited. Experiments on weather insurance with private sector leadership are also encouraged. Information is a major ingredient of a successful insurance. Timely availability of crop estimates or forecasts can be extremely useful for insurance. While India's data protocols have gained impressive efficiency and transparency in recent times, use of technology like remote sensing and drones and quick reactions of banks and agriculture ministry to contingencies are projected to be helpful. Water is a main input to agriculture and given that rainfall is variable and likely to be more unpredictable with CC, irrigation as a means to controlled water supply and the effective use of rainwater are critical input for adaptation also. Investment on irrigation will have to take account of the potentially conjunctive role of ground and surface water and the mutual interaction between rainfall and irrigation as also the precision of water application based on micro-irrigation. Given the character of water as a common property and especially in river basins, water use efficiency as adaptation will take account of the knowledge of spatial interactions of water use (Elhorst 2003, 2011; Ping et al. 2004).

CAN AGRICULTURE INFLUENCE CLIMATE CHANGE?

Agriculture accounts for 16 per cent of world's GHG emission net of land use change effects of which crop cultivation has a 61 per cent share, the remaining accounted for by animal husbandry and manure management (Fischer et al. 2005). It is, however, not a major source of carbon emission except at the preliminary stage when land use changes reduce forest cover. Although forests and forest soils sequester carbon, cereals, pulses, grass, and other crops and their residues too act as sink albeit at a lesser degree. Fallowing is a way to recover carbon but multi-cropping, suitable choices of cover crops like grass in case of monocropping, mulching, careful planning of crop rotation, zero tillage with plowing back of residues, and even application of nitrogenous fertilizers are practices advocated by agronomists for preventing loss of SOM for sustainability of agriculture but which are beneficial also for CC (Agboola 1981; AL-Kaisi et al. 2008; Detwiler, Hall, and Bogdonoff 1985; Lee and Yu 2010; Lemon 1977; Loomis 1979; Mayers 1984; Rustogi, Singh, and Pathak 2002). Emission of methane and nitrous oxide are associated more closely with agriculture (USEPA 2006) though both these elements of GHG are also emitted by many natural processes also. Rice cultivation in wetland through transplantation is a source of methane (WRI 2008; USEPA 2006), known for its powerful global heating potential and long atmospheric life. Nitrous oxide is an intermediate output of a chain of chemical reactions to which nitrogenous fertilizers are subjected to before they become harmless. Measuring this emission is not easy.

Mitigating emission of GHG is perceived to contradict with agricultural development given that rice is the backbone of food security in the country and nitrogenous fertilizer has been the key to India's green revolution and food security. However, mitigation can be consistent and complementary with sustainable agricultural development if agriculture is managed prudently. Fertilizer use is associated with resource scarcity, leaching, and pollution of air and water. Thus, judicious use of fertilizer is beneficial for public budget, agriculture, and environment as well as for mitigation. Methane emission is a natural process in oxygen-free wetlands regardless of rice cultivation and thus cannot necessarily be attributed to agriculture. Puddling soil for transplantation which is associated with methanogenic microbes has several advantages like nitrogen fixation, and redundancy of weedicides, and even emission may be inhibited by the submerged ambience (Neue 1993). Direct seeding of rice, mid-season drainage, and SRI in areas not prone to waterlogging can reduce water use and methane emission without compromising production. Methane described as ghost light in parts of the country can be captured to energize farm operations. Precision based water and fertilizer use will help to reduce emission while enhancing sustainability. Organic fertilizers being sources of GHG cannot be treated as solution but arguably, by making soil less liable to erosion, can check carbon loss. The net effect of manuring on CC may also be contested if it is seen as a transformation rather than addition to GHG emission (Rustogi, Singh, and Pathak 2002). Managing wastes and crop residues and producing biofuels are other ways agriculture can contribute to mitigation.

POLICY SUGGESTIONS

To discharge the nation's fundamental obligations of food security and sustainable livelihoods, adaptation to climatic events needs to be ingrained in agricultural policy. Also as a responsible member of the global community suitable mitigation initiatives through agriculture compatible with the same obligations need to be critically and cautiously appraised, taking into view that mitigation can go hand in hand with conservation of resources and sustainable development. Planning and financing of adaptation will reduce farm distress and ensure food sufficiency. Cautious planning of rotations, crops and fallows, use of cover crops, and mulching can make farming conducive to carbon sequestration turning farm lands to sinks. Taking advantage of the carbon market may bring higher financial gains to farmers. Mitigation and adaptation, if pursued with prudence, will create new sectors with economic opportunity. India's future agricultural policy needs to be designed seeking balances and synergies among multiple objectives like mitigation, adaptation, food security, sustainability, and livelihood.

REFERENCES

Agboola, A.A. 1981. 'The Effect of Different Soil Tillage and Management Practices on the Physical and Chemical Properties of Soils and Maize Yield in a Rainforest Zone of Western Nigeria', *Agronomy Journal* 73(2): 247–51.

Aggarwal, P.K. and S.K. Sinha. 1993. 'Effects of Probable Increase in Carbon Dioxide and Temperature on Wheat Yield in India', *Journal of Agro-meteorology*, 48(5): 811–14.

Ainsworth, E.A and S.P Long. 2005. 'What We Have Learned from 15 Years of Free-air CO_2 Enrichment (FACE)? A Meta-analytic Review of the Responses of Photosynthesis, Canopy Properties and Plant Production to rising CO_2', *New Phytologist* 165(2), February: 351–72.

AL-Kaisi, M. Mahdi, Marc L. Kruse, and John E. Sawyer. 2008. 'Effect of Nitrogen Fertilizer Application on Growing Season Soil Carbon Dioxide Emission in a Corn-Soybean Rotation', *Journal of Environmental Quality* 37(2): 325–32.

Detwiler R.P., C.A.S. Hall, and P. Bogdonoff. 1985. 'Land Use Change and Carbon Exchange in the Tropics', *Environment Management* 9(4): 335–44.

Easterling, W.E., P.K. Aggarwal, P. Batima, K.M. Brander, L. Erda, S.M. Howden, A. Kirilenko, J. Morton, J.F. Soussana, J. Schmidhuber, and F.N. Tubiello. 2007. *Food, Fibre and Forest Products. Climate Change 2007: Impacts, Adaptation and Vulnerability.* Contribution of Working Group II to the Fourth Assessment Report of the Intergovernmental Panel on Climate Change.

Elhorst, J.P. 2003. 'Specification and Estimation of Spatial Panel Data Models', *International Regional Science Review* 26(3): 244–68.

———. 2011. 'Spatial Panel Models,' working paper, University of Groningen, Department of Economics, Econometrics and Finance.

Fischer, Gunther, Mahendra Shah, Francesco N. Tumeillo, and Harrij van Velhuizen. 2005. 'Socio-economic and Climate Change Impacts on Agriculture: An Integrated Assessment 1990–2080', *Philosophical Transactions of the Royal Society* B 360 (1463): 2067–83.

Ghosh, Nilabja, Sangeeta Chakravarty, and Yogesh Bhatt. 2008. 'Crop Supply Forecasting for Food Management A Review of Practices', *Agricultural Situation in India*, Ministry of Agriculture.

Ghosh, Nilabja, Yogesh Bhatt, and S.S. Yadav. 2016. 'What Restrains Demand for Crop Insurance in India?', Forthcoming Book Chapters in *Glimpse of Indian Agriculture* (Same volume).

Hundal, S.S. and Prabhjyot Kaur. 2007. 'Climatic Variability and its Impact on Cereal Productivity in Indian Punjab: A Simulation Study', *Current Science* 92(4): 506–12.

Kalra, Naveen, D. Chakraborty, Anil Sharma, and H.K. Rai. 2008. 'Effect of Increasing Temperature on Yield of Some Winter Crops in Northwest India', *Current Science* 94(1): 82–8.

Kathy, B., D.P. Nicholas, and P. Gianfranco. 2011. 'Spatial Approaches to Panel Data in Agricultural Economics: A Climate Change Application', *Journal of Agricultural and Applied Economics* 43(3): 325–38.

Kimball B.A., K. Kobayashi, M. Bindi. 2002. 'Responses of Agricultural Crops to Free-air CO_2 Enrichment', *Advances in Agronomy* 77: 293–368.

Klien, Richard J.T., E. Lisa F. Schipper, and Suraje Desai. 2005. 'Integrating Mitigation and Adaptation into Climate and Development Policy: Three Research Questions', *Environmental Science and Policy* 8(6): 579.

Kumar, K.S. 2013. 'Climate Sensitivity of Indian Agriculture', In *The Future of Indian Agriculture: Technology and Institutions*, edited by Nilabja Ghosh and C.S.C. Sekhar. New Delhi: Academic Foundation, 156–69.

Kumari, Anita and Nilabja Ghosh. 2014. 'The Trade in Temperature and Rainfall and the Possible Vulnerability of Indian Agriculture to Climate Change', *Food Studies: An Interdisciplinary Journal* 2 (4): 42–59.

Lahiri, A.K. and P. Roy. 1985. 'Rainfall and Supply Response: A Study of Rice in India', *Journal of Development Economics* 18.

Lee, Jeffery., C. Vernon Cole., Klause Flach, Sauerbeck, and Bobby Steward. 1993. 'Sources and Sinks of Carbon', *Water, Air, and Soil Pollution* 70(1–4): 111–22.

Lee, L.F. and J. Yu. 2010. 'Estimation of Spatial Autoregressive Panel Data Models With Fixed Effects', *Journal of Econometrics* 154(2): 165–85.

Lemon, E. 1977. 'The Land's Response To More Carbon Dioxide', in *The Fate of Fossil Fuel CO_2 in the Oceans*, edited by N.R. Andersen and Malahoff, 97–130. New York USA: Plenum Press.

Long, P. Stephen, Elizabeth. A. Ainsworth, Andrew D.B. Leakey, Josef Nosberger, Donald R. Ort. 2006. 'Food for Thought: Lower-Than-Expected Crop Yield Stimulation with Rising Carbon Dioxide Concentration', *Science* 312, 30 June: 1918–21.

Loomis, R.S. 1979. 'Carbondioxide and the Biosphere', In *Workshop on the Global Effects of Carbon Dioxide from Fossil Fuels*, edited by W.P. Elliott and L. Machta, 51–62. Department of Energy CONF-77o358, National Technical Information Service, Springfield, Virginia USA.

Mayer, N. 1984. *The Primary Source: Tropical Forests and Our Future*. New York, USA: W.W. Norton and Company.

Mitchell, T. and T.M. Tanner, eds. 2007. 'Embedding Climate Change Adaptation in Development Processes', in *Focus*: Issue 2. Institute of Development Studies, University of Sussex, UK. www.ids.ac.uk/climatechangeadaptation.

Mythili, G. 2008. *Acreage and Yield Response for Major Crops in the Pre-and Post-reform Periods in India: A Dynamic Panel Data Approach*, Mumbai: India Gandhi Institute of Development Research (Report prepared for IGIDR-ERS/USDA Project: Agricultural Markets and Policy).

Nadolnyak, D., D. Vedenov, and James Novak. 2008. 'Information Value of Weather-Based Yield Forecasts in Selecting Optimal Crop Insurance Coverage', *American Journal of Agricultural Economics* 90(5): 1245–55.

Neue Heinz-Ulrich. 1993. 'Methane Emission from Rice fields', *Bio-Science* 43(7): 466–74.

Nishiyama I, Satake T. 1981. 'High Temperature Damage in Rice Plants', *Japan Journal of Tropological Agriculture* 25: 14–19.

Parry, M., C. Rosenzweig, and M. Livermore. 2005. 'Climate Change, Global Food Supply and Risk of Hunger', *Phil. Trans. R. Soc.* 360(1463): 2125–38.

Rasul Golam. 2014. 'Food, Water and Energy Security in South Asia: A Nexus Perspective from Hindu Kush Himalayan Region', *Environmental Science and Policy* 39 35–48.

Rustogi, Monika, Shalini Singh, and H. Pathak. 2002. 'Emission of Carbon Dioxide from Soil', *Current Science* 82(5), 10 March: 510–17.

Satake, T. and S. Yoshida. 1978. 'High Temperature-induced Sterility in Indica Rices at Flowering', *Japan Journal of Crop Science* 47: 6–17.

Tol, R.S.J. 2002. 'New Estimates of the Damage Costs of Climate Change, Part I: Benchmark Estimates', *Environmental and Resource Economics* 21(1): 47–73.

US Environmental Protection Agency (USEPA). 2006. *Global Mitigation of Non-CO_2 Green House Gases*, Office of Atmospheric Programs, Washington DC, USA.

Welch R. Jarrod, Jeffrey R. Vincent, Maximilan Auffhammer, Piedad F. Moya, Achim Dobermann and David Dawe. 2010. 'Rice Yields in Tropical/Subtropical Asia Exhibit Large but Opposing Sensitivities to Minimum and Maximum Temperature', *Proceedings of the Academy of Sciences of United States of America*. Edited by Gurdev S. Khush, Early Edition, 107(33), 17 August, 14562–7.

World Resources Institute (WRI). 2008. *Climate Analysis Indicators Toolkit* (CAIT), accessed January 2008, available at: http://cait.wri.org/.

Yoshida S., F.T. Parao. 1976. 'Climatic Influence on Yield and Yield Components of Lowland Rice in the Tropics', *International Rice Research Institute*, Climate and rice: 471–94.

59

Determinants of Stagnation in Productivity of Important Crops in India

Elumalai Kannan

Agriculture plays an important role in the Indian economy by providing employment and livelihood to rural people even though its contribution to gross domestic product (GDP) has declined. While the share of agriculture in GDP declined from 36.4 per cent in 1982–83 to 14.0 per cent in 2012–13, dependence of rural workforce on agriculture did not decline at the same pace. According to 2011 Population Census, about 58 per cent of total workforce depended on agriculture for employment. Notwithstanding, India's agricultural sector witnessed appreciable growth in production of various crops over time. The main source of long-run growth was technological augmentation of yields, which could be realized through concerted efforts of farmers, agricultural technologists, and government departments.

However, the technological gains were not evenly shared across different parts of the country. The nature of technology was such that it benefitted only regions endowed with water resources, so the rainfed areas remain neglected (Chand and Chauhan 1999). Even with these problems, India's agriculture had registered impressive growth in 1980s. This seemed to have set complacency among policy planners and state extension machineries resulting in deceleration in output growth and reduction in public investment during 1990s (Bhalla and Singh 2009). In fact, the decade of 1990s showed mixed performance for the agricultural sector (Bhalla and Singh 2001, 2009; Rao 2003). Initially, there was a positive sign of growth but soon under the pressure of inter-sectoral growth pulls the investment trends in the agricultural sector decelerated.

Agricultural growth decelerated across states from the decade of 1990s to early 2000s. This deceleration, although most marked in rainfed areas, occurred in almost all the states and covered all major sub-sectors. The growth in productivity of all crops was showing declining trend across the regions by the end of 2004–05. Wheat, an important component of National Food Security, showed a declining trend in productivity, while rice and ragi seem to have reached yield plateau after 1992–93. These results clearly show the signs of acute distress and stagnation in productivity in the sector. Against this backdrop, the present study looks at the trends in agricultural productivity and its determinants in selected Indian states, namely Maharashtra, Karnataka, Punjab, and Himachal Pradesh.

The specific objectives of the study are:

1. to analyse the growth pattern of production and productivity of important crops across major states;

2. to trace the determinants for changes in productivity and stagnation of important crops; and
3. to suggest suitable interventions to overcome the problems of stagnation.

DATA AND METHODOLOGY

The study covered four major states—Maharashtra, Karnataka, Punjab, and Himachal Pradesh—representing different geographical settings of the country. The study used secondary data published in various sources such as Season and Crop Report of Maharashtra State, Districtwise Agricultural Statistical Information of Maharashtra State, Economic Survey of Maharashtra, Statistical Abstract of Karnataka, Karnataka Agriculture: A Profile, Statistical Abstract of Punjab, Annual Seasons and Crop Report of Himachal Pradesh, Cost of Cultivation of Principal Crops in India, Fertilizer Statistics, Agricultural Census, Livestock Census, and Finance Accounts.

The study employed growth accounting (index number method) technique to measure the Total Factor Productivity (TFP) of select crops in these four states. Tornqvist-Theil index was used to construct aggregate output index and aggregate input index for major crops across selected states.

Output Index

$$\ln\left(Q_t / Q_{t-1}\right) = 1/2\sum_j \left(S_{jt} + S_{jt-1}\right)\ln\left(Q_{jt} / Q_{jt-1}\right)$$

Input Index

$$\ln\left(X_t / X_{t-1}\right) = 1/2\sum_i \left(S_{it} + S_{it-1}\right)\ln\left(X_{it} / X_{it-1}\right)$$

TFP Index

$$\ln\left(TFP_t / TFP_{t-1}\right) = \ln\left(Q_t / Q_{t-1}\right) - \ln\left(X_t / X_{t-1}\right)$$

where, S_{jt} is the share of output j in total revenue, Q_{jt} is output j, S_{it} is the share of input i in total input cost, X_{it} is input i, and all specified in time t.

To analyse the determinants of total factor productivity, explanatory variables like government expenditure on research and education, extension and farmers training, rural literacy, canal irrigation, rainfall, fertilizer consumption, and regulated markets were regressed against TFP.

RESULTS AND DISCUSSION

Recent Developments in Agriculture

Maharashtra

Monsoon rain plays a critical role in the development of agriculture in Maharashtra, as about 83 per cent of cropped area is cultivated under rainfed condition. During 1960–61 to 1964–65, out of 26 districts, four districts were in high rainfall zone, 10 districts in medium rainfall group and remaining 12 districts in low rainfall category. The corresponding numbers during 2000–01 to 2004–05 were 6 districts, 10 districts, and 17 districts (out of 33 reported districts). Most districts in Pune, Latur, and Amravati divisions came under low rainfall group. This wide variation in rainfall across regions is a major constraint in affecting agricultural production. Despite huge spending on the irrigation projects, the proportion of gross area irrigated to gross cropped area in the state was around 18 per cent as against about 43 per cent at the national level in 2006–07. In fact, proportion of irrigated area was less than 15 per cent in 18 out of 33 districts in 2002–03. The low level of irrigation as well as the erratic rainfall pattern often discourages the farmers to take up intensive cultivation in many regions in the state.

Land use pattern in Maharashtra had slightly changed over time with marginal decline in the proportion of forest area and increase in area under non agricultural use. Net sown area (NSA) marginally declined between triennium ending (TE) 1962–63 and 2004–05. However, area cultivated more than once increased nearly three times between TE 1982–83 and TE 2004–05. Intensive cultivation of annual crops such as sugarcane and horticulture was responsible for this sharp increase. Area under cultivable waste, permanent pastures, land under tree crops and grooves and, current as well as other fallows accounted for about 49.33 lakh ha. With a proper wasteland development programme, these lands can be brought under productive use, which may help to reduce the rural poverty in the state. As per Agricultural Census 2000–01, marginal and small holdings together accounted for about 73.4 per cent of the total holdings and 38.7 per cent of operated area in Maharashtra. There is a sharp reduction in the number of large holdings over time. The average size of the operational holdings declined to 1.66 ha in 2001 from 2.21 ha in 1990–91. A similar trend has also been recorded in all the districts of the state.

Cropping pattern of the Maharashtra state had changed considerably in the last four decades. The area under cereals declined from 55.60 per cent during TE 1962–63 to 39.74 per cent in TE 2004–05, mainly due to substantial reduction in area under jowar. Though the productivity of pulses is low, their total area has increased from 12.60 per cent in TE 1962–63 to 15.26 per cent in TE 2004–05. Among oilseeds, proportion of the area under soybean increased considerably since the early 1990s. Cotton is an important crop predominantly cultivated under rainfed condition. Area under cotton increased from 2.6 million ha in TE 1962–63 to 3.2 million ha during TE 2000–01, but then declined to 2.8 million ha in TE 2004–05. One interesting observation is that despite significant increase in productivity of cotton due to introduction of Bt cotton in 2002, area under cotton was almost stable. The share area under sugarcane area increased from just 0.78 per cent in TE 1962–63 to 3.00 per cent of total cropped area in TE 2000–01 and then drastically declined to 1.98 per cent in TE 2004–05.

The growth analysis revealed that during the pre-green revolution period (1960–61 to 1966–67), production of almost all the crops was significantly low. But, growth in foodgrains production increased at the rate of 4.97 per cent per annum, despite a low growth in area (0.88 per cent) during 1967–68 to 1979–80. Productivity growth was instrumental for increasing production of foodgrains. The production of groundnut declined due to drastic decline in area. During post-green revolution period (1980–81 to 1989–90), except rice, wheat, and sugarcane, all other crops recorded increase in production. The decrease in production of wheat and sugarcane was due to decrease in area and, however, in case of rice, it was due to decline in productivity. During this period, the production of oilseed and pulses increased significantly, which may be due to implementation of Technology Mission in the country during 1986–87. During post reform period (1990–91 to 2004–05), growth in production of almost all the crops has declined due to decline in area and productivity. Production of all the major cereals decreased either due to stagnation in yield or reduction in area or both. Except soybean, all other oilseed crop production decelerated due to negative productivity growth. As soybean emerged as an alternative crop in rainfed areas, area under cereals and cotton has been increasingly shifting towards soybean cultivation in recent years.

Increase in productivity depends on factors like technology, use of quality seeds, fertilizers, irrigation, pesticides, and micronutrients. The area under high yielding varieties (HYV) significantly increased from 1.08 lakh ha in 1966–67 to 82.32 lakh ha in 2003 implying that farmers are fully aware of the importance of HYVs in crop cultivation. Credit availability is of crucial importance for agricultural development. The ground level credit flow to agriculture sector in Maharashtra increased significantly from Rs 859 crore in 1980–81 to Rs 1,283 crore in 1990–91, Rs 4,576 crore in 2000–01, and Rs 12,113 crore in 2007–08.

The number of plant protection implements such as sprayer and dusters has increased from 0.46 lakh in 1972 to 7.59 lakh in 2003. The use of four wheel tractors increased significantly to 1.05 lakh in 2003 from just 0.01 lakh in 1961. Further, number of electric pump sets has increased tremendously over time due to subsidy schemes operated by state government. Though there is significant increase in number of pump sets in the state, wide variation among the regions exist with 54 per cent of pump sets being located in Western Maharashtra. Despite poor irrigation facility, consumption of fertilizers increased from 1.62 kg/ha in 1960–61 to 103.1kg/ha in 2007–08. Out of the total irrigated area (3.86 million ha), paddy, wheat, jowar, and bajra together accounted for only 38.05 per cent, while sugarcane alone accounted for 16.87 per cent. Pulses and oilseeds together have accounted for only about 14.0 per cent of irrigated area.

The analysis of trends in cost of cultivation of major crops in the last three decades showed that the cost of cultivation of major crops increased substantially. Agricultural prices play a critical role in influencing cropping pattern, farmers' income, and welfare. The average annual growth rates of Wholesale Prices Index (WPI), Farm Harvest Prices (FHP), and Minimum Support Prices (MSP) of major crops shows that during the period 1980–81 to 1989–90, the growth rate of MSP of all the crops was higher than the FHP. During 1990–91 to 2004–05, except wheat and cotton, all other crops have recorded higher growth in MSP as compared to WPI and FHP. Thus, there was a significant increase in minimum support prices of agricultural commodities during 1990–91 to 2004–05.

Karnataka

Rainfall plays an important role in maintaining the crop productivity in Karnataka. Five year average annual rainfall in Karnataka was 1322.9 mm in 1964 bit it declined continuously to reach 1164.6 mm in 1989. Though rainfall activities picked up during 1990s, but again decelerated during 2000s. In Karnataka, over three-fourth of cultivated land depend on rainfall. The net irrigated area increased from 9.1 lakh ha in 1962–63 to 25.5 lakh ha in 2004–05. The gross irrigated area also increased from 10.0 lakh ha in 1962–63 to 18.6 lakh ha in 1982–83 and then to 29.6 lakh ha in 2004–05. The proportion of gross irrigated area to gross cropped area stood at 24.8 per cent. Among sources of irrigation, tanks were predominant source of supply of irrigation water during 1960s. Overtime, canal and tube wells emerged as the major sources of irrigation. Drying up of tanks due to poor maintenance, encroachment, conflicts among users, and vagaries of rainfall has forced the farmers to resort to ground water. As a result, area irrigated through tube wells increased remarkably from 0.3 per cent in 1972–73 to 31.3 per cent in 2004–05. Open wells were another important source of irrigation till early 1990s constituting about 25 per cent of net irrigated area. However, its share has declined to 16.5 per cent in 2004–05.

The land use pattern has witnessed some changes over time. Land put to non-agricultural uses increased from 4.4 per cent of total geographical area in 1962–63 to 7.0 per cent in 2004–05. Increase in area for non-agricultural uses might have come from the reduction in area under cultivable waste and permanent pastures and other grazing land. Net sown area in total geographical area has almost remained constant at 10 million ha. Considerable increase in area sown more than once has led to rise in the gross cropped area from 56.8 per cent in 1962–63 to 62.6 per cent in 2004–05. This has resulted in increasing cropping intensity from 103.6 in 1962–93 to 114.7 in 1992–93 and further to 118.5 in 2004–05.

The number of landholdings has more than doubled to 75.81 lakhs in 2005–06 from the level recorded in 1970–71. While the number of marginal, small, and semi-medium holdings increased substantially over time, the number of medium and large land holdings has declined considerably. The changing structure of landholdings clearly indicates that farm sizes are increasingly becoming marginal and small. The huge increase in number of holdings overtime has reduced the average size of area operated per holding. The area operated per holding was 3.2 ha in 1970–71, which declined to 1.63 ha in 2005–06. Size of the operated area per landholding influences the choice of crops grown. In general, marginal and small farmers practise intensive cultivation of crops when compared to medium and large farmers.

Food grain crops dominate the cropping pattern accounting for over two-third of total gross cropped area (GCA) in Karnataka. Surprisingly, per cent area under food grains has declined from 71.9 per cent in 1962–63 to 60.0 per cent in 2004–05. Among food grains, coarse cereals occupy prominent place in the cropping pattern. In 2004–05, jowar accounted for about 14.6 per cent of total area followed by rice (9.9 per cent), sunflower (9.2 per cent), ragi (7.4 per cent), and groundnut (7.3 per cent). However, per cent area under certain food crops like jowar, bajra, ragi, and small millets has come down drastically since 1962–63. Area under maize has increased considerably from meagre 0.1 per cent of GCA in 1962–63 to 5.9 per cent in 2004–05. Similarly, per cent area under arhar in total cropped area has increased from 2.5 per cent in 1972–73 to 4.5 per cent in 2004–05.

The per cent area under groundnut has declined sharply since 2000s due to persistent drought like conditions prevailing in the State. However, share of area under sunflower has registered sharp increase from 1.0 per cent in 1982–83 to 9.2 per cent in 2004–05. Among cash crops, area under cotton has declined drastically over time. Sugarcane area has increased considerably from 1960s to 2000s, but has showed declining trend since 2001–02. The analysis of cropping pattern revealed that there is marked shift in area from cereals to pulses, oilseeds and high value crops like vegetables and plantation crops.

The compound annual growth in area under food grains was 0.3 per cent during pre-green revolution and it declined to -0.1 per cent during green revolution period. However, growth in food grains production was high at of 3.5 per cent during the green revolution period. This high growth rate has largely come from growth in yield (3.8 per cent). During post-green revolution period growth in area under food grains was positive at 0.4 per cent, but growth in production and yield has decelerated. During the period of economic reforms, growth in area under food grains decelerated but yield has registered positive growth rate of 0.4 per cent. These results broadly indicate that growth in yield of food grains has decelerated during period of post-green revolution and economic reforms. However, some reversal of growth in yield of rice, ragi, arhar, and gram was evident during economic reforms period. The growth in yield of cotton has increased from -6.2 per cent in pre-green revolution period to 3.9 per cent and 9.7 per cent in green-revolution and post-green revolution periods, respectively.

Area under HYVs has increased remarkably from 0.7 lakhs ha in 1966 to 21.2 lakhs ha in 1982 and then to 38.4 lakhs ha in 2003. Among crops, area under HYVs of rice was at 88.3 per cent, jowar at 65.1 per cent, and wheat at 50.6 per cent. It indicates that there exists greater scope to increase production of these crops through cultivation of HYVs. Further, the extent of use of machineries like tractors and pump sets indicate the level of mechanization in agriculture. In 2003, the number of tractors per thousand hectare of GCA worked out to be only 9.3, which is lower than all India average of 16.7. Similarly, the number of pump sets per thousand hectare of GCA was 67.7 as against all India average of 111. The number of plant protection equipment per thousand hectares of cropped area was found to be low at 7.7.

The consumption of NPK per hectare of cropped area was 3.4 kg in 1961 and has increased continuously to reach 76.6 kg in 1992 and then to 90 Kg. However,

NPK consumption in Karnataka is low when compared with all India average consumption of 136 kg per ha. The low consumption may be attributed to dry land based and coarse cereals dominated cropping pattern. Karnataka was one of the states had witnessed spate of farmers' suicide during 2000s. Both the Central and State Governments have attempted to increase the flow of agricultural credit. Amount of agricultural credit advanced in the State has risen from Rs 1,886 crores in 1997 to Rs 4,039 crores in 2003.

Cost structure of important crops was studied in terms of share of traditional and modern inputs. Traditional inputs like land and human labour have accounted for over 50 per cent of total cost of paddy cultivation in Karnataka. The cost share of modern inputs like pesticides and machine labour, by and large, has increased during study period. Of the total cost of jowar cultivation, traditional inputs accounted for about three-fourth of total cost. In case of arhar, traditional inputs accounted for about two-third of the total cost. Though use of machine labour has increased, but animal labour continues to dominate operations involved in its cultivation. As groundnut is cultivated largely under dry land conditions, the share of improved inputs like pesticides, irrigation, and machine labour is found to be low. In case of cotton cultivation, traditional inputs constituted about 70 per cent of total cost. Pesticides and fertilizrs accounted for 3.8 per cent and 2.4 per cent, respectively in 2004–05.

In case of prices, after slump in growth during 1967–68 to 1979–80, wholesale and farm harvest prices have increased during 1980–81 to 1989–90. Among all commodities, the increase in growth of farm harvest price was relatively high for rice. During 1990–91 to 2004–05 growth in wholesale and farm harvest prices has moderated. However, growth in minimum support price for most of commodities has jumped during this period, registering over 7 per cent per annum. This has led to improving the terms of trade in favour of agriculture.

Punjab

In the state of Punjab, almost 97 per cent of the cultivated area is under assured irrigation which is the major reason for higher productivity and input use in agriculture. More than 75 per cent of annual rainfall is received during the southwest monsoon. There has been continuous increase in the net sown area in the state since 1960–61 and the proportion of net sown area to total geographical area stood at 83.14 per cent by the triennium ending 2006–07.

Data from the 2000–01 agriculture census indicated that the average holding size in the state had improved to nearly 4.03 ha, but declined marginally to 3.95 ha in the 2005–06. Cereals, particularly rice and wheat dominate the cropping pattern scenario in the state as about 70 per cent of gross cropped area in the state in is occupied by these two crops. There was tremendous increase in area, production, and yield under paddy for all the periods under study. Wheat also showed the same trend but the increase was at lesser pace than for the paddy. Increase in area and productivity of these crops are the main movers for the increase in production of these crops. All other crops showed either decrease in area or the insignificant increase in area during this period. The paddy-wheat rotation became predominant at the cost of area under maize, other cereals, oilseed, and pulses in the state.

The total availability of agricultural credit has increased from Rs 945 million during 1971–75 to Rs 88,838 million in 2001–05. Total consumption of NPK in Punjab, which was merely 276 thousand nutrient tonnes during 1971–75, had increased over time and reached to a level of 14.52 lakh nutrient tonnes by the period 2001–05. In terms of per ha consumption, it was the highest at about 184 kgs/ha of fertilizers in the period 2001–05. The Punjab agriculture is highly mechanized in nature. The density of tractors per thousand ha is 64 in Punjab, which is the highest in India. Similarly, electric tube wells are increasing rapidly in Punjab and it accounted for 8.27 lakh during 2001–05. The number of diesel tube wells has reached 2.42 lakh in 2001–05.

The area under HYVs increased from about 22 lakh ha in the period 1971–75 to about 61 lakh ha during 2001–01. Accordingly, the level of use of major inputs has increased significantly during 2000–01 to 2005–06 for paddy, wheat, and cotton crops. Although the procurement price, wholesale price, and farm harvest price showed significantly consistent growth over the years but for paddy and maize crops, the growth of MSP in the period 1980–81 to 1989–90 was higher than for the growth in wholesale price and FHP, while it had reversed during 1990–91 to 2006–07.

Himachal Pradesh

In Himachal Pradesh, rainfall pattern is varied and unequal across districts. The agricultural production in Himachal Pradesh is more or less dependent upon rains. During 2003–04, out of 540 thousand ha of net area sown, only 19.4 per cent is irrigated. Kuhls is the main source of irrigation in the state and accounted for 84 per cent of irrigated area in 2003–04. The contribution of tube wells and canals constituted 10 and 6 per cent, respectively.

The land use pattern shows that the area under forest has decreased from 32 per cent in 1962–63 to 24 per cent in the year 2003–04. The land put to non agricultural uses increased from about 4 per cent to about 10 per cent during this period. Permanent pasture and other grazing land accounted for 33 per cent of the reported area during 2003–04. Land under current fallow and other fallow land slightly increased from 1.0 per cent to 1.60 per cent during the study period. Meanwhile, the proportion of marginal and small holdings accounts for 86 per cent of the total holdings in the state. The percentage of marginal holdings had gone up to 67 per cent in 2000–01 with an area of 28 per cent. The average size of holdings in the state has declined from 1.53 ha in 1970–71 to 1.07 ha in 2000–01.

Foodgrains accounted for about 85 per cent of the GCA in 2003–04. Among cereals, the area under maize and wheat has increased from 27 per cent and 33 per cent in TE 1962–63 to 31 per cent and 38 per cent in 2003–04, respectively. However, the area under two principle cereals paddy and barley has decreased. The area under fruit and vegetable crops registered the highest increase from 4 per cent in 1962–63 to 12 per cent in 2003–04.

Growth analysis revealed that the area under food crops has registered a positive growth through 1960s to 1980s, whereas in the later period of 1990–91 to 2003–04 negative growth has been observed at the rate of 0.43 per cent per annum. The growth in production of food crops was positive in all the study periods but it was at the decreasing rate. In the case of rice, growth in area has resulted in positive growth in production during 1990–91 to 2003–04 when compared to other periods. Only during 1967–68 to 1979–80, growth in yield was responsible to increase the production. As far as maize is concerned, positive growth in both area and yield resulting positive growth in production was observed during 1960–61 to 1966–67 and 1967–68 to 1979–80.

The increase in production of wheat was due to increase in both area and yield in the 1960–61 to 1966–67. In the period 1966–67 to 1979–80, the growth in yield was negative but production increased positively because of expansion of area. While in the period of 1980–81 to 1989–90, both area and yield are responsible for the increase in production, in the period 1990–91 to 2003–04, the growth in production was negative due to the negative growth in area and yield. In the case of barley, growth in production had decreased despite increase in area, but negative growth in the yield during 1980–81 to 1989–90. Similar pattern also was observed during 1990–91 to 2003–04.

The total area brought under HYV's in the period of 1976–80 was only 379 thousand ha which increased to 555 thousand ha in 1991–95. After that the area was started to decline and reached 118 thousand ha in 2001–04. The minimum support price for wheat rice and maize has increased continuously over time. Similarly, farm harvest prices of these crops have increased substantially during the same period. Farm harvest prices for wheat, rice, and maize were high as compared to their respective minimum support prices during the study period.

The amount of agricultural credit disbursed stood at Rs 45613 lakh in 2001–04. The number of hand and power operated sprayer and duster are 56,142 and 4,053, respectively in 2001–04 as against of 32,948 and 1,791 in 1991–95. The number of tractors has increased from four in 1961–65 to 4,222 in 2001–04. Similarly, the number of electric and diesel pump sets has increased from two and six in 1961–65 to 1,571 and 1,048 in 2001–04, respectively. The use of chemical fertilizers is low in Himachal Pradesh. The consumption of NPK has increased from 9 kg in 1971–75 to 44 kg per ha in 2001–04.

During the period of 1996–97 to 2004–05, for maize the overall total cost per ha was worked out to Rs 9,262. Out of total cost per ha cost of traditional inputs accounted for 80 per cent. As far as wheat is concerned, per ha total cost was estimated to Rs 11155 during the period of 1996–97 to 2005–06. Out of this traditional inputs accounted for 65 per cent.

The wholesale price index of rice and wheat has registered growth rate of 3.96 and 2.10 per cent, respectively during the period 1990–91 to 2004–05 as compared to 1.98 per cent and 1.69 per cent during 1980–81 to 1989–90. The growth of WPI of maize was high at 2.95 per cent in the period of 1980–81 to 1989–90 but declined to 2.15 per cent per annum in 1990–91 to 2004–05. The MSP of rice, wheat, and maize has increased at the rate of 7.72 per cent, 9.10 per cent, and 7.94 per cent per, respectively during 1990–91 to 2004–05.

Total Factor Productivity (TFP) and its Determinants

Maharashtra

The growth in TFP indices was positive for jowar, moong, urad, sunflower, safflower, and sugarcane, but negative for bajra, groundnut, and cotton during the overall period. The significant deceleration in TFP growth for major crops during the recent period has serious implications for the agricultural development of Maharashtra. There is an urgent need to increase TFP growth in all the major crops especially pulses, oilseeds, and cotton to make their cultivation profitable and promote crop diversification.

The analysis of determinants of TFP growth in agriculture in Maharashtra showed that number of regulated markets and annual rainfall are the most important sources of growth. The effect of research, literacy, and rainfall during June and August was found to be negative.

Karnataka

Output, input, and TFP indices were constructed for major crops at different periods depending up on availability of data. Annual growth in TFP of paddy was high at 1.48 per cent during 1990s. Higher output growth triggered by technological change has resulted in positive TFP growth. During entire period of analysis, that is, 1973–74 to 2004–05 TFP has risen at 0.75 per cent. The contribution of TFP to output growth was 35.6 per cent. For jowar, TFP registered negative growth during 1980s and 1990s. However, TFP had risen positively during 1970s and early 2000s. TFP growth in maize was negative during the 1997–98 to 2004–05 and also during the sub-period 2000–01 to 2004–05. TFP of ragi recorded annual growth rate of 0.42 per cent during 1984–85 to 2004–05, which contributed about 37.72 per cent of total output growth. For Arhar, output growth was mainly driven by growth in input during 1980s and 1990s.

During 1976–77 to 2004–05, TFP contributed about 77.72 per cent to groundnut output growth indicating that technology has played greater role in augmenting its production of groundnut. In sunflower production, growth in inputs was the main driver of output growth. Annual growth in input, output and TFP of cotton was 1.37 per cent, 5.03 per cent, and 3.66 per cent, respectively during 1974–75 to 2004–05. Technical change played an important role in increasing output growth with the contribution of 72.73 per cent. Sugarcane output grew positively across all the periods under study. However, higher input growth than output growth has resulted in negative TFP growth of 5.27 per cent in 1990–91 to 1999–2000. For Karnataka as whole, input and output indices have registered growth rate of 0.85 per cent and 1.34 per cent, respectively. TFP has risen at 0.49 per cent per annum and it has contributed about 36.53 per cent to total output growth.

To analyse determinants of TFP, multiple regression analysis was carried out. Results indicated that that government expenditure on research, education, and extension, canal irrigation, rainfall, balanced use of fertilizers, and markets were the important drivers of crop productivity in Karnataka. Effect of rural literacy was found to be negative and significant. The possible explanation for such result is migration of rural literates to urban areas due to increased availability of non-farm employment opportunities and distress like conditions prevailing in the agriculture sector. Thus, they may not contribute directly to increasing agricultural productivity.

Punjab

For paddy, input index has registered negative growth in both the periods. The growth in output is high at 6.55 per cent in 1981–82 to 1989–90 and has nearly doubled to 12.67 per cent during 1990–91 to 2004–05. Higher growth in output has resulted in positive and high growth in TFP. In fact, technical change seems to have played a greater role in registering impressive paddy output growth.

In case of wheat, output and TFP have registered respective growth rate of 9.44 per cent and 9.11 per cent in 1981–82 to 1989–90 and they have continued to maintain the growth momentum during recent period also. The TFP growth during 1990–91 to 2004–05 was high at 13.33 per cent. With respect to cotton, the introduction of Bt cotton technology was reflected in higher TFP growth during the period 1990–91 to 2004–05 as compared to that of the period 1981–82 to 1989–90. However, cotton output growth seems to have decelerated during the recent period.

Himachal Pradesh

TFP was calculated for two major crops namely, maize and wheat. The input and output index of maize grew at 1.02 and 1.16 per cent, respectively. The annual growth in TFP was low at 0.15 per cent, indicating that there is huge scope for accelerating output growth through introduction of new technologies in the cultivation of maize. In case of wheat, both the output and input indices grew almost at the same rate during the period under study. Shockingly, growth in TFP was stagnant and thus it is necessary that proper policy interventions should be made for increasing investment in research, infrastructure, and extension services.

CONCLUSIONS AND POLICY RECOMMENDATIONS

The broader conclusions and policy recommendations for the selected states are presented in this section.

Maharashtra

Instability in crop production is one of the characteristics of Maharashtra agriculture with considerable inter-district disparities in the growth rates observed among different crops. Ensuring adequate and timely supply of quality inputs to farmers can address this problem. Public investment on irrigation increase the utilization of the unutilized potential will also help. At the same time, private investment in research and development activities in agriculture needs to be promoted for accelerating growth.

Dry land farming technology development played a significant role in promoting productivity and output growth in Maharashtra. Research efforts, therefore, should be intensified further to develop high yielding varieties of the crops suitable to different agro-climatic conditions of the region. Drought tolerant and pest resistant varieties should also be developed.

Over exploitation of groundwater should be restricted through legal, technological, and institutional options. Cultivation of crops using surface method of irrigation should not be allowed and micro-irrigation (drip and sprinkler) should be made compulsory in all those areas. Digging new wells should not be allowed in the areas that are classified as 'dark' block. At the same time volumetric pricing of water should be introduced to enhance water use efficiency.

Crop insurance should be expanded to cover all districts under agrarian distress. Commercial banks, Regional Rural Banks, and cooperative banks should make crop insurance mandatory for all agricultural loanees. Short duration cum high yielding varieties of pulse crops should be made available to improve the cultivation of these crops. Government should give top priority to bring technological improvement particularly on varietal development by allocating more resources for research activities on pulses. An appropriate network of extension services will have to be created to stimulate and encourage both top-down and bottom-up flows of information among farmers, extension workers, and research scientists to promote the generation, adoption, and evaluation of location specific farm technologies.

Karnataka

With limited scope for improving irrigation facilities, dry land farming offers potential source of agricultural growth in Karnataka. Concerted efforts should be made to develop and promote dry land forming through appropriate policy interventions and investment decisions. As farm sizes are becoming increasingly marginal and small, farmers of these groups should be given a centre stage in all policy decision making. For instance, whatever technological interventions is made to enhance the agricultural production and farm income, they should be applicable to marginal and small holdings also.

Extension services should be invigorated through appropriate policy mechanism. In this direction, coordination between various government departments, training, and technology disseminating centres and research organizations should be strengthened for effective transfer of technology to farmers' field. The concept of 'on farm school' should be promoted in every village to impart training and education on agricultural practices, new technology, pests and diseases control, and other related aspects. In order to popularize it, attendance should be made compulsory for those farmers who receive subsidy on any agricultural inputs/machinery.

As in Karnataka experiences vary in agro-climatic conditions, natural resources, and irrigation facilities, prioritization of extension activities should be undertaken at the district level. Low level of TFP growth implies that there is huge scope to increase agricultural production through new technological breakthrough by enhancing investments in agricultural research and technology. More private investment should be attracted by providing incentives and favourable policy environment.

Punjab

Predominance of paddy-wheat rotation has serious repercussions for soil health due to toxicity, alkalinity, salinity, micronutrient deficiency, and depletion of groundwater. Thus, paddy-wheat rotation needs diversification towards more sustainable cropping system. Area under paddy should be diverted to other crops like oilseeds, cotton, maize, pulses, and fruits and vegetables with proper incentives to farmers.

The excess utilization of water, especially for rice cultivation, has created water-logged conditions in the canal command areas (Southern Punjab) and a sizable area is getting saline for want of adequate drainage. Free electricity and highly subsidized water have compounded problems. However, dissemination of knowledge on proper field preparation, timely planting, and irrigation techniques can lead to higher water-use efficiency.

The importance of green manuring to enrich the soil with organic-carbon is well established, but this practice does not seem to have been adopted by the farmers. It could, however, be successfully adopted with little adjustment and timely planting of rice in mid-June instead of May. Besides, the crop residue management practice has to be improved for which research would be required.

Public investment in agriculture needs to be expanded by augmenting agricultural credit and increasing allocation for agriculture. Since the centre and state governments are facing a resource crunch, public investment can be increased only if the present level of subsidy on agricultural inputs such as power, water, and fertilizers is readjusted.

The extension system has weakened over time, as a result educating farmers about the latest developments in agriculture has almost come to a standstill. However, with far-reaching changes in communication technology and breakthroughs in space technology, remote sensing, satellite broadcasting, and the media spread revolution, extension workers will have to be totally reoriented and retrained to adapt themselves to these developments and emerging opportunities.

Himachal Pradesh

Increase in number of small land holdings overburdens agricultural sector in terms of sustainability of livelihoods. In future, the focus of agricultural development programmes has to be on small and marginal farmers, which will, besides boosting economic growth, also reduce poverty and rural inequalities.

The land use data of the state reveals that the area under barren and uncultivable land is increasing over time which resulted in marginal land being brought under the plough and overgrazing of pastures. Thus, the process of land degradation in the state needs to be contained and it is necessary to examine sustainable approaches to agricultural development. The changing cropping pattern is creating regional imbalances in terms of high concentration of income and employment in mid and high hill regions. Appropriate research, extension, and policy strategies are required to take full advantage of the agro-climatic opportunities of the region for balanced growth of commercial and traditional crops.

REFERENCES

Bhalla, G.S. and Gurmail Singh. 2001. *Indian Agriculture: Four Decades of Development*, New Delhi: Sage Publications.

———. 2009. 'Economic Liberalisation and Indian Agriculture: A Statewise Analysis', *Economic and Political Weekly*, 44(52), 26 December: 34–44.

Chand, Ramesh and Sonia Chauhan. 1999. 'Are Disparities in Indian Agriculture Growing?', Policy Brief No. 8. New Delhi: National Centre for Agricultural Economics and Policy Research.

Charyulu, D.K. and S. Biswas. 2010. 'Organic Input Production and Marketing in India—Efficiency, Issues and Policies', CMA Publication No. 239, India.

Rao, C.H. Hanumantha. 2003. 'Reform Agenda for Agriculture', *Economic and Political Weekly*, 33(29): 615–20.

Rekha, Nail and S.N. Prasad. 2006. 'Pesticide residue in organic and conventional food-risk analysis', *Journal of Chemical and Health safety*, 13(6): 12–19.

ANNEXURE 59A

Table 59A The Details of AERCs and the Lead Person Involved in the State Report

S. No.	States	AERC who Carried the Study in the State	The Lead Person Who Carried Out the Study
1.	Himachal Pradesh	AERC Shimla	Ranveer Singh
2.	Punjab	AERC Ludhiana	Sanjay Kumar
3.	Maharashtra	AERC Pune	S.S. Kalamkar
4.	Karnataka	ADRTC Bangalore	Elumalai Kanan and Khalil Shah

60

Agriculture Diversification in India

Brajesh Jha

Agriculture diversification in the recent period is increasingly being considered as a panacea for many ills in the agricultural development of the country. As the level of farm diversification is supposed to increase the farm income; the utility of diversification as risk management practices remains important. At the country level, diversification is supposed to increase the extent of self-sufficiency for the country. At the regional level, diversification is being promoted to mitigate negative externalities associated with monocropping.[1] Some of the above expectation from diversification is also rooted in different interpretations of agricultural diversification in the country. While diversification was historically construed as the opposite of concentration; increase in area under the high value commodities is being referred as agricultural diversification in the recent period.

The high value commodities refer to a group of commodities wherein trade was liberalized in the 1990s; and difference between domestic and international prices was very high during the initial period of trade liberalization in the country. The above difference in price tapered-off for some commodities and the concept/term 'high value' was not very relevant for these commodities. The high value usually refers to fruits, vegetables, and many agricultural exportable commodities. The fruit and vegetable-led diversification in the recent period have also been presumed as a precondition for achieving the 4 per cent rate of growth in agriculture. Considering the multi-dimensional importance of agricultural diversification, it becomes pertinent to understand the pattern, process, problems, and prospects of agricultural diversification in the country. The present study is an attempt with following specific objectives:

1. to understand pattern of agricultural diversification in India with special reference to Haryana;
2. to study the drivers of agricultural diversification in India and Haryana;
3. to examine various problems of existing crop-enterprise mix in Haryana; and
4. to study the prospects of crop diversification in the state of Haryana.

METHODOLOGY AND DATA

As is apparent from the above discussion, there are two broad approaches to agricultural diversification. In the first approach, diversification is measured with the concentration ratio; while in the second approach, diversification is measured with the per cent of non-food crops in the gross cropped area (NFCP). The present study has used various

[1] Mono cropping is about cultivation of the same set of crops in a region over a long period of time.

concentration indices: Harfindhal and Entropy to work out agricultural diversification. The Harfindhal index (DHI) is a sum of the square of the proportion of individual activities in a portfolio. With an increase in diversification, sum of the square of the proportion of activities decreases and so also the DHI. This is a measure of concentration, alternately, an inverse measure of diversification since the DHI decreases with an increase in diversification. The DHI is bound by zero (complete diversification) to one (complete specialization).

$$\text{Harfindhal index } (D_h) = \Sigma\, P_i^2,$$

where, $P_i = A_i / \Sigma_1 A_i$ is the proportion of the *i*th activity in acreage/income.

The above DHI is a measure of concentration and the index decreases with diversification, while Entropy indices discussed further is a positive measure of diversification. In order to make the DHI comparable with the Entropy index, the Simpson index that is (1-DHI) has been worked out. The Entropy index is a direct measure of diversification having a logarithmic character. This index increases with an increase of diversification. It approaches zero when the farm is specialized and takes a maximum value when there is perfect diversification. The upper limit of the Entropy Index is determined by the base chosen for taking logarithms and the number of crops. The upper value of the index can exceed one, when the number of total crops is higher than the value of logarithm's base, and it is less than one when the number of crops is lower than the base of logarithm. Thus, the major limitation of the Entropy Index is that it does not give a standard scale for assessing the degree of diversification.

$$\text{Entropy index (EI)} = \Sigma_i\, P_i * \log (1/P_i)$$

The modified Entropy index (MEI) is used to overcome the limitation of the EI by using a variable base of logarithm instead of a fixed base of logarithm. The EI lies between zero (complete specialization) to one (perfect diversification). The EI is bound by zero and one. It can be computed as:

$$\text{MEI} = -\Sigma_i\, (P_i * \log_N P_i)$$

The MEI is equal to EI/logN, it is worth mentioning that the base of the logarithm is shifted to 'N' number of crops. This index has a lower limit equal to zero when there is complete specialization or concentration and it assumes an upper limit of one in the case of perfect diversification, that is, it is bounded by zero and one.

$$\text{Maximum MEI (when Pi approaches 1/N)} = \Sigma\, 1/N * \log_N N = \Sigma\, 1/N = 1$$

Since the MEI imparts uniformity and fixity to the scale used as a norm to examine the extent of diversification; the index is quite useful. The MEI, however, measures deviations from equal distribution among existing activities, that is, the number of crops only, and does not incorporate the number of activities in it. This index measures diversification given the number of crops and the index is not sensitive to the change in the number of crops (Shiyani and Pandya 1998).

In order to understand the drivers of agricultural diversification, different variants of agricultural diversification, concentration ratios and changes in the per cent of NFCP have been regressed on the structure of land holdings, irrigation intensity, institutional credit, road network, and urbanization. The above regression analysis has been undertaken at the level of country and also for the state of Haryana. Linear and double-log equations were estimated with the ordinary least square technique (OLS) for the year 2003–04, 1993–94, and 1983–84. The results from the log-based OLS estimates were more suitable and were, therefore, presented in Table 60.2, given later in the chapter. The Model and Specification of variables are as:

$$AGDIV1/2 = \oint(PCI, AOH, SMH, IRIP, RDEN, URB, ICD, MKTP)$$

where,

AGDIV1 = Agricultural diversification as measured with Simpson Index

AGDIV2 = Per cent of cropped area under NFCP

PCI = Per capita net state domestic product at 1993/94 prices, used in aggregate level analysis

SMH = Per cent of small and marginal holdings in total agricultural holdings, used in the aggregate level analysis

AOH = Average size of operational holdings in hectare in state-level analysis

IRIP = Intensity of irrigation is per cent of gross irrigated to gross sown area

RDEN = Road density is the length of road (in km) per thousand square km of geographical area in the country level analysis while road density in state-level analysis is per cent of villages connected with metal road

URB = Urbanization and road density is highly correlated; URB is the per cent of urban to total population in the district and states. URB has been used for the state level analysis.

ICD = Institutional Credit is the ground-level credit disbursed for agricultural and allied activities per unit of gross cropped area

MTPI = Market Penetration is the net sown area per unit of regulated market. This is an adverse measure of market penetration.

Though the role of trade on agricultural diversification is clear, a suitable variable that represent trade and can also be incorporated in the multiple regression analysis to study drivers of agricultural diversification in the present framework is difficult to obtain. The study recognizes role of price in agricultural diversification; in fact, trade, both (external and internal) affects domestic prices of agricultural commodities; price subsequently affects acreage under a crop (food or non-food crops); a suitable representative price of crop that can explain diversification of agriculture is difficult to find; therefore, ignored from the multiple regression analysis to determine factors responsible for agricultural diversification in country and states.

Some of the possible consequences of agricultural diversification are measured with productivity and vulnerability of farmers. Productivity can be defined as a ratio of the output to inputs used in production. Partial productivity indices, such as output per worker, are the ratio of output to a single factor of production. However, since the amounts of other inputs may vary, increases in output per worker can result from either increase in the use of other inputs or to changes in technology. An index of total factor productivity (TFP) compares changes in output with changes in aggregate inputs. Changes in TFP provide a better picture of changes in technology and other unmeasured improvement in means of production (such as improvements in input quality). The present study uses Divisia index, this is based on a translog production function, characterized by constant return to scale. It allows for variable elasticity of substitution and does not require assumption of Hicks neutrality. The translog index of TFP growth in a three-input framework is given by the following equation:

$$\Delta lnTFP_t = \Delta lnQ_t - ASL\Delta lnL_t - ASF\Delta lnF_t - ASC\Delta lnC_t \quad (1)$$

In this equation, Q denotes gross output, L labour, F fertilizer, C other input costs. The Δ indicates change over previous year, for instance, ($\Delta lnQ_t = lnQ_t - lnQ_{t-1}$). ΔlnL_t, ΔlnF_t, and ΔlnC_t are defined in similar way. ASL is the average cost share of labour, this average is based on present (t) and previous (t − 1) years. Similarly, ASF and ASC are average cost shares of fertilizer and other inputs in the cost of production of wheat and paddy. $\Delta lnTFP_t$ is the rate of growth of total factor productivity. The growth rate in TFP has been computed for each year using equation 1. These have been used to obtain an index of TFP in the following way. Let A denote the index of TFP. The index for the base year, A(0), is taken as 100. Then, the index for subsequent year is computed using the following equation:

$$A_t/A_{t-1} = \exp[\Delta lnTFP_t] \quad (2)$$

Following the above procedure, present study obtained output and input indices. After obtaining TFP, output, and input indices for different years, the growth rate for entire period is worked out by fitting an exponential (or semi-log or log-lin) trend equation to the estimates of TFP, output, and input indices.

RESULTS AND DISCUSSION

An enquiry into Central Statistical Organization (CSO) income series for different sectors and sub-sectors of economy shows decline in the share of agriculture in the aggregate economy; the pace of this transformation has, however, been slower in some of the Indian states like West Bengal, Kerala, and Bihar. In agriculture and allied sector there has been significant structural changes; the share of horticulture in crop, bovine in livestock, cross-bred in bovine, fisheries in allied sector, and inland in total fisheries has increased in the recent period. There is also realignment of commodities within a commodity group. The role of trade on the above structural changes in agriculture and allied sector is evident. The study argues that within horticulture, vegetables as compared to fruits hold better promises since the productivity of vegetables has increased significantly in the recent decade. In the livestock sector the structural changes in favour of buffaloes and cross-bred cows have been very encouraging, especially in the light of the competition for food and fodder in the country. Though milk production after 1970s has been increasing consistently at a rate of more than 4 per cent; relatively narrow spatial base of milk production, kind of linkage between crop and livestock sector, increase of stall-fed cattle rearing, and similar other factors raises doubt on continuance of the similar rate of growth of milk production in the country.

Agriculture production data shows that the share of leading producing state of particular crop has increased between the year 1983 and 2004, suggesting increased specialization of agriculture in the country. This specialization is not necessarily in accordance with the natural resource endowment of the region; favourable institutions and incentive structures have emerged important. The changes in the per cent of gross cropped area (GCA) in the recent decade show that the area under fruits and vegetables has increased significantly; while the per cent of GCA under fine cereals, oilseeds, and commercial crops has stagnated. The per cent of GCA under coarse cereals and pulses have been decreasing since 1970s; the decreasing trend in the per cent of GCA has however ceased in the 1990s.

Temporal changes in the per cent of GCA across states clearly show periodic shift in the acreage of some crops towards the specific regions of the country. The data on agriculture income, agriculture production, and land use for agriculture as measured with the GCA provides enough evidences of specialization. The state of Haryana, for example, is being specialized under paddy and wheat crops. Again in Haryana, specialization has been the maximum in the Kurukshetra district. A comparative account of crop diversification trends in Haryana and the country shows that the small crop-specific region is being created in Haryana and so for other parts of the country as well. Though many of the above changes are influenced by natural resource endowment of the region; price and non-price incentive structure have also provided impetus to the above process of specialization in agriculture.

A temporal and spatial comparison of alternate measurements of diversification: Simpson index and per cent of non food crops is presented in Table 60.1. At the all-India level there is no significant change in diversification indices during the reference period (1983–84 to 2003–04). Simpson indices are generally high for the larger states. A large state, in fact, consists of diverse agro-climatic regions and diversity provides scope of cultivating a variety of crops in different regions of the state. The increase in diversification indices is significant in the state of Goa, West Bengal, Maharashtra, Andhra Pradesh, Tamil Nadu. The states showing a significant decline in diversification indices during the reference period are Haryana, Meghalaya, and Odisha. The per cent of cropped area under non-food crops, another measurement of agricultural diversification specifically resource diversification, has increased significantly during the reference period. Increase in the per cent of cropped area is observed in many states; some states that showed trend dissimilar to the above are Bihar, Haryana, Karnataka, Punjab, and Rajasthan.

The determinants of agricultural diversification were studied with different variants of agricultural diversification: Simpson index and change in per cent of NFCP. Alternate variants of agricultural diversification was regressed on a set of independent variables like the per capita income (PCI), concentration of small and marginal farmers (SMH), average size of operational holding (AOH), irrigation intensity (IRIP), institutional credit (ICD), road density (RDEN), and urbanization (URB). The regression analysis was undertaken at the level of country and also for the state of Haryana. In the country level regression analysis individual state was an observation while districts are observations in the state level analysis. Since the per capita income was not available for districts, income as explanatory variable was

Table 60.1 Agricultural Diversification in India

State	Simpson Index			Per cent of Non-Food Crops		
	1983–84	1993–94	2003–04	1983–84	1993–94	2003–04
Andhra Pradesh	0.83	0.83	0.87	31.16	45.86	46.51
Assam	0.45	0.42	0.42	32.58	31.76	34.30
Arunachal Pradesh	0.07	0.08	0.10	38.98	50.89	53.96
Bihar	0.70	0.68	0.67	10.57	11.67	10.17
Haryana	0.80	0.79	0.77	26.82	32.94	32.09
Jammu and Kashmir	0.70	0.69	0.69	19.59	21.00	21.35
Himachal Pradesh	0.67	0.65	0.64	16.9	16.69	18.28
Gujrat	0.87	0.88	0.88	52.47	62.52	62.36
Karnataka	0.89	0.9	0.92	33.82	43.87	40.84
Kerala	0.71	0.71	0.68	74.13	83.23	90.31
Maharashtra	0.84	0.86	0.88	31.12	33.61	45.54
Madhya Pradesh	0.87	0.87	0.86	18.44	28.63	33.41
Odisha	0.66	0.5	0.41	28.36	40.46	38.8
Punjab	0.64	0.63	0.61	28.26	24.31	21.85
Rajasthan	0.83	0.85	0.82	29.86	39.59	32.88
Tamil Nadu	0.81	0.81	0.85	32.06	43.70	53.58
Uttar Pradesh	0.82	0.79	0.77	18.07	20.65	21.06
West Bangal	0.45	0.44	0.50	20.86	24.66	32.51
All India	0.88	0.88	0.88	26.68	34.14	35.19

Source: Author's computation.

considered for the state level analysis only. The variable related to road density was incorporated at the country level analysis whereas urbanization instead of road density was considered for the state-level analysis of agricultural diversification in the present study.

The linear and double-log equations were estimated with the OLS for the year 2003–04, 1993–94, and 1983–84. The log-based OLS estimates were found to be more robust than the linear estimates. The signs and significance of estimates varied over the years. In the year 2003–04 diversification, as measured with the per cent area under NFCPs, was affected positively by the per capita income. Road density had emerged important in the recent period. The irrigation intensity affected the NFCPs adversely; the NFCPs was, however, indifferent to the farm sizes. Though the adjusted R-square shows that the above set of independent variables together explain the variation of simpson indices better than the NFCPs; the estimated results from simpson indices contradict many of the established findings on the determinants of agricultural diversification in the country. This further encourages study of determinants of agricultural diversification at the level of state-Haryana.

The regression analysis for the state of Haryana (Table 60.2) clearly shows that irrigation had led to specialization under fine cereals. Infrastructure and market intensity further contributed to the above trend towards specialization. The above process of diversification increasingly became indifferent to the size of holding and the availability of institutional credit. Diversification as measured with the per cent of GCA under non-food crops increased with the increase of urbanization in the state. Analysis of determinants of crop diversification at the level of state was more discernible than the country level analysis. In this background the study further analysed farm-level diversification.

The farm level diversification studied with the sample households from Kurukshetra district of Haryana found that small farm was the least diversified in the region. The study argues that in the light of the established literature on farmer's attitude towards risk and farm size, a positive relationship between diversification and farm size was difficult to accept if diversification is a risk management practice. The study found that diversification with crops was not a risk-reducing proposition in the northwest India since crop incomes on an average farm is not negatively correlated amongst themselves; an essential condition for diversification to reduce risk. Whereas diversification with dairy enterprises reduces risk in farm portfolio since crop and dairy incomes were negatively correlated amongst themselves. Farm level evidences suggested that crops like basmati paddy, potato, vegetables were as remunerative as paddy-wheat crop combinations; profitable levels of these crops were however constrained as these crops involved more risk than the latter (paddy-wheat).

With the commercialization subsistence type of crop production is being replaced by the specialized farms in the

Table 60.2 Regression Estimates for Determinants of Crop Diversification in Haryana

Variables	Simpson Index			Per cen of Non-Food Crops		
	2003–04	1993–94	1983–84	2003–04	1993–94	1983–84
AOH	−.02	.26	.14	.06	.73	1.02***
	(−0.14)	(1.6)	(0.89)	(0.10	(1.12)	(2.49)
IRI	−0.29***	−.32***	−.04	−1.05***	−.98***	−.32
	(−3.69)	(−2.78)	(−.30)	(−3.19)	(2.18)	(1.09)
MPTI	.08	.09*	−.02	.23	.35	0.16
	(1.46)	(1.73)	(−0.11)	(1.02)	(1.61)	(0.47)
URB	.12*	.16	.06	.48	.37	0.58*
	(1.68)	(1.66)	(0.48)	(1.62)	(0.99)	(1.88)
ICD	−.15	.06	−.08	−.68	.11	.06
	(1.37)	(0.50)	(−0.60)	(−1.56)	(0.26)	(0.17)
No. of observation	19	16	12	19	16	12
R-squared	0.649	0.606	0.266	0.619	0.544	0.619
Adjusted R^2	0.514	0.408	−0.00	0.473	0.316	0.301
F – statistics	4.80	3.07	0.44	4.23	2.39	1.95

Note: Asterisk shows level of significance, (*) shows significance at 10 per cent level, (**) shows significance at the level of 5 per cent and, (***) shows significance at 1 per cent level. Values in parentheses show t-statistics.
Source: Author's computation.

Kurukshetra district of Haryana, India. Assured irrigation further contributed to the above process of specialization. The continuation of the same or similar set of enterprises (specific crop-combination) over the years has led to certain kind of externalities in a region and accumulation of negative externalities over a long period of time is often referred as problems. These problems have been studied separately as economic, ecological, and social problems. The economic problem discusses the issue of decreasing profitability and changing nature of vulnerability of farmers. Though profitability and vulnerability is often used in a micro-sense; in dearth of suitable statistics at the level of farm the present study has discussed the above issue with the productivity and prices of agricultural commodities for the country in general and the state of Haryana in particular. Farmers' vulnerability is discussed with the district (Kurukshetra) level information.

The profitability of a crop as obtained from the CACP data varies across states; profitability of bajra in Gujarat, rapeseed-mustard in Rajasthan has increased significantly during the reference period (1989–2005); interestingly profitability of the same crop (bajra, rapeseed-mustard) has decreased in Haryana. Again, profitability of paddy and wheat has not decreased in Haryana. The difference of profit between the crop growing states provides basis for region-specific specialization under a crop. The above difference in profit is not merely on account of natural resource endowment of the region, favourable institution for crop must have played important role. The cultivation of same crops over a long period of time has, however, resulted in decline of profitability; primarily on account of decreasing productivity especially total factor productivity.

The partial productivity with respect to land was worked out for many crops during the period 1983–2004. The number of states registering decline in productivity has increased during the reference period (1983–84 and 2003–04). Many of the states showing decline in productivity of a crop were actually cultivating the said crop for a relatively longer period of time. Examples of crops and states in this category are West Bengal, Punjab, and Haryana in paddy whereas Haryana and Punjab in wheat. The above examples to some extent suggested the effect of specialization and monocropping on the productivity of a crop. In commodities like wheat, rape-mustard, soybean, and sugarcane, most of the states are showing decline of productivity in the recent period.

Further analysis of total factor productivity of paddy and wheat shows higher fluctuation in productivity indices towards the late 1990s. The factor productivity indices were less than 100 in many years, suggesting that increase in the value of output is less than the increase in the cost of farm inputs during the year. A significant increase in the cost of production has been on account of the cost of pesticides; this increase in the cost of pesticide is particularly reported from the district which has been early adopters of the biochemical technology.

The changing nature of vulnerability is further contributing to the economic woes of a farmer. The vulnerability of farmers is discussed with risk, more specifically downside risk, in the yield and the price of agricultural commodities. Risk is measured with the instability indices and the same is computed for the important states of the country. Instability indices of yield show that risk in many of the crops have increased over the years. While the yield-induced risk is supposed to decrease over the years as investment in irrigation has increased. The market-induced risk in agriculture has also increased with trade liberalization in agriculture. The wheat and paddy presents a different case in Haryana; these crops are cultivated in the assured irrigated area, therefore, production related vulnerability is less. The price-related vulnerability is also less for these crops since price of the above crops following government procurement are assured in the region. Estimates for downside risk clearly indicate that farmers cultivating crops other than fine cereals are vulnerable to production as well as market forces; the market-induced vulnerability has increased in the recent period.

The growing specialization of agriculture is expected to have some implications for market and market price as well. The spatial correlation between market prices of agricultural commodities suggests that disparity in prices across states has increased for most of the commodities. The widening of disparity in agricultural prices in the 1990s, following liberalization, is surprising since disparity in prices across states are supposed to decrease with the removal of restrictions on domestic and external market. In many agricultural commodities domestic prices are also integrating with the international market. In spite of it widening of prices of agricultural commodities has been an enigma. The study argues that a high transport cost in domestic market and initiation of limited external trade are probably contributing to the disparity of agricultural prices in the 1990s. The widening disparity in the prices of a commodity across states provides the basis for region-specific specialization under particular crop.

The ecological consequences of the above specialization in Haryana have been studied with respect to soil and groundwater resources. The present study found that macro- and micronutrient status of soil in the most of the districts of Haryana deteriorated suggesting that the rate of use of natural resources was greater than the combined effect of the regeneration of resources through the natural process and application of additional resources to soil. Though degradation of macronutrient status depends on a variety of factors, intensive agricultural practice is the most important. The deterioration in the status of particular soil micronutrients from certain regions of country, to large extent, is on account of crop-specific specialization in different regions of the country. As a matter of fact, macro- and micronutrient requirements of crop varies, a continuous cultivation of the same crop depending on its specificity for nutrients draws particular resources more than the natural regeneration and artificial application of nutrients in soil; consequently deficiency of particular nutrients emerges. Examples of zinc deficiency in the paddy growing regions, sulphur deficiency in the rape-mustard growing regions are noteworthy.

The spatial and temporal distribution of the levels of groundwater table suggested that districts in the north and south of Haryana recorded fall in groundwater table whereas regions in the west and the middle of Haryana recorded increase in the groundwater table. Since the changes in the water table represent the net effect of recharge and draft of groundwater during the reference period, increase in the groundwater table indicates that draft of groundwater is less than the recharge of groundwater. Bulk of draft of groundwater is for irrigation purposes; a lower draft of groundwater in the particular region of the state raises doubt about the irrigation quality of the ground water. The groundwater table in the large part of the state, especially in the north and south of Haryana, has gone down rapidly. This is also the region where irrigation quality of groundwater is good.

The degradation of natural resources has wider ramification for the society. The deterioration in the status of soil nutrients not only increases production cost but also causes decline in the productivity of soil. The decline in the organic matter content of soil as is evident with the low nitrogen status has encouraged heavy application of synthetic nitrogenous fertilizers; a significant part of which leaches down to increase nitrate content of groundwater, this reduces potable quality of ground water. A continuous decline in the groundwater table has led to introduction of submersible technology in the region; its differential access to small and large farmers has emerged as a new source of inequity in the semi-arid regions of rural India. Following specialization agricultural operations in Haryana has become dependent on migratory workforce; while family labour on the small farm remains unemployed for a large part of the year. In order to get rid of uncertainty of migrant labour, increasing number of large farmers is harvesting wheat with harvester. Such mechanization besides reducing employment in agriculture has several other implications. The present study argues for cultivation of a variety of crops with different sowing and maturity period on different farms of a region; this will reduce skewed demand for labour and utilize labour efficiently in the region.

The ongoing pattern of growth in agriculture is specialization. The economic, ecological, and social implication of specialization in agriculture is amply clear with a case of the north-west India. Since the above pattern of growth in agriculture is being pursued in other regions of the country there is need to minimize the potential danger of the above growth in agriculture. Farmers in the north-west of India are also aware of many of the negative externalities associated with rice and wheat cropping system. The region also has potential to diversify into other crops. The existing price and institutional frameworks were, however, such that an average farmer was discouraged to grow crops other then paddy and wheat. Only in the recent period urbanization and income induced agricultural diversification in favour of horticulture had started showing its impact; some of the small farmers have started growing vegetables. The study found that even with the existing crop-combinations there is possibility of minimizing the negative externalities of intensive agricultural practices by adopting alternate practices such as integrated nutrients and pest management (IN&PM). The existing price policy for agricultural inputs has been such that chemical/synthetic inputs are preferred by farmers over the farm-based organic inputs. The present study argues that a region can be collection of specialized farms; these farms are specialized under different crops and the same has been referred here as mosaic-kind of agriculture diversification. Encouraging such diversification requires a suitable mix of price, technology, and institution related policies in Indian agriculture.

POLICY SUGGESTIONS

Considering different definitions and notions of agricultural diversification the present study discusses pattern, process, problems, and possibilities of agricultural diversification in India in general and Haryana in particular. Though the share of agriculture in overall economy has been decreasing, share of livestock and fisheries in agriculture has increased. There have been significant structural changes in the livestock and fisheries sectors of the economy as well. In agricultural crops there has been significant increase in the per cent of GCA under fruits and vegetables. A threat to the availability of fine cereals on the above account is however a misplaced one; since crop diversification trend from states like Haryana are not supportive to the trend of agricultural diversification at the aggregate level. This suggests that market-induced specialization in horticulture and public procurement-induced specialization in fine cereals can coexist at the regional and national level. A synthesis/combination of crop diversification related evidences from macro-, meso-, and micro-level suggests that certain crops are more remunerative in the given resource endowments and institutional framework; farms are getting specialized under these crops. Such specialization has not necessarily increased risk on farm since covariance in crop income is increasing with the commercialization of agriculture in the country. The specialization of agriculture is evident from other sets of data as well. The consequences of specialization as is evident from the state of Haryana and Punjab are too strong to be ignored. The present study, therefore, argues for a mosaic-kind of agricultural diversification in the country.

REFERENCES

Jha, Brajesh. 1996. 'Farm-level Diversification: Some Disconcerting Evidences from the Greenbelt of India', *Agricultural Economics Research Review*, 9(1): 49–56.

———. 2000a. 'Implications of Intensive Agriculture on Soil and Ground Water Resources in Kurukshetra District', *Indian Jl of Agricultural Economics*, 55(2): 182–93.

———. 2000b. *Indian Agriculture and the Multilateral Trading System*. New Delhi: Bookwell.

———. 2009. 'Evaluating Agricultural Policy in a Farming System Framework: A Case Study from North West India', Institute of Economic Growth (IEG) Working Paper Series No. E/299/2009.

Jha B., N. Kumar, and B. Mohanty. 2009. 'Pattern of Agricultural Diversification in India', IEG Working Paper Series No. E/302/2009.

Jha B. and A. Tripathy. 2010. 'Towards Understanding the Process of Agricultural Diversification in India', *Indian Economic Journal*, 58(2): 101–20.

Joshi P.K., A. Gulati, and Ralph Cummings Jr, (eds). 2007. *Agricultural Diversification and Small Holders in South Asia*. New Delhi: Academic Foundation.

Shiyani R.L. and H.R. Pandya. 1998. 'Diversification of Agriculture in Gujarat: A Spatio-Temporal Analysis', *Indian Jl of Agricultural Economics*, 53(4): 627–39.

61

Macro Management of Agriculture Schemes in India

An Assessment

Komol Singha

Agriculture has played a significant role in the overall development of India since pre-independence period. At present, the agriculture sector accounts for 13 per cent of the country's gross domestic product (GDP) and employs around 50 per cent of the total workforce. So, the development of this sector is paramount for the achievement of rapid growth of the country's economy. Agricultural growth is necessary for not only attaining high overall growth but also for food security, eradication of poverty, ensuring better employment, and so on.

The Five Year Plans initiated programmes for agricultural development in India. However, despite the concerted efforts, the sector remains weak on all fronts. To bring about all-round development of agriculture, a scheme called 'Macro Management of Agriculture' (MMA hereafter) was approved by the Cabinet Committee on Economic Affairs (CCEA) on 4 October 2000. It became operational from 2001 in all the states and union territories (UTs) by integrating the existing 27 centrally-sponsored schemes of agriculture and its related activities.

The centrally-sponsored MMA scheme has been formulated to ensure that central assistance is spent on focused areas and specific interventions are made for the development of agriculture across the states in the country. The scheme provides sufficient flexibility for the states to pursue the development programmes on the basis of their regional priorities. Thus, states have been given a free hand to finalize their sector-wise allocation as per requirements of their developmental priorities. However, with the launching of the National Horticulture Mission (NHM) in 2005–06, altogether 10 schemes pertaining to horticulture development were taken out of the purview of the MMA scheme.

Further, the scheme was revised in July 2008 to improve its efficacy in supplementing and complementing the efforts of the states for enhancement of agricultural production and productivity, and provide opportunity to draw upon their agricultural development programmes relating to crop production and natural resource management, with the flexibility to use 20 per cent of resources for innovative components. The revised MMA scheme has formula-based allocation criteria and provides assistance in the form of grants to the states UTs on 90:10 ratios except in the north-eastern states and UTs where the central share is 100 per cent.

Ever since the implementation of the MMA, no concrete study on the impact of its schemes and sub-schemes has been carried out. The present chapter serves as a consolidated report of five sub-schemes evaluated by nine Agro Economic Research Centres (AERCs) all over the country. The specific objectives of the study are:

1. to assess the impact of states' interventions under the five sub-schemes, namely, (i) Integrated Cereal Development Programme (ICDP) of rice, wheat, and coarse cereals, (ii) foundation/certified seed production of vegetable crops, (iii) special jute developmental programme, (iv) sustainable development of sugarcane-based cropping system, and (v) integrated nutrient management/balanced integrated use of fertilizers;
2. to analyse the impact of efforts made by the states and UTs in ensuring timely availability of sufficient quantity and quality seeds; and
3. to give a macro-level picture of the five sub-schemes evaluated by nine AERCs in the country.

In addition to the above objectives, there is always a need to assess the impact of interventions made under the specific sub-schemes of the MMA scheme, so as to examine the impact of such a decentralised approach at the grassroots level and to verify whether or not local needs have been fulfilled. Hence, the chapter also attempts:

1. to examine the public expenditure pattern of MMA;
2. to review the distribution of various benefits, like seeds distribution, bio-agents, micronutrients, agricultural implements, soil testing, mini-kits, demonstrations, training, and so on, under each scheme, across different sizes of farming;
3. to capture the changes under each of the schemes in terms of the area under cultivation, yield, and income before and after the implementation of each scheme; and
4. to highlight the problems and difficulties in getting access to various facilities under each scheme by farming sizes.

METHODOLOGY AND DATA SOURCE

The study is based on both the primary and secondary data collected from published and unpublished sources. The primary data were collected with the help of a three-stage stratified sampling (block, village, and farmers) survey using tested questionnaires for each scheme. The secondary data were also collected from the official sources as well as the different nodal agencies at the state, district, and block levels. The study concentrates on the few selected sub-schemes (three to five schemes depending on AERCs) in the ten states evaluated by nine AERCs in the country. The schemes covered under MMA are: (i) ICDP of rice, wheat and coarse cereals, (ii) foundation/certified seed production of vegetable crops, (iii) special jute developmental programme, (iv) sustainable development of sugarcane-based cropping system, and (v) balanced integrated use of fertilizers/integrated nutrient management. The year 2007–08 is considered as the year of reference for the study and hereafter, throughout the text, the year 2007–08 is understood as 'after the scheme' and 2000–01 as 'before the scheme'.

TRENDS IN PUBLIC EXPENDITURE UNDER MMA

It is mandatory to consider the budget estimates and expenditure made under a scheme to assess its progress. As such, it is evident that the Government of India has been assigning more importance to the MMA scheme because its budget estimates more than doubled only after nine years of its commencement (2009). This underlines the growing importance of the agricultural sector and its budget expenses. It is evident from the study that the actual expenditure has almost doubled from Rs 381 crore in 2000–01 to Rs 678 crore in 2001–02. But, very interestingly, the growth trend was not as smooth as Adolf Wagner's (1883; 1893) theory suggests. With constant growth, it reached the highest level at Rs 1,186 crore in 2004–05, but declined to Rs 706 crore in 2009–10.

To understand and view the inter-state disparities of expenditure pattern, the public expenditure made for various MMA schemes is presented in the following Table 61.1. For the purpose of analysis, the states have been classified under seven zones, namely, northern, southern, eastern, western, central, north-eastern (NE) and Union territories. The funds for the schemes are made available from two streams—central allocation and states' share. In fact, according to the guidelines of the MMA scheme, the contribution of the centre and the states is made at the ratio of 90:10, except for the NE zone.[1]

During the period under review (2005–06 to 2009–10), the total allocation of funds under MMA schemes was Rs 4,97,649 lakh. Out of which, Rs 4,56,082 lakh (91.65 per cent) was made available by the Central government and Rs 41,567 lakh (8.35 per cent) was the share of the state governments. It is evident that the central zone, which consists of Chhattisgarh, Madhya Pradesh, Uttar Pradesh, and Uttarakhand, received the highest allocation

[1] In the NE zone, 100 per cent of the fund is provided by the centre as it is considered to comprise 'Special Category' states.

of funds with Rs 1,04,272 lakh (20.95 per cent) followed by the southern zone with Rs 95,056 lakh (19.10 per cent). Karnataka tops within the southern zone with a total fund of Rs 30,833 lakh. Further, the northern zone, which consists of Haryana, Himachal Pradesh, Punjab, Jammu and Kashmir, and Rajasthan, stands third in terms of allocation of funds. Its allocation of Rs 85,511 lakh works out to be 17.18 per cent of the total allocated funds under MMA schemes. North-eastern, eastern, and western zones stand at fourth, fifth, and sixth positions respectively with their percentage shares of 16.30 per cent, 14.33 per cent, and 12 per cent of funds allocation respectively. Union territories have been allocated a nominal amount of Rs 1,093 lakh under MMA schemes which is only 0.21 per cent of the total allocated funds. As for the state-wise compound annual growth rate of fund allocation under the MMA over the years (2005–06 to 2009–10), Bihar is at the top with 50.38 per cent followed by Madhya Pradesh (24.04 per cent), Mizoram (23.16 per cent), and Uttar Pradesh (17.63 per cent). At the bottom are the Andaman and Nicobar Islands with –33.58 per cent, followed by Kerala, Goa, and Rajasthan with –29.38 per cent, –28.94 per cent, and –12.28 per cent, respectively.

Table 61.1 Fund Allocation and Expenditure under MMA (2005–06 to 2009–10) (Rs in Lakh)

States	Total funds Available	Total Expenditure
Andhra Pradesh	39461.08	22833.71
Arunachal Pradesh	11567.09	10901.18
Assam	8650.91	5287.00
Bihar	17157.31	14176.93
Chhattisgarh	15939.59	13699.83
Goa	1740.23	1428.23
Gujarat	28521.53	23173.88
Haryana	11710.79	11572.53
Himachal Pradesh	12190.46	11292.42
Jammu and Kashmir	18104.65	14638.95
Jharkhand	6891.71	5037.12
Karnataka	36217.33	32524.65
Kerala	20117.55	12621.10
Madhya Pradesh	33460.27	26561.83
Maharashtra	59058.98	54006.57
Manipur	12258.32	12103.41
Meghalaya	12502.20	10874.71
Mizoram	7677.55	7189.25
Nagaland	11560.54	11560.54
Odisha	18951.39	16150.88
Punjab	11928.04	7322.26
Rajasthan	45101.31	35207.36
Sikkim	10228.70	9295.63
Tamil Nadu	26587.31	24025.31
Tripura	12953.55	9198.15
Uttar Pradesh	51345.67	44973.24
Uttarakhand	13021.05	11892.44
West Bengal	21236.29	16807.28
Delhi	397.61	21.13
Puducherry	141.53	62.09
Andaman and Nicobar Islands	126.52	75.62
Chandigarh	0.00	0.00
Dadra & Nagar Haveli	40.46	16.03
Daman & Diu	0.00	3.13
Lakshadweep	58.83	52.31

Source: Ministry of Agriculture and Farmers Welfare, Govt. of India.

DEMOGRAPHIC PROFILE OF THE SAMPLE FARMERS IN THE STUDY

The demographic profile of the sample farmers is presented in Table 61.2 (caste-wise) and Table 61.3 (farm size-wise). Table 61.2 shows that a total of 1,320 sample farmers of different categories (caste-wise) are found to be involved in five different schemes/components of MMA. The five different schemes/components are: Sustainable Development of Sugarcane-Based Cropping System (SUBACS), Integrated Cereal Development Programme (ICDP) (rice, wheat, and coarse cereal), Integrated Nutrient Management (INM), Foundation and Certified Seed Production

Table 61.2 Demographic Profile of the Sample Farmers (by Category)

Schemes	SC	ST	OBC	Gen	Total
SUBACS	25	28	55	122	230
	(10.9)	(12.2)	(23.9)	(53.0)	(100.0)
ICDP	75	30	156	269	530
	(14.2)	(5.7)	(29.4)	(50.8)	(100.0)
INM	32	7	38	163	240
	(13.3)	(2.9)	(15.8)	(67.9)	(100.0)
FCSPVC	41	14	77	3	135
	(30.4)	(10.4)	(57.0)	(2.2)	(100.0)
SJDP	27	1	66	91	185
	(14.6)	(0.5)	(35.7)	(49.2)	(100.0)
Grand Total	**200**	**80**	**392**	**648**	**1320**
	(15.2)	**(6.1)**	**(29.7)**	**(49.1)**	**(100.0)**

Note: Figures in the parentheses are percentages to the total.
Source: Compiled from the different Schemes/sub-scheme of MMA.

Table 61.3 Socio-economic Profile of the Sample Farmers (by Farm Size)

Schemes	Marginal	Small	Semi Medium	Medium	Large	Total
SUBACS	66	76	45	41	2	230
	(28.7)	(33.0)	(19.6)	(17.8)	(0.9)	(100)
ICDP	153	142	113	86	36	530
	(28.9)	(26.8)	(21.3)	(16.2)	(6.8)	(100)
INM	45	56	61	51	27	240
	(18.8)	(23.3)	(25.4)	(21.3)	(11.3)	(100)
FCSPVC	77	30	15	9	4	135
	(57.0)	(22.2)	(11.1)	(6.7)	(3.0)	(100)
SJDP	83	62	33	7	0	185
	(44.9)	(33.5)	(17.8)	(3.8)	(0.0)	(100)

Note: Figures in the parentheses are percentages to the total.
Source: Compiled from the different schemes/sub-schemes of MMA.

of Vegetable Crops (FCSPVC), and Special Jute Development Programme (SJDP); and the categories of farmers (communities of farmers) involved in the five different schemes are: schedule caste (SC), schedule tribe (ST), other backward caste (OBC), and general category.

In Table 61.3, under the five different schemes mentioned above, the general category comprises the highest percentage of farmers with 648 sample farmers: 49.1 per cent of the total sample farmers of the MMA schemes. It is followed by OBC with 29.7 per cent and SC with 15.2 per cent. The ST category of farmers is at the bottom with 6.1 per cent (80 sample farmers only) of the total 1,320 sample farmers. A major portion of the SC farmers (30.4 per cent) are involved in FCSPVC sub-scheme and the rest are more or less uniformly involved in other four sub-schemes, ranging from 11 per cent to 15 per cent. Similarly, most of the ST farmers (12.2 per cent) are found in SUBACS and the least number of this community is found with SJDP, consisting of only 0.5 per cent.

A majority of the OBC farmers (57 per cent) got involved in the FCSPVC sub-scheme, while the least number of this category is with INM (15.8 per cent). However, in case of the general category farmers, 68 per cent of them are with INM, followed by ICDP (51 per cent) and SUBACS (53 per cent). At the bottom is the FCSPVC component with only 2.2 per cent of this category. More interestingly, the general category farmers are being benefited more (49.1 per cent) when all the sub-schemes or components of MMA are taken together. Even component-wise, the general category farmers are the most-benefited: 53 per cent of the total farmers involved in SUBACS are general category farmers, and at the bottom is the SC category with 11 per cent. Similarly, ICDP, INM, and SJDP components are benefiting mostly the general category farmers with 51 per cent, 68 per cent, and 49.2 per cent, respectively. However, in the FCSPVC sub-scheme, only 2.2 per cent of the general category of farmers got benefited (the lowest percentage in this sub-scheme). The ST category of farmers is at the bottom in the three sub-schemes of ICDP, INM, and SJDP with 6 per cent, 3 per cent, and 0.5 per cent, respectively.

As regards the scheme-wise classification (see Table 61.2), the highest number of 530 sample farmers (40.1 per cent of the total sample) are involved in the ICDP component. It is followed by INM with 240 farmers (18.1 per cent of the total). SUBACS is in the third position with 230 sample farmers (17.4 per cent of the total sample). The fourth and fifth positions are occupied by SJDP and FCSPVC with 185 sample farmers (14 per cent) and 135 sample farmers (10.2 per cent), respectively.

A majority of the sample farmers involved in SUBACS is the small category farmers with 33 per cent, followed by marginal farmers with 28.7 per cent. In case of ICDP, a majority is the marginal category with 28.9 per cent, followed very closely by the small farmers with 26.8 per cent. Similarly, FCSPVC and SJDP sub-schemes are also dominated by the marginal farmers with 57 per cent and 44.9 per cent, respectively. The small farmers follow the marginal category in both the schemes with 22.2 per cent and 33.5 per cent, respectively. However, INM is led by semi-marginal farmers with 23.3 per cent of the total sample farmers, followed by small farmers with 23.3 per cent. This indicates that the MMA scheme is, by and large, reaching out to the lower-rung farmers.

FINANCIAL TARGETS AND ACHIEVEMENT OF MMA SCHEME

To determine the progress and development of the MMA scheme, an assessment of the financial assistance is very important. The trend and pattern of financial assistance and achievements of various sub-schemes/components of MMA scheme from 2000–01 to 2008–09 are presented in Table 61.4.

It is clear from Table 61.4 that the overall percentage of financial targets of different sub-schemes of the MMA scheme is quite impressive and it has increased gradually from the inception of the scheme till 2008–09. As for the performance of the scheme, only 57.6 per cent of the financial achievement was registered when the scheme was introduced in 2000–01. Almost all the sub-schemes recorded very poor achievement rate during the initial stages.

Table 61.4 Financial Targets and Achievement under Selected MMA Schemes (Rs in Lakh)

Schemes		SUBACS	ICDP	INM	FCSPVC	SJDP	Total	Per cent of Ach./Targ.
2000–01	Tar.	N A	993.6	274.0	1223.0	0.0	2490.6	**57.6**
	Ach.	N A	415.9	17.0	1002.0	0.0	1434.9	
2001–02	Tar.	32.0	858.1	289.0	10600.0	173.6	11952.7	**76.9**
	Ach.	32.0	631.8	30.0	8364.0	130.0	9187.8	
2002–03	Tar.	853.3	901.6	193.0	N A	69.2	2017.1	**81.9**
	Ach.	835.5	712.4	43.0	N A	61.6	1652.5	
2003–04	Tar.	892.5	938.7	136.0	7637.0	16.0	9620.2	**93.8**
	Ach.	838.9	864.2	43.0	7270.0	11.5	9027.6	
2004–05	Tar.	996.0	1186.1	228.0	16745.0	119.5	19274.6	**95.5**
	Ach.	994.0	936.2	100.0	16273.0	102.6	18405.8	
2005–06	Tar.	917.2	1453.2	204.0	5261.0	10.9	7846.3	**91.0**
	Ach.	876.8	1281.3	134.0	4843.0	4.3	7139.4	
2006–07	Tar.	1222.8	2336.3	396.0	N A	21.3	1873.7	**90.9**
	Ach.	1199.9	2044.2	372.0	N A	N A	3636.9	
2007–08	Tar.	1996.0	1790.0	416.0	N A	N A	4202.0	**92.0**
	Ach.	1894.3	1617.7	355.0	N A	N A	3867.0	
2008–09	Tar.	6909.8	714.3	N A	N A	N A	7624.1	**96.6**
	Ach.	6671.5	693.3	N A	N A	N A	7364.8	
Total	**Tar.**	**13819.6**	**9069.2**	**2136.0**	**41466.0**	**410.5**	**69004.0**	**89.4**
	Ach.	**13342.9**	**9197.0**	**1094.0**	**37752.0**	**310.0**	**61695.9**	

Note: NA implies as Not Available. Tar.: Target, Ach.: Achievement.
Source: Compiled from the different schemes/sub-schemes of MMA.

It gradually improved over the years and reached 96.6 per cent achievement rate during 2008–09. In fact, the improvement level from 2003–04 onwards registered above 90 per cent. The percentage achievement rate against the target was 89.4 per cent during the period (2000–01 till 2008–09).

As regards the component-wise achievement level of MMA till 2008–09, ICDP registered the highest performance with 101 per cent achievement level over the year. It is followed by SUBACS with 96.5 per cent and FCSPVC with 91 per cent. SJDP and INM fall in the fourth and fifth positions with 80.5 per cent and 51.2 per cent, respectively.

SEED PROCUREMENT UNDER MMA

Timely availability of seed is one of the important factors for successful implementation of the MMA scheme. Sources of seed for the sample farmers of the four sub-schemes (excluding INM) under MMA are given in Table 61.5. When we see the scheme-wise procurement of seeds by sample farmers, ICDP registered 73.2 per cent of the total sample farmers of all the four schemes. SUBACS is a distant second with 12.5 per cent of the total sample farmers, followed by SJDP with 10 per cent. Only 4 per cent of the total sample farmers (excluding INM) procured seed of the FCSPVC sub-scheme.

Table 61.5 Sources of Seed Procurement under MMA (No. of Farmers)

Schemes	Government Outlets	Retail Shops	Open Market	Domestic	Others	Total
SUBACS	136	2	8	55	29	230
	(59.1)	(0.9)	(3.5)	(23.9)	(12.6)	(100)
ICDP*	295	40	87	921	0	1343
	(22.0)	(3.0)	(6.5)	(68.6)	(0.0)	(100)
FCSPVC	53	0	15	7	0	75
	(70.7)	(0.0)	(20.0)	(9.3)	(0.0)	(100)
SJDP	140	6	24	15	0	185
	(75.7)	(3.2)	(13.0)	(8.1)	(0.0)	(100)
Total	**624**	**48**	**134**	**998**	**29**	**1833**
	(34.0)	(2.6)	(7.3)	(54.4)	(1.6)	(100)

Note: *It is in terms of quantity in quintals.
Figures in the parentheses are percentages to the total.
Source: Compiled from the different schemes/sub-schemes of MMA.

As for the sources of seed procurement of the farmers of all the four sub-schemes, most of the farmers, that is, 54.4 per cent, got their seeds from domestic sources. The government outlets with 34 per cent are in the second position. Open market provides 7.3 per cent of the total seeds to the sample farmers. Retail shops provide a negligible 2.6 per cent of the total seed requirement while the other sources (unspecified sources) supply 1.6 per cent.

From Table 61.5 it is clear that 59.1 per cent of the sample farmers of SUBACS component procured seed from the government outlets, followed by domestic source with 23.9 per cent. Retail shops provided the least share (0.9 per cent) of seeds of this sub-scheme. In case of ICDP, 68.6 per cent of the farmers got seeds through the domestic sources while retail shops recorded only 3 per cent. However, for FCSPVC and SJDP sub-schemes, government outlets provide 70.7 per cent and 75.7 per cent of the total sample farmers, respectively. At the bottom are the domestic sources and retail shops with only 9.3 per cent and 3.2 per cent, respectively.

DISTRIBUTION OF AGRICULTURAL IMPLEMENTS UNDER MMA

Similar to seed procurement, distribution of agricultural implements seeks to bring about technological change under the MMA scheme. Table 61.6 shows the number of beneficiary farmers under four different sub-schemes (excluding INM).

Of the total 191 implements distributed, cono-weeder recorded the biggest number with 60 and all the 60 cono-weeders (100 per cent) were distributed under the ICDP sub-scheme. It is followed by seed-cum-fertilizer with 45 numbers (23.5 per cent of the total implements distributed), and all of them (100 per cent) were distributed under the ICDP sub-scheme. Out of the 191 beneficiaries, 152 farmers (80 per cent) benefited under the ICDP sub-scheme alone. Of the 191 implements, 60 were cono-weeders and 45 were seed-cum-fertilizers (39.4 per cent and 29.6 per cent, respectively). Under the SUBACS sub-scheme, 39 implements (20.4 per cent) were distributed. Of the total beneficiaries under this sub-scheme, 51.2 per cent were sprayers and 43.5 per cent were cultivators. All the sprayers (100 per cent) and 42.5 per cent of the total cultivators were distributed under the SUBACS sub-scheme. No implement was distributed under the FCSVC and SJDP sub-schemes.

AREA, YIELD, AND PRODUCTION UNDER MMA

Table 61.7 shows the overall changes in the production, productivity, and areas brought under cultivation under

Table 61.6 Distribution of the Agricultural Implements under MMA Schemes (Numbers)

Implements	SUBACS	ICDP	FCSVC	SJDP	Total
Disk Blade	2	0	0	0	2
	(100)	(0.0)	(0.0)	(0.0)	(100)
Sprayer	20	0	0	0	20
	(100)	(0.0)	(0.0)	(0.0)	(100)
Peddler	0	8	0	0	8
	(0.0)	(100.0)	(0.0)	(0.0)	(100)
Seed cum Fertiliser	0	45	0	0	45
	(0.0)	(100)	(0.0)	(0.0)	(100)
Cultivator	17	23	0	0	40
	(42.5)	(57.5)	(0.0)	(0.0)	(100)
Multipurpose tool bar	0	1	0	0	1
	(0.0)	(100)	(0.0)	(0.0)	(100)
Thresher	0	2	0	0	2
	(0.0)	(100)	(0.0)	(0.0)	(100)
Power–Driven	0	8	0	0	8
	(0.0)	(100)	(0.0)	(0.0)	(100)
Sprinkler or Drip	0	5	0	0	5
	(0.0)	(100)	(0.0)	(0.0)	(100)
Other (Cono-weeder)	0	60	0	0	60
	(0.0)	(100)	(0.0)	(0.0)	(100)
Total	**39**	**152**	**0**	**0**	**191**
	(20.4)	**(80)**	**(0.0)**	**(0.0)**	**(100)**

Note: Figures in the brackets are the percentages to the total.
Source: Compiled from the different schemes/sub-schemes of MMA.

Table 61.7 Area Yield and Production under the Selected MMA

Schemes	Area (Acres)		Production (Qtls)		Yield (Qtls/Acre)	
	Before	After	Before	After	Before	After
SUBACS*	629.0	651.0	15547.0	18553.0	25.0	29.0
ICDP**	3830.6	4062.9	63791.7	75178.9	16.7	18.5
FCSPVC***	29.5	40.0	1307.5	2624.0	44.3	65.6
SJDP	121.6	135.5	694.1	812.3	5.7	6.0

Notes: *For Sugarcane crop, production is presented in terms of tonnes.
** It includes all the three crops (rice, wheat, and coarse cereal).
*** For this particular sub-scheme, only two states (Karnataka and Uttar Pradesh) are included, and the area, yield, and production of vegetable crops for Uttar Pradesh are given in terms of money value. Since the quantity of the crop is missing in Uttar Pradesh, the evaluation is done for Karnataka only.
Source: Compiled from the different schemes/sub-schemes of MMA.

the four sub-schemes of MMA. It has brought positive changes in the area of cultivation after the implementation of the scheme (2007–08). Similarly, its impact is visible even in productivity and production.

As far as the component-wise analysis is concerned, FCSPVC registered the highest growth after the implementation of the scheme with 65.6 quintals of output per acre. Similarly, the area brought under cultivation increased from 29.5 acres to 40 acres after the scheme, a change of 35.5 per cent. It is followed by SUBACS with a change of productivity from 25 quintals per acre to 29 quintals per acre after the implementation of the scheme. The change in the area of cultivation under this sub-scheme is 3.4 per cent after the scheme, that is, from 629 acres to 651 acres.

In the third position is the ICDP sub-scheme that brought about a change of 1.8 quintals in productivity from 16.7 quintals per acre before the scheme to 18.5 quintals per acre after the scheme. The total area brought under cultivation increased by 232.3 acres, from 3,830.6 acres to 4,062.9 acres after the scheme. At the bottom, the SJDP sub-scheme registered a growth of productivity from 5.7 quintals of output per acre to 6 quintals per acre after the implementation of the scheme. The area under cultivation has increased by 13.9 acres under this sub-scheme, from 121.6 acres to 135.5 acres.

CONCLUDING REMARKS AND RECOMMENDATION

Obviously, the tasks are many and performing these tasks requires coordinated efforts among different departments of the government. Nevertheless, considering the broader objectives of the MMA scheme, the aforesaid policy implications boil down only to minor corrections in the strategies for effective implementation of the schemes concerned, so as to sustain the success of the macro management mode of agriculture. Much more coordination is needed between the farmers and the government or it's implementing agencies.

In all, the MMA scheme has done a commendable job in the field of agriculture. For making the agricultural development programmes much more successful in the states, institutional and infrastructural supports need to be developed. Also, efficient planning, monitoring, and sincere execution of the policies by the government agencies is essential to make the schemes viable and successful. MMA is a significant scheme introduced for the development of the agricultural sector in the country. This scheme has brought about tangible benefits for farmers, especially for the lower-rung farmers. In totality, the sub-schemes under MMA and their objectives have made a tremendous impact on agriculture in the states.

Having analysed the main features of the scheme, some policy implications of the study may be enumerated as follows. As the success and development of the MMA scheme depends on the involvement of the targeted farmers, more attention should be given to the participation and training of the lower-rung farmers. For disseminating advanced technology to the grassroots farmers under this scheme, their participation in the training provided by the agencies is essential. Unfortunately, the rate of farmers' participation under the scheme is very poor. Proper information has to be disseminated to the farmers and they should be mobilized on time. The study reveals that the seed procurement and the supply of agricultural implements have not met the targets of the scheme satisfactorily. Similarly, most of the implements are received by the higher-rung farmers. More attention needs to be given to the distribution system and delivery mechanism of the scheme.

REFERENCES

Arashiro, Z. 2010. *Financial Inclusion in Australia: Towards Transformative Policy*. Social Policy Working Paper No.13, Brotherhood of St. Laurence, and University of Melbourne Centre for Public Policy, Melbourne.

Desai, S. and Dubey, A. 2011. 'Caste in 21st Century in India: Competing Narratives', *Economic and Political Weekly*, 46(11): 40–49.

Dev, S.M. 2006. 'Financial Inclusion: Issues and Challenges', *Economic and Political Weekly*, 41(41): 4310–13.

Drèze, J. and Khera, R. 2010. 'The BPL Census and a Possible Alternative', *Economic and Political Weekly*, 45(9): 54–63.

Jakab, M. Alexander S. Preker, Chitra Krishnan, Pia Schneider, Francois Diop, Johannes Jutting, Anil Gumber, Kent Ranson, and Siripen Supakankunti. 2001. *Social Inclusion and Financial Protection through Community Financing-Initials Results from Five Household Surveys*. HNP Discussion Paper: The World Bank, Washington DC.

Kamath, R. 2007. 'Financial Inclusion Vis-à-vis Social Banking', *Economic and Political Weekly*, 42(15): 1334–35.

Leeladhar, V. 2005. 'Taking Banking Services to the Common Man: Financial Inclusion', Commemorative lecture of Fedbank Hormis Memorial Foundation at Ernakulam 2 December.

Midgley, J. 2005. 'Financial Inclusion, Universal Banking and Post Offices in Britain', *Area*, 37(3): 277–85.

Ministry of Agriculture. 2011. *Macro Management of Agriculture*, Govt. of India, New Delhi.

Singha, K. and Murthy, K. 2011. *Macro Management of Agricultural Schemes: Consolidated Report*. Bangalore: ADRTC, Institute for Social and Economic Change.

Thorat, U. 2007. *Financial Inclusion: The Indian Experience*. HMT-DFID Financial Inclusion Conference (19 June) at London.

Wagner, Adolph. 1883. *Finanzwissenschaft*. Leipzig: C.F. Winter'sche Verlags-handlung.

———. 1893 *Grundlegung der politischen Ökonomie*. Leipzig: C.F. Winter'sche Verlagshandlung.

——— 1958. *Three Extracts of Public Finance*. London: MacMillan and Co. Ltd.

Wilson, L. 2006. 'Developing a Model for the Measurement of Social Inclusion and Social Capital in Regional Australia', *Social Indicators Research*, 75(3): 335–60.

62

What Restrains Demand for Crop Insurance in India

Nilabja Ghosh, Yogesh C. Bhatt, and S.S. Yadav

Farmers customarily undertake various indigenous measures to cope with risk. Unfortunately, the ex ante methods relating to productions decisions such as the use of safe varieties of seeds, staggering of planting time, spatial scattering of plots, crop diversification, share cropping, and avoidance of purchased inputs all have negative implications for efficiency. Methods adopted after the occurrence of the unfavourable event on the contrary are extremely distressful (Mellor 1969). While despair even leads to suicides in extreme cases, dissaving, borrowing at exorbitant interest rates, sale or mortgage of assets, search for off farm employment, and migration to cities are common recourses. Further, the farmers would compromise on their household expenditure on nutrition, health, and children's education with long term effects. Migration disrupts normal social life, degrades human development, and inflicts social problems in urban areas.

No government can be blind to the distress of large section of people involved. Agricultural risk, therefore, translates to budgetary support, which has developmental and political costs. Worse, since lending to farmers becomes risky, the banks are hesitant to extend credit to agriculture, which becomes resource starved and stagnated. The apprehensions about climate change and recent events of repeated weather adversities of different types create further compulsion for the government and the larger public to create systems to protect farmers' welfare and their incentives towards higher productivity. The atmosphere of uncertainty in which crop cultivation takes place and the implications of risk for welfare and efficiency are reasons enough to suppose that crop insurance could play an extremely useful role in India.

Insurance being the trading of risk, its market is governed by demand and supply forces because it has a utility for those who are affected by risk. Market-driven insurance has proved successful in many cases but for agriculture the market for insurance is not readily formed due to adverse selection, moral hazard, covariate risk, high administrative cost, and poor purchasing power of the farmers. Public support has been a compulsion in many countries.

The Government of India has therefore taken up the responsibilities of facilitating the formation of an agricultural insurance market. With this end the National Agricultural Insurance Scheme (NAIS) was launched in 1999–2000 by which farmers pay a price as a 'premium' for coverage of their risk to the insurer and in turn is liable to claim an indemnity as damage in case the crop fails to deliver a guaranteed yield. The success of insurance lies largely in the ability of a scheme to draw participation and create a large and varied pool of risk profiles so that the revenue and cost even out over a period of time across the purchasers of insurance. The demand for insurance, however, depends on how meaningful the insurance appears in the perception of the farmers. The design of the scheme is extremely important.

OBJECTIVE

This chapter reviews the performance of the NAIS for crop insurance in India and looks for the shortcomings that restrict its demand. Participation in NAIS is assessed by states, irrigation endowments, crops grown, farm classes, and participation in the credit market. The presumptions made by the scheme about the character of risk covered are assessed by examining the normality of crop yield using linear time trends. The chapter makes a comparison between the simulated yield losses as based on the NAIS formula of the threshold yield and as calculated using the observed variability of yield. A CI scheme designed with due care towards the specification of the threshold can create utility and appeal among farmers for the insurance product and for their adopting advanced new technologies with greater productive potential.

CROP INSURANCE IN INDIA AND ITS PARAMETERS

In India the crop insurance programme has a long history (Mishra 1996) and the evolution has been far from smooth.[1] The experiences of other countries also discouraged the venture (Hazell and Valdes 1985; Hazell, Pomareda, and Valdes 1986). The NAIS was initially implemented by a general insurer already in the market but in 2003 the task was handed over to an autonomous implementing agency called the Agriculture Insurance Company (AIC) of India Limited. Projected to become viable and commercially independent within a short span of time, the NAIS started with limited government support. The scheme is multiple-peril type but offers comprehensive insurance to yield losses due to a few named risks mentioned in the scheme. The NAIS aims to stabilize farm income and help agriculture to be progressive. As with other agricultural programmes in India, the interests of the small farmers who constitute over 80 per cent of holdings and are the primary victims of crop risk are central to the objective.

Area-Yield Insurance

Conceptually, the area-yield insurance (AYI), as a design, has advantageous features that can circumvent many of the difficulties that foil the insurance market (Skees and Reed 1986, Miranda 1991). Risk is pooled not from individual farmers but from various groups of farmers or areas and the indemnity is assessed uniformly at the area level. The concept of an area is based on the possibility of sufficient homogeneity existing within the unit so that the majority of the farmers in the unit are likely to encounter a loss simultaneously and the risk exposure is expected to be similar across all the farmers in the area. The farmers insured in the area receive indemnities at the same rate when the yield at the area level falls short of what is considered its normal regardless of their own individual losses. The indemnity is calculated based on the individual contract size and the area level yield shortfall. No payment is made when the area level yield is above its normal level.

Administrative costs and chances of moral hazard are reduced in the AYI. Ideally, the individual's yield shortfall should be attended to, but AYI offers only a second best solution to the individual based scheme. That homogeneity is an elusive idea and *basis risk* is not addressed are dominant criticism in NAIS. The size of the area is ideally small but in practice, the specification of an area actually depends on the availability of regular and past yield data. Only improved information by statistical, administrative, and technological means can reduce the size of the area over time.

The area yield insurance can do little to overcome the problem of adverse selection for which compulsion is usually prescribed. Compulsion, however, is a contradiction to market principle and should in principle be restricted if not avoided. The NAIS in India is, however, compulsory but only for institutional borrowers in those states that have opted to participate in the scheme. For farmers who have not borrowed from banks or cooperative societies participation is optional. This mandate makes the demands for insurance and institutional credit nearly indistinct. To avoid the burden of paying premiums, farmers obtain finance from moneylenders or mortgage or sell family gold to raise finance for farming. Also farmers who have not been deemed suitable or eligible for institutional loans due to technical reasons are necessarily outside the pool. Even states have the choice not to participate under certain conditions. All these factors limit participation.

The NAIS is largely an area based scheme although, on an experimental basis, insurance for localized calamities is given to individuals. Crops differ in their vulnerability to different events and the NAIS covers a large number of

[1] The (Expert) Committee under the Chairmanship of Prof. Dharm Narain advised against launching the scheme in view of the cost involved and the availability of simpler ways of achieving the purpose.

crops to diversify exposure as well as to serve the farmers. To determine the specified minimum yield below which a loss is deemed to have been incurred, a threshold needs to be calculated based on past information. The area based scheme is parsimonious in information requirement since actual yield information at the farmer level is irrelevant for its functioning.

Threshold Yield

Areas are assigned certain levels of indemnity (LOI) at 90 per cent and 80 per cent and 60 per cent based on last ten years data of crop yields and thereby classified into three risk groups namely, low risk, medium risk, and high risk, respectively. The threshold yield (TY) of a crop in the unit is the moving average based on the past three years' yield in case of rice and wheat, and five years' on case of others, multiplied by the state LOI. This formula is a subject of criticism both from farmers and from review committees. If the actual yield per hectare of the insured crop in the area as obtained from the crop cutting experiments (CCE) of the government falls short of the specified TY, then all the insured farmers growing that crop in the defined area are deemed to have suffered a loss. They are eligible for indemnity where,

Indemnity = [(TY–actual yield)/TY] of 'area' × Sum insured of farmer (1)

Premium

The premium rates are basically the prices for insurance of different crops in different areas. In a competitive market they should reflect the marginal utility from coverage to insurance buyers whereas in the interest of supplier they ideally reflect the cost of insurance which includes the cost of both risk and administration.

However, for food grains and oilseeds the premium rates in India are far from market determined. Indeed, because premiums add to the cost of cultivation and to the price if the crop is insured, maximum bounds are fixed by the government based on welfare and political criterion. Part of this imposition can be justified because coverage is mandatory for farmers who borrow from institutional lenders. Premium rates are fixed at 3.5 per cent for bajra and oilseeds, 2.5 per cent for other kharif crops, 1.5 per cent for wheat, and 2 per cent for other rabi crops. For commercial crops the rates are necessarily actuarial and are much higher. The scheme foresees a transition to the actuarial regime but the actual rates shall be applied at the option of the states. A subsidy of 50 per cent on premium was initially allowed in respect of small and marginal farmers but this was to be phased out and at present 10 per cent subsidy is given on the premiums paid by them, borne by the central and state governments jointly.

DATA AND METHOD

State-wise data on insurance as provided by the implementing agency covering the states: Andhra Pradesh, Assam, Bihar, Gujarat, Haryana, Karnataka, Madhya Pradesh, Maharashtra, Odisha, Rajasthan, Tamil Nadu, Uttar Pradesh, and West Bengal for major kharif[2] crops: cotton, groundnut, maize, and rice is used. Rice and maize together account for 64.1 per cent of kharif and 41.9 per cent of combined (rabi and kharif) acreage under total foodgrains. Groundnut accounts for 29.6 and 23.6, respectively of area under total oilseeds. Coverage, premiums collected, claims disbursed, and the parameters of the scheme comprise the insurance data. Besides insurance data, data on cropped and irrigated acreages and yield rates of crops (Ministry of Agriculture and Farmers Welfare [MoAFW], website) will be used. The riskiness of the crops is assessed statistically using time series data on crop yield.

Classification of States by Irrigation Endowment

Apart from insurance there are other ways in which risk is formally managed by the market or by policy. Irrigation is a key instrument of risk management of ancient vintage in that less irrigated regions may be described as more risk prone to weather events than more irrigated ones. We have classified the thirteen major participating states into three categories Highly irrigated (HI), Medium irrigated (MI), and Low irrigated (LI). HI is represented by states with irrigation intensity exceeding 50 per cent, MI by states with irrigation intensity between 30 per cent and 50 per cent and states with irrigation intensity up to 30 per cent are classified as LI. Official data 2002–03 are used for classification. States are found to be quite varied in irrigation endowment even within a geographic region. For example, Uttar Pradesh in north, Bihar in east, and Tamil Nadu in the south of India are all in the HI category. Similarly, the other irrigation-based categories also draw states from varied regions. (Table 62.1).

The three categories making up 94 per cent of territory are compared in Table 62.2. The MI category of states which claims 29.6 per cent of cropped area of the

[2] Kharif is a crop season, roughly June to September, Rabi is another season roughly October to December.

Table 62.1 Irrigation Endowments of States with Crop Insurance in Three Categories (2002–03)

S.No.	High (>50%)		Medium (30%–50%)		Low (<30%)	
	Name	Irrigation	Name	Irrigation	Name	Irrigation
1	Uttar Pradesh	70.3	Rajasthan	39.9	Madhya Pradesh	25.6
2	Bihar	57.5	Andhra Pradesh	39.2	Karnataka	24.5
3	Tamil Nadu	50.5	West Bengal	36.7	Odisha	21.8
4			Gujarat	31.4	Himachal Pradesh	0.00
5					Maharashtra	18.1
6					Assam	5.5

Note: Irrigation = Net irrigated area/Net sown area (per cent).
Source: Ministry of Agriculture (2005).

Table 62.2 Crop Insurance and Shares of State-groups in All India Totals (2003–04)

State groups	Sum Insured		Small/Marginal Holdings	Cropped Area	Irrigated Area	FGOLS Area	COM Area
	Rs/Hect.	%	%	%	%	%	%
High Irrigated	266.76	10.9	43.8	23.8	39.5	25.2	19.1
Medium Irrigated	861.59	43.5	21.0	29.6	27.2	28.2	34.2
Low Irrigated	651.47	45.6	32.8	41.0	22.3	39.9	44.8

Note: FGOLS = Foodgrains-oilseeds, COM = Commercial.
Source: AIC, Agricultural Census 1995, MOA (2005).

total accounts for 44 per cent of the sum insured and has the highest sum insured per cropped hectare. The HI category has a small 11 per cent share in sum insured compared to its 24 per cent share in area and its 44 per cent share of small farmers in the country but holds 40 per cent of the irrigated area. The largest share of cropped area and sum insured are in the LI category which is more commercialized.

The Measurement of Risk

The coefficient of variation (CV) of yield is widely considered as a measure of risk. Associated with this is a tacit assumption that yield rates are normally distributed but the presumption of normality has been widely questioned in literature (Day 1976). One rationale provided for conjecturing a skewed distribution is that too much or too little rain or heat during any of the critical periods of plant growth such as sowing, germination, flowering, and harvest, is sufficient to reduce yield drastically though ideal weather prevails in the other periods. Thus, common sense suggested that less than average yields cases are more likely than greater than average yields cases. That similar reasoning can apply to favour a reverse argument too complicates the argument. Although much investigation transpired around this issue (Moss and Shonkwiler 1993; Nelson and Preckel 1989), empirical studies have not generally been able to reject the normality hypothesis in practice.

The normality test is complicated by the usual nature of agricultural data. A time-series data of yield rates often used for analysis usually incorporates a time trend reflective of the secular progress of technology. Without suitable detrending, this trend will be nested in the distribution. Tests such as the Wald-Wolfwitz run method Day, 1976, Chi-square test (Dandekar 1976, Rustagi 1988) have been used to establish normality but not all of them corrected for a trend. Just and Weninger (1999) emphasized that for a normality test it is critical to deal with the random component only and so elimination of the deterministic component is a prerequisite. They used a polynomial time function to isolate the random elements but using rigorous statistical tests, they could not rule out normality. Day's own experiment results were also said to have been weak in rejecting normality.

To study the yield distributions, we have, therefore, detrended the yield rates of major kharif crops using linear time trend equations estimated over the period the years 1973 to 2005 for each irrigation based categories and for the aggregate sample. Since this period overlaps with a period when the effect of the agrarian technology generating the so called green revolution faded out, the trend is likely to encounter a structural change.

The presence of structural breaks was examined using the Chow tests in a search model. Since the trend curves have indeed shifted downwards in most cases between 1996 to 2001 and such shifts are not likely to be unobserved by the operating farmers, the trend equations need to incorporate these structural changes in order to obtain the random and unpredictable components around the expected values. As an example, we can consider an upwardly moving series that has shifted downwards at a given point of break. If we fail to take account of the shift and measure the deviations around the linear and unbroken trend the distribution, even when normal, would tend to include a high proportion of large negative values typically projecting a positively skewed yield distribution which would be actually misleading.

$$Y_t = a_0 + a_1\, time + a_2\, time \times dummy + u_i \qquad (2)$$

$$\text{where Dummy} = \begin{cases} 1 \text{ if time} > 23 \\ 0 \quad \text{otherwise} \end{cases}, \quad \text{time} = 1, 2, \ldots, 33$$

RESULTS AND DISCUSSIONS

The following discussion begins by assessing the performance of NAIS in terms of its acceptance by the risk victims, reflecting also the risk pooling effect and its financial outcome indicative of its viability. The shortcoming is then further investigated by working out representative threshold yields using the scheme parameters and yield data and comparing them with notional lower-band yields of the corresponding time period using estimated time trends of yield.

Penetration, Financial Success and Participation, Irrigation Management Risk

Crop insurance constituted only 2.7 per cent of the GDP from agriculture of India where 7.6 per cent of the gross cropped area and 10 per cent of the farmers in the country are covered (GOI 2004). The achievements are modest and Table 62.3 also indicates serious seasonal bias towards the kharif (monsoon) season and a high share (93 per cent) of loanee farmers in the pool. This makes insurance demand almost synonymous to credit demand. The claim to premium ratio is high at nearly 4 and loss accounts for 8 per cent of the sum insured. Despite the subsidies the share of small farmers is about 65 per cent, much less than their 80 per cent share in total farmer's population.

The financial viability of the scheme can be indicated by the success of crop insurance as a market driven product. The financial performance is measured by the loss which

Table 62.3 Performance of Crop Insurance in India

YEAR	C/P (Cum)	SI/ GDPA (%)	AI/ GCA (%)	Non-loanee (%)	KH/ RABI (Ratio)	Loss (%)	SMF/ TF (%)
2000	5.47	1.74	5.67	1.95	4.31	12.31	66.38
2001	3.5	1.74	5.66	4.02	5.01	2.96	67.14
2002	4.33	2.32	6.89	11.88	5.13	14.63	67.69
2003	4.04	2.10	6.52	14.4	2.66	7.19	61.99
2004	3.5	3.16	8.47	6.98	3.49	3.92	63.93
2005	3.27	3.28	8.66	5.56	2.67	4.54	61.48
2006	3.37	3.60	9.27	NA	2.26	7.89	NA
2007	3.21	3.95	9.42	NA	2.28	4.25	NA
2008	3.44	NA	NA	NA	1.42	10.53	NA
Average	3.79	2.74	7.57	7.47	3.25	7.58	64.77

Note: C/P = Claim/premium is cumulated (cum) over years. SI = Sum Insured, SMF = Small and Marginal Farmers insured, TF = Total Farmer insured, AI = Area Insured, GDPA = Gross Domestic Product of Agriculture, GCA = Gross Cropped Area, KH = Kharif. NA-Not Available, Loss = Claims paid less premiums collected, Non-loanee = farmers who did not borrow from institutional lenders.
Sources: Ministry of Agriculture, NAIS.

is the excess of payouts in claims over the revenue from premiums. If insurance can be viewed as temporal pooling of risk, that is, over the years losses in some years are made up by surpluses[3] then as the claim to premium ratios in Table 62.3 shows, the NAIS has incurred losses in all the years of its existence in 2000–01 to 2005–06. Till now the losses are borne by government.

Figure 62.1 shows that majority of the crops have generated losses and groundnut is found to have generated the largest loss to the NAIS followed by kharif paddy. Ideally one would like to see both loss making and surplus generating crops but the basket of crops insured has not contributed towards risk pooling at the given parameters of the scheme. Similarly, we also find that there is little effective pooling across states as none of the states has generated a surplus. This is probably related to adverse selection but since it is fairly easy to classify the states by their risk profile, this can be corrected by appropriate pricing. It may be noted that Punjab, the most agriculturally advanced state and possibly one of the safe ones, has not yet agreed to participate and Haryana, another progressive state, was a late entrant and its participation rate is low. Gujarat,

[3] The receipt will, in general, be expected to exceed the payments in good years and fall short in poor years when a large number of farmers claim damages. In principle, over a number of years the average receipt will ideally converge with the average expenditure.

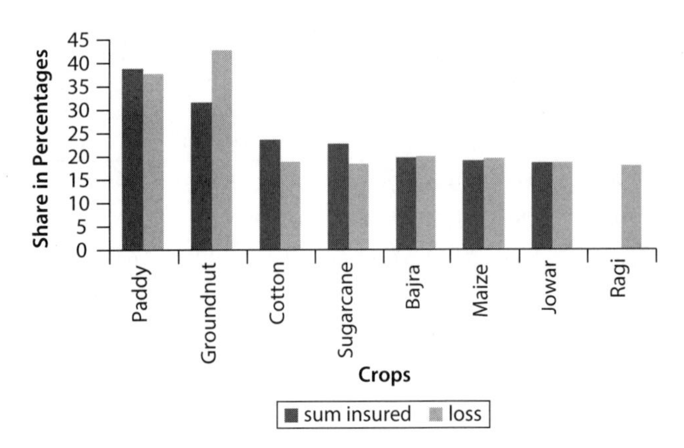

Figure 62.1 Share (%) of Sum Insured and Loss (2000–05) for Kharif Crops
Sources: Ministry of Agriculture, NAIS.

Table 62.4 All India Time Trend of Crop Yields with a Break

Variable	Con.	Time	Time* Dummy	Adj-R2	DW	F-stat	Break
Rice kharif	914.3	32.5	−7.4	0.86	2.5	6.1	2000
	(21.9)	(12.5)	(−3.5)				
Maize	818.5	35.6		0.86	2.2		–
	(16.7)	(14.2)					
Groundnut	683.3	9.2		0.23	2.7		–
	(12.3)	(3.3)					
Cotton	104.8	4.34		0.86	1.7	19.2	2000
	(9.0)	(7.3)					

Sources: Ministry of Agriculture, NAIS.

Andhra Pradesh, and Karnataka lead in terms of their contribution to losses (Figure 62.2).

Yield Trends

The trend equations estimated at the aggregate level over all the states using flexible break points, based on data of crop yields (equation 2) over 1973–74 to 2005–06 are presented at the all India level in Table 62.4. Similar trend equations are estimated at the level of irrigation-based categories (Annexure Table 62A.2).

A positive time trend is noted for all the crops and a structural slow down detected in rice at the all India level. Cotton shows an accelerating trend from year 2000.[4] Slowdown is also evident in the irrigated regions. The detrended series are shown to be stationary by a simple ADF technique (Table 62A.1). A common way of determining the degree of skewness of a data is to compare the numerical value for 'Skewness' with twice the 'Standard Error of Skewness' and the distribution is marked as significantly skewed only if the skewness is high enough. The risk measured both in coefficient of variation and skewness along with the Jarque-Bera test of normality are summarized in Table 62.5. The signs of skewness of crop yield distribution were found to be both positive and negative (Table 62A.2 in the Annexure). However, there is only one case out of the 16 estimated relations in which the skewness is found statistically significant. When the normality test is done only this same case (groundnut in HI) is identified as non-normal. However, the coefficient of variation exceeds 0.1 in as many as 11 of the cases. Thus, although it appears that there is a case for examining the distribution on a case by case basis, a broad association of risk with the coefficient of measure may not be unjustified.

[4] This probably captures the effect of a technological change in the form of Bt. Cotton (Biotechnology) grown is some of the states since early 2000s.

Threshold Yield Under Yield Dynamics

The determination of the threshold yield (TY) is critical in the crop insurance design. A TY so low that rarely does

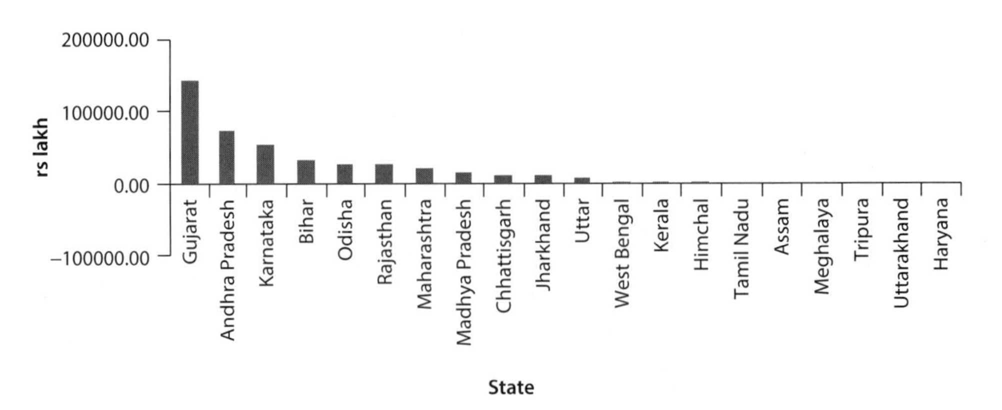

Figure 62.2 State-wise Loss (2000–05): Kharif

Sources: Ministry of Agriculture, NAIS.

Table 62.5 High Variation and Non-Normality of Crop Yield (DE-trended)

	High	Medium	Low	All
Coefficient of Variation (>10%)	Maize, Cotton, Groundnut	Maize, Cotton, Goundnut	Rice, Kharif, Cotton, Groundnut	Cotton, Groundnut
Skewness+ (>2 Std. Dev)	Groundnut			
Skewness− (<2 Std. Dev)				
J.B Statistics (Non-Normality)	Groundnut			

Sources: Ministry of Agriculture, NAIS.

Table 62.6 Comparison between Estimated Threshold and a Lower Band Yield Rates (average of 2000–01 to 2005–06)

Region	Proportion	High	Medium	Low	All
Rice kharif	$TY/(Y^t - n.\sigma)$ (where n = 1)	0.94	0.7	0.78	0.8
Maize	$TY/(Y^t - n.\sigma)$ (where n = 1)	0.79	0.75	0.75	0.74
Groundnut	$TY/(Y^t - n.\sigma)$ (where n = 1)	1.1	0.78	0.73	0.74
Cotton	$TY/(Y^t - n.\sigma)$ (where n = 1)	0.79	0.74	0.62	0.65

Note: TY = threshold yield, Y^t = Trend yield, σ = standard errors of estimate
Sources: Ministry of Agriculture, NAIS.

the actual yield touch it in reality would imply that farmers would not be able to claim indemnity (see equation 1) for successive years despite the continuity of premium payment and such a scheme would be thoroughly unattractive in the market. The meaningfulness of the TY for the farmer's decision to participate would depend greatly on what the farmer considers as a meaningful threshold. The TY determined in India (as given earlier) is usually criticized to be not high enough to mean any advantage for the farmers and it has been recommended even by a review committee to revise the formula in order to arrive at a high enough threshold yield to make participation rewarding. The threshold is built upon the premises that a normal yield can be calculated purely by averaging the past yield rates. In a case of yield dynamics, it is normal to project the yield into the unknown future, especially when the progressive farmer incessantly perseveres to stretch the limits.

In practice, a moving average of the past yields is inadequate to capture the movement. For example, a normal yield of 200 obtained by averaging three consecutive realizations of 100, 200, and 300 would be no different from one obtained from unchanging realizations of 100, 100, and 100. The formula makes no distinction between a stagnant and a dynamic situation and, in fact, can prove to be a disincentive to progress. Formulas such as the average of a few best yield realizations of the past and exclusion of drought years from the calculation may be closer to the notional normal in practice but yet be theoretically inadequate in a dynamic situation of progress. They also do not conform to the objective of risk insurance. One adverse consequence of this inadequacy has been a popular pressure on relevant statistical departments to report lower yield rates than the true yield which is extremely inimical to the interests of development.

To make an assessment we have generated TY of four major kharif crops at the state level using the threshold formula as given earlier and the LOI values specified officially by the NAIS. The values so obtained are compared with a notional worst case scenario given by the lower band value in our trend equations (equation 2). When a deviation of one standard error is considered around the estimated trend value. In general, the TY is found to be lower than the lower band. The only exception is in cases of groundnut in HI region. In the case of rice, the threshold is only 80 per cent of the lower band and is also low in the case of maize and groundnut and lowest for cotton. Although farmers' assessment is subjective, the threshold level by this demonstration appears to be too low in relation to the probability of occurrence of the event. The figures given in Figure 62.3 are computed averages of the two low points for kharif rice over the years 2000–01 to 2005–06.

CONCLUSIONS: CROP INSURANCE FOR A PROGRESSIVE FUTURE

Crop insurance, despite its significance for Indian agriculture and the government's active support has proved

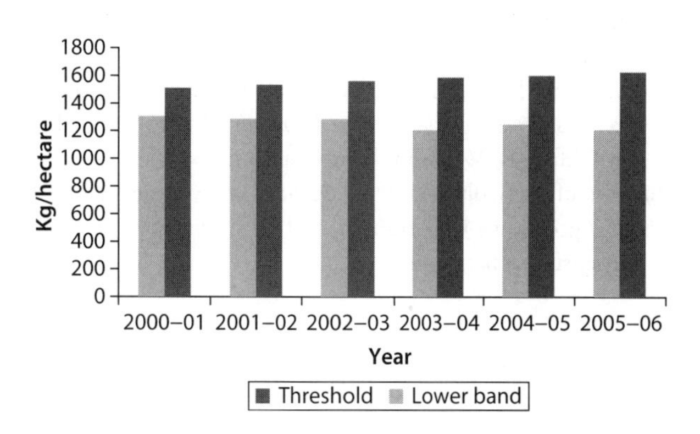

Figure 62.3 A Comparison of the Threshold Yield with the Lower Band Yield Using One Standard Deviation for Kharif Rice
Sources: Ministry of Agriculture, NAIS.

to be a losing proposition even after several years of the launch of the national level scheme. The reasons would probably be poor risk pooling due to inadequate participation linked to the design. This study suggests that much is wrong in the design of the scheme. Besides the uneven ground created by the rigidly constrained uniform prices of insurance for cereal and oilseeds crops despite their vulnerability to risk, the regressive characters of the TY formulation is a serious weakness. The TY, that in principle is linked to the actually revealed uncertainties in nature, also serves little purpose by failing to recognize that progressive farmers have to invest to achieve higher yields than experienced before and would take coverage of insurance with the hope that their probable losses will be realistically indemnified. Indeed, it appears that the applied formula could be fixing thresholds at levels way too low to make any meaning for the farmer to pay a price for the insurance. Indeed, they pay for coverage merely mandated for credit.

REFERENCES

Dandekar, V.M. 1976. 'Crop Insurance in India', *Economic and Political Weekly*, 11(26) June: A61–A80.

Day, N. 1976. 'A New Measure of Age Standardized Incidence, The Cumulative Rate', in *Cancer Incidence in Five Continents, Vol. 111*, edited by J.A.H. Waterhouse, C.S. Muir, P. Correa, and J. Powell, 443–52. Lyon, International Agency for Research on Cancer (IARC Scientific Publications No. 151).

Government of India (GOI). 2004. *Report of the Joint Group on Crop Insurance*. New Delhi: Ministry of Agriculture.

Hazell, P. and A. Valdes.1985. 'Is there a Role for Crop Insurance in Agricultural Development?', *International Food Policy Research Institute*, Food Policy Statement No. 5, IFPRI, Washington D.C.

Hazell, P., C. Pomareda, and A. Valdes. 1986. *Crop Insurance for Agricultural Development: Issues and Experience*. Baltimore MD: Johns Hopkins University Press.

Just, Richard E. and Quinn Weninger. 1999. 'Are Crop Yields Normally Distributed?', *American Journal of Agricultural Economics*, 81(2): 287–304.

Mellor. 1969. 'The Subsistence Farmer in Traditional Economics', in *Subsistence Agriculture and Economic Development*, edited by Clifton R. Wharton. Frank Cass and Co.

Ministry of Agriculture. 1995. Agriculture Census. Ministry of Agriculture Government of India. Available at http://agcensus.dacnet.nic.in/.

———. 2005. Agricultural Statistics at a Glance. Government of India Ministry of Agriculture Department of Agriculture & Cooperation Directorate of Economics & Statistics.

Ministry of Agriculture and Farmers Welfare (MoAFW). N.d. Website. Directorate of Economics and Statistics, Department of Agriculture, Cooperation and Farmers Welfare, Government of India. http://eands.dacnet.nic.in/.

Miranda, M.J. 1991. 'Area Yield Crop Insurance Reconsidered', *American Journal of Agricultural Economics*, 73(2) May: 233–42.

Mishra, Pramod K. 1996. *Agricultural Risk, Insurance and Income: A Study of the Impact and Design of India's Comprehensive Crop Insurance Scheme*. Hong Kong: Avebury.

Moss, Charles B. and J.S. Shonkwiler. 1993. 'Estimating Yield Distributions with a Stochastic Tend and Non-normal Errors', *American Journal of Agricultural Economics*, 75(4) November: 1056–62.

Nelson, C.H. and P.V. Preckel. 1989. 'The Condition Beta Distribution as a Stochastic Production Function', *American Journal of Agricultural Economics*, 71(2), 1 May: 370–8.

Rustagi, Narendra K. 1988. *Crop Insurance in India*. B.R. Publishing Corporation.

Skees, J.R. and M.R. Reed. 1986. 'Rate Making for Farm Level Crop Insurance: Implications for Adverse Selection', *American Journal of Agricultural Economics* 68 (3): 653–59.

ANNEXURE 62A

Table 62A.1 Dickey-Fuller Test Statistics for Stationarity of Yield Rates (de-trended)

Crops	HI	MI	LI	Pooled
Kharif Rice	3.40	4.97	4.95	4.18
Rabi rice	2.68	4.13	3.59	2.93
Wheat	3.63	3.28	5.71	3.38
Maize	3.73	4.19	3.75	4.31
Groundnut	3.18	5.80	3.35	
Soyabean	5.08	3.54	3.12	3.55
Tur	5.46	2.00*	3.30	6.25
Cotton	4.44	4.38	3.92	4.48
Sugarcane	5.50	3.22	3.60	5.17
Potato	4.75	3.53	3.64	4.71

Note: ADF test with lag one is conducted. Critical values are 10 per cent-2.6, 5 per cent- 2.9.
Sources: Ministry of Agriculture, NAIS.

Table 62A.2 Time Trend Equations of Crop Yields Rates with a Break in Trend for the Three Regions

Regions	Crops	Cotton	Groundnut	Maize	Rice
	Constant	213.5 (12.2)	616.2 (7.1)	652.1 (10.3)	869.2 (16.7)
High	Time	3.09 (2.8)	35.4 (6.6)	39.2 (10)	42.4 (13.0)
(Hi)	T*D	−1.89 (−2.1)	−14.9 (−3.4)	−1.09 (−0.3)	−11.4 (−4.4)
	R2	0.16	0.57	0.84	0.86
	D.W.	1.20	11.70	1.80	1.70
	Break	2000 [2.5]	1998 [14.9]		2000 [12.2]
	Ditribution of Detrend Yield				
	CV	16.70	18.61	11.76	8.36
	Skewness	0.53	1.91*	0.29	−0.34
	Kurtosis	3.72	8.99	2.66	2.31
	JB-stat	2.29 (0.32)	69.42 (0.00)	0.61 (0.73)	1.29 (0.52)
Medium	C	146.9 (6.4)	729.7 (7.7)	800.7 (7.9)	1129.5 (26.6)
(MI)	T	5.85 (4.1)	3.01 (0.5)	30.3 (4.8)	38.9 (14.7)
	T*D	−0.5 (0.4)	5.63 (1.2)	4.31 (0.8)	0.5 (0.2)
	R2	0.44	0.77	0.61	0.92
	D.W.	0.97	2.50	2.30	2.10
	Break	2000 [20.4]	1999 [3.1]		
	Distribution of Detrend Yield				
	CV	22.90	28.41	18.64	5.80
	Skewness	−0.13	0.53	−0.19	0.07
	Kurtosis	4.83	1.44	2.59	2.59
	JB-stat	4.68 (0.096)	0.09 (0.96)	0.43 (0.81)	0.26 (0.88)
Low	C	70.8 (7.9)	618.1 (15.2)	1167.98 (18.1)	843.2 (15.8)
(LI)	T	3.4 (6.1)	10.84 (4.3)	27.5 (6.8)	20.3 (6.1)
	T*D	−0.26 (−0.6)	4.69 (−2.3)	0.67 (0.21)	−7.73 (−2.9)
	R2	0.64	0.34	0.73	0.54
	D.W.	1.83	2.30	1.60	2.80
	Break		1999 [4.8]	1992 [2.8]	1999 [6.0]
	Distribution of Detrend Yield				
	CV	17.15	12.85	9.63	11.43
	Skewness	−0.23	0.50	−0.09	−0.31
	Kurtosis	2.66	2.61	2.33	2.24
	JB-stat	0.45 [0.80*]	1.60 [0.45*]	0.67 [0.71*]	1.34 [0.51*]
All	Distribution of Detrend Yield				
(POOLED)	CV	17.95	18.20	9.47	7.17
	Skewness	−0.02	−0.11	−0.23	−0.55
	Kurtosis	3.70	3.23	3.00	2.51
	JB-stat	0.61 (0.74)	0.15 (0.93)	0.30 (0.86)	2.00 (0.37)

Note: Figures in parentheses are T-Statistics. JB-test is Jarque–Bera test.
Sources: Ministry of Agriculture, NAIS.

63

End-Term Evaluation Study in Respect of the Implementation of Bringing Green Revolution to Eastern India (BGREI) Programme*

Debanshu Majumder, Debajit Roy, and Ranjan Kumar Biswas

The spread of high yielding variety (HYV) technology resulting in the 'Green Revolution in India' since the mid-1960s had been successful in enhancing the crop productivity and achieving self-sufficiency in foodgrains production in the country. However, the most widely debated issue about this 'Green Revolution' was the growing income disparities between different regions and between different categories of farmers. Therefore, it becomes particularly important to address regional imbalances in growth, imparting stability to agricultural output, and bringing the benefits of agricultural research technology to the resource poor farmers across all the regions of the country to ensure economic equity.

A new technology based on hybrid variety of rice and wheat (the two staple crops in eastern region) seeds were thought of to make a dent in the existing level of productivity. Furthermore, it is worth noting in this regard that the Green Revolution technology that was propagated in the mid-1960s depended heavily on assured and controlled irrigation that was catered mostly by the tube wells. With the passage of time indiscriminate and over use of tube well, irrigation has resulted in an acute depletion of sub-soil water table in the country. Hence, there had been a need for an alternative technology that could address the environmental issues in the process of pushing up the productivity frontier. The program of Bringing Green Revolution to Eastern India (BGREI) is intended to address the underlying constraints for enhancing productivity of rice and wheat in seven states of eastern India (Assam, Bihar, Chhattisgarh, Jharkhand, Eastern Uttar Pradesh, Odisha, and West Bengal) so that agricultural productivity is reasonably enhanced in these areas.

The programme takes care of needed technology in terms of assured provision for incentivized supply of recommended agricultural inputs to the farmers adopting cluster approach in order to ensure equity amongst farmers across selected locations in the BGREI states. The process of input inducement under BGREI programme

* The present chapter is based on the study done by AERC Allahabad, Assam, Bihar and Jharkhand, Chhattisgarh, Waltair, and West Bengal.

differs from other crop development programs in respect of the provision of cash doles for deep ploughing in rainfed areas/land preparation & line sowing/transplanting for all ecologies and making provision of improved seed supply. Besides this, the programme intended to enhance supply of agriculture credit and procurement of agriculture commodities by the public sector agencies at the minimum support prices.

The programme of Bringing Green Revolution in Eastern India was launched in the year 2010–11 to enhance the agriculture production in the states of Assam, Bihar, Chhattisgarh, Jharkhand, Odisha, Eastern Uttar Pradesh, and West Bengal. It was conceived as a lateral to Rashtriya Krishi Vikas Yojna (RKVY). The programme included a bouquet of activities including three broad categories of interventions namely, organizing block demonstrations of rice and wheat in different rice and wheat ecologies; asset building for water management such as construction of shallow tube wells/dug wells/bore wells, and distribution of pump sets, drum seeders, zero till seed drills, and site-specific activities such as construction/renovation of field/irrigation channels/electric power supply for agricultural purposes and institution building for inputs supply. The programme envisaged adopting both medium and long term strategies for asset building activities relating to water conservation and utilization in combination with short term strategies pertaining to transfer of technology through block demonstration.

The programme was implemented in a cluster approach. The size of cluster for the interventions was determined as 1000 ha. Selection of villages/blocks was made based on ecology. From the ecologies beneficiary farmers were selected for each cluster. In each Block Demonstration one Progressive Farmer for every 100 ha of area was selected for providing handholding support to the beneficiary farmers. In order to ensure effective implementation of the programme, district-wise scientific resources drawn from ICAR-SAU system were roped besides 3-tier monitoring system put in place at national, state, and district levels. Institutional support for technical backstopping has been arranged through Central Rice Research Institute (CRRI) besides provision of honorarium to progressive farmers and field staff of State Department of Agriculture concerned as a stop gap arrangement for extension support at ground level. There was overwhelming response to the BGREI programme at all the levels in the BGREI states and crop production prospects were reported to have made a breakthrough. Enthused with these reports, Department of Agriculture and Cooperation decided for conducting an 'End Term Evaluation of BGREI program'. This chapter focuses on evaluation of Block Demonstrations of rice and wheat to the extent possible besides understanding the planning and implementation strategies adopted by the BGREI states.

DATABASE AND METHODOLOGY

The sample units of demonstrations, for each of the BGREI states were selected from five rice ecologies namely; rainfed uplands, rainfed shallow low land, rainfed medium deep water, rainfed deep water, and irrigated. At the first stage of sampling, for each state, one district was selected from each of the ecologies considering the concentration of demonstrations in the district. In the second stage, one representative block from one Block Demonstration under each of the different ecologies was selected following the same procedure. In the third stage, a total number of ten beneficiaries and five non-beneficiaries were selected at random from each selected block. In sum, a total number of 450 beneficiaries and 225 non-beneficiaries spread over 34 selected districts across all the seven BGREI states were covered in the study.

For secondary data on different aspects of BGREI programme—financial allocation and utilization, we had to depend on various government sources including state directorate of agriculture in each BGREI states. Data on area, production, and yield for rice and wheat at the state level (both NFSM and BGREI districts) were made available to us by the BGREI Cell, New Delhi.

A homogeneity test of the respondent farmers (both beneficiaries and non-beneficiaries) in respect of land holding size and level of education was carried out separately to probe into the characteristics of the respondents in respect of their position in economic and social ladder The results reveal that the respondents were more or less homogenous with little variations across ecologies and household characteristics. However, homogeneity test for the beneficiaries was not conducted in respect of Bihar, Jharkhand, and eastern Uttar Pradesh.

It is to be noted that the result of the test for homogeneity signifies that the two sections of respondent namely, beneficiaries and non-beneficiaries, are alike in terms of their land holding sizes and educational attainments. Hence, it is possible to get an impression of the impact of an intervention like BGREI comparing the two groups.

RESULT AND DISCUSSIONS

As mentioned earlier, programme of BGREI had major focus on technology transfer with assured technical backstopping, water asset building, and site specific needs. Accordingly, the entire programme was sub-divided in the following three projects backed with the provision of their monitoring:

1. Block Demonstrations of rice and wheat
2. Water asset building
3. Site-specific needs

Adoption of Bringing Green Revolution to Eastern India Programme

The composition of the BGREI programme in 2011–12 included lion's share for short term interventions namely technology promotion through block demonstrations to the tune of 64.5 per cent of total allocation. The site specific need, however, was allotted about 19.1 per cent of total outlay while water asset building activities comprised of about 16 per cent (Table 63.1).

It appears from the data on fund allocation in the BGREI states that allocation of funds among these interventions within the state did not maintain a strict compliance with the prescribed norm. However, the proportions of allocation among the three interventions on the whole for all BGREI states had been rather successful in maintaining a near proximity to the prescribed norm.

Block Demonstration

It is revealed that in Assam, Bihar, Odisha, and West Bengal the expenditure in block demonstration were over 60 per cent. However, in Chhattisgarh, Jharkhand, and eastern Uttar Pradesh expenditure in block demonstration were found less than 60 per cent (the proportion in Jharkhand was 30.9 per cent). However, for all the BGREI states taken together the proportion of expenditure in block demonstration was to the tune of 60 per cent of total outlay.

This study revealed that the beneficiaries have not used entire recommended input package. In many cases, beneficiary farmers have not undertaken seed treatment; weed control through weedicides, application of micronutrients, and plant protection measures. The farmers did not receive the inputs package specified in the BGREI guidelines uniformly across all the BGREI States. Deep ploughing and line sowing has not been adopted in several cases. This gets reflected from the primary survey across all ecologies.

There was mixed response of beneficiaries of block demonstrations of rice and wheat regarding adequacy of input packs for block demonstrations. The farmers' opinion was solicited with regard to the overall rating of the BGREI programme. There was mixed response of beneficiaries of block demonstrations of rice and wheat in this regard. The overall 74 per cent beneficiaries rated the programme as 'Good' and 26 per cent rates it as 'Average'.

Table 63.1 Component-specific Allocation under BGREI during 2010–11 and 2011–12

Activities	2010–11	2011–12
	Allocation (%)	Allocation (%)
Block Demonstration	51.70	64.50
Water Asset Building	35.60	16.14
Site-Specific Needs	12.00	19.11
Program Management	0.30	0.19
Monitoring	0.40	0.03
Evaluation	0.00	0.08
Total BGREI	100.00	100.00

Source: BGREI Cell, DAC, GOI.

Water Asset Building

The composition of the programme in 2011–12 also included a separate provision for water asset building at farmers' level for on-farm water harvesting. Provisions were for dug wells in rainfed areas and shallow tube wells and bore wells in the areas with high water table for assured irrigation. Among the BGREI states Chhattisgarh and West Bengal seemed to have made fewer attempts in this respect. Expenditure towards water asset building is found to be quite high in Bihar and eastern Uttar Pradesh (registering over 30 per cent) in comparison with the other BGREI states. However, no water asset building activities were carried out in Jharkhand.

It is observed that in Assam, Chhattisgarh, and eastern Uttar Pradesh installation of shallow tube wells and pump sets had been widespread. In Bihar the achievement was substantially low in this regard. In Jharkhand, however, no target was set as to physical water asset building activities and no work has been done in this respect.

Site-specific Activities

Proportion of expenditure towards site specific activities was very high in Jharkhand (over 69 per cent) followed by Chhattisgarh. In Bihar, however, no site specific activities were taken up. In the other states, namely, Assam, Odisha, eastern Uttar Pradesh, and West Bengal, the proportion of expenditure varied around 14 per cent to 19 per cent.

Technical Backstopping

As far as implementation of BGREI is concerned, there had been progressive farmers, state extension workers, KVKs, and SAUs, who had been entrusted to provide technical backstopping to the farmers. Performance index are percentages computed on the basis of responses from farmers as regards to their access to technical knowhow from sources mentioned above. Results indicate that 47 per cent beneficiaries accessed technical know-how from the local extension worker of State Department of Agriculture followed by 36 per cent from progressive farmers, 11 per cent from *Krishi Vigyan Kendras*, and 6 per cent from State Agricultural University.

Assam: 51 per cent beneficiaries accessed technical support from the progressive farmers followed by 43 per cent from the local extension worker and 6 per cent from the Krishi Vigyan Kendras.

Bihar: 11 per cent beneficiaries accessed technical support from the progressive farmers (*Krishi Salahkars* appointed on contractual basis under RKVY) followed by 70 per cent from the local extension worker and 19 per cent from the Krishi Vigyan Kendras.

Chhattisgarh: 17 per cent beneficiaries accessed technical support from the progressive farmers followed by 67 per cent from the local extension worker and 16 per cent from the Krishi Vigyan Kendras.

Jharkhand: 62 per cent beneficiaries accessed technical support from the progressive farmers followed by 28 per cent from the local extension worker and 10 per cent from the Krishi Vigyan Kendras.

Odisha: 28 per cent beneficiaries accessed technical support from the progressive farmers followed by 26 per cent from the local extension workers, 23 per cent from the Krishi Vigyan Kendras, and 23 per cent from State Agricultural University.

Eastern Uttar Pradesh: 45 per cent beneficiaries accessed technical support from the progressive farmers followed by 46 per cent from the local extension worker, 6 per cent from the Krishi Vigyan Kendras, and 3 per cent from State Agricultural University.

West Bengal: 60 per cent beneficiaries accessed technical support from the progressive farmers followed by 31 per cent from the local extension worker and 9 per cent from the Krishi Vigyan Kendras. From the primary data (sample survey) for West Bengal it is revealed 18 beneficiary farmers and five progressive farmers had acquired the soft skill and 27 of the extension workers from the state departments had regular contact with the beneficiary farmers.

The general opinion among the beneficiary farmers was that the provision of technical backstopping had been adequate. On the whole 73 per cent beneficiaries reported adequacy in technical backstopping.

In Assam, 60 per cent beneficiary farmers reported that technical backstopping under BGREI program was 'adequate'. It was reported adequate by 72 per cent in Bihar. In Chhattisgarh the corresponding figure was 100 per cent, in Jharkhand 80 per cent, in Odisha 52 per cent, in eastern Uttar Pradesh 100 per cent, and in West Bengal 52 per cent. On the whole, 73 per cent beneficiaries reported adequacy in technical backstopping.

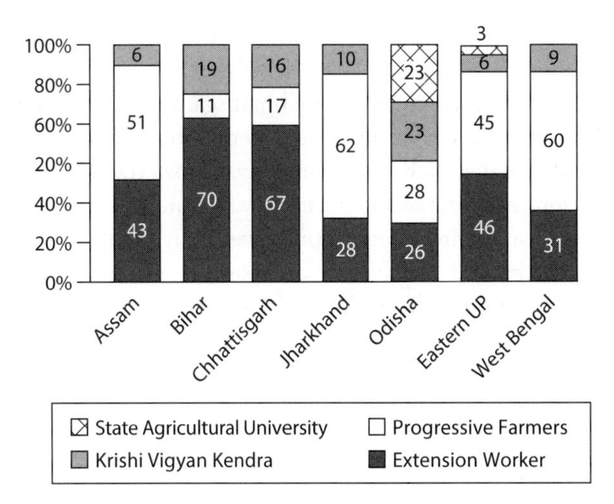

Figure 63.1 Performance Index of Agency Specific access to Technical Backstopping under BGREI in 2011–12

Source: Field Survey 2012.

It might be mentioned that the scientists of SAUs and ICAR (ICAR-SAU system) were identified for providing technical support to the BGREI beneficiaries during 2011–12 with the help of KVKs and extension workers from state department of agriculture. A sizeable majority of the respondents (68 per cent) reported that extension workers of state department of agriculture provided the best technical support followed by progressive farmers (19 per cent).

Monitoring

A three-tier monitoring structure has been put in place at national, state, and district levels. CRRI is the nodal agency for monitoring the programme.

It appears from the official statistics that were made available; CRRI scientists have carried out the awareness meetings regarding implementation of BGREI programme in general and provided necessary technical backstopping.

The staff of BGREI cell has visited the 61 BGREI districts out of 114 districts during kharif -2011 and 14 districts during rabi 2011–12 out of 54 districts. All the states stood by the programme and accomplished task of programme formulation and implementation on time.

In Assam 12 State Level Monitoring Team (SLMT) meetings were held in 2010–11 and six meetings in 2011–12. Fifteen such meetings were held in Bihar during

2011–12, though only one meeting at the state level was held in Jharkhand in 2011–12. Only two meetings of SLMTs were conducted in Chhattisgarh while five and three meetings were held in eastern Uttar Pradesh and Odisha, respectively. In West Bengal, however, no information regarding SLMT meeting was available from State Agricultural Directorate despite repeated requests.

No detailed account of composition of the District Level Monitoring Team (DLMT) was available for most of the states. Neither the numbers of meetings, discussions, and resolutions taken in such meeting was available from the states barring Odisha.

IMPACT OF BRINGING GREEN REVOLUTION TO EASTERN INDIA PROGRAMME

Bringing Green Revolution to Eastern India Programme and Changes in Cropping Intensity

The BGREI program was conceived in a manner to take into account the varying ecologies within the state. Moreover, the programme had the strategic emphasis on increasing yield rates resulting from technology dissemination particularly in rainfed areas. On the other hand, increase in cropping intensity (CI) depends on assured and controlled irrigation, the source in most of the cases is the shallow tube wells. But the experience of Green Revolution propagated in the mid-1960s, which depended heavily on sub soil water, had raised scepticism among agricultural scientists and environmentalists regarding the technology itself. Heavy dependence on sub soil water had been contributing in rapid depletion of sub soil water table. A new strategy was thus conceived that could thrive on surface water, rainfall, and water conservation.

The state wise changes in CI on the farms of BGREI beneficiaries' vis-à-vis non-beneficiaries during 2011–12 over 2010–11 are:

Assam: There has been marginal change (up to 3 per cent) in the CI of BGREI beneficiaries (2.09 per cent) and non-beneficiaries (2.99 per cent) for the state as a whole.

Bihar: A marginal change (up to 3 per cent) in the CI is observed among BGREI beneficiaries (2.09 per cent) and non-beneficiaries (1.13 per cent) in the state.

Chhattisgarh: The average CI for all ecologies in aggregate has shown significant increase in respect of BGREI beneficiaries (9.6 per cent) as compared to non-beneficiaries (1.52 per cent).

Jharkhand: There has been marginal change (up to 3 per cent) in the CI of BGREI beneficiaries (2.6 per cent) whereas CI has shown declining trend amongst non-beneficiaries (–1.2 per cent) in Jharkhand.

Odisha: The pooled average of CI for all ecologies has shown significant decrease in respect of BGREI beneficiaries (−19.8 per cent) as compared to non-beneficiaries (−12.7 per cent).

Eastern Uttar Pradesh: The average CI for the state for rice block demonstrations was less in respect of BGREI beneficiaries (201 per cent) as compared to non-beneficiaries (221 per cent). In case of wheat also the average CI was lower among beneficiaries (169 per cent) as compared to non-beneficiaries (179 per cent).

West Bengal: Average CI for all ecologies taken together shows marginal increase for both BGREI beneficiaries and non-beneficiaries.

On the whole, it can be said that there has been marginal changes over two years in cropping intensity for both beneficiary and non-beneficiary farmers with variations across states. Moreover, no substantial difference is observed among beneficiary and non-beneficiary farms in terms of their cropping intensity. Hence, the change in CI in the states (as derived from sample survey results) cannot be attributed to the program of BGREI. There may have been some other factors (for example, rainfall) influencing the cropping intensity in the states in the years of reference. Over and above, the BGREI programme as conceived had focused on increasing the yield of crops of which we shall be discussing presently.

Bringing Green Revolution to Eastern India Programme and Increase in Grain Yields

In all the states data was collected as to the yield of crops among the sample beneficiaries and non-beneficiary farms. Land size was conceived as one of the main determinants of household's position in the economic hierarchy within the village and level of education had been thought as an important attribute that could have an impact on adoption of the new technology under BGREI programme. And both sections namely treatment and control were found homogeneous.

It is revealed from the mean yield achieved by the beneficiaries and non-beneficiaries that there exists a difference in grain yield between them (Table 63.2). In most of the states the average yield of crops among beneficiaries was substantially higher than their counterparts (that is, non-beneficiaries). For kharif paddy the difference is more pronounced in the states under consideration. In Assam, however, for rabi pulses, the difference between the yield rate of beneficiaries and non-beneficiaries seems to be less prominent.

The results of the mean difference test also reveal a similar pattern of differences in average grain yield between

Table 63.2 Mean Difference Test of Grain Yield of Paddy, Wheat and Pulses between BGREI Beneficiaries and Non-beneficiaries in 2011–12.

State	Farmer Groups	Yield in Kg/ha		
		N	Mean Yield (Kg/Ha)	t values
Kharif-2011: Paddy				
Assam*	Beneficiary	50	4708.85	8.014
	Non-beneficiary	25	3769.10	
Bihar*	Beneficiary	50	3874.30	8.468
	Non-beneficiary	25	3448.60	
Chhattisgarh*	Beneficiary	40	4287.30	3.097
	Non-beneficiary	20	3740.00	
Jharkhand*	Beneficiary	50	2977.30	6.751
	Non-beneficiary	25	2691.20	
Odisha*	Beneficiary	50	5576.86	31.353
	Non-beneficiary	25	3880.92	
UP*	Beneficiary	50	7164.80	2.554
	Non-beneficiary	25	3884.00	
WB*	Beneficiary	50	5059.25	3.125
	Non-beneficiary	25	4743.00	
***Rabi:* 2011–12: Pulses**				
Assam	Beneficiary	40	695.76	1.489
	Non-beneficiary	20	614.57	
***Rabi:* 2011–12: Wheat**				
Eastern UP*	Beneficiary	20	7564.25	1.767
	Non-beneficiary	10	4269.10	
Summer-2012: Paddy				
Assam*	Beneficiary	50	5733.75	9.878
	Non-beneficiary	25	4594.21	

Note: * Mean Difference Significant at 0.01 level.
Source: Field Survey-2012.

beneficiaries and non-beneficiaries, with better performance for the former group:

Assam: The test results clearly indicates that the difference in yield rates for the kharif paddy, summer paddy, and rabi pulses in Assam across beneficiary and non-beneficiary farmers is found to be statistically significant at 1 per cent level, having bias in favour of the former group.

Bihar, Chhattisgarh, Jharkhand, Odisha, and West Bengal: The test results clearly indicates that the difference in yield rates for kharif paddy in these states across beneficiary and non-beneficiary farmers is found to be statistically significant at 1 per cent level, again having bias in favour of the former group.

Eastern Uttar Pradesh: There is clear indication from the results that the difference in yield rates of kharif paddy and of wheat in eastern Uttar Pradesh across beneficiary and non-beneficiary farmers is found to be statistically significant at 1 per cent level; beneficiaries reaping the benefits of the program.

Bringing Green Revolution to Eastern India Programme and Yield Gap

Normally yield gap is the difference between yield obtained at the farm level and the potential yield of a particular variety on the experiment station. Differences in yield gap between beneficiary and non-beneficiary farmers would suggest the impact of changes brought about in terms of yield enhancement. It also suggests the scope of yield enhancement across ecologies.

In case of Assam, no yield gap was witnessed in kharif paddy in respect of BGREI beneficiaries as well as non-beneficiaries in all the rice ecologies except rainfed medium deep water low land and irrigated land in which there is wide yield gap in the range from 15 per cent to 34 per cent. The reason, in case of Assam, for not showing the yield gap in general may be that farmers' yield were compared with the quinquennial mean yield fixed at the preceding year. In fact, the farmers' yield should have been compared with *potential* yield of the varieties used by the farmers.

In Bihar, the yield gap among the beneficiary farms is 44.71 per cent and 50.73 per cent in respect of non-beneficiary farms.

The yield gap reported in Chhattisgarh is in the range of 4 to 47 per cent across ecologies. The extent of yield gap of paddy in Chhattisgarh was found to be comparatively low (12.8 per cent) for beneficiary farmers as compared to the non-beneficiary farmers (31.8 per cent). The actual yield of paddy in the state was found to be 4148 kg/ha and 3239 kg/ha, respectively for beneficiary and non-beneficiary farmers as against its potential yield of 4750 kg/ha.

In Jharkhand, the yield gap among the beneficiary farms is 42.71 per cent and 58.13 per cent in respect of non-beneficiary farms. This signifies that beneficiaries enjoy higher yield rate than their counterpart.

In case of Odisha, the yield gap of paddy is compared with potential yield of paddy across kharif and summer seasons amongst the selected BGREI beneficiary and non-beneficiary farmers. Accordingly, the yield gap in Odisha was in the range from 0.3 per cent in irrigated ecology in Rayagada district in respect of BGREI beneficiaries to 76 per cent in rainfed ecology in Ganjam district in respect of non-beneficiaries during kharif 2010. The yield gap of paddy in kharif 2011 was in the range from 0.6 per cent in irrigated ecology in respect of BGREI beneficiaries in Rayagada district to 78 per cent in respect of non-beneficiaries of rainfed ecology in Ganjam district. Thus, rainfed systems are more vulnerable to yield fluctuations than the irrigated areas among other things. However, in most of the district the yield gap for beneficiaries was substantially lower than their non-beneficiary counterparts.

In West Bengal, yield gap was calculated by comparing the potential yield with the farmers' yield. It shows that yield gap varied from 12 per cent to 27 per cent across ecologies in respect of BGREI beneficiaries. In respect of non-beneficiary farmers, the yield gap is 15 per cent to 31 per cent across all five rice ecologies. Thus, there is almost same trend in respect of yield gap in rice amongst BGREI beneficiaries as well as non-beneficiaries.

Hence, it can be said that the beneficiary farmers in general in all the BGREI states had an edge over the non-beneficiaries in enhancing the yield of crop.

FARMERS AND PROBLEMS IN MARKETING OF AGRICULTURE PRODUCE

In course of our study in West Bengal as well as in other states, one had to come across repeated complain from the respondents regarding the problems they are faced with as regards to marketing of agricultural produce. Two main problems as identified by the respondents were 'low price of the agricultural output in the market' and 'problem of transporting the output to the market'.

The opinion of the beneficiary farmers of the BGREI programme was secured relating to problems faced in marketing of agriculture produce. The arrangement of assured procurement of agriculture produce is as essential as promotion of technology.

A sizeable proportion of respondents (72 per cent) reported that farm gate prices are always lower than minimum support price (MSP) due to non-existence of the provision of market intervention for cereals. Rest of the farmers (28 per cent) reported that there is problem of transportation of harvested produce to the markets due to poor rural roads, remotely located markets, and lack of transport facility.

CONCLUSIONS AND POLICY IMPLICATIONS

The study revealed that there are certain gaps in varying extents between recommended, promoted, and implemented strategies across different states due to lack of uniformity in input package and mode of implementation across the states. After a detailed analysis of yield rates across beneficiary and non-beneficiary farmers across different states, the study reveals a positive crop response to promoted technology under BGREI programme. Though it seems too early to conclude strongly as to the definite impact of the programme, nonetheless there are signs towards a positive change. In course of the study, the impact of various interventions of block demonstrations to drive growth in rice and wheat is reflected in changes in yield rates. The BGREI programme, as conceived, addressed towards increasing the yield rather than the cropping intensity. Hence, the impact of intervention under block demonstration programmes under BGREI is more prominent in increasing the yield rates for the beneficiary farms as compared to non-beneficiaries. In case of technical backstopping, the scientists of SAUs, KVKs, and ICAR (ICAR-SAU system) were identified for providing technical support to the BGREI beneficiaries during 2011–12. Through a regular contact technology dissemination had been quite successful in the BGREI states.

Efforts should be made to reduce the gaps between recommended, promoted, and implemented strategies. In course of dissemination of technology, provision of progressive farmers and regular monitoring from state agriculture departments can play vital role. As such, such links between the beneficiaries and state machineries should be encouraged. Interventions through crop demonstrations has helped decline the gap between ecology specific potential and actual yields across beneficiary farms. Hence, such demonstration programmes should be encouraged. Eastern India covered under the BGREI programme has exhibited a glimpse of a high potential for yield enhancement of rice, wheat, and rabi pulses through a favourable positive crop response. There is a huge scope to exploit this potential through scientific and technological intervention like BGREI, and hence, the programme should continue with greater effort and coordination. An all-round effort should be made to ensure the timeliness of input delivery system prescribed under the recommended technology.

REFERENCES

Birsa Agriculture University. 2012. Birsa Kisan Diary-2012, Birsa Agriculture University, Ranchi. (http://www.baujharkhand.org).

Boswell, E.P., R.T. Koide, D.L. Shumway, and H.D. Addy. 1998. 'Winer Wheat Cover *Cropping*, Va Mycorrhizal Fungi and Maize Growth and Yield', *Agriculture, Ecosystems and Environment*, 67(1): 55–65.

Bravo-Ureta, Boris E. and E. Antonio Pinheiro. 1997. 'Technical, Economic and Allocative Efficiency in Peasant Farming: Evidence from the Dominican Republic', *The Developing Economics*, 35(1): 48–67.

Campbell, J.A. and M.E. Akhtar. 2009. Impact of Tillage on Soil Water Regimes in the Rainfed Areas of Pakistan. http://www.cab.org/GARA/Full Text/Pdf/2009/2009.

Chaudhary, M.R, P.R. Gajri, S.S. Parihar, and R. Khera. 1985. 'Effect of Deep Tillage on Soil Physical Properties and Maize Yield on Coarse Textured Soils', *Soil and Tillage Research*, 6(1): 31–44.

Compendium of Work Breakdown Structure with Timelines, Records and Reporting Formats and documentation thereof-BGREI; 26 April 2011 published by CRRI-Cuttack (India).

Farrell, J.M. 1957. 'The Measurement of Productive Efficiency', *Journal Royal Stats 506* 120 (3): 253–90.

Gaud, William S. 1968. 'The Green Revolution: Accomplishments and Apprehensions', AgBioWorld, 8 March. Available at http://www.agbioworld.org/biotech-info/topics/borlaug/borlaug-green.html. Retrieved 8 August 2011.

Government of India. 2007. Serving Farmers and Saving Farming—Towards Faster and More Inclusive Growth of Farmers' Welfare, Reports of the National Commission on Farmers, Ministry of Agriculture, Govt. of India.

———. 2011. *Agricultural Statistics At A Glance 2011*. New Delhi: Directorate of Economics and Statistics, Ministry of Agriculture, Government of India.

Grigg, D.B. 1974. The Agricultural Systems of the World: An Evolutionary Approach. Cambridge Geograp/hic Studies, p. 358. Cambridge: Cambridge University Press.

Guidelines for Extending Green Revolution to Eastern India, Department of Agriculture and Cooperation, Ministry of Agriculture, Government of India, March 2011.

Jill, Lenne. 1997. *Pests and Poverty: The Continuing Need for Crop Protection Research. The White Paper: Eliminating World Poverty: A Challenge for the 21st Century*. DFID, 29(4), London, UK: 235–50.

Jondrow, J., C.A. Knot Lovell, vans Materov and P. Schmidt. 1982. 'On the Estimation of Technical Efficiency in the Stochastic Frontier Production Function Model', *Journal of Econometrics*, 19(2/3): 233–8.

Kabir, Z; I.P. O'Halloran, and C. Hamel. 1999. 'Combined Effects of Soil Disturbance and Fallowing on Plant and Fungal Components of Mycorrhizal Corn (*Zea mays* L.)', *Soil Biology and Biochemistry*, 31: 307–14.

Kabir, Z. and R.T. Koide. 2000. 'The Effects of Dandelion or a Cover Crop on Mycorrhizal Inoculum Potential, Soil Aggregation and Yield of Maize', *Agriculture, Ecosystems and Environment*, 78: 167–74.

———. 2002. 'Mixed Cover Crops, Mycorrhizal Fungi, Soil Properties and Sweet Corn Yield', *Plant Soil* 238: 205–15.

National Policy for Farmers. 2007. New Delhi: Ministry of Agriculture, Government of India.

National Sample Survey Office (NSSO). 2005a. *Situation Assessment Survey of Farmers—Consumption Expenditure of Farmers Households*, NSS Report No. 495, New Delhi: Ministry of Statistics and Program Implementation, Government of India.

———. 2005b. *Situation Assessment Survey of Farmers—Some Aspects of Farming*, NSS Report No. 496. New Delhi: Ministry of Statistics and Program Implementation. Government of India.

———. 2005c. *Situation Assessment Survey of Farmers—Income, Expenditure and Productive Assets of Farmer Households*, NSS Report No. 497. New Delhi: Ministry of Statistics and Programme Implementation. Government of India.

———. 2005d. *Situation Assessment Survey of Farmers—Indebtedness of Farmer Households*, NSS Report No. 498. New Delhi: Ministry of Statistics and Programme Implementation, Government of India.

———. 2005e. *Situation Assessment Survey of Farmers—Access to Modern Technology for Farming-2003*, NSS Report No. 499. New Delhi: Ministry of Statistics and Programme Implementation, Government of India.

Pandey, S., M. Mortimer, L. Wade, T.P. Tuong, K. Lopez, and B. Hardy. 2002. 'Direct Seeding: Research Strategies and Opportunities', Proceedings of the International Workshop on Direct Seeding in Asian Rice Systems: Strategic Research Issues and Opportunities, 25–28 January 2000, Bangkok, Thailand.

Pandey, S. and L. Velasco. 2002. Economics of Direct Seeding in Asia: Patterns of Adoption and Research Priorities. Proceedings of the International Workshop on Direct Seeding in Asian Rice Systems: Strategic Research Issues and Opportunities, 25–28 January 2000, 383. Bangkok, Thailand. Los Baños (Philippines): International Rice Research Institute..

Mehta, Rajeev. 2009. Situation Assessment Survey for Farm Sector Policy Formulation in India-Expert Consultation on Statistics in Support of Policies to Empower Small Farmers Sponsored by FAO at Bangkok-Thailand, STAT-EMPOWER-6, September.

Ramesh Chand and S.S. Raju. 2008. 'Instability in Indian Agriculture', Discussion Paper-NPP-01/2008, National Professor Project, National Centre for Agricultural Economics & Policy Research, New Delhi.

Reddy, D.S., D.R. Reddy, and G.V. Chary. 1983. 'A Note on the Effect of Deep Ploughing on Basic Filtration Rate of Soil, Root Growth and Grain Yield Under Rainfed Agriculture at Anantpur', *Annals of Arid Zone*, 16(1): 149–52.

Reddy, P., Bala Hussain, S. Sreenivasulu, and C. Manohar. 'Direct Seeding with Drum Seeder—Future Prospects', Acharya Ranga Krishi Vigyan Kendra, Tirupati, Andhra Pradesh.

Rossi, P.H., W.M. Lipsey, and H.E. Freeman. 2004. *Evaluation: A Systematic Approach*, 7th edition. Thousand Oaks, CA: SAGE.

Sharma, P.D. and C.L. Acharya. 1987. 'Effect of Soil Management on Rainfed Wheat in Northern India, Nutrient Uptake, Plant Growth & Yield', *Soil and Tillage Research*, 9(1): 79–89.

Singh, A.K, B.P. Bhatt, and P.S. Minhas. 2011. *Technical Output and Recommendations—Strategies for Agricultural Transformation of Eastern Region-Brainstorming Session on Second Green Revolution*. Patna: ICAR Research Complex for Eastern Region.

Singh, M.V. 2009. 'Micronutrient Nutritional problems in Soils of India and Improvement for human and Animal Health', *Indian Journal of Fertilizer*, 5(4): 11–16, 19–26, and 56.

Sprague, M.A. and G.B. Triplette. 1986. *No Tillage Surface Tillage Agriculture—The Tillage Revolution*. Somerset, N.J.: John Wiley & Sons Inc.

Swaminathan, M.S. 2000. 'An Evergreen Revolution', *Biologist*, 47(2): 85–89.

Zahangir, Kabir. 2005. 'Tillage or No-tillage: Impact on Mycorrhizae', *Canadian Journal of Plant Science*, 85(1): 23–9.

64

Organic Input Production and Marketing in India

Efficiency, Issues, and Policies

D. Kumara Charyulu and Subho Biswas

India had developed a vast and rich traditional agricultural knowledge since ancient times and presently finding solutions to problems created by over use of agrochemicals. Today's modern farming is not sustainable in consonance with economics, ecology, equity, energy, and socio-cultural dimensions. Indiscriminate use of chemical fertilizers, weedicides, and pesticides has resulted in various environmental and health hazards along with socio-economic problems. Though agricultural production has continued to increase, but productivity rate per unit area has started to decline. The entire agricultural community is trying to find out an alternative sustainable farming system, which is ecologically sound, economically and socially acceptable. Sustainable agriculture is unifying concept, which considers ecological, environmental, philosophical, ethical, and social impacts, balanced with cost effectiveness. The answer to the problem probably lies in returning to our own roots. Traditional agricultural practices, which are, based on natural and organic methods of farming offer several effective, feasible, and cost effective solutions to most of the basic problems being faced in conventional farming system. Many long-term studies have reported that soil under organic farming conditions had lower bulk density, higher water holding capacity, higher microbial biomass carbon and nitrogen, and higher soil respiration activities compared to the conventional farms. This indicates that sufficiently higher amounts of nutrients are made available to the crops due to enhanced microbial activity under organic farming.

India is bestowed with lot of potential to produce all varieties of organic products due to its agro-climatic regions. In several parts of the country, the inherited tradition of organic farming is an added advantage. This holds promise for the organic producers to tap the market which is growing steadily in the domestic market related to the export market. Currently, India ranks 33rd in terms of total land under organic cultivation and 88th position for agriculture land under organic crops to total farming area. The cultivated land under certification is around 2.8 million ha (2007–08, 1.9 per cent of GCA). This includes one million ha under cultivation and the rest is under forest area (wild collection) (Agricultural and Processed Food Products Export Development Authority [APEDA] 2010). India exported 86 items during 2007–08 with the total volume of 37,533 MT. The export realization was around 100.4 million USD registering a 30 per cent growth over the previous year (APEDA 2010).

With having such a due importance of organic farming in India, the important event in the history of the modern nascent organic farming was the unveiling of the National Programme for Organic Production (NPOP) in 2000. The subsequent accreditation and certification program was started in 2001. The implementation of NPOP is ensured by the formulation of the National Accredited Policy and Programme (NAPP). Later, the Department of Agriculture and Cooperation, Ministry of Agriculture has also launched a central sectoral scheme entitled 'National Project on Organic Farming (NPOF)' during 10th five-year plan. The main objectives of the programme are: capacity building through service providers; financial support to different production units engaged in production of bio-fertilizers, fruit and vegetable waste compost and vermi-hatchery compost; human resource development through training on certification and inspection, production technology, and so on.

Setting up of organic input units with capital investment subsidy is one of major component under NPOF for encouraging the organic inputs production since 2004. Availability of quality organic inputs is critical for success of organic farming in India. To promote organic farming in the country and to increase the agricultural productivity while maintaining the soil health and environmental safety; organic input units are being financed as credit-linked and back-ended subsidy through National Bank for Agriculture and Rural Development (NABARD) and NCDC. These units will not only reduce the dependency on chemical fertilizers but also efficiently convert the organic waste in to plant nutrient resources. Three types of organic input production units namely; Fruit/vegetable waste units, bio-fertilizer unit, and vermi-hatchery units are being supported at 25 per cent of their total project costs respectively. Around 455 vermi-hatchery units, 31 bio-fertilizer units, and 10 fruit and vegetable waste units were sanctioned across different states by NABARD till May 2009. But, NCDC has so far sanctioned only two bio-fertilizer units in Maharashtra state.

OBJECTIVES

At this juncture, it is very interesting to know what is the present status of these units, what is the production and capacity utilization of each unit, and suggestions for enhancing capacity utilization, and so on. It is also very important to get the feedback from promoters for further improving in the implementation of the scheme. However, very little effort has been made so far to find out the performance of organic input units in terms of its capacity utilization, cost of production and efficiency. Very few attempts were also made till now to assess the economics and efficiency of organic farming in India. Such analysis can provide valuable insights for undertaking appropriate measures for faster expansion organic farming in the country.

DATA AND METHODOLOGY

The selection of sample organic input units for the study was purposively chosen from four states, namely Punjab, Uttar Pradesh, Gujarat, and Maharashtra. One of the reasons for choosing these four states was to see the comparison in efficiency of organic input units between the northern states (where irrigation and fertilization application is default in states like Punjab and Uttar Pradesh) and western states (where irrigation and fertilizer application is optional in states like Maharashtra and Gujarat) of India. A total sample of 40 vermi-hatchery units were chosen in clusters under four states for the study. The criterion used for selection of sample units from each state was on their respective weights in the population. Due to their extremely scattered nature in each state, vermi-hatchery sample units were chosen in two to three groups/clusters in order to minimize the travel costs and time. Similarly, four bio-fertilizer units (three NABARD and one NCDC sanctioned) and two fruit and vegetable units were also covered in the study. A well-structured and pre-tested questionnaire was administered to extract some common quantitative parameters of each type of unit; with utmost emphasis was placed on qualitative case analysis through interaction with the organic input units. The study also covered a sample of 60 (15 each from one state) organic farmers to canvass a structured questionnaire to extract constraints in procuring and usage of organic inputs, and so on. Another random sample of 60 (15 each from one state) conventional farmers was also selected to compare the crop economics and productivity of crops with organic cultivation in the respective study areas.

This study used a model based non-parametric DEA approach for efficiency analysis of organic input units. Multiple regression models are also used to estimate the drivers for efficiency in input units. The same DEA approach is also used for estimating the efficiency between organic and conventional farming systems. Similarly, the determinants for efficiency in organic farming are also identified.

RESULTS AND DISCUSSION

Capacity Utilization of Sample Units

The details of capacity utilization of sample units are presented in Table 64.1. Capacity utilization is a concept that refers to the extent to which an enterprise actually uses its

Table 64.1 Capacity Utilization of Sample Units (TPA)

Item	Gujarat	Maharashtra	Punjab	Uttar Pradesh	Over all
Average installed capacity	150	150	150	150	150
Current capacity utilization	24.2	187	33	105	76.2
Capacity utilization rate (%)	16.1	124.6	22.0	70.0	50.8
Average recovery rate (%)†	48.0	52.5	33.3	39.7	42.7
Gestation period per cycle (days)†	46.5	35	60	50	48.8
Avg no. of cycles per year (range)†	5–7	10–15	3–5	6–8	7–9

Note:† reviewed based on farmers' past experiences.

installed productive capacity. The results presented in the table were referring to the capacity utilization of organic input units in the last one year.

The average installed capacity of the sample units was 150 tonnes per annum (TPA). Overall, the average capacity utilization was around 76.2 TPA. The average capacity utilization rate was 50.8 per cent which indicates nearly half of its full potential. Across different states, the average capacity utilization was the highest in Maharashtra followed by Uttar Pradesh, Punjab, and Gujarat. The actual production in Maharashtra units was more than its installed capacity. The lowest capacity utilization was observed in Gujarat at the rate of 24.2 TPA. This capacity utilization rate was one sixth of the actual potential (16.1 per cent). The reasons for low capacity utilization are lack of demand, poor production skills, and insufficient infrastructure. Even though the units in Punjab were well equipped, their productivity levels were also low. This is because of absence of market demand for vermi-compost. In case of Uttar Pradesh, the average capacity utilization rate was 70.0 per cent. The demand is slowly picking up due to its nearness to different export channels exist in and around New Delhi.

The average recovery rate per unit was 42.7 per cent. Across different states, the highest recovery rate was noticed in case of Maharashtra (52.5 per cent) followed by Gujarat, Uttar Pradesh, and Punjab. The high recovery rate in Maharashtra may be one of the reasons for its high productivity. Even though, the rate of recovery was high in Gujarat, the productivity was low because of lack of production skills and influence of climatic parameters (high temperatures, heavy rains, and so on). The average gestation period per cycle for the entire sample was 48.8 days. It is dependent on various parameters like number of worms per cubic meter, age of the worms, raw material type, and production season, and so on. The time period was the lowest in Maharashtra due to their higher efficiency levels while it was the highest in Punjab. Overall, the average number of cycles per annum produced by the organic inputs was seven to nine. This number was very low in case of Punjab because of high gestation period.

India has enough potential for production of sufficient quantities of organic inputs. Substantial capacities have been generated for production of different organic manures through diverse state and central financial assistance schemes. As per National Centre of Organic Farming (NCOF) (2007–08), the total compost/vermicompost production (includes rural, urban, FYM, and other sources) at all India was 3830.9 lakh tonnes and area covered by these units was 1694.8 lakh ha. Similarly, the total green manure production in the country was 133.5 lakh tonnes with 13.0 lakh ha area coverage. The total installed capacity created for production of different bio-fertilizers in the country was 67,162 tonnes. But, their actual production of different bio-fertilizers was 38932.6 tonnes. This data clearly indicates that only 58 per cent of their capacity was utilized. However, the growth in production of bio-fertilizers was quite significant when compared to 2004–05. The results also showed that the total production of bio-fertilizers was the highest in case of Tamil Nadu followed by Karanataka, Andhra Pradesh, and Kerala. It also concludes that the awareness and usage of bio-fertilizers was higher in south zone than other zones in India. Among different types of bio-fertilizers, the share of Phosphorous Solubilizing Bacteria (PSB) production was higher. The status of biopesticides production in India is still in infant stage. The production is slowly gaining momentum with the increased awareness of the farmers.

Economics of Vermi-Compost Production

The summary of economics of vermi-compost production across different states is presented in Table 64.2. The results clearly reveal that the production of vermi-compost was a profitable venture in India. The weighted average cost of production per quintal was Rs 286 and price realization for the same was Rs 506. The net margin per quintal of vermicompost production was Rs 220. This is a quite significant margin in agri-business sector. Among different states, the

Table 64.2 Summary of Economics of Vermi-Compost Production in India (Rs)

Item	Gujarat	Maharashtra	Punjab	Uttar Pradesh	Weighted Average
Cost of production per quintal (Rs)	453	218	433	324	286
Price realization per quintal (Rs)*	233	447	488	678	506
Net margin per quintal (Rs)	−220	229	55	354	220

*Note:** Including the sale of worms.

cost of production was the highest in Gujarat followed by Punjab, Uttar Pradesh, and Maharashtra. Good production skills, higher market demand, and economies of scale of production are, may be, the reasons for higher productivity and low cost of production in Maharashtra. Per quintal price realization was the highest in Uttar Pradesh followed by Punjab, Maharashtra, and Gujarat. Proximity to Delhi metropolitan and presence of vermi-compost export channels have helped Uttar Pradesh state to realize more price per quintal. Even though, the productivity and market demand were relatively lower in Punjab, existence of green houses and nurseries in Chandigarh facilitating to reap reasonable price for vermi-compost. The average net margin per quintal was the highest in Uttar Pradesh while it was the lowest and in negative value in Gujarat state. By administering proper training to promoters and providing technical know-how in vermi-compost production would yield good results in Gujarat state as well.

Efficiency of Organic Inputs

The frequency distribution of technical, allocative, and economic efficiencies of sample organic input units both under CRS and VRS models of DEA approach is presented in Table 64.3. The estimated mean technical, allocative, and economic efficiencies under DEA-CRS model were 63.7, 50.95, and 32.95 per cent, respectively. Similarly, these values under DEA-VRS model were 83.39, 59.42, and 50.24 per cent, respectively. In terms of technical efficiency, about 45 per cent of the sample units have more than 90 per cent efficiency under the VRS model. Under the CRS model, only 20 per cent of the sample units have more than 90 per cent efficiency. In case of allocative efficiency, majority of sample units (40 per cent) fell under less than 50 per cent category under VRS model while 47.5 per cent of the same belonged to less than

Table 64.3 Frequency Distribution of Efficiency of Organic Input Units (n = 40)

Efficiency (%)	CRS			VRS		
	TE	AE	EE	TE	AE	EE
1–50	47.5	47.5	85.0	12.5	40.0	57.5
51–60	5.0	17.5	2.5	2.5	7.5	10.0
61–70	5.0	10.0	0.0	5.0	20.0	10.0
71–80	12.5	7.5	5.0	10.0	2.5	2.5
81–90	10.0	10.0	2.5	25.0	20.0	10.0
91–100	20.0	7.5	5.0	45.0	10.0	10.0
Max (%)	100	100	100	100	100	100
Min (%)	25.4	12.8	8.8	44.6	16.3	13.4
Mean (%)	63.7	50.95	32.95	83.39	59.42	50.24
Standard deviation (%)	24.0	25.7	24.1	18.8	25.2	26.4

50 per cent category under CRS assumption. 85 per cent of sample units exhibited less than 50 per cent of economic efficiency under CRS assumption. Correspondingly under VRS model, the large share of sample (57.5 per cent) was also belonging to the same class.

It is concluded from the table that majority of the sample organic units (47.5 per cent) showed less than 50 per cent technical efficiency under CRS assumption, indicating that most of the organic production units were inefficient. In other words, the inputs under CRS model (VRS) can be reduced by 36 per cent (16 per cent) to attain the same level of output. To supplement the above statement, the most frequent interval of allocative and economic efficiency was 1 to 50 per cent under both CRS and VRS assumptions. Further, it reveals that the organic production units were suffering from both technical inefficiency in using resources as well as unable to allocate inputs in the cost minimizing way. The scale efficiency index among sample varied from 32.7 per cent to 100 per cent, with a mean value of 77.7 per cent.

Due to very little accessible information on economics and efficiency of organic farming in India, an attempt is made to assess it in different crops and states. The results showed mixed response. Overall, crop economics results concluded that the unit cost of production is lower in organic farming in case of cotton (both in Gujarat and Punjab) and sugarcane (both in Uttar Pradesh and Maharashtra) crops whereas the same is lower in conventional farming for Paddy and Wheat (both in Punjab and Uttar Pradesh) crops. The DEA efficiency analysis conducted on different crops indicated that the efficiency levels are lower in organic farming when compared to conventional farming, relative to their production frontiers.

These results conclude that there is ample scope for increasing the efficiency under organic farms. The determinants of the efficiency in organic farming are education of the farmer and formal participation in training programmes.

The broad suggestions for promotion of organic input units are also collected from the sample respondents. The major issues are: prompt and timely conduct of JMC visits; quick and timely disbursement of subsidies to promoters; inclusion of buffaloes in the scheme, inclusion of training on vermi-compost production, and insurance components in the going scheme; assist farmer producers in obtaining licensing and certification of compost; supply of quality seed stock at cheaper rates; intervention of NABARD/state government/SAUs in marketing of compost; encouragement of organic inputs usage by subsidies; creation of market demand by promoting more awareness programmes; and finally further increase in subsidy upper limit in the establishment of input units.

Finally, the nagging key policy decisions are hindering the growth of organic farming in the country. In India, APEDA is the highest controlling body for organic certification for export. Till date there are no domestic standards for organic produce within India. Although there is no system for monitoring and labelling of organic produce sold within India, which particularly affects the retail market. An innovative cost effective certification method uniform in standards across various countries need be developed to connect numerous small and marginal farmers in the country.

CONCLUSIONS AND POLICY IMPLICATIONS

The study brought out the following conclusions and policy implications after thorough understanding about organic farming in India as well as analysis of primary data:

1. Out of 40 vermi-hatchery units covered in the study, only 22 units are functioning on the day of visit. The main reasons for not functioning are: lack of demand for vermi-compost, neither JMC visit nor no subsidy release from NABARD, death of worms in high temperatures, heavy rains, and floods. The number of non-functioning units were maximum (100 per cent) in case of Gujarat.
2. NABARD has finished the conduct of JMC visits only in case of 70 per cent units. The remaining 30 per cent units are still waiting for JMC visits and final subsidy. This indicates a huge delay in the process of subsidy release. Out of the 28 units (70 per cent) which completed JMC visits, only 19 units have received the final subsidy amount. Almost 32 per cent of units are waiting for release of final subsidy. This was another bottleneck in the scheme where lot of time was consuming for processing.
3. On an average, the total financial out lay per unit was Rs 5.9 lakh. The outlay was the highest in case of Maharashtra whereas it was the lowest in Punjab. The results conclude that there is a huge gap between subsidy released till now (0.93 lakh) and eligible subsidy (1.5 lakh) per unit. This gap is the highest in case of Gujarat (1.23 lakh) followed by Uttar Pradesh (0.27 lakh) and Punjab (0.25 lakh).
4. The socio-economic characters of promoters were regressed against efficiency values to determine the drivers for efficiency in vermi-hatchery units. The results concluded that the size of the unit, contribution of family labour have shown positive relation with technical as well as scale-efficiencies. Participation in the training programmes is also enhancing technical efficiency of the corresponding units. The age of the unit and subsidies discouraged the scale-efficiency.
5. Majority of the sample promoters did not face any problem in establishment of vermi-hatchery units. Very few expressed some difficulties while establishing them. The major problems are: non-availability of quality worms in the vicinity, lack of sufficient raw materials, wild boar attacks on compost units, no proper guidance from NABARD, heavy rains, and delay in release of bank loan amounts, and so on.
6. Almost all vermi-hatchery units are following direct sales method rather than depending upon any other intermediary. The quantity of total sales is very high in direct sales. Nearly half of the sample promoters expressed that they are facing severe marketing problems in marketing of their compost.

Policy Implications

The Ministry of Agriculture should introduce favourable governmental policies and strategies for the promotion of organic farming in India. These should include:

1. A single authority at national level with a well-defined role should be responsible for the organic sector in the country. This includes the responsibility for regulating and supervising the organic sector at both domestic and outside the country. With regard to export, the national authority should act as counterpart to the authorities of the importing countries and could, thus, strengthen the organic sector's export potential.

2. Current market demand is considerably higher than the supply, a situation which creates potential opportunities for countries in the short and medium term. So, India should use this opportunity timely to tap the national and international markets by framing a well-defined strategy on organic farming sector at the national level. The development of international markets will also stimulate domestic as well as regional market opportunities.
3. The quality organic input production (compost, bio-fertilizers, and biopesticides) in the country should be further encouraged with latest technologies and improved way of financial assistance so as to reduce the high dependency on inorganic fertilizers in a phase manner and to save our domestic subsidies. It not only protects our soil health but also sustains the environmental and natural resources.
4. The organic input units established under various schemes in the country should be linked up with suitable market channels to improve their capacity utilization or to make use of entire installed capacities. NABARD/state agriculture department/ IFFCO should intervene in providing necessary support for their marketing of organic inputs. Establishment of organic input marketing channels is the need of the hour for expansion of organic farming in the country.
5. The technical efficiency of organic input production should also be enhanced by imparting more production skills to the promoters. The economic and scale efficiency of the units should also be improved by providing more technical guidance, quality seed stock, and training programmes.
6. A comprehensive programme/scheme should be developed to assist the farmers that who want to convert their lands from conventional to organic farming. It includes some conversion or input subsidies, providing technical guidance, and finally certification of farm. It will dramatically expand the organic farming in the country and ultimately sustains our food production.
7. Finally, the most important task would be to ensure consistency of government policies on organic sector. Through focusing of policies and activities, the organic sector can be developed more quickly and more effectively. Institutional barriers to the development of the organic sector are considered greater than the technical and trade barriers. So, most relevant institutions and partners should be prepared to competently involve in the promotion of the organic sector in the country.

REFERENCES

Coelli, T.J. 1996. 'A Guide to DEAP 2.1: A Data Envelopment Analysis Computer Program', *CEPA working paper No.8/96, ISBN 1863894969*, Department of Econometrics, University of New England: 1–49.

Coelli, T., P.D.S Rao, and G.E Battese. 2002. *An Introduction to Efficiency and Productivity Analysis*. London: Kluwer Academic Publishers.

Ghosh, N. 2004a. 'Promoting Bio-fertilizers in Indian Agriculture', *Economic and Political Weekly*, 25 December: 5617–25.

———. 2004b. 'Reducing Dependence on Chemical Fertilizers and its Financial Implications for Farmers in India', *Ecological Economics* 49: 149–62.

Kumara, Charyulu D. and Subho Biswas. 2011. *Organic Input Production and Marketing in India: Efficiency, Issues and Policies*. CMA Publication, Allied Publishers Pvt Ltd.

Lampkin, N.H. 1994. 'Organic Farming: Sustainable Agriculture in Practice', in *The Economics of Organic Farming—An International Perspective*, edited by N.H. Lampkin and S. Padel. Oxon (UK): CAB International Publishers.

National Centre of Organic Farming (NCOF), Ghaziabad (http://dacnet.nic.in/ncof/, accessed on 10th January 2010).

Rekha et al. 2006. 'Pesticide Residue in Organic and Conventional Food-risk Analysis', *Journal of Chemical and Health Safety*, 13(6), November/December.

Singh, S. 2009. 'Organic Produce Supply Chains in India: Organization and Governance', *CMA publication* 222, IIM, Ahmedabad. Ahmedabad: Allied Publishers Pvt Ltd.

65

Impacts and Constraints of Organic Farming in West Bengal

Ranjan Kumar Biswas, Debanshu Majumder, and Ashok Sinha

Use of chemical inputs in farming operation like fertilizers, pesticides, and others have played a positive role in increasing agricultural productivity for making India self-sufficient in foodgrain production. This results an increased yield of foodgrain production in India from 644 kg per hectare in 1966–67 to 1636 kg per hectare in 2000–01 (www.indiastat.com), that is, yield of foodgrains increased by around two and half times within the period of thirty-four years and at the same period there happen a more than twelve-fold increase in the consumption of chemical fertilizers: from 1.1 million tonnes to 13.56 million tonnes (Pawan Wadhwa 2001). Besides the increased foodgrain production inorganic inputs have also contributed towards increasing productivity of cash crops.

After continuing that type of farming practices, based on the use of inorganic and mineral components for a long time, there has been an increasing demand for rethinking agricultural growth strategy. Agricultural sustainability, degradation of soil, biodiversity, impact on human health, and environment as a whole are important criteria nowadays. In the 1990s of the twentieth century, a focus on long-term sustainability of agriculture has been enhanced as an alternative to inorganic farming. Organic farming with the usage of bio-fertilizers and biopesticides are being espoused not only in developed countries but in developing countries also.

Organic farming is one of several approaches to sustainable agriculture and many of the techniques used (for example, inter-cropping, crop rotation, mulching, integration of crops, and livestock) are practiced under various agricultural systems. What makes organic agriculture unique, as regulated under various laws and certification programmes, is that (i) almost all synthetic inputs are prohibited, and (ii) 'soil building' crop rotations are mandated. The basic rules of organic production are that natural inputs are approved and synthetic inputs are prohibited.

In reality, organic farming is a system of farming which devoid of chemical inputs and in which the biological potential of the soil and underground water resources are conserved and protected from the natural and human induced degradation or depletion. Organic farming allows the powerful laws of nature to increase both agricultural yields and disease resistance. Organic farming is also a rule based agricultural system in which the operator has to follow the standards of organic farming set by the certification organization.

It is observed that under organic farming practice, yield of crops does not decrease. If equal quantity of nutrient is applied through organic manure, then the question of decrease in yield does not arise. Secondly, fertilizer use efficiency will be much higher under organic conditions, the leaching and evaporation losses will be lesser. Furthermore, the moisture retention capacity of the soil increases which helps to grow crops even under drought condition. In view of the above, the present study has been conducted to examine the impacts and constraints of organic farming in West Bengal. The reference period of the study is 2009–10. The specific objectives of the study are:

1. to study the status of organic farming in West Bengal;
2. to study the comparative economics of crop production under organic and inorganic farming;
3. to study the impact of organic farming in relation to quality of produce and price premium;
4. to study the farmers' awareness regarding organic farm practices; and
5. to study the constraints in adoption of organic farming.

DATA AND METHODOLOGY

Both primary and secondary data have been used in this chapter. The primary data has been collected by personal interview using pre-tested survey schedule, specially prepared for this purpose. The reference period of the study is 2009–10. Different aspects of farm operation have been obtained for both organic and inorganic farming systems. These aspects are (i) record of organic farmers indicating the number of years engaged in organic practices, (ii) season-wise record of crops both in organic and inorganic farms, (iii) input and output record of both organic and inorganic farms, (iv) cost of cultivation as well as cost of production record for different crops of both group of farmers, (v) record of price received from sale of products in market, and (vi) input uses record both in organic and inorganic farms. The secondary data have been collected from respective Project Implementing Authority (PIA) of the study area.

Selection of Areas

The study has been confined to two districts, that is, one from southern part and another from northern part of West Bengal. Emphasis has been made in selection of districts where both government and non-government organizations are working for organic farming. Thus, north 24-Parganas district and Jalpaiguri district from southern and northern part respectively of the state have been selected purposively. In the second stage, four blocks two from each district have been selected purposively. Among the selected four blocks, government agency has been working in two blocks, called Bio-Village and non-government organizations (NGOs) are working in other two blocks. In the next stage, two bio-villages namely, Babpur village of Barasat-I block of north 24-Parganas district and Ghughudanga village of Jalpaiguri Sadar block of Jalpaiguri district have been selected purposively. Among the NGOs activity areas, two villages namely, Panji village of Baduria block and Purba Satali village of Kalchini block of north 24-Parganas and Jalpaiguri district respectively have been selected randomly.

Selection of Farmers

In the first stage, the list of the farmers along with their size of holdings has been collected. In the second stage, all the farmers have been sub-divided into five categories based on size of land holdings: (i) sub-marginal (below 0.50 ha), (ii) marginal (0.51 ha to 1.00 ha), (iii) small (1.01 ha to 2.00 ha), (iv) medium (2.01 ha to 4.00 ha), and (v) big (4.01 ha and above). In the next stage, 30 farmers, that is, 15 each from organic and inorganic farms have been selected from each village based on simple random sampling with proportional allocation. Thus, all total 120 farm households have been selected for in-depth study.

Measurement of Variables

On the basis of extensive review of studies, the relevant variables associated with the adoption and non-adoption of organic farming was identified. Eight variables related to adoption of organic farming are measured on the basis of 5-point scale following the scoring method as very strong = 5, strong = 4, medium = 3, low = 2, and nil = 1. Similarly, seventeen variables related to non-adoption of organic farming are measured as very strong = 1, strong = 2, medium = 3, low = 4, and nil = 5.

Analytical Framework

To achieve the first object simple tabular analysis such as averages, percentages, and so on have been used. Tabular presentation for understanding the comparative economics of crop production under organic and inorganic farming through the estimation of Cost of Cultivation, Cost of Production and Return-Cost Ratio has been done on the basis of standard cost concepts, which are as follows:

Cost A_1: (Hired human labour wage + Bullock labour wage + Hired machinery charges + Cost of seeds/seedlings + Cost of fertilizers + Cost of manures + Cost of insecticides and pesticides + Cost of biopesticides + Irrigation charges + Interest on working capital (at the rate of 4 per cent pa, for example KCC) + Land revenue and taxes + Depreciation on farm implements and machinery + Miscellaneous expenses).

Cost A_2: (*Cost A_1* + Rent for leased in land)

Cost B_1: (*Cost A_2* + Interest on fixed capital: It has been calculated as per duration of a specific crop, on the basis of an assumption of at the rate of Rs 0.20 per day, that is, Rs 6.00 per month (30 days).

Cost B_2: (Cost B_1 + Rent for own land: It has been calculated on the basis of rent for leased in land prevailing at the study area during the period of 2007–08 to 2009 –10.

Cost C: (*Cost B_2* + Imputed value of family labour)

The tabular analysis for studying the impact of organic farming in relation to quality of produce and price premium based on perception of consumers' of various income groups has been done. An exercise has been carried out to get an estimate of the degree of association between consumer's level of monthly income and his willingness to pay higher price for organic products. For the purpose we used the χ^2 statistic that tests the independence of attributes. The result was tested against the null hypothesis:

H_0 = Consumer's monthly income level and consumer's willingness to pay higher price are independent, with alternative hypothesis being

H_1 = Consumer's monthly income level and consumer's willingness to pay higher price are associated.

$$\chi^2 = \Sigma \left\{ \frac{(f_0 - fe)^2}{fe} \right\}$$

where, f_0= Observed frequencies of respective cells; f_e= Expected frequencies of respective cells This approximately follows a chi-square distribution with d.f. = (number of rows – 1) × (number of columns – 1).

To measure the awareness about organic farming, the sample organic farmers have been interviewed with a structured questionnaire following five points ranking scale. The awareness of the farmers in the field of organic farming has been ranked through rank score method as, 5 (very strong), 4 (strong), 3 (medium), 2 (low), and 1 (nil). The weighted mean of rank score is presented in Tabular form.

The constraints faced by farmers to practice organic farming in the study area have been ranked with the help of Rank Based Quotient (RBQ) analysis as follows and the results are presented in tabular form.

$$RBQ = \Sigma_{i=1}^{n} \frac{\mathrm{fi}(\mathrm{n}+1-\mathrm{i}) * 100}{\mathrm{N} * \mathrm{n}}$$

where,

N = Total number of farmers

n = Total number of ranks (there are five ranks altogether, so, n = 5)

i = The rank for which the RBQ is calculated (for a constraint)

f = Number of farmers reporting the rank (for the constraint)

The constraints faced by inorganic farmers are listed in seventeen groups. These constraints have been incorporated in the survey schedule using a five-point scoring pattern, that is, 'very strong', 'strong', 'moderate', 'low', and 'nil' giving numeral scores 1, 2, 3, 4, and 5, respectively.

RESULTS AND DISCUSSIONS

The facts and findings along with discussions have been presented in this section under the heads of the specific objectives:

Status of Organic Farming in West Bengal

As per the information collected from respective PIA regarding total number of organic farmers and total area under organic farming in the study area, it has been observed that 184 farmers in Panji village and 119 farmers in Babpur bio-village in north 24-Parganas district are practicing organic farming. These farmers represent 37.32 per cent and 24.34 per cent to the total farmers of Panji and Babpur bio-villages, respectively (Table 65.1). It has also been noticed that the farmers of Panji village are practicing organic farming for last seventeen years and the farmers of Babpur bio-village are practicing the same for last five years. It also reveals that 6.57 per cent and 6.14 per cent of total cultivable area in Panji and Babpur bio-villages, respectively have come under organic farming practices. Here, there is a clear indication that the performance measured in terms of area under organic practice, is better in government activity area than NGO activity area in 24-Parganas district.

In Jalpaiguri district, however, there were 437 farmers in Purba Satali village and 111 farmers in Ghughudanga bio-village representing 47.24 per cent and 18.59 per cent to the total farmers in Purba Satali and Ghughudanga bio-village,

Table 65.1 Status of Organic Farming in Respect to Number of Farms and Land Area in Hectare

Particulars	Status			
District →	North 24 Parganas		Jalpaiguri	
Village →	Panji	Babpur	Purba Satali	Ghughudanga
PIA	NGO	Govt. of WB	NGO	Govt. of WB
Total farmers	493	489	925	597
Organic farmers (Number)	184 (37.32)	119 (24.34)	437 (47.24)	111 (18.59)
Total area (ha)	121.31	182.08	474.53	350.93
Organic farms area (ha)	7.97 (6.57)	11.18 (6.14)	10.06 (2.12)	13.23 (3.77)
Adoption of organic farming in the vill. (yrs)	17	5	11	5

Note: Figure in parentheses indicates the percentage of organic farmers and organic land area to total farmers and total land area of the village.
Source: Farmers' register and Land register of PIA.

respectively which had been practicing organic farming. In terms of area under operation, only 10.06 ha of land out of 474.53 ha (2.12 per cent) of cultivable land could be brought under organic practices in eleven years time span in Purba Satali village. In Ghughudanga bio-village, however, the picture of organic practices is a bit encouraging in comparison with Purba Satali village. Out of 350.93 ha of total cultivable area, organic farms cover 13.23 ha (3.77 per cent) in Ghughudanga (Table 65.1).

However, despite more number of organic farmers in NGO areas of both north 24-Parganas and Jalpaiguri districts, it should be worthwhile to mention that the area under organic farming is more encouraging in both NGO and government activity areas in Jalpaiguri district than north 24-Parganas district. The probable reason for this may be comparatively easy accessibility of organic manures as well as organic inputs in Jalpaiguri district than that of north 24-Parganas district.

Comparative Economics of Crop Production under Organic and Inorganic Farming

Attempts to examine the economic viability of organic farming with return/cost analysis have been undertaken with the following crops. It is obvious that the Cost A_1 does play a vital role for variation in cost of cultivation between organic and inorganic farming system. So in this section, despite the analysis of total cost of cultivation, an in depth analysis of Cost A_1 has been undertaken for identifying the actual factor(s) that are responsible for variation in cost of cultivation between organic and inorganic farming system.

Economics of Lady's Finger

Lady's finger is one of the important vegetable crops grown commercially during summer. This is well preferred by the local people. Majority of growers of the state follows inorganic system of cultivation for lady's finger. However, per hectare cost of cultivation of lady's finger is calculated as Rs 74,019.47 and Rs 65,257.62 for organic and inorganic farming system in NGO area, respectively. Turning to government area, the per hectare cost of cultivation of lady's finger for organic and inorganic system have been found as Rs 73,427.67 and Rs 65,328.81, respectively. The return/cost ratio of organic lady's finger is higher in NGO area (1.75) than government area (1.56). This ratio is more or less same in inorganic farming system for both NGO and government area and the ratio have been calculated as 1.88 and 1.87 for NGO and government area, respectively (Table 65.2).

It is observed that the total cost of cultivation of lady's finger with organic technology is substantially higher (approximately Rs 8.00 per ha) than the total cost of cultivation under inorganic farming practices of the crop. Analysis of Cost A_1 indicates that this is due to higher cost of organic manures that are being used for cultivation of the crop. On the other hand, cost for plant protection material is higher in inorganic system than organic system. The estimated cost of irrigation in inorganic farm is also higher than organic farm.

Moreover, human labour component also seems to be higher in organic method of cropping. As a matter of fact, the total cost remains higher in organic than inorganic system. So, even if the price per quintal of lady's finger from organic farms are higher than that of its inorganic

Table 65.2 Comparative Cost of Cultivation of Lady's Finger

Cost Items	NGO Area		Government Bio-village Area	
	OFS	IFS	OFS	IFS
Cost A_1	49,060.92	41,250.61	52,989.14	45,600.71
Cost A_2	49,060.92	41,250.61	52,989.14	45,600.71
Cost B_1	49,078.92	41,268.61	53,007.14	45,618.71
Cost B_2	51,428.92	43,618.61	55,357.14	47,968.71
Cost C	74,019.47	65,257.62	73,427.67	65,328.81
Yield (qtl/ha)	111.77	122.65	100.56	124.21
Price(Rs/qtl)	1,153.21	993.38	1,135.45	981.62
By product	–	–	–	–
Price of by-product	–	–	–	–
Gross return(Rs)	1,29,847.46	1,22,767.68	1,15,067.91	1,22,838.07
Net return(Rs)	55,827.99	57,510.06	41,640.25	57,509.25
R/C ratio	1.75	1.88	1.56	1.87
Total cost/ha	74,019.47	65,257.62	73,427.67	65,328.81
Total cost/qtl	662.25	532.06	730.19	525.95

Note: OFS = Organic Farming System, IFS = Inorganic Farming System.
Source: Field survey.

counterpart and hence the gross return per hectare, the net return per hectare and return-cost ratio remains favourable towards the inorganic farms. Besides, the higher return/cost ratio of organic lady's finger in NGO area may be the impact of practicing organic farming for a longer duration in NGO area than government area.

Economics of Potato

Normally, farmers grow potato in winter season as commercial venture. The production of potato is higher in winter season with intensive use of synthetic fertilizers in West Bengal. There is a significant difference between organic and inorganic system of potato cultivation in terms of productivity, total cost, gross return, net return, and return/cost ratio. The total cost of cultivation is higher in organic potato (Rs 91,621.17/ha) than inorganic potato (Rs 73,686.54) in the NGO area. The cost of organic potato (Rs 1,05,762.21) is also higher than inorganic potato (Rs 73,019.59) in government area. The profitability in terms of gross return is quite encouraging under inorganic potato (Rs 1,63,857.58) than the organic potato (Rs 1,49,414.75) in NGO area and in government area it is Rs 1,64,018.19 under inorganic system and Rs 1,37,589.18 under organic system. The net return is also higher (Rs 90,171.05/ha) in inorganic potato than organic potato (Rs 57,793.58) in NGO area. In the government area, net return is Rs 90,998.60 and Rs 31,826.96 for inorganic and organic system, respectively. There is a difference between price premium received for organic potato and prevailing market price of inorganic potato in both the area. The prevailing market price is observed to be Rs 768.77 per qtl and Rs 768.76 per qtl for inorganic potato in NGO and government area, respectively. Whereas, the premium price of organic potato is Rs 859.54 per qtl and Rs 846.65 per qtl in NGO and government area, respectively. The return/cost ratio is estimated at 1.62 and 2.21 in NGO area and 1.29 and 2.23 in government area, for organic and inorganic potato, respectively. This indicates inorganic potato is more profitable than the organic potato. Potato is a highly soil exhaustive crops and requires supplementation of high dose of nutrient which is only possible through the application of inorganic inputs (Table 65.3).

Analysis of Cost A_1 explains the reason behind the higher cost of cultivation of organic potato. It is due to more cost of organic manures and biopesticides applied in potato cultivation. The cost of seed under both the systems and both the areas is more or less same. However, the use of higher doses of synthetic fertilizers in inorganic potato induces higher productivity than organic potato. Therefore, to sustain the present productivity as well as to enhance the potato productivity, potato cultivation practices may continue to be inorganic.

So far we have discussed about the vegetable lady's finger and potato those are being grown by the farmers with organic technology. There are also a number of other

Table 65.3 Comparative Cost of Cultivation of Potato

Cost Items	NGO Area		Government Bio-village Area	
	OFS	IFS	OFS	IFS
Cost A_1	63,525.16	46,185.00	82,617.50	50,486.86
Cost A_2	63,525.16	46,185.00	82,617.50	50,486.86
Cost B_1	63,549.16	46,209.00	82,641.50	50,510.86
Cost B_2	65,899.16	48,559.00	85,011.50	52,860.86
Cost C	91,621.17	73,686.54	1,05,762.21	73,019.59
Yield (qtl/ha)	172.54	211.56	161.30	211.76
Price (Rs/qtl)	859.54	768.77	846.65	768.76
By product	–	–	–	–
Price of by-product	–	–	–	–
Gross return (Rs)	1,49,414.75	1,63,857.58	1,37,589.18	1,64,018.19
Net return (Rs)	57,793.58	90,171.05	31,826.96	90,998.60
R/C ratio	1.62	2.21	1.29	2.23
Total cost/ha	91,621.17	73,686.54	1,05,762.21	73,019.59
Total cost/qtl	531.01	348.30	655.69	344.82

Note: OFS = Organic Farming System, IFS = Inorganic Farming System.
Source: Field survey.

vegetables like cowpea, brinjal, cauliflower, chilli, and so on, those are also grown by the farmers with alternative technology. In all cases the total cost of organic production process remains higher in comparison with the output produced by chemical technology. But the produce of organic farms has an edge over others in terms of market price and hence gross return.

It is important at this juncture to mention that the crops, of which we are concerned, are vegetable crops. These vegetables have a good market opportunity. But perishable nature of the produce necessitates a well-knit network of market access and transportation. Moreover, warehousing facilities for vegetables are still meagre in the state.

Impact of Organic Farming in Relation to Quality of Produces and Price Premium

To measure the impact on quality of organic farm product and its price, consumers' perception has been studied in eight selected markets, where organic vegetables are sold by the organic farmers. These output markets were chosen purposively for the study for the fact that agricultural produces with both organic and inorganic technology flow to these markets. To assess the consumers' preference in this regard, a sample of 126 buyers from different income groups (up to Rs 10,000 pm, Rs >10,000 to <20,000 pm, and Rs 20,000 and above pm) were selected. The selection of consumers is made purposively those who purchase produce from such outlets where both organic as well as inorganic products are available.

It is expected that the level of income would be an important factor in determining the consumer demand towards organic foods. To have an idea about consumers' attitude towards organic vegetables we had to rely on a proxy variable namely 'consumers' willingness to pay higher price for organic produce'. The sublime assumption being more willing the consumer is to pay higher price for organic product, the higher is his/her preference towards the product, it is expected to vary with the income level of the consumer. Hence, consumer of higher income group would prefer organic products more. Field level data get corroborated with our expectation (Table 65.4).

We carried out an exercise to get an estimate of the degree of association between consumer's level of monthly income and his willingness to pay higher price for organic products. For the purpose we used the χ^2 statistic that tests the independence of attributes. From the Table 65.4, a 3×3 contingency table was prepared to test the degree of association between monthly income and consumer's response. The result was tested against the null hypothesis and the estimated value of χ^2 was:

$\chi^2 = 31.989$ with 4 degrees of freedom which was significant at 0.99 level.

From the result we get a clear indication of positive association between consumer's monthly income and his

Table 65.4 Price Premium that Consumers' Willing to Pay for Organic Products (in per cent)

Willingness to Pay Price Premium	Monthly Income (in Rs)			
	Up to 10000	10000–20000	Above 20000	Total
Up to 20%	62 (95.4)	37 (77.1)	4 (30.8)	103 (81.7)
21% to 30%	3 (4.6)	9 (18.8)	8 (61.5)	20 (15.9)
31% to 40%	0 (0.0)	2 (4.2)	1 (7.7)	3 (2.4)
Total	65 (100.0)	48 (100.0)	13 (100.0)	126 (100.0)

Note: Figures in parenthesis indicate percentage.
Source: Market survey.

willingness to pay a higher price for organically produced crops. Hence, the null hypothesis was rejected.

The important points are to be noted here that lady's finger, cowpea, brinjal, and potato in NGO area of both north 24-Parganas and Jalpaiguri district are sold over 10 per cent higher price as compared to inorganic farm product though cauliflower and chilli are sold by a price of less than 10 per cent higher price as compared to inorganic farm product in the same market. Perhaps the small size of organic cauliflower and chilli is the factor for disliking of the consumers. However, the price premium in the study area of both the districts pushed up the farm income of the organic farmers.

Farmers' Awareness Regarding Organic Farm Practices

To measure the awareness about organic farming, the sample organic farmers have been interviewed with a structured questionnaire following 5 points ranking scale. The weighted mean of rank score reveals that the organic farmers' awareness is maximum in relation to the good quality of the product in both NGO and government area followed by beneficial attributes of the organic farm product for the human health. The interesting point may be noted here that the level of awareness of the farmers of NGO area regarding 'high profitable' is placed third by rank, that is, the farmers of the said area believe that organic farming system is high profitable than any other system of farming, but the farmers of the government area consider this phenomenon by ranking eighth. This may be the cause of practicing organic farming for a longer duration by the farmers of NGO area. The fact is that during the conversion period from inorganic to organic farming, the yield of crops is reduced in organic farm and at the initial years application of higher quantity of organic manures is required for maintaining the status of nutrients in the soil. As a result, higher cost involvement for manuring the soil and lower yield leads lower profit from farm operation. Farmers of Government area are practicing organic farming for last five years, so the level of their profit

Table 65.5 Ranking of Organic Farmers' Awareness

Questionnaire	NGO Area		Government Bio-village Area	
	Score	Rank	Score	Rank
High profitable	3.30	3	2.23	8
Minimum production risk	2.77	7	2.47	6
Higher employment potentiality	2.80	6	3.33	4
Lower recurring cost for inputs	2.97	5	3.50	3
Beneficial for health	4.17	2	4.33	2
Increasing consumer demand	3.20	4	2.50	5
Higher price of organic product	2.53	8	2.34	7
Good quality	4.20	1	4.63	1

Source: Field survey.

of organic farming is not equal to the level of profit that are perceived by the farmers of NGOs area, who are practicing organic farming for over a decade (Table 65.5).

Constraints in Adoption of Organic Farming

Despite having potential the organic farming in West Bengal is almost at nascent stage and several issues have to be resolved for its promotion. Systematic constraints analysis from the perspective of farmers is an important step to resolve these issues. The inorganic farmers of the study area of the state express their opinion about the constraint encountered with organic production; the farmers' opinion is collected through focus group discussion by structured questionnaire. Constraints are analysed through using RBQ technique for ranking. It has been found that seventeen constraints are dominating for non-adoption of organic farming in these study areas.

However, it is also a fact that among these constraints, the constraints like high cost of organic inputs, lack of market for organic product, non-availability of organic inputs, lower yield, and lacking of price advantage for organic product are found to be the major constraints. The other constraints appear to be lack of consumers demand for organic product and lower profitability. Small holding size, inconvenience of organic techniques, no scope, higher production risk, non-availability of suitable land for organic farming are also posing hindrance towards farmers' willingness to go for organic farming (Table 65.6).

Table 65.6 Field Level Constraints of Organic Farming as Perceived by the Sample Farms

Constraints	RBQ	Rank
Not aware	24.67	15
No scope	28.33	10
Small holding size	30.67	8
Lower profitability	36.26	7
Lower yield	47.56	4
High cost of organic inputs	78.73	1
Higher production risk	27.72	11
Lacking of price advantage	47.49	5
Lack of market	74.71	2
Lower employment potentiality	26.03	16
More recurring cost for inputs	24.74	14
Non-availability of suitable land	27.00	12
Non-availability of organic inputs	51.50	3
Lack of consumers demand	41.86	6
Inconvenience of organic techniques	29.67	9
Lack of experience on organic farming	21.67	17
Lack of training on organic practices	26.33	13

Source: Field survey.

CONCLUSIONS AND POLICY IMPLICATIONS

It has been found that the area under organic farming of the sample farmers is more in Jalpaiguri district than that of North 24-Parganas district. The probable reason for this may be due to the fact of comparatively easy accessibility of organic manures as well as organic inputs in Jalpaiguri district than that of North 24-Parganas district. So, we may conclude that organic inputs are mainly responsible for expansion of organic farm area.

Economics of organic vis-à-vis inorganic farm practices indicates that the cost of cultivation was higher and production was lower in organic than inorganic system but price of the organic product was higher than inorganic in the study area. This was resulted a favourable return/cost ratio for organic farming system. So, it may be concluded that price premium is too important for keeping the organic farming profitable.

The higher cost of production in organic farming is mainly the reason for higher price of organic produce at present. And there exists a definite positive association between consumer's monthly income and his willingness to pay higher price for organic products. So, low cost production technology is required for easy accessibility of organic produces to all the people.

Though the organic produces are of good quality and play beneficial role for human health, there are seventeen constraints found to be dominating in non-adoption of organic farming in the study areas. Among these, no local market for organic product, unavailability of organic inputs, no consumers demand for organic product, lack of training of organic practices, and so on are important. So, strong and appropriate market structure with training facility for organic farming should be given priority for larger adoption of this farming system.

The main policy implications of the study include formation of Farmers' Organization for a reasonable price premium, use of recommended doses of plant nutrients, interlinking credit with output for organic farm production to facilitate export in this section, and to encourage organic farmers. The government should provide start-up funding as subsidy for a broad scale farmer conversion programme through kinds, that is, inputs of organic in nature. Marketing co-operatives by pooling the small and scattered produce of the producers' can improve the bargaining strength of organic growers and can, thus, effectively eliminate the margin appropriated by the market intermediaries. Organic food products should be integrated into public procurement, such as in schools, hospitals, and so on, through the requirement of at least a certain percentage of organic foods, if these are available, to stimulate both a base market demand and improve the public information and consumer exposure to organics.

REFERENCES

Alvares, C. (ed.). 2002. *The Organic Farming Reader*. Goa: Other India Press.

Birthal, P. 2005. *Agriculture Diversification Opportunities for Small Farmers*. NCAP. New Delhi.

Chatterjee, A.S. 2005. *Ecological Farming and NRM. Food and Nutrition Security Community (FAO)*. New Delhi.

Daniel, A. 2001. Institute for Integrated Rural Development (IIRD). *Concepts. Principles and Basic Standards of Indian Organic Agriculture*. Aurangabad: Kanchannagar.

Deb, Debal. 2004. *Industrial vs Ecological Agriculture*. New Delhi: Navdanya.

Food and Agriculture Organization. 1997. *Bulletin for Organic Agriculture*. FAO. UN.

Ghosh, A.K. and Debal Deb (eds). 2000. *Manual for Sustainable Agriculture & Biodiversity Conservation*. New Delhi: Navdanya.

International Federation of Organic Agriculture Movements. 1998. *Guideline for Organic Agriculture*. IFOAM. Germany.

Joshi, Mukund and T.K. Prabhakarasetty. 200). *Sustainability Through Organic Farming*. New Delhi: Kalyani Publishers.

Maiti, R.G. 2007. 'Organic Horticulture in India—Its Past, Present and Future', National Workshop on 'Organic Horticulture' at BCKV. Mohanpur. West Bengal: 53–4.

Palaniappan, S.P. and K. Annadurai. 2003. *Organic Farming (Theory & Practice)*. Jodhpur: Scientific Publishers.

Shiva. V., P. Pande, J. Singh. 2004. *Principles of Organic Farming*. New Delhi: Navdanya.

66

Rural Diversification in India

Brajesh Jha

The traditional vision of rural economies as purely agricultural has become increasingly obsolete. The structure of economy changes with the growth of economy. As an economy grows proportion of non-farm sector in total income and employment increases at aggregate and also at disaggregate levels. Farm households across the developing world earn an increasing share of their income from non-farm sources. In India rural non-farm sector (RNFS) has been essential part of a village life since ages. It primarily supplies non-agricultural goods and services to agriculture-based households in a relatively closed economy. The village's demand for such items was, by and large, fulfilled with consumer goods conveniently produced in and around village. Of late, trade became important and with trade, commodities and services were increasingly specialized. The specialization is often associated with scale economies and consumer goods industries are increasingly shifting towards consumption centre (the city). Government of India attempts to regulate the above shift of industries through its policies; such policies range from protection to promotion of the above industries has achieved limited success. From world over, there are not many evidences against the above shift of consumer goods industries. Though some of the East Asian countries in the recent decades have recorded successes of consumer goods industries in the conundrum of village and cities often referred as peri-urban.

In India also peri-urban region at times attracts manufacturing industries; however, percolation of benefits of such industry in rural vicinity requires adequate integration of such industries with rural hinterland. In dearth of such integration the RNFS continues to languish in the large part of country. The bulk of small and marginal farmers depend on off-farm income, and adequate off-farm income requires growth of productive employment in RNFS. The sector has therefore drawn attention of researchers. The specific objectives of the chapter are:

1. to present review of studies related to RNFS;
2. to examine the kind of transition of rural economy in India;
3. to study the process and policies for vibrant growth in rural sector; and
4. to bring out various issues of income and employment diversification in rural India.

METHODOLOGY AND DATA

The present chapter has been based on secondary information collected from various secondary sources like National Sample Survey Organization (NSSO), Central Statistical Organization (CSO), Directorate of Labour, Shimla, and depicts of secondary information in per cent, average in tables of chapters.

RESULTS AND DISCUSSION

Rural transition referred as structural change in rural economy is presumed as rural diversification in the present study. The structural changes require the increasing role of non-farm sector in rural employment and income. The studies related to growth of rural non-farm employment (RNFE) can be categorized on different basis. The sources of data, issues related to employment can be some basis of categorization. The majority of studies on RNFE are based on information from secondary sources, the studies based on primary information are also important. The primary survey based studies range from the study of a village to the study of region consisting of villages, districts and states (Reddy 2005; Wilson 1999). These studies are often designed to disentangle certain issues related to RNFS, which are otherwise not possible to discuss with data from secondary sources. Examples of such issues are the role of non-farm sector in the household income of an average farmer, association of non-farm employment with income groups, and poverty of rural households.

The household resource allocation pattern in different non-farm enterprises promises a richer understanding of accumulation patterns and pathways out of poverty (Bharadwaj 1993). The role assets, non-farm opportunities, and public policies play in reduction of poverty has been assessed at micro-level (Awasthi 2005). A poor household's welfare depends on productivity of labour especially unskilled labour (Papola 1987). The comparison of poor and non-poor households in a region may address questions like how aggregate trends in farm and non-farm productivity, changing labour allocations and income composition, and the fluidity of rural-urban interactions drives dynamic transformation in a variety of rural settings (Harris 1987).

Some of primary information based studies include thousands of samples spread across the country (Lanjouw and Shariff 2001). The study showed that non-farm income contributed about one third of total household income in rural India in 1993–94 and that the share of income from the non-farm sector is much higher than that of employment. The study also suggests that the share of rural non-farm income was higher for middle three quintals as compared to the lowest and the highest quintals. Further the share of casual non-farm labour is higher for the poor quintals while that of non-farm regular employment is higher for the richer ones. The NSS Survey results on farmer's condition published as Situation Assessment of Farmers in 2004 provides some of the above information, such surveys are, however, not very frequent.

The studies based on information from secondary sources attempt to look into the trends and pattern of employment in sectors (rural vis-à-vis urban), industries (agriculture and non-agriculture) and collate these trends with the performances of economy at aggregate and disaggregate level. Many of such studies used employment data as available from the NSS quinquennial survey results. The NSS quinquennial employment survey results have encouraged researchers to delineate pattern of employment in historical years. Examples of such studies are Sen 2003; Sundaram 2001; and Visaria and Basant 1994. In each of this study researchers have compared current year information with the historical data and arrive at certain policy conclusions. Some of such policy conclusions are jobless growth of economy during 1990s, a significant shift of rural employment from agriculture to non-agriculture sector, prevalence of working poor in economy. Some studies on rural non-farm employment also look into the patterns of employment across gender, social groups, and castes (Unni and Rani 1999; Newman and Thorat 2007). Some researchers have used data from different secondary sources to arrive at certain policy conclusions like village industries are dying (Saith 2001).

These policy conclusions are often debated. Few of them raise finger about compatibility of these data across reference years. Some of such limitations also crop up following evolution of certain concepts in the particular quinquennial rounds of employment. The periodic changes in sample questions designed to aim at improvement in survey results are another reason for non-compatibility of data over the survey years. Example of such change in survey design for question is shift from yearly to monthly recall period in the NSS Survey of Consumption expenditure (Sen 2003).

The NSS survey results on employment as published in various reports are for 32 states and Union territories of India. Some states in India are, however, too big and diverse to present homogeneous situation, researchers interested in micro-pictures of different regions of a state, therefore, utilize unit-level data from NSSO survey results generally available in soft version. The extraction of unit level information is often complicated and arriving at a regional picture is often not very easy. Therefore, researchers interested in regional analysis of a state often use Decennial Census results on employment as it provides district level information (Bhalla 1994). Analysis of RNFE at the level of state frequently uses census data (Mahendra Dev 2000). The census results on employment have several limitations; first, the employment figures are based on usual status of employment that includes both principal and subsidiary status of employment; while the NSSO employment results provide estimates on the basis of three concepts of employment: current, weekly, and usual status of employment. In census data, categorization of workers into cultivators, agriculture labour, household, and non-household industry workers and workers other than the above have changed frequently to make these data difficult to compare over a long period of time.

Many studies on RNFE discuss determinants of growth of rural non-farm employment in country, states, and regions. Factors associated with the growth of rural non-farm employment are broadly categorized as development- or distress-related. In development related factors agriculture has been important; Mellor (1978) illustrates production and consumption linkages between agriculture and non-farm sector. Later, Hossain (1988) found a third important link between agriculture and rural non-farm sector, and termed it as labour market interaction effect. In the post-green revolution period several studies from India have found significant positive effect of agriculture on the growth of rural non-farm employment (Chadha 2001; Hazell 1991). Researchers in the subsequent period are, however, reporting the decreased role of agriculture on rural transformation (Bhalla 1994; Jha 2010).

Some research findings suggest that development of urban centres provide impetus to non-farm employment in the adjoining rural areas because of low factor prices in rural areas (Visaria and Basant 1994). There are also few studies which contradict the above results; Kundu, Sarangi, and Das (2005) did not find any significant relationship between the growth of rural non-farm employment and either of towns: large (class I–III), medium, and small (class IV–VI). Manufacturing at the level of town shows a strong relationship with the availability of infrastructure and services. The urban centres with adequate infrastructure can thus promote rural non-farm employment, provided it is integrated with the nearest rural town. The growth of rural non-farm employment is in industries and services and each of these requires its own set of infrastructure.

Industrial activities around town is believed to have been promoting rural non-farm activities. Das (2005), however, reports that industry-led development strategy has favoured establishment of large modern units in rural pockets resulting in the rise of new industrial centres in the medium-sized towns and decline of rural industries. Analysis of cluster development programme ascribes failure of small industry programmes to neglect of villages rather modern industry and obsession with certain old techniques. The right kind of industrial clusters deeply interwoven with the natural resource endowment of the region need to be prioritized to promote rural non-farm employment in the country (Jha 2010). Some researchers argue infrastructure as the most important determinants of rural transformation (Acharya and Mitra 2000; Bhalla 1994). Infrastructure includes overheads (social, physical) and also industry-specific hardwares/infrastructure. Different kinds of infrastructure may not have similar effects on all kinds of rural non-farm activities. Dearth of assured electricity in rural area with adequate physical infrastructural facilities like road has also led to flight of rural industries to urban areas in certain regions (Jha 2006). Islam (1997) highlights importance of human resource-related parameters like education and skill on rural transformation in India.

The growth of rural non-farm sector also requires supportive institutions. Many researchers including Fisher and Mahajan (1996) emphasizes on institutional framework. They highlight enabling environment for the growth of RNFS. They conclude that investments in public goods, rather than on ineffective state run promotional institutions and subsidies promote growth of RNFS. The governance of such institutions should be overhauled with a strong representation of users and they should be made to earn their revenue through the levy of user fees. In the list of supportive institutions some researchers highlight the role of institutional credit; others emphasize the role of voluntary organizations in facilitating rural activities (Rath 2005). Government is also creator, protector, and facilitator of the institutions. Institutions are often created in certain context, the context at times changes and institutions also need to be re-innovated. The instance of rural non-farm development agency (RUDA) in Rajasthan is an example of such institutional innovations. The multiplicity of government agencies and public institutions is also reported to have constrained promotion of rural activities in India.

The residual sector hypotheses as proposed by Vaidyanathan (1986) argues that in a situation where the labour absorptive capacity of agriculture becomes limited and the urban industrial sector is not able to accommodate the ever-growing labour force, the RNFS tend to act as a *sponge* for the surplus labour. The above discussions, thus, suggest that not only pull but also push related factors promote the growth of RNFS in India. The evidences on determinants of growth of RNFE often vary on the basis of sample of analysis. Many of above studies are based on state level information; where, small numbers of poorer regions actually co-exist with the prosperous regions. Therefore, Jha (2006) has studied factors of rural transformation by combining information on factors from different levels (states and districts).

The literature on migration has made important contributions to our understanding of rural-urban labour flows; only few have focused on the rural side of non-farm and farm labour market (Srivastava 2011). Industry, business, and services the non-agriculture sector in rural economy is often termed as RNFS. Studies show that RNFS consists of heterogeneous group of industries. Growth in each of these depends on a set of variables development and distress related. Agriculture, infrastructure are some development related factors. Studies show that growth of RNFS and its determinants varies over years and also across regions. Though secondary information on employment provides the kind of transition in sectors, industries and gender over years; the methodology used for arriving at certain trend are often debated. With these review of studies the subsequent section discusses transition of RNFS in the post MGNREGS phase.

EXTENT OF RURAL DIVERSIFICATION

Review of studies related to rural diversification shows dearth of desired transition from agriculture to non-agriculture sector in rural economy. The present chapter, after reviewing rural diversification, attempts to understand the kind of diversification in rural economy in the post-MGNREGS period. Pattern of rural diversification with secondary information discusses diversification in income and employment of rural economy. This also discusses rural wages in the current years.

The extent of diversification of income can be seen from Table 66.1. Table illustrates distribution of net domestic product (NDP) in different industries in aggregate and rural economy in India. The country level distribution in different industries is presented for selected years (Table 66.1). Interestingly, 2004–05 is the most recent year for which income data at country level is available from the CSO Income Accounts. Table clearly shows that the share of agriculture in total income was around half in the year 1972–73, the corresponding figure was less than one fifth of total income in 2004–05. Interestingly industries/sectors other than agriculture accounts for more than 60 per cent of rural income in 2004–05. Trade and services were accounting for around one-third of rural income in 2004–05, they were accounting for less than 10 per cent of rural income in 1970–71. Rural income from manufacturing increased during the year 1981 has stagnated thereafter at around 10 per cent. The extent of rural diversification is often discussed with the data on employment.

Table 66.2 presents distribution of workers into nine major industries during the selected years of reference 1973, 1983, 1993–94, 1999–2000, 2004–05, and 2009–10. The above table shows per cent distribution of male and female workers separately for rural and urban sector of India. The details of workers in the present table are based on usual status of employment. Table clearly shows a sharp decline in the share of workers in agriculture in the recent period (1999–2000 and 2009–10). Though this decline was noticed for both male and female, the rate of decline was significantly lower for female workers. Manufacturing traditionally the most important source of growth of rural non-farm employment has yielded its prime position to construction. In the recent period employment growth appears to have been propelled by construction and transport-storage-communication group of industries in rural sector. Though these industries together account for less than 15 per cent of male workers in the rural sector; the trend is encouraging since these industries are associated with development of infrastructure in rural sector.

Table 66.1 Diversity of Rural Income in Major Industries: Distribution of Rural vis-à-vis Total Income (NDP) in Major Industries in Historical Years

Industry	Year									
	2004–05		1999–2000		1993–94		1980–81		1970–71	
	Rural	Total	Rural	Total	Rural	Total	Rural	Total	Rural	Total
Agriculture	38.9	19.9	51.8	26.9	56.1	32.1	64.4	40.0	72.4	46.9
Mining	3.7	2.7	1.8	2.1	2.5	2.0	1.2	1.3	0.8	0.9
Manufacturing	11.5	13.1	10.9	12.7	8.1	14.7	9.2	16.9	5.9	14.2
Utilities	0.9	1.3	1.4	1.5	0.8	1.2	0.6	0.8	0.4	0.6
Construction	7.8	8.2	5.7	6.4	4.5	5.4	4.1	5.2	3.5	5.0
Trade etc	14.9	17.5	8.6	15.4	8.7	15.1	6.7	13.0	2.7	9.3
Transport etc.	5.7	8.3	4.0	6.6	4.0	6.2	1.3	3.4	1.3	3.4
Business services	8.4	15.0	6.5	13.2	0.0	0.0	0.0	0.0	0.0	0.0
Public services	8.2	14.1	9.1	15.2	9.3	12.1	7.3	11.0	6.4	10.0

Note: In the above table Trade consists of trade, hotels, and restaurant; Transport consists of transport, storage, and communication; Private services consists of finance, insurance, real estate, and business services; while Public services consists of Community, Social, and personal services.
Source: CSO.

Table 66.2 Diversity of Rural Workforce: Distribution of Usual Status Workers (principal + subsidiary status) in India by Sectors, Sex, and Residence over the Years

Sectors	Locale	Male						Female						Person
		1972–73	1983	1993–94	1999–2000	2004–05	2009–10	1972–73	1983	1993–94	1999–2000	2004–05	2009–10	2009–10
Agriculture etc.	Rural	83.2	77.5	74.1	71.4	66.5	62.8	89.7	87.5	86.2	85.3	83.3	79.4	67.9
	Urban	10.8	10.6	9	6.5	6.1	6.	32	31.5	24.7	17.6	18.1	13.9	7.5
Mining & quarrying	Rural	0.4	0.6	0.7	0.6	0.6	0.8	0.2	0.3	0.4	0.3	0.3	0.3	0.6
	Urban	1	1.2	1.3	0.9	0.9	0.7	0.7	0.7	0.6	0.4	0.2	0.3	0.6
Manufacturing	Rural	5.7	7	7	7.3	7.9	7.0	4.7	6.4	7	7.6	8.4	7.5	7.2
	Urban	26.9	26.8	23.5	22.4	23.5	21.8	26.2	26.7	24.1	24	28.2	27.9	23.0
Utilities	Rural	0.1	0.2	0.3	0.2	0.2	0.2	–	–	0.1	0	0	0.0	0.2
	Urban	0.8	1.1	1.2	0.8	0.8	0.7	0.1	0.2	0.3	0.2	0.2	0.4	0.6
Construction	Rural	1.6	2.2	3.2	4.5	6.8	11.3	1.1	0.7	0.9	1.1	1.5	5.2	9.4
	Urban	4.3	5.1	6.9	8.7	9.2	11.4	3.3	3.2	4.1	4.8	3.8	4.7	10.2
Trade, hotels and restraunt	Rural	3.1	4.4	5.5	6.8	8.3	8.2	1.5	1.9	2.1	2	2.5	2.8	6.4
	Urban	20.2	20.4	21.9	29.4	28	27.0	9.5	9.5	10	16.9	12.2	12.1	24.3
Transport, storage and communications	Rural	1	1.7	2.2	3.2	3.8	4.1	-	0.1	0.1	0.1	0.2	0.2	2.9
	Urban	9	10	9.7	10.4	10.7	10.4	1	0.6	1.3	1.8	1.4	1.4	8.7
Services	Rural	4.8	6.1	7	6.2	5.9	5.5	2.8	2.8	3.4	3.6	3.7	4.6	5.4
	Urban	27.1	24.7	26.4	19	20.8	21.9	27.2	26.7	35	34.2	35.9	39.3	25.5

Source: Computed from Relevant quinquennial survey results on Employment (NSSO).

Government schemes such as MGNREGS appear to have strengthened casual work in construction.

The RNFS actually consists of several heterogeneous industries and employment growth in each of these industries is influenced by a host of separate factors (Jha 2006). These factors have different kind of implications for welfare of rural society. Table 66.3 presents states with different levels of rural diversification in employment. The RNFE per cent in some states is less than 20 per cent; the corresponding figure in some of the other states is more than 35 per cent. A higher per cent of RNFE in states do not necessarily reflect favourable situation (per capita income, rural poverty) in rural society. The growth in rural non-farm employment is often associated with distress related factors. Different kind of development and distress related factors appears to have been associated with the growth of rural non-farm employment in the country.

The above discussions in brief suggest that agriculture's share in rural income has reduced significantly but share of rural workers engaged in agriculture remains higher. The share of male workers in agriculture has reduced significantly after 1999. Manufacturing has lost its pivotal position in RNFS, transport, and associated industries in 2004–05, construction in 2009–10 has emerged as engine of growth in rural non-farm sector. The MGNREGS appears to have helped employment in construction. The growth of employment is meaningless

Table 66.3 States with different levels of Employment Diversification in Rural India, 2009–10

RNFE (per cent)	States
Less than 22.0	Chhattisgarh (15.1), Gujarat (21.8), Madhya Pradesh (20.6), Mizoram (19.4)
22.0–28.9	Arunachal Pradesh (24.5), Karnataka (24.3), Nagaland (25.8)
29.0–35.9	Andhra Pradesh (31.4), Assam (29.5), Bihar (33.2), Meghalaya (29.2), Odisha (32.4), Uttarakhand (30.4), Uttar Pradesh (33.3), All India (32.1)
36.0–42.9	Haryana (40.2), Himachal Pradesh (37.3), J&K (40.0), Punjab (37.9), Rajasthan (36.6), Tamil Nadu (36.3),
43.0 and more	Delhi (99.9), Goa (76.2), Jharkhand (45.3), Kerala (64.3), Manipur (46.7), Sikkim (46.2), Tripura (69.6), West Bengal (43.8), Andaman and Nicobar Islands (56.8), Chandigarh (97.1), Dadar & Nagar Haveli (45.0), Daman & Diu (45.0), Lakshadweep (54.8), Puducherry (54.0)

Source: Computed from relevant quinquennial Survey Results on Employment from NSSO.

in the light of significant numbers of *working poor* in the country. Therefore, employment is discussed with the quality of employment (wages) of rural workers in the next subsection.

Though discussion on quality of rural employment may include levels of disguised and seasonal unemployment of rural workers, also employment in organized sector (NCEUS 2007); the present chapter, however discusses quality of employment with wages and salary of rural workers in the country. Wage to some extent reflects productivity of labour in the sector. The wage of an average illiterate, adult person, employed as regular/salaried worker and casual worker is available in different volumes of NSS survey results on employment. The real wage in rural and urban sector is obtained by adjusting nominal wage with consumer price indices of agricultural and industrial workers (CPIAL and CPIIW), respectively. The values of real wage for different type of workers during reference period are presented in the study for Ministry of Agriculture. One of important trend in real wages is the significantly lower real wages for rural workers, the same has stagnated during 1999–2004, but the rural wages has increased in 2009–10. Another important trend in wage is gender-wise difference in wages, the same has increased after 1999–2000, the gender-wise difference has, however, reduced in the year 2009–10. This was also the year (2009–10) after adoption of MGNREGS in the country.

To further discern effect of MGNREGS on rural wages, trend in real wages for different occupation groups in rural sector is presented for the period between 2000–01 and 2009–10. The nominal wages for the present analysis is obtained from Rural Wages in India, published by Labour Bureau, Shimla. The real wage is obtained by dividing the nominal wage of workers with the consumer price indices of rural workers and the same for years 2000–01, 2004–05, and 2009–10 is presented in Figure 66.1. These years represent three different situations with respect to implementation of NREGS in India.[1] The pictorial presentation is for rural workers engaged in different kind of activities like harvesting, well digging, carpenter, blacksmith, cobbler, mason, tractor driver, and unskilled workers. The wages for workers are often for different categories of workers: male, female, and child. Sometimes it for unskilled male workers only while other times it is also for other categories of workers. For example, in well digging wages are for male and female workers only.

[1] If 2000–01 represents year before NREGS; the years 2004–05 and 2009–10 reflect nascent and advance stage of NREGS, respectively.

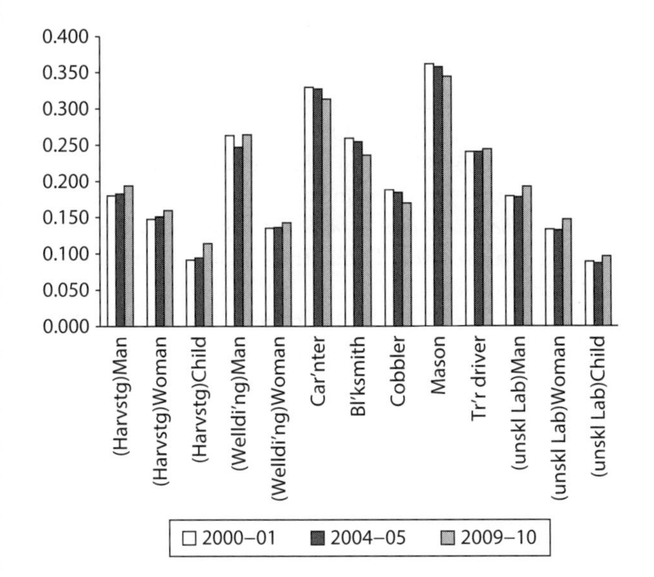

Figure 66.1 Rural Wages for Different Occupations and Sex over Years

Source: Computed from Rural wages in India, GOI Publication from Labour Bureau, Shimla.

The occupations compared in the above figure ranges from unskilled to semi-skilled and skilled type of works. It is interesting to note that real wages for unskilled works have increased for all categories of workers (male, female, and child), the increase is particularly sharp in the year 2009–10. Similar increase in real wage is reported from other semi-skilled occupations like harvesting, well digging. However, real wages for skilled workers like carpenter, blacksmith, cobbler, and mason continue to decrease during the reference period. The MGNREGS could not check fall of real wages of skilled workers. This is discouraging as country requires skilled workers for development of rural sector. The real wages for workers associated with certain agricultural activities like harvesting, tractor driving has not decreased. In fact, real wages for workers in harvesting has increased significantly. The real wages for tractor drivers has also increased marginally, though real wages for other skilled workers has declined in the country. The minimum wage for agriculture once decided administratively, causes upward pressure on rural labour market especially when workers have other options (employment in MGNREGS); the wages for agricultural workers has, therefore, increased (Jha 2006).

The above analysis of wages suggests that wedge between real wages in rural vis-à-vis urban sector and female vis-à-vis male have started reducing after initiation of MGNREGS. The real wages have increased for all categories of rural workers especially unskilled and semi-skilled workers after MGNREGS.

POLICY SUGGESTIONS

The review of studies and analysis shows that employment in trade, transport, and business services has increased in the recent period. These activities are dominated by private players. Therefore, studies to examine structure and dynamics of key privately delivered rural services will be important. Though transport and commerce are typically integrated into supply chain studies; the services such as private schools, health clinics, communications, finance, and personal services remain under-explored. The potential effects of penetration of large-scale retailers, supplying non-farm goods and services, also merits attention to track potential effects on RNFE. There have been efforts to make supply chain for agriculture commodities inclusive. How private sector has contributed to equity-enhancing supply chain development is important? Such studies may illustrate prospects for alternate interventions and policies. Bulk of rural workers is illiterate and unorganized. Their engagement with multiple activities to sustain their livelihood provides them limited scope of interacting and bargaining with other players in economy. Rural institutions are, therefore, important and the country provides wide and varied experiences with each of different kinds of rural institutions. Institutional experimentation in private and public arenas can help channel future promotional efforts more constructively. In this context, careful documentations of alternative models and their efficacy in meeting the stated objectives will be very useful.

The study that deals with the opportunities and constraints of labour selling households at different gradient of rural-urban continuum will improve our understanding of rural labour market and how such households interact spatially with regional and urban migration opportunities. What proportion of earnings of a migrant is remitted back to their family member? How remittances are influencing different rural activities and also welfare of the rural households are important questions to investigate? A better integration of rural-urban continuum often restricts migration, but how it improves welfare of rural labour is also important to investigate? On the similar context analysis of trends and forces affecting rural-urban integration merits attention. Any information on spatial distribution of price of essential commodities and wages of workers in rural-urban continuums has implications for welfare of workers and merits attention of researchers.

Government expenditure for rural sector has increased in the recent years (after 2004–05). These expenditures are in different kind of programmes for rural labour, studies to evaluate relative efficiency of expenditure in each of the programme will be very useful. Governance of these programmes has also been important. Union Government in the recent years is encouraging decentralization at the level of district (panchayat). In a decentralized environment how different rural programmes are working, how workers of different social groups are participating in these programmes, are important to investigate.

The analysis of data shows that rural non-farm economy now accounts for more than 60 per cent of rural income though it accounts for only 32 per cent of rural workforce in country. There was a sharp fall in the share of agriculture in total rural work force after 1999–2000; this fall was more in the case of male workers. Bulk of female workers (85 per cent rural female workers) is in agriculture, feminization of agriculture and rural sector is, therefore, reported. Though manufacturing has traditionally been the most important industry in the rural non-farm sector; its share in employment and income is stagnating. Now construction, transport, trade, and services lead the rural non-farm sector. Women account for around 30 per cent of rural workforce, the disguised unemployment among female workers appears to have reduced marginally following MGNREGS programme. The MGNREGS has also helped in increasing rural wages for unskilled workers and also female workers in rural sector. The rural wages for skilled workers continue to decline in rural sector.

REFERENCES

Acharya, Sarathi and Mitra Arup. 2000. 'The Potential for Rural Industries and Trade to Provide Decent Work Conditions: A Data Reconnaissance in India', SAAT Working Papers, ILO-SAAT, New Delhi.

Awasthi, Dinesh. 2005. 'Entrepreneurship, Technology and Rural Non-farm Sector: Field Experiences and Policy Imperatives', In *Rural Transformation in India: The Role of Non Farm Sector*, edited by R. Nayyar and A.N. Sharma. New Delhi: IHD-Manohar publisher and Distributor, 474–88.

Bhalla, Sheila. 1994. 'Poverty, Workforce Development and Rural Labour Market in India', *Indian J of Labour Economics* 37(4): 609–22.

Bharadwaj, Krishna. 1993. 'Regional Differentiation in India', in *Agrarian Relation, Institutional Change and Colonial Legacy, Vol. III*, edited by Binoy N. Verma. New Delhi. Ashish Publishing House.

Central Statistical Organisation (CSO). *Annual Series of Income*, several issues, New Delhi.

———. 2008. *Economic Census 2005-All India Report*. New Delhi.

Chadha, G.K. 2001. 'Impact of Economic Reforms on Rural Employment: No Smooth Sailing Anticipated', *Indian J of Agricultural Economics* 56(3): 491–97.

Cox, M. 2008. 'Chile Case study', A study prepared for the project, 'Applying historical precedent to New Conventional Wisdom on public sector roles in Agriculture and rural development', Rome: FAO.

Das Keshab. 2005. 'Can Firm Clusters Foster Non-farm Jobs? Policy Issues for Rural India', in *Rural Transformation in India: The Role of Non Farm Sector*, edited by R. Nayyar and A.N. Sharma. New Delhi: IHD-Manohar publisher and Distributor, 415–28.

Fisher, Thomas and Vijay Mahajan.1996. *The Forgotten Sector: Rural Non Farm Employment and Enterprises in Rural India*. New Delhi and ITDG, U.K.: Oxford and IBH.

Harris, B. 1987. 'Regional Growth Linkages from Agriculture', *Journal of Development Studies* 23 (2): 275–89.

Hazell, P.B.R. and S. Haggblade. 1991. 'Rural-Urban Growth Linkages in India', *Indian J of Agricultural Economics* 46(4): 515–29.

Hossain M. 1988. *Nature and Impact of the Green Revolution in Bangladesh*. Research Report no. 67, International Food Policy Research Institute, Washington, D.C. USA.

Islam Nurul. 1997. *The Non-farm Sector and Rural Development: Review of Issues and Evidences*, FAED paper no. 22, IFPRI, Washington DC.

Jha, B. 2006. *Rural Non-farm Employment in India: Macro Trends, Micro Evidences and Policy Options*, IEG Working Paper Series No. E/272/2006.

———. 2010. 'Policy Analysis for Increasing Rural Non Farm Employment for Farm Households in India', An unpublished report submitted to Ministry of Agriculture and FAO Regional Office for Asia and Pacific, Bangkok.

Kundu, Amitabh, Niranjan Sarangi, and Bal Paritosh Das. 2005. 'Economic Growth, Poverty and Non Farm Employment: An Analysis of Rural–Urban Inter-linkages', in *Rural Transformation in India: The Role of Non Farm Sector*, edited by R. Nayyar and A.N. Sharma. New Delhi: IHD-Manohar, 137–54.

Lanjouw, P. and A. Shariff. 2002. 'Rural Non-farm Employment in India: Access, Impact and Poverty Impact', Working Paper Series No. 81, National Council of Applied Economic Research (NCAER), New Delhi.

Mahendra Dev, S. 2000. 'Economic Reforms, Poverty, Income Distribution and Employment', *Economic and Political Weekly* 35 10, 4 March: 823–35.

Mellor, John W. 1978. *The New Economics of Growth—A Strategy for India and the Developing World.* Ithaca: A twentieth Century Fund Study, Cornell University Press.

National Sample Survey Organisation (NSSO). 1990. *Results of the Fourth Quinquennial Survey on Employment and Unemployment (All India)*, NSS Forty-third Round (July 1987- June 1988), NSS Report no. 409, New Delhi.

———. (NSSO, 1997). *Employment and Unemployment in India, 1993–94, NSS Fiftieth Round (July 1993–June 1994)*, NSS Report no. 409, New Delhi.

———. (NSSO 2001). *Employment and Unemployment Situation in India, 1999–2000, NSS Fifty-fifth Round (July 1999–June*

2000), NSS Report no. 458, Key Results, Part-I, Part-II, New Delhi.

———. (NSSO 2003). *Income Expenditure and Productive Assets of Farm Households 2003*, NSS Report No. 497, New Delhi

———. (NSSO 2011). *Employment and Unemployment Situation in India, 2009–10, NSS 66th Round (July 2009–June 2010)*, NSS Report no. 537(66/10/1), New Delhi.

NCEUS. 2007. 'Conditions of Work and Promotion of Livelihood in the Unorganized Sector', National Commission for Enterprises in the Unorganized Sector, GOI, New Delhi.

Newman, Katherine S. and Sukhdeo Thorat. 2007. 'Caste and Economic Discrimination: Causes, Consequences and Remedies', *Economic and Political Weekly*, 42(41), 13 October: 4121–4.

Papola, T.S. 1987. 'Rural Industrialization and Agricultural Growth: A Case Study in India', in Islam Rizwanul (ed.) *Rural Industrialization and Employment in Asia*. New Delhi: ILO/ARTEP.

Rath, Binayak. 2005. 'Self-help Groups and Rural Non-Farm Employment Opportunities', in *Rural Transformation in India: The Role of Non Farm Sector*, edited by R. Nayyar and A.N. Sharma. New Delhi: IHD-Manohar.

Reddy, D.N. 2005. 'Sustainable Rural Livelihoods and Rural Non Farm Employment: Observations Based on Village Studies from South India', in R. Nayyar and A.N. Sharma (eds) *Rural Transformation in India: The Role of Non Farm Sector*. New Delhi: IHD-Manohar.

Saith, Ashwani. 2001. 'From Village Artisans to Industrial Clusters: Agendas and Policy Gaps in Indian Rural Industrialization', *Journal of Agrarian Change*, 1(1): 81–123.

Sen, Abhijit. 2003. 'Globalization Growth and Inequality in South Asia—The Evidence from Rural India', in Jayati Ghosh and C. P. Chandrasekhar (eds) *Work and Well being in the Age of Finance*. New Delhi: Tulika Books, 469–507.

Srivastava, Ravi. 2011. 'Labour Migration in India: Recent Trends, Patterns and Policy Issues', *The Indian Journal of Labour Economics*, 54(3): 411–40.

Sundaram, K. 2001b. 'Employment and Poverty in 1990s: Further Results from NSS 55th Round Employment-Unemployment Survey, 1999–2000', *Economic and Political Weekly* 36(32), 11 August: 3039–49.

Unni, Jeemol and Uma Rani. 1999. 'Informal Sector: Women in the Emerging Labour Market', *Indian Journal of Labour Economics* 42(4): 625–40.

Vaidyanathan, A. 1986. 'Labour Use in Rural India: A Study of Spatial and Temporal Variations', *Economic and Political Weekly*, 21(52): A130–A146.

Visaria, P. and R. Basant. 1994. 'Non Agricultural Employment in India: Problems and Perspective', in P. Visaria and R. Basant (eds) *Non Agricultural Employment in India*. New Delhi: Sage Publications

Wilson, Kalpana. 1999. 'Patterns of Accumulation and Struggles of Rural Labour: Some Aspects of Agrarian Change in Central Bihar', *Journal of Peasant Studies*, 26(2–30): 316–54.

ANNEXURE 66A

Table 66A.1 Distribution of States on the Basis of Average No. of Days Worked in MGNREG

Average No. of Days Worked in MGNREG	States
Less than 30 man days	Bihar (24), Goa (14), Gujarat (25), Jharkhand (23), Kerala (26), MP (29), Orissa (26), Uttarakhand (34.3), West Bengal (17), Pudducherry (12)
30–44 man days	Chandigarh (35), Haryana (39), J&K (34), Karnataka (30), Maharashtra (34), Punjab (30), Tamil Nadu (43), UP (31), Andaman and Nicobar island (31), Dadar and Nagar Haveli (32)
More than 44 man days	HP (47), Manipur (57), Meghalaya (50), Mizoram (76), Nagaland (40), Rajasthan (71), Sikkim (59), Tripura (61), Lakshadweep (54)

Note: Rural non-farm economy includes all rural economic activity outside of agriculture. Non-farm activity may take place at home or in factories or be performed by itinerant traders. Some agriculturally focused studies measure off-farm' income or employment; this refers to 'off the owners own farm'. Consequently, off-farm income includes wage employment in agriculture earned on other people's farms along with non-farm earnings from the owner's non-farm enterprises or from non-farm wage earnings. Rural non-farm income is, thus, smaller than total 'off-farm income' by the amount of wage earnings in agriculture.

67

Jhum Cycle and Current Situation of Jhum Cultivation in North-Eastern States of India

K.C. Talukdar

Jhum cultivation is a primitive method of farming which is practiced by the tribal by slashing and burning the vegetation and then keeping it abandoned for gaining soil fertility during which the jhumias shift their homestead in search of new plot of land for cultivation. After few years the same farm family arrives at the same plot of land creating a jhum cycle. The jhumias practice mixed cropping in the jhum field to produce their food and fibre for their domestic use during the year so that the families are self-sustained and not so market dependent. With the increase in population and shrinkage of culturable land for jhuming the period of jhum cycle has declined in the recent years and the issue of jhum cultivation is under scanning from view point of ecology, environment, economy, and social considerations and is in search of settled or semi-settled cultivation to its safeguard. As the jhumias shift their homestead from place to place, it is also known as shifting cultivation. About 85 per cent of total cultivation in North-eastern (NE) India is done by shifting cultivation and 0.45 million hectares are affected in this region and which has direct bearing on the jhum cycle. The system of farming through jhuming is self-sustained producing their own food and fibre and also generate marketed surplus for meeting their household expenditures. In the recent years the policy goals have been to safeguard the flora and fauna and to improve livelihood of the jhumias through alternative farming practices to improve livelihood by adopting various measures. This chapter, therefore, is an attempt to examine the following objectives:

1. to study the current pattern of crop cultivation in the hill areas;
2. to study the economics of jhum cultivation in terms of generation of employment, income, and profitability in jhum cultivation as compared to competing crops cultivated in the hill areas;
3. to study the impact of government sponsored jhum controlled measures in the NE region to wean away the jhumias from jhuming; and
4. to suggest policy implications.

METHODOLOGY AND DATA

The study was conducted during the year 2007–08 in Mizoram and Meghalaya states of North-east India. The study was based on both primary and secondary data published in various sources. Primary data were collected following Multistage Stratified Random Sampling technique. Two districts, namely, Ri-Bhoi and West Garo hills from Meghalaya and Aizawl and Kolasib from Mizoram were selected at the first stage. One block from each district was selected at random at the second stage. From each block, 50 samples were drawn from the jhum areas purposively. Thus, a total of 200 jhumia farm families from Meghalaya and Mizoram were selected to collect the primary data. The data were subject to tabular and functional analyses to generate parameters for interpretation. Cost and returns were also examined by using the cost concepts. A multiple regression analysis was also carried out to examine the effect of family size, level of consumption, density of population, per capita culturable land, education level, and distance to the market and the parameters generated were tested for interpretation.

RESULTS AND DISCUSSION

Shifting Cultivation in the Hills of North-eastern India

The area under shifting cultivation in NE India was found to vary from state to state (NEC, Basic statistics for NE India 2000). It varied from 190 sq. km in Nagaland to 900 sq. km in Manipur followed by 700 sq. km in Arunachal Pradesh and 630 sq. km in Mizoram engaging 70,000 to 1,16,046 families in jhum cultivation. Fallow period varied from three to six years. In Tripura quite a less number of families (43,000) were engaged in shifting cultivation. The length of fallow period varied from 2.10 years in Assam to 5.90 years in Tripura followed by Nagaland (5.80 years) and Meghalaya (5.70 years). About 90 per cent of area under shifting cultivation in NE India was in Arunachal Pradesh, Assam, Manipur, and Mizoram.

Cropping Pattern and Area Shift in North Eastern Hills

Cropping pattern in a region indicates the pre dominance of crop based on soil and climate and also on pattern of livelihood. It was observed that during 1990–91 the cropping pattern in the NE hills was dominated by cereals which was highest in Manipur (90 per cent), and lowest in Meghalaya (54.70 per cent) followed by Sikkim and Tripura. In most of the states of NE India, proportionate area under food grains was higher than the country average. In Meghalaya and Tripura it was 7.50 per cent in case of oilseeds. Fruits and vegetables shared 7.90 per cent of gross cropped area (GCA) which was predominant in Arunachal Pradesh, Meghalaya, Tripura, and Mizoram. The area under fruits and vegetables was more than double over all India average.

A marginal shift in cropping pattern in the hills of NE India was observed during 2003–04. Foodgrains still predominated the cropping pattern, but there was marginal shift from foodgrains to fruits and vegetables in most of the states. Oilseeds area shifted up in Arunachal Pradesh (1.20 per cent) and Nagaland (4.60 per cent) and rice area shifted up in Manipur (16 per cent) and in Nagaland (20.90 per cent). Maize area was found to shift in Tripura (0.40 per cent) and Sikkim (1.20 per cent). It was found that area under foodgrains declined in all the states and the highest decline was in Nagaland (19.50 per cent), Mizoram (16.50 per cent), and in Manipur (14 per cent). In the states, the area under fruits and vegetables gained from 1.80 per cent in Arunachal Pradesh to 5.40 per cent in Tripura. The region observed a shift of about 2.40 per cent against 0.70 per cent for fruits and vegetables in the country.

Shift in net and GCA was also observed in the region from 1970–71 to 2002–03. The region experienced a shift of net cropped area of 35.15 per cent while GCA was found to shift at 47.91 per cent. Thus, cropping intensity was also found to shift at 8.87 per cent which was higher than the shift of cropping intensity in the country during the period. It was found to decline in Sikkim, Mizoram, Meghalaya, and also in Manipur due to marginal decline and stagnation of GCA in the states.

Shift in area under foodgrains and non-food grain crops indicated that annual growth of area under food grains during 1970–71 to 2002–03 increased in Arunachal Pradesh (34.57 per cent), Mizoram (36.37 per cent), Sikkim (56.67 per cent), Manipur (2.44 per cent) and was found to decline in Assam (–5.53 per cent), Meghalaya (–5.08 per cent), Tripura (–1.38 per cent) in the region. In comparison to the country (–10.28 per cent), the decline in area under foodgrains in the region was much less (–3.03 per cent). Similarly, area under non-foodgrain crops also declined in Arunachal Pradesh (–61.71 per cent), Manipur (10.83 per cent), Mizoram (–61.12 per cent), and in Meghalaya (–20.11 per cent). In the whole region, the area under non-foodgrains was found to gain by 11.00 per cent over the shift in the country. It was also observed that the area under non-foodgrain crops was the highest in Meghalaya (45.97 per cent of GCA) and the lowest in Manipur (20.55 per cent) in the region.

MICRO-LEVEL ANALYSIS

Cropping Pattern and Cropping Intensity in Jhuming

Attempt was made to substantiate the parameters obtained through published macro sources with the micro level data. It was observed that the principal foodgrain crops grown in the jhum areas were rice followed by maize and millet while non-foodgrain crops were sesamum, chillies, ginger, and tapioca. Cotton was the important non-food crop grown by the jhumias both in Meghalaya and Mizoram. It was observed that out of total GCA, about 87.08 per cent were occupied by foodgrains and about 10.40 per cent were occupied by non-foodgrain crops. Area under non-food crops was 2.02 per cent in Meghalaya. Rice was the main crop under foodgrains followed by chilli under non-foodgrain crops. West Garo hills district of Meghalaya practiced more of jhum cultivation with higher GCA under paddy, maize, chilli, and ginger in the district. It was observed that about 98 per cent of GCA was shared by food crops in jhum cultivation. It was also found that at lower level of holding GCA was more under food crops in both the districts of Meghalaya and the area under non-food crops was higher at higher level of holding.

In Mizoram the cropping pattern was almost similar to Meghalaya. Rice was the principal foodgrain crop followed by maize and millets. Ginger and chillies were the important non-foodgrain crops in both the districts of Mizoram. Food crops occupied more than 98 per cent of GCA and it was higher at lower level of holding. GCA was relatively higher in Meghalaya than Mizoram. Cotton cultivation in Meghalaya was more as commercial cash crop. In Kolasib district of Mizoram GCA was higher at marginal and small holdings due to higher area put under rice cultivation in the jhum land. Large farmers in the Kolasib district of Mizoram did not cultivate crops under jhum.

It was observed that cropping intensity in both the districts of Meghalaya decreased with the increase in farm size. The average cropping intensity was 111.99 and 113.99 per cent, respectively in Ri-Bhoi and West Garo hills districts, respectively. In Aizawl and Kolasib district of Mizoram it was 109.05 and 112.00 per cent, respectively.

Cropping Pattern in Settled Cultivation

Ri-Bhoi District of Meghalaya

The overall net cropped area was 4.40 hectares out of which 1.80 hectare was in 2–4 ha size group followed by 1.50 ha in 1–2 ha size group of farms. Under terrace cultivation, the GCA was also higher in these two groups and paddy shared 45.94 per cent and 42.40 per cent, respectively. It was followed by potato and vegetables. There was more variation in paddy, potato, and vegetables and the area of chilli was almost constant.

Wet rice cultivation covered 6.80 hectares out of which 44.12 per cent were shared by the size group of 4 ha and above followed by 23.53 per cent and 17.64 per cent in 1–2 and 0–1 size groups, respectively.

Horticultural crops were also cultivated in Ri-Bhoi district under terrace. Pineapple was found as the main crop grown followed by squash. Out of total area of 3.90 hectares under horticultural crops, pineapple occupied 74.36 per cent while squash and other vegetables occupied 25.64 per cent. Both marginal and large holdings put more area under horticultural crops. Out of total cropped area during survey 38.29 per cent were shared by terrace cultivation while 39.22 per cent were shared by wet rice cultivation and 22.49 per cent by horticultural crops. More area was put under terrace cultivation by small and medium groups while wet rice area was more in case of large groups. Area under horticultural crops was more in the marginal groups followed by medium and large groups.

West Garo Hills District of Meghalaya

In this district net cropped area was 2.50 hectares and it was found in medium and large size of farms. Under terrace cultivation GCA was 3.68 hectares and large and medium farms shared 57.06 and 42.93 per cent, respectively. Out of total GCA paddy occupied 55.71 per cent followed by potato (24.46 per cent) and vegetables (13.59 per cent). In terrace cultivation West Garo Hills district of Meghalaya was not so extensive.

In the district 5.80 hectares were put under wet rice cultivation and the highest share was in the large group of farms followed by marginal (31.03 per cent) and small (20.69 per cent) farms. Horticultural crops like pineapple, orange, and vegetables occupied 8.25 hectares out of which share of pineapple and orange had substantial area. Its cultivation was relatively higher at lower level of holdings. The district also started tea cultivation in 2.20 hectares of land. It was found to be started in small and large group of farms. Including horticultural crops total cropped area in this district was 19.93 hectares and was found to be more in large farms. Out of total gross cropped hectares terrace and wet rice cultivation shared 18.46 and 29.10 per cent while horticulture and tea cultivation shared 41.39 and 11.04 per cent, respectively. Terrace cultivation was more extensive in medium and large farms while cultivation of horticultural crops was extensive in lower holdings. It was found that more than 88 per cent of area was put under terrace, wet rice cultivation (WRC), and horticultural crops in the West Garo Hills district of Meghalaya.

Aizawl District of Mizoram

There was no terrace cultivation in Mizoram. WRC was done in 4.75 hectares of land and its share was more in large farms. Area under total horticultural crops was 6.60 hectares out of which 5.94 hectares were put under squash and vegetables which were cultivated extensively by the marginal group of farms. In Aizawl district of Mizoram, GCA including horticultural crops was relatively lower than Meghalaya. It was 11.35 hectares only. In this district the share of area under horticultural crops was higher

(58.15 per cent) than the WRC (41.85 per cent). The district did not start tea cultivation.

Kalasib District of Mizoram

In this district also there was no terrace cultivation and WRC was more popular in larger holdings. The farmers used to cultivate only squash and vegetables. There was no tea cultivation and the GCA was 10.50 hectares out of which 60 per cent were under WRC and 40.00 per cent were under horticultural crops. Rice cultivation was practiced more by the marginal and small farmers while horticultural crops were cultivated more by the medium farmers. Farmers of this district did not cultivate tea as plantation crop.

Share of Jhum Area to Non-Jhum Area

Share of jhum and non-jhum area was also examined. It was observed that in all the sample farms, 31.20 hectares were put under jhum cultivation while 15.10 hectares were put under settled cultivation which were 67.39 and 32.61 per cent, respectively for Ri-Bhoi district of Meghalaya. In West Garo Hills district of Meghalaya, the corresponding figures were 63.49 and 36.51 per cent, respectively. It was also observed that with the increase in size of holdings, jhum area declined while the area under settled cultivation was found to increase in both the states. This might be due to food and income security of the farmers at low farm sizes as the farmers at higher income groups could secure food and income and need not to depend more on land for frequent seasonal use.

In Mizoram, area under jhum and settled cultivation was relatively lower and more emphasis was put on jhum cultivation in both the districts. Proportionate area under jhum was higher in Kolasib district. It was also observed that like Meghalaya, jhum area was higher at lower level of holding while area under settled cultivation was higher at higher level of holding in both the districts of Mizoram. In Mizoram also farmers at lower level of holdings practiced jhum cultivation in more areas and the farmers at counter parts practiced more of settled cultivation.

Farm Size and Productivity of Crops under Jhum Cultivation

It was observed that farmers of both the states cultivated rice, maize, and millets under foodgrains while crops like sesam, chillies, ginger, tapioca, and cotton under non-foodgrain crops in the states. The size of holding varied from 0.79 hectare to 4.17 hectares with an average size of 1.54 hectare in Ri-Bhoi district and it varied from 0.82 hectare to 5.24 hectares with an average size of 1.71 hectare in West Garo hills district of Meghalaya. District-wise analysis indicated that productivity of foodgrain crops was quite low under jhum cultivation. Rice was more productive in Meghalaya while maize was more productive in Mizoram. Productivity of rice varied from 8.48 qtl/ha in large farms to 9.69 qtl/ha in marginal farms in Ri-Bhoi district of Meghalaya. In west Garo hills district the productivity of rice was found to vary from 8.87 qtl/ha in marginal farms to 9.84 qtl/ha in large farms. Productivity was found to be higher at lower level of holding. Productivity of maize was lower than rice in both the districts of Meghalaya. However, West Garo hills district was marginally more productive and the marginal and small farms were more productive. Productivity of millets was quite low. It varied from 1.67 qtl/ha to 2.69 qtl/ha in Ri-Bhoi district and 1.88 qtl/ha to 2.54 qtl/ha in West Garo hills. Productivity of millets was higher in lower level of holdings.

In Mizoram, land holding varied from 0.71 ha to 4.15 hectares with an average size of 1.22 ha in Aizawl district while it varied from 0.88 to 3.75 hectares in Kolasib district with an average size of 1.42 hectares. Productivity of rice was quite low, and was about one third of Meghalaya. It was mainly because of rat menace. In case of maize, productivity was about double of Meghalaya. In Aizawl district productivity of maize varied from 16.50 qtl/ha to 17.19 qtl/ha while it varied from 13.25 qtl/ha to 13.85 qtl/ha in Kalasib district. Aizawl district was found to be more productive for maize in Mizoram. It was also found that productivity of millet was lower in Mizoram. Its productivity was relatively higher in Kalasib district. It was found that productivity of rice and millets was more in Meghalaya while maize was more productive in Mizoram.

Productivity of non-food crops indicated that cash crops like sesamum was more productive in the districts of Meghalaya and varied from 1.86 qtl/ha to 2.92 qtl/ha and the average productivity was higher in Ri-Bhoi district. Productivity of sesamum was relatively lower in Mizoram. Its average productivity was 2.00 qtl/ha in Aizawl district and 2.13 qtl/ha in Kolasib district. Mizoram was highly productive for chilli. The average productivities of chilli in Ri-Bhoi and West Garo Hills districts were 4.64 qtl/ha and 4.27 qtl/ha, respectively while it were 5.09 qtl/ha and 5.03 qtl/ha in Aizawl and Kalasib districts of Mizoram. Productivity of chilli was higher at lower level of holdings in both the states. In case of ginger productivity was marginally higher in Mizoram than Meghalaya. The average productivity of ginger was 19.24 qtl/ha and 18.86 qtl/ha in Ri-Bhoi and west Garo hills districts while it were 20.89 qtl/ha and 19.91 qtl/ha in Aizawl and Kolasib districts of Mizoram, respectively. Marginal and small farms were more productive in both the states. Tapioca was a popular non-foodgrain crop in Meghalaya and also in Mizoram. The medium and large farmers of Mizoram did not cultivate tapioca. It was found to be more productive in Meghalaya than Mizoram. The average productivities of tapioca in Ri-Bhoi and West Garo hills districts were 5.93 qtl/ha and 6.25 qtl/ha, respectively while it were found to be 5.44 qtl/ha and 4.75 qtl/ha in Aizawl and Kalasib districts, respectively.

Under non-food crop, cotton was the only crop grown in both the states. Productivity was lower in both the states. Mizoram was more productive in cotton cultivation. The average productivities in Aizawl and Kalasib districts were 3.21 qtl/ha and 4.79 qtl/ha, respectively while it were 2.14 qtl/ha and 2.37 qtl/ha in Ri-Bhoi and West Garo hills districts of Meghalaya. The marginal and small farms were more productive in the states. It was observed that large size of farms over 4 ha did not cultivate these crops in Kalasib district of Mizoram

Generation of Income and Employment per Hectare

Study of cost and return indicated that net return per hectare in jhum cultivation was marginal to negative in both the states. It was found higher in lower holdings of Meghalaya. It was negative in all the districts of Mizoram for all classes of holdings. Employment per hectare was marginally higher than 70 Man days (MDs) and the return per MD was higher in Ri-Bhoi and West Garo hills districts of Meghalaya. In Mizoram it was about half of Meghalaya. Low productivity and low farm harvest price in the hills led to low gross return per unit of land. In Mizoram low productivity was found mainly due to rat menace in the districts.

In Meghalaya, net returns and employment were higher in lower level of holdings. Labour productivity was higher in Meghalaya than in Mizoram.

Per Capita Food and Non-Foodgrain Production and Marketed Surplus under Jhum Cultivation

Per Capita Foodgrain Production

Per capita food grain production in terms of kg per man equivalent was found to be lower at lower level of holdings. It was relatively higher in Meghalaya than in Mizoram. It was mainly due to high population size and lower size of holding. The average per capita foodgrain production was 139.85 kg/man equivalent in Ri-Bhoi district and 166.26 kg/man equivalent in West Garo Hills district of Meghalaya. It was found much lower in Mizoram. In Aizawl and Kolasib districts, the average per capita food grain production was 62.04 and 74.28 kg/ man equivalent, respectively. This was mainly due to low productivity in Mizoram for rat menace.

Per Capita Non-Foodgrain Production

Non-foodgrain production per capita was found to be quite low. Its production was low at lower level of holding in both the states. The average per capita production of non-foodgrain varied from 16.35 to 19.05 kgs in Ri-Bhoi and West Garo Hills districts of Meghalaya, respectively. In Mizoram, it was 15.61 and 18.23 kgs/man equivalents in Aizawl and Kalasib districts, respectively. Per capita production in Mizoram was marginally lower than Meghalaya as main emphasis was put for foodgrain production by the farmers.

Value Productivity

Value of productivity of foodgrains per hectare in Meghalaya was relatively higher than Mizoram. In both the districts value of productivity was higher on marginal and small farms and the average value of productivity in Ri-Bhoi district was Rs 6203.13 while in West Garo hills district it was Rs 6327.03. The average value productivity of foodgrains in Mizoram was about half of Meghalaya due to rat menace. It was marginally higher on marginal, small, and medium farms. The value of productivity of food grains was Rs 3318.65 in Aizawl district while it was Rs 3433.09 in Kolasib district of Mizoram. Value of productivity was low because of low farm harvest price and low productivity per hectare.

Value of Marketed Surplus

Value of marketed surplus of foodgrains in total value of production was quite low. It was 22.64 per cent in Ri-Bhoi district and 21.16 per cent in West Garo Hills district of Meghalaya. Share of marketed surplus of food grains to value of productivity in Mizoram was marginally higher than Meghalaya. It was 29.19 per cent in Aizawl district and 28.52 per cent in Kalasib district of Mizoram. It indicated that the farmers of Mizoram depended more on grains for family living keeping less quantity for the market.

Food Requirement and Production from Jhum

Meghalaya

Foodgrain production from jhum in the Ri-Bhoi district of Meghalaya was much lower due to low productivity.

The marginal and small farms were relatively more productive and it varied from 52.05 qtl/ha in large farms to 93.05.05 qtl/ha in marginal farms. Share of foodgrains from jhum was very low in the district. Out of total foodgrain requirements, share of foodgrains from jhum cultivation was 48.87 and 73.53 per cent, respectively for marginal and small farms while it was 87.81 and 114.20 per cent in medium and large farms. It was found to be surplus in large farms.

In West Garo Hills district, average production was higher than Ri-Bhoi district. Total production was higher in small farms and requirement of food grains in marginal and small farms was higher. However, 40.62 and 95.03 per cent of total requirement of foodgrains were met from jhum cultivation. Foodgrain production of medium and large farms was surplus over requirements.

Mizoram

In Aizawl district of Mizoram, total production of foodgrains was almost half of Meghalya because of low productivity. It varied from 13.30 qtl/ha in medium farms to 33.6 qtl/ha in small farms. Requirement of foodgrains was the highest in marginal farms due to high population size. It was observed that 23.42 and 28.64 per cent of total requirement of foodgrains were met from jhum cultivation. There was no surplus production from jhum cultivation in this district of Mizoram.

In Kalasib district, there were no large farms and the average production of foodgrains was marginally higher. Total requirement of foodgrains was relatively higher in this district, especially in marginal and small farms. The share of foodgrain productions in total requirement was marginally higher. In both the districts, jhum cultivation could not meet the total requirements of foodgrains.

Relative Economics of Jhum and Settled Cultivation

Comparative economics of jhum and settled cultivation in Meghalaya and Mizoram indicated that income and employment generated in settled cultivation was higher in both the states. Net income was several times higher in settled cultivation over jhum and it was found to be negative in Mizoram. Settled cultivation was more capital intensive but more paying than jhum cultivation. Return per Man day was much higher than prevailing wage rate in the states. The Benefit–Cost Ratio (BCR) for jhum cultivation was higher for settled cultivation in both the states. However, it was found to be much lower in Mizoram for jhum cultivation. Average economic farm situation was better in Meghalaya than in Mizoram.

Economics of Settled Cultivation for Horticultural Crops

Ri-Bhoi District of Meghalaya

Economic analysis of few horticultural crops, namely, pineapple, oranges, squash, and other vegetables indicated that cultivation of horticultural crops was much profitable in the settled cultivation over the cereal crops. Net returns per hectare were higher in marginal farms with the highest BCR of 1.58. The average net returns per hectare were Rs 8490.21 and generated 152 man days of employment per hectare. Return per man day varied from Rs 175.57 in large farms to Rs 193.62 in marginal farms.

West Garo Hills District of Meghalaya

Relative to Ri-Bhoi district settled cultivation of vegetables in this district was less remunerative. Average net returns per hectare were Rs 6160.88 and it varied from Rs 4508.12 to Rs 7362.48 per hectare. It generated 187 man days per hectare and the highest employment of 196 Man Days was found in the marginal farms. Returns per Man Day generated were higher than the Ri-Bhoi district. It varied from Rs 160.20 to Rs 168.55 in large and small farms, respectively. The average BCR was 1.25 and it varied from 1.18 in large farms to 1.30 in marginal farms. This district was found to be costlier to cultivate vegetables than the Ri-Bhoi district of Meghalaya.

Aizawl District of Mizoram

In Aizawl district of Mizoram gross returns from cultivation of horticultural crops were higher than Meghalaya. However, average cost of production was also higher. The average net returns per hectare were Rs 8553.98 and it was the lowest in large farms (Rs 5458.96) and was the highest in the marginal farms (Rs 9194.77). Per hectare generation of employment was 237 Man days and the average returns per man day were Rs 168.23. The BCR varied from 1.17 in large farms to 1.31 in small farms. It was found to be more remunerative in lower holdings.

Kalasib District of Mizoram

Profitability of horticultural crops in settled cultivation in this district was relatively lower than Aizawl district. The average net returns were Rs 5399.52 and it was the highest in the marginal farms. Employment generated was 234 man days per hectare and a worker could earn Rs 153.80 per man day. It was marginally higher in the marginal farms. The BCR varied from 1.18 in large farms to 1.21 in marginal farms.

It was observed that cultivation of horticultural crops in settled cultivation was more profitable over rice and other cereals. Profitability was relatively higher in Meghalaya and the marginal and small farms were more viable through cultivation of horticultural crops.

Jhum Cycle and Current Situation of Jhum Cultivation in North Eastern States of India

West Garo Hills District of Meghalaya

Tea cultivation was started in West Garo Hills district of Meghalaya as a plantation crop under settled cultivation in the recent years. These cases were found under small and large farms. The farms sold green leaves and the average net returns per hectare were Rs 38112.54 and employment generation was 393.18 man days per hectare. The average return per Man day was Rs 268.14 with the BCR of 1.57 in the district. Cultivation of tea in the marginal holdings was found to be more economic in the district.

Factors Affecting Jhum Cycle in Meghalaya

The length of a jhum cycle depends on various socio-economic and physical factors. Attempt was made to examine the effects of these factors on length of the jhum cycle. The factors like education level of the family measured in terms of indices (X_1), Per capita culturable land (X_2) in hectares, family size (X_3) in numbers, population density (X_4) as a ratio and distance to the city (X_5), and consumption level (X_6) as independent variables were regressed with the length of the jhum cycle (C_t) in years. The following results were obtained.

Ri-Bhoi District

$$C_t = 1.9206 - 0.1441X_1 - 0.3613X_2 + 0.0412X_5^{*} + 0.0526X_6$$

$$R^2 = 0.11$$

(0.2533) (0.6294) (1.5042) (0.5963)
(Figures in parentheses indicate t values)

*significant at 10 per cent probability level

In Ri-Bhoi district, the length of jhum cycle was affected significantly only by the distance to the city with desired sign. This indicated that the jhum cycle was shorter at nearness to the city. The effects of other factor were insignificant. The effect of consumption was indecisive. Low value the coefficient of determination indicated that some other exogenous factors influenced the jhum cycle. The variable like family size and population density were highly correlated.

West Garo Hills District

$$C_t = 9.4078 - 0.2719X_1^{*} - 0.07477X_2 - 0.0612X_5^{**} + 0.0272\,X_6$$

$$R^2 = 0.43$$

(1.0668) (0.1966) (4.3082) (0.6025)
(Figures in parentheses indicate t values)

** Significant at 5 per cent level of probability,
* Significant at 10 per cent level of probability

In West Garo Hills district, the variables like education level of the family, per capita cultivable land, distance to the city and consumption level could explain about 43 per cent of the total variation of length of jhum cycle. Desired sign was observed for education and distance to the city. Impact of education was higher in affecting the jhum cycle. The effect of distance to the market was higher than the other variables. The value of coefficient of multiple determination was 43 per cent.

Aizawl District

$$C_t = 4.3830 + 0.2811X_1^{*} + 0.16148X_2 - 0.0065X_3 - 0.04461X_4^{**}$$

$$R^2 = 0.5023$$

(0.8295) (0.41184) (0.1092) (3.7086)
(Figures in parentheses indicate t values)

** Significant at 5 per cent level of probability, * Significant at 10 per cent level of probability.

In Aizawl district, the effects of education and per capita cultivable land were indeterminate. Family size and population density had adverse effects on jhum cycle. However, except education and population density the effects of all other variables were insignificant. The value of coefficient of multiple determination was 50.23 per cent.

Kalasib District

$$C_t = 7.0351 + 0.0615X_1 - 0.1290X_2 + 0.0358X_3 - 0.0837X_4^{**}$$

$$R^2 = 0.7339$$

(0.3124) (0.3311) (0.8761) (7.9915)
(Figures in parentheses indicate t values)

** Significant at 5 per cent level of probability

In Kalasib district, the effects of education and Population density had significant effect on jhum cycle. However, the effects of all other variables were insignificant. The value of coefficient of multiple determination was 50.23 per cent.

Measures for Controlling Jhum Cultivation in Meghalaya in Different Plan Periods

The Department of Soil and Water Conservation, Government of Meghalaya has implemented various jhum control measures in the state since fifth five-year plan.

Terracing

Terracing was considered as a measure to control jhum in the state. During fifth to tenth five year plans, 4965.82 hectares were achieved under terracing. Achievement was quite low in the state. It was 39.34 per cent in the fifth five year plan and decreased to 6.59 per cent in tenth five year plan. Physical target varied from 557.10 hectares in tenth five year plan to 1250.00 hectares in eighth five year plan.

Reclamation of Valley Bottom Land

During fifth to tenth five-year plans, 2178.82 hectares of valley land were reclaimed and out of this the highest achievement (74.76 per cent) was observed during fifth five year plan and it was the lowest in seventh five year plan period. The achievement exceeded the target during eighth five year plan period while achievement was far below the target during ninth five year plan.

Follow-up Programme

Under the follow-up programme, 4452.02 hectares were achieved during fifth to tenth plan period. Achievement was the highest (27.79 per cent) during the fifth plan and declined to 21.15 per cent during tenth plan. Achievement of physical target was up to the extent of 67.27 per cent during eighth five-year plan

Afforestation

Afforestation programme in Meghalaya was not so encouraging. During fifth five year plan it achieved 1570.79 hectares under afforestation and it declined to mere 41.08 hectares which shared 62.01 and 1.62 per cent of total achievement, respectively. The physical target was quite high (4000.00 ha) during ninth five-year plan out of which achievement was only 7.33 per cent. In the state, 2533.17 hectares were put under afforestation by the Department of Soil and Water Conservation.

Water Conservation and Distribution Works

Under this measure, 5315.22 hectares were achieved in the state. Highest achievement of this measure was in the fifth five year plan (33.04 per cent) and the lowest achievement was in the ninth five year plan (2.05 per cent). During the eighth five year plan it exceeded the target. Out of all the measures adopted by the Government of Meghalaya, this measure gained much importance in the state.

Camp and Camp Equipment

Numbers of camps were organized in the jhum areas for encouraging settled cultivation so that the farmers need not shift their farm stead. In fifth five year plan about 113 such camps were organized by the Department. In subsequent plan periods it was found to decline faster.

Dwelling House and Drinking Water

Permanent settlement was also encouraged by the Department of Soil Conservation and it constructed dwelling houses. It was observed that during fifth five-year plan 1,665 numbers of dwelling houses were constructed by the Department. However, this measure was not followed in the subsequent years. Similarly, 235 drinking water points were established by the Department in the jhum areas for settlement. It was higher in the fifth, seventh, and eighth five-year plans. It was quite negligible in the tenth five-year plan period.

Link Roads

Development of link roads in the jhuming areas was also encouraged for permanent settlement. It was observed that from fifth to ninth five year plan, 532.27 kms of link roads were constructed in the state. Achievement was higher in the sixth five year plan which was about 33.97 per cent of the total. It declined very fast in the subsequent plan periods. In the tenth five year plan there was no construction of link roads in the state.

Development of Horticultural Cash Crops

Department of Soil and Water Conservation of Meghalaya put a continuous effort to develop horticulture in the jhum land as a controlling measure. During fifth to tenth plan periods, 6982.98 hectares of land were converted to horticultural crops. Out of total area, 43.21 per cent were put during fifth five-year plan which declined in the sixth and seventh plan periods and was found to increase again in the eighth and ninth plan periods. During the tenth five-year plan, development of horticultural cash crops in the jhum area was quite negligible. It was observed that out of all the measures, the Department put more emphasis on development of horticultural cash crops in the jhum areas as control measures of jhum.

Development of Nursery

This measure was adopted by the department in the recent years. During tenth five-year plan about nine lakh such nurseries were targeted and 6,60,813 numbers of nurseries (73.42 per cent) were established in the state.

Thus, it was seen that the Department of Soil and Water Conservation in Meghalaya followed different control measures for jhum cultivation. Out of all the measures terracing, water conservation, link roads, area converted to horticultural cash crops, and nursery bed preparations were more emphasized by the Department in the state.

Measures for Controlling Jhum Cultivation in Mizoram

Watershed Development Programme Implemented by the Department of Agriculture

The Department of Agriculture, Government of Mizoram implemented watershed development programmes under central scheme. The two such scheme were Integrated Wasteland Development Programs (IWDP) and National Watershed Development Project for Rainfed Areas (NWDPRA).

Watershed Development Project in Shifting Cultivation Areas (WDPSCA)

This programme was launched in 1994–95 in the seven NE states and continued up to tenth plan. The Department of Agriculture in Mizoram has so far completed 33 projects covering an area of 65774 hectares with the expenditures of Rs 2185.87 lakhs. During the eighth and ninth plan period, 2877 numbers of drainage line structures, 3991 numbers of water harvesting structures, 36440 numbers of field bunds have been executed.

Control of Shifting Cultivation

During 2002–03, a separate fund of Rs 480 lakh was earmarked for controlling shifting cultivation in the state. The fund was spent under different components like Administrative Cost (project management), Agricultural Extension and Training, Development Component (natural resource management), Production System (production of crops) Land Use Incentive, and Rehabilitation (community organization). Under the Development Component, 213 ha of potential area has been developed for terraced cultivation. Besides these, a total length of 63 kms of Potential Area connectivity (Agril link road) has been constructed at different locations wherever necessary. Under the component of Household Production System, 450 numbers of families were also assisted.

Demonstrations on commercial crops and vegetable crops were also conducted in areas of 200 ha and 500 ha, respectively under the component of Land Use Incentive. Agro forestry and Drainage Line treatment like construction of brushwood check-dam, water harvesting structure/dam and fabion structure were also done under the scheme.

During 2004–05, a fund of Rs 350 lakhs was earmarked to be spent under different components. Under Development Component, 49 hectares of potential area was developed for WRC and 39 hectares of hillside slopes was constructed for terrace cultivation. Besides, a total length of 11 kms of Potential Area Connectivity (Agril. Link Roads) was constructed at different locations.

Under the component of land use, incentive demonstrations on commercial and vegetable crops were conducted in 1,200 and 500 hectares, respectively. Agroforestry and drainage line treatment like construction of brushwood check-dam, loose bolder check-dam, water harvesting structure/dam, and Gabian structure was done under the scheme.

During 2005–06, Rs 550 lakhs were earmarked. In this year under Development component 320 hectares of potential area for WRC, terraces of hill slopes, has been developed for cultivation during the year. Besides, a total length of 15 kms of Potential Area Connectivity (Agril. Link Road) were constructed at different locations.

Construction of rainwater harvesting structure, individual, and community has been developed with physical outlay of 214 numbers during 2005–06. About 200 hectares for cultivation of medicinal and aromatic plants will be converted as an alternative to jhuming and 1000 hectares for innovative management practices of Integrated Nutrient Management, respectively. Drainage Line treatment of upper, middle, and lower reaches of Gabian structures has to be done under the scheme.

Mizoram Intodelhna Project

Mizoram Intodelhna project (MIP) funded by the Government of India for the year 2002–03 was launched from March 2002. The main purpose of MIP is the upliftment of rural poor especially the shifting cultivators. Its main concern is attainment of self-sufficiency, food security, and better livelihood for the cultivators as a whole. This project introduced by the Government of Mizoram is practically known as 'Project for self-sufficiency in Mizoram'.

Government of Mizoram introduced MIP with an allocation of total fund of Rs 2000 lakhs. Under the component of assistance to individual farmers for Land Development, 9366 numbers of families were assisted providing Rs 4000 per family. Training, Awareness campaign, workshop, and so on were also conducted at the state, district, and (Partners in Agriculture) PIA levels. During 2002–03 the following programme components were implemented in different style.

1. Land development in low land area and terraces in the upland areas.
2. Development of piggery for diversification of farming system.

Supervision of Works in the Field

Supervision of field was done by the project officers (MIP) and Assisstant Project Officers (MIP) covering all selected villages under overall control of the Deputy Commissioner who is also the Chairman of the District Level Monitoring Cell (DLMC) and District Level Project Committee (DLPC).

Monitoring and Evaluation

A comprehensive monitoring system has been designed for MIP creating a separate cell at village, district, and state level. During 2003–04, Rs 2,000 lakhs were allocated for this project. The project was continued for 2004–05, as well.

CONCLUSION AND POLICY IMPLICATIONS

It can be concluded from the study that area under jhum cultivation was higher than that of settled cultivation in both the states. Jhumias raise their food crops in the jhum field and the commercial cash crops were raised in the settled cultivation. It was less productive and less remunerative than settled cultivation. It is a continuous socio-economic process linked up with religion of the tribals. Jhum cycle was affected mainly due to paucity of land and high population growth. The system of jhuming has not been abolished so far after so much of government plan and programmes. Jhumias must be attracted towards the government programmes which are alternative to jhuming so that these are feasible and acceptable without affecting their religious rites, quality of life of the jhumias can be improved through various measures like improvement of land tenure system, use of common property resources, market development, introduction of small entrepreneurship, food and healthcare facilities, good communications, and so on.

Initiative and Strategies for Controlling Jhum Cultivation

It is a primitive practice to sustain the tribal. Different integrated tribal development programmes have been initiated, but these programmes are not yielding much response due to ignorance of the authorities about the socio-economic and agro ecological conditions of shifting cultivation and less involvement of officials to make the people more conscious about the adverse effect of jhum cultivation. In the recent years, International Development Agency through India-Canada Environment Facility, a project has been launched in Nagaland known as Nagaland Environment Protection for Economic Development to make Nagaland self-sufficient in agroforestry. This project involves rural people for jhum as well as agroforestry under supervision along with the village team. Through such efforts, gradually jhum is made more stable and profitable and has started tackling much of the tribal-forest conflict.

The Strategies

1. Through equitable distribution of wasteland among the tribals more income and employment can be generated through development of agroforestry and silvipasture practices.
2. Encouragement of forest based activities through formation of cooperatives.
3. Forming Village Forest Committee can protect the communities by giving suitable incentives after harvest. This will also encourage increase the employment opportunities through different rural development schemes.
4. Literacy campaign should be launched in the remote areas educating the tribal women and children.

REFERENCES

Anonymous. 2000. *Basic Statistics for North East India*. North East Council, Shillong.

Barlett H.H. 1956. 'Fire, Primitive Agriculture and Grazing in the Tropics,' in *Mans's Role in Changing the Face of the Earth*, edited by W.L. Thomas, 692–720. The University of Chicago and London.

Bhowmick, P.K. 1976. 'Shifting Cultivation: A Plea for New Strategy.' North East India Council for Social Science Research, Shillong: 5–10.

Bora, D. and N.R. Goswami. 1977. *A Comparative Study of Crop Production Under Shifting and Terrace Cultivation (A Case Study in Garo Hills, Meghalaya).* Agro Economic Research Centre for North East India, Jorhat.

FAI. 2005. http://agricoop.nic.in.

FAO. 1957. 'Shifting Cultivation.' *Tropical Agriculture* 34(3). FAO staff paper.

Goswami, P.C and N. Saha. 1969. *A Study on the Problems of Agricultural Development in the Hills of N.E India.* Agro economic Research Centre for N.E India, Jorhat-13.

Goswami S.N. 1997. 'Economic Appraisal of Indigenous Farming System of West Garo Hills District of Meghalaya (India).' *Indian Journal of Agricultural Economics* 52(2): 252–59.

Government of India, CMIE, various issues pertaining to 1998, 2005, 2006.

Gupta T.R. and S. Sabrani. 1978. 'Control of Shifting Cultivation: The Need for an Integrative Approach and Systematic Appraisal.' *Indian Journal of Agricultural Economics* 33(4): 1–8.

Mandal, R.K. 2006. 'Control of Jhum Cultivation in Arunachal Pradesh, Retrospect and Prospect.' *Kurukshetra* 55(2): 16–19.

Saha, N. 1973. *The Economics of Shifting Cultivation in North East India.* Ad-hoc study, No. 22. Agro economic Research Centre for North East India, Jorhat.

Saikia, P.D. and D. Bora. 1978. 'Pattern of Crop Production Under Shifting and Terrace Cultivation—A Case Study in Garo Hills.' *Indian Journal of Agricultural Economics* 33(4): 43.

68

Participation of Plains Tribal Women in Non-Agricultural Development Activities

Runjun Savapandit

Women constitute half of the human resources and in that sense, half the economic wealth of the country. If half of the nation's human resources are neglected, the overall progress of the country would obviously be hampered. In the third world countries, especially where agriculture still continues to be the main source of livelihood and the main area of work for most people, women's contribution is quite high. Recognizing the need for involving women in various development activities, the Government of India has initiated several affirmative measures by way of launching different programmes and schemes to bring them in to the mainstream of development.

India is the habitat of 121.06 crore people comprising different castes and sub-castes along with 573 tribes and sub-tribes. They are broadly classified into different classes and groups depending upon their present state of development and some other significant factors.

It is evident from various Census reports that the work participation of men and women in different economic activities have a wide disparity in India as well as in Assam. In 1951 Census, the female participation rate was 16.50 per cent of the total work force of the country, that is, less than one-fifth of the male counterpart (83.50 per cent), while in Assam less than one-fourth (21.17 per cent) of the male counterparts were engaged. In 1961 Census, the work participation of women improved considerably both at the national and state level. Thus, the female participation was almost half of the male counterparts at the national level (31.53 per cent) and at the state level, it was more than half of the male counterparts (33.36 per cent). In 1971 Census, the rate of work participation of women in the country again declined to the extent of 17.35 per cent and at the state level it was only 7.87 per cent of total work force. At the National level there was a slight improvement in the female participation rate to the extent of 20.21 per cent (that is, less than one-fourth of the male counterparts) during 1981 census. In 1991 Census, again there was some improvements both at national and state levels, as the work participation rate of women were recorded to be 24.85 per cent at the national level and 28.74 per cent at the state level. In 2001 Census at national level there was marginal improvement in the women participation rate, that is, 31.63 per cent while in Assam, it slightly declined to 27.97 per cent (that is, less than one-third of the male counterparts).

Assam is the habitation of various tribal communities having different social structure and cultural heritage. The total population of Assam as per 2011 census is 3.12 cores consisting of 22.31 lakh scheduled caste (SC) and 38.84 lakhs scheduled tribe (ST) population. The tribes in the plain districts of Assam include nine social groups, namely Barman in Cachar, Boro, Boro-Kachari, Deori, Hojai, Kachari including Sonowal, Lalung, Mech, Mising, and Rabha.

Generally, women are lagging behind in the pace of all types of development and so is their status in the society. In backward societies they are in more disadvantageous position than that of developed societies. But it is a well-known fact, that no society or nation can be developed and civilized without developing the women folk. Thus, time has come to analyse the role of women in different agriculture based and non-agriculture based activities.

The Constitution of India provides not only equal rights and privileges for all the citizens, but also specially mentions the need for making special provisions for women. Accordingly, from the beginning of the Indian planning, the planners and policymakers have introduced various welfare and development programmes for women to improve their social and economic status in the society.

The necessity of improving the status of women has now been recognized by all the sections of the society all over the world. With a view to improve the economic condition of the female population in India, various development schemes are introduced during the plan periods with provisions of active participation of women at par with their male counterpart.

The socio-economic and educational upliftment of the tribal people in the state of Assam is a major thrust area of the Government of Assam. For the welfare of the plain tribes and backward classes, the Directorate of Welfare of Plains Tribes and Backward Classes was set up in 1976. Since inception the Directorate has been trying hard to uplift the socio-economic condition of plains tribes and backward classes in Assam by implementing different schemes and programmes. The various schemes taken up by the Government of Assam for the welfare of the plains tribal are:

Educational Development Programme: This includes:

1. For educational development of ST students, Government has awarded post-matriculation scholarship to 1,02,700 numbers of students and pre-matriculation scholarship to 25,410 numbers of students with financial involvement of Rs 999.98 lakh and Rs 40 lakh, respectively during the year 2005–06.
2. Government provides stipend for Craftsmen Training to the plains tribal students and they may undergo training in various trades in Assam. Till recently, stipend has been provided to 592 numbers of students with a financial involvement of Rs 6.60 lakh.
3. Under Infrastructure Development Scheme, several educational institutions and schools in rural areas have been provided financial assistance for renovation and repairing which include development of science laboratories, playgrounds, and so on. During the period from 2003–04 to 2005–06 the financial assistance was provided to 122 number of schools amounting to Rs 165.73 lakh.

Economic Development Scheme: This programme includes:

1. Family-Oriented Income Generating Schemes (FOIGS). Under FOIGS the Directorate of Welfare of Plain Tribes and Backward Classes provides assistance in the form of providing tractor, power tiller, sewing machine, yarn, eri, muga kit, and so on, to the plain tribal people living below poverty line to remove or reduce poverty in the tribal society.

Infrastructure Development Scheme (IDS): This programme includes:

1. Renovation and repairing of educational institutions
2. Construction of roads/bridges
3. Construction of community centre
4. Drinking water facilities
5. Youth welfare and cultural activities and
6. Vocational Training Institute (VTI) Scheme

Besides FOIGS and IDS, the Assam Tribal Development Authority (ATDA) has been given financial assistance for tribal women for upliftment of their economic conditions.

There is a growing consciousness all over the world about the role of women in economic development. It has now been realized that economic progress in any sphere of economic activity is integrally related with the economic well-being and social status of women. The United Nations proclaimed 1975 as the 'International Women Year' and the period between 1976 to 1985 was the United Nations 'Decade for Women' which demanded need based policy measures to improve the quality of life of women.

The objectives of the present study are:

1. to examine the nature and types of assistance and the economic viability of the non-agricultural development schemes designed for the plains tribal women;
2. to assess the impact of the schemes for generation of employment and improving the economic conditions of the plains tribal women;

3. to examine the problems faced by the sample women entrepreneurs as well as implementing agencies in implementing the schemes for the economic benefit of plains tribal women; and
4. to suggest policy implications.

METHODOLOGY AND DATA

In order to draw sample a complete list of women beneficiaries under different non-agricultural development schemes of government agencies were collected from the concerned authorities of the sample districts. Jorhat district of upper Brahmaputra Valley Zone and Dhemaji district of North Bank Plains were selected for the study on the basis of the concentration of plains tribal women entrepreneurs in consultation with the concerned state departments. From each selected district 50 women beneficiaries were selected by adopting random sampling method. Thus, a total of 100 plains tribal women beneficiaries were covered by the study. Of the total samples, 40 numbers from each cotton and 'muga' weavers and 20 numbers from 'eri' weavers were taken for the study. To make a comparative analysis of the economic conditions of the beneficiary women, both pre and post benefit years were taken into consideration.

The field level data were collected with the help of a set of specially designed schedules through personal interview method. Some secondary-level information regarding the participation of women in different non-agricultural economic activities were collected from various published and unpublished sources of the concerned districts and the state.

RESULTS AND DISCUSSION

Total population of the Jorhat district sample was 276 comprising 52.90 per cent males and 47.10 per cent females and in Dhemaji district total number of family members of the sample households was 332, comprising 55.42 per cent males and 44.58 per cent females.

The literacy rate in the sample districts of Jorhat and Dhemaji were 94.59 per cent and 97.24 per cent, respectively.

In Jorhat district 36.23 per cent of the total population were worker comprising 22.10 per cent male and 14.13 per cent female workers and in Dhemaji district, 37.05 per cent of the total population were workers, of which 23.49 per cent were male workers and 13.60 per cent female workers.

The main occupation of male members of the working population was agriculture and that of female members was weaving in both the districts.

The total owned land of the sample families were 59.34 hectares and 72.39 hectares in Jorhat and Dhemaji district, respectively.

In Jorhat district, of the total operational holding of 67.26 hectares, 90.45 per cent (60.84 hectares) were under field crops and 9.55 per cent (6.42 hectares) were under other miscellaneous crops. In Dhemaji district, out of the total operational holdings of 78.21 hectares, 88.35 per cent were under field crops and 11.65 per cent were under miscellaneous crops.

In Jorhat district, average yield of high yielding variety (HYV) *Ahu* paddy under irrigated condition was 3,590 kg/ha and 3,316 kg/ha in unirrigated situation. The average yield of HYV *Sali* paddy in irrigated condition was 3,089 kg/ha and 2,772 kg/ha in un-irrigated condition.

In Dhemaji district average yield of HYV *Ahu* paddy was 3,780 kg/ha (in irrigated condition). The average yield of HYV *Sali* paddy was 3,165 kg/ha (irrigated) and 3,097 kg/ha (unirrigated).

The present study was concentrated with the programmes under Tribal Development Authority and Social Welfare Department of Assam. The schemes are (i) Cotton Yarn and looms, (ii) Muga Kit, and (iii) Eri Kit.

Out of 100 sample beneficiary weavers, 40 households were provided 139.60 kg of cotton yarn and eight numbers of looms. Forty households were provided Muga kit ('Chalani', 'Dola', and 'Net') and the remaining 20 household were provided Eri kit (Hand lamp, 'Dola', 'Dabri', and Sprayer) by the state government.

Although various tools and implements are required for cotton weaving, only cotton yarn and looms were supplied by the government. However, the other implements required for cotton weavers such as-loom shed, 'ugha', 'chereki', loom, 'rash', 'puttal', 'kathi', 'holi', and 'charkha', and so on, were arranged by the households themselves. Total cost of home produced implements was Rs 288,819 and the value of purchased items was Rs 151,651. On the other hand, the value of items given by the government department was Rs 18,080. The total cost of looms, both home produced and purchased was Rs 243,880, whereas the cost of looms provided by the government department was only Rs 14,560.

As per report of the sample households, 3101.85 kg of yarn was required to meet their demand and its value was of Rs 471,161. Of the total requirement, 55 per cent of yarn was purchased from the market and the rest 45 per cent were supplied by the government department.

The total quantity of Muga yarn used by the sample households was 184.73 kg and the value was Rs 6,56,393, of which 72.97 per cent were home produced and 27.03 per cent was purchased from the market.

The total quantity of Eri yarn woven by the sample household was 306.41 kg, of which 82 per cent was home produced and 18 per cent was purchased from the market. Total number of various cotton clothes prepared by the sample families was 20,319 and its value was Rs 29,21,706. Of the total products, they sold 71.81 per cent and the rest 28.19 per cent was kept for use of family members.

Muga weaver prepares 'riha', 'mekhala', 'chadar', shirting, and 'saree' in their looms. The total number of different clothes woven by the sample household was 1,414 and the value was Rs 14,48,050. Of the total products they sold 87.27 per cent and rest 12.73 per cent was kept for family members. Total number of various Eri clothes prepared by the sample families was 2,201 and its value was Rs 819,400. Of the total products, they sold 89.87 per cent and the remaining 10.13 per cent was kept for family use.

Total number of man days required for cotton clothes production was 20,914, of which 93.25 per cent were family labour and 6.75 per cent were hired labour. The total wage requirement for family labour was Rs 8,93,893 and Rs 63,463 was spent on hired labour.

Total number of mandays required for Muga clothes production was 5,453 of which 93.86 per cent were family labour and 6.14 per cent hired labour. Total wages (both family and hired) required for 'Muga' clothes production were Rs 2,45,385, of which 93.86 per cent was for family labour and 6.14 per cent for hired labour.

In Eri clothes production, the requirement of total number of man days was 9,382 with a wage bill of Rs 4,28,778. In Eri clothes production, hired labour was not engaged by the sample households. Family members were sufficient for this purpose.

The total annual cost of production of cotton clothes was Rs 15,52,911 of which 30.34 per cent was attributed to raw material costs and 57.82 per cent labour costs (both family and hired labour). The per loom cost was found at Rs 10,936. Total annual return from cotton clothe was found at Rs 2,921,706. The per loom net return was recorded at Rs 9,639 and the B.C.R. was found to be 1.88:1. This indicates that cloth production is economically viable in case of sample cotton weavers. The per loom cost was found at Rs 10,234 and the per loom net return stood at Rs 4850 and the benefit–cost ratio (BCR) was found at 1.47:1. Higher BCR indicated profitability of Muga cloth production.

The per loom cost on Eri cloth production was Rs 14,339 and the per loom net return stood at Rs 4,717 and the BCR was found at 1.33:1. A higher BCR indicated higher profitability amongst the Eri cloth weavers. Though the main source of income of the sample households was weaving, some of the sample households managed to earn some additional income from other sources as well. Tables 68.1 and 68.2 represent the per-household distribution of annual income by sources in Jorhat and Dhemaji districts, respectively. The total per-household income from various sources in Jorhat district was found at Rs 1,01,011.98 of which 47.31 per cent came from weaving. In Dhemaji district, the total per household income from various sources was found to be Rs 97,970.40, of which 57.15 per cent came from weaving. Thus, it may be inferred that weaving was the principal source of income of the sample households in both the districts under reference.

Table 68.1 Per-Household Distribution of Annual Income by Sources of Sample Households in Jorhat District
(Value in Rs)

Farm size (Ha)	HHs	Cultivation	Wage Labour	Livestock, Forestry, & Fishery	Weaving	Other HH Cottage Industry	Service	Trade, Com. & Transport	Total	Annual HH Income
Below 1.00	24	19489	8400	5,100	46,501.42	1733.33	336,600	11,220	2,297,190	90,476
		(20.37)	(8.78)	(5.33)	(48.57)	(1.81)	(14.66)	(0.49)	(100.00)	
1.00–2.00	18	20568	10,000	7,673	48,803.72	2,568.61	230	14,400	1,857,846	103,214
		(19.93)	(9.69)	(7.43)	(47.28)	(2.49)	(12.40)	(0.78)	(100.00)	
2.00–4.00	7	21836	0	12,920	50,265.57	5,123.57	136,800	12,870	780,686	118,669
		(19.58)		(11.58)	(45.07)	(4.59)	(17.52)	(1.65)	(100.00)	
4.00 &	1	32,989	0	30,030	43,528	8,630	0	0	115,177	140,177
above		(28.64)		26.07	37.79	7.49			(100.00)	
Total	50	20474.24	7631.64	7618.80	47,794.18	2,646.28	14,075.10	769.74	1,01,011.98	100
		(20.27)	(7.56)	(7.54)	(47.31)	(2.62)	(13.94)	(0.76)	(100.00)	

Note: Figures in parenthesis indicate percentage to total annual Household Income.
Source: Field survey data.

Table 68.2 Distribution of Annual Income by Sources of Sample Households in Dhemaji District
(Value in Rs)

Farm size (Ha)	HHs	Cultivation	Wage Labour	Livestock, Forestry, & Fishery	Weaving	Other HH Cottage Industry	Service	Trade, Com. & Transport	Total	Annual HH Income
Below 1.00	21	18,489	6,857.14	4,571.43	53,673.38	428.57	4,371.43	340.00	88,730.95	88,731
		(20.84)	(7.73)	(5.15)	(60.49)	(0.48)	(4.93)	(0.38)	(100)	
1.00–2.00	14	12,568	6,171.43	10,285.71	65,803.71	450.00	7,200.00	257.14	1,02,736.00	102,736
		(12.23)	(6.01)	(10.01)	(64.05)	(0.44)	(7.01)	(0.25)	(100)	
2.00–4.00	13	20,836	0	14,520.00	54,250.00	1120.00	13,984.62	1,405.38	1,06,116.00	106,116
		(19.64)		(13.68)	(51.12)	(1.06)	(13.18)	(1.32)	(100)	
4.00 &	2	42,489	0	10,762.50	23,000.00	0	32,615.00	0	1,08,866.50	108,867
above		(39.03)		(9.89)	(21.13)		(4.73)		(100)	
Total	50	18,399.50	4,607.72	9,004.92	55,988.92	597.16	8,792.00	580.16	97,970.40	97,985
		(18.78)	(4.70)	(9.19)	(57.15)	(0.61)	(8.97)	(0.59)	(100)	

Note: Figures in parentheses indicate percentage to total annual Household Income.
Source: Field survey data.

From the findings, it is revealed that the contribution of tribal women in economic upliftment of their family is very much satisfactory. The total income from various sources (from both the districts) was found at Rs 99,49,019, of which 52.16 per cent came from weaving.

CONCLUSION AND POLICY IMPLICATIONS

The basic objective of Tribal Development Authorities is to encourage the tribal women in their income generating non-agricultural activities by providing them various aid by the concerned department. Based on field observations and findings of the study, following policy suggestions have been put forward.

Creation of basic infrastructural facilities such as connecting roads, bridges, transport, marketing facility, electricity, and so on are considered as most essential for development of a region. Investment strategy should be broad-based and the assets and benefits provided by the different departments should be sufficient enough to generate income and employment to help the women in raising their standard of living.

Sometimes, fair and exhibitions, and so on should be arranged to promote local products through display or advertisement, as the consumers are less aware of locally manufactured products. This will give an opportunity to the consumers to assess the local products in terms of quality and value.

The benefits under the government sponsored schemes should be offered to economically weaker families and needy lot.

Training programme should be arranged in rural areas to encourage the women to adopt modern system of weaving. Training should be of short duration because women in rural areas cannot stay away from home for a long period because of other commitments.

Adoption of modern technologies, especially use of sophisticated tools and equipments as well as machineries in weaving requires reasonable amount of money. Lack of finance in case of the poor weaver families should be mitigated by providing institutional finance to the willing weaver families of the study area.

The financial institutions should come forward to lend credit at lower rate of interest with simplification of procedural formalities for fast and easy disposal of loans, so that the women could get the financial help on time.

Appropriate steps should be taken to safeguard the golden yarn, Muga. With its limited production and expensive prices, there is every possibility that the entire muga culture may be at peril by way of replacing it by low cost Tasar yarn.

A harmonious combination of traditional designs with new ideas taking the help of computer technologies should be worked out to evolve more appealing patterns in the Muga and cotton clothes woven by the weavers. This will help in attracting more customers both in local and outside markets. Establishment of a research cum training centre in this regard may cater to a long pending necessity of the sample areas.

The weavers of the study area should visit industrially developed states of the country to gather fair knowledge on latest development in the process of production and marketing which can help them in enhancing their profits and making the production units viable one. In this regard, government may organize educational tour which will help the enthusiastic weavers to a great extent.

Though tribal women play an important role in various economic activities in the study area, it is not sufficient enough to stand boldly in their day to day life. Due to inadequate infrastructural supports, lack of proper education, lack of capital, inadequate government aids, inadequate marketing facilities, and so on, most of the women beneficiaries are unable to expand their existing ventures. In this regard, development of infrastructure is a must. At the same time, training programmes could be of much help to tribal women for development of self-confidence and communication skill to interact with various marketing agencies and financial organizations. Proper steps should be taken by the government to develop socio-economic support for women which include adequate facilities for caring of children, healthcare, insurance, and peaceful working atmosphere so that the tribal women can contribute to their family income which in turn can help in economic upliftment of the tribal society.

REFERENCES

Ahmed, Jaynal Uddin. 2012. 'Constraints Experienced by Women Entrepreneurs in South Assam', *Journal of the North-Eastern Council,* 22(2–3), April and July, August, 2012

Baishya, P. 1989. *Small and Cottage Industries: A Study in Assam.* Assam: Manas Publication.

Baruah, M. et. al. 2006. 'Socio-Economic Status of Handloom Weavers of Sualkuchi of Assam', *Rural India,* Oct–Nov 2006.

Borah, R. 2007. *Viable Entrepreneurial Trades for Women in Agriculture—A Study in Assam, Agro-Economic Research Centre for North-East.* Assam Agricultural University, 10–11.

Borah, Ruplekha. 1998. 'Perceived Drudgery and Factors Influencing Time Spent in Agricultural Operation by Farm Women in Assam', *Rural India,* November.

Choudhary, Robindra Kumar. 1994. 'Work Participation and Economic Status of Women', in *Women of Assam,* edited by Renu Devi. New Delhi: Omsons publications.

Dutta, P.C. 1998. *Problems and Prospects of Muga Silk Production in Assam.* Agro-Economic Research Centre for North-East India, Assam Agricultural University.

Government of Assam. 2007. *Statistical Hand Book of Assam, 2007.* Directorate of Economics and Statistics, Guwahati, Government of Assam.

———. 2007–08. *Economic Survey of Assam.* Directorate of Economics and Statistics, Guwahati, Government of Assam.

Kumar, D. and V. Himachalan. 1991. 'Women Entrepreneurship Development in India-Problem and Prospects', *Monthly Commentary,* June 1991.

Mazumdar, Bina. 1978. 'Towards Equality, Status of Women in India', In *Women in World, Illusion and Reality,* edited by Urmila Phadnis and Indian Malani. New Delhi: Vikas Publishing House.

Medhi, D.N. 1956. *Handloom Weaving and Designing.* Guwhati: Lawyers Book Stall.

Phukan, Ranjita. 1998. 'Women Entrepreneurs Development in Assam with Special Reference to Dibrugarh and Tinsukia Districts', Unpublished Ph.D. Thesis submitted to the Dibrugarh University, 95–100.

Saikia, Anuva. 1992. 'Economic Status of Women in Rural Areas of Assam', in *Status of Women in Assam,* edited by S.L. Boruah. New Delhi: Indian Society of Agricultural Economics.

———. 2004. 'Employment Pattern of Rural Women and their Involvement in Decision Making, A Study in Jorhat District of Assam', *Women in Agriculture and Rural Development, Indian Journal of Agricultural Economics,* New Delhi.

Saikia, P.D. and D. Borah. 1979. *Tribal Women of North-East India.* Agro-Economic Research Centre for North-East India. Assam Agricultural University.

Sharma, A. 1994. 'Impact of Economic Development on Women with Particular Reference to Assam', in *Women of Assam,* edited by Renu Devi. New Delhi: Omsons publications.

About the Editors

GENERAL EDITORS

Sangeeta Verma is Principal Adviser in the Department of Consumer Affairs, Ministry of Consumer Affairs, Food and Public Distribution, Government of India. She has served in various capacities for the Central Government and for the State of Uttar Pradesh, advising on public policy in both the economic and social spheres. She has vigorously pursued the cause of gender mainstreaming and gender sensitivity in government policy and programmes. She has worked as a member of several Working Groups set up by the erstwhile Planning Commission for the Twelth Five Year Plan and has represented India at numerous International forums including meetings of the G-20 Agricultural Deputies. Her articles on various subjects have been published in magazines and newspapers of national repute.

P.C. Bodh is Adviser to the Government of India in the Ministry of Agriculture and Farmers Welfare (MOAFW), New Delhi. He has worked as a Director in Planning Commission, Ministry of Water Resources; as a Deputy Director in the Directorate General of Foreign Trade; and as an Assistant Director in the Ministry of Labour, before the current assignment as Adviser to the Ministry of Agriculture since 2014. His writings include four books: *Mount Kailash and the Wonder Lake* (2000), *Tales from the Buddha's Life* (2004), *Global Digital Economy and the 11th Plan ICT Equity Issues* (2009), and *Zeroing in on Happiness* (2013).

VOLUME EDITORS

Vijay Paul Sharma is currently Chairman, Commission for Agricultural Costs and Prices (CACP), MOAFW. He was a Professor in Centre for Management in Agriculture (CMA), Indian Institute of Management, Ahmedabad (IIMA) when this study was completed. He has served as a consultant to many national and international organizations in the agri-food and development sector and is a member/chairman of several national and state-level committees.

Parmod Kumar is Professor, Agricultural Development and Rural Transformation Centre, Institute for Social and Economic Change, Bengaluru, India. He has authored nine research volumes and published more than 40 research articles in reputed national and international journals. He is leading several research projects sponsored by the Government of India and various international organizations. Kumar was conferred the IDRC India Social Science Research Award for his work on the public distribution system.

Nilabja Ghosh is a faculty member of Institute of Economic Growth, New Delhi. She has taught economics for over 10 years and served in reputed institutes as a researcher in agricultural economics and rural development, supporting public policy with research input, improving the statistical database, conducting studies to support the MOAFW for short-term policymaking and participating in international cooperative research.

D.K. Grover has been the visiting fellow/scientist at World Vegetable Centre, Taiwan; International Rice Research Institute (IRRI), Philippines; and International Food Policy Research Institute (IFPRI), Washington, DC, USA. He has worked as World Bank Consultant of Agricultural Economics in Ethiopia. He has authored one research book and contributed several research papers. He has also been a member of several expert panels/committees/editorial boards and interdisciplinary teams both at national and international levels along with serving as session chairman/discussant in several international conferences.

Usha Tuteja was the former Director (Acting) of the Agricultural Economics Research Centre, University of Delhi. She has authored a book and several research papers and articles for reputed journals and newspapers on the agrarian problems in India. She has also conducted around 25 primary data-based studies on problems of pulse cultivation, agricultural input subsidies, contribution of rural women in work and income, rural women entrepreneurs, rural non-farm employment, evaluation of National Horticulture Mission (NHM) and National Food Security Mission (NFSM), pulses, crop diversification, and on wide range of aspects related to agriculture.